Algorithms and Combinatorics 20

Springer
Berlin
Heidelberg
New York
Barcelona
Hong Kong
London
Milan
Paris
Singapore
Tokyo

Kazuo Murota

Matrices and Matroids for Systems Analysis

With 64 Figures

Springer

Kazuo Murota
Kyoto University
Research Institute for Mathematical Sciences
606-8502 Kyoto, Japan
e-mail: murota@kurims.kyoto-u.ac.jp

Cataloging-in-Publication Data applied for.

Die Deutsche Bibliothek - CIP-Einheitsaufnahme
Murota, Kazuo: Matrices and matroids for systems analysis / Kazuo Murota. – Berlin; Heidelberg; New York; Barcelona; Hong Kong; London; Milan; Paris; Singapore; Tokyo: Springer, 2000
(Algorithms and combinatorics; 20)
ISBN 3-540-66024-0

Mathematics Subject Classification (1991):
05B35, 15A21, 68R05, 93B99

ISSN 0937-5511
ISBN 3-540-66024-0 Springer-Verlag Berlin Heidelberg New York

Printed in Germany

Production: PRO EDIT GmbH, 69126 Heidelberg, Germany
Typesetting: Camera-ready copy produced from the authors' output file
Printed on acid-free paper SPIN 10723294 46/3143hs - 5 4 3 2 1 0

Preface

Interplay between matrix theory and matroid theory is the main theme of this book, which offers a matroid-theoretic approach to linear algebra and, reciprocally, a linear-algebraic approach to matroid theory. The book serves also as the first comprehensive presentation of the theory and application of mixed matrices and mixed polynomial matrices.

A matroid is an abstract mathematical structure that captures combinatorial properties of matrices, and combinatorial properties of matrices, in turn, can be stated and analyzed successfully with the aid of matroid theory. The most important result in matroid theory, deepest in mathematical content and most useful in application, is the intersection theorem, a duality theorem for a pair of matroids. Similarly, combinatorial properties of polynomial matrices can be formulated in the language of valuated matroids, and moreover, the intersection theorem can be generalized for a pair of valuated matroids.

The concept of a mixed matrix was formulated in the early eighties as a mathematical tool for systems analysis by means of matroid-theoretic combinatorial methods. A matrix is called a mixed matrix if it is expressed as the sum of a "constant" matrix and a "generic" matrix having algebraically independent nonzero entries. This concept is motivated by the physical observation that two different kinds of numbers, fixed constants and system parameters, are to be distinguished in the description of engineering systems. Mathematical analysis of a mixed matrix can be streamlined by the intersection theorem applied to the pair of matroids associated with the "constant" and "generic" matrices. This approach can be extended further to a mixed polynomial matrix on the basis of the intersection theorem for valuated matroids.

The present volume grew out of an attempted revision of my previous monograph, "Systems Analysis by Graphs and Matroids — Structural Solvability and Controllability" (Algorithms and Combinatorics, Vol. 3, Springer-Verlag, Berlin, 1987), which was an improved presentation of my doctoral thesis written in 1983. It was realized, however, that the progress made in the last decade was so remarkable that even a major revision was inadequate. The present volume, sharing the same approach initiated in the above monograph, offers more advanced results obtained since then. For developments in the neighboring areas the reader is encouraged to consult:

- A. Recski: "Matroid Theory and Its Applications in Electric Network Theory and in Statics" (Algorithms and Combinatorics, Vol. 6, Springer-Verlag, Berlin, 1989),
- R. A. Brualdi and H. J. Ryser: "Combinatorial Matrix Theory" (Encyclopedia of Mathematics and Its Applications, Vol. 39, Cambridge University Press, London, 1991),
- H. Narayanan: "Submodular Functions and Electrical Networks" (Annals of Discrete Mathematics, Vol. 54, Elsevier, Amsterdam, 1997).

The present book is intended to be read profitably by graduate students in engineering, mathematics, and computer science, and also by mathematics-oriented engineers and application-oriented mathematicians. Self-contained presentation is envisaged. In particular, no familiarity with matroid theory is assumed. Instead, the book is written in the hope that the reader will acquire familiarity with matroids through matrices, which should certainly be more familiar to the majority of the readers. Abstract theory is always accompanied by small examples of concrete matrices.

Chapter 1 is a brief introduction to the central ideas of our combinatorial method for the structural analysis of engineering systems. Emphasis is laid on relevant physical observations that are crucial to successful mathematical modeling for structural analysis.

Chapter 2 explains fundamental facts about matrices, graphs, and matroids. A decomposition principle based on submodularity is described and the Dulmage–Mendelsohn decomposition is derived as its application.

Chapter 3 discusses the physical motivation of the concepts of mixed matrix and mixed polynomial matrix. The dual viewpoint from structural analysis and dimensional analysis is explained by way of examples.

Chapter 4 develops the theory of mixed matrices. Particular emphasis is put on the combinatorial canonical form (CCF) of layered mixed matrices and related decompositions, which generalize the Dulmage–Mendelsohn decomposition. Applications to the structural solvability of systems of equations are also discussed.

Chapter 5 is mostly devoted to an exposition of the theory of valuated matroids, preceded by a concise account of canonical forms of polynomial/rational matrices.

Chapter 6 investigates mathematical properties of mixed polynomial matrices using the CCF and valuated matroids as main tools of analysis. Control theoretic problems are treated by means of mixed polynomial matrices.

Chapter 7 presents three supplementary topics: the combinatorial relaxation algorithm, combinatorial system theory, and mixed skew-symmetric matrices.

Expressions are referred to by their numbers; for example, (2.1) designates the expression (2.1), which is the first numbered expression in Chap. 2.

Similarly for figures and tables. Major symbols used in this book are listed in Notation Table.

The ideas and results presented in this book have been developed with the help, guidance, encouragement, support, and criticisms offered by many people. My deepest gratitude is expressed to Professor Masao Iri, who introduced me to the field of mathematical engineering and guided me as the thesis supervisor. I appreciate the generous hospitality of Professor Bernhardt Korte during my repeated stays at the University of Bonn, where a considerable part of the theoretical development was done. I benefited substantially from discussions and collaborations with Pawel Bujakiewicz, François Cellier, Andreas Dress, Jim Geelen, András Frank, Hisashi Ito, Satoru Iwata, András Recski, Mark Scharbrodt, András Sebő, Masaaki Sugihara, and Jacob van der Woude. Several friends helped me in writing this book. Most notable among these were Akiyoshi Shioura and Akihisa Tamura who went through all the text and provided comments. I am also indebted to Daisuke Furihata, Koichi Kubota, Tomomi Matsui, and Reiko Tanaka. Finally, I thank the editors of Springer-Verlag, Joachim Heinze and Martin Peters, for their support in the production of this book, and Erich Goldstein for English editing.

Kyoto, June 1999 *Kazuo Murota*

Contents

1. Introduction to Structural Approach — Overview of the Book

This chapter is a brief introduction to the central ideas of the combinatorial method of this book for the structural analysis of engineering systems. We explain the motivations and the general framework by referring, as a specific example, to the problem of computing the index of a system of differential-algebraic equations (DAEs). In this approach, engineering systems are described by mixed polynomial matrices. A kind of dimensional analysis is also invoked. It is emphasized that relevant physical observations are crucial to successful mathematical modeling for structural analysis. Though the DAE-index problem is considered as an example, the methodology introduced here is more general in scope and is applied to other problems in subsequent chapters.

1.1 Structural Approach to Index of DAE

1.1.1 Index of Differential-algebraic Equations

Let us start with a simple electrical network[1] of Fig. 1.1 to introduce the concept of an index of a system of *differential-algebraic equations* (*DAEs*) and to explain a graph-theoretic method.

The network consists of a voltage source V (branch 1), two ohmic resistors R_1 and R_2 (branch 2 and branch 3), an inductor L (branch 4), and a capacitor C (branch 5). A state of this network is described by a 10 dimensional vector $\boldsymbol{x} = (\xi^1, \cdots, \xi^5, \eta_1, \cdots, \eta_5)^{\mathrm{T}}$ representing currents ξ^i in and the voltage η_i across branch i $(i = 1, \cdots, 5)$ with reference to the directions indicated in Fig. 1.1. The governing equations in the frequency domain are given by a system of equations $A^{(1)}\boldsymbol{x} = \boldsymbol{b}$, where $\boldsymbol{b} = (0, 0, 0, 0, 0; V, 0, 0, 0, 0)^{\mathrm{T}}$ is another 10 dimensional vector representing the source, and $A^{(1)}$ is a 10×10 matrix defined by

[1] This example, described in Cellier [28, §3.7], was communicated to the author by P. Bujakiewicz, F. Cellier, and R. Huber.

$$
A^{(1)} = \begin{array}{c}
\begin{array}{ccccc|ccccc}
\xi^1 & \xi^2 & \xi^3 & \xi^4 & \xi^5 & \eta_1 & \eta_2 & \eta_3 & \eta_4 & \eta_5
\end{array} \\
\left[\begin{array}{ccccc|ccccc}
1 & -1 & 0 & 0 & -1 & & & & & \\
-1 & 0 & 1 & 1 & 1 & & & & & \\ \hline
 & & & & & -1 & 0 & 0 & 0 & -1 \\
 & & & & & 0 & 1 & 1 & 0 & -1 \\
 & & & & & 0 & 0 & -1 & 1 & 0 \\ \hline
0 & 0 & 0 & 0 & 0 & -1 & 0 & 0 & 0 & 0 \\
0 & R_1 & 0 & 0 & 0 & 0 & -1 & 0 & 0 & 0 \\
0 & 0 & R_2 & 0 & 0 & 0 & 0 & -1 & 0 & 0 \\
0 & 0 & 0 & sL & 0 & 0 & 0 & 0 & -1 & 0 \\
0 & 0 & 0 & 0 & -1 & 0 & 0 & 0 & 0 & sC
\end{array}\right]
\end{array} . \tag{1.1}
$$

As usual, s is the variable for the Laplace transformation that corresponds to $\mathrm{d}/\mathrm{d}t$, the differentiation with respect to time (see Remark 1.1.1 for the Laplace transformation). The first two equations, corresponding to the 1st and 2nd rows of $A^{(1)}$, represent *Kirchhoff's current law* (*KCL*), while the following three equations *Kirchhoff's voltage law* (*KVL*). The last five equations express the element characteristics (*constitutive equations*). The system of equations, $A^{(1)}\boldsymbol{x} = \boldsymbol{b}$, represents a mixture of differential equations and algebraic equations (i.e., a linear time-invariant DAE), since the coefficient matrix $A^{(1)}$ contains the variable s.

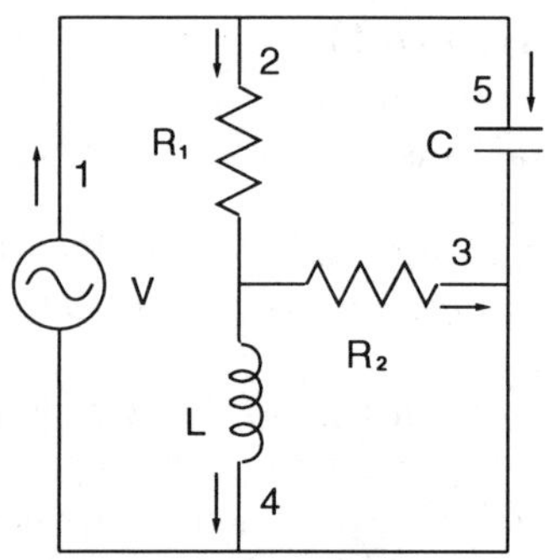

Fig. 1.1. An electrical network

For a linear time-invariant DAE in general, say $A\boldsymbol{x} = \boldsymbol{b}$ with $A = A(s)$ being a nonsingular polynomial matrix in s, the *index* is defined (see Remark 1.1.2) by

$$\nu(A) = \max_{i,j} \deg_s (A^{-1})_{ji} + 1. \tag{1.2}$$

Here it should be clear that each entry $(A^{-1})_{ji}$ of A^{-1} is a rational function in s and the degree of a rational function p/q (with p and q being polynomials) is defined by $\deg_s(p/q) = \deg_s p - \deg_s q$. An alternative expression for $\nu(A)$ is

$$\nu(A) = \max_{i,j} \deg_s((i,j)\text{-cofactor of } A) - \deg_s \det A + 1. \tag{1.3}$$

For the matrix $A^{(1)}$ of (1.1), we see

$$\max_{i,j} \deg_s((i,j)\text{-cofactor of } A^{(1)}) = \deg_s((6,5)\text{-cofactor of } A^{(1)}) = 2,$$

$$\det A^{(1)} = R_1 R_2 + sL \cdot R_1 + sL \cdot R_2 \tag{1.4}$$

by direct calculation and therefore $\nu(A^{(1)}) = 2 - 1 + 1 = 2$ by the formula (1.3).

The solution to $A\boldsymbol{x} = \boldsymbol{b}$ is of course given by $\boldsymbol{x} = A^{-1}\boldsymbol{b}$, and therefore $\nu(A) - 1$ equals the highest order of the derivatives of the input $\boldsymbol{b}$ that can possibly appear in the solution $\boldsymbol{x}$. As such, a high index indicates difficulty in the numerical solution of the DAE, and sometimes even inadequacy in the mathematical modeling. Note that the index is equal to one for a system of purely algebraic equations (where $A(s)$ is free from s), and to zero for a system of ordinary differential equations in the normal form ($\mathrm{d}\boldsymbol{x}/\mathrm{d}t = A_0\boldsymbol{x}$ with a constant matrix A_0, represented by $A(s) = sI - A_0$).

Remark 1.1.1. For a function $x(t)$, $t \in [0, \infty)$, the *Laplace transform* is defined by $\hat{x}(s) = \int_0^\infty x(t)\mathrm{e}^{-st}\mathrm{d}t$, $s \in \mathbf{C}$. The Laplace transform of $\mathrm{d}x(t)/\mathrm{d}t$ is given by $s\hat{x}(s)$ if $x(0) = 0$. See Doetsch [49] and Widder [341] for precise mathematical accounts and Chen [33], Kailath [152] and Zadeh–Desoer [350] for system theoretic aspects of the Laplace transformation. □

Remark 1.1.2. The definition of the index given in (1.2) applies only to linear time-invariant DAE systems. An index can be defined for more general systems and two kinds are distinguished in the literature, a differential index and a perturbation index, which coincide with each other for linear time-invariant DAE systems. See Brenan–Campbell–Petzold [21], Hairer–Lubich–Roche [100], and Hairer–Wanner [101] for details. □

Remark 1.1.3. Extensive study has been made recently on the DAE index in the literature of numerical computation and system modeling. See, e.g., Brenan–Campbell–Petzold [21], Bujakiewicz [26], Bujakiewicz–van den Bosch [27], Cellier–Elmqvist [29], Duff–Gear [60], Elmqvist–Otter–Cellier [72], Gani–Cameron [86], Gear [88, 89], Günther–Feldmann [98], Günther–Rentrop [99], Hairer–Wanner [101], Mattsson–Söderlind [188], Pantelides [264], Ponton–Gawthrop [272], and Ungar–Kröner–Marquardt [324]. □

1.1.2 Graph-theoretic Structural Approach

Structural considerations turn out to be useful in computing the index of DAE. This section describes the basic idea of the graph-theoretic structural methods.

In the graph-theoretic structural approach we extract the information about the degree of the entries of the matrix, ignoring the numerical values of the coefficients. Associated with the matrix $A^{(1)}$ of (1.1), for example, we consider

$$A^{(1)}_{\rm str} = \begin{array}{c} \begin{array}{cccccccccc} \xi^1 & \xi^2 & \xi^3 & \xi^4 & \xi^5 & \eta_1 & \eta_2 & \eta_3 & \eta_4 & \eta_5 \end{array} \\ \left[\begin{array}{ccccc|ccccc} t_1 & t_2 & 0 & 0 & t_3 & & & & & \\ t_4 & 0 & t_5 & t_6 & t_7 & & & & & \\ \hline & & & & & t_8 & 0 & 0 & 0 & t_9 \\ & & & & & 0 & t_{10} & t_{11} & 0 & t_{12} \\ & & & & & 0 & 0 & t_{13} & t_{14} & 0 \\ \hline 0 & 0 & 0 & 0 & 0 & t_{15} & 0 & 0 & 0 & 0 \\ 0 & t_{16} & 0 & 0 & 0 & 0 & t_{17} & 0 & 0 & 0 \\ 0 & 0 & t_{18} & 0 & 0 & 0 & 0 & t_{19} & 0 & 0 \\ 0 & 0 & 0 & s\,t_{20} & 0 & 0 & 0 & 0 & t_{21} & 0 \\ 0 & 0 & 0 & 0 & t_{22} & 0 & 0 & 0 & 0 & s\,t_{23} \end{array}\right] \end{array}$$

where $t_1, \cdots, t_{23}$ are assumed to be independent parameters.

For a polynomial matrix $A = A(s) = (A_{ij})$ in general, we consider a matrix $A_{\rm str} = A_{\rm str}(s)$, called the *structured matrix* associated with A, in a similar manner. For a nonzero entry A_{ij}, let $\alpha_{ij} s^{w_{ij}}$ be its leading term, where $\alpha_{ij} \in \mathbf{R} \setminus \{0\}$ and $w_{ij} = \deg_s A_{ij}$. Then $(A_{\rm str})_{ij}$ is defined to be equal to $s^{w_{ij}}$ multiplied by an independent parameter t_{ij}. Note that the numerical information about the leading coefficient α_{ij} is discarded with the replacement by t_{ij}. Namely, we define the (i,j) entry of $A_{\rm str}$ by

$$(A_{\rm str})_{ij} = \begin{cases} t_{ij} s^{\deg_s A_{ij}} & (\text{if } A_{ij} \neq 0) \\ 0 & (\text{if } A_{ij} = 0) \end{cases} \tag{1.5}$$

where t_{ij} is an independent parameter. We refer to the index of $A_{\rm str}$ in the sense of (1.2) or (1.3) as the *structural index* of A and denote it by $\nu_{\rm str}(A)$, namely,

$$\nu_{\rm str}(A) = \nu(A_{\rm str}). \tag{1.6}$$

Two different matrices, say A and A', are associated with the same structured matrix, $A_{\rm str} = A'_{\rm str}$, if $\deg_s A_{ij} = \deg_s A'_{ij}$ for all (i,j). In other words, a structured matrix is associated with a family of matrices that have a common structure with respect to the degrees of the entries. Though there is no guarantee that the structural index $\nu_{\rm str}(A)$ coincides with the true index $\nu(A)$ for a particular (numerically specified) matrix A, it is true that $\nu_{\rm str}(A') = \nu(A')$ for "almost all" matrices A' that have the same structure as A in the sense of $A'_{\rm str} = A_{\rm str}$. That is, the equality $\nu_{\rm str}(A') = \nu(A')$ holds true for "almost all" values of t_{ij}'s, or, in mathematical terms, "generically" with respect to the parameter set $\{t_{ij} \mid A_{ij} \neq 0\}$. (The precise definition of "generically" is given in §2.1.)

The structural index has the advantage that it can be computed by an efficient combinatorial algorithm free from numerical difficulties. This is based on a close relationship between subdeterminants of a structured matrix and matchings in a bipartite graph.

Specifically, we consider a bipartite graph $G(A) = (\mathrm{Row}(A), \mathrm{Col}(A); E)$ with the left vertex set corresponding to the row set $\mathrm{Row}(A)$ of the matrix A, the right vertex set corresponding to the column set $\mathrm{Col}(A)$, and the edge set corresponding to the set of nonzero entries of $A = (A_{ij})$, i.e.,

$$E = \{(i,j) \mid i \in \mathrm{Row}(A), j \in \mathrm{Col}(A), A_{ij} \neq 0\}.$$

Each edge $(i,j) \in E$ is given a weight $w_{ij} = \deg_s A_{ij}$.

For instance, the bipartite graph $G(A^{(1)})$ associated with our example matrix $A^{(1)}$ of (1.1) is given in Fig. 1.2(a). The thin lines indicate edges (i,j) of weight $w_{ij} = 0$ and the thick lines designate two edges, $(i,j) = (9,4), (10,10)$, of weight $w_{ij} = 1$.

A matching M in $G(A)$ is, by definition, a set of edges (i.e. $M \subseteq E$) such that no two members of M have an end-vertex in common. The weight of M, denoted $w(M)$, is defined by

$$w(M) = \sum_{(i,j)\in M} w_{ij},$$

while the size of M means $|M|$, the number of edges contained in M. We denote by $\mathcal{M}_k$ the family of all the matchings of size k in $G(A)$ for $k = 1, 2, \cdots$, and by $\mathcal{M}$ the family of all the matchings of any size (i.e., $\mathcal{M} = \cup_k \mathcal{M}_k$).

For example, the thick lines in Fig. 1.2(b) show a matching M of weight $w(M) = 1$ and of size $|M| = 10$, and $M' = (M \setminus \{(3,10), (10,5)\}) \cup \{(10,10)\}$ is a matching of weight $w(M') = 2$ and of size $|M'| = 9$.

Assuming that A_{str} is an $n \times n$ matrix, we consider the defining expansion of its determinant:

$$\det A_{\mathrm{str}} = \sum_{\pi \in \mathcal{S}_n} \mathrm{sgn}\,\pi \cdot \prod_{i=1}^{n} (A_{\mathrm{str}})_{i\pi(i)} = \sum_{\pi \in \mathcal{S}_n} \mathrm{sgn}\,\pi \cdot \prod_{i=1}^{n} t_{i\pi(i)} \cdot s^{\sum_{i=1}^{n} w_{i\pi(i)}},$$

where $\mathcal{S}_n$ denotes the set of all the permutations of order n, and $\mathrm{sgn}\,\pi = \pm 1$ is the signature of a permutation π. We observe the following facts:

1. Nonzero terms in this expansion correspond to matchings of size n in $G(A)$;
2. There is no cancellation among different nonzero terms in this expansion by virtue of the independence among t_{ij}'s.

These two facts imply the following:

1. The structured matrix A_{str} is nonsingular (i.e., $\det A_{\mathrm{str}} \neq 0$) if and only if there exists a matching of size n in $G(A)$;
2. In the case of a nonsingular A_{str}, it holds that

$$\deg_s \det A_{\mathrm{str}} = \max_{M_n \in \mathcal{M}_n} w(M_n). \tag{1.7}$$

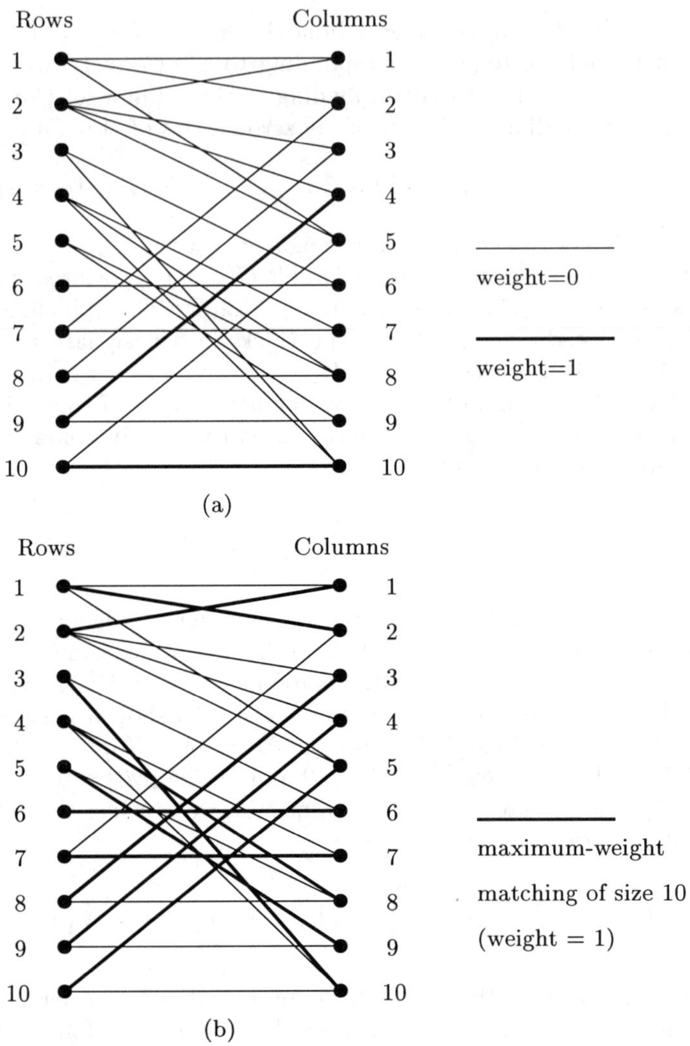

Fig. 1.2. Graph $G(A^{(1)})$ and the maximum-weight matching

A similar argument applied to submatrices of A_{str} leads to more general formulas:

$$\operatorname{rank} A_{\mathrm{str}} = \max_{M \in \mathcal{M}} |M|,$$

$$\max_{|I|=|J|=k} \deg_s \det A_{\mathrm{str}}[I,J] = \max_{M_k \in \mathcal{M}_k} w(M_k) \quad (k = 1, \cdots, r_{\mathrm{str}}), \qquad (1.8)$$

where $A_{\mathrm{str}}[I,J]$ means the submatrix of A_{str} having row set I and column set J, and $r_{\mathrm{str}} = \operatorname{rank} A_{\mathrm{str}}$. It should be clear that the left-hand side of (1.8) designates the maximum degree of a minor (subdeterminant) of order k.

A combination of the formulas (1.3) and (1.8) yields

$$\nu_{\mathrm{str}}(A) = \max_{M_{n-1} \in \mathcal{M}_{n-1}} w(M_{n-1}) - \max_{M_n \in \mathcal{M}_n} w(M_n) + 1 \tag{1.9}$$

for a nonsingular $n \times n$ polynomial matrix A. Thus we have arrived at a combinatorial expression of the structural index.

For the matrix $A^{(1)}$ we have (cf. Fig. 1.2)

$$\max_{M_{n-1}^{(1)} \in \mathcal{M}_{n-1}^{(1)}} w(M_{n-1}^{(1)}) = 2, \qquad \max_{M_n^{(1)} \in \mathcal{M}_n^{(1)}} w(M_n^{(1)}) = 1$$

and therefore $\nu_{\mathrm{str}}(A^{(1)}) = 2 - 1 + 1 = 2$, in agreement with $\nu(A^{(1)}) = 2$.

It is important from the computational point of view that efficient combinatorial algorithms are available for checking the existence of a matching of a specified size and also for finding a maximum-weight matching of a specified size. Thus the structural index ν_{str}, with the expression (1.9), can be computed efficiently by solving weighted bipartite matching problems utilizing those efficient combinatorial algorithms.

A number of graph-theoretic techniques (which may be considered variants of the above idea) have been proposed as "structural algorithms" (Bujakiewicz [26], Bujakiewicz–van den Bosch [27], Duff–Gear [60], Pantelides [264], Ungar–Kröner–Marquardt [324]). It is accepted that structural considerations should be useful and effective in practice for the DAE-index problem and that the generic values computed by graph-theoretic "structural algorithms" have practical significance.

1.1.3 An Embarrassing Phenomenon

While the structural approach is accepted fairly favorably, its limitation has also been realized in the literature. A graph-theoretic structural algorithm, ignoring numerical data, may well fail to render the correct answer if numerical cancellations do occur for some reason or other. So the failure of a graph-theoretic algorithm itself should not be a surprise. The aim of this section is to demonstrate a further embarrassing phenomenon that the structural index of our electrical network varies with how KVL is described.

Recall first that the 3rd row of the matrix $A^{(1)}$ represents the conservation of voltage along the loop 1–5 (V–C). In place of this we now take another loop 1–2–4 (V–R_1–L) to obtain a second description of the same electrical network. The coefficient matrix of the second description is given by

$$
A^{(2)} = \begin{array}{c}
\begin{array}{ccccc|ccccc}
\xi^1 & \xi^2 & \xi^3 & \xi^4 & \xi^5 & \eta_1 & \eta_2 & \eta_3 & \eta_4 & \eta_5
\end{array} \\
\left|\begin{array}{ccccc|ccccc}
1 & -1 & 0 & 0 & -1 & & & & & \\
-1 & 0 & 1 & 1 & 1 & & & & & \\ \hline
 & & & & & -1 & -1 & 0 & -1 & 0 \\
 & & & & & 0 & 1 & 1 & 0 & -1 \\
 & & & & & 0 & 0 & -1 & 1 & 0 \\ \hline
0 & 0 & 0 & 0 & 0 & -1 & 0 & 0 & 0 & 0 \\
0 & R_1 & 0 & 0 & 0 & 0 & -1 & 0 & 0 & 0 \\
0 & 0 & R_2 & 0 & 0 & 0 & 0 & -1 & 0 & 0 \\
0 & 0 & 0 & sL & 0 & 0 & 0 & 0 & -1 & 0 \\
0 & 0 & 0 & 0 & -1 & 0 & 0 & 0 & 0 & sC
\end{array}\right|
\end{array}, \qquad (1.10)
$$

which differs from $A^{(1)}$ in the 3rd row. The associated structured matrix $A^{(2)}_{\rm str}$ differs from $A^{(1)}_{\rm str}$ also in the 3rd row, and is given by

$$
A^{(2)}_{\rm str} = \begin{array}{c}
\begin{array}{ccccc|ccccc}
\xi^1 & \xi^2 & \xi^3 & \xi^4 & \xi^5 & \eta_1 & \eta_2 & \eta_3 & \eta_4 & \eta_5
\end{array} \\
\left|\begin{array}{ccccc|ccccc}
t_1 & t_2 & 0 & 0 & t_3 & & & & & \\
t_4 & 0 & t_5 & t_6 & t_7 & & & & & \\ \hline
 & & & & & t_{24} & t_{25} & 0 & t_{26} & 0 \\
 & & & & & 0 & t_{10} & t_{11} & 0 & t_{12} \\
 & & & & & 0 & 0 & t_{13} & t_{14} & 0 \\ \hline
0 & 0 & 0 & 0 & 0 & t_{15} & 0 & 0 & 0 & 0 \\
0 & t_{16} & 0 & 0 & 0 & 0 & t_{17} & 0 & 0 & 0 \\
0 & 0 & t_{18} & 0 & 0 & 0 & 0 & t_{19} & 0 & 0 \\
0 & 0 & 0 & s\,t_{20} & 0 & 0 & 0 & 0 & t_{21} & 0 \\
0 & 0 & 0 & 0 & t_{22} & 0 & 0 & 0 & 0 & s\,t_{23}
\end{array}\right|
\end{array},
$$

where $\{t_i \mid i = 1, \cdots, 7, 10, \cdots, 26\}$ is the set of independent parameters.

Naturally, the index should remain invariant against this trivial change in the description of KVL, and in fact we have

$$\nu(A^{(1)}) = \nu(A^{(2)}) = 2.$$

It turns out, however, that the structural index does change, namely,

$$\nu_{\rm str}(A^{(1)}) = 2, \qquad \nu_{\rm str}(A^{(2)}) = 1,$$

where the latter is computed from the graph $G(A^{(2)})$ in Fig. 1.3; we have

$$\max_{M^{(2)}_{n-1} \in \mathcal{M}^{(2)}_{n-1}} w(M^{(2)}_{n-1}) = 2, \qquad \max_{M^{(2)}_{n} \in \mathcal{M}^{(2)}_{n}} w(M^{(2)}_{n}) = 2$$

and therefore

$$\nu_{\rm str}(A^{(2)}) = \nu(A^{(2)}_{\rm str}) = 2 - 2 + 1 = 1$$

according to the expression (1.9).

The discrepancy between the structural index $\nu_{\rm str}(A^{(2)})$ and the true index $\nu(A^{(2)})$ is ascribed to the discrepancy between $\deg_s \det A^{(2)}_{\rm str} = 2$ and

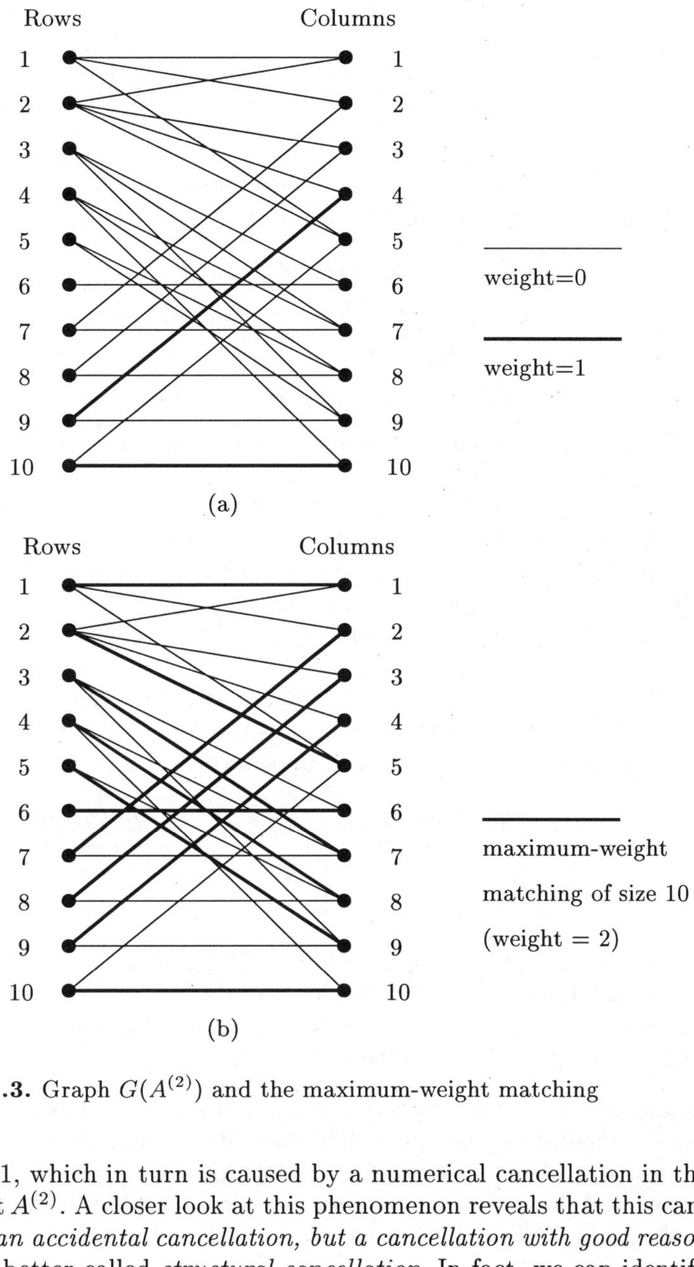

Fig. 1.3. Graph $G(A^{(2)})$ and the maximum-weight matching

$\deg_s \det A^{(2)} = 1$, which in turn is caused by a numerical cancellation in the expansion of $\det A^{(2)}$. A closer look at this phenomenon reveals that this cancellation is *not an accidental cancellation, but a cancellation with good reason* which could be better called *structural cancellation.* In fact, we can identify a 2×2 singular submatrix of the coefficient matrix for the KCL and a 3×3 singular submatrix of the coefficient matrix for the KVL:

$$
\begin{array}{cc} \xi^1 & \xi^5 \\ \hline 1 & -1 \\ -1 & 1 \end{array}, \qquad \begin{array}{ccc} \eta_2 & \eta_3 & \eta_4 \\ \hline -1 & 0 & -1 \\ 1 & 1 & 0 \\ 0 & -1 & 1 \end{array}
$$

as the reason for this cancellation. More specifically, the expansion of det $A^{(2)}_{\mathrm{str}}$ contains four "spurious" quadratic terms

$$\underline{t_1 \cdot t_7} \cdot \underline{t_{25} \cdot t_{11} \cdot t_{14}} \cdot t_{15} \cdot t_{16} \cdot t_{18} \cdot st_{20} \cdot st_{23}, \tag{1.11}$$

$$\underline{t_1 \cdot t_7} \cdot \underline{t_{26} \cdot t_{10} \cdot t_{13}} \cdot t_{15} \cdot t_{16} \cdot t_{18} \cdot st_{20} \cdot st_{23}, \tag{1.12}$$

$$\underline{t_3 \cdot t_4} \cdot \underline{t_{25} \cdot t_{11} \cdot t_{14}} \cdot t_{15} \cdot t_{16} \cdot t_{18} \cdot st_{20} \cdot st_{23}, \tag{1.13}$$

$$\underline{t_3 \cdot t_4} \cdot \underline{t_{26} \cdot t_{10} \cdot t_{13}} \cdot t_{15} \cdot t_{16} \cdot t_{18} \cdot st_{20} \cdot st_{23}, \tag{1.14}$$

which cancel one another when the numerical values as well as the system parameters are given to t_{ij}'s ($t_1 = t_7 = t_{10} = t_{11} = t_{14} = 1$, $t_3 = t_4 = t_{13} = t_{15} = t_{25} = t_{26} = -1$, $t_{16} = R_1$, $t_{18} = R_2$, $t_{20} = L$, $t_{23} = C$). In fact, det $A^{(2)}$, which is equal to det $A^{(1)} = R_1 R_2 + sL \cdot R_1 + sL \cdot R_2$ given in (1.4), does not contain those terms. Note that the term (1.11) corresponds to the matching in Fig. 1.3(b), and recall that the system parameters R_1, R_2, L, C are treated as mutually independent parameters, which cannot be cancelled out among themselves.

This example demonstrates that the structural index is not determined uniquely by a physical/engineering system, but it depends on its mathematical description. It is emphasized that both

$$
A^{(1)} : \begin{array}{ccccc} \eta_1 & \eta_2 & \eta_3 & \eta_4 & \eta_5 \\ \hline -1 & 0 & 0 & 0 & -1 \\ 0 & 1 & 1 & 0 & -1 \\ 0 & 0 & -1 & 1 & 0 \end{array} \quad \text{and} \quad A^{(2)} : \begin{array}{ccccc} \eta_1 & \eta_2 & \eta_3 & \eta_4 & \eta_5 \\ \hline -1 & -1 & 0 & -1 & 0 \\ 0 & 1 & 1 & 0 & -1 \\ 0 & 0 & -1 & 1 & 0 \end{array}
$$

are equally a legitimate description of KVL and there is nothing inherent to distinguish between the two. In this way the structural index is vulnerable to our innocent choice. This makes us reconsider the meaning of the structural index, which will be discussed in the next section.

Remark 1.1.4. The limitation of the graph-theoretic structural approach, as explained above, is now widely understood. Already Pantelides [264] recognized this phenomenon and more recently Ungar–Kröner–Marquardt [324] expounded this point with reference to an example problem arising from an analysis of distillation columns in chemical engineering. □

1.2 What Is Combinatorial Structure?

In view of the "embarrassing phenomenon" above we have to question the physical relevance of the structural index (1.6) and reconsider how we should recognize the combinatorial structure of physical systems. The objective of this section is to discuss this issue and to introduce an advanced framework of structural analysis that uses mixed (polynomial) matrices as the main mathematical tool. The framework realizes a reasonable balance between physical faith and mathematical convenience in mathematical modeling of physical/engineering systems. As for physical faith, it is based on two different observations; the one is the distinction between "accurate" numbers (fixed constants) and "inaccurate" numbers (independent system parameters), and the other is the consistency with respect to physical dimensions. As for mathematical convenience, the analysis of mixed (polynomial) matrices and the design of efficient algorithms for them can be done successfully by means of matroid theory. Hence the name of "matroid-theoretic approach" for the advanced framework based on mixed matrices, as opposed to the conventional graph-theoretic approach to structural analysis.

1.2.1 Two Kinds of Numbers

Let us continue with our electrical network. The matrix $A^{(2)}$ of (1.10) can be written as

$$A^{(2)}(s) = A_0^{(2)} + sA_1^{(2)}$$

with

$$A_0^{(2)} = \begin{array}{|ccccc|ccccc|}\hline 1 & -1 & 0 & 0 & -1 & & & & & \\ -1 & 0 & 1 & 1 & 1 & & & & & \\ \hline & & & & & -1 & -1 & 0 & -1 & 0 \\ & & & & & 0 & 1 & 1 & 0 & -1 \\ & & & & & 0 & 0 & -1 & 1 & 0 \\ \hline 0 & 0 & 0 & 0 & 0 & -1 & 0 & 0 & 0 & 0 \\ 0 & R_1 & 0 & 0 & 0 & 0 & -1 & 0 & 0 & 0 \\ 0 & 0 & R_2 & 0 & 0 & 0 & 0 & -1 & 0 & 0 \\ 0 & 0 & 0 & 0 & 0 & 0 & 0 & 0 & -1 & 0 \\ 0 & 0 & 0 & 0 & -1 & 0 & 0 & 0 & 0 & 0 \\ \hline \end{array}\,, \quad A_1^{(2)} = \begin{array}{|ccccc|ccccc|}\hline 0 & 0 & 0 & 0 & 0 & & & & & \\ 0 & 0 & 0 & 0 & 0 & & & & & \\ \hline & & & & & 0 & 0 & 0 & 0 & 0 \\ & & & & & 0 & 0 & 0 & 0 & 0 \\ & & & & & 0 & 0 & 0 & 0 & 0 \\ \hline 0 & 0 & 0 & 0 & 0 & 0 & 0 & 0 & 0 & 0 \\ 0 & 0 & 0 & 0 & 0 & 0 & 0 & 0 & 0 & 0 \\ 0 & 0 & 0 & 0 & 0 & 0 & 0 & 0 & 0 & 0 \\ 0 & 0 & 0 & L & 0 & 0 & 0 & 0 & 0 & 0 \\ 0 & 0 & 0 & 0 & 0 & 0 & 0 & 0 & 0 & C \\ \hline \end{array}\,. \tag{1.15}$$

We observe here that the nonzero entries of the coefficient matrices $A_k^{(2)}$ $(k = 0, 1)$ are classified into two groups: one group of fixed constants (± 1) and the other group of system parameters R_1, R_2, L and C. Accordingly, we can split $A_k^{(2)}$ $(k = 0, 1)$ into two parts:

$$A_k^{(2)} = Q_k^{(2)} + T_k^{(2)} \qquad (k = 0, 1)$$

with

$$Q_0^{(2)} = \left[\begin{array}{ccccc|ccccc}
1 & -1 & 0 & 0 & -1 & & & & & \\
-1 & 0 & 1 & 1 & 1 & & & & & \\
\hline
 & & & & & -1 & -1 & 0 & -1 & 0 \\
 & & & & & 0 & 1 & 1 & 0 & -1 \\
 & & & & & 0 & 0 & -1 & 1 & 0 \\
\hline
0 & 0 & 0 & 0 & 0 & -1 & 0 & 0 & 0 & 0 \\
0 & 0 & 0 & 0 & 0 & 0 & -1 & 0 & 0 & 0 \\
0 & 0 & 0 & 0 & 0 & 0 & 0 & -1 & 0 & 0 \\
0 & 0 & 0 & 0 & 0 & 0 & 0 & 0 & -1 & 0 \\
0 & 0 & 0 & 0 & -1 & 0 & 0 & 0 & 0 & 0
\end{array}\right], \quad
T_0^{(2)} = \left[\begin{array}{ccccc|ccccc}
0 & 0 & 0 & 0 & 0 & & & & & \\
0 & 0 & 0 & 0 & 0 & & & & & \\
\hline
 & & & & & 0 & 0 & 0 & 0 & 0 \\
 & & & & & 0 & 0 & 0 & 0 & 0 \\
 & & & & & 0 & 0 & 0 & 0 & 0 \\
\hline
0 & 0 & 0 & 0 & 0 & 0 & 0 & 0 & 0 & 0 \\
0 & R_1 & 0 & 0 & 0 & 0 & 0 & 0 & 0 & 0 \\
0 & 0 & R_2 & 0 & 0 & 0 & 0 & 0 & 0 & 0 \\
0 & 0 & 0 & 0 & 0 & 0 & 0 & 0 & 0 & 0 \\
0 & 0 & 0 & 0 & 0 & 0 & 0 & 0 & 0 & 0
\end{array}\right],$$

$$Q_1^{(2)} = \left[\begin{array}{ccccc|ccccc}
0 & 0 & 0 & 0 & 0 & & & & & \\
0 & 0 & 0 & 0 & 0 & & & & & \\
\hline
 & & & & & 0 & 0 & 0 & 0 & 0 \\
 & & & & & 0 & 0 & 0 & 0 & 0 \\
 & & & & & 0 & 0 & 0 & 0 & 0 \\
\hline
0 & 0 & 0 & 0 & 0 & 0 & 0 & 0 & 0 & 0 \\
0 & 0 & 0 & 0 & 0 & 0 & 0 & 0 & 0 & 0 \\
0 & 0 & 0 & 0 & 0 & 0 & 0 & 0 & 0 & 0 \\
0 & 0 & 0 & 0 & 0 & 0 & 0 & 0 & 0 & 0 \\
0 & 0 & 0 & 0 & 0 & 0 & 0 & 0 & 0 & 0
\end{array}\right], \quad
T_1^{(2)} = \left[\begin{array}{ccccc|ccccc}
0 & 0 & 0 & 0 & 0 & & & & & \\
0 & 0 & 0 & 0 & 0 & & & & & \\
\hline
 & & & & & 0 & 0 & 0 & 0 & 0 \\
 & & & & & 0 & 0 & 0 & 0 & 0 \\
 & & & & & 0 & 0 & 0 & 0 & 0 \\
\hline
0 & 0 & 0 & 0 & 0 & 0 & 0 & 0 & 0 & 0 \\
0 & 0 & 0 & 0 & 0 & 0 & 0 & 0 & 0 & 0 \\
0 & 0 & 0 & 0 & 0 & 0 & 0 & 0 & 0 & 0 \\
0 & 0 & 0 & L & 0 & 0 & 0 & 0 & 0 & 0 \\
0 & 0 & 0 & 0 & 0 & 0 & 0 & 0 & 0 & C
\end{array}\right].$$

It is assumed that the system parameters, R_1, R_2, L, C, are independent parameters. Even when concrete numbers are given to R_1, R_2, L, C, those numbers are not expected to be exactly equal to their nominal values, but they lie in certain intervals of real numbers of engineering tolerance. Even in the extreme case where both R_1 and R_2 are specified to be 1Ω, for example, their actual values will be something like $R_1 = 1.02\Omega$ and $R_2 = 0.99\Omega$.

Generally, when a physical system is described by a polynomial matrix

$$A(s) = \sum_{k=0}^{N} s^k A_k, \tag{1.16}$$

it is often justified (see §1.2.2) to assume that the nonzero entries of the coefficient matrices A_k $(k = 0, 1, \cdots, N)$ are classified similarly into two groups. In other words, we can distinguish the following *two kinds of numbers*, together characterizing a physical system. We may refer to the numbers of the first kind as "fixed constants" and to those of the second kind as "system parameters."

Accurate numbers (fixed constants): Numbers accounting for various sorts of conservation laws such as Kirchhoff's laws which, stemming from topological incidence relations, are precise in value (often ± 1), and therefore cause no serious numerical difficulty in arithmetic operations on them.

Inaccurate numbers (system parameters): Numbers representing independent system parameters such as resistances in electrical networks and masses in mechanical systems which, being contaminated with noise and other errors, take values independent of one another, and therefore can be modeled as algebraically independent numbers.[2]

Accurate numbers often appear in equations for conservation laws such as Kirchhoff's laws, the law of conservation of mass, energy, or momentum, and the principle of action and reaction, where the nonvanishing coefficients are either 1 or -1, representing the underlying topological incidence relations. Integer coefficients in chemical reactions (*stoichiometric coefficients*), such as "2" and "1" in $2 \cdot H_2O = 2 \cdot H_2 + 1 \cdot O_2$, are also accurate numbers. Another example of accurate numbers appears in the defining relation $\mathrm{d}x/\mathrm{d}t = 1 \cdot v$ between velocity v and position x. Typical accurate numbers are illustrated in Fig. 1.4.

The above observation leads to the assumption that the coefficient matrices A_k $(k = 0, 1, \cdots, N)$ in (1.16) are expressed as

$$A_k = Q_k + T_k \qquad (k = 0, 1, \cdots, N), \tag{1.17}$$

where

(A-Q1): Q_k $(k = 0, 1, \cdots, N)$ are matrices over $\mathbf{Q}$ (the field of rational numbers), and

(A-T): The collection $\mathcal{T}$ of nonzero entries of T_k $(k = 0, 1, \cdots, N)$ is algebraically independent over $\mathbf{Q}$.

Namely, each A_k may be assumed to be a *mixed matrix*, in the terminology to be introduced formally in §1.3. Then $A(s)$ is split accordingly into two parts:

$$A(s) = Q(s) + T(s) \tag{1.18}$$

with

$$Q(s) = \sum_{k=0}^{N} s^k Q_k, \qquad T(s) = \sum_{k=0}^{N} s^k T_k. \tag{1.19}$$

Namely, $A(s)$ is a *mixed polynomial matrix* in the terminology of §1.3.

Our intention in the splitting (1.17) or (1.18) is to extract a more meaningful combinatorial structure from the matrix $A(s)$ by treating the Q-part numerically and the T-part symbolically. This is based on the following observations.

Q-part: The nonzero pattern of the Q-matrices is subject to our arbitrary choice in the mathematical description, as we have seen in our electrical network, and hence the structure of the Q-part should be treated numerically, or linear-algebraically. In fact, this is feasible in practice, since the entries of the Q-matrices are usually small integers, causing no serious numerical difficulty in arithmetic operations.

---

[2] Informally, "algebraically independent numbers" are tantamount to "independent parameters," whereas a rigorous definition of algebraic independence will be given in §2.1.1.

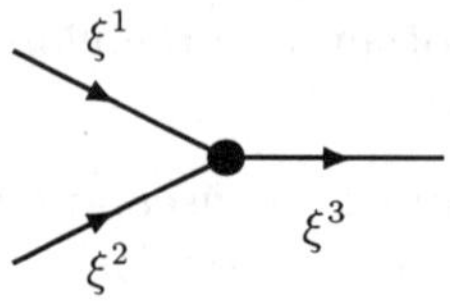

$$-1 \cdot \xi^1 - 1 \cdot \xi^2 + 1 \cdot \xi^3 = 0$$

KCL

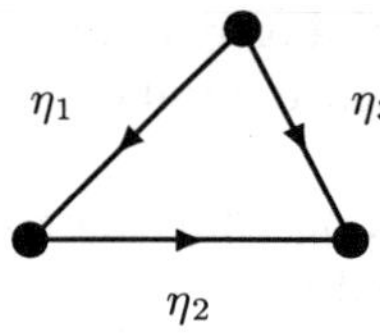

$$-1 \cdot \eta_1 - 1 \cdot \eta_2 + 1 \cdot \eta_3 = 0$$

KVL

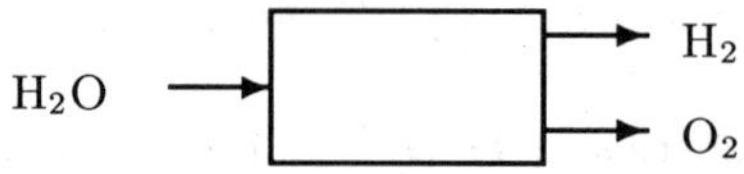

$$2 \cdot H_2O = 2 \cdot H_2 + 1 \cdot O_2$$

Stoichiometry

Velocity v – displacement x $\qquad$ $v = 1 \cdot \dot{x} \ (= s \cdot x)$

Current ξ – charge Q $\qquad$ $\xi = 1 \cdot \dot{Q} \ (= s \cdot Q)$

Fig. 1.4. Accurate numbers

T-part: The nonzero pattern of the T-matrices is relatively stable against our arbitrary choice in the mathematical description of constitutive equations and therefore it can be regarded as representing some aspect of the combinatorial structure of the system. It can be treated properly by graph-theoretic concepts and algorithms.

Combination: The structural information from the Q-part and the T-part can be combined properly and efficiently by virtue of the fact that each part defines a well-behaved and well-studied combinatorial structure called matroid. Mathematical and algorithmic results from matroid theory afford effective methods of system analysis.

We may summarize the above as follows:

Q-part	by	linear algebra
T-part	by	graph theory
Combination	by	matroid theory

In §1.3 we shall take a glimpse at how the DAE-index problem can be treated using mixed polynomial matrices and how the embarrassing phenomenon of §1.1.3 can be resolved properly.

1.2.2 Descriptor Form Rather than Standard Form

In introducing mixed polynomial matrices we have assumed that the nonzero entries of the coefficient matrices are either fixed constants or independent parameters. This is an assumption on a description of a physical system, and not an assumption on the system itself. For a system in question there can be many different descriptions, but some of them may satisfy the assumption and others may fail to meet it. In this section we discuss this issue by comparing the state-space equations (Kalman [153]) and the descriptor equations (Luenberger [182, 183]).

Let us consider another example, a simple mechanical system (Fig. 1.5) which consists of two masses m_1, m_2, two springs k_1, k_2, and a damper f; u is the force exerted from outside.

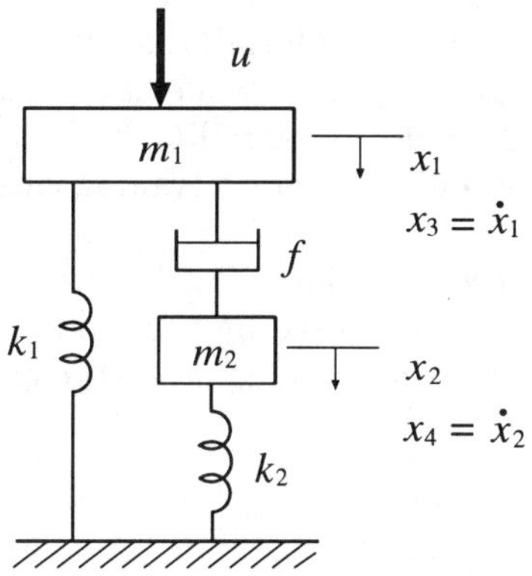

Fig. 1.5. A mechanical system

We may describe the system in the form of *state-space equations*:

$$\dot{\boldsymbol{x}}(t) = \hat{A}\boldsymbol{x}(t) + \hat{B}\boldsymbol{u}(t) \tag{1.20}$$

in terms of $\boldsymbol{x} = (x_1, x_2, x_3, x_4)$ and $\boldsymbol{u} = (u)$, where x_1 and x_2 are vertical displacements (downwards, as indicated in Fig. 1.5) of masses m_1 and m_2, respectively, and x_3 and x_4 are their velocities, and

$$\hat{A} = \begin{array}{c} \begin{array}{cccc} x_1 & x_2 & x_3 & x_4 \end{array} \\ \left[\begin{array}{cccc} 0 & 0 & 1 & 0 \\ 0 & 0 & 0 & 1 \\ -k_1/m_1 & 0 & -f/m_1 & f/m_1 \\ 0 & -k_2/m_2 & f/m_2 & -f/m_2 \end{array}\right] \end{array}, \quad \hat{B} = \begin{array}{c} u \\ \left[\begin{array}{c} 0 \\ 0 \\ 1/m_1 \\ 0 \end{array}\right] \end{array}. \tag{1.21}$$

It should be clear that $\dot{\boldsymbol{x}}$ is a short-hand notation for $\mathrm{d}\boldsymbol{x}/\mathrm{d}t$, the time derivative of $\boldsymbol{x}$.

The state-space equations (1.20) have been useful for investigating analytic and algebraic properties of a dynamical system, and the structural or combinatorial analysis at the early stage[3] was based on it. It is gradually recognized, however, that the state-space equations are not very suitable for representing the combinatorial structure of a system in that the entries of matrices $\hat{A}$ and $\hat{B}$ of (1.20) are usually not independent but interrelated to one another, being subject to algebraic relations. For instance, we have $\hat{A}_{33} + \hat{A}_{34} = 0$ in (1.21), and consequently $\hat{A}$ of (1.21) does not admit a splitting into Q-part and T-part satisfying (A-Q1) and (A-T).

In this respect, the so-called *descriptor form*

$$\bar{F}\dot{\boldsymbol{x}}(t) = \bar{A}\boldsymbol{x}(t) + \bar{B}\boldsymbol{u}(t) \tag{1.22}$$

is more promising, having more flexibility to avoid complicated algebraic relations among entries of the coefficient matrices. Here $\boldsymbol{x}$ is called the descriptor-vector and $\boldsymbol{u}$ is the input-vector. The matrix $\bar{F}$ is not necessarily nonsingular, so that the reduction of (1.22) to the standard state-space form (1.20) is not straightforward. Even when $\bar{F}$ is nonsingular, the reduction to the standard state-space form (1.20) with $\hat{A} = \bar{F}^{-1}\bar{A}$ and $\hat{B} = \bar{F}^{-1}\bar{B}$ entailing complicated algebraic relations among the entries of $\hat{A}$ and $\hat{B}$, is not advantageous from the combinatorial point of view.

To describe our mechanical system in the descriptor form (1.22), it may be natural to introduce two additional variables x_5 (= force by the damper f) and x_6 (= relative velocity of the two masses). Additional equations (constraints) for these variables are given by[4]

$$x_5 = f x_6, \qquad x_6 = \dot{x}_1 - \dot{x}_2.$$

Then the coefficient matrices in (1.22) are given by

$$\bar{F} = \left[\begin{array}{cccccc} 1 & 0 & 0 & 0 & 0 & 0 \\ 0 & 1 & 0 & 0 & 0 & 0 \\ 0 & 0 & m_1 & 0 & 0 & 0 \\ 0 & 0 & 0 & m_2 & 0 & 0 \\ 0 & 0 & 0 & 0 & 0 & 0 \\ 1 & -1 & 0 & 0 & 0 & 0 \end{array}\right], \quad \bar{A} = \left[\begin{array}{cccccc} 0 & 0 & 1 & 0 & 0 & 0 \\ 0 & 0 & 0 & 1 & 0 & 0 \\ -k_1 & 0 & 0 & 0 & -1 & 0 \\ 0 & -k_2 & 0 & 0 & 1 & 0 \\ 0 & 0 & 0 & 0 & -1 & f \\ 0 & 0 & 0 & 0 & 0 & 1 \end{array}\right], \quad \bar{B} = \left[\begin{array}{c} 0 \\ 0 \\ 1 \\ 0 \\ 0 \\ 0 \end{array}\right]. \tag{1.23}$$

---

[3] Structural approach in the literature of control theory was initiated by Lin [173] in the mid-seventies.

[4] We could replace the equation $x_6 = \dot{x}_1 - \dot{x}_2$ by $x_6 = x_3 - x_4$, which may be more natural. Our choice is to make the example less trivial.

The Laplace transform of the equation (1.22) gives a *frequency domain description*:

$$s\bar{F}\hat{\boldsymbol{x}}(s) = \bar{A}\hat{\boldsymbol{x}}(s) + \bar{B}\hat{\boldsymbol{u}}(s), \qquad \text{or} \qquad \left[\bar{A} - s\bar{F} | \bar{B}\right] \begin{bmatrix} \hat{\boldsymbol{x}}(s) \\ \hat{\boldsymbol{u}}(s) \end{bmatrix} = \mathbf{0},$$

where $\boldsymbol{x}(0) = \mathbf{0}$, $\boldsymbol{u}(0) = \mathbf{0}$ is assumed (see Remark 1.1.1 for the Laplace transform). Then the system is described by a polynomial matrix

$$A(s) = \left[\bar{A} - s\bar{F} | \bar{B}\right]. \tag{1.24}$$

For our mechanical system we have

$$A(s) = \begin{array}{c} \begin{array}{ccccccc} x_1 & x_2 & x_3 & x_4 & x_5 & x_6 & u \end{array} \\ \left[\begin{array}{cccccc|c} -s & 0 & 1 & 0 & 0 & 0 & 0 \\ 0 & -s & 0 & 1 & 0 & 0 & 0 \\ -k_1 & 0 & -sm_1 & 0 & -1 & 0 & 1 \\ 0 & -k_2 & 0 & -sm_2 & 1 & 0 & 0 \\ 0 & 0 & 0 & 0 & -1 & f & 0 \\ -s & s & 0 & 0 & 0 & 1 & 0 \end{array}\right] \end{array} \tag{1.25}$$

as the matrix of (1.24). Note that no complicated algebraic expressions are involved in this matrix, for which it is reasonable to assume (A-Q1) and (A-T) above. Consequently, $A(s)$ of (1.25) is expressed as $A(s) = Q(s) + T(s)$ with

$$Q(s) = \begin{array}{c} \begin{array}{ccccccc} x_1 & x_2 & x_3 & x_4 & x_5 & x_6 & u \end{array} \\ \left[\begin{array}{cccccc|c} -s & 0 & 1 & 0 & 0 & 0 & 0 \\ 0 & -s & 0 & 1 & 0 & 0 & 0 \\ 0 & 0 & 0 & 0 & -1 & 0 & 1 \\ 0 & 0 & 0 & 0 & 1 & 0 & 0 \\ 0 & 0 & 0 & 0 & -1 & 0 & 0 \\ -s & s & 0 & 0 & 0 & 1 & 0 \end{array}\right] \end{array}, \quad T(s) = \begin{array}{c} \begin{array}{ccccccc} x_1 & x_2 & x_3 & x_4 & x_5 & x_6 & u \end{array} \\ \left[\begin{array}{cccccc|c} 0 & 0 & 0 & 0 & 0 & 0 & 0 \\ 0 & 0 & 0 & 0 & 0 & 0 & 0 \\ -k_1 & 0 & -sm_1 & 0 & 0 & 0 & 0 \\ 0 & -k_2 & 0 & -sm_2 & 0 & 0 & 0 \\ 0 & 0 & 0 & 0 & 0 & f & 0 \\ 0 & 0 & 0 & 0 & 0 & 0 & 0 \end{array}\right] \end{array}. \tag{1.26}$$

Here we have $\mathcal{T} = \{m_1, m_2, k_1, k_2, f\}$ as the set of system parameters.

It is emphasized again that the coefficient matrices $\hat{A}$ and $\hat{B}$ in the standard state-space form do not admit such natural splitting into two parts. Thus we may conclude that the descriptor form is more suitable for representing the combinatorial structure than the standard state-space form.

1.2.3 Dimensional Analysis

Here is a kind of *dimensional analysis* concerning "accurate numbers," i.e., concerning the constant part $Q(s) = \sum_{k=0}^{N} s^k Q_k$ of the matrix $A(s)$ in (1.18).

First we consider the physical dimensional consistency in the system of equations $A(s)\boldsymbol{x} = \boldsymbol{b}$, where $A(s)$ is assumed to be an $m \times n$ matrix. Since this system is to represent a physical system, relevant physical dimensions are associated with both the variables (corresponding to the components of $\boldsymbol{x}$) and the equations (corresponding to the components of $\boldsymbol{b}$), or alternatively, with both columns and rows of the matrix $A(s)$. Also the entries of $A(s)$ have physical dimensions.

In our mechanical system, for instance, we may choose time T, length L and mass M as the *fundamental quantities* in the dimensional analysis. Then the dimensions of velocity and force are given by $T^{-1}L$ and $T^{-2}LM$, respectively. The physical dimensions associated with the equations, i.e., with the rows of $A(s)$ of (1.25), are

row 1	row 2	row 3	row 4	row 5	row 6
velocity	velocity	force	force	force	velocity
$T^{-1}L$	$T^{-1}L$	$T^{-2}LM$	$T^{-2}LM$	$T^{-2}LM$	$T^{-1}L$

(1.27)

whereas those with the variables (x_i and u), i.e., with the columns of $A(s)$, are

col 1	col 2	col 3	col 4	col 5	col 6	col 7
length	length	velocity	velocity	force	velocity	force
L	L	$T^{-1}L$	$T^{-1}L$	$T^{-2}LM$	$T^{-1}L$	$T^{-2}LM$

(1.28)

The $(3,1)$-entry "$-k_1$" of $A(s)$, for example, has a dimension of $T^{-2}M$.

The *principle of dimensional homogeneity* demands that

$$\begin{aligned}&[\text{Dimension of } i\text{th row}]\\ &= [\text{Dimension of } (i,j) \text{ entry}] \times [\text{Dimension of } j\text{th column}]\end{aligned} \tag{1.29}$$

for each (i,j) with $A_{ij} \neq 0$. For instance, this identity reads

$$T^{-2}LM = T^{-2}M \times L$$

for $(i,j) = (3,1)$ in our mechanical system.

Choosing time as one of the fundamental dimensions, we denote by $-r_i$ and $-c_j$ the exponent to the dimension of time associated respectively with the ith row and the jth column. Then the (i,j) entry of $A(s)$ should have the dimension of time with exponent $c_j - r_i$.

In our mechanical system we have

$$\begin{aligned}r_1 &= r_2 = 1, \ r_3 = r_4 = r_5 = 2, \ r_6 = 1;\\ c_1 &= c_2 = 0, \ c_3 = c_4 = 1, \ c_5 = 2, \ c_6 = 1, \ c_7 = 2\end{aligned}$$

from (1.27) and (1.28).

The "accurate numbers" usually represent topological and/or geometrical incidence coefficients (cf. Fig. 1.4), which have no physical dimensions, so that it is natural to expect that the entries of Q_k in (1.19) are dimensionless constants. On the other hand, the variable (indeterminate) "s" should have the physical dimension of the inverse of time, since it corresponds to $\mathrm{d}/\mathrm{d}t$, the differentiation with respect to time. This implies, in particular, that each entry of the term $s^k Q_k$ has the physical dimension of time with exponent $-k$. On the other hand, the (i,j) entry of $A(s)$, and hence the (i,j) entry of $Q(s)$, should have the dimension of time with exponent $c_j - r_i$, as pointed out above.

Combining these two facts we obtain

$$r_i - c_j = k \qquad \text{if} \quad (Q_k)_{ij} \neq 0, \tag{1.30}$$

or in matrix form:

$$Q(s) = \mathrm{diag}\,[s^{r_1}, \cdots, s^{r_m}] \cdot Q(1) \cdot \mathrm{diag}\,[s^{-c_1}, \cdots, s^{-c_n}], \tag{1.31}$$

where $\mathrm{diag}\,[d_1, d_2, \cdots]$ means a diagonal matrix having diagonal entries $d_1, d_2, \cdots$. It follows from this decomposition that every nonvanishing subdeterminant of $Q(s)$ is a monomial in s over $\mathbf{Q}$, i.e., of the form αs^p with a nonvanishing rational number α and a nonnegative integer p.

In our mechanical system, it can be verified that $Q(s)$ of (1.26) admits an expression of the form (1.31):

$$\left[\begin{array}{cccccc|c} -s & 0 & 1 & 0 & 0 & 0 & 0 \\ 0 & -s & 0 & 1 & 0 & 0 & 0 \\ 0 & 0 & 0 & 0 & -1 & 0 & 1 \\ 0 & 0 & 0 & 0 & 1 & 0 & 0 \\ 0 & 0 & 0 & 0 & -1 & 0 & 0 \\ -s & s & 0 & 0 & 0 & 1 & 0 \end{array}\right]$$

$$= \left[\begin{array}{cccccc} s & 0 & 0 & 0 & 0 & 0 \\ 0 & s & 0 & 0 & 0 & 0 \\ 0 & 0 & s^2 & 0 & 0 & 0 \\ 0 & 0 & 0 & s^2 & 0 & 0 \\ 0 & 0 & 0 & 0 & s^2 & 0 \\ 0 & 0 & 0 & 0 & 0 & s \end{array}\right] \cdot \left[\begin{array}{cccccc|c} -1 & 0 & 1 & 0 & 0 & 0 & 0 \\ 0 & -1 & 0 & 1 & 0 & 0 & 0 \\ 0 & 0 & 0 & 0 & -1 & 0 & 1 \\ 0 & 0 & 0 & 0 & 1 & 0 & 0 \\ 0 & 0 & 0 & 0 & -1 & 0 & 0 \\ -1 & 1 & 0 & 0 & 0 & 1 & 0 \end{array}\right] \cdot \left[\begin{array}{cccccc|c} 1 & 0 & 0 & 0 & 0 & 0 & 0 \\ 0 & 1 & 0 & 0 & 0 & 0 & 0 \\ 0 & 0 & s^{-1} & 0 & 0 & 0 & 0 \\ 0 & 0 & 0 & s^{-1} & 0 & 0 & 0 \\ 0 & 0 & 0 & 0 & s^{-2} & 0 & 0 \\ 0 & 0 & 0 & 0 & 0 & s^{-1} & 0 \\ \hline 0 & 0 & 0 & 0 & 0 & 0 & s^{-2} \end{array}\right].$$

Note that the diagonal entries s^{r_i} and s^{-c_j} are determined from the negative of the exponents to T (time) in (1.27) and (1.28).

We have thus arrived at a subclass of mixed polynomial matrices suitable for representing the structure of linear time-invariant dynamical systems. Namely, we are to consider the class of polynomial matrices $A(s)$ in indeterminate s with rational coefficients which are represented as

$$A(s) = Q(s) + T(s),$$

where

(A-Q2): Every nonvanishing subdeterminant of $Q(s)$ is a monomial in s over $\mathbf{Q}$, and

(A-T): The collection $\mathcal{T}$ of the nonzero coefficients of the entries of $T(s)$ is algebraically independent over $\mathbf{Q}$.

The dual viewpoint of dimensional analysis and structural analysis constitutes the physical foundation of the mathematical development explained in this book. Chapter 3 will be devoted to a full discussion about this issue.

1.3 Mathematics on Mixed Polynomial Matrices

While the previous section is devoted to physical motivations for mixed polynomial matrices, this section offers an informal introduction to their mathematical aspects through a successful treatment of the DAE-index problem left unanswered in §1.1.3.

1.3.1 Formal Definitions

The concept of a mixed matrix is defined formally as follows. Let $\boldsymbol{K}$ be a subfield of a field $\boldsymbol{F}$. A matrix $A = (A_{ij})$ over $\boldsymbol{F}$ (i.e., $A_{ij} \in \boldsymbol{F}$) is called a *mixed matrix* with respect to $(\boldsymbol{K}, \boldsymbol{F})$ if

$$A = Q + T, \tag{1.32}$$

where

(M-Q) $Q = (Q_{ij})$ is a matrix over $\boldsymbol{K}$ (i.e., $Q_{ij} \in \boldsymbol{K}$), and
(M-T) $T = (T_{ij})$ is a matrix over $\boldsymbol{F}$ (i.e., $T_{ij} \in \boldsymbol{F}$) such that the set of its nonzero entries is algebraically independent over $\boldsymbol{K}$.

For example, $A_0^{(2)}$ in (1.15) is a mixed matrix with respect to $(\boldsymbol{K}, \boldsymbol{F}) = (\mathbf{Q}, \mathbf{Q}(\mathcal{T}))$, where $\mathcal{T} = \{R_1, R_2\}$ and $\mathbf{Q}(\mathcal{T})$ is the field of rational functions in $\mathcal{T}$ with rational coefficients.

Similarly, a polynomial matrix $A(s)$ over $\boldsymbol{F}$ (i.e., $A_{ij} \in \boldsymbol{F}[s]$) is called a *mixed polynomial matrix* with respect to $(\boldsymbol{K}, \boldsymbol{F})$ if

$$A(s) = Q(s) + T(s) = \sum_{k=0}^{N} s^k Q_k + \sum_{k=0}^{N} s^k T_k \tag{1.33}$$

for some integer $N \geq 0$, where

(MP-Q1) Q_k $(k = 0, 1, \cdots, N)$ are matrices over $\boldsymbol{K}$, and
(MP-T) T_k $(k = 0, 1, \cdots, N)$ are matrices over $\boldsymbol{F}$ such that the set of their nonzero entries is algebraically independent over $\boldsymbol{K}$.

A mixed polynomial matrix with respect to $(\boldsymbol{K}, \boldsymbol{F})$ is a mixed matrix with respect to $(\boldsymbol{K}(s), \boldsymbol{F}(s))$. It should be obvious that (MP-Q1) and (MP-T) generalize (A-Q1) and (A-T), respectively. Corresponding to (A-Q2) we consider

(MP-Q2) Every nonvanishing subdeterminant of $Q(s)$ is a monomial in s over $\boldsymbol{K}$.

The major part of this book is devoted to the development of a theory of mixed (polynomial) matrices. Mixed matrices can be treated successfully by means of the standard results in the theory of matroids and submodular functions. For mixed polynomial matrices, on the other hand, a quantitative generalization of matroid theory, the theory of valuated matroids, is needed.

Matroid	(Chap. 2)	Mixed matrix	(Chap. 4)
Valuated matroid	(Chap. 5)	Mixed polynomial matrix	(Chap. 6)

Remark 1.3.1. The concept of a mixed matrix is useful also in solving a system of linear/nonlinear equations $\boldsymbol{f}(\boldsymbol{x}) = \boldsymbol{0}$. Suppose we solve this equation numerically by the *Newton method.* This amounts to solving

$$J(\boldsymbol{x})\Delta\boldsymbol{x} = -\boldsymbol{f}(\boldsymbol{x})$$

for a correction $\Delta\boldsymbol{x}$, where $J(\boldsymbol{x})$ is the Jacobian matrix of $\boldsymbol{f}(\boldsymbol{x})$. The equations may be divided into linear and nonlinear parts as

$$\boldsymbol{f}(\boldsymbol{x}) = Q\boldsymbol{x} + \boldsymbol{g}(\boldsymbol{x}),$$

where Q is a constant matrix representing the linear part and $\boldsymbol{g}(\boldsymbol{x})$ is the nonlinear part. Accordingly, the Jacobian matrix $J(\boldsymbol{x})$ takes the form of

$$J(\boldsymbol{x}) = Q + T(\boldsymbol{x}),$$

where $T(\boldsymbol{x})$ is the Jacobian matrix of the nonlinear part $\boldsymbol{g}(\boldsymbol{x})$. This expression suggests that we may treat $J(\boldsymbol{x}) = Q + T(\boldsymbol{x})$ as a mixed matrix by regarding (or modeling) the nonvanishing entries of $T(\boldsymbol{x})$ as being algebraically independent over $\boldsymbol{K} = \mathbf{R}$, where $\boldsymbol{F}$ is a certain field consisting of functions (e.g., the field of rational functions). This direction will be pursued further in §4.4.5. □

1.3.2 Resolution of the Index Problem

In this section we describe how the DAE-index problem can be treated successfully by means of mixed polynomial matrices. In so doing we intend to convey a general idea of the mathematical issues around mixed polynomial matrices without entering into technical details. Precise definitions and statements are given in subsequent chapters (§6.2 in particular).

For the DAE-index problem, we consider the degree of the determinant of an $n \times n$ mixed polynomial matrix $A(s) = Q(s) + T(s)$. The row set and the column set of $A(s)$ are denoted by R and C, respectively, and the submatrices of Q and T with row set I and column set J are written as $Q[I, J]$ and $T[I, J]$.

The following identity (cf. Theorem 6.2.4) shows how to combine the structural information from Q-part and T-part to obtain the information about A.

Theorem 1.3.2. *For a nonsingular mixed polynomial matrix* $A(s) = Q(s) + T(s)$,

$$\deg_s \det A = \max_{\substack{|I|=|J| \\ I\subseteq R, J\subseteq C}} \{\deg_s \det Q[I,J] + \deg_s \det T[R\setminus I, C\setminus J]\}. \quad (1.34)$$

(We adopt the convention that $\deg_s 0 = -\infty$, *so that the maximum in* (1.34) *is taken effectively over all* (I, J) *such that both* $Q[I, J]$ *and* $T[R\setminus I, C\setminus J]$ *are nonsingular.)* □

It is mentioned that the assumptions (MP-Q1) and (MP-T) are crucial for this formula, whereas generally the right-hand side of (1.34) is only an upper bound on $\deg_s \det A$.

The right-hand side of the identity (1.34) involves a maximization over all pairs (I, J), the number of which is almost as large as $2^{|R|+|C|}$, too large for an exhaustive search for maximization. Fortunately, however, it is possible to design an efficient algorithm to compute this maximum on the basis of the facts that each of the functions

$$f_Q(I,J) = \deg_s \det Q[I,J], \qquad f_T(I,J) = \deg_s \det T[I,J]$$

can be evaluated easily, and that the maximization ("combination of Q-part and T-part") can be done efficiently, as follows.

Q-part: The matrix $Q(s)$ is a polynomial matrix with coefficients in $\boldsymbol{K}$, and the function $f_Q(I, J)$ may be computed in many different ways (see §7.1). If the stronger condition (MP-Q2) on $Q(s)$ is satisfied with the expression (1.31), it holds that

$$f_Q(I,J) = \begin{cases} \sum_{i\in I} r_i - \sum_{j\in J} c_j & (\text{if } \det Q(1)[I,J] \neq 0) \\ -\infty & (\text{otherwise}) \end{cases}$$

where $Q(1)$ is a matrix over $\boldsymbol{K}$ and therefore $\det Q(1)[I, J]$ can be computed by means of arithmetic operations over $\boldsymbol{K}$.

T-part: The matrix $T(s)$ is a structured matrix in that the nonzero coefficients are algebraically independent. Therefore the function $f_T(I, J)$ can be evaluated efficiently by means of bipartite matchings, as has been explained in §1.1.2 (recall (1.7) in particular).

Combination: Each of $f_Q(I, J)$ and $f_T(I, J)$ enjoys a combinatorially nice property, being a variant of a valuated matroid. Therefore, the maximum can be computed by a straightforward application of an algorithmic scheme for the valuated matroid intersection problem, where the functions $f_Q(I, J)$ and $f_T(I, J)$ are evaluated polynomially many times (polynomial in n). If the stronger condition (MP-Q2) on $Q(s)$ is satisfied, the maximization can be reduced to an easier problem (the ordinary weighted matroid intersection problem).

For easy reference, let us give a temporary name, say "Algorithm D," to the above algorithm for computing the degree of determinant, while deferring a concrete description to Chap. 6 (cf. Remark 6.2.17).

With "Algorithm D" for the degree of determinant we can compute the index $\nu(A)$ of $A(s)$ based on the formula (1.3):

$$\nu(A) = \max_{i,j} \deg_s((i,j)\text{-cofactor of } A) - \deg_s \det A + 1.$$

A simplest way is to apply "Algorithm D" repeatedly to the whole matrix A and all the submatrices of order $n-1$ (which are n^2 in number) to obtain $\deg_s \det A$ and $\deg_s((i,j)\text{-cofactor of } A)$ for all (i,j). This naive method already gives a polynomial-time algorithm for $\nu(A)$, though an improvement for efficiency is possible.

Let us illustrate the above method for the matrix $A^{(2)}$ of (1.10), the second coefficient matrix of our electrical network. We regard it as a mixed polynomial matrix $A^{(2)}(s) = Q^{(2)}(s) + T^{(2)}(s)$ with

$$Q^{(2)}(s) = \begin{array}{|ccccc|ccccc|} \multicolumn{1}{c}{\xi^1} & \xi^2 & \xi^3 & \xi^4 & \multicolumn{1}{c}{\xi^5} & \eta_1 & \eta_2 & \eta_3 & \eta_4 & \multicolumn{1}{c}{\eta_5} \\ \hline 1 & -1 & 0 & 0 & -1 & & & & & \\ -1 & 0 & 1 & 1 & 1 & & & & & \\ \hline & & & & & -1 & -1 & 0 & -1 & 0 \\ & & & & & 0 & 1 & 1 & 0 & -1 \\ & & & & & 0 & 0 & -1 & 1 & 0 \\ \hline 0 & 0 & 0 & 0 & 0 & -1 & 0 & 0 & 0 & 0 \\ 0 & 0 & 0 & 0 & 0 & 0 & -1 & 0 & 0 & 0 \\ 0 & 0 & 0 & 0 & 0 & 0 & 0 & -1 & 0 & 0 \\ 0 & 0 & 0 & 0 & 0 & 0 & 0 & 0 & -1 & 0 \\ 0 & 0 & 0 & 0 & -1 & 0 & 0 & 0 & 0 & 0 \\ \hline \end{array}, \quad T^{(2)}(s) = \begin{array}{|ccccc|ccccc|} \multicolumn{1}{c}{\xi^1} & \xi^2 & \xi^3 & \xi^4 & \multicolumn{1}{c}{\xi^5} & \eta_1 & \eta_2 & \eta_3 & \eta_4 & \multicolumn{1}{c}{\eta_5} \\ \hline 0 & 0 & 0 & 0 & 0 & & & & & \\ 0 & 0 & 0 & 0 & 0 & & & & & \\ \hline & & & & & 0 & 0 & 0 & 0 & 0 \\ & & & & & 0 & 0 & 0 & 0 & 0 \\ & & & & & 0 & 0 & 0 & 0 & 0 \\ \hline 0 & 0 & 0 & 0 & 0 & 0 & 0 & 0 & 0 & 0 \\ 0 & R_1 & 0 & 0 & 0 & 0 & 0 & 0 & 0 & 0 \\ 0 & 0 & R_2 & 0 & 0 & 0 & 0 & 0 & 0 & 0 \\ 0 & 0 & 0 & sL & 0 & 0 & 0 & 0 & 0 & 0 \\ 0 & 0 & 0 & 0 & 0 & 0 & 0 & 0 & 0 & sC \\ \hline \end{array}$$

under the assumption that R_1, R_2, L, and C are independent parameters. The matrix $Q^{(2)}(s)$, being free from s, satisfies the stronger condition (MP-Q2) trivially. On the right-hand side of (1.34) in Theorem 1.3.2 we can take

$$I = R \setminus \{\text{row } 8, \text{row } 9\}, \qquad J = C \setminus \{\text{column } 3, \text{column } 4\}$$

as a maximizer,[5] for which

$$\deg_s \det Q^{(2)}[I,J] = 0, \qquad \deg_s \det T^{(2)}[R \setminus I, C \setminus J] = 1.$$

Hence we obtain $\deg_s \det A^{(2)} = 1$. In a similar manner we obtain

[5] It can be checked easily that, if $\deg_s \det T^{(2)}[R \setminus I', C \setminus J'] = 2$, then $\deg_s \det Q^{(2)}[I', J'] = -\infty$ (i.e., $\det Q^{(2)}[I', J'] = 0$). In particular, this is the case with $I' = R \setminus \{\text{row } 7, \text{row } 8, \text{row } 9, \text{row } 10\}$ and $J' = C \setminus \{\text{column } 2, \text{column } 3, \text{column } 4, \text{column } 10\}$, which corresponds to the "spurious" quadratic terms (1.11)–(1.14).

$$\max_{i,j} \deg_s((i,j)\text{-cofactor of } A^{(2)}) = 2,$$

and therefore $\nu(A^{(2)}) = 2 - 1 + 1 = 2$ by (1.3). Thus, the present method gives the correct answer for the matrix $A^{(2)}$ for which the graph-theoretic method fails.

Finally, we mention a duality theorem which forms a basis for "Algorithm D" for computing the maximum in (1.34), and which provides a systematic method for index reduction. This theorem is a concrete manifestation of a general duality theorem for the valuated matroid intersection problem.

Consider a transformation of $A(s)$ to

$$\tilde{A}(s) = \operatorname{diag}(s; -p_R) \cdot A(s) \cdot \operatorname{diag}(s; p_C)$$

with two integer vectors $p_R \in \mathbf{Z}^R$ and $p_C \in \mathbf{Z}^C$, where $\operatorname{diag}(s;p)$ for a vector $p = (p_i)$ means a diagonal matrix $\operatorname{diag}[s^{p_1}, s^{p_2}, \cdots]$. This is quite a natural transformation that corresponds to rewriting the system of equations $A\boldsymbol{x} = \boldsymbol{b}$ into $\tilde{A}\tilde{\boldsymbol{x}} = \tilde{\boldsymbol{b}}$ in terms of

$$\tilde{\boldsymbol{x}} = \operatorname{diag}(s; -p_C) \cdot \boldsymbol{x}, \qquad \tilde{\boldsymbol{b}} = \operatorname{diag}(s; -p_R) \cdot \boldsymbol{b}.$$

The transformation yields another mixed polynomial[6] matrix $\tilde{A}(s) = \tilde{Q}(s) + \tilde{T}(s)$ with

$$\begin{aligned} \tilde{Q}(s) &= \operatorname{diag}(s; -p_R) \cdot Q(s) \cdot \operatorname{diag}(s; p_C), \\ \tilde{T}(s) &= \operatorname{diag}(s; -p_R) \cdot T(s) \cdot \operatorname{diag}(s; p_C). \end{aligned}$$

For any choice of $p_R \in \mathbf{Z}^R$ and $p_C \in \mathbf{Z}^C$ we obviously have

$$\begin{aligned} &\max_{\substack{|I|=|J| \\ I \subseteq R, J \subseteq C}} \{\deg_s \det \tilde{Q}[I,J] + \deg_s \det \tilde{T}[R \setminus I, C \setminus J]\} \\ &\leq \max_{\substack{|I|=|J| \\ I \subseteq R, J \subseteq C}} \deg_s \det \tilde{Q}[I,J] + \max_{\substack{|I|=|J| \\ I \subseteq R, J \subseteq C}} \deg_s \det \tilde{T}[R \setminus I, C \setminus J]. \end{aligned} \tag{1.35}$$

A natural question here is whether the inequality in (1.35) turns into an equality for an appropriate choice of p_R and p_C.

The duality theorem asserts the existence of a pair of vectors $p_R \in \mathbf{Z}^R$ and $p_C \in \mathbf{Z}^C$ ("dual variables") which makes the inequality of (1.35) into an equality. Combining this with Theorem 1.3.2 we obtain the following theorem (cf. Theorem 6.2.15).

Theorem 1.3.3. *For a nonsingular mixed polynomial matrix $A(s) = Q(s) + T(s)$, there exist $p_R \in \mathbf{Z}^R$ and $p_C \in \mathbf{Z}^C$ such that*

$$\tilde{A}(s) = \operatorname{diag}(s; -p_R) \cdot A(s) \cdot \operatorname{diag}(s; p_C) = \tilde{Q}(s) + \tilde{T}(s)$$

[6] Strictly speaking, "polynomial" is to be replaced by "Laurent polynomial" (cf. §2.1.1), since negative powers of s can appear in $\tilde{A}(s)$.

satisfies

$$\deg_s \det \tilde{A} = \max_{\substack{|I|=|J| \\ I\subseteq R, J\subseteq C}} \deg_s \det \tilde{Q}[I,J] + \max_{\substack{|I|=|J| \\ I\subseteq R, J\subseteq C}} \deg_s \det \tilde{T}[R\backslash I, C\backslash J]. \quad (1.36)$$

A further condition (i) $p_R \geq \mathbf{0}$, $p_C \geq \mathbf{0}$, *or* (ii) $p_R \leq \mathbf{0}$, $p_C \leq \mathbf{0}$, *may be imposed on* p_R *and* p_C. □

The "Algorithm D" for computing $\deg_s \det A$ consists of finding such a pair of vectors (p_R, p_C) to transform $A(s)$ to $\tilde{A}(s)$ as well as an optimal pair of subsets (I, J) in the right-hand side of (1.34).

In addition to the algorithmic significance, the transformation to $\tilde{A}(s)$ has another meaning of index reduction transformation. Recall that the index is equal to one for a system of purely algebraic equations and to zero for a system of ordinary differential equations in the normal form.

Corollary 1.3.4. *The index of* $\tilde{A}(s)$ *is at most one:* $\nu(\tilde{A}) \leq 1$.

Proof. For any $I \subseteq R$ and $J \subseteq C$, consider the expression of $\deg_s \det \tilde{A}[I,J]$ as in Theorem 1.3.2. This shows that $\deg_s \det \tilde{A}[I,J]$ is bounded by the right-hand side of (1.36), which is equal to $\deg_s \det \tilde{A}$. Then (1.3) establishes the claim. ■

Theorem 1.3.3 is illustrated below for the matrix $A^{(2)}$ of (1.10) of our electrical network. By "Algorithm D" we can find

$$p_R = (0,0,0,0,0;0,0,0,0,1), \qquad p_C = (1,0,0,0,1;0,0,0,0,0)$$

as the "dual variables," which yields

$$\tilde{A}^{(2)}(s) = \begin{array}{c} \begin{array}{cccccccccc} Q^1 & \xi^2 & \xi^3 & \xi^4 & Q^5 & \eta_1 & \eta_2 & \eta_3 & \eta_4 & \eta_5 \end{array} \\ \left[\begin{array}{ccccc|ccccc} s & -1 & 0 & 0 & -s & & & & & \\ -s & 0 & 1 & 1 & s & & & & & \\ \hline & & & & & -1 & -1 & 0 & -1 & 0 \\ & & & & & 0 & 1 & 1 & 0 & -1 \\ & & & & & 0 & 0 & -1 & 1 & 0 \\ \hline 0 & 0 & 0 & 0 & 0 & -1 & 0 & 0 & 0 & 0 \\ 0 & R_1 & 0 & 0 & 0 & 0 & -1 & 0 & 0 & 0 \\ 0 & 0 & R_2 & 0 & 0 & 0 & 0 & -1 & 0 & 0 \\ 0 & 0 & 0 & sL & 0 & 0 & 0 & 0 & -1 & 0 \\ 0 & 0 & 0 & 0 & -1 & 0 & 0 & 0 & 0 & C \end{array}\right] \end{array}. \quad (1.37)$$

As asserted in (1.36), we have

$$\deg_s \det \tilde{A}^{(2)} = 2 = 1 + 1 = \max_{I,J} \deg_s \det \tilde{Q}^{(2)}[I,J] + \max_{I,J} \deg_s \det \tilde{T}^{(2)}[I,J]$$

in contrast to

$$\deg_s \det A^{(2)} = 1 < 0 + 2 = \max_{I,J} \deg_s \det Q^{(2)}[I,J] + \max_{I,J} \deg_s \det T^{(2)}[I,J].$$

In accordance with the change of variables $\tilde{\boldsymbol{x}} = \mathrm{diag}\,(s; -p_C) \cdot \boldsymbol{x}$, the 1st and 5th columns of $\tilde{A}^{(2)}$ are indexed by

$$Q^1 = s^{-(p_C)_1}\xi^1 = \int \xi^1 \mathrm{d}t \quad \text{(charge supplied by the source)},$$

$$Q^5 = s^{-(p_C)_5}\xi^5 = \int \xi^5 \mathrm{d}t \quad \text{(charge stored in the capacitor)},$$

respectively. The last row of $\tilde{A}^{(2)}$ represents the constitutive equation $Q^5 = C\eta_5$ ("charge is proportional to voltage") rather than $\xi^5 = C\dot{\eta}_5$ ("current is proportional to the time derivative of voltage"), which corresponds to the last row of $A^{(2)}$. The "dual variables" p_R and p_C admit such a physical interpretation.

Remark 1.3.5. The structural index computed by a "structural algorithm," whether graph-theoretic or matroid-theoretic, is equal to the true index value only in the generic case ("almost surely") and there is no guarantee of equality for a particular (numerically specified) matrix. "Combinatorial relaxation algorithms," to be described in Chap. 7, are refinements of the "structural algorithms" such that they first compute the generic values by combinatorial algorithms and then invoke a minimum amount of numerical arithmetic operations to always guarantee the true index value. □

Remark 1.3.6. Theorem 1.3.2 implies as an immediate corollary that A is nonsingular if and only if there exist $I \subseteq R$ and $J \subseteq C$ such that both $Q[I,J]$ and $T[R \setminus I, C \setminus J]$ are nonsingular. Application of this result to submatrices yields a rank identity:

$$\mathrm{rank}\, A = \max_{\substack{|I|=|J| \\ I \subseteq R, J \subseteq C}} \{\mathrm{rank}\, Q[I,J] + \mathrm{rank}\, T[R \setminus I, C \setminus J]\},$$

which is true for a mixed matrix $A = Q + T$ in general. This identity acts as a first bridge between mixed matrices and matroids, as will be seen in Chap. 4 (cf. Theorem 4.2.8, in particular). □

1.3.3 Block-triangular Decomposition

Kirchhoff's laws can be written in many different ways, on which the success and failure of the graph-theoretic method depends. This is what we have seen with the coefficient matrices $A^{(1)}$ and $A^{(2)}$ of our electrical network of Fig. 1.1; $A^{(1)}$ is good for the graph-theoretic method, while $A^{(2)}$ is not.

The matroid-theoretic method, i.e., the method of mixed polynomial matrices, works for both $A^{(1)}$ and $A^{(2)}$. It is stable against arbitrary choices in the representation of KCL and KVL, giving a correct answer for any legitimate representation of Kirchhoff's laws. Though the "embarrassing phenomenon" has been resolved by the matroid-theoretic method, no answer has yet been given to the question: Why is $A^{(1)}$ better than $A^{(2)}$?

In this section we shall answer this question by considering *block-triangular decompositions* of the coefficient matrices. In so doing we intend to introduce another mathematical aspect of mixed (polynomial) matrices.

Recall from §1.1.3 that the discrepancy $\nu_{\rm str}(A^{(2)}) \neq \nu(A^{(2)})$ is caused by the "spurious" quadratic terms (1.11)–(1.14) in $\det A^{(2)}_{\rm str}$, whereas those terms do not appear in $\det A^{(1)}_{\rm str}$. The reason for this difference is apparent from the following block-triangular forms of $A^{(1)}$ and $A^{(2)}$ obtained through reorderings of rows and columns:

$$
\bar{A}^{(1)} = \begin{array}{r|cccccccccc|}
 & \xi^1 & \xi^2 & \xi^3 & \xi^4 & \eta_2 & \eta_3 & \eta_4 & \xi^5 & \eta_5 & \eta_1 \\ \hline
\mathrm{KCL} & 1 & -1 & 0 & 0 & & & & -1 & & \\
\mathrm{KCL} & -1 & 0 & 1 & 1 & & & & 1 & & \\
\mathrm{KVL} & & & & & 1 & 1 & 0 & & -1 & \\
\mathrm{KVL} & & & & & 0 & -1 & 1 & & & \\
\mathrm{branch\ 2} & 0 & R_1 & 0 & 0 & -1 & 0 & 0 & & & \\
\mathrm{branch\ 3} & 0 & 0 & R_2 & 0 & 0 & -1 & 0 & & & \\
\mathrm{branch\ 4} & 0 & 0 & 0 & sL & 0 & 0 & -1 & & & \\
\mathrm{branch\ 5} & & & & & & & & -1 & sC & \\
\mathrm{KVL} & & & & & & & & & -1 & -1 \\
\mathrm{branch\ 1} & & & & & & & & & & -1 \\ \hline
\end{array},
$$

$$
\bar{A}^{(2)} = \begin{array}{r|cccccccccc|}
 & \xi^1 & \xi^2 & \xi^3 & \xi^4 & \eta_2 & \eta_3 & \eta_4 & \xi^5 & \eta_5 & \eta_1 \\ \hline
\mathrm{KCL} & 1 & -1 & 0 & 0 & & & & -1 & & \\
\mathrm{KCL} & -1 & 0 & 1 & 1 & & & & 1 & & \\
\mathrm{KVL} & & & & & 1 & 1 & 0 & & -1 & \\
\mathrm{KVL} & & & & & 0 & -1 & 1 & & 0 & \\
\mathrm{branch\ 2} & 0 & R_1 & 0 & 0 & -1 & 0 & 0 & 0 & 0 & \\
\mathrm{branch\ 3} & 0 & 0 & R_2 & 0 & 0 & -1 & 0 & 0 & 0 & \\
\mathrm{branch\ 4} & 0 & 0 & 0 & sL & 0 & 0 & -1 & 0 & 0 & \\
\mathrm{branch\ 5} & 0 & 0 & 0 & 0 & 0 & 0 & 0 & -1 & sC & \\
\mathrm{KVL} & & & & & -1 & 0 & -1 & & 0 & -1 \\
\mathrm{branch\ 1} & & & & & & & & & & -1 \\ \hline
\end{array}.
$$

In $\bar{A}^{(1)}$ the entry "sC" cannot contribute to $\det \bar{A}^{(1)}$ for a combinatorial reason that it lies in an off-diagonal block, whereas "sC" disappears in $\det \bar{A}^{(2)}$ as a result of cancellation. This is why "sC" does not appear in $\det A^{(1)}_{\rm str}$ but does survive in $\det A^{(2)}_{\rm str}$.

The above observation leads naturally to a further question: Which description of Kirchhoff's laws gives a finest block-triangular decomposition? This may be rephrased to the following problem: Given a coefficient matrix A of the form

$$A = \begin{array}{|c|c|} \hline \text{K C L} & O \\ \hline O & \text{K V L} \\ \hline \multicolumn{2}{|c|}{\text{constitutive eqns}} \\ \hline \end{array} \, ,$$

such as $A^{(1)}$ and $A^{(2)}$, find a finest block-triangular decomposition using a transformation of the form

$$\bar{A} = P_{\mathrm{r}} \cdot \begin{array}{|c|c|c|} \hline S_{\mathrm{KCL}} & O & O \\ \hline O & S_{\mathrm{KVL}} & O \\ \hline O & O & I \\ \hline \end{array} \cdot \begin{array}{|c|c|} \hline \text{K C L} & O \\ \hline O & \text{K V L} \\ \hline \multicolumn{2}{|c|}{\text{constitutive eqns}} \\ \hline \end{array} \cdot P_{\mathrm{c}}, \tag{1.38}$$

where S_{KCL} and S_{KVL} are nonsingular matrices, I is an identity matrix, and P_{r} and P_{c} are permutation matrices.

It will be shown in Chap. 4 that there exists a canonical (finest) block-triangular decomposition under the transformation (1.38) and it can be computed by an efficient algorithm. The canonical form of $A = A^{(2)}$ is given by

$$\bar{A}^{(3)} = \begin{array}{r|cccccccccc|} & \xi^1 & \xi^2 & \xi^3 & \xi^4 & \eta_2 & \eta_3 & \eta_4 & \xi^5 & \eta_5 & \eta_1 \\ \hline \text{KCL} & 1 & -1 & & & & & & -1 & & \\ \text{KCL} & & -1 & 1 & 1 & & & & & & \\ \text{KVL} & & & & & 1 & 1 & 0 & & -1 & \\ \text{KVL} & & & & & 0 & -1 & 1 & & & \\ \text{branch 2} & & R_1 & 0 & 0 & -1 & 0 & 0 & & & \\ \text{branch 3} & & 0 & R_2 & 0 & 0 & -1 & 0 & & & \\ \text{branch 4} & & 0 & 0 & sL & 0 & 0 & -1 & & & \\ \text{branch 5} & & & & & & & & -1 & sC & \\ \text{KVL} & & & & & & & & & -1 & -1 \\ \text{branch 1} & & & & & & & & & & -1 \\ \hline \end{array} \tag{1.39}$$

that is obtained from $A^{(2)}$ through (1.38) with

$$S_{\mathrm{KCL}} = \begin{array}{|cc|} \hline 1 & 0 \\ 1 & 1 \\ \hline \end{array} \, , \qquad S_{\mathrm{KVL}} = \begin{array}{|ccc|} \hline -1 & 0 & 0 \\ -1 & -1 & -1 \\ -1 & 0 & -1 \\ \hline \end{array} \, .$$

The canonical form does not depend on how Kirchhoff's laws are initially written (in particular, $A^{(1)}$ and $A^{(2)}$ have a common canonical form (1.39)), and therefore, it can be regarded as representing a certain combinatorial structure inherent in our electrical network.

Returning to the determinant, note first that the transformation (1.38) changes the determinant only by a constant factor (i.e., $\det \bar{A} = \pm \det S_{\mathrm{KCL}} \cdot \det S_{\mathrm{KVL}} \cdot \det A$). The determinant of $\bar{A}$ is obviously the product of the determinants of the diagonal blocks. Hence the off-diagonal entries, such as "sC" in (1.39), do not appear in the determinant. Furthermore, a theorem in Chap. 4 (Theorem 4.5.4) states that any system parameter contained in a diagonal block appears in $\det \bar{A}$. This explains, in our example, why R_1, R_2 and sL appear in

$$\det A^{(1)} = \det A^{(2)} = -\det \bar{A}^{(3)} = R_1 R_2 + sL \cdot R_1 + sL \cdot R_2.$$

For the degree of $\det \bar{A}^{(3)}$, we may apply the method of §1.3.2 to each diagonal block and sum up the degrees thus obtained. This will be more efficient than to apply the same method to the whole matrix.

Block-triangular decompositions of the above kind are one of the major topics studied in this book; among which are the Dulmage–Mendelsohn decomposition in Chap. 2 and the combinatorial canonical form (CCF) in Chap. 4.

Notes. The concept of a mixed matrix was introduced in Murota–Iri [238] with the observation on two kinds of numbers explained in §1.2.1, while the dimensional analysis in §1.2.3 is due to Murota [200]. This chapter is an improved version of a presentation (Murota [223]) at ICIAM 95.

2. Matrix, Graph, and Matroid

This chapter lays the mathematical foundation for combinatorial methods of systems analysis. Combinatorial properties of numerical matrices can be stated and analyzed with the aid of matroid theory, whereas those of polynomial matrices are formulated in the language of valuated matroids in Chap. 5. Emphasis is laid also on the general decomposition principle based on submodularity, and accordingly the Dulmage–Mendelsohn decomposition, which serves as a fundamental tool for the generic-case analysis of matrices, is presented in a systematic manner.

2.1 Matrix

2.1.1 Polynomial and Algebraic Independence

Let $\boldsymbol{K}$ be a field (typically $\boldsymbol{K} = \mathbf{Q}$ (rational numbers) or $\boldsymbol{K} = \mathbf{R}$ (real numbers)) and X be an indeterminate (independent variable). We denote by $\boldsymbol{K}[X]$ the *ring of polynomials* in X over $\boldsymbol{K}$, i.e.,

$$\boldsymbol{K}[X] = \{\sum_{k=0}^{N} \alpha_k X^k \mid 0 \leq N \in \mathbf{Z}, \alpha_k \in \boldsymbol{K} \ (0 \leq k \leq N)\}.$$

For $p(X) = \sum_{k=0}^{N} \alpha_k X^k$ with $\alpha_N \neq 0$, the highest index N is called the *degree* of $p(X)$, denoted $\deg p$, whereas α_N is the *leading coefficient*. A polynomial is called *monic* if the leading coefficient is equal to one. We denote by $\boldsymbol{K}(X)$ the field of rational functions in X over $\boldsymbol{K}$, i.e.,

$$\boldsymbol{K}(X) = \{p(X)/q(X) \mid p(X), q(X) \in \boldsymbol{K}[X], q(X) \neq 0\}.$$

For a rational function $f(X) \in \boldsymbol{K}(X)$, the degree of $f(X)$, denoted $\deg f$, is defined by $\deg f = \deg p - \deg q$ with reference to an expression $f(X) = p(X)/q(X)$ with $p(X), q(X) \in \boldsymbol{K}[X]$, where $\deg f$ is well-defined, independent of the expression. A rational function $f(X) \in \boldsymbol{K}(X)$ is said to be *proper* if $\deg f \leq 0$, and *strictly proper* if $\deg f < 0$. A rational function $f(X) \in \boldsymbol{K}(X)$ is called a *Laurent polynomial* if $X^N f(X) \in \boldsymbol{K}[X]$ for some integer $N \in \mathbf{Z}$; in other words, $f(X)$ is a Laurent polynomial if and only if $f(X) \in \boldsymbol{K}[X, 1/X]$, where

$$\boldsymbol{K}[X, 1/X] = \{ \sum_{k=-N_1}^{N_2} \alpha_k X^k \mid 0 \leq N_1, N_2 \in \mathbf{Z}, \alpha_k \in \boldsymbol{K} \ (-N_1 \leq k \leq N_2)\}.$$

For a Laurent polynomial $f(X)$ we define

$$\mathrm{ord} f = -\min\{N \in \mathbf{Z} \mid X^N f(X) \in \boldsymbol{K}[X]\}. \tag{2.1}$$

Obviously, $f(X)$ is a polynomial if and only if $\mathrm{ord} f \geq 0$.

Let $\boldsymbol{F}$ be an extension field of $\boldsymbol{K}$ (i.e., $\boldsymbol{F} \supseteq \boldsymbol{K}$). For a subset $\mathcal{Y}$ of $\boldsymbol{F}$, we denote by $\boldsymbol{K}(\mathcal{Y})$ and $\boldsymbol{K}[\mathcal{Y}]$ the *field adjunction* and the *ring adjunction*, respectively; that is, $\boldsymbol{K}(\mathcal{Y})$ is the extension field of $\boldsymbol{K}$ generated by $\mathcal{Y}$ over $\boldsymbol{K}$ while $\boldsymbol{K}[\mathcal{Y}]$ the ring generated by $\mathcal{Y}$ over $\boldsymbol{K}$.

An element y of $\boldsymbol{F}$ is called *algebraic* over $\boldsymbol{K}$ if there exists a nontrivial polynomial $p(X)$ in one indeterminate X over $\boldsymbol{K}$ such that $p(y) = 0$, where $p(X)$ is called nontrivial if some of its coefficients are distinct from zero. An element of $\boldsymbol{F}$ is called *transcendental* over $\boldsymbol{K}$ if it is not algebraic over $\boldsymbol{K}$. A subset (more precisely, multiset) $\mathcal{Y} = \{y_1, \cdots, y_q\}$ of $\boldsymbol{F}$ is called *algebraically independent* over $\boldsymbol{K}$ if either of the following equivalent conditions holds:

(i) For any i, y_i is transcendental over $\boldsymbol{K}(\mathcal{Y} \setminus \{y_i\})$,
(ii) There exists no nontrivial polynomial $p(X_1, \cdots, X_q)$ in q indeterminates over $\boldsymbol{K}$ such that $p(y_1, \cdots, y_q) = 0$,

where $p(X_1, \cdots, X_q)$ is called nontrivial if some of its coefficients are distinct from zero. We call $\mathcal{Y}$ *algebraically dependent* if it is not algebraically independent. In this book we often use an informal expression like "$y_1, \cdots, y_q$ are algebraically independent over $\boldsymbol{K}$" to mean the set $\{y_1, \cdots, y_q\}$ is algebraically independent over $\boldsymbol{K}$, which is a stronger assertion than the elementwise transcendency of each y_i over $\boldsymbol{K}$.

The *degree of transcendency* of $\boldsymbol{F}$ over $\boldsymbol{K}$, denoted $\dim_{\boldsymbol{K}} \boldsymbol{F}$, means the maximum cardinality of a subset of $\boldsymbol{F}$ that is algebraically independent over $\boldsymbol{K}$, where $\dim_{\boldsymbol{K}} \boldsymbol{F}$ can be infinite. For instance, $\dim_{\mathbf{Q}} \mathbf{R} = +\infty$.

Example 2.1.1. Let t_1, t_2, t_3 be independent parameters (indeterminates) over $\mathbf{Q}$. For $z_1 = t_1 + t_2$, $z_2 = (t_2 + t_3)^2$, $z_3 = t_1 - t_3$ we can find a nontrivial polynomial $p(X_1, X_2, X_3) = (X_1 - X_3)^2 - X_2$ for which $p(z_1, z_2, z_3) = 0$. Hence $\{z_1, z_2, z_3\}$ is algebraically dependent over $\mathbf{Q}$, whereas each z_i is transcendental over $\mathbf{Q}$. We see $\dim_{\mathbf{Q}} \mathbf{Q}(z_1, z_2, z_3) = 2$. On the other hand, for $y_1 = t_1 t_2$, $y_2 = t_2 + t_3$, $y_3 = 2t_3/t_1$, there is no nontrivial polynomial $q(X_1, X_2, X_3)$ over $\mathbf{Q}$ such that $q(y_1, y_2, y_3) = 0$, and therefore, $\{y_1, y_2, y_3\}$ is algebraically independent over $\mathbf{Q}$. Accordingly, $\dim_{\mathbf{Q}} \mathbf{Q}(y_1, y_2, y_3) = 3$. □

We refer to Jacobson [149, 150] and van der Waerden [325] as general references for algebraic concepts.

2.1.2 Determinant

We consider matrices over a field $\boldsymbol{F}$. For a matrix A, the *row set* and the *column set* are denoted by $\mathrm{Row}(A)$ and $\mathrm{Col}(A)$, i.e., $A = (A_{ij} \mid i \in \mathrm{Row}(A), j \in \mathrm{Col}(A))$, where A_{ij} is the (i,j)-entry. For $I \subseteq \mathrm{Row}(A)$ and $J \subseteq \mathrm{Col}(A)$, $A[I,J] = (A_{ij} \mid i \in I, j \in J)$ means the submatrix of A with row set I and column set J.

For a square matrix A, its *determinant* is denoted by $\det A$. Namely, for an $n \times n$ matrix A, we define

$$\det A = \sum_{\pi \in \mathcal{S}_n} \mathrm{sgn}\, \pi \cdot \prod_{i=1}^{n} A_{i\pi(i)}, \tag{2.2}$$

where $\mathcal{S}_n$ denotes the set of all the permutations of order n, and $\mathrm{sgn}\, \pi = \pm 1$ is the signature of a permutation π. A matrix is *nonsingular* if it is square and its determinant is distinct from zero. The set of nonsingular matrices of order n over $\boldsymbol{F}$ will be denoted by $\mathrm{GL}(n, \boldsymbol{F})$.

For $I \subseteq R \equiv \mathrm{Row}(A)$ and $J \subseteq C \equiv \mathrm{Col}(A)$, we denote by $\det A[I,J]$ the determinant of the submatrix $A[I,J]$. Since the sign of a determinant depends on the orderings of rows and columns, we always assume that the orderings of the elements of I and J are compatible with those of R and C, unless otherwise stated. The determinant of a submatrix is sometimes referred to as a *minor* or as a *subdeterminant.*

Proposition 2.1.2 (Laplace expansion). *For a square matrix A and $i \in R$, it holds that*

$$\det A = \sum_{j \in C} (-1)^{i+j} \cdot A_{ij} \cdot \det A[R \setminus \{i\}, C \setminus \{j\}]. \tag{2.3}$$

Proof. In the summation of (2.2) classify π according to the value of $\pi(i)$ to obtain

$$\det A = \sum_{j=1}^{n} A_{ij} \left[\sum_{\pi : \pi(i) = j} \mathrm{sgn}\, \pi \cdot \prod_{k \neq i} A_{k\pi(k)} \right].$$

The expression in [ ] is equal to $(-1)^{i+j} \det A[R \setminus \{i\}, C \setminus \{j\}]$. ■

The formula (2.3) is called the Laplace expansion with respect to row i. A more general form of this formula is an expansion with respect to a subset $I \subseteq R$. The following expression (2.4) is called the generalized Laplace expansion with respect to row set $I \subseteq R$.

Proposition 2.1.3 (Generalized Laplace expansion). *For a square matrix A and $I \subseteq R$, it holds that*

$$\det A = \sum_{J \subseteq C, |J| = |I|} \mathrm{sgn}\,(I,J) \cdot \det A[I,J] \cdot \det A[R \setminus I, C \setminus J]. \tag{2.4}$$

Here, $\mathrm{sgn}\,(I, J) = \pm 1$ *denotes the signature of the permutation*

$$\begin{pmatrix} i_1, \cdots, i_k; i_{k+1}, \cdots, i_n \\ j_1, \cdots, j_k; j_{k+1}, \cdots, j_n \end{pmatrix},$$

where $I = \{i_1, \cdots, i_k\}$, $R \setminus I = \{i_{k+1}, \cdots, i_n\}$ *with* $i_1 < \cdots < i_k$; $i_{k+1} < \cdots < i_n$, *and* $J = \{j_1, \cdots, j_k\}$, $C \setminus J = \{j_{k+1}, \cdots, j_n\}$ *with* $j_1 < \cdots < j_k$; $j_{k+1} < \cdots < j_n$.

Proof. The proof is similar to the one for Proposition 2.1.2 ■

We now derive the Grassmann–Plücker identity, which would be the most important identity for determinants in the context of the combinatorial study of matrices, to be explained in Remark 2.1.8.

Proposition 2.1.4 (Grassmann–Plücker identity). *Let* A *be a matrix with* $|R| \le |C|$, *where* $R = \mathrm{Row}(A)$ *and* $C = \mathrm{Col}(A)$. *For* $J, J' \subseteq C$ *with* $|J| = |J'| = |R|$ *and* $i \in J \setminus J'$, *it holds that*

$$\det A[R, J] \cdot \det A[R, J'] = \sum_{j \in J' \setminus J} \det A[R, J - i + j] \cdot \det A[R, J' + i - j], \tag{2.5}$$

where $J - i + j$ *is a short-hand notation for* $(J \setminus \{i\}) \cup \{j\}$ *in which the column* j *is put at the position of column* i *in* J; *similarly for* $J' + i - j = (J' \cup \{i\}) \setminus \{j\}$.

Proof. To avoid complication in notation, we present the idea for a 4×6 matrix $A = \begin{bmatrix} \boldsymbol{a}_1 & \boldsymbol{a}_2 & \boldsymbol{a}_3 & \boldsymbol{a}_4 & \boldsymbol{a}_5 & \boldsymbol{a}_6 \end{bmatrix}$ with $J = \{1, 2, 3, 4\}$, $J' = \{3, 4, 5, 6\}$ and $i = 1$, where $\boldsymbol{a}_j$ ($j = 1, \cdots, 6$) are 4-dimensional column vectors. Consider a 8×8 matrix

$$\tilde{A} = \left[\begin{array}{cccc|cccc} \boldsymbol{a}_1 & \boldsymbol{a}_2 & \boldsymbol{a}_3 & \boldsymbol{a}_4 & \boldsymbol{a}_3 & \boldsymbol{a}_4 & \boldsymbol{a}_5 & \boldsymbol{a}_6 \\ \hline \boldsymbol{a}_1 & \boldsymbol{0} & \boldsymbol{0} & \boldsymbol{0} & \boldsymbol{a}_3 & \boldsymbol{a}_4 & \boldsymbol{a}_5 & \boldsymbol{a}_6 \end{array}\right],$$

which is singular. The generalized Laplace expansion (2.4) applied to $\tilde{A}$ yields (2.5). ■

Example 2.1.5. The Grassmann–Plücker identity is illustrated here. For

$$A = \boxed{\begin{array}{cccccc} 1 & 0 & 0 & a_{14} & a_{15} & a_{16} \\ 0 & 1 & 0 & a_{24} & a_{25} & a_{26} \\ 0 & 0 & 1 & a_{34} & a_{35} & a_{36} \end{array}}$$

with $C = \{1, 2, 3, 4, 5, 6\}$, $J = \{1, 2, 3\}$, $J' = \{4, 5, 6\}$, $i = 1$, we have the following identity:

$$\begin{vmatrix} 1 & 0 & 0 \\ 0 & 1 & 0 \\ 0 & 0 & 1 \end{vmatrix} \begin{vmatrix} a_{14} & a_{15} & a_{16} \\ a_{24} & a_{25} & a_{26} \\ a_{34} & a_{35} & a_{36} \end{vmatrix} = \begin{vmatrix} a_{14} & 0 & 0 \\ a_{24} & 1 & 0 \\ a_{34} & 0 & 1 \end{vmatrix} \begin{vmatrix} 1 & a_{15} & a_{16} \\ 0 & a_{25} & a_{26} \\ 0 & a_{35} & a_{36} \end{vmatrix}$$
$$+ \begin{vmatrix} a_{15} & 0 & 0 \\ a_{25} & 1 & 0 \\ a_{35} & 0 & 1 \end{vmatrix} \begin{vmatrix} a_{14} & 1 & a_{16} \\ a_{24} & 0 & a_{26} \\ a_{34} & 0 & a_{36} \end{vmatrix} + \begin{vmatrix} a_{16} & 0 & 0 \\ a_{26} & 1 & 0 \\ a_{36} & 0 & 1 \end{vmatrix} \begin{vmatrix} a_{14} & a_{15} & 1 \\ a_{24} & a_{25} & 0 \\ a_{34} & a_{35} & 0 \end{vmatrix},$$

where $|\cdot|$ means a determinant. Note how the columns are ordered. □

The determinant of a matrix product AB admits an expansion in terms of the minors of A and B, called the Cauchy–Binet formula.

Proposition 2.1.6 (Cauchy–Binet formula). *For an $m \times n$ matrix A and an $n \times m$ matrix B, where $m \leq n$, it holds that*

$$\det(AB) = \sum_{|J|=m} \det A[R, J] \cdot \det B[J, C],$$

where $R = \mathrm{Row}(A)$, $C = \mathrm{Col}(B)$, and J runs over all subsets of size m of $\mathrm{Col}(A) = \mathrm{Row}(B)$.

Proof. This formula can be derived from the generalized Laplace expansion (2.4) applied to the $(m+n) \times (m+n)$ matrix

$$M = \begin{bmatrix} O_m & A \\ B & -I_n \end{bmatrix}.$$

In fact, the expansion (2.4) with respect to the row set $I = R$ yields

$$\det M = \sum_{|J|=m} (-1)^n \det A[R, J] \cdot \det B[J, C],$$

a summation over $J \subseteq \mathrm{Col}(A)$ with $|J| = m$, whereas

$$\begin{bmatrix} I_m & A \\ O & I_n \end{bmatrix} \cdot \begin{bmatrix} O_m & A \\ B & -I_n \end{bmatrix} = \begin{bmatrix} AB & O \\ B & -I_n \end{bmatrix}$$

implies $\det M = (-1)^n \det(AB)$. ■

Proposition 2.1.7 (Schur complement). *If A is square and D is nonsingular, then*

$$\det \begin{bmatrix} A & B \\ C & D \end{bmatrix} = \det D \cdot \det \left[A - BD^{-1}C \right].$$

Proof. This follows from

$$\begin{bmatrix} I & -B \\ O & I \end{bmatrix} \cdot \begin{bmatrix} I & O \\ O & D^{-1} \end{bmatrix} \cdot \begin{bmatrix} A & B \\ C & D \end{bmatrix} = \begin{bmatrix} A - BD^{-1}C & O \\ D^{-1}C & I \end{bmatrix}.$$

■

The submatrix $A - BD^{-1}C$ is often called the Schur complement.

Remark 2.1.8. The Grassmann–Plücker identity has important combinatorial implications, which play significant roles in this book.

In the Grassmann–Plücker identity, if the left-hand side of (2.5) is distinct from zero, then there exists at least one nonzero term in the summation on the right-hand side. This means that $\mathcal{B} = \{J \subseteq C \mid \det A[R, J] \neq 0\}$ (= the family of column bases of A) has the following property:

(BM$_\pm$) For $J, J' \in \mathcal{B}$ and for $i \in J \setminus J'$, there exists $j \in J' \setminus J$ such that $J - i + j \in \mathcal{B}$ and $J' + i - j \in \mathcal{B}$.

This property is formulated in §2.3.4 as the simultaneous exchange property for matroids.

In case the entries of the matrix A are rational functions (more generally, when $\boldsymbol{F}$ is a non-Archimedean valuated field), we can talk of the consistency of (2.5) with respect to degree (additive valuation). Let $\omega : \mathcal{B} \to \mathbf{Z}$ denote the degree of $\det A[R, J]$ for $J \in \mathcal{B}$, i.e., $\omega(J) = \deg \det A[R, J]$. The degree of the left-hand side of (2.5) is equal to $\omega(J) + \omega(J')$, and there exists, in the summation on the right-hand side, at least one term of degree greater than or equal to this. This means the function $\omega : \mathcal{B} \to \mathbf{Z}$ has the following property:

(VM) For $J, J' \in \mathcal{B}$ and for $i \in J \setminus J'$, there exists $j \in J' \setminus J$ such that $J - i + j \in \mathcal{B}$ and $J' + i - j \in \mathcal{B}$, and $\omega(J) + \omega(J') \leq \omega(J - i + j) + \omega(J' + i - j)$.

This property is formulated in §5.2 as an axiom of valuated matroids.

In case the entries of the matrix are real numbers (more generally, when $\boldsymbol{F}$ is an ordered field), we can talk of the consistency of (2.5) with respect to sign (± 1). Let $\sigma : \mathcal{B} \to \{1, -1\}$ denote the sign of $\det A[R, J]$ for $J \in \mathcal{B}$, i.e., $\sigma(J) = \text{sgn} \det A[R, J]$, where J is considered an ordered set and assume $\sigma(\pi J) = \text{sgn}\,\pi \cdot \sigma(J)$ for a permutation π of J. If the left-hand side of (2.5) is positive, then there exists at least one positive term in the summation on the right-hand side. This means the function $\sigma : \mathcal{B} \to \{1, -1\}$ has the following property:

(OM) For $J, J' \in \mathcal{B}$ and for $i \in J \setminus J'$, there exists $j \in J' \setminus J$ such that $J - i + j \in \mathcal{B}$ and $J' + i - j \in \mathcal{B}$, and $\sigma(J) \cdot \sigma(J') = \sigma(J - i + j) \cdot \sigma(J' + i - j)$,

where we still assume the convention in Proposition 2.1.4 that $J - i + j$ denotes the ordered set in which the column j is put at the position of column i in J. This property is formulated as an axiom of oriented matroids, though not treated in this book (see Björner–Las Vergnas–Sturmfels–White–Ziegler [14]). □

2.1.3 Rank, Term-rank and Generic-rank

The *rank* of A, as defined in linear algebra, is equal to (i) the maximum number of linearly independent column vectors of A, (ii) the maximum number of linearly independent row vectors of A, and (iii) the maximum size of a nonsingular submatrix of A. We denote the rank of A by $\text{rank}\, A$. A maximum-sized set of linearly independent column vectors is called a column basis of A. A row basis is defined in a similar manner.

Let A be a matrix with $R = \text{Row}(A)$ and $C = \text{Col}(A)$, and define $\rho : 2^C \to \mathbf{Z}$ and $\lambda : 2^R \times 2^C \to \mathbf{Z}$ by

$$\rho(J) = \operatorname{rank} A[R, J], \qquad J \subseteq C, \tag{2.6}$$
$$\lambda(I, J) = \operatorname{rank} A[I, J], \qquad I \subseteq R,\ J \subseteq C. \tag{2.7}$$

These functions satisfy the following inequalities, of which the first is called the *submodular inequality*. These inequalities are most fundamental in the combinatorial study of matrices, as will be explained later.

Proposition 2.1.9.

(1) $\rho(J_1) + \rho(J_2) \geq \rho(J_1 \cap J_2) + \rho(J_1 \cup J_2), \qquad J_1, J_2 \subseteq C.$

(2) $\lambda(I_1, J_1) + \lambda(I_2, J_2) \geq \lambda(I_1 \cup I_2, J_1 \cap J_2) + \lambda(I_1 \cap I_2, J_1 \cup J_2),$
$I_1, I_2 \subseteq R;\ J_1, J_2 \subseteq C.$

Proof. (1) We denote by $\boldsymbol{a}_j$ the column vector of A at column $j \in C$. For the submatrix $A[R, J_1 \cap J_2]$, take a column basis, say $\{\boldsymbol{a}_j \mid j \in B_{12}\}$, where $B_{12} \subseteq J_1 \cap J_2$ and $|B_{12}| = \rho(J_1 \cap J_2)$. It is possible to make a column basis of $A[R, J_1]$ by adding some vectors from among $\{\boldsymbol{a}_j \mid j \in J_1 \setminus J_2\}$ to the already chosen set $\{\boldsymbol{a}_j \mid j \in B_{12}\}$. Let $\{\boldsymbol{a}_j \mid j \in B_1\}$ be the added vectors, where $B_1 \subseteq J_1 \setminus J_2$ and $|B_{12}| + |B_1| = \rho(J_1)$. Similarly, we can make a column basis of $A[R, J_1 \cup J_2]$ by augmenting $\{\boldsymbol{a}_j \mid j \in B_{12} \cup B_1\}$ with some vectors of $\{\boldsymbol{a}_j \mid j \in J_2 \setminus J_1\}$. Let $\{\boldsymbol{a}_j \mid j \in B_2\}$ be the added vectors, where $B_2 \subseteq J_2 \setminus J_1$ and $|B_{12}| + |B_1| + |B_2| = \rho(J_1 \cup J_2)$. Since $\{\boldsymbol{a}_j \mid j \in B_{12} \cup B_2\}$ is a set of linearly independent vectors and $B_{12} \cup B_2 \subseteq J_2$, we have $|B_{12}| + |B_2| \leq \rho(J_2)$. This establishes the desired inequality.

(2) Consider $\tilde{A} = (I_m \mid A)$, where I_m is the identity matrix of order $m = |R|$. We have $\mathrm{Col}(\tilde{A}) = R \cup C$. Putting $\tilde{\rho}(I \cup J) = \operatorname{rank} \tilde{A}[\mathrm{Row}(\tilde{A}), I \cup J]$, we see from

$\tilde{A} = (I_m \mid A) =$

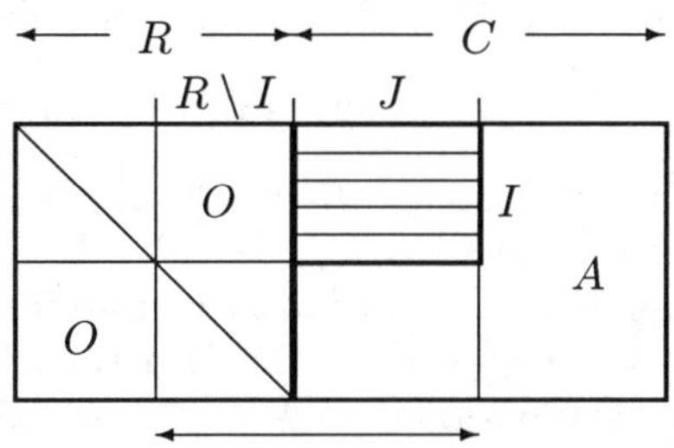

that $\lambda(I, J) = \tilde{\rho}((R \setminus I) \cup J) + |I| - m$. Then the submodularity of $\tilde{\rho}$ established in (1) is equivalent to the claimed inequality for λ. ∎

The concept of the term-rank of a matrix, introduced by Ore [256], is a combinatorial version of the rank and plays a significant role in the combinatorial analysis of matrices. As already mentioned, the rank of a matrix is equal to the maximum size of a nonsingular submatrix, i.e.,

$$\operatorname{rank} A = \max\{|I| \mid A[I, J] \text{ is nonsingular},\ I \subseteq R, J \subseteq C\}.$$

As a combinatorial version of nonsingularity, let us say that a matrix A is *term-nonsingular* if the defining expansion (2.2) of the determinant contains at least one nonvanishing term, that is, if $A_{i\pi(i)} \neq 0$ ($\forall\, i \in R$) for some bijection $\pi : R \to C$. Obviously, nonsingularity implies term-nonsingularity, since (2.2) is distinct from zero only if the summation contains a nonzero term. The *term-rank* of A is then defined by

$$\text{term-rank}\, A = \max\{|I| \mid A[I, J] \text{ is term-nonsingular, } I \subseteq R, J \subseteq C\}.$$

In other words, the term-rank of A is defined as the maximum of k such that $A_{i_1 j_1} \neq 0$, $A_{i_2 j_2} \neq 0$, $\cdots$, $A_{i_k j_k} \neq 0$ for some suitably chosen distinct rows $i_1, i_2, \cdots, i_k$ and distinct columns $j_1, j_2, \cdots, j_k$. A set of pairs $\{(i_1, j_1), (i_2, j_2), \cdots, (i_k, j_k)\}$ with the property that $A_{i_1 j_1} \neq 0$, $A_{i_2 j_2} \neq 0$, $\cdots$, $A_{i_k j_k} \neq 0$ is sometimes called a *partial transversal.* It holds that

$$\text{rank}\, A \leq \text{term-rank}\, A, \tag{2.8}$$

since nonsingularity implies term-nonsingularity.

Another related concept, called the generic-rank, is defined for a matrix containing parameters, as follows. Let the entries A_{ij} of A be rational functions over a field $\boldsymbol{K}$ in q independent parameters, or indeterminates, $X_1, \cdots, X_q$; i.e., $A_{ij} \in \boldsymbol{K}(X_1, \cdots, X_q)$ (= the field of rational functions in $(X_1, \cdots, X_q)$ over $\boldsymbol{K}$). Then any subdeterminant is a rational function in $X_1, \cdots, X_q$ over $\boldsymbol{K}$. The rank of A viewed as a matrix over $\boldsymbol{K}(X_1, \cdots, X_q)$ is defined to be the maximum size of a submatrix whose determinant is a nonzero rational function. We call this rank the *generic rank* of A, and denote it by generic-rank A. The generic-rank of A is equal to the rank of A with parameters $X = (X_1, \cdots, X_q)$ fixed to a set of numbers $t = (t_1, \cdots, t_q)$ (in some extension field of $\boldsymbol{K}$) which is algebraically independent over $\boldsymbol{K}$:

$$\text{generic-rank}\, A = \text{rank}\, A(t). \tag{2.9}$$

This means, in the case of $\boldsymbol{K} = \mathbf{Q}$, that (2.9) holds true for "almost all" choices of real numbers $t = (t_1, \cdots, t_q) \in \mathbf{R}^q$.

Let $\boldsymbol{F}$ be an extension field of $\boldsymbol{K}$ containing an infinite number of elements (typically, $\boldsymbol{K} = \mathbf{Q}$ and $\boldsymbol{F} = \mathbf{R}$). If a set of numerical values $a \in \boldsymbol{F}^q$ are substituted for the parameters $X = (X_1, \cdots, X_q)$, each entry of $A(a)$ belongs to $\boldsymbol{F}$, and therefore the rank of $A(a)$ as a matrix over $\boldsymbol{F}$ can be defined. This rank is uniquely determined for those parameter values $a \in \boldsymbol{F}^q$ which lie outside some proper algebraic variety[1] ($\subset \boldsymbol{F}^q$). The uniquely determined rank is equal to the maximum of rank $A(a)$ over $a \in \boldsymbol{F}^q$, and also to the generic-rank of A:

$$\text{generic-rank}\, A = \max_{a \in \boldsymbol{F}^q} \text{rank}\, A(a).$$

[1] By a proper algebraic variety is meant a proper subset of $\boldsymbol{F}^q$ that can be represented as $\{(x_1, \cdots, x_q) \in \boldsymbol{F}^q \mid p(x_1, \cdots, x_q) = 0\}$ for some $p(X) \in \boldsymbol{K}[X_1, \cdots, X_q]$.

Example 2.1.10. For a matrix $A = \begin{bmatrix} X^2 & X \\ X & 1 \end{bmatrix}$, we have generic-rank $A = 1$ and term-rank $A = 2$. □

Example 2.1.11. For a matrix $A = \begin{bmatrix} X & X \\ 1 & X \end{bmatrix}$ over $\boldsymbol{K} = \mathrm{GF}(2)$ (= the field consisting of 0 and 1), we have generic-rank A = term-rank $A = 2$, whereas $\max_{a \in \boldsymbol{K}} \operatorname{rank} A(a) = 1$. □

We are often interested in the cases where the generic-rank and the term-rank coincide. A matrix A is called a *generic matrix* if the set of its nonvanishing entries is algebraically independent over some field. This means that each of the nonvanishing entries of A can be regarded as an independent parameter by itself. For example, among three matrices

$$A_1 = \begin{bmatrix} X_1 & X_2 \\ X_3 & 0 \end{bmatrix}, \; A_2 = \begin{bmatrix} X_1X_2 & X_2 + X_3 \\ 2X_3/X_1 & 0 \end{bmatrix}, \; A_3 = \begin{bmatrix} X_1 + X_2 & (X_2 + X_3)^2 \\ X_1 - X_3 & 0 \end{bmatrix},$$

where X_1, X_2, X_3 are algebraically independent numbers (or independent parameters), A_1 and A_2 are generic matrices and A_3 is not. Note that, in A_2, there is no algebraic relation among $Y_1 = X_1X_2$, $Y_2 = X_2 + X_3$, $Y_3 = 2X_3/X_1$, whereas, in A_3, we have a relation $(Z_1 - Z_3)^2 = Z_2$ for $Z_1 = X_1 + X_2$, $Z_2 = (X_2 + X_3)^2$, $Z_3 = X_1 - X_3$ (cf. Example 2.1.1).

The following fact is obvious, but fundamental (Edmonds [67]).

Proposition 2.1.12. *For a generic matrix A, it holds that*

$$\text{generic-rank}\, A = \text{term-rank}\, A. \tag{2.10}$$

Proof. Consider a $k \times k$ term-nonsingular submatrix of A, where k = term-rank A. The defining expansion of its determinant contains a term, say, $A_{i_1 j_1} A_{i_2 j_2} \cdots A_{i_k j_k}$, which cannot be cancelled out. Hence generic-rank $A \geq$ term-rank A. The reverse inequality is true in general by (2.8), in which rank A = generic-rank A. ■

Remark 2.1.13. The term-rank of A is in fact a graph-theoretic concept (see §2.2.3 for terminology). Consider a bipartite graph $G = (V^+, V^-; \tilde{A})$ that has the vertex bipartition $(V^+, V^-) = (C, R)$ corresponding to the column set C and the row set R, and the arc set $\tilde{A}$ defined by

$$\tilde{A} = \{(j, i) \mid i \in R, j \in C, A_{ij} \neq 0\}.$$

That is, an arc represents a nonvanishing entry of A. Then term-rank of A is equal to the maximum size of a matching in G, which can be computed efficiently in polynomial time (cf. §2.2.3). □

2.1.4 Block-triangular Forms

We distinguish two kinds of block-triangular decompositions for matrices. The one employs a simultaneous permutation of rows and columns and the other uses two independent permutations.

The first block-triangular decomposition is defined for a square matrix A such that $\mathrm{Row}(A)$ and $\mathrm{Col}(A)$ have a natural one-to-one correspondence $\psi : \mathrm{Col}(A) \to \mathrm{Row}(A)$. Let $(C_1, \cdots, C_b)$ be a partition of $C = \mathrm{Col}(A)$ into disjoint blocks and $(R_1, \cdots, R_b)$ the corresponding partition of $R = \mathrm{Row}(A)$ with $R_k = \psi(C_k)$ for $k = 1, \cdots, b$. We say that A is *block-triangularized* with respect to $(R_1, \cdots, R_b)$ and $(C_1, \cdots, C_b)$ if

$$A[R_k, C_l] = O \quad \text{for} \quad 1 \le l < k \le b.$$

If this is the case, we can bring A into an explicit upper block-triangular form $\bar{A} = PAP^{\mathrm{T}}$ in the ordinary sense by using a permutation matrix P, where it is tacitly assumed that $\mathrm{Row}(A) = \mathrm{Col}(A) = \{1, 2, \cdots, n\}$ and $\psi(j) = j$ for $j = 1, 2, \cdots, n$. For a general ψ, however, $\bar{A} = PAP^{\mathrm{T}}$ should be replaced by $\bar{A} = PA\Psi^{-1}P^{\mathrm{T}}$ with another permutation matrix Ψ representing ψ.

A partial order is induced among the blocks $\{C_k \mid k = 1, \cdots, b\}$ in a natural manner by the zero/nonzero structure of a block-triangular matrix $\bar{A}$. The partial order $\preceq$ is the reflexive and transitive closure of the relation defined by: $C_k \preceq C_l$ if $A[R_k, C_l] \neq O$.

Usually we want to find a finest partition of C as well as the corresponding one of R for which a given matrix A is block-triangularized. This problem can be treated successfully by means of a graph-theoretic method, as will be explained in §2.2.1.

Example 2.1.14. For a 6×6 matrix

$$A = \begin{array}{r|cccccc|} & 1 & 2 & 3 & 4 & 5 & 6 \\ \hline 1 & & a_{12} & a_{13} & & & \\ 2 & & a_{22} & & & a_{25} & \\ 3 & & & a_{33} & & & \\ 4 & a_{41} & & & a_{44} & & a_{46} \\ 5 & a_{51} & & a_{53} & & & \\ 6 & & & a_{63} & a_{64} & & a_{66} \\ \hline \end{array} \tag{2.11}$$

the finest block-triangular decomposition is given by

$$PAP^{\mathrm{T}} = \begin{array}{rr|cc|ccc|c|} & & \multicolumn{2}{c}{C_1} & \multicolumn{3}{c}{C_2} & \multicolumn{1}{c}{C_3} \\ & & 4 & 6 & 1 & 2 & 5 & 3 \\ \hline R_1 & 4 & a_{44} & a_{46} & a_{41} & & & \\ & 6 & a_{64} & a_{66} & & & & a_{63} \\ \hline & 1 & & & & a_{12} & & a_{13} \\ R_2 & 2 & & & & a_{22} & a_{25} & \\ & 5 & & & a_{51} & & & a_{53} \\ \hline R_3 & 3 & & & & & & a_{33} \\ \hline \end{array} \tag{2.12}$$

with $R_1 = C_1 = \{4, 6\} \preceq R_2 = C_2 = \{1, 2, 5\} \preceq R_3 = C_3 = \{3\}$. □

The second block-triangular decomposition is defined for a matrix A of any size, where no correspondence between $\mathrm{Col}(A)$ and $\mathrm{Row}(A)$ is assumed. Let $(C_0; C_1, \cdots, C_b; C_\infty)$ and $(R_0; R_1, \cdots, R_b; R_\infty)$, where $b \geq 0$, be partitions of $C = \mathrm{Col}(A)$ and $R = \mathrm{Row}(A)$, respectively, into disjoint blocks such that

$$\begin{array}{lll} |R_0| < |C_0| & \text{or} & |R_0| = |C_0| = 0, \\ |R_k| = |C_k| > 0 & \text{for} & k = 1, \cdots, b, \\ |R_\infty| > |C_\infty| & \text{or} & |R_\infty| = |C_\infty| = 0. \end{array} \tag{2.13}$$

We say that A is *block-triangularized* with respect to $(R_0; R_1, \cdots, R_b; R_\infty)$ and $(C_0; C_1, \cdots, C_b; C_\infty)$ if

$$A[R_k, C_l] = O \quad \text{for} \quad 0 \leq l < k \leq \infty. \tag{2.14}$$

The submatrices $A[R_0, C_0]$ and $A[R_\infty, C_\infty]$ are called the *horizontal tail* and the *vertical tail*, respectively. Clearly, if A is block-triangularized in this sense, we can put it into an explicit upper block-triangular form $\bar{A} = P_\mathrm{r} A P_\mathrm{c}$ in the ordinary sense by using certain permutation matrices P_r and P_c.

A partial order is induced among the blocks $\{C_k \mid k = 1, \cdots, b\}$ in a similar manner by the zero/nonzero structure of a block-triangular matrix A. The partial order $\preceq$ is the reflexive and transitive closure of the relation defined by: $C_k \preceq C_l$ if $A[R_k, C_l] \neq O$. It is often convenient to extend the partial order onto $\{C_0, C_\infty\} \cup \{C_k \mid k = 1, \cdots, b\}$ by defining

$$C_0 \preceq C_k \preceq C_\infty \qquad (\forall k). \tag{2.15}$$

We adopt this convention unless otherwise stated.

Usually we want to find finest partitions of R and C for which a given matrix A is block-triangularized. This problem can be treated successfully by another graph-theoretic method, called the Dulmage–Mendelsohn decomposition, to be explained in §2.2.3.

Example 2.1.15. If a transformation of the form $\bar{A} = P_\mathrm{r} A P_\mathrm{c}$ is applicable to the matrix A of Example 2.1.14, the finest block-triangular decomposition using two permutation matrices is given by

$$P_\mathrm{r} A P_\mathrm{c} = \begin{array}{ll|cccccc|} & & C_1 & & C_2 & C_3 & C_4 & C_5 \\ & & 4 & 6 & 5 & 2 & 1 & 3 \\ \hline R_1 & 4' & a_{44} & a_{46} & a_{41} & & & \\ & 6' & a_{64} & a_{66} & & & & a_{63} \\ R_2 & 2' & & & a_{25} & a_{22} & & \\ R_3 & 1' & & & & a_{12} & & a_{13} \\ R_4 & 5' & & & & & a_{51} & a_{53} \\ R_5 & 3' & & & & & & a_{33} \\ \hline \end{array} . \tag{2.16}$$

This consists of five blocks $(R_1, C_1) = (\{4', 6'\}, \{4, 6\})$, $(R_2, C_2) = (\{2'\}, \{5\})$, $(R_3, C_3) = (\{1'\}, \{2\})$, $(R_4, C_4) = (\{5'\}, \{1\})$, $(R_5, C_5) = (\{3'\}, \{3\})$ with partial order $C_1 \preceq C_2 \preceq C_3 \preceq C_5$, $C_4 \preceq C_5$. □

Remark 2.1.16. The two kinds of decompositions above are closely related as follows, and this fact seems to cause complications and confusions in the literature. A considerable number of papers propose or describe a "two-stage method," so to speak, that first chooses $\psi : \mathrm{Col}(A) \to \mathrm{Row}(A)$ such that $A_{\psi(j),j} \neq 0$ ($j \in \mathrm{Col}(A)$) and then finds the finest decomposition of the first kind with respect to the chosen ψ. This amounts to a decomposition under a transformation $P_{\mathrm{r}}AP_{\mathrm{c}} = PA(\Psi^{-1}P^{\mathrm{T}})$, where Ψ is the permutation matrix representing ψ. The following points are emphasized here concerning this "two-stage method."

(1) The decomposition produced by the "two-stage method" depends apparently on the choice of ψ. The resulting decomposition, however, is not affected by the nonuniqueness of ψ, but coincides with the finest decomposition under a transformation of the form $P_{\mathrm{r}}AP_{\mathrm{c}}$ (see the algorithm for the DM-decomposition in §2.2.3). In this sense, the "two-stage method" is fully justified from the mathematical point of view.

(2) Still, the "two-stage method" seems to lack in philosophical soundness. The invariance (or insensitivity) of the resulting decomposition to the choice of ψ indicates that the "two-stage method" based on $PA\Psi^{-1}P^{\mathrm{T}}$ should be recognized in a different manner, more intrinsically without reference to ψ. It can be said that the "two-stage method" is not so much a decomposition concept as an algorithmic procedure for computing the (finest) decomposition under transformations of the form $P_{\mathrm{r}}AP_{\mathrm{c}}$. □

In applications of the second block-triangularization technique it is often required to impose an additional condition

$$\mathrm{rank}\, A[R_k, C_k] = \min(|R_k|, |C_k|) \quad \text{for} \quad k = 0, 1, \cdots, b, \infty \tag{2.17}$$

on the diagonal blocks in the decomposition. If this is the case, A is said to be *properly block-triangularized* with respect to $(R_0; R_1, \cdots, R_b; R_\infty)$ and $(C_0; C_1, \cdots, C_b; C_\infty)$. Note that the additional condition (2.17) is of numerical nature, while the condition (2.14) refers to the zero/nonzero structure only.

Not every matrix has a proper block-triangular form. Consider, for example, $A = \begin{bmatrix} 1 & 1 \\ 1 & 1 \end{bmatrix}$, which can never be properly block-triangularized for any partitions. The term-rank is the key concept for the statement of a necessary and sufficient condition for the existence of a proper block-triangular form.

Proposition 2.1.17. *A matrix A can be put in a proper block-triangular form with a suitable choice of partitions of R and C, if and only if* $\mathrm{rank}\, A = \text{term-rank}\, A$.

Proof. This will be proven later as an immediate corollary of Proposition 2.2.26. ■

In this book we often encounter questions of the following type:

> A class of matrices and a class of "admissible transformations" for the class of matrices are specified. Given a matrix A belonging to the class, can we transform it to a proper block-triangular matrix $\bar{A}$ by means of an admissible transformation?

By Proposition 2.1.17 this question is equivalent to:

> Given a matrix A, can we transform it by means of an admissible transformation to a matrix $\bar{A}$ such that $\mathrm{rank}\,\bar{A} = \text{term-rank}\,\bar{A}$?

The simplest problem of this kind is the case where the class of matrices comprises all the matrices and any equivalence transformation ($\bar{A} = S_{\mathrm{r}} A S_{\mathrm{c}}$ with nonsingular S_{r} and S_{c}) is admissible. In this case the answer is in the affirmative and the proper block-triangular matrix $\bar{A}$ is given by the rank normal form of A, i.e., $\bar{A} = \begin{bmatrix} O & I_r \\ O & O \end{bmatrix}$ with I_r denoting the identity matrix of size $r = \mathrm{rank}\,A$. Other instances of such questions, more of combinatorial nature, include: the Dulmage–Mendelsohn decomposition of generic matrices (§2.2.3), the combinatorial canonical form of layered mixed matrices (§4.4), the combinatorial canonical form of matrices with respect to pivotal transformations (Remark 4.7.10), and the decomposition of generic partitioned matrices (§4.8.4).

2.2 Graph

Graphs are convenient tools to represent the structures of matrices and systems. Decompositions of graphs, when combined with appropriate physical interpretations, lead to effective decomposition methods for matrices and systems.

2.2.1 Directed Graph and Bipartite Graph

Let $G = (V, A)$ be a *directed graph* with vertex set V and arc set A. For an arc $a \in A$, $\partial^+ a$ denotes the *initial vertex* of a, $\partial^- a$ the *terminal vertex* of a, and $\partial a = \{\partial^+ a, \partial^- a\}$ the set of *vertices incident* to a. For a vertex $v \in V$, $\delta^+ v$ means the set of arcs going out of v, $\delta^- v$ the set of arcs coming into v, and $\delta v = \delta^+ v \cup \delta^- v$ the set of arcs incident to v. The *incidence matrix* of G is a matrix with row set indexed by V and column set by A such that, for $v \in V$ and $a \in A$, the (v, a) entry is equal to 1 if $v = \partial^+ a$, to -1 if $v = \partial^- a$, and to 0 otherwise, where, for an arc a with $\partial^+ a = \partial^- a$, the corresponding column is set to zero.

For V' ($\subseteq V$) the *(vertex-)induced subgraph*, or the *section graph*, on V' is a graph $G' = (V', A')$ with $A' = \{a \in A \mid \partial^+ a \in V', \partial^- a \in V'\}$. We also denote G' by $G \setminus V''$, where $V'' = V \setminus V'$, saying that G' is obtained from G by deleting the vertices in V''.

For two vertices u and v, we say that v is *reachable* from u on G, which we denote as $u \overset{*}{\longrightarrow} v$, if there exists a directed path from u to v on G. Based on the reachability we define an equivalence relation $\sim$ on V by: $u \sim v \iff [u \overset{*}{\longrightarrow} v$ and $v \overset{*}{\longrightarrow} u]$. In fact it is straightforward to verify (i) [reflexivity] $v \sim v$, (ii) [symmetry] $u \sim v \Rightarrow v \sim u$, and (iii) [transitivity] $u \sim v, v \sim w \Rightarrow u \sim w$. Accordingly, the vertex set V is partitioned into equivalence classes $\{V_k\}_k$, called *strongly connected components* (or *strong components*, in short). Namely, two vertices u and v belong, by definition, to the same strong component if and only if $u \overset{*}{\longrightarrow} v$ and $v \overset{*}{\longrightarrow} u$. A partial order $\preceq$ can be defined on the family $\{V_k\}_k$ of strong components by

$$V_k \preceq V_l \iff v_l \overset{*}{\longrightarrow} v_k \quad \text{on } G \text{ for some } v_k \in V_k \text{ and } v_l \in V_l.$$

Each strong component V_k determines a vertex-induced subgraph $G_k = (V_k, A_k)$ of G, also called a strong component of G. The *partial order* $\preceq$ is induced naturally on the family of strong components $\{G_k\}_k$ by: $G_k \preceq G_l \iff V_k \preceq V_l$. The decomposition of G into partially ordered strong components $\{G_k\}_k$ is referred to as the *strong component decomposition* of G. An efficient algorithm of complexity $\mathrm{O}(|A|)$ is known for the strong component decomposition (see Aho–Hopcroft–Ullman [1, 2], Tarjan [310]).

For an $n \times n$ matrix $A = (A_{ij} \mid i,j = 1,\cdots,n)$ we can represent the zero/nonzero pattern of the matrix in terms of a directed graph[2] $G = (V, \tilde{A})$ with $V = \{1,\cdots,n\}$ and $\tilde{A} = \{(j,i) \mid A_{ij} \neq 0\}$. The strong component decomposition of the graph G corresponds to a (finest possible) block-triangular decomposition of the matrix A by means of a simultaneous permutation of the rows and the columns.

Example 2.2.1. For the matrix A of (2.11) the zero/nonzero structure can be represented by a directed graph G of Fig. 2.1. The graph has three strong components, $V_1 = \{4,6\}$, $V_2 = \{1,2,5\}$, $V_3 = \{3\}$, with $V_1 \preceq V_2 \preceq V_3$. The strong component decomposition of G yields a block-triangular form given in (2.12), where V_k corresponds to $R_k = C_k$ for $k = 1,2,3$. □

In applications of linear algebra it is often crucial to recognize the relevant transformations associated with a matrix. For example, we can talk of the Jordan canonical form of A only if A is subject to *similarity transformations*, SAS^{-1} with nonsingular S. This is the case when A corresponds to a mapping in a single vector space. If a matrix A corresponds to a mapping between a pair of different vector spaces, it is subject to *equivalence transformations*, $S_{\mathrm{r}}AS_{\mathrm{c}}$ with nonsingular matrices S_{r} and S_{c}. In this case, it is meaningless to consider the Jordan canonical form of A, whereas it is still sound to talk about the rank of A. Consider, for instance, a state-space equation $\dot{\boldsymbol{x}} = A\boldsymbol{x} + B\boldsymbol{u}$ for a dynamical system. The matrix A here is subject to similarity transformations, and B to equivalence transformations. It should be clear

[2] This graph is called the *Coates graph* in Chen [34].

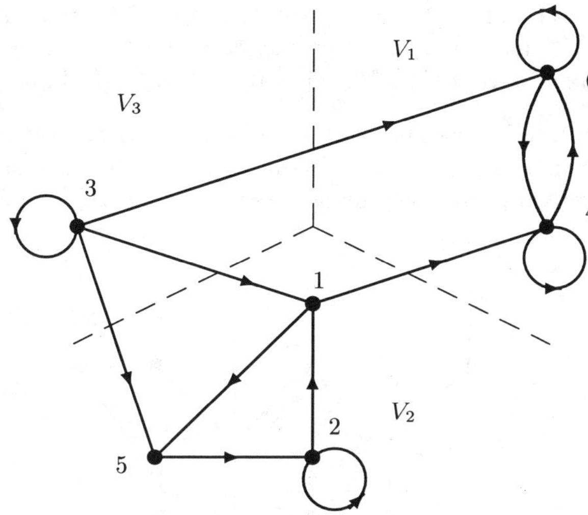

Fig. 2.1. Strong component decomposition (Example 2.2.1)

that even if B happens to be square, having as many rows as columns, it is meaningless to consider the Jordan canonical form of B.

Such distinctions in the nature of matrices should be respected also in the combinatorial analysis of matrices. As we have observed, the decomposition of a matrix A through the strong component decomposition of the associated graph gives the finest block-triangularization under a transformation of the form $PAP^{\mathrm{T}} = PAP^{-1}$ with a permutation matrix P. For this decomposition method to be applicable it is assumed tacitly that the matrix in question represents a mapping in a single vector space and is subject to similarity transformations, so that the structure of the matrix can in turn be represented by the associated graph defined above.

For a matrix A under equivalence transformations, on the other hand, a natural transformation of a combinatorial nature will be given by $P_{\mathrm{r}}AP_{\mathrm{c}}$ with two permutation matrices P_{r} and P_{c}. For such a matrix there is no reason for restricting P_{c} to be the inverse of P_{r}, and accordingly the strong component decomposition does not make much sense. Note that the associated graph itself is not very meaningful, since the associated graph does not remain isomorphic when the matrix A changes to $P_{\mathrm{r}}AP_{\mathrm{c}}$.

The structure of such a matrix A (subject to equivalence transformations) can be better represented by another graph $G = (V, \tilde{A})$ with $V = \mathrm{Col}(A) \cup \mathrm{Row}(A)$ and $\tilde{A} = \{(j,i) \mid A_{ij} \neq 0\}$. By definition, each arc has the initial vertex in $\mathrm{Col}(A)$ and the terminal vertex in $\mathrm{Row}(A)$, and therefore this graph is a bipartite graph. Recall that, in general, a graph $G = (V, \tilde{A})$ is called a *bipartite graph* if the vertex set V can be partitioned into two disjoint parts, say V^+ and V^-, in such a way that $|\partial a \cap V^+| = |\partial a \cap V^-| = 1$ for all $a \in \tilde{A}$.

We write $G = (V^+, V^-; \tilde{A})$ for a bipartite graph and often assume $\partial^+ a \in V^+$ and $\partial^- a \in V^-$ for $a \in \tilde{A}$. In this notation the bipartite graph associated with a matrix A is $G = (V^+, V^-; \tilde{A})$ with $V^+ = \mathrm{Col}(A)$ and $V^- = \mathrm{Row}(A)$.

Example 2.2.2. In case the matrix A of (2.11) in Example 2.1.14 represents a mapping between a pair of different vector spaces, the structure of A is expressed more appropriately by a bipartite graph, shown in Fig. 2.2. □

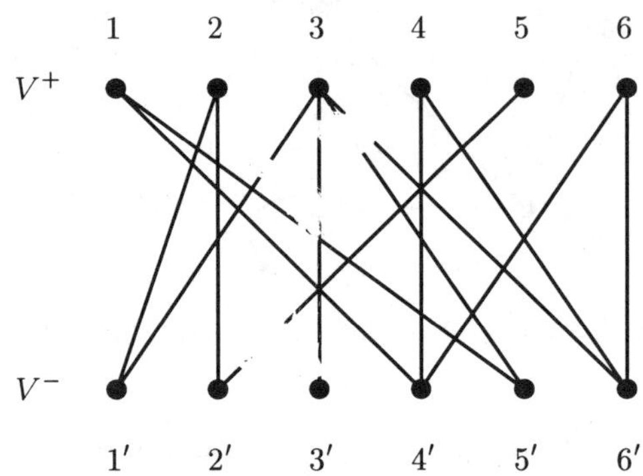

Fig. 2.2. Bipartite graph representation (Example 2.2.2)

The decomposition under a transformation of the form $P_{\mathrm{r}} A P_{\mathrm{c}}$ will be treated in §2.2.3 as the Dulmage–Mendelsohn decomposition.

Remark 2.2.3. For a nonzero entry A_{ij} of a matrix A, the associated bipartite graph, as defined above, has an arc (j, i) directed from column j to row i. This convention makes sense when we consider signal-flow graphs and is often found in the literature of engineering. It is also legitimate to direct an arc from row i to column j. In this book we use whichever convention is more convenient in the context. □

Let us dwell on the distinction of the two kinds of graph representations by referring to the two kinds of descriptions of dynamical systems introduced in §1.2.2, namely, the standard form of state-space equations (1.20):

$$\dot{\boldsymbol{x}} = \hat{A}\boldsymbol{x} + \hat{B}\boldsymbol{u}$$

and the descriptor form (1.22):

$$\bar{F}\dot{\boldsymbol{x}} = \bar{A}\boldsymbol{x} + \bar{B}\boldsymbol{u}.$$

The matrix $\hat{A}$ in the standard form has a natural one-to-one correspondence between $\mathrm{Row}(\hat{A})$ and $\mathrm{Col}(\hat{A})$, since the ith equation describes the dynamics of the ith variable. The matrix $\bar{A}$ in the descriptor form, on the other hand, has no such natural correspondence between $\mathrm{Row}(\bar{A})$ and $\mathrm{Col}(\bar{A})$. In other words, the concept of "diagonal" is meaningful for the matrix $\hat{A}$ and not for the matrix $\bar{A}$. Mathematically, $\hat{A}$ is subject to similarity transformations, $S\hat{A}S^{-1}$, and $\bar{A}$ to equivalence transformations, $S_{\mathrm{r}}\bar{A}S_{\mathrm{c}}$.

Accordingly, the standard form (1.20) is represented by a directed graph $G = (V, A)$ called the *signal-flow graph*.[3] The vertex set V and the arc set A are defined by

$$V = X \cup U, \qquad X = \{x_1, \cdots, x_n\}, \quad U = \{u_1, \cdots, u_m\},$$
$$A = \{(x_j, x_i) \mid \hat{A}_{ij} \neq 0\} \cup \{(u_j, x_i) \mid \hat{B}_{ij} \neq 0\}.$$

The natural graphical representation of the descriptor form (1.22), on the other hand, is the bipartite graph $G = (V^+, V^-; \tilde{A})$ associated with the matrix $D(s) = [\bar{A} - s\bar{F} \mid \bar{B}]$, where s is an indeterminate. Namely, $V^+ = \mathrm{Col}(D) = X \cup U$ stands for the set of variables and $V^- = \mathrm{Row}(D)$ for the set of equations, say $\{e_1, \cdots, e_n\}$, and the arcs correspond to the nonvanishing entries of $D(s)$, i.e.,

$$\tilde{A} = \{(x_j, e_i) \mid \bar{A}_{ij} \neq 0 \text{ or } \bar{F}_{ij} \neq 0\} \cup \{(u_j, e_i) \mid \bar{B}_{ij} \neq 0\}.$$

It is sometimes convenient to assign weight 1 to arc (x_j, e_i) with $\bar{F}_{ij} \neq 0$ and weight 0 to the other arcs.

The above distinction between standard form and the descriptor form implies, in particular, that the finest decomposition of $\hat{A}$ is obtained through the strong component decomposition, whereas that of $\bar{A}$ is through the Dulmage–Mendelsohn decomposition.

For the standard form (1.20) another graph representation is sometimes useful. For $k \geq 1$ the *dynamic graph* of time-span k is defined to be $G_0^k = (X_0^k \cup U_0^{k-1}, A_0^{k-1})$ with

$$X_0^k = \bigcup_{t=0}^{k} X^t, \quad X^t = \{x_i^t \mid i = 1, \cdots, n\} \quad (t = 0, 1, \cdots, k),$$
$$U_0^{k-1} = \bigcup_{t=0}^{k-1} U^t, \quad U^t = \{u_j^t \mid j = 1, \cdots, m\} \quad (t = 0, 1, \cdots, k-1),$$
$$A_0^{k-1} = \{(x_j^t, x_i^{t+1}) \mid \hat{A}_{ij} \neq 0; t = 0, 1, \cdots, k-1\}$$
$$\cup \{(u_j^t, x_i^{t+1}) \mid \hat{B}_{ij} \neq 0; t = 0, 1, \cdots, k-1\}.$$

Example 2.2.4. The graph representations are illustrated for the mechanical system treated in §1.2.2 (see also Fig. 1.5). The signal-flow graph representing the standard form (1.21) and the bipartite graph associated with

[3] The signal-flow graph defined here is different from the *Mason graph* (Chen [34]), which is sometimes called the signal-flow graph, too.

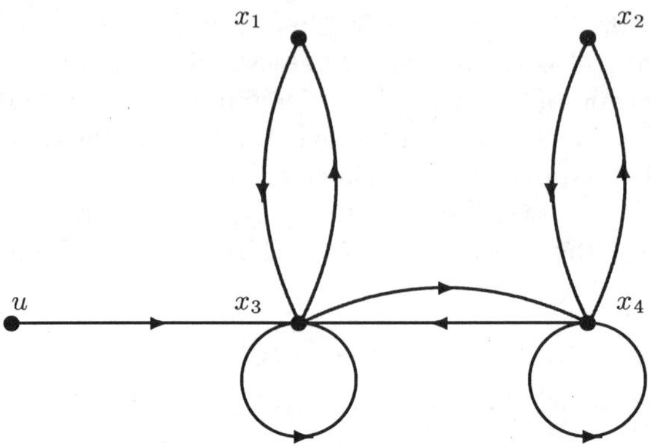

Fig. 2.3. Signal-flow graph of the mechanical system of Fig. 1.5

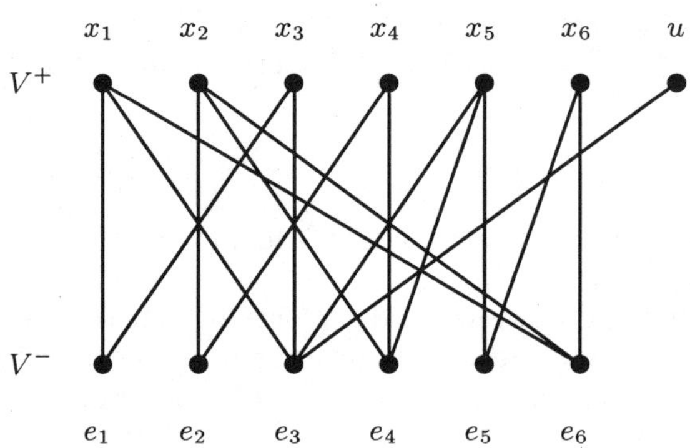

Fig. 2.4. Bipartite graph of the mechanical system of Fig. 1.5

the descriptor form (1.23) are given in Figs. 2.3 and 2.4, respectively. The dynamic graph of time-span $k = 4$ for (1.21) is also depicted in Fig. 2.5. □

2.2.2 Jordan–Hölder-type Theorem for Submodular Functions

We describe here a general decomposition principle of submodular functions, known as the Jordan–Hölder-type theorem for submodular functions. We shall make essential use of this general framework in a number of different places in this book. Recall that we have already encountered in §2.1.3 a typical submodular function, the rank function ρ of (2.6) associated with a matrix.

Let V be a finite set, $\mathcal{L}$ ($\neq \emptyset$) be a *sublattice* of the boolean lattice 2^V:

$$X, Y \in \mathcal{L} \Rightarrow X \cup Y, X \cap Y \in \mathcal{L}, \tag{2.18}$$

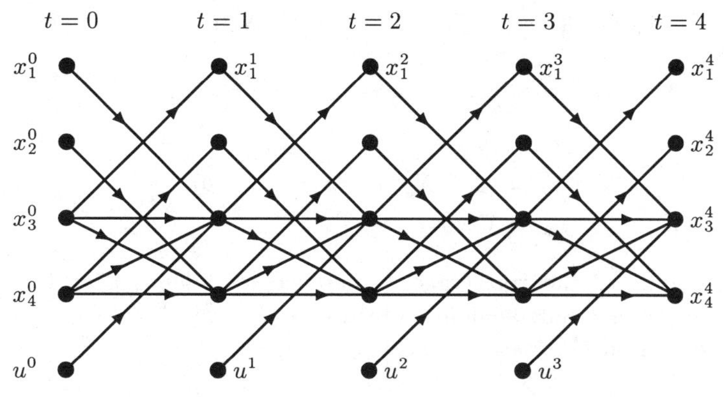

Fig. 2.5. Dynamic graph G_0^4 of the mechanical system of Fig. 1.5

and $f : 2^V \to \mathbf{R}$ be a *submodular function*:

$$f(X) + f(Y) \geq f(X \cup Y) + f(X \cap Y), \qquad X, Y \subseteq V. \tag{2.19}$$

We say that $\mathcal{L}$ is an *f-skeleton* if, in addition, f is *modular* on $\mathcal{L}$:

$$f(X) + f(Y) = f(X \cup Y) + f(X \cap Y), \qquad X, Y \in \mathcal{L}. \tag{2.20}$$

The decomposition principle applies to a pair of a submodular function f and an f-skeleton $\mathcal{L}$. In principle, an f-skeleton $\mathcal{L}$ can be specified quite generally, but in our subsequent applications it is often derived from f itself as the family of the minimizers, as follows (Ore [257]).

Theorem 2.2.5. *For a submodular function $f : 2^V \to \mathbf{R}$, the family of the minimizers:*

$$\mathcal{L}_{\min}(f) = \{X \subseteq V \mid f(X) \leq f(Y), \forall Y \subseteq V\} \tag{2.21}$$

forms a sublattice of 2^V, and moreover it is an f-skeleton.

Proof. Let α denote the minimum value of f. For $X, Y \in \mathcal{L}_{\min}(f)$ we have

$$2\alpha = f(X) + f(Y) \geq f(X \cup Y) + f(X \cap Y) \geq 2\alpha,$$

which shows $f(X \cup Y) = f(X \cap Y) = \alpha$, i.e., $X \cup Y, X \cap Y \in \mathcal{L}_{\min}(f)$. ■

First we consider a representation of a sublattice $\mathcal{L}$ of 2^V, independent of a submodular function f. This is a fundamental result from lattice theory, called Birkhoff's representation theorem, which shows a one-to-one correspondence between sublattices of 2^V and pairs of a partition of V into blocks with a partial order among the blocks. This correspondence is given as follows.

Given a sublattice $\mathcal{L}$ of 2^V, take any *maximal ascending chain* of $\mathcal{L}$:

$$X_0 \ (= \min \mathcal{L}) \subsetneqq X_1 \subsetneqq X_2 \subsetneqq \cdots \subsetneqq X_b \ (= \max \mathcal{L}), \tag{2.22}$$

where $X_k \in \mathcal{L}$ $(k = 0, 1, \cdots, b)$, and put

$$\begin{aligned} V_0 &= X_0, \\ V_k &= X_k \setminus X_{k-1} \quad (k = 1, \cdots, b), \\ V_\infty &= V \setminus X_b. \end{aligned} \tag{2.23}$$

Then the family of the "intervals" (difference sets) $\{V_k \mid k = 1, \cdots, b\}$ is uniquely determined independently of the choice of the chain. A partial order $\preceq$ is introduced on $\{V_k \mid k = 1, \cdots, b\}$ by

$$V_k \preceq V_l \quad \iff \quad [V_l \subseteq X \in \mathcal{L} \ \Rightarrow \ V_k \subseteq X]. \tag{2.24}$$

In this way, a sublattice $\mathcal{L}$ determines a pair of a *partition* $\{V_0; V_1, \cdots, V_b; V_\infty\}$ of V and a *partial order* $\preceq$ on $\{V_1, \cdots, V_b\}$, which will be denoted by

$$\mathcal{P}(\mathcal{L}) = (\{V_0; V_1, \cdots, V_b; V_\infty\}, \preceq). \tag{2.25}$$

Note that $V_k \neq \emptyset$ for $k = 1, \cdots, b$, whereas V_0 and V_∞ are distinguished blocks that can be empty. By (2.24) the indexing of the blocks is consistent with the partial order in the sense that $V_k \preceq V_l \Rightarrow k \leq l$.

It is often convenient to extend the partial order onto $\{V_0, V_\infty\} \cup \{V_k \mid k = 1, \cdots, b\}$ by defining

$$V_0 \preceq V_k \preceq V_\infty \qquad (\forall k). \tag{2.26}$$

We adopt this convention unless otherwise stated.

Remark 2.2.6. In the above argument we have constructed the partition $\{V_0; V_1, \cdots, V_b; V_\infty\}$ with reference to a particular maximal chain of $\mathcal{L}$. There is another way of construction of $\mathcal{P}(\mathcal{L})$ from $\mathcal{L}$ that refers to join-irreducible elements instead of a maximal chain. An element $Z \in \mathcal{L}$ is said to be *join-irreducible* if $Z = Z' \cup Z''$ with $Z', Z'' \in \mathcal{L}$ means $Z' = Z$ or $Z'' = Z$. Let $\{Z_k \mid k = 1, \cdots, b\}$ be the family of all the join-irreducible elements of $\mathcal{L}$ distinct from $\min \mathcal{L}$. For each Z_k there exists a unique element of $\mathcal{L}$, say Y_k, that lies immediately below Z_k (i.e., $Y_k \subset Z_k$ and $\nexists X \in \mathcal{L}$ such that $Y_k \subset X \subset Z_k$). Define $V_k = Z_k \setminus Y_k$ for $k = 1, \cdots, b$, $V_0 = \min \mathcal{L}$ and $V_\infty = V \setminus \max \mathcal{L}$. Then $\{V_0; V_1, \cdots, V_b; V_\infty\}$ forms a partition of V. A partial order $\preceq$ is induced on $\{V_k \mid k = 1, \cdots, b\}$ by $[V_k \preceq V_l \iff Z_k \subseteq Z_l]$. It is known that this construction coincides with the one defined by (2.23) and (2.24). □

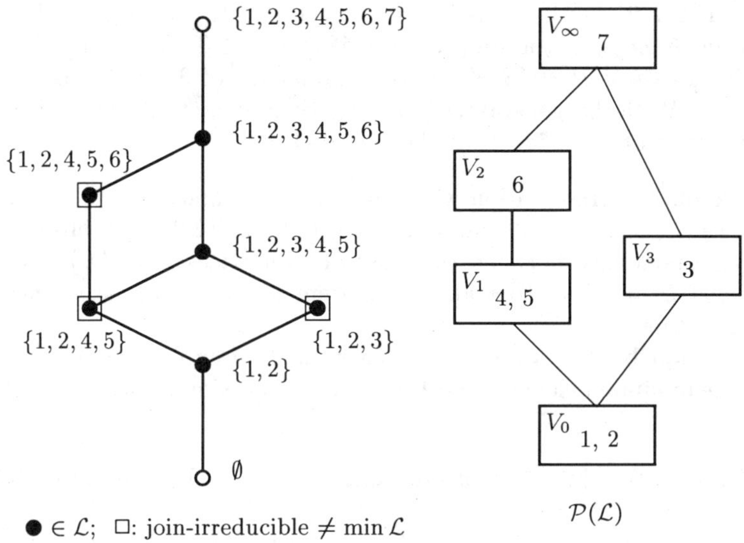

Fig. 2.6. Representation of a sublattice by partially ordered blocks

Example 2.2.7. Let $\mathcal{L}$ be a sublattice of 2^V indicated by • in Fig. 2.6, where $V = \{1,2,3,4,5,6,7\}$. We have $\min \mathcal{L} = \{1,2\}$ and $\max \mathcal{L} = \{1,2,3,4,5,6\}$. A maximal chain (2.22) with

$$X_0 = \{1,2\}, \quad X_1 = \{1,2,4,5\}, \quad X_2 = \{1,2,4,5,6\}, \quad X_3 = \{1,2,3,4,5,6\}$$

yields $V_0 = \{1,2\}$, $V_1 = \{4,5\}$, $V_2 = \{6\}$, $V_3 = \{3\}$, $V_\infty = \{7\}$. The partial order $\mathcal{P}(\mathcal{L})$ is depicted also in Fig. 2.6. There are three join-irreducible elements distinct from $\min \mathcal{L}$, i.e., $Z_1 = \{1,2,4,5\}$, $Z_2 = \{1,2,4,5,6\}$, $Z_3 = \{1,2,3\}$, indicated by □ in Fig. 2.6, and the immediately-below elements Y_k are $Y_1 = \{1,2\}$, $Y_2 = \{1,2,4,5\}$, $Y_3 = \{1,2\}$. Note that Z_k corresponds to V_k as $V_k = Z_k \setminus Y_k$ for $k = 1,2,3$. □

Conversely, suppose we are given $\mathcal{P} = (\{V_0; V_1, \cdots, V_b; V_\infty\}, \preceq)$, a pair of a *partition* of V with two distinguished subsets V_0 and V_∞ and a *partial order* $\preceq$ on $\{V_1, \cdots, V_b\}$. Define $\mathcal{L}(\mathcal{P}) \subseteq 2^V$ by

$$\begin{aligned}\mathcal{L}(\mathcal{P}) = \{X \subseteq V \mid &\text{(i) } V_0 \subseteq X \subseteq V \setminus V_\infty;\\ &\text{(ii) } X \cap V_l \neq \emptyset,\ V_k \preceq V_l\ (1 \leq k, l \leq b) \Rightarrow V_k \subseteq X\},\end{aligned} \tag{2.27}$$

which implies that $X \in \mathcal{L}(\mathcal{P})$ can be expressed as $X = \bigcup_{k \in I \cup \{0\}} V_k$ for some $I \subseteq \{1, \cdots, b\}$. Then $\mathcal{L} = \mathcal{L}(\mathcal{P})$ forms a sublattice of 2^V with $\min \mathcal{L} = V_0$ and $\max \mathcal{L} = V \setminus V_\infty$.

Example 2.2.8. For the $\mathcal{P}$ in Fig. 2.6, $\mathcal{L}(\mathcal{P})$ consists of six elements: $V_0 = \{1,2\}$, $V_0 \cup V_1 = \{1,2,4,5\}$, $V_0 \cup V_3 = \{1,2,3\}$, $V_0 \cup V_1 \cup V_2 = \{1,2,4,5,6\}$, $V_0 \cup V_1 \cup V_3 = \{1,2,3,4,5\}$, $V_0 \cup V_1 \cup V_2 \cup V_3 = \{1,2,3,4,5,6\}$. □

Remark 2.2.9. For a partially ordered set $\mathcal{P}=(S,\preceq)$ in general, $T\subseteq S$ is called an *order ideal* (or simply *ideal*) if $[s\preceq t\in T\Rightarrow s\in T]$. The family of order ideals of $\mathcal{P}=(S,\preceq)$ forms a sublattice of 2^S (with respect to set inclusion). With this general terminology, we may say that $\mathcal{L}(\mathcal{P})$ is isomorphic to the lattice of order ideals of $(\{V_1,\cdots,V_b\},\preceq)$. □

Birkhoff's representation theorem states, roughly, that the mappings $\Phi:\mathcal{L}\mapsto\mathcal{P}(\mathcal{L})$ of (2.25) and $\Psi:\mathcal{P}\mapsto\mathcal{L}(\mathcal{P})$ of (2.27) establish a one-to-one correspondence between the class of sublattices $\Lambda=\{\mathcal{L}\}$ and that of partitions $\Pi=\{\mathcal{P}\}$. To make the statement more precise we need some more notation.

We denote by $\Lambda(V;V_0,V_\infty)$ the collection of the sublattices of 2^V that have V_0 as the minimum element and $V\setminus V_\infty$ as the maximum, where $V_0\cap V_\infty=\emptyset$, i.e.,

$$\Lambda(V;V_0,V_\infty)=\{\mathcal{L}\mid\mathcal{L}\text{: sublattice of }2^V,\ \min\mathcal{L}=V_0,\max\mathcal{L}=V\setminus V_\infty\}. \tag{2.28}$$

For $\mathcal{L}_1,\mathcal{L}_2\in\Lambda(V;V_0,V_\infty)$, $\mathcal{L}_1\wedge\mathcal{L}_2$ will mean the sublattice $\mathcal{L}_1\cap\mathcal{L}_2$, and $\mathcal{L}_1\vee\mathcal{L}_2$ the sublattice generated by $\mathcal{L}_1\cup\mathcal{L}_2$. The family $\Lambda(V;V_0,V_\infty)$ forms a lattice $(\Lambda(V;V_0,V_\infty),\vee,\wedge)$ (in the sense of Remark 2.2.14) with respect to $\wedge$ and $\vee$ thus defined.

We denote by

$$\Pi(V;V_0,V_\infty)=\{\mathcal{P}\mid\mathcal{P}=(\{V_0;V_1,\cdots,V_b;V_\infty\},\preceq)\} \tag{2.29}$$

the collection of the pairs of a partition $\{V_0;V_1,\cdots,V_b;V_\infty\}$ of V with two distinguished subsets V_0 and V_∞ and a partial order $\preceq$ on $\{V_1,\cdots,V_b\}$. A partial order, denoted also as $\preceq$, can be introduced on $\Pi(V;V_0,V_\infty)$ with respect to the refinement relation as follows. For $\mathcal{P}_1=(\{V_0;\{V_k^{(1)}\};V_\infty\},\preceq_1)$, $\mathcal{P}_2=(\{V_0;\{V_l^{(2)}\};V_\infty\},\preceq_2)\in\Pi(V;V_0,V_\infty)$, we say $\mathcal{P}_1\preceq\mathcal{P}_2$ if and only if

(i) $\{V_k^{(1)}\}$ is a refinement of $\{V_l^{(2)}\}$ as a partition of $V\setminus(V_0\cup V_\infty)$, that is, any $V_k^{(1)}$ is contained in some $V_l^{(2)}$, and
(ii) $V_{k_1}^{(1)}\subseteq V_{l_1}^{(2)},\ V_{k_2}^{(1)}\subseteq V_{l_2}^{(2)},\ V_{k_1}^{(1)}\preceq_1 V_{k_2}^{(1)}\Longrightarrow V_{l_1}^{(2)}\preceq_2 V_{l_2}^{(2)}$.

It is easy to see that the partially ordered set $(\Pi(V;V_0,V_\infty),\preceq)$ thus defined forms a lattice $(\Pi(V;V_0,V_\infty),\vee,\wedge)$, in which $\mathcal{P}_1\vee\mathcal{P}_2$ is the finest *common aggregation* of $\mathcal{P}_1$ and $\mathcal{P}_2$ and $\mathcal{P}_1\wedge\mathcal{P}_2$ is the coarsest *common refinement* of $\mathcal{P}_1$ and $\mathcal{P}_2$.

We are now ready to state Birkhoff's representation theorem.

Theorem 2.2.10 (Birkhoff's representation theorem). *Let V_0 and V_∞ be disjoint subsets of V. The two families $\Lambda(V;V_0,V_\infty)$ and $\Pi(V;V_0,V_\infty)$ are in one-to-one correspondence to each other through mutually inverse mappings, $\Phi:\Lambda(V;V_0,V_\infty)\to\Pi(V;V_0,V_\infty)$ and $\Psi:\Pi(V;V_0,V_\infty)\to\Lambda(V;V_0,V_\infty)$, defined by $\Phi:\mathcal{L}\mapsto\mathcal{P}(\mathcal{L})$ of (2.25) and $\Psi:\mathcal{P}\mapsto\mathcal{L}(\mathcal{P})$ of (2.27),* *respectively. Moreover, for* $\mathcal{L}_1, \mathcal{L}_2 \in \Lambda(V; V_0, V_\infty)$ *and* $\mathcal{P}_1, \mathcal{P}_2 \in \Pi(V; V_0, V_\infty)$ *it holds that*

$$\mathcal{P}(\mathcal{L}_1 \wedge \mathcal{L}_2) = \mathcal{P}(\mathcal{L}_1) \vee \mathcal{P}(\mathcal{L}_2), \qquad \mathcal{P}(\mathcal{L}_1 \vee \mathcal{L}_2) = \mathcal{P}(\mathcal{L}_1) \wedge \mathcal{P}(\mathcal{L}_2),$$
$$\mathcal{L}(\mathcal{P}_1 \wedge \mathcal{P}_2) = \mathcal{L}(\mathcal{P}_1) \vee \mathcal{L}(\mathcal{P}_2), \qquad \mathcal{L}(\mathcal{P}_1 \vee \mathcal{P}_2) = \mathcal{L}(\mathcal{P}_1) \wedge \mathcal{L}(\mathcal{P}_2).$$

□

Remark 2.2.11. Theorem 2.2.10 above, referring to two distinguished subsets V_0 and V_∞, is slightly different from the standard statement of Birkhoff's representation theorem. This is for convenience in our subsequent applications. The essence, however, lies in the case of $V_0 = V_\infty = \emptyset$. □

Concerning the partial order, we introduce the following additional notation:

$$V_k \prec V_l \iff V_k \preceq V_l \ \text{ and } V_k \neq V_l; \tag{2.30}$$

$$V_k \prec\!\!\cdot\, V_l \iff \begin{cases} \text{(i) } V_k \prec V_l \quad \text{and} \\ \text{(ii) } \nexists\, V_j \text{ such that } \ V_k \prec V_j \prec V_l; \end{cases} \tag{2.31}$$

$$\langle V_l \rangle = \bigcup_{V_k \prec V_l} V_k. \tag{2.32}$$

Example 2.2.12. In Example 2.2.7 (see Fig. 2.6) it holds that $V_0 \prec\!\!\cdot\, V_1$ and $V_1 \prec\!\!\cdot\, V_2$, while $V_0 \prec\!\!\cdot\, V_2$ is not true. We have $\langle V_0 \rangle = \emptyset$, $\langle V_1 \rangle = V_0 = \{1, 2\}$, $\langle V_2 \rangle = V_0 \cup V_1 = \{1, 2, 4, 5\}$, $\langle V_3 \rangle = V_0 = \{1, 2\}$, $\langle V_\infty \rangle = V_0 \cup V_1 \cup V_2 \cup V_3 = \{1, 2, 3, 4, 5, 6\}$. Note that $\langle V_k \rangle = Y_k$ for $k = 1, 2, 3$. □

So far we have considered how a sublattice $\mathcal{L}$ of 2^V induces a decomposition of the ground set V into partially ordered blocks. We now go on to explain how a submodular function f is decomposed into minors on those blocks if it is modular on $\mathcal{L}$. With reference to the maximal chain (2.22) we define $f_k : 2^{V_k} \to \mathbf{R}$ by

$$\begin{aligned} f_0(Y) &= f(Y), \qquad Y \subseteq V_0, \\ f_k(Y) &= f(X_{k-1} \cup Y) - f(X_{k-1}), \qquad Y \subseteq V_k \quad (k = 1, \cdots, b), \\ f_\infty(Y) &= f(X_b \cup Y) - f(X_b), \qquad Y \subseteq V_\infty. \end{aligned} \tag{2.33}$$

Obviously, each $f_k : 2^{V_k} \to \mathbf{R}$ is a submodular function.

The following theorem is sometimes called the *Jordan–Hölder-type theorem for submodular functions*, after an analogous statement in module theory.

Theorem 2.2.13. *Assume that* $f : 2^V \to \mathbf{R}$ *is submodular and a sublattice* $\mathcal{L} \subseteq 2^V$ *is an* f*-skeleton, as in* (2.19) *and* (2.20)*. Let* f_k $(k = 1, \cdots, b)$ *be defined by* (2.33) *with reference to a maximal chain of* $\mathcal{L}$*. For* $k = 1, \cdots, b$*, it holds that*

$$f_k(Y) = f(\langle V_k \rangle \cup Y) - f(\langle V_k \rangle), \qquad Y \subseteq V_k.$$

In particular, the family $\{(V_k, f_k) \mid k = 1, \cdots, b\}$ *is determined independently of the choice of a maximal chain.*

Proof. Noting $X_{k-1} \supseteq \langle V_k \rangle$, we put $W = X_{k-1} \setminus \langle V_k \rangle$. Then $f_k(Y) = f(\langle V_k \rangle \cup Y \cup W) - f(\langle V_k \rangle \cup W)$. It follows from the submodularity (2.20) of f that

$$\begin{aligned} f(\langle V_k \rangle \cup V_k) - f(\langle V_k \rangle \cup Y) &\geq f(\langle V_k \rangle \cup V_k \cup W) - f(\langle V_k \rangle \cup Y \cup W), \\ f(\langle V_k \rangle \cup Y) - f(\langle V_k \rangle) &\geq f(\langle V_k \rangle \cup Y \cup W) - f(\langle V_k \rangle \cup W). \end{aligned} \tag{2.34}$$

Addition of these yields

$$f(\langle V_k \rangle \cup V_k) - f(\langle V_k \rangle) \geq f(\langle V_k \rangle \cup V_k \cup W) - f(\langle V_k \rangle \cup W).$$

The inequality here is an equality by the modularity of f on $\mathcal{L}$, since $\langle V_k \rangle \cup V_k$, $\langle V_k \rangle$, $\langle V_k \rangle \cup V_k \cup W$ $(= X_k)$ and $\langle V_k \rangle \cup W$ $(= X_{k-1})$ all belong to $\mathcal{L}$ by Birkhoff's representation theorem. Therefore, we have an equality also in (2.34). ■

Remark 2.2.14. In abstract terms a *lattice* is a triple $\mathcal{L} = (S, \vee, \wedge)$ of a nonempty set S and two binary operations $\vee$ and $\wedge$ on S (called "join" and "meet" respectively) such that $x \vee x = x$, $x \wedge x = x$; $x \vee y = y \vee x$, $x \wedge y = y \wedge x$; $x \vee (y \vee z) = (x \vee y) \vee z$, $x \wedge (y \wedge z) = (x \wedge y) \wedge z$; $x \wedge (x \vee y) = x$, $x \vee (x \wedge y) = x$ for $x, y, z \in S$. A lattice $\mathcal{L} = (S, \vee, \wedge)$ gives rise to a partially ordered set $\mathcal{P} = (S, \preceq)$ with $\preceq$ defined by $[x \preceq y \iff x \vee y = y]$. Such partially ordered set $\mathcal{P} = (S, \preceq)$ enjoys a nice property that for $x, y \in S$ there exist a (unique) minimum element among $\{z \in S \mid x \preceq z, y \preceq z\}$ (denoted as $\sup\{x, y\}$) and a (unique) maximum element among $\{z \in S \mid z \preceq x, z \preceq y\}$ (denoted as $\inf\{x, y\}$). Conversely, a partially ordered set $\mathcal{P} = (S, \preceq)$ such that $\sup\{x, y\}$ and $\inf\{x, y\}$ exist for any $x, y \in S$ induces a lattice $\mathcal{L} = (S, \vee, \wedge)$ with $\vee$ and $\wedge$ defined by $x \vee y = \sup\{x, y\}$ and $x \wedge y = \inf\{x, y\}$. A lattice $\mathcal{L} = (S, \vee, \wedge)$ is called *distributive* if $x \wedge (y \vee z) = (x \wedge y) \vee (x \wedge z)$, $x \vee (y \wedge z) = (x \vee y) \wedge (x \vee z)$. See Birkhoff [12] and Aigner [4] for lattice theory. □

Notes. The general decomposition principle given as Theorem 2.2.13 is due to Iri [129], Nakamura [245], and Nakamura–Iri [247]. This is an outcome of a series of successive generalizations of the concept of *principal partitions* for graphs and matroids. See also Kishi–Kajitani [157, 158, 159], Ohtsuki–Ishizaki–Watanabe [254], Ozawa [260, 262] for principal partitions of graphs, and Bruno–Weinberg [25], Iri [126], Nakamura–Iri [246], Narayanan–Vartak [249], Tomizawa [313], Tomizawa–Fujishige [316] for principal partitions of matroids. In this book we shall make use of this decomposition principle in a number of places. For example, it underlies the Dulmage–Mendelsohn decomposition of bipartite graphs (§2.2.3), the min-cut decomposition for independent matching problems (§2.3.5), the M-decomposition of graphs (§4.3.2), and the combinatorial canonical form of layered mixed matrices (§4.4). Another general decomposition principle for submodular functions, called the principal structure, is described in §4.9.2.

2.2.3 Dulmage–Mendelsohn Decomposition

This section is devoted to a comprehensive account of the Dulmage–Mendelsohn decomposition (or the DM-decomposition for short), a unique decomposition of a bipartite graph with respect to maximum matchings due to Dulmage–Mendelsohn [63, 64, 65, 66]. A standard reference for matching theory, with emphasis on structures rather than algorithms, is Lovász–Plummer [181].

Let $G = (V^+, V^-; A)$ be a *bipartite graph* with vertex set consisting of two disjoint parts V^+ and V^- and with arc set A, where arcs are directed from V^+ to V^-. For M ($\subseteq A$) in general, we denote by $\partial^+ M$ (resp., $\partial^- M$) the set of vertices in V^+ (resp., V^-) incident to arcs in M. Also we put $\partial M = \partial^+ M \cup \partial^- M$.

A *matching* M is a subset of A such that no two arcs in M share a common vertex incident to them. In other words, M is a matching if and only if $|M| = |\partial^+ M| = |\partial^- M|$. A matching of maximum size (cardinality) is called a *maximum matching*. The size of a maximum matching in G is denoted by $\nu(G)$. A matching with $\partial^+ M = V^+$ and $\partial^- M = V^-$ is called a *perfect matching*. An arc of G is said to be *admissible* if it is contained in some maximum matching in G.

A *cover* is a pair (U^+, U^-) of $U^+ \subseteq V^+$ and $U^- \subseteq V^-$ such that no arcs exist between $V^+ \setminus U^+$ and $V^- \setminus U^-$. The size of a cover (U^+, U^-) is defined to be $|U^+| + |U^-|$ and a cover of minimum size is called a *minimum cover*. We denote by $\mathcal{C}(G)$ the family of minimum covers of G.

A duality relation exists between the maximum matchings and the minimum covers.

Theorem 2.2.15 (König–Egerváry). *For a bipartite graph we have*

$$\max\{|M| \mid M : \text{matching}\} = \min\{|U^+| + |U^-| \mid (U^+, U^-) : \text{cover}\}. \quad (2.35)$$

Proof. This is a special case of Theorem 2.3.27. ∎

To rewrite Theorem 2.2.15 into another form we define $\Gamma : 2^{V^+} \to 2^{V^-}$ and $\gamma : 2^{V^+} \to \mathbf{Z}$ by

$$\Gamma(X) = \{v \in V^- \mid \exists u \in X : (u, v) \in A\}, \qquad X \subseteq V^+, \quad (2.36)$$

$$\gamma(X) = |\Gamma(X)|, \qquad X \subseteq V^+, \quad (2.37)$$

where $\Gamma(X)$ denotes the set of vertices in V^- adjacent to some vertex in X ($\subseteq V^+$). In passing we note the following fundamental properties.

Lemma 2.2.16. *For $X, Y \subseteq V^+$ it holds that*

$$\Gamma(X \cup Y) = \Gamma(X) \cup \Gamma(Y), \quad \Gamma(X \cap Y) \subseteq \Gamma(X) \cap \Gamma(Y),$$
$$\gamma(X) + \gamma(Y) \geq \gamma(X \cup Y) + \gamma(X \cap Y).$$

Proof. The first claim is easy to verify, while the second follows from

$$|\Gamma(X\cup Y)|+|\Gamma(X\cap Y)| \le |\Gamma(X)\cup\Gamma(Y)|+|\Gamma(X)\cap\Gamma(Y)| = |\Gamma(X)|+|\Gamma(Y)|.$$

■

The second expression of the duality is given in terms of γ as follows.

Theorem 2.2.17 (Hall–Ore).

$$\max\{|M| \mid M\text{: matching}\} = \min\{\gamma(X) - |X| \mid X \subseteq V^+\} + |V^+|. \quad (2.38)$$

Proof. This follows from Theorem 2.2.15 and the fact that (U^+, U^-) is a cover if and only if $\Gamma(V^+ \setminus U^+) \subseteq U^-$. ■

The function

$$p_0(X) = \gamma(X) - |X|, \qquad X \subseteq V^+, \quad (2.39)$$

appearing in the above identity is called the *surplus function* in Lovász–Plummer [181], and $-p_0$ is the *deficiency* according to Ore [256].

Lemma 2.2.18. *The surplus function* $p_0(X)$ *of* (2.39) *is submodular, i.e.,*

$$p_0(X) + p_0(Y) \ge p_0(X \cup Y) + p_0(X \cap Y).$$

Proof. This is immediate from the submodularity of γ in Lemma 2.2.16. ■

We may say that the second expression (2.38) for the duality reveals the submodularity inherent in the problem at the sacrifice of the symmetry apparent in the first expression (2.35). On the basis of the submodularity of p_0 we shall derive the DM-decomposition.

Example 2.2.19. The above theorems are illustrated here for the bipartite graph $G = (V^+, V^-; A)$ in Fig. 2.7, where $V^+ = \{u_1, \cdots, u_7\}$ and $V^- = \{v_1, \cdots, v_7\}$. In Theorem 2.2.15,

$$M = \{(u_2, v_1), (u_3, v_2), (u_4, v_3), (u_5, v_4), (u_6, v_5), (u_7, v_6)\}$$

is a maximum matching of size $|M| = 6$, and

$$(U^+, U^-) = (\{u_4, u_5, u_6, u_7\}, \{v_1, v_2\})$$

is a minimum cover of size $|U^+| + |U^-| = 6$. In Theorem 2.2.17, the surplus function $p_0(X) = \gamma(X) - |X|$ takes the minimum value -1 at $X = \{u_1, u_2, u_3\}$, for example, and hence the right-hand side of (2.38) is equal to 6. It is mentioned in advance that the four subgraphs G_0, G_1, G_2, G_∞, indicated by vertical broken lines in Fig. 2.7, are the components of the DM-decomposition to be derived below. □

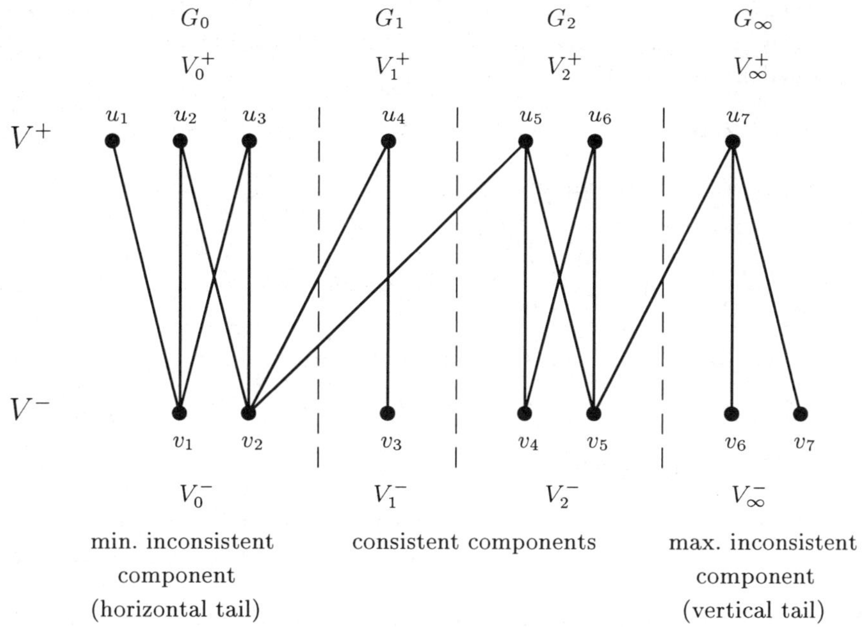

Fig. 2.7. DM-decomposition

The family of subgraphs $G_k = (V_k^+, V_k^-; A_k)$ in the DM-decomposition is constructed as follows. In view of the minimax relation (2.38) it is natural to look at the family of the minimizers of surplus function p_0:

$$\mathcal{L}_{\min}(p_0) = \{X \subseteq V^+ \mid p_0(X) \leq p_0(Y), \forall Y \subseteq V^+\}, \tag{2.40}$$

which forms a sublattice of 2^{V^+} by virtue of the submodularity of p_0 (cf. Lemma 2.2.18 and Theorem 2.2.5). According to the Jordan–Hölder-type theorem for submodular functions (§2.2.2), the sublattice $\mathcal{L}_{\min}(p_0)$ determines

$$\mathcal{P}(\mathcal{L}_{\min}(p_0)) = (\{V_0^+; V_1^+, \cdots, V_b^+; V_\infty^+\}, \preceq), \tag{2.41}$$

a pair of a partition of V^+ and a partial order $\preceq$. Here $V_k^+ \neq \emptyset$ for $k = 1, \cdots, b$, whereas V_0^+ and V_∞^+ are distinguished blocks that can be empty. In accordance with (2.23) we may assume $V_0^+ = X_0$, $V_k^+ = X_k \setminus X_{k-1}$ $(k = 1, \cdots, b)$, $V_\infty^+ = V^+ \setminus X_b$ for a maximal chain $(X_k \mid k = 0, 1, \cdots, b)$ of $\mathcal{L} = \mathcal{L}_{\min}(p_0)$ (cf. (2.22)). Define

$$\begin{aligned} V_0^- &= \Gamma(X_0), \\ V_k^- &= \Gamma(X_k) \setminus \Gamma(X_{k-1}) \qquad (k = 1, \cdots, b), \\ V_\infty^- &= V^- \setminus \Gamma(X_b) \end{aligned} \tag{2.42}$$

to obtain a partition $(V_0^-; V_1^-, \cdots, V_b^-; V_\infty^-)$ of V^-, which is determined independently of the chosen maximal chain by the following lemma. The notation $\langle V_k^+ \rangle$ is defined in (2.32).

Lemma 2.2.20. $V_k^- = \Gamma(V_k^+) \setminus \Gamma(\langle V_k^+ \rangle) \qquad (k = 1, \cdots, b).$

Proof. Theorem 2.2.13 implies $|V_k^-| = \gamma(X_{k-1} \cup V_k^+) - \gamma(X_{k-1}) = \gamma(\langle V_k^+ \rangle \cup V_k^+) - \gamma(\langle V_k^+ \rangle)$. Since $V_k^- = \Gamma(X_{k-1} \cup V_k^+) \setminus \Gamma(X_{k-1}) = \Gamma(V_k^+) \setminus \Gamma(X_{k-1})$ and $\Gamma(\langle V_k^+ \rangle \cup V_k^+) \setminus \Gamma(\langle V_k^+ \rangle) = \Gamma(V_k^+) \setminus \Gamma(\langle V_k^+ \rangle)$, it follows that $|V_k^-| = |\Gamma(V_k^+) \setminus \Gamma(X_{k-1})| = |\Gamma(V_k^+) \setminus \Gamma(\langle V_k^+ \rangle)|$, in which $\Gamma(X_{k-1}) \supseteq \Gamma(\langle V_k^+ \rangle)$. Therefore, $V_k^- = \Gamma(V_k^+) \setminus \Gamma(\langle V_k^+ \rangle)$. ■

The arc set A of G is partitioned accordingly as

$$A = \left(\bigcup_{k=0}^{\infty} A_k \right) \cup \left(\bigcup_{k \neq l} A_{kl} \right),$$

$$A_k = \{a \in A \mid \partial^+ a \in V_k^+, \ \partial^- a \in V_k^-\} \qquad (k = 0, 1, \cdots, b, \infty),$$

$$A_{kl} = \{a \in A \mid \partial^+ a \in V_l^+, \ \partial^- a \in V_k^-\} \qquad (k \neq l; \ k, l = 0, 1, \cdots, b, \infty).$$

Thus we have obtained the family of subgraphs $G_k = (V_k^+, V_k^-; A_k)$ $(k = 0, 1, \cdots, b, \infty)$, which we call the *DM-components*. Furthermore we define $G_k \preceq G_l$ if and only if $V_k^+ \preceq V_l^+$ in $\mathcal{P}(\mathcal{L}_{\min}(p_0))$, where it is emphasized that $G_0 \preceq G_k \preceq G_\infty$ for any k. We call G_0 the *horizontal tail* (or the *minimal inconsistent component*), G_∞ the *vertical tail* (or the *maximal inconsistent component*) and the others G_k $(k = 1, \cdots, b)$ the *consistent components*.

Example 2.2.21. For the graph in Fig. 2.7 it can be verified that

$$\mathcal{L}_{\min}(p_0) = \{\{u_1, u_2, u_3\}, \{u_1, u_2, u_3, u_4\}, \{u_1, u_2, u_3, u_5, u_6\}, \{u_1, u_2, u_3, u_4, u_5, u_6\}\}.$$

This yields (2.41) with $b = 2$, $V_0^+ = \{u_1, u_2, u_3\}$, $V_1^+ = \{u_4\}$, $V_2^+ = \{u_5, u_6\}$, $V_\infty^+ = \{u_7\}$ and the partial order: $V_0^+ \preceq V_k^+ \preceq V_\infty^+$ for $k = 1, 2$. Hence the four subgraphs G_0, G_1, G_2, G_∞, indicated by vertical broken lines in Fig. 2.7 with the partial order $G_0 \preceq G_k \preceq G_\infty$ for $k = 1, 2$. The identity in Lemma 2.2.20 for $k = 2$ reads $\{v_4, v_5\} = \{v_2, v_4, v_5\} \setminus \{v_1, v_2\}$. □

The DM-decomposition reveals the structure of a bipartite graph with respect to maximum matchings and minimum covers, as follows. Recall the notation $\nu(G)$ for the size of a maximum matching in G and $\mathcal{C}(G)$ for the family of minimum covers of G.

Theorem 2.2.22 (Dulmage–Mendelsohn decomposition). *Let $G_k = (V_k^+, V_k^-; A_k)$ $(k = 0, 1, \cdots, b, \infty)$ be the DM-components of a bipartite graph $G = (V^+, V^-; A)$.*

(1) *For* $1 \leq k \leq b$ *(consistent components):* $\nu(G_k) = |V_k^+| = |V_k^-|$, $\mathcal{C}(G_k) = \{(V_k^+, \emptyset), (\emptyset, V_k^-)\}$, *and each* $a \in A_k$ *is admissible in* G_k*;*
For $k = 0$ *(horizontal tail):* $\nu(G_0) = |V_0^-|$, $|V_0^-| < |V_0^+|$ *if* $V_0^- \neq \emptyset$, $\mathcal{C}(G_0) = \{(\emptyset, V_0^-)\}$, *and each* $a \in A_0$ *is admissible in* G_0*;*
For $k = \infty$ *(vertical tail):* $\nu(G_\infty) = |V_\infty^+|$, $|V_\infty^+| < |V_\infty^-|$ *if* $V_\infty^+ \neq \emptyset$, $\mathcal{C}(G_\infty) = \{(V_\infty^+, \emptyset)\}$, *and each* $a \in A_\infty$ *is admissible in* G_∞.

(2) *The partial order* $\preceq$ *among the components* G_k *is represented by the existence of arcs:*

$$A_{kl} = \emptyset \quad \text{unless} \quad G_k \preceq G_l \quad (1 \leq k, l \leq b); \tag{2.43}$$

$$A_{kl} \neq \emptyset \quad \text{if} \quad G_k \prec\cdot\, G_l \quad (1 \leq k, l \leq b). \tag{2.44}$$

(3) *The minimum covers of* G *are in one-to-one correspondence with the order ideals:*

$$\mathcal{C}(G) = \{(\bigcup_{k \in \overline{I}} V_k^+, \bigcup_{k \in I} V_k^-) \mid \{G_k \mid k \in I\} : \text{order ideal }\},$$

where $\overline{I} = \{0, 1, \cdots, b, \infty\} \setminus I$, *and it is emphasized that* $0 \in I$ *and* $\infty \notin I$ *if* I *corresponds to an order ideal.*

(4) M $(\subseteq A)$ *is a maximum matching of* G *if and only if* $M \subseteq \bigcup_{k=0}^{\infty} A_k$ *and* $M \cap A_k$ *is a maximum matching of* G_k *for* $k = 0, 1, \cdots, b, \infty$.

Proof. We prove (1), (3), (2) and (4). First note from (2.42) that

$$A_{kl} = \emptyset \qquad (k > l). \tag{2.45}$$

(1) [Size of vertex sets] Since $V_0^+ \in \mathcal{L}_{\min}(p_0)$, we have

$$0 = p_0(\emptyset) \geq \min p_0 = p_0(V_0^+) = \gamma(V_0^+) - |V_0^+| = |V_0^-| - |V_0^+|.$$

If the equality holds here, then $p_0(\emptyset) = \min p_0$, which implies $V_0^+ = \emptyset$ and therefore $V_0^- = \emptyset$. For $k = 1, \cdots, b$, we have $\min p_0 = \gamma(X_{k-1}) - |X_{k-1}| = \gamma(X_k) - |X_k|$, which means $|V_k^+| = |V_k^-|$ by (2.42). If $V_\infty^+ \neq \emptyset$, then $p_0(V^+) > \min p_0 = p_0(X_b)$, i.e., $\gamma(V^+) - |V^+| > \gamma(X_b) - |X_b|$. Combination of this with $|V^-| \geq \gamma(V^+)$, $|V_\infty^-| = |V^-| - \gamma(X_b)$, $|V_\infty^+| = |V^+| - |X_b|$ yields $|V_\infty^-| > |V_\infty^+|$.

[Size of maximum matchings] For $k = 0, 1, \cdots, b$, put $Y_k = \bigcup_{l=0}^{k} V_l^-$ and let $G^{(k)}$ be the subgraph of G induced on $X_k \cup Y_k$. It follows from (2.45) and Theorem 2.2.17 that

$$\nu(G^{(k)}) = \min\{p_0(X) \mid X \subseteq X_k\} + |X_k| = |Y_k|,$$

which implies $\nu(G_k) = |V_k^-|$ for $k = 0, 1, \cdots, b$. Since $A_{\infty k} = \emptyset$ for $k = 0, 1, \cdots, b$ by (2.45) and $\nu(G^{(b)}) = |Y_b|$, we have

$$|V_\infty^+| \geq \nu(G_\infty) \geq \nu(G) - |Y_b| = \min p_0 + |V^+| - |Y_b| = |V^+| - |X_b| = |V_\infty^+|.$$

[Minimum covers] For $k = 0, 1, \cdots, b, \infty$, the surplus function $p^{(k)} : 2^{V_k^+} \to \mathbf{Z}$ of G_k is given by

$$p^{(k)}(X) = |\Gamma(X) \cap V_k^-| - |X| = p_0(X_{k-1} \cup X) - p_0(X_{k-1}), \qquad X \subseteq V_k^+,$$

where $X_{k-1} = \emptyset$ for $k = 0$ and $X_{k-1} = X_b$ for $k = \infty$. This shows that

$$\mathcal{L}_{\min}(p^{(k)}) = \begin{cases} \{V_0^+\} & (k = 0) \\ \{\emptyset, V_k^+\} & (1 \le k \le b) \\ \{\emptyset\} & (k = \infty). \end{cases} \tag{2.46}$$

[Admissibility of each arc] For $a = (u, v) \in A_k$ consider a cover (W^+, W^-) of $G_k \setminus \{u, v\}$. Since $(W^+ \cup \{u\}, W^- \cup \{v\})$ is a cover of G_k but not a minimum cover by (2.46), we have $|W^+| + |W^-| + 2 \ge \nu(G_k) + 1$. Therefore, $\nu(G_k \setminus \{u, v\}) = \min\{|W^+| + |W^-|\} \ge \nu(G_k) - 1$. A maximum matching of $G_k \setminus \{u, v\}$ augmented with (u, v) yields a maximum matching of G_k.

(3) This follows from the facts that (U^+, U^-) is a cover if and only if $\Gamma(V^+ \setminus U^+) \subseteq U^-$, and that $X \in \mathcal{L}_{\min}(p_0)$ if and only if X corresponds to an order ideal.

(2) For the proof of (2.43) suppose that $G_k \not\preceq G_l$. Then there exists an order ideal I such that $k \notin I$ and $l \in I$. By (3), $(\bigcup_{j \in \overline{I}} V_j^+, \bigcup_{j \in I} V_j^-)$ is a minimum cover. This means in particular that there exists no arc between V_l^+ and V_k^-, that is, $A_{kl} = \emptyset$.

For the proof of (2.44) suppose that $G_k \prec\cdot\, G_l$, where $k < l$. Put

$$I = \{i \mid k < i < l, G_k \prec G_i\}, \qquad I^* = I \cup \{k\},$$
$$J = \{j \mid k < j < l\} \setminus I, \qquad J^* = J \cup \{l\}.$$

We have (i) $i \in I^*, j \in J \Rightarrow G_i \not\preceq G_j$, and (ii) $i \in I \Rightarrow G_i \not\preceq G_l$. The statement (i) is due to the transitivity of the partial order and the statement (ii) is by the assumption $G_k \prec\cdot\, G_l$. It then follows that

$$i \in I^*, j \in J^*, (i, j) \ne (k, l) \Rightarrow G_i \not\preceq G_j \Rightarrow A_{ij} = \emptyset,$$

where (2.43) is used. If $A_{kl} = \emptyset$ is the case, we have $A_{ij} = \emptyset$ for $i \in I^*$ and $j \in J^*$. This implies $X = X_{k-1} \cup \left(\bigcup_{j \in J^*} V_j^+\right)$ belongs to $\mathcal{L}_{\min}(p_0)$, since

$$p_0(X) = (|Y_{k-1}| + \sum_{j \in J^*} |V_j^-|) - (|X_{k-1}| + \sum_{j \in J^*} |V_j^+|) = |Y_{k-1}| - |X_{k-1}| = \min p_0.$$

Hence we have $X \in \mathcal{L}_{\min}(p_0)$, $V_k^+ \not\subseteq X$ and $V_l^+ \subseteq X$. This contradicts the definition (2.24) of $V_k^+ \preceq V_l^+$. Therefore, $A_{kl} \ne \emptyset$.

(4) Let (U^+, U^-) be a minimum cover. By Theorem 2.2.15, a matching M is maximum if and only if there exists no $a \in M$ such that $\partial^+ a \in U^+$ and $\partial^- a \in U^-$. Then the assertion follows from (1)–(3) above. ■

Corollary 2.2.23.

(1) $a \in A$ *is admissible in* G *if and only if* $a \in \bigcup_{k=0}^{\infty} A_k$.

(2) $V_0^+ = \{v \in V^+ \mid \nu(G) = \nu(G \setminus \{v\})\}$,
$V_\infty^- = \{v \in V^- \mid \nu(G) = \nu(G \setminus \{v\})\}$.

Proof. (1) This follows from (1) and (4) of Theorem 2.2.22.

(2) Note $V^+ \setminus V_0^+ = \{v \in V^+ \mid \exists (U^+, U^-) \in \mathcal{C}(G), v \in U^+\}$. If $(U^+, U^-) \in \mathcal{C}(G)$ and $v \in U^+$, then $(U^+ \setminus \{v\}, U^-)$ is a cover of $G \setminus \{v\}$, and hence $\nu(G \setminus \{v\}) \leq |U^+| + |U^-| - 1 = \nu(G) - 1$. Conversely, for $v \in V_0^+$ and any cover (W^+, W^-) of $G \setminus \{v\}$, $(W^+ \cup \{v\}, W^-)$ is a cover of G but not a minimum cover. Hence $|W^+ \cup \{v\}| + |W^-| \geq \nu(G) + 1$. Therefore, $\nu(G \setminus \{v\}) = \min\{|W^+| + |W^-|\} \geq \nu(G)$. Similarly for V_∞^-. ■

An algorithm for the DM-decomposition is given below. For a matching M an auxiliary graph $\tilde{G}_M = (V^+ \cup V^-, \tilde{A}; S^+, S^-)$ is defined by

$$\tilde{A} = \{(u,v) \mid (u,v) \in A \text{ or } (v,u) \in M\}, \quad S^+ = V^+ \setminus \partial^+ M, \quad S^- = V^- \setminus \partial^- M.$$

Recall the notation $\xrightarrow{*}$ for the reachability by a directed path.

Algorithm for the DM-decomposition of $G = (V^+, V^-; A)$

1. Find[4] a maximum matching M on $G = (V^+, V^-; A)$.
2. Let $V_0 = \{v \in V^+ \cup V^- \mid u \xrightarrow{*} v \text{ on } \tilde{G}_M \text{ for some } u \in S^+\}$.
3. Let $V_\infty = \{v \in V^+ \cup V^- \mid v \xrightarrow{*} u \text{ on } \tilde{G}_M \text{ for some } u \in S^-\}$.
4. Let G' denote the graph obtained from $\tilde{G}_M$ by deleting the vertices $V_0 \cup V_\infty$ (and arcs incident thereto).
5. Let V_k $(k = 1, \cdots, b)$ be the strong components of G'.
6. Let $G_k = (V_k^+, V_k^-; A_k)$ be the subgraph of G induced on V_k $(k = 0, 1, \cdots, b, \infty)$.
7. Define a partial order $\preceq$ on $\{G_k \mid k = 1, \cdots, b\}$ as follows:

$$G_k \preceq G_l \iff v_l \xrightarrow{*} v_k \text{ on } \tilde{G}_M \text{ for some } v_k \in V_k \text{ and } v_l \in V_l.$$

Also define $G_0 \preceq G_k \preceq G_\infty$ for any k. □

The validity of the above algorithm will be established in §2.3.5 as a special case of a more general algorithm (the algorithm for the min-cut decomposition of an independent matching problem; see Lemma 2.3.35, to be specific). This implies, in particular, that the decomposition constructed by the above algorithm does not depend on the initially chosen maximal matching M.

---

[4] A maximum matching can be found in $O(|A| \, (\min(|V^+|, |V^-|))^{1/2})$ time by an augmenting path method using layered networks; see Lawler [171] and Papadimitriou–Steiglitz [265]. More recent algorithms can be found in Ahuja–Magnanti–Orlin [3].

A bipartite graph $G = (V^+, V^-; A)$ with $V^+ \cup V^- \neq \emptyset$ is said to be *DM-irreducible* if it cannot be decomposed into more than one nonempty component in the DM-decomposition. Otherwise, it is called *DM-reducible.* A graph G with $V^+ = \emptyset$ or $V^- = \emptyset$ is DM-irreducible, as the whole graph is a (vertical or horizontal) tail. Note that G with $A = \emptyset$, $V^+ \neq \emptyset$ and $V^- \neq \emptyset$ is DM-reducible, as it can be decomposed into two nonempty components, the horizontal tail $G_0 = (V^+, \emptyset; \emptyset)$ and the vertical tail $G_\infty = (\emptyset, V^-; \emptyset)$.

The following theorem is a reformulation of the result due to Marcus–Minc [186] and Brualdi [22] (see also Brualdi–Ryser [24, Theorem 4.2.2]).

Theorem 2.2.24. *For a bipartite graph* $G = (V^+, V^-; A)$ *with* $|V^+| = |V^-|$ *the following three conditions are equivalent:*

(i) *G is DM-irreducible,*

(ii) $\mathcal{C}(G) = \{(V^+, \emptyset), (\emptyset, V^-)\}$,

(iii) $\nu(G \setminus \{u, v\}) = \nu(G) - 1$ *for* $\forall u \in V^+$, $\forall v \in V^-$.

Proof. The equivalence between (i) and (ii) follows from Theorem 2.2.22(1).

For (ii) $\Rightarrow$ (iii), first note (ii) implies $\nu(G) = |V^+|$, and hence $\nu(G \setminus \{u, v\}) \leq |V^+| - 1 = \nu(G) - 1$. Take a minimum cover (W^+, W^-) of $G \setminus \{u, v\}$. Since $(W^+ \cup \{u\}, W^- \cup \{v\})$ is a cover of G but not a minimum cover, we see $\nu(G) + 1 \leq |W^+ \cup \{u\}| + |W^- \cup \{v\}| = \nu(G \setminus \{u, v\}) + 2$.

For (iii) $\Rightarrow$ (ii) suppose that (ii) fails. We divide into two cases: (a) $\nu(G) = |V^+|$ and (b) $\nu(G) < |V^+|$. In case (a), there exists $(U^+, U^-) \in \mathcal{C}(G)$ such that $U^+ \neq \emptyset$ and $U^- \neq \emptyset$. For $u \in U^+$ and $v \in U^-$, $(U^+ \setminus \{u\}, U^- \setminus \{v\})$ is a cover of $G \setminus \{u, v\}$. Hence $\nu(G \setminus \{u, v\}) \leq |U^+ \setminus \{u\}| + |U^- \setminus \{v\}| = \nu(G) - 2$. In case (b), both horizontal and vertical tails are nonempty. For $u \in V_0^+$ and $v \in V_\infty^-$ we have $\nu(G \setminus \{u, v\}) = \nu(G)$ by Corollary 2.2.23(2). ■

The concept of DM-decomposition may be extended to matrices by means of the DM-decomposition of associated bipartite graphs. Recall from §2.2.1 that for a matrix $A = (A_{ij})$ the associated bipartite graph is defined by $G = (V^+, V^-; \tilde{A})$ with $V^+ = \mathrm{Col}(A)$, $V^- = \mathrm{Row}(A)$ and $\tilde{A} = \{(j, i) \mid A_{ij} \neq 0\}$. A DM-component $G_k = (V_k^+, V_k^-; A_k)$ of G corresponds to the submatrix $A[V_k^-, V_k^+]$, which will be referred to as a *DM-component* of A.

The following relation is obvious from the definitions, but provides the DM-decomposition with a linear algebraic significance.

Proposition 2.2.25. $\text{term-rank}\, A = \nu(G)$. □

The *DM-decomposition* of A gives the finest block-triangularization of a matrix by means of a transformation $P_{\mathrm{r}} A P_{\mathrm{c}}$ using two permutation matrices P_{r} and P_{c}, where it is imposed that each diagonal block in a block-triangularization has full term-rank. In fact, Theorem 2.2.22(1) combined with Proposition 2.2.25 above guarantees this term-rank condition for the diagonal blocks produced by the DM-decomposition. Furthermore, term-rank coincides with rank for a generic matrix, in which all nonzero entries are independent parameters (cf. Proposition 2.1.12). Hence, for a generic matrix, the DM-decomposition gives the finest proper block-triangularization in the sense of §2.1.4.

For instance, the matrix version of the DM-decomposition of the graph in Fig. 2.7 is given by

$$
A = \begin{array}{c|ccc|c|cc|c|}
 & & V_0^+ & & V_1^+ & V_2^+ & & V_\infty^+ \\ \hline
V_0^- & t_1 & t_2 & t_3 & & & & \\
 & & t_4 & t_5 & t_6 & t_7 & & \\ \hline
V_1^- & & & & t_8 & & & \\
V_2^- & & & & & t_9 & t_{10} & \\
 & & & & & t_{11} & t_{12} & t_{13} \\ \hline
V_\infty^- & & & & & & & t_{14} \\
 & & & & & & & t_{15} \\ \hline
\end{array} .
$$

Note that term-rank $A[V_k^-, V_k^+] = \min(|V_k^+|, |V_k^-|)$ for $k = 0, 1, 2, \infty$. The matrix $P_{\mathrm{r}} A P_{\mathrm{c}}$ of (2.16) in Example 2.1.15 gives an instance of the DM-decomposition of a term-nonsingular matrix.

Though term-rank is a natural combinatorial counterpart, it is not the same as rank, which is undoubtedly more important in applications. A numerical (nongeneric) matrix may or may not have the same rank as term-rank, and accordingly, the DM-decomposition may or may not be a proper block-triangular form. The present argument shows the following.

Proposition 2.2.26. *For a matrix A the following three conditions* (i)–(iii) *are equivalent.*

(i) rank A = term-rank A.

(ii) *The DM-decomposition is a proper block-triangularization, i.e., the DM-components $A[V_k^-, V_k^+]$ $(k = 0, 1, \cdots, b, \infty)$ satisfy*

$$\operatorname{rank} A[V_k^-, V_k^+] = \min(|V_k^+|, |V_k^-|) \qquad (k = 0, 1, \cdots, b, \infty).$$

(iii) *There exist $I \subseteq \mathrm{Row}(A)$ and $J \subseteq \mathrm{Col}(A)$ such that* $\operatorname{rank} A[I, J] = 0$, $\operatorname{rank} A[\mathrm{Row}(A) \setminus I, J] = |\mathrm{Row}(A) \setminus I|$, *and* $\operatorname{rank} A[I, \mathrm{Col}(A) \setminus J] = |\mathrm{Col}(A) \setminus J|$.

Proof. The equivalence of (i) and (ii) is immediate from Theorem 2.2.22. For (iii) take $I = \mathrm{Row}(A) \setminus V_0^-$ and $J = \mathrm{Col}(A) \setminus V_0^+$. ■

The concept of DM-irreducibility can be naturally defined for matrices, and it coincides with the well-studied concept of full indecomposability (cf. Brualdi–Ryser [24] and Schneider [288]). To see this, first recall that a matrix A is said to be *fully indecomposable* if it does not contain a zero submatrix $A[I, J] = O$ with $I \neq \emptyset$, $J \neq \emptyset$ and $|I| + |J| = \max(|\mathrm{Row}(A)|, |\mathrm{Col}(A)|)$. Since $A[I, J] = O$ if and only if $(V^- \setminus I, V^+ \setminus J)$ is a cover of the associated graph $G = (V^+, V^-; \tilde{A})$, matrix A is fully indecomposable if and only if G has no cover (U^+, U^-) such that $U^+ \neq V^+$, $U^- \neq V^-$ and

$|U^+| + |U^-| = \min(|V^+|, |V^-|)$. The latter condition is equivalent to the DM-irreducibility (cf. Theorem 2.2.24). Henceforth we use DM-irreducibility as a synonym of full indecomposability.

The DM-irreducibility for square generic matrices admits two further characterizations in addition to those given in Theorem 2.2.24. The first says that the DM-irreducibility for a generic matrix is equivalent to the inverse matrix having a completely dense nonzero pattern.

Theorem 2.2.27. *A square generic matrix A is DM-irreducible if and only if A is nonsingular and $(A^{-1})_{ji} \neq 0$ for all (j, i).*

Proof. Since $(A^{-1})_{ji} = \det A[R \setminus \{i\}, C \setminus \{j\}] / \det A$, where $R = \mathrm{Row}(A)$ and $C = \mathrm{Col}(A)$, the claim here reduces to the equivalence of (i) and (iii) in Theorem 2.2.24. ∎

The determinant of a generic matrix A can be regarded as a polynomial in the nonzero entries. Specifically, let $\mathcal{T}$ denote the set of nonzero entries of A, which is algebraically independent over a ground field $\boldsymbol{K}$. Then $\det A \in \boldsymbol{K}[\mathcal{T}]$, where $\boldsymbol{K}[\mathcal{T}]$ means the ring of polynomials in $\mathcal{T}$ over $\boldsymbol{K}$.

The following theorem gives an algebraic characterization of the DM-irreducibility in terms of the irreducibility of the determinant as a multivariate polynomial. This is proven in Ryser [285] and credited essentially to Frobenius [78] in Ryser [286].

Theorem 2.2.28. *A square generic matrix A is DM-irreducible if and only if $\det A$ is an irreducible (nonzero) polynomial in $\boldsymbol{K}[\mathcal{T}]$, where $\mathcal{T}$ denotes the set of nonzero entries of A.*

Proof. The "if" part is obvious, since, for a DM-reducible A with no tails, $\det A$ is equal to the product of the determinants of the diagonal blocks of the DM-decomposition of A (and $\det A = 0$ if a nonempty tail exists). For the "only if" part assume that $\det A$ is factored as $\det A = f_1 \cdot f_2$ with $f_1, f_2 \in \boldsymbol{K}[\mathcal{T}] \setminus \boldsymbol{K}$. For $k = 1, 2$, let $\mathcal{T}_k$ denote the set of the variables of $\mathcal{T}$ that appear in f_k. Put

$$R_k = \{i \in R \mid A_{ij} \in \mathcal{T}_k\}, \quad C_k = \{j \in C \mid A_{ij} \in \mathcal{T}_k\} \qquad (k = 1, 2),$$

where $R = \mathrm{Row}(A)$ and $C = \mathrm{Col}(A)$. Then $R_1 \cap R_2 = \emptyset$, $R_1 \cup R_2 = R$, $C_1 \cap C_2 = \emptyset$, $C_1 \cup C_2 = C$ for $k = 1, 2$, since for each pair of terms in f_1 and f_2 their product remains in $f_1 \cdot f_2 = \det A$ as a nonvanishing term, which in turn corresponds to a perfect matching in the associated bipartite graph. We may assume $|R_1| \geq |C_1| \geq 1$ without loss of generality. If $A[R_1, C_2] = O$, A is DM-reducible. If $A_{ij} \neq 0$ for some $i \in R_1$ and $j \in C_2$, the variable $t = A_{ij}$ cannot appear in $\det A$, since otherwise t must be contained in f_k for $k = 1$ or 2, which implies $i \in R_k$ and $j \in C_k$, a contradiction to $R_1 \cap R_2 = \emptyset$ and $C_1 \cap C_2 = \emptyset$. The disappearance of t in $\det A$ implies $\det A[R \setminus \{i\}, C \setminus \{j\}] = 0$, or equivalently, $\nu(G \setminus \{i, j\}) < \nu(G) - 1$ in terms of the associated bipartite graph G. This shows the DM-reducibility by Theorem 2.2.24. ∎

Remark 2.2.29. We have derived the DM-decomposition in a systematic manner on the basis of the Jordan–Hölder-type theorem for submodular functions, though alternative quicker derivations would have been possible. Our systematic derivation here enables us to generalize the DM-decomposition to a more sophisticated decomposition in §4.4, called the CCF (combinatorial canonical form) of layered mixed matrices. The DM-decomposition serves as one of the main tools for the graph-theoretic methods for systems analysis (see Duff–Erisman–Reid [59], Murota [204, Chaps. 2 and 3]), whereas the CCF is for the matroid-theoretic methods to be developed in Chap. 4 and Chap. 6. Applications of the DM-decomposition can be found in Ashcraft–Liu [8], Erisman–Grimes–Lewis–Poole–Simon [73], Hellerman–Rarick [109, 110], O'Neil–Szyld [255], and Pothen–Fan [273]. □

2.2.4 Maximum Flow and Menger-type Linking

In §2.2.4 and §2.2.5 we describe some fundamental results from network flow theory that are needed in this book; maximum flow in §2.2.4 and minimum cost flow in §2.2.5. For systematic and comprehensive expositions of network flow theory, the reader is referred to standard textbooks such as Ahuja–Magnanti–Orlin [3], Cook–Cunningham–Pulleyblank–Schrijver [40], Ford–Fulkerson [75], Iri [123], Lawler [171], Nemhauser–Rinnooy Kan–Todd [250], Nemhauser–Wolsey [251], and Papadimitriou–Steiglitz [265].

Consider a *network* $N = (V, A, c; s^+, s^-)$, where V is the vertex set, A is the arc set, $c : A \to \mathbf{R}_+ \cup \{+\infty\}$ is a function defining the capacity of arcs ($\mathbf{R}_+$: set of nonnegative reals), and s^+ and s^- are two distinct vertices called the *source* and the *sink* ($s^+, s^- \in V$ and $s^+ \neq s^-$). A *flow* in N is a function $\varphi : A \to \mathbf{R}$. A function $\partial\varphi : V \to \mathbf{R}$ derived from φ by

$$\partial\varphi(v) = \sum\{\varphi(a) \mid a \in \delta^+ v\} - \sum\{\varphi(a) \mid a \in \delta^- v\}, \qquad v \in V, \tag{2.47}$$

is called the *boundary* of φ. Recall that for $v \in V$, $\delta^+ v$ means the set of arcs going out of v and $\delta^- v$ the set of arcs coming into v. A *feasible flow* in N is a flow $\varphi : A \to \mathbf{R}$ that satisfies

$$\begin{aligned} \text{capacity condition} : \ & 0 \le \varphi(a) \le c(a), \qquad a \in A, \\ \text{conservation condition} : \ & \partial\varphi(v) = 0, \qquad v \in V \setminus \{s^+, s^-\}. \end{aligned}$$

We call $\mathrm{val}(\varphi) = \partial\varphi(s^+)$ $(= -\partial\varphi(s^-))$ the *value* of φ. The *maximum flow problem* is to find a feasible flow φ that maximizes $\mathrm{val}(\varphi)$.

Suppose we tear the network N into two parts in such a way that s^+ and s^- belong to different parts. Such tearing is specified by a set $S \subseteq V$ such that $s^+ \in S$ and $s^- \in V \setminus S$; we put

$$\mathcal{S} = \{S \subseteq V \mid s^+ \in S, s^- \in V \setminus S\}. \tag{2.48}$$

The set of arcs going from S to $V \setminus S$ is denoted as

$$C(S) = \{a \in A \mid \partial^+ a \in S,\ \partial^- a \in V \setminus S\}, \tag{2.49}$$

which is referred to as the *cut* corresponding to S. Total amount of flow from s^+ to s^- is obviously bounded by the total capacity of the arcs in $C(S)$. That is, denoting the capacity of the cut by

$$\kappa(S) = \sum\{c(a) \mid a \in C(S)\}, \tag{2.50}$$

we have an inequality

$$\mathrm{val}(\varphi) \leq \kappa(S) \tag{2.51}$$

for any flow φ and any $S \in \mathcal{S}$.

The celebrated max-flow min-cut theorem asserts that the inequality (2.51) is satisfied with equality for a suitable choice of φ and S.

Theorem 2.2.30 (Max-flow min-cut theorem). *The maximum value of a flow is equal to the minimum capacity of a cut:*

$$\max\{\mathrm{val}(\varphi) \mid \varphi : \text{feasible flow}\} = \min\{\kappa(S) \mid S \in \mathcal{S}\}.$$

If the capacity function is integer-valued, there exists an integer-valued maximum flow. □

In passing it is mentioned that the function $\kappa : \mathcal{S} \to \mathbf{R}$ satisfies the submodular inequality:

$$\kappa(S) + \kappa(T) \geq \kappa(S \cup T) + \kappa(S \cap T), \qquad S, T \in \mathcal{S}. \tag{2.52}$$

This can be proven easily by the nonnegativity of c.

Next we turn to Menger-type linkings in a graph. Let $G = (V, A; X, Y)$ be a graph with vertex set V composed of three disjoint parts as $V = X \cup U \cup Y$. We call X the *entrance* and Y the *exit* (including the case where $X = \emptyset$ or $Y = \emptyset$). By a *Menger-type linking*[5] from X to Y is meant a set of pairwise vertex-disjoint directed paths, each from a vertex in X to a vertex in Y. The size of a linking is defined to be the number of directed paths from X to Y contained in the linking. A linking of the maximum size is called a *maximum linking* and, in case $|X| = |Y|$, a linking of size $|X|$ is called a *perfect linking.* By a *separator* of (X, Y) is meant such a subset of V that intersects any directed path from a vertex in X to a vertex in Y. A separator of minimum cardinality is called a *minimum separator.*

Menger-type linkings can be treated as a special case of network flow. Assuming, for simplicity of presentation, that there is no arc entering $x \in X$

[5] Four variants of Menger-type linking are considered in the literature according to (i) whether arcs are directed or undirected, and (ii) whether paths are vertex-disjoint or arc-disjoint. Only the version of directed arcs and vertex-disjoint paths is used in this book.

or leaving $y \in Y$, we consider a network $N_G = (\tilde{V}, \tilde{A}, c; s^+, s^-)$ using copies of X, Y and U and new vertices s^+ and s^- as follows:

$$
\begin{aligned}
\tilde{V} &= \{s^+, s^-\} \cup X_* \cup U_* \cup U^* \cup Y^*, \\
&\quad X_* = \{x_* \mid x \in X\}, \quad U_* = \{u_* \mid u \in U\}, \\
&\quad U^* = \{u^* \mid u \in U\}, \quad Y^* = \{y^* \mid y \in Y\}, \\
\tilde{A} &= \tilde{A}_{\rm o} \cup \tilde{A}_{\rm d}, \\
&\quad \tilde{A}_{\rm o} = \{(v_*, w^*) \mid (v, w) \in A\}, \\
&\quad \tilde{A}_{\rm d} = \{(s^+, x_*) \mid x \in X\} \cup \{(u^*, u_*) \mid u \in U\} \cup \{(y^*, s^-) \mid y \in Y\}, \\
c(a) &= \begin{cases} 1 & (a \in \tilde{A}_{\rm d}) \\ +\infty & (a \in \tilde{A}_{\rm o}). \end{cases}
\end{aligned}
$$

Note that U_* and U^* are disjoint copies of U.

There exists a one-to-one correspondence between Menger-type maximum linkings in G from X to Y and integral maximum flows in N_G from s^+ to s^- which have no circulation (flow along a cycle). On the other hand, minimum separators of (X, Y) in G correspond to minimum cuts with respect to (s^+, s^-) in N_G. The max-flow min-cut theorem for N_G implies the following relationship between linkings and separators.

Theorem 2.2.31 (Menger's theorem). *Let $G = (V, A; X, Y)$ be a graph with entrance X and exit Y. The maximum size of a Menger-type vertex-disjoint linking from X to Y is equal to the minimum cardinality of a separator of (X, Y).* □

Remark 2.2.32. Based on the submodularity (2.52) of the cut capacity function κ, a unique decomposition of a network into subnetworks can be defined (see Picard–Queyranne [268] and Murota [204, §8.2]). This decomposition can be tailored readily to a decomposition of a graph into subgraphs with respect to Menger-type maximum linkings and minimum separators. The decomposition thus obtained is named "Menger-decomposition" (or "M-decomposition" for short) by Murota [196, 205]. See also Murota [204, §8.3] for details. □

2.2.5 Minimum Cost Flow and Weighted Matching

The minimum cost flow problem can be described as follows. Let $G = (V, A)$ be a graph with vertex set V and arc set A. Suppose we are also given an upper capacity function $\overline{c} : A \to \mathbf{R} \cup \{+\infty\}$, a lower capacity function $\underline{c} : A \to \mathbf{R} \cup \{-\infty\}$, and a cost function $\gamma : A \to \mathbf{R}$. Namely, we are given a network $N = (V, A, \overline{c}, \underline{c}, \gamma)$. A flow in N is a function $\varphi : A \to \mathbf{R}$, and a feasible flow (circulation) in N is a flow $\varphi : A \to \mathbf{R}$ that satisfies

$$
\begin{aligned}
\text{capacity condition}: &\quad \underline{c}(a) \le \varphi(a) \le \overline{c}(a), \qquad a \in A, \\
\text{conservation condition}: &\quad \partial\varphi(v) = 0, \qquad v \in V,
\end{aligned}
$$

where $\partial\varphi : V \to \mathbf{R}$ is the boundary of φ defined in (2.47). The *cost* of flow φ is defined to be

$$\mathrm{cost}(\varphi) = \sum_{a \in A} \gamma(a)\varphi(a).$$

The *minimum cost flow problem* is to find a feasible flow φ that minimizes $\mathrm{cost}(\varphi)$. A feasible flow that attains the minimum cost is called an *optimal flow* or a *minimum cost flow.*

The optimality of a feasible flow can be characterized in a manner suitable for algorithmic verification. With a feasible flow $\varphi : A \to \mathbf{R}$ we associate an auxiliary network $\tilde{N}_\varphi = (V, \tilde{A}, \tilde{\gamma})$. The arc set of $\tilde{N}_\varphi$ is given by $\tilde{A} = A^* \cup B^*$ with

$$A^* = \{a \mid a \in A, \varphi(a) < \overline{c}(a)\},$$
$$B^* = \{\overline{a} \mid a \in A, \underline{c}(a) < \varphi(a)\} \qquad (\overline{a}\text{: reorientation of } a)$$

and the arc length function $\tilde{\gamma} : \tilde{A} \to \mathbf{R}$ is defined by

$$\tilde{\gamma}(a) = \begin{cases} \gamma(a) & (a \in A^*) \\ -\gamma(\overline{a}) & (a = (u,v) \in B^*, \overline{a} = (v,u) \in A). \end{cases} \tag{2.53}$$

Then optimality of a feasible flow can be characterized as follows.

Theorem 2.2.33. *For a feasible flow $\varphi : A \to \mathbf{R}$, the following three conditions* (i)–(iii) *are equivalent.*

(i) *φ is optimal, i.e., φ minimizes* $\mathrm{cost}(\varphi)$.

(ii) *There exists a "potential" function $p : V \to \mathbf{R}$ such that, for each $a \in A$,*

$$\gamma(a) + p(\partial^+ a) - p(\partial^- a) > 0 \implies \varphi(a) = \underline{c}(a), \tag{2.54}$$
$$\gamma(a) + p(\partial^+ a) - p(\partial^- a) < 0 \implies \varphi(a) = \overline{c}(a). \tag{2.55}$$

(iii) *There exists no cycle of negative length with respect to $\tilde{\gamma}$ in the auxiliary network $\tilde{N}_\varphi$.*

Moreover, if the cost γ is integer-valued, we can choose p to be integer-valued in (ii).

Proof. Put $\gamma_p(a) = \gamma(a) + p(\partial^+ a) - p(\partial^- a)$ for $a \in A$.

(ii) ⇒ (i): For any feasible $\varphi' : A \to \mathbf{R}$ we have

$$\begin{aligned}\mathrm{cost}(\varphi') = \sum_{a \in A} \gamma_p(a)\varphi'(a) &= \sum_{a:\gamma_p(a)>0} \gamma_p(a)\varphi'(a) + \sum_{a:\gamma_p(a)<0} \gamma_p(a)\varphi'(a) \\ &\geq \sum_{a:\gamma_p(a)>0} \gamma_p(a)\underline{c}(a) + \sum_{a:\gamma_p(a)<0} \gamma_p(a)\overline{c}(a) = \mathrm{cost}(\varphi).\end{aligned}$$

(i) ⇒ (iii): Suppose there exists a cycle of negative length in $\tilde{N}_\varphi$. Let $\tilde{C} \subseteq \tilde{A}$ be the arc set of such a cycle of minimum cardinality. Then φ' defined by

$$\varphi'(a) = \begin{cases} \varphi(a) + \varepsilon & (a \in A^* \cap \tilde{C}) \\ \varphi(a) - \varepsilon & (\overline{a} \in B^* \cap \tilde{C}) \\ \varphi(a) & (\text{otherwise}) \end{cases} \tag{2.56}$$

with a sufficiently small $\varepsilon > 0$ is feasible and

$$\text{cost}(\varphi') - \text{cost}(\varphi) = \varepsilon \sum_{a \in \tilde{C}} \tilde{\gamma}(a) < 0.$$

(iii) ⇒ (ii): Since there exists no cycle of negative length in $\tilde{N}_\varphi$, there exists $p : V \to \mathbf{R}$ such that $\tilde{\gamma}(a) + p(\partial^+ a) - p(\partial^- a) \geq 0$ $(\forall\, a \in \tilde{A})$ (see Theorem 2.2.35 below). This condition is equivalent to the condition in (ii). ■

As to the existence of optimal flows, the following theorem states fundamental facts. The second statement below refers to another auxiliary network $\tilde{N}_\infty = (V, A^* \cup B^*, \tilde{\gamma})$ defined by

$$A^* = \{a \mid a \in A, \overline{c}(a) = +\infty\},$$
$$B^* = \{\overline{a} \mid a \in A, \underline{c}(a) = -\infty\} \qquad (\overline{a}\text{: reorientation of } a)$$

and $\tilde{\gamma} : A^* \cup B^* \to \mathbf{R}$ of (2.53). It is noted that the existence of an optimal flow is equivalent to the boundedness of the cost ($\inf_\varphi \text{cost}(\varphi) > -\infty$) under the assumption of feasibility.

Theorem 2.2.34. (1) *A feasible flow exists if and only if*

$$\sum\{\overline{c}(a) \mid a \in C(S)\} \geq \sum\{\underline{c}(a) \mid a \in C(V \setminus S)\}$$

for each $S \subseteq V$, where $C(S)$ is defined by (2.49). *Moreover, if the capacity functions $\overline{c}$ and $\underline{c}$ are integer-valued and there exists a (real-valued) feasible flow, then there exists an integer-valued feasible flow.*

(2) *Assume that a feasible flow exists. An optimal flow exists if and only if there exists no cycle of negative length with respect to $\tilde{\gamma}$ in the auxiliary network $\tilde{N}_\infty = (V, A^* \cup B^*, \tilde{\gamma})$.*

(3) *If the capacity functions $\overline{c}$ and $\underline{c}$ are integer-valued and there exists a (real-valued) optimal flow, then there exists an integer-valued optimal flow.*

Proof. (1) This is a theorem due to Hoffman [111], which may be regarded as a variant of Theorem 2.2.30.

(2) The "only if" part is immediate from Theorem 2.2.33 ((i) ⇒ (iii) to be specific). The "if" part can be shown by repeated modifications of feasible flows as in (2.56).

(3) This can be shown by repeated modifications of integral feasible flows as in (2.56), in which we can take $\varepsilon = 1$. Note that an initial integral feasible flow exists by (1). ■

The following basic facts are stated here for the convenience of reference. The function $\gamma : A \to \mathbf{R}$ is now interpreted as the arc-length function of a network $N = (V, A, \gamma)$.

Theorem 2.2.35. *Let $N = (V, A, \gamma)$ be a network with $\gamma : A \to \mathbf{R}$.*

(1) *There exists $p : V \to \mathbf{R}$ such that*

$$\gamma(a) + p(\partial^+ a) - p(\partial^- a) \geq 0 \qquad (\forall\ a \in A) \tag{2.57}$$

if and only if there exists no (directed) cycle of negative length with respect to γ. Moreover, if γ is integer-valued, we can take integer-valued p.

(2) *There exists $p : V \to \mathbf{R}$ such that*

$$\gamma(a) + p(\partial^+ a) - p(\partial^- a) = 0 \qquad (\forall\ a \in A) \tag{2.58}$$

if and only if

$$\sum_{a \in A} \chi_C(a)\gamma(a) = 0 \qquad (\forall\ C : \text{circuit in } N), \tag{2.59}$$

where $\chi_C(a) = 1$ or -1 according to whether $a \in A$ is contained in C in the positive or negative direction[6] *(and $\chi_C(a) = 0$ if $a \in A$ is not contained in C). Moreover, if γ is integer-valued, we can take integer-valued p.* □

Next we turn to the *weighted bipartite matching problem.* Let $G = (V^+, V^-; A)$ be a bipartite graph and $w : A \to \mathbf{R}$ be a weight function. The weight of a matching M is defined to be

$$w(M) = \sum_{a \in M} w(a).$$

Given a nonnegative integer k, the *maximum weight k-matching problem* is to find a k-matching M (i.e., a matching M of size k) that maximizes the weight $w(M)$. A k-matching that attains the maximum weight is called an *optimal k-matching.*

This problem can be formulated as the minimum cost flow problem in a network $N_G = (\tilde{V}, \tilde{A}, \overline{c}, \underline{c}, \gamma)$ with

$$\tilde{V} = V^+ \cup V^- \cup \{s^+, s^-\},$$
$$\tilde{A} = A \cup \{(s^+, u) \mid u \in V^+\} \cup \{(v, s^-) \mid v \in V^-\} \cup \{(s^-, s^+)\},$$

and $\overline{c}$, $\underline{c}$, γ given by

	$\underline{c}(a)$	$\overline{c}(a)$	$\gamma(a)$
$a \in A$	0	$+\infty$	$-w(a)$
$a = (s^+, u)$ $(u \in V^+)$	$-\infty$	1	0
$a = (v, s^-)$ $(v \in V^-)$	$-\infty$	1	0
$a = (s^-, s^+)$	k	k	0

Then the optimality criterion for N_G is translated into the following theorem for the weighted matching problem on $G = (V^+, V^-; A)$.

[6] It is understood that a circuit in N (i.e., a circuit of the underlying undirected graph) is endowed with a direction.

Theorem 2.2.36. *A k-matching M in $G = (V^+, V^-; A)$ is optimal (maximum) with respect to $w : A \to \mathbf{R}$ if and only if there exist a "potential" function $p : V^+ \cup V^- \to \mathbf{R}$ and a scalar $q \in \mathbf{R}$ such that $p(v) \geq 0$ $(v \in V^+ \cup V^-)$, $\{v \in V^+ \cup V^- \mid p(v) > 0\} \subseteq \partial M$, and*

$$w(u,v) - p(u) - p(v) - q \begin{cases} = 0 & ((u,v) \in M) \\ \leq 0 & ((u,v) \in A) \end{cases} \tag{2.60}$$

where $w(u,v)$ means $w(a)$ for $a = (u,v) \in A$. Moreover, if the weight w is integer-valued, we can choose p and q to be integer-valued.

Proof. Note first that an integral feasible flow in N_G corresponds to a k-matching in G. Let $\tilde{p} : \tilde{V} \to \mathbf{R}$ be the potential function in Theorem 2.2.33, and put $p(u) = \tilde{p}(u) - \tilde{p}(s^+)$ $(u \in V^+)$, $p(v) = \tilde{p}(s^-) - \tilde{p}(v)$ $(v \in V^-)$, and $q = \tilde{p}(s^+) - \tilde{p}(s^-)$. Then (2.54) and (2.55) for $\tilde{p}$ are equivalent to the conditions on p and q above. ■

2.3 Matroid

2.3.1 From Matrix to Matroid

As the name shows, the concept of a matroid is a combinatorial abstraction of matrices with respect to linear independence. The abstract definition of a matroid, to be given in §2.3.2, is preceded here by linear algebraic motivations explained by means of a concrete example.

Take a 3×5 matrix

$$A = \begin{array}{c} \begin{matrix} 1 & 2 & 3 & 4 & 5 \end{matrix} \\ \boxed{\begin{matrix} 1 & 0 & 0 & 1 & 0 \\ 0 & 1 & 0 & 1 & 1 \\ 0 & 0 & 1 & 0 & 1 \end{matrix}} \end{array},$$

of which the columns are indexed by $V = \mathrm{Col}(A) = \{1, \cdots, 5\}$, and consider linear dependence/independence among column vectors, say $\{\boldsymbol{a}_v \mid v \in V\}$. A subset $I \subseteq V$ is said to be independent if the corresponding subfamily $\{\boldsymbol{a}_v \mid v \in I\}$ of column vectors is linearly independent. Denote by $\mathcal{I} \subseteq 2^V$ the family of independent subsets, i.e.,

$$\mathcal{I} = \{I \subseteq V \mid \{\boldsymbol{a}_v \mid v \in I\} \text{ is linearly independent}\}, \tag{2.61}$$

where

(I-1) $\emptyset \in \mathcal{I}$

by convention. For the matrix A above an inspection shows

$$\mathcal{I} = \{\emptyset, \{1\}, \{2\}, \{3\}, \{4\}, \{5\}, \{1,2\}, \{1,3\}, \{1,4\}, \{1,5\}, \{2,3\}, \{2,4\}, \{2,5\}, \{3,4\}, \{3,5\}, \{4,5\}, \{1,2,3\}, \{1,2,5\}, \{1,3,4\}, \{1,3,5\}, \{1,4,5\}, \{2,3,4\}, \{2,4,5\}, \{3,4,5\}\}.$$

Obviously, it holds that

(I-2) $I \subseteq J \in \mathcal{I} \implies I \in \mathcal{I}$,

since a subset of an independent subset is also independent. A nontrivial property of $\mathcal{I}$ is described by

(I-3) $I, J \in \mathcal{I}, |I| < |J| \implies I \cup \{v\} \in \mathcal{I}$ for some $v \in J \setminus I$.

For $I = \{1,2\}$ and $J = \{2,3,4\}$, for instance, we can take $v = 3$ to obtain $I \cup \{v\} = \{1,2,3\} \in \mathcal{I}$, whereas $v = 4$ leads to $I \cup \{v\} = \{1,2,4\} \notin \mathcal{I}$. It is a good exercise in linear algebra (and hence left to the reader) to prove (I-3) in general, where $\mathcal{I}$ is defined by (2.61) for a given matrix $A = (\boldsymbol{a}_v \mid v \in V)$. In this way a matrix gives rise to a pair $(V, \mathcal{I})$ with the properties (I-1), (I-2), (I-3).

Since $\mathcal{I}$ satisfies (I-2), it is redundant to enumerate all the members of $\mathcal{I}$, and only the maximal members of $\mathcal{I}$ (maximal with respect to set inclusion) suffice. In our example, the family $\mathcal{B}$ of the maximal members of $\mathcal{I}$ is given by

$$\mathcal{B} = \{\{1,2,3\}, \{1,2,5\}, \{1,3,4\}, \{1,3,5\}, \{1,4,5\}, \{2,3,4\}, \{2,4,5\}, \{3,4,5\}\},$$

which is the family of column bases of the matrix A. The family $\mathcal{B}$ satisfies

(BM$_-$) For $B, B' \in \mathcal{B}$ and for $u \in B \setminus B'$, there exists $v \in B' \setminus B$ such that $B - u + v \in \mathcal{B}$,

where $B - u + v$ is a short-hand notation for $(B \setminus \{u\}) \cup \{v\}$. This is a consequence of the Grassmann–Plücker identity (see Remark 2.1.8 for (BM$_\pm$), which implies (BM$_-$)). For $B = \{1,2,3\}$, $B' = \{3,4,5\}$ and $u = 1$, for example, we can take $v = 4$ to obtain $B - u + v = \{2,3,4\} \in \mathcal{B}$, whereas $v = 5$ yields $B - u + v = \{2,3,5\} \notin \mathcal{B}$. Thus a matrix gives rise to a pair $(V, \mathcal{B})$ with the property (BM$_-$).

The linear independence structure of column vectors can be represented also by the rank function $\rho : 2^V \to \mathbf{Z}$ defined by

$$\rho(X) = \operatorname{rank} A[\operatorname{Row}(A), X], \qquad X \subseteq V.$$

The following two properties of ρ are obvious:

(R-1) $0 \le \rho(X) \le |X|$,
(R-2) $X \subseteq Y \implies \rho(X) \le \rho(Y)$.

The key property of ρ is the submodularity:

(R-3) $\rho(X) + \rho(Y) \ge \rho(X \cup Y) + \rho(X \cap Y), \quad X, Y \subseteq V,$

which has been shown in Proposition 2.1.9(1). Thus a matrix yields a pair (V, ρ) with the properties (R-1), (R-2), (R-3).

To sum up, a matrix gives rise to $(V, \mathcal{I})$, $(V, \mathcal{B})$ and (V, ρ), each representing (some aspects of) the linear independence structure of column vectors. The conditions (I-1), (I-2), (I-3) for $\mathcal{I}$, (BM$_-$) for $\mathcal{B}$, and (R-1), (R-2), (R-3) for ρ are stated without reference to the original matrix, and are meaningful by themselves as conditions on a family $\mathcal{I} \subseteq 2^V$, a family $\mathcal{B} \subseteq 2^V$, and a set function $\rho : 2^V \to \mathbf{Z}$, respectively. It turns out that these three abstract structures, $(V, \mathcal{I})$, $(V, \mathcal{B})$ and (V, ρ), are equivalent (in an appropriate sense), and therefore define one and the same combinatorial structure underlying linear independence. This structure is named a matroid.

We are now ready to formally define a matroid in terms of abstract axioms.

2.3.2 Basic Concepts

A matroid can be defined in many different ways. For our purpose it is convenient to feature independent sets, bases, and the rank function. Statements marked with $^{((\mathrm{P}))}$ are given proofs at the end of §2.3.2.

A *matroid* is a pair $\mathbf{M} = (V, \mathcal{I})$ of a finite set V and a collection $\mathcal{I}$ of subsets of V such that

(I-1) $\emptyset \in \mathcal{I}$,
(I-2) $I \subseteq J \in \mathcal{I} \implies I \in \mathcal{I}$,
(I-3) $I, J \in \mathcal{I}$, $|I| < |J| \implies I \cup \{v\} \in \mathcal{I}$ for some $v \in J \setminus I$.

The set V is called the *ground set* and $I \in \mathcal{I}$ an *independent set*; accordingly, $\mathcal{I}$ is the *family of independent sets.*

Denote by $\mathcal{B}$ the family of the maximal members of $\mathcal{I}$ (maximal with respect to set inclusion); namely, $\mathcal{B} = \max \mathcal{I}$ in short. The family $\mathcal{B}$ satisfies$^{((\mathrm{P1}))}$

(BM$_-$) For $B, B' \in \mathcal{B}$ and for $u \in B \setminus B'$, there exists $v \in B' \setminus B$ such that $B - u + v \in \mathcal{B}$.

We call (BM$_-$) the *(one-sided) basis exchange property.* A member of $\mathcal{B}$ is called a *base*. The size of a base is uniquely determined$^{((\mathrm{P2}))}$ and is called the *rank* of $\mathbf{M}$, denoted as $\mathrm{rank}\,\mathbf{M}$; i.e.,

$$\mathrm{rank}\,\mathbf{M} = |B| = \max\{|I| \mid I \in \mathcal{I}\} \quad \text{for } B \in \mathcal{B}.$$

Conversely$^{((\mathrm{P3}))}$, a nonempty family $\mathcal{B}$ of subsets of V forms the basis family of a matroid if it satisfies the axiom (BM$_-$); the matroid $\mathbf{M} = (V, \mathcal{I})$ is given by

$$\mathcal{I} = \{I \subseteq V \mid I \subseteq B \in \mathcal{B}\}. \tag{2.62}$$

For $X \subseteq V$, the *rank* $\rho(X)$ of X is defined as the uniquely determined$^{((\mathrm{P4}))}$ cardinality of a maximal independent set contained in X. That is,

$$\rho(X) = \max\{|I| \mid I \subseteq X,\ I \in \mathcal{I}\}.$$

The rank function $\rho : 2^V \to \mathbf{Z}$ satisfies[((P5))] the conditions:

(R-1) $0 \leq \rho(X) \leq |X|$,
(R-2) $X \subseteq Y \implies \rho(X) \leq \rho(Y)$,
(R-3) $\rho(X) + \rho(Y) \geq \rho(X \cup Y) + \rho(X \cap Y), \quad X, Y \subseteq V.$

The property (R-3) is the submodularity. Conversely[((P6))], a function $\rho : 2^V \to \mathbf{Z}$ satisfying these properties is the rank function of a matroid; the matroid $\mathbf{M} = (V, \mathcal{I})$ is given by

$$\mathcal{I} = \{I \subseteq V \mid \rho(I) = |I|\}.$$

To sum up, a matroid can be defined as $(V, \mathcal{I})$ with (I-1)–(I-3), $(V, \mathcal{B})$ with (BM$_-$), or (V, ρ) with (R-1)–(R-3). Given one of these, we can derive the other two as follows:

Given	Define	
$(V, \mathcal{I}) \Rightarrow$	$\mathcal{B} = \max \mathcal{I}$,	$\rho(X) = \max\{|I| \mid I \subseteq X,\ I \in \mathcal{I}\}$
$(V, \mathcal{B}) \Rightarrow$	$\mathcal{I} = \{I \mid I \subseteq B \in \mathcal{B}\}$,	$\rho(X) = \max\{|X \cap B| \mid B \in \mathcal{B}\}$
$(V, \rho) \Rightarrow$	$\mathcal{I} = \{I \subseteq V \mid \rho(I) = |I|\}$,	$\mathcal{B} = \{B \subseteq V \mid \rho(B) = |B| = \rho(V)\}$

We use notations $\mathbf{M} = (V, \mathcal{B}, \mathcal{I}, \rho)$, $\mathbf{M} = (V, \mathcal{B}, \rho)$, etc., whenever convenient.

A subset $X \subseteq V$ not belonging to $\mathcal{I}$ is called a *dependent set*, and a minimal dependent set (minimal with respect to set inclusion) is a *circuit*. We call $X \subseteq V$ a *spanning set* if it contains a base. For $X \subseteq V$, the *closure* $\mathrm{cl}(X)$ of X is defined by

$$\mathrm{cl}(X) = \{v \in V \mid \rho(X \cup \{v\}) = \rho(X)\}. \tag{2.63}$$

It is also possible to characterize a matroid in terms of the family of dependent sets, the family of circuits, the family of spanning sets, or the closure function.

An element of V not contained in any base is called a *loop*, whereas an element contained in every base is a *coloop*. A pair of elements of V, neither of which is a loop, are said to be *in parallel*, if there exists no base that contains both of them. A pair of elements of V, neither of which is a coloop, are said to be *in series*, if there exists no base that is disjoint from them. The relation of being in parallel is transitive in that if both $\{u, v\}$ and $\{v, w\}$ are parallel pairs, then $\{u, w\}$ is also a parallel pair. The relation of being in series is also transitive.

Given a matroid $\mathbf{M} = (V, \mathcal{B}, \mathcal{I}, \rho)$, we can derive a number of matroids, as follows.

The *dual* of $\mathbf{M}$, denoted $\mathbf{M}^*$, is a matroid on V of which the basis family $\mathcal{B}^*$ is given by

$$\mathcal{B}^* = \{V \setminus B \mid B \in \mathcal{B}\}.$$

In fact, $\mathcal{B}^*$ satisfies the exchange property (BM$_-$) because (BM$_-$) for $\mathcal{B}^*$ is tantamount to the condition

(BM$_+$) For $B, B' \in \mathcal{B}$ and for $v \in B' \setminus B$, there exists $u \in B \setminus B'$ such that $B - u + v \in \mathcal{B}$,

for $\mathcal{B}$, and it can be shown (cf. Theorem 2.3.14) that (BM$_+$) for $\mathcal{B}$ is equivalent to (BM$_-$) for $\mathcal{B}$. Note that (BM$_+$) is not identical with (BM$_-$). The rank function ρ^* of $\mathbf{M}^*$ is given by

$$\rho^*(X) = |X| + \rho(V \setminus X) - \rho(V), \qquad X \subseteq V. \tag{2.64}$$

The *restriction* of $\mathbf{M}$ to U ($\subseteq V$), denoted as $\mathbf{M}^U$, is a matroid on U in which X ($\subseteq U$) is independent if and only if X is independent in $\mathbf{M}$. We also say that $\mathbf{M}^U$ is obtained from $\mathbf{M}$ by deleting the elements of $V \setminus U$. The rank function ρ^U of $\mathbf{M}^U$ is simply the restriction of ρ to U, i.e., $\rho^U(X) = \rho(X)$ for $X \subseteq U$.

The *contraction* of $\mathbf{M}$ to $U(\subseteq V)$, denoted as $\mathbf{M}_U$, is a matroid on U in which X ($\subseteq U$) is independent if and only if $X \cup B$ is independent in $\mathbf{M}$ for a base B of $\mathbf{M}^{V\setminus U}$. We also say that $\mathbf{M}_U$ is obtained by contracting the elements of $V \setminus U$. The rank function ρ_U of $\mathbf{M}_U$ is given by

$$\rho_U(X) = \rho(X \cup (V \setminus U)) - \rho(V \setminus U), \qquad X \subseteq U.$$

We have $(\mathbf{M}_U)^* = (\mathbf{M}^*)^U$.

The *truncation* of $\mathbf{M}$ to k, where $k \leq \operatorname{rank} \mathbf{M}$, is a matroid on V in which X ($\subseteq V$) is a base if and only if $|X| = k$ and X is independent in $\mathbf{M}$. The rank function is given by $\min(\rho(X), k)$.

The *elongation* of $\mathbf{M}$ to l, where $l \geq \operatorname{rank} \mathbf{M}$, is a matroid on V in which X ($\subseteq V$) is a base if and only if $|X| = l$ and X is spanning in $\mathbf{M}$. The dual of the truncation of $\mathbf{M}$ to k coincides with the elongation of $\mathbf{M}^*$ to $|V| - k$, where $k \leq \operatorname{rank} \mathbf{M}$.

For two matroids $\mathbf{M}_1$ and $\mathbf{M}_2$ on disjoint ground sets V_1 and V_2, respectively, their *direct sum*, denoted as $\mathbf{M}_1 \oplus \mathbf{M}_2$, is a matroid on $V_1 \cup V_2$ in which X ($\subseteq V_1 \cup V_2$) is independent if and only if $X \cap V_i$ is independent in $\mathbf{M}_i$ for $i = 1, 2$. Besides this rather trivial operation, there is another operation of "adding" two matroids, called union operation, which will be treated in §2.3.6.

A matroid $\mathbf{M}_1 = (V, \rho_1)$ is said to be a *strong quotient* of another matroid $\mathbf{M}_2 = (V, \rho_2)$ if

$$\rho_2(X) - \rho_2(Y) \geq \rho_1(X) - \rho_1(Y), \qquad X \supseteq Y. \tag{2.65}$$

If this is the case, we write $\mathbf{M}_2 \to \mathbf{M}_1$ and say that $\mathbf{M}_2 \to \mathbf{M}_1$ is a *strong map*. Obviously, $\operatorname{rank} \mathbf{M}_2 \geq \operatorname{rank} \mathbf{M}_1$ if $\mathbf{M}_2 \to \mathbf{M}_1$.

Lemma 2.3.1. *If* $\mathbf{M}_2 \to \mathbf{M}_1$ *and* $\operatorname{rank} \mathbf{M}_2 = \operatorname{rank} \mathbf{M}_1$, *then* $\mathbf{M}_2 = \mathbf{M}_1$.

Proof. The inequality (2.65) with $Y = \emptyset$ gives $\rho_2(X) \geq \rho_1(X)$, whereas (2.65) with $X = V$ yields $\rho_2(Y) \leq \rho_1(Y)$ since $\rho_2(V) = \rho_1(V)$. Hence $\rho_2 = \rho_1$. ∎

Proofs of the Marked Statements. Proofs of the marked statements are sketched below.

((P1)) For $B, B' \in \mathcal{B} = \max \mathcal{I}$, we have $|B| = |B'|$ since $|B| < |B'|$ would imply by (I-3) the existence of $v \in B' \setminus B$ such that $B \cup \{v\} \in \mathcal{I}$, a contradiction to the maximality of B. Then application of (I-3) to $I = B - u \in \mathcal{I}$ and $J = B' \in \mathcal{I}$ yields (BM$_-$).

((P2)) We show (BM$_-$) $\Rightarrow |B| = |B'|$. For $B, B' \in \mathcal{B}$, put $B_1 = B - u + v \in \mathcal{B}$ using $u \in B \setminus B'$ and $v \in B' \setminus B$ in (BM$_-$). Then $|B_1| = |B|$ and $|B_1 \setminus B'| = |B \setminus B'| - 1$. Applying the same argument to (B_1, B') we obtain $B_2 \in \mathcal{B}$ with $|B_2| = |B_1| = |B|$ and $|B_2 \setminus B'| = |B_1 \setminus B'| - 1 = |B \setminus B'| - 2$. Repeating this we can prove $|B'| = |B|$.

((P3)) (I-1) and (I-2) are obviously satisfied by $\mathcal{I}$ of (2.62). To show (I-3) suppose that $I \subseteq B_I \in \mathcal{B}$, $J \subseteq B_J \in \mathcal{B}$ and $|I| < |J|$, where $|B_I \cap B_J|$ is maximized over such B_I and B_J. Then $B_I \setminus I \subseteq B_J$, since otherwise there exist $u \in (B_I \setminus B_J) \setminus I$ and $v \in B_J \setminus B_I$, for which $B'_I = B_I - u + v \in \mathcal{B}$, $I \subseteq B'_I$ and $|B'_I \cap B_J| = |B_I \cap B_J| + 1$ (a contradiction). Since $|B_I| = |B_J|$ by ((P2)), it follows from $|B_I| = |I| + |B_I \setminus I| = |I| - |J| + |(B_I \setminus I) \cap J| + |(B_I \setminus I) \cup J| \le |I| - |J| + |(B_I \setminus I) \cap J| + |B_J|$ that $|(B_I \setminus I) \cap J| \ge |J| - |I| > 0$, i.e., $\exists v \in (B_I \setminus I) \cap J$. For this v we have $I \cup \{v\} \subseteq B_I$, and hence $I \cup \{v\} \in \mathcal{I}$.

((P4)) Let I and J be two maximal independent sets contained in X. If $|I| < |J|$, then (I-3) implies $I \cup \{v\}$ is also a maximal independent set contained in X, a contradiction.

((P5)) (R-1) and (R-2) are obviously satisfied. To show (R-3) first observe from ((P4)) that for $X_1 \subseteq X_2 \subseteq X_3$, there exist independent sets $I_i \subseteq X_i$ $(i = 1, 2, 3)$ such that $I_1 \subseteq I_2 \subseteq I_3$ and $|I_i| = \rho(X_i)$ $(i = 1, 2, 3)$. Applying this fact to $X_1 = X \cap Y$, $X_2 = X$, $X_3 = X \cup Y$ we obtain J_1, J_2, J_3 such that $J_1 \subseteq X \cap Y$, $J_2 \subseteq X \setminus Y$, $J_3 \subseteq Y \setminus X$, $J_1 \in \mathcal{I}$, $J_1 \cup J_2 \in \mathcal{I}$, $J_1 \cup J_2 \cup J_3 \in \mathcal{I}$, $|J_1| = \rho(X \cap Y)$, $|J_1| + |J_2| = \rho(X)$, $|J_1| + |J_2| + |J_3| = \rho(X \cup Y)$. Combination of these equalities with an inequality $\rho(Y) \ge |J_1| + |J_3|$ yields the submodularity (R-3).

((P6)) (I-1) is obvious. For (I-2) suppose that $I \subseteq J$ and $\rho(J) = |J|$. By (R-1) we have $|I| \ge \rho(I)$ and $|J \setminus I| \ge \rho(J \setminus I)$. Addition of these, combined by (R-3), yields $|J| \ge \rho(I) + \rho(J \setminus I) \ge \rho(\emptyset) + \rho(J) = |J|$. Hence $|I| = \rho(I)$. For (I-3) it suffices to show $[\rho(I \cup \{v\}) = \rho(I)$ for all $v \in J \setminus I \;\Rightarrow\; \rho(J) \le \rho(I)]$ for $I, J \in \mathcal{I}$. Let $v_1, \cdots, v_m$ be the elements of $J \setminus I$. It follows from (R-2) and (R-3) that $\rho(I) \le \rho(I \cup \{v_1, v_2\}) \le \rho(I \cup \{v_1\}) + \rho(I \cup \{v_2\}) - \rho(I) = \rho(I)$. Hence $\rho(I \cup \{v_1, v_2\}) = \rho(I)$. Similarly, $\rho(I) \le \rho(I \cup \{v_1, v_2, v_3\}) \le \rho(I \cup \{v_1, v_2\}) + \rho(I \cup \{v_3\}) - \rho(I) = \rho(I)$. Continuing in this way, we arrive at $\rho(J) \le \rho(I \cup \{v_1, \cdots, v_m\}) = \rho(I)$.

Notes. The concept of a matroid was introduced by Whitney [340] and the earlier key papers are compiled in Kung [166]. For topics on matroids and submodular functions not covered in this book, the reader is referred to Bixby–Cunningham [13], Edmonds [68], Fujishige [82], Lawler [171], Oxley [259], Topkis [319], Welsh [333], and White [336, 337, 338] for theory, and to

Iri [127, 128], Iri–Fujishige [130], Lee–Ryan [172], Murota [204], and Recski [277] for applications.

2.3.3 Examples

Example 2.3.2 (Free matroid). Let V be a finite set. Put $\mathcal{I} = 2^V$ and $\mathcal{B} = \{V\}$. This is called the *free matroid* on V, in which every subset of V is independent. We have $\rho(X) = |X|$ for $X \subseteq V$. □

Example 2.3.3 (Uniform matroid). Let V be a finite set and $r \leq |V|$ be an integer. Put $\mathcal{I} = \{I \subseteq V \mid |I| \leq r\}$ and $\mathcal{B} = \{B \subseteq V \mid |B| = r\}$. This is called the *uniform matroid* of rank r. The uniform matroid of rank $|V|$ is the free matroid, and that of rank zero is called the *trivial matroid.* □

Example 2.3.4 (Partition matroid). Let $(V_i \mid i \in P)$ be a partition of a finite set V, i.e., $\bigcup_{i \in P} V_i = V$ and $V_i \cap V_j = \emptyset$ for $i \neq j$. Then $\mathcal{I} = \{I \subseteq V \mid |I \cap V_i| \leq 1 \ (\forall i \in P)\}$ forms the family of independent sets of a matroid on V, called a *partition matroid.* □

Example 2.3.5 (Transversal matroid). For a bipartite graph $G = (V^+, V^-; A)$ let $\mathcal{I}$ be the family of subsets of V^+ that can be matched into V^-; i.e., a subset I of V^+ belongs to $\mathcal{I}$ if and only if there exists a matching that covers I (see §2.2.3 for matchings). Then $\mathcal{I}$ satisfies (I-1)–(I-3), and defines a matroid on V^+. A matroid obtained in this manner is called a *transversal matroid.* The uniform matroid of rank r on V is a transversal matroid defined by a complete bipartite graph with $V^+ = V$, $|V^-| = r$ and $A = \{(v, i) \mid v \in V, i \in V^-\}$. The partition matroid on V defined by $(V_i \mid i \in P)$ is a transversal matroid with $V^+ = V$, $V^- = P$ and $A = \{(v, i) \mid v \in V_i, i \in P\}$. □

Example 2.3.6 (Gammoid). Let $G = (W, A; S, T)$ be a directed graph with vertex set W, arc set A and disjoint entrance S and exit T ($S \subseteq W$, $T \subseteq W$). Denote by $\mathcal{I}$ the family of subsets of S that can be linked into T; i.e., a subset I of S belongs to $\mathcal{I}$ if and only if there exists a Menger-type (vertex-disjoint) linking of size $|I|$ from I to a subset of T (see §2.2.4 for linkings). Then $\mathcal{I}$ satisfies (I-1)–(I-3), and defines a matroid on S. A matroid obtained in this manner is called a *gammoid.* If G is a bipartite graph, the gammoid defined by G is a transversal matroid. □

Example 2.3.7 (Matching matroid). Let $G = (V, A)$ be a graph with vertex set V and arc set A. Denote by $\mathcal{I}$ the family of subsets of V that can be covered by some matching.[7] That is, $\mathcal{I} = \{I \subseteq V \mid I \subseteq \partial M$ for some matching $M\}$. Then $\mathcal{I}$ satisfies (I-1)–(I-3), and defines a matroid on V. A matroid obtained in this manner is called a *matching matroid.* □

[7] A matching in a nonbipartite graph is a subset of arcs no two of which share a common vertex incident to them.

Example 2.3.8 (Linear matroid). Let A be a matrix over a field $\boldsymbol{F}$, and V be the set of the column vectors of A. Linear independence among the column vectors of A defines a matroid on V, which will be denoted by $\mathbf{M}(A)$. Namely, $I \subseteq V$ is independent in $\mathbf{M}(A)$ if and only if the column vectors in I are linearly independent. The rank function ρ of $\mathbf{M}(A)$ is given by (2.6). A matroid that can be obtained in this way is called a *linear matroid representable* over $\boldsymbol{F}$. If the matrix A is given explicitly, the matroid is said to be *represented* over $\boldsymbol{F}$.

A linear subspace U of $\boldsymbol{F}^V$ defines a matroid, denoted as $\mathbf{M}\{U\}$, in which $I(\subseteq V)$ is independent if and only if there exists no vector $\boldsymbol{x} = (x(v) \mid v \in V) \in U \setminus \{\mathbf{0}\}$ such that $\{v \in V \mid x(v) \neq 0\} \subseteq I$. This matroid can be linearly represented by a matrix A such that $U = \ker A$; namely, $\mathbf{M}\{U\} = \mathbf{M}(A)$ if $U = \ker A$. This shows

$$\mathrm{rank}\,[\mathbf{M}\{U\}] + \dim U = |V|.$$

The orthogonal complement $U^\perp$ of U (or, more precisely, the subspace of the dual space of $\boldsymbol{F}^V$ consisting of elements that annihilate on U) corresponds to the matroid dual to $\mathbf{M}\{U\}$, i.e., $\mathbf{M}\{U^\perp\} = \mathbf{M}\{U\}^*$. For two nested subspaces $U_1 \subseteq U_2$ it holds that $\mathbf{M}\{U_1\} \to \mathbf{M}\{U_2\}$, where "$\to$" denotes a strong map. To see this we may assume $U_1 = \ker A[I_1, V]$ and $U_2 = \ker A[I_2, V]$ for a matrix A and $I_1 \supseteq I_2$. Then Proposition 2.1.9(2) implies (2.65) for $\rho_i(J) = \mathrm{rank}\, A[I_i, J]$, $i = 1, 2$. Hence we may interpret the strong map relation as a combinatorial abstraction of the nesting of linear subspaces. □

Example 2.3.9 (Graphic matroid). Let $G = (V, A)$ be a graph with vertex set V and arc set A. Define $\mathcal{I}$ to be the family of subsets of A that contain no (undirected) cycles in G. Then $\mathcal{I}$ satisfies (I-1)–(I-3) and defines a matroid on A. This matroid coincides with the linear matroid defined by the incidence matrix of G. A matroid obtained in this way is called a *graphic matroid.* □

Example 2.3.10 (Algebraic matroid). Let $\boldsymbol{F}$ be an extension field of a field $\boldsymbol{K}$, and V a finite subset of $\boldsymbol{F}$. Define $\mathcal{I}$ to be the family of subsets of V that are algebraically independent over $\boldsymbol{K}$ (see §2.1.1 for algebraic independence). Then $\mathcal{I}$ satisfies (I-1)–(I-3) and defines a matroid on V. A matroid obtained in this way is called an *algebraic matroid.* The rank function is given by the degree of transcendency of the extension field $\boldsymbol{K}(X)$ over $\boldsymbol{K}$:

$$\rho(X) = \dim_{\boldsymbol{K}} \boldsymbol{K}(X), \qquad X \subseteq V.$$

□

2.3.4 Basis Exchange Properties

We consider a number of basis exchange properties in a matroid. First recall

(BM$_-$) For $B, B' \in \mathcal{B}$ and for $u \in B \setminus B'$, there exists $v \in B' \setminus B$ such that $B - u + v \in \mathcal{B}$,

which characterizes the basis family $\mathcal{B}$ of a matroid (see §2.3.2).

We have observed in Remark 2.1.8 that the Grassmann–Plücker identity (Proposition 2.1.4) implies a stronger exchange property:

(BM$_\pm$) For $B, B' \in \mathcal{B}$ and for $u \in B \setminus B'$, there exists $v \in B' \setminus B$ such that $B - u + v \in \mathcal{B}$ and $B' + u - v \in \mathcal{B}$,

for a linear matroid (cf. Example 2.3.8). The following lemma (Brualdi [23]) claims that (BM$_\pm$) is implied by (BM$_-$) in general, and therefore satisfied by any matroid. (BM$_\pm$) is often called the *simultaneous exchange property*.

Lemma 2.3.11. *For $\mathcal{B} \subseteq 2^V$, (BM$_-$) $\Longrightarrow$ (BM$_\pm$).*

Proof. First note that (BM$_-$) implies

(BM$_{+\mathrm{loc}}$) For $B, B' \in \mathcal{B}$ with $|B \setminus B'| = 2$ and for $v \in B' \setminus B$, there exists $u \in B \setminus B'$ such that $B - u + v \in \mathcal{B}$.

Define

$$\mathcal{D} = \{(B, B') \mid B, B' \in \mathcal{B},\ \exists u_* \in B \setminus B', \forall v \in B' \setminus B : \\ B - u_* + v \notin \mathcal{B}, \text{ or } B' + u_* - v \notin \mathcal{B}\},$$

which denotes the set of pairs (B, B') for which the simultaneous exchange (BM$_\pm$) fails. We want to show $\mathcal{D} = \emptyset$.

To the contrary suppose that $\mathcal{D} \neq \emptyset$. Take $(B, B') \in \mathcal{D}$ such that $|B \setminus B'|$ is minimum, and let $u_* \in B \setminus B'$ be as in the definition of $\mathcal{D}$. Take any $u_0 \in (B \setminus B') \setminus \{u_*\}$, which is possible since $|B \setminus B'| \geq 2$. Define

$$X = \{v \in B' \setminus B \mid B' + u_* - v \in \mathcal{B}\}, \quad Y = \{v \in B' \setminus B \mid B - u_0 + v \in \mathcal{B}\},$$

where $Y \neq \emptyset$ by (BM$_-$). If $X \cap Y \neq \emptyset$, take any $v_0 \in X \cap Y$; otherwise take any $v_0 \in Y$. We have $B_1 \equiv B - u_0 + v_0 \in \mathcal{B}$ by $v_0 \in Y$.

Claim: $(B_1, B') \in \mathcal{D}$.

To prove this claim it suffices to show

$$B' + u_* - v \in \mathcal{B}, v \in B' \setminus B_1 \Longrightarrow B_2 \equiv B_1 - u_* + v \notin \mathcal{B}.$$

Note that $B - u_* + v \notin \mathcal{B}$ by the choice of u_*. In case of $X \cap Y \neq \emptyset$, we have $B' + u_* - v_0 \in \mathcal{B}$ by $v_0 \in X$, and therefore $B - u_* + v_0 \notin \mathcal{B}$. Then the contraposition of (BM$_-$) applied to (B, B_2) shows $B_2 \notin \mathcal{B}$, since $B \in \mathcal{B}$, $B - u_* + v_0 \notin \mathcal{B}$ and $B - u_* + v \notin \mathcal{B}$. In the remaining case of $X \cap Y = \emptyset$, since $v \in X$, we have $v \notin Y$, i.e., $B - u_0 + v \notin \mathcal{B}$. Then the contraposition of (BM$_{+\mathrm{loc}}$) applied to (B, B_2) shows $B_2 \notin \mathcal{B}$, since $B \in \mathcal{B}$, $B - u_0 + v \notin \mathcal{B}$ and $B - u_* + v \notin \mathcal{B}$. Thus the above claim has been proven.

Since $|B_1 \setminus B'| = |B \setminus B'| - 1$, the above claim contradicts our choice of $(B, B') \in \mathcal{D}$. Therefore we conclude $\mathcal{D} = \emptyset$, completing the proof of the theorem. (See also Brualdi [23], Kung [167], Welsh [333] for alternative proofs.) ■

In connection to dual matroids we have seen in §2.3.2 another exchange property[8]

(BM$_+$) For $B, B' \in \mathcal{B}$ and for $u \in B \setminus B'$, there exists $v \in B' \setminus B$ such that $B' + u - v \in \mathcal{B}$.

The next lemma shows that (BM$_+$) also implies the simultaneous exchange property (BM$_\pm$).

Lemma 2.3.12. *For* $\mathcal{B} \subseteq 2^V$, (BM$_+$) $\Longrightarrow$ (BM$_\pm$).

Proof. This is an immediate corollary to Lemma 2.3.11, since (BM$_+$) for $\mathcal{B}$ is equivalent to (BM$_-$) for $\mathcal{B}^* = \{V \setminus B \mid B \in \mathcal{B}\}$, and (BM$_\pm$) for $\mathcal{B}$ is equivalent to (BM$_\pm$) for $\mathcal{B}^*$. ■

We mention a weaker statement on simultaneous exchange:

(BM$_{\pm\mathrm{w}}$) For distinct $B, B' \in \mathcal{B}$, there exist $u \in B \setminus B'$ and $v \in B' \setminus B$ such that $B - u + v \in \mathcal{B}$ and $B' + u - v \in \mathcal{B}$.

The following fact was observed by Kelmans [156] (according to White [339]) and independently by Tomizawa [314].

Lemma 2.3.13. *For* $\mathcal{B} \subseteq 2^V$, (BM$_{\pm\mathrm{w}}$) $\Longrightarrow$ (BM$_-$).

Proof. Take $B, B' \in \mathcal{B}$ and $u \in B \setminus B'$. By (BM$_{\pm\mathrm{w}}$) there exist $u_1 \in B \setminus B'$ and $v_1 \in B' \setminus B$ such that $B - u_1 + v_1 \in \mathcal{B}$ and $B' + u_1 - v_1 \in \mathcal{B}$. If $u_1 = u$, (BM$_-$) is satisfied with $v = v_1$. Otherwise, put $B'' = B' + u_1 - v_1 \in \mathcal{B}$, for which $u \in B \setminus B''$ and $|B'' \setminus B| = |B' \setminus B| - 1$. Again by (BM$_{\pm\mathrm{w}}$) there exist $u_2 \in B \setminus B''$ and $v_2 \in B'' \setminus B$ such that $B - u_2 + v_2 \in \mathcal{B}$ and $B'' + u_2 - v_2 \in \mathcal{B}$. If $u_2 = u$, (BM$_-$) is satisfied with $v = v_2$. Otherwise, continue the above argument to eventually obtain a valid $v \in B \setminus B'$. ■

The above lemmas imply the following theorem, stating the equivalence among exchange properties. This means that any one of (BM$_+$), (BM$_-$), (BM$_\pm$), (BM$_{\pm\mathrm{w}}$) serves as an axiom of the basis family of a matroid.

Theorem 2.3.14. *For* $\mathcal{B} \subseteq 2^V$, *the exchange properties* (BM$_+$), (BM$_-$), (BM$_\pm$), (BM$_{\pm\mathrm{w}}$) *are equivalent.*

Proof. Obviously, (BM$_\pm$) $\Rightarrow$ (BM$_-$), (BM$_\pm$) $\Rightarrow$ (BM$_+$) and (BM$_\pm$) $\Rightarrow$ (BM$_{\pm\mathrm{w}}$). We have (BM$_-$) $\Rightarrow$ (BM$_\pm$) by Lemma 2.3.11, and (BM$_+$) $\Rightarrow$ (BM$_\pm$) by Lemma 2.3.12. Finally, (BM$_{\pm\mathrm{w}}$) $\Rightarrow$ (BM$_-$) by Lemma 2.3.13. ■

[8] At first sight (BM$_+$) here may appear different from the one in §2.3.2, but they are identical through a change of notation $B \leftrightarrow B'$ and $u \leftrightarrow v$.

Remark 2.3.15. In the proof of Lemmas 2.3.11 and 2.3.12, we intentionally avoided referring to rank functions. This is because the exchange properties play the major role in the argument of valuated matroids in §5.2. The equivalence of ($\mathrm{BM}_{\pm}$) and ($\mathrm{BM}_{\pm \mathrm{w}}$) remains valid for their generalizations in valuated matroids (see Theorem 5.2.25) and the present proof technique is generalized to prove it. □

The simultaneous exchange property has an important consequence, as observed by Brualdi [23]. For $B \in \mathcal{B}$ and $B' \subseteq V$ we define the *exchangeability graph*, denoted $G(B, B')$, as a bipartite graph $(B \setminus B', B' \setminus B; A)$ having the vertex bipartition $(B \setminus B', B' \setminus B)$ and the arc set

$$A = \{(u, v) \mid u \in B \setminus B', v \in B' \setminus B, B - u + v \in \mathcal{B}\}. \tag{2.66}$$

Lemma 2.3.16 (Perfect-matching lemma). *Let $B \in \mathcal{B}$. If B' is also a base, then $G(B, B')$ has a perfect matching.*

Proof. For any $u_1 \in B \setminus B'$ there exists $v_1 \in B' \setminus B$ such that $B - u_1 + v_1 \in \mathcal{B}$ and $B_2' := B' + u_1 - v_1 \in \mathcal{B}$. By the same argument applied to (B, B_2'), there exist $u_2 \in (B \setminus B') \setminus \{u_1\}$ and $v_2 \in (B' \setminus B) \setminus \{v_1\}$ such that $B - u_2 + v_2 \in \mathcal{B}$ and $B_3' := B_2' + u_2 - v_2 = B' + \{u_1, u_2\} - \{v_1, v_2\} \in \mathcal{B}$. Repeating this process we obtain $B - u_i + v_i \in \mathcal{B}$ $(i = 1, \cdots, m)$, where $m = |B \setminus B'| = |B' \setminus B|$, $B \setminus B' = \{u_1, \cdots, u_m\}$ and $B' \setminus B = \{v_1, \cdots, v_m\}$. ■

The converse of the above statement is not always true, as follows.

Example 2.3.17. Consider the matroid $\mathbf{M} = (V, \mathcal{B})$ defined by a matrix

u_1	u_2	v_1	v_2
1	0	1	1
0	1	1	1

on the column set $V = \{u_1, u_2, v_1, v_2\}$. Take $B = \{u_1, u_2\}$ and $B' = \{v_1, v_2\}$. Then $B \in \mathcal{B}$ and $B' \notin \mathcal{B}$, whereas $G(B, B')$ is a complete bipartite graph, admitting two perfect matchings, $M_1 = \{(u_1, v_1), (u_2, v_2)\}$ and $M_2 = \{(u_1, v_2), (u_2, v_1)\}$. □

A partial converse of Lemma 2.3.16 holds in the following form.

Lemma 2.3.18 (Unique-matching lemma). *Let $B \in \mathcal{B}$ and $B' \subseteq V$ with $|B'| = |B|$. If there exists exactly one perfect matching in $G(B, B')$, then $B' \in \mathcal{B}$.*

Proof. The proof is given later based on a series of lemmas below. ■

First we note the following fact, rephrasing the existence of a unique perfect matching with reference to suitable orderings of the elements of $B \setminus B'$ and $B' \setminus B$.

Lemma 2.3.19. *For $B \in \mathcal{B}$ and $B' \subseteq V$ with $|B' \setminus B| = |B \setminus B'| = m$, the graph $G(B, B')$ has a unique perfect matching if and only if there exist some indexings of the elements of $B \setminus B'$ and $B' \setminus B$, say $B \setminus B' = \{u_1, \cdots, u_m\}$ and $B' \setminus B = \{v_1, \cdots, v_m\}$, such that $B - u_i + v_i \in \mathcal{B}$ $(1 \leq i \leq m)$ and $B - u_i + v_j \notin \mathcal{B}$ $(1 \leq i < j \leq m)$.*

Proof. This is immediate from the properties of the DM-decomposition described in Theorem 2.2.22. Note that there exists a unique perfect matching if and only if the tails are empty and each consistent DM-component is composed of a single arc. ∎

Lemma 2.3.20. *Let $B \in \mathcal{B}$ and u, u°, v, v° be four distinct elements with $\{u, u^\circ\} \subseteq B$, $\{v, v^\circ\} \subseteq V \setminus B$, and put $B' = B - \{u, u^\circ\} + \{v, v^\circ\}$. If $M = \{(u, v), (u^\circ, v^\circ)\}$ is the unique perfect matching in $G(B, B')$, then $B' \in \mathcal{B}$.*

Proof. Put $B^\circ = B - u^\circ + v^\circ$ and $B^* = B - u + v$. By applying the simultaneous exchange axiom to (B°, B^*) with $u \in B^\circ \setminus B^*$ we obtain $B^* - v' + u \in \mathcal{B}$ and $B^\circ + v' - u \in \mathcal{B}$ for some $v' \in B^* \setminus B^\circ = \{u^\circ, v\}$. If $v' = u^\circ$, we have $B^* - u^\circ + u = B - u^\circ + v \in \mathcal{B}$ and $B^\circ + u^\circ - u = B - u + v^\circ \in \mathcal{B}$, which means that $M' = \{(u^\circ, v), (u, v^\circ)\}$ is another perfect matching in $G(B, B')$, a contradiction to the uniqueness of M. Therefore we must have $v' = v$, and then $B' = B^\circ + v - u \in \mathcal{B}$. ∎

Lemma 2.3.21. *Let $B \in \mathcal{B}$ and $B' \subseteq V$ with $|B'| = |B|$. If there exists exactly one perfect matching M in $G(B, B')$, then for any $(u^\circ, v^\circ) \in M$ it holds that $B^\circ \equiv B - u^\circ + v^\circ \in \mathcal{B}$ and there exists exactly one perfect matching in $G(B^\circ, B')$.*

Proof. The first assertion, $B^\circ \in \mathcal{B}$, is obvious. Using the notation in Lemma 2.3.19 we have $M = \{(u_i, v_i) \mid i = 1, \cdots, m\}$ and $(u^\circ, v^\circ) = (u_k, v_k)$ for some k. For $i \neq k$, $j \neq k$, put

$$B_{ij} = B^\circ - u_i + v_j = B - \{u_i, u^\circ\} + \{v_j, v^\circ\}.$$

Since $G(B, B_{ii})$ has a unique perfect matching $\{(u_i, v_i), (u^\circ, v^\circ)\}$, we have $B_{ii} \in \mathcal{B}$ by Lemma 2.3.20. We also claim that $B_{ij} \notin \mathcal{B}$ if $i < j$. To see this, suppose $B_{ij} \in \mathcal{B}$. Then Lemma 2.3.16 implies the existence of a perfect matching in $G(B, B_{ij})$, which is either $M_1 = \{(u_i, v_j), (u^\circ, v^\circ)\}$ or $M_2 = \{(u_i, v^\circ), (u^\circ, v_j)\}$. But M_1 is possible only if $i \geq j$ and M_2 is possible only if $i \geq k \geq j$. Hence $G(B^\circ, B')$ meets the condition in Lemma 2.3.19. ∎

We are now in the position to prove the unique-matching lemma.

(Proof of Lemma 2.3.18) The proof is by induction on $m = |B \setminus B'|$. The case of $m = 1$ is obvious. So assume $m \geq 2$. Take any (u°, v°) contained in the unique perfect matching, and put $B^\circ = B - u^\circ + v^\circ$. Lemma 2.3.21

shows that $B^\circ \in \mathcal{B}$ and $G(B^\circ, B')$ has a unique perfect matching. Then the induction hypothesis yields $B' \in \mathcal{B}$. □

As a corollary to the unique-matching lemma we obtain an exchange-augmentation property for independent sets. Recall that $\mathcal{I}$ denotes the family of independent sets of the matroid $\mathbf{M} = (V, \mathcal{B}, \mathcal{I}, \rho)$.

Lemma 2.3.22. *Suppose that* $\{u_1, \cdots, u_m\} \subseteq I \in \mathcal{I}$ *and* $\{v_0, v_1, \cdots, v_m\} \subseteq V \setminus I$, *where* u_i $(1 \le i \le m)$ *and* v_j $(0 \le j \le m)$ *are distinct. If* $I + v_0 \in \mathcal{I}$, $I + v_i \notin \mathcal{I}$ $(1 \le i \le m)$, $I - u_i + v_i \in \mathcal{I}$ $(1 \le i \le m)$ *and* $I - u_i + v_j \notin \mathcal{I}$ $(1 \le i < j \le m)$, *then* $I - \{u_1, \cdots, u_m\} + \{v_0, v_1, \cdots, v_m\} \in \mathcal{I}$.

Proof. Put $B = I + v_0$ and $B' = I - \{u_1, \cdots, u_m\} + \{v_0, v_1, \cdots, v_m\}$. Then B is a base of the truncation of $\mathbf{M}$, say $\mathbf{M}' = (V, \mathcal{B}')$, with $\operatorname{rank} \mathbf{M}' = |I| + 1$. We claim that $B - u_i + v_i \in \mathcal{B}'$ $(1 \le i \le m)$ and $B - u_i + v_j \notin \mathcal{B}'$ $(1 \le i < j \le m)$. The former follows from

$$\begin{aligned}\rho(B - u_i + v_i) &\ge \rho(I + v_0 + v_i) + \rho(I - u_i + v_i) - \rho(I + v_i) \\ &= (|I| + 1) + |I| - |I| = |I| + 1,\end{aligned}$$

whereas the latter is obvious from $I - u_i + v_j \notin \mathcal{I}$ $(1 \le i < j \le m)$. Then Lemma 2.3.18 together with Lemma 2.3.19 implies that $B' = B - \{u_1, \cdots, u_m\} + \{v_1, \cdots, v_m\}$ belongs to $\mathcal{B}'$, and hence to $\mathcal{I}$. ■

Remark 2.3.23. In the case where the matroid is defined by a matrix, the unique-matching lemma is a restatement of an obvious fact that a triangular matrix having nonzero diagonal elements is nonsingular. Let $\mathbf{M} = (V, \mathcal{B})$ be a matroid defined by a matrix A with $V = \operatorname{Col}(A)$ and $\operatorname{rank} A = |R|$, where $R = \operatorname{Row}(A)$. For $B \in \mathcal{B}$ define $\tilde{A} = A[R, B]^{-1} A$, where it is noted that $\operatorname{Row}(\tilde{A})$ can be identified with B while $\operatorname{Col}(\tilde{A}) = V$. Then $B - u + v \in \mathcal{B}$ if and only if (u, v) entry of $\tilde{A}$ is distinct from zero. For $B' \subseteq V$ with $|B'| = |B| = |R|$, the graph $G(B, B')$ has a unique perfect matching if and only if the rows $(B \setminus B')$ and the columns $(B' \setminus B)$ of the submatrix $\tilde{A}[B \setminus B', B' \setminus B]$ can be rearranged so that the resulting matrix may be a triangular matrix with nonzero diagonal entries. If this is the case, the submatrix $\tilde{A}[B \setminus B', B' \setminus B]$ is nonsingular, which corresponds to the nonsingularity of $A[R, B']$, i.e., the condition $B' \in \mathcal{B}$. □

Remark 2.3.24. The unique-matching lemma reveals a key property underlying the (unweighted or linear-weighted) matroid intersection algorithm, to be explained later. In the literature (e.g., Iri–Tomizawa [133, Lemma 2], Krogdahl [164], Lawler [171, Lemma 3.1 of Chap. 8], and Schrijver [291, Theorem 4.3]) this fact is stated often with an explicit reference to the orderings of the elements just as in Lemma 2.3.19 and Lemma 2.3.22, and is accordingly referred to as "no-shortcut lemma" (cf. Kung [167] for this name). We have adopted the present form, referring to the uniqueness of a perfect matching, because this is suitable for its extension to valuated matroids in §5.2. □

2.3.5 Independent Matching Problem

The matroid intersection problem and its extensions will play a major role in this book. The problem may be described as follows:

> **[Matroid intersection problem]**
> Given a pair of matroids $\mathbf{M}_1$ and $\mathbf{M}_2$ defined on a common ground set V, find a common independent set of maximum size.

In this section we feature an equivalent variant of the matroid intersection problem called the independent matching problem, which is defined as follows. Suppose we are given a bipartite graph $G = (V^+, V^-; A)$ and two matroids $\mathbf{M}^+ = (V^+, \mathcal{B}^+, \mathcal{I}^+, \rho^+)$ and $\mathbf{M}^- = (V^-, \mathcal{B}^-, \mathcal{I}^-, \rho^-)$. Here, (V^+, V^-) is the bipartition of the vertex set of G, A is the arc set of G; $\mathbf{M}^+$ is a matroid on V^+ with the family of bases $\mathcal{B}^+$, the family of independent sets $\mathcal{I}^+$, and the rank function ρ^+; and similarly for $\mathbf{M}^-$. Arcs are directed from V^+ to V^- and therefore the initial vertex $\partial^+ a \in V^+$ and the terminal vertex $\partial^- a \in V^-$ for each $a \in A$.

A matching $M (\subseteq A)$ is called an *independent matching* if

$$\partial^+ M \in \mathcal{I}^+, \qquad \partial^- M \in \mathcal{I}^-, \tag{2.67}$$

where $\partial^+ M$ (resp., $\partial^- M$) denotes the set of vertices in V^+ (resp., V^-) incident to M. That is, $M (\subseteq A)$ is an independent matching if and only if $|M| = |\partial^+ M| = |\partial^- M|$ and the sets of end-vertices of M, i.e., $\partial^+ M$ and $\partial^- M$, are independent in $\mathbf{M}^+$ and $\mathbf{M}^-$, respectively. The independent matching problem is to find an independent matching M of maximum cardinality:

> **[Independent matching problem]**
> Find a matching $M (\subseteq A)$ that maximizes $|M|$ subject to the constraint that $\partial^+ M \in \mathcal{I}^+$ and $\partial^- M \in \mathcal{I}^-$.

The matroid intersection problem above is a special case of the independent matching problem, in which V^+ and V^- are disjoint copies of V, $\mathbf{M}^+ \simeq \mathbf{M}_1$, $\mathbf{M}^- \simeq \mathbf{M}_2$, and $A = \{(v^+, v^-) \mid v \in V\}$, where $v^+ \in V^+$ and $v^- \in V^-$ denote the copies of $v \in V$.

Example 2.3.25. Here is an example of the independent matching problem. Consider a bipartite graph $G = (V^+, V^-; A)$, shown in Fig. 2.8(a), with $V^+ = \{x_1, x_2, x_3, x_4\}$, $V^- = \{y_1, y_2, y_3, y_4, y_5\}$, and $A = \{(x_1, y_1), (x_2, y_1), (x_2, y_2), (x_3, y_2), (x_3, y_3), (x_4, y_4), (x_4, y_5)\}$. The matroid $\mathbf{M}^+$ is assumed to be a free matroid on V^+, whereas $\mathbf{M}^-$ is a linear matroid defined by the matrix

y_1	y_2	y_3	y_4	y_5
1	1	1	0	0
1	2	0	1	0
0	0	0	0	1

.

Then $M = \{(x_3, y_3), (x_4, y_4)\}$ is an independent matching with $\partial^+ M = \{x_3, x_4\}$ and $\partial^- M = \{y_3, y_4\}$ being independent in $\mathbf{M}^+$ and $\mathbf{M}^-$, respectively. Another matching $M' = \{(x_2, y_2), (x_3, y_3), (x_4, y_4)\}$ is not an independent matching, since $\partial^- M = \{y_2, y_3, y_4\}$ is not independent in $\mathbf{M}^-$. □

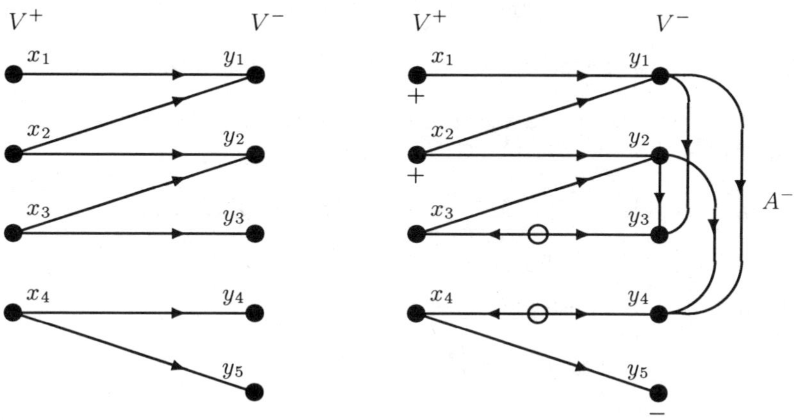

(a) Graph G (b) Auxiliary graph $\tilde{G}_M$

Fig. 2.8. Graph G and auxiliary graph $\tilde{G}_M$ (○: arc in M; +: vertex in S^+; −: vertex in S^-)

The objective of this section is twofold:

1. To establish an extension of the min-max duality for bipartite matchings, which has been formulated as the König–Egerváry theorem and the Hall–Ore theorem, to that for independent matchings.
2. To give an efficient algorithm for finding a maximum independent matching.

We shall prove the min-max duality for independent matchings by showing the validity of the algorithm.

Let us recall the König–Egerváry theorem (Theorem 2.2.15). It states that the maximum size of a matching is equal to the minimum size of a cover, where a cover means a pair (U^+, U^-) with $U^+ \subseteq V^+$ and $U^- \subseteq V^-$ such that $\partial^+ a \in U^+$ or $\partial^- a \in U^-$ for each $a \in A$, and the size of (U^+, U^-) is defined to be $|U^+| + |U^-|$. We also recall that the inequality

$$|M| \leq |U^+| + |U^-| \tag{2.68}$$

is an obvious relation valid for any matching M and any cover (U^+, U^-).

For a cover (U^+, U^-) we define the *rank* of (U^+, U^-) by

$$\rho^+(U^+) + \rho^-(U^-),$$

with reference to the rank functions of the respective matroids $\mathbf{M}^+$ and $\mathbf{M}^-$. The obvious relation (2.68) is extended as follows.

Lemma 2.3.26. *The inequality*

$$|M| \le \rho^+(U^+) + \rho^-(U^-) \tag{2.69}$$

holds true for any independent matching M and any cover (U^+, U^-).

Proof. Since (U^+, U^-) is a cover, M can be expressed as $M = M^+ \cup M^-$ with $\partial^+ M^+ \subseteq U^+$ and $\partial^- M^- \subseteq U^-$. Then

$$|M^+| = |\partial^+ M^+| = \rho^+(\partial^+ M^+) \le \rho^+(U^+),$$
$$|M^-| = |\partial^- M^-| = \rho^-(\partial^- M^-) \le \rho^-(U^-),$$

and the addition of these yields

$$|M| \le |M^+| + |M^-| \le \rho^+(U^+) + \rho^-(U^-).$$

■

The duality in the independent matching problem consists in the assertion that the equality holds in (2.69) for some M and (U^+, U^-). We say (U^+, U^-) is a minimum cover if it attains the minimum in (2.70) below.

Theorem 2.3.27.

$$\begin{aligned} &\max\{|M| \mid M : \text{independent matching}\} \\ &= \min\{\rho^+(U^+) + \rho^-(U^-) \mid (U^+, U^-) : \text{cover}\}. \end{aligned} \tag{2.70}$$

Proof. Lemma 2.3.26 shows $\max |M| \le \min\{\rho^+(U^+) + \rho^-(U^-)\}$. The equality is proven later in Lemma 2.3.32 along with the validity of an algorithm for computing this common value. ■

It is sometimes convenient to recast Theorem 2.3.27 into different forms. For $U \subseteq V^+ \cup V^-$ define the *cut capacity* of U by

$$\kappa(U) = \begin{cases} \rho^+(V^+ \setminus U) + \rho^-(V^- \cap U) & (\not\exists a \in A : \partial^+ a \in U, \partial^- a \notin U) \\ +\infty & (\exists a \in A : \partial^+ a \in U, \partial^- a \notin U). \end{cases} \tag{2.71}$$

Noting that $\kappa(U)$ is finite if and only if $(V^+ \setminus U, V^- \cap U)$ is a cover, we can rewrite Theorem 2.3.27 to

$$\max\{|M| \mid M\text{: independent matching}\} = \min\{\kappa(U) \mid U \subseteq V^+ \cup V^-\}. \tag{2.72}$$

The function κ to be minimized is submodular in U.

Another form of Theorem 2.3.27 refers to a function $\Gamma : 2^{V^-} \to 2^{V^+}$ defined by

$$\Gamma(Y) = \{u \in V^+ \mid \exists v \in Y : (u,v) \in A\}, \qquad Y \subseteq V^-. \tag{2.73}$$

Since (U^+, U^-) is a cover if and only if $\Gamma(V^- \setminus U^-) \subseteq U^+$, the right-hand side of (2.70) can be rewritten as

$$\begin{aligned} &\min\{\rho^+(U^+) + \rho^-(U^-) \mid (U^+,U^-)\text{: cover}\} \\ &= \min\{\rho^+(\Gamma(V^- \setminus U^-)) + \rho^-(U^-) \mid U^- \subseteq V^-\} \\ &= \min\{\rho^+(\Gamma(Y)) + \rho^-(V^- \setminus Y) \mid Y \subseteq V^-\}. \end{aligned}$$

Therefore Theorem 2.3.27 can be expressed as

$$\begin{aligned} &\max\{|M| \mid M\text{: independent matching}\} \\ &= \min\{\rho^+(\Gamma(Y)) + \rho^-(V^- \setminus Y) \mid Y \subseteq V^-\}. \end{aligned} \tag{2.74}$$

Again the function $\rho^+(\Gamma(Y)) + \rho^-(V^- \setminus Y)$ to be minimized is submodular in Y. These alternative expressions (2.71) and (2.74) reveal the submodularity inherent in the problem at the sacrifice of the symmetry apparent in the original expression (2.70).

The duality result above implies a number of important consequences. First of all, if both $\mathbf{M}^+$ and $\mathbf{M}^-$ are free matroids, for which $\rho^+(U^+) = |U^+|$ and $\rho^-(U^-) = |U^-|$, the identity (2.70) reduces to the König–Egerváry theorem (Theorem 2.2.15), whereas (2.74) reduces to the Hall–Ore theorem (Theorem 2.2.17).

Next, consider the case where $\mathbf{M}^-$ is free (and $\mathbf{M}^+$ is general). The expression (2.74) in this case takes the form

$$\begin{aligned} &\max\{|M| \mid M\text{: matching with } \partial^+ M \in \mathcal{I}^+\} \\ &= \min\{\rho^+(\Gamma(Y)) + |V^- \setminus Y| \mid Y \subseteq V^-\}. \end{aligned} \tag{2.75}$$

We call this the *Rado–Perfect theorem* (cf. Rado [274], Perfect [266]).

Thirdly, the matroid intersection theorem of Edmonds [68, 70] can also be derived from Theorem 2.3.27.

Theorem 2.3.28 (Matroid intersection theorem). *For two matroids* $\mathbf{M}_i = (V, \mathcal{I}_i, \rho_i)$ $(i = 1, 2)$ *it holds that*

$$\max\{|I| \mid I \in \mathcal{I}_1 \cap \mathcal{I}_2\} = \min\{\rho_1(X) + \rho_2(V \setminus X) \mid X \subseteq V\}.$$

Proof. Let V^+ and V^- be disjoint copies of V and put $A = \{(v^+, v^-) \mid v \in V\}$, where $v^+ \in V^+$ and $v^- \in V^-$ denote the copies of $v \in V$. Consider an independent matching problem on $(V^+, V^-; A)$ with $\mathbf{M}^+ \simeq \mathbf{M}_1$ and $\mathbf{M}^- \simeq \mathbf{M}_2$. Then the assertion follows from Theorem 2.3.27. ■

We now turn to the algorithm for computing an independent matching of the maximum size. The algorithm, starting with the empty matching M, finds a sequence of independent matchings M with $|M| = 0, 1, 2, \cdots$ with the

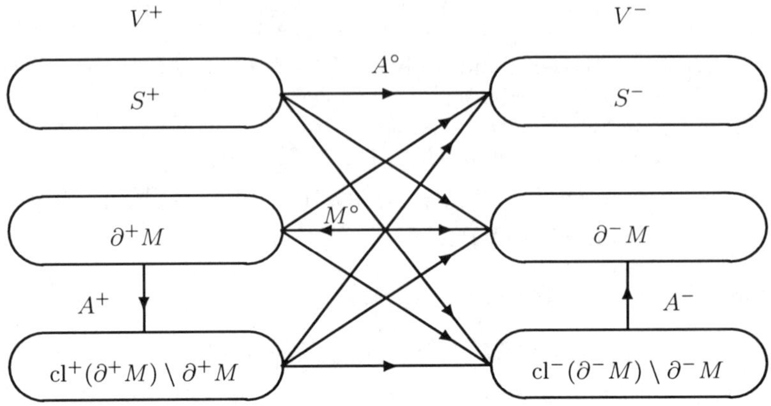

Fig. 2.9. Auxiliary graph for the independent matching problem

aid of an auxiliary graph $\tilde{G}_M = (\tilde{V}, \tilde{A}; S^+, S^-)$ that has vertex set $\tilde{V}$, arc set $\tilde{A}$, entrance vertex set S^+ and exit vertex set S^- (see Fig. 2.9). The vertex set $\tilde{V}$ is given by

$$\tilde{V} = V^+ \cup V^-,$$

whereas S^+, S^- and $\tilde{A}$ are defined with reference to independent matching M, as follows. The entrance S^+ and the exit S^- are defined by

$$S^+ = V^+ \setminus \mathrm{cl}^+(\partial^+ M), \qquad S^- = V^- \setminus \mathrm{cl}^-(\partial^- M),$$

where

$$\begin{aligned} \mathrm{cl}^+(X) &= \{v \in V^+ \mid \rho^+(X \cup \{v\}) = \rho^+(X)\}, \qquad X \subseteq V^+, \\ \mathrm{cl}^-(X) &= \{v \in V^- \mid \rho^-(X \cup \{v\}) = \rho^-(X)\}, \qquad X \subseteq V^-, \end{aligned}$$

in accordance with (2.63). The arc set $\tilde{A}$ consists of four disjoint parts:

$$\tilde{A} = A^\circ \cup M^\circ \cup A^+ \cup A^-,$$

where

$$\begin{aligned} A^\circ &= \{a \mid a \in A\} \qquad \text{(copy of } A\text{)}, \\ M^\circ &= \{\overline{a} \mid a \in M\} \qquad (\overline{a}\text{: reorientation of } a), \\ A^+ &= \{(u,v) \mid u \in \partial^+ M, v \in \mathrm{cl}^+(\partial^+ M) \setminus \partial^+ M, \partial^+ M - u + v \in \mathcal{I}^+\}, \\ A^- &= \{(v,u) \mid u \in \partial^- M, v \in \mathrm{cl}^-(\partial^- M) \setminus \partial^- M, \partial^- M - u + v \in \mathcal{I}^-\}. \end{aligned}$$

Note that $\tilde{V} = V^+ \cup V^-$ is partitioned into six disjoint parts with possible additional connections by $M^\circ \cup A^+ \cup A^-$.

Example 2.3.29 (Continued from Example 2.3.25). The auxiliary graph $\tilde{G}_M = (\tilde{V}, \tilde{A}; S^+, S^-)$ for the independent matching $M = \{(x_3, y_3), (x_4, y_4)\}$ is depicted in Fig. 2.8(b), where $S^+ = \{x_1, x_2\}$, $S^- = \{y_5\}$, $A^\circ = A$, $M^\circ = \{(y_3, x_3), (y_4, x_4)\}$, $A^+ = \emptyset$, and $A^- = \{(y_1, y_3), (y_1, y_4), (y_2, y_3), (y_2, y_4)\}$. □

The algorithm for the independent matching problem reads as follows.

Algorithm for independent matching problem

Starting with the empty matching M, repeat (i)–(ii) below:
(i) Find a shortest path P (in the number of arcs) from S^+ to S^- in $\tilde{G}_M$. [Stop if there is no path from S^+ to S^-.]
(ii) Update M to

$$\overline{M} = (M \setminus \{a \in M \mid \overline{a} \in P \cap M^\circ\}) \cup (P \cap A^\circ).$$

□

The validity of the algorithm is shown by the lemmas after the example below.

Example 2.3.30. In the auxiliary graph $\tilde{G}_M$ in Fig. 2.8(b), we can take $P = \{(x_1, y_1), (y_1, y_4), (y_4, x_4), (x_4, y_5)\}$ as a shortest path from S^+ to S^-. Then the matching is updated to $\overline{M} = \{(x_1, y_1), (x_3, y_3), (x_4, y_5)\}$ in Step (ii) of the algorithm, and the auxiliary graph changes to $\tilde{G}_{\overline{M}}$ shown in Fig. 2.10. The entrance is $S^+ = \{x_2\}$, whereas the exit S^- is empty, and therefore, no path exists from S^+ to S^-. Then the algorithm terminates in Step (i). It will be shown in Example 2.3.34 that $\overline{M}$ is a maximum independent matching. □

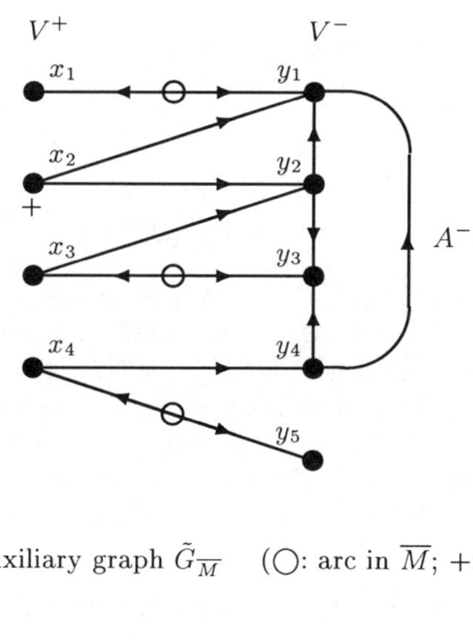

Fig. 2.10. Auxiliary graph $\tilde{G}_{\overline{M}}$ (○: arc in $\overline{M}$; +: vertex in S^+)

The following two lemmas justify the above algorithm. The former shows that the independence of $\partial^+ M$ and $\partial^- M$ in the respective matroids is maintained, whereas the latter guarantees that the independent matching at the termination of the algorithm is of the maximum size.

Lemma 2.3.31. *$\partial^+ \overline{M} \in \mathcal{I}^+$ and $\partial^- \overline{M} \in \mathcal{I}^-$.*

Proof. Let $v_0 \in S^+$ be the starting vertex of P and put $\{(u_i, v_i) \mid i = 1, \cdots, l\} = P \cap A^+$, where $l = |P \cap A^+|$ and the indices are chosen so that $v_0, u_1, v_1, u_2, v_2, \cdots, u_l, v_l$ represents the order in which they appear on P. Then

$$\partial^+ \overline{M} = \partial^+ M - \{u_1, \cdots, u_l\} + \{v_0, v_1, \cdots, v_l\}.$$

Since P is a shortest path, Lemma 2.3.22 guarantees $\partial^+ \overline{M} \in \mathcal{I}^+$. The other claim can be proven similarly. ■

Lemma 2.3.32. *Let $U \subseteq \tilde{V}$ be the set of vertices reachable from S^+ in $\tilde{G}_M$ at the termination of the algorithm, and put $U^+ = V^+ \setminus U$ and $U^- = V^- \cap U$. Then (U^+, U^-) is a cover for the independent matching problem and $|M| = \rho^+(U^+) + \rho^-(U^-)$. Therefore, M is a maximum independent matching.*

Proof. By the definition there is no $a \in \tilde{A}$ with $\partial^+ a \in U$ and $\partial^- a \in \tilde{V} \setminus U$. In particular, there is no $a \in A$ with $\partial^+ a \in V^+ \setminus U^+$ and $\partial^- a \in V^- \setminus U^-$, namely, (U^+, U^-) is a cover.

Put $I^+ = \partial^+ M$, $J^+ = I^+ \cap U^+$, and $I^+ \setminus J^+ = \{u_1, \cdots, u_m\}$. For each $v \in U^+ \setminus J^+ \subseteq \mathrm{cl}^+(I^+) \setminus I^+$, we have $\rho^+(I^+ + v - u_i) \leq |I^+| - 1$ for $i = 1, \cdots, m$, since there is no arc going out of U. The submodularity of ρ^+ implies that

$$\rho^+(I^+ + v - \{u_1, u_2\}) \leq \rho^+(I^+ + v - u_1) + \rho^+(I^+ + v - u_2) - \rho^+(I^+ + v) \leq |I^+| - 2.$$

Repeating such process, we obtain

$$\rho^+(J^+ + v) = \rho^+(I^+ + v - \{u_1, \cdots, u_m\}) \leq |I^+| - m = |J^+|.$$

Hence, for $v, v' \in U^+ \setminus J^+$ with $v \neq v'$,

$$\rho^+(J^+ + \{v, v'\}) \leq \rho^+(J^+ + v) + \rho^+(J^+ + v') - \rho^+(J^+) \leq |J^+|.$$

Repeating this we obtain $\rho^+(U^+) \leq |J^+|$, and hence $\rho^+(U^+) = |J^+|$.

Symmetrically, put $I^- = \partial^- M$, $J^- = I^- \cap U^-$. For each $v \in U^- \setminus J^-$ and $u \in I^- \setminus J^-$ there is no arc (v, u), and hence $\rho^-(I^- + v - u) \leq |I^-| - 1$. By a similar argument using the submodularity of ρ^- we obtain $\rho^-(U^-) = |J^-|$.

Then $|M| = |J^+| + |J^-| = \rho^+(U^+) + \rho^-(U^-)$, where the first equality follows from the fact that $\{\partial^+ a, \partial^- a\} \subseteq U$ or $\{\partial^+ a, \partial^- a\} \subseteq \tilde{V} \setminus U$ for each $a \in M$. Finally, we recall Lemma 2.3.26 to conclude that M is a maximum independent matching. ■

To sum up, we obtain the following optimality criterion in terms of the auxiliary graph.

Theorem 2.3.33. *An independent matching M is maximum if and only if there exists no directed path from S^+ to S^- in $\tilde{G}_M = (\tilde{V}, \tilde{A}; S^+, S^-)$.* □

Example 2.3.34 (Continued from Example 2.3.30). In Fig. 2.10, there exists no path from S^+ to S^-. This means, by Theorem 2.3.33, that $\overline{M}$ is a maximum independent matching. A minimum cover (U^+, U^-) can be constructed as in Lemma 2.3.32. The set of vertices reachable from S^+ is given by $U = \{x_1, x_2, x_3, y_1, y_2, y_3\}$, and $U^+ = V^+ \setminus U = \{x_4\}$ and $U^- = V^- \cap U = \{y_1, y_2, y_3\}$ satisfy $\rho^+(U^+) = 1$ and $\rho^-(U^-) = 2$, adding up to $|\overline{M}| = 3$. □

The auxiliary graph $\tilde{G}_M$ is useful to capture the family of all minimum covers. A pair (U^+, U^-) is a minimum cover if and only if $U = (V^+ \setminus U^+) \cup U^-$ gives the minimum cut capacity κ defined in (2.71). Namely, the family of all minimum covers is expressed as

$$\{(U^+, U^-) \mid U^+ = V^+ \setminus U, U^- = V^- \cap U, U \in \mathcal{L}_{\min}(\kappa)\} \tag{2.76}$$

in terms of the family of all minimum cuts

$$\mathcal{L}_{\min}(\kappa) = \{U \subseteq V^+ \cup V^- \mid \kappa(U) \leq \kappa(W), \ \forall W \subseteq V^+ \cup V^-\}.$$

The family $\mathcal{L}_{\min}(\kappa)$ forms a lattice due to the submodularity of κ (cf. Theorem 2.2.5), and by Birkhoff's representation theorem (Theorem 2.2.10) it can be represented as a pair of a partition $\{V_0; V_1, \cdots, V_b; V_\infty\}$ of $V^+ \cup V^-$ and a partial order $\preceq$ on $\{V_1, \cdots, V_b\}$. Let us call $(\{V_0; V_1, \cdots, V_b; V_\infty\}, \preceq)$ the *min-cut decomposition for the independent matching problem.* This decomposition can be computed from the auxiliary graph $\tilde{G}_M$ on the basis of the following fact.

Lemma 2.3.35. *Let $\tilde{G}_M = (\tilde{V}, \tilde{A}; S^+, S^-)$ be the auxiliary graph associated with a maximum independent matching M. The family of the minimum cuts $\mathcal{L}_{\min}(\kappa)$ is represented in terms of $\tilde{G}_M$ as*

$$\mathcal{L}_{\min}(\kappa) = \{U \subseteq \tilde{V} \mid S^+ \subseteq U \subseteq \tilde{V} \setminus S^-; \nexists a \in \tilde{A} : \partial^+ a \in U, \partial^- a \notin U\}. \tag{2.77}$$

Proof. For $U \subseteq \tilde{V}$ define $(U^+, U^-) = (V^+ \setminus U, V^- \cap U)$, $(I^+, I^-) = (\partial^+ M, \partial^- M)$, and $(J^+, J^-) = (I^+ \cap U^+, I^- \cap U^-)$. By definition, $\kappa(U) < +\infty \iff \nexists a \in A : \partial^+ a \in U, \partial^- a \notin U$. For such U, we have $U \in \mathcal{L}_{\min}(\kappa) \iff$ (i) $|M| = |J^+| + |J^-|$, (ii) $|J^+| = \rho^+(U^+)$, and (iii) $|J^-| = \rho^-(U^-)$, since $|M| \leq |J^+| + |J^-| \leq \rho^+(U^+) + \rho^-(U^-)$. Denote by $\mathcal{L}'$ the right hand side of (2.77). The proof of Lemma 2.3.32 shows $\mathcal{L}_{\min}(\kappa) \supseteq \mathcal{L}'$. Conversely, suppose $U \in \mathcal{L}_{\min}(\kappa)$. The following claims show $U \in \mathcal{L}'$.

Claim 1: $S^+ \subseteq U \subseteq \tilde{V} \setminus S^-$. If $S^+ \not\subseteq U$, there exists $v \in U^+ \cap S^+$. Then $J^+ + v \in \mathcal{I}^+$, which implies $\rho^+(U^+) \geq \rho^+(J^+ + v) = |J^+| + 1$, a contradiction to (ii) above. Similarly, $S^- \cap U \neq \emptyset$ contradicts (iii).

Claim 2: $\nexists a \in M^\circ : \partial^+ a \in U, \partial^- a \notin U$. This follows from (i).

Claim 3: $\nexists a \in A^+ : \partial^+ a \in U, \partial^- a \notin U$. If there is $a = (u,v) \in A^+$ with $u \in U$, $v \notin U$, then $J^+ + v \subseteq I^+ + v - u \in \mathcal{I}^+$, leading to a contradiction to (ii).

Claim 4: $\nexists a \in A^- : \partial^+ a \in U, \partial^- a \notin U$. This is proven similarly. ∎

The min-cut decomposition for the independent matching problem can be found by the following procedure. It differs from the one for the DM-decomposition only in the first step. Recall the notation $\stackrel{*}{\longrightarrow}$ for the reachability by a directed path.

Algorithm for min-cut decomposition of an independent matching problem

1. Find a maximum independent matching M.
2. Let $V_0 = \{v \in V^+ \cup V^- \mid w \stackrel{*}{\longrightarrow} v \text{ on } \tilde{G}_M \text{ for some } w \in S^+\}$.
3. Let $V_\infty = \{v \in V^+ \cup V^- \mid v \stackrel{*}{\longrightarrow} w \text{ on } \tilde{G}_M \text{ for some } w \in S^-\}$.
4. Let $\tilde{G}'$ denote the graph obtained from $\tilde{G}_M$ by deleting the vertices $V_0 \cup V_\infty$ (and arcs incident thereto).
5. Let V_k $(k = 1, \cdots, b)$ be the strong components of $\tilde{G}'$.
6. Define a partial order $\preceq$ on $\{V_k \mid k = 1, \cdots, b\}$ as follows:

$$V_k \preceq V_l \iff v_l \stackrel{*}{\longrightarrow} v_k \text{ on } \tilde{G}' \text{ for some } v_k \in V_k \text{ and } v_l \in V_l.$$

□

Example 2.3.36. The min-cut decomposition for the independent matching problem in Example 2.3.25 is obtained from the auxiliary graph $\tilde{G}_{\overline{M}}$ in Fig. 2.10. According to the notation in the algorithm, we have $V_0 = \{x_1, x_2, x_3, y_1, y_2, y_3\}$ and $V_\infty = \emptyset$. The subgraph $\tilde{G}'$, having vertex set $\{x_4, y_4, y_5\}$ and arc set $\{(x_4, y_4), (x_4, y_5), (y_5, x_4)\}$, is decomposed into two strong components, $V_1 = \{y_4\}$ and $V_2 = \{x_4, y_5\}$ with $V_1 \preceq V_2$. The min-cut decomposition is given by $(\{V_0; V_1, V_2; V_\infty\}, \preceq)$. We have $\mathcal{L}_{\min}(\kappa) = \{V_0, V_0 \cup V_1, V_0 \cup V_1 \cup V_2\}$, with which the family of all minimum covers is obtained as (2.76). Thus we can enumerate all the minimum covers as $(U^+, U^-) = (\{x_4\}, \{y_1, y_2, y_3\}), (\{x_4\}, \{y_1, y_2, y_3, y_4\}), (\emptyset, \{y_1, y_2, y_3, y_4, y_5\})$. □

Remark 2.3.37. The independent matching problem is closely related to the rank of a *triple matrix product*, as is pointed out by Tomizawa–Iri [317]. Consider a triple matrix product $P = Q_1 T Q_2$, where T is a generic matrix (cf. §2.1.3) and Q_i $(i = 1, 2)$ are numerical matrices. Put $R_1 = \mathrm{Row}(Q_1)$, $C_2 = \mathrm{Col}(Q_2)$, and first suppose $|R_1| = |C_2| = k$. By the Cauchy–Binet formula (Proposition 2.1.6) we have

$$\det P = \sum_{|I|=|J|=k} \pm \det Q_1[R_1, I] \cdot \det T[I, J] \cdot \det Q_2[J, C_2].$$

There is no numerical cancellation in the summation above by virtue of the assumed algebraic independence of the nonzero entries of T, and hence P is nonsingular if and only if $Q_1[R_1, I]$, $T[I, J]$, and $Q_2[J, C_2]$ are all nonsingular for some I and J. By applying the above argument to square submatrices of P for a general $P = Q_1 T Q_2$, we see that

$$\begin{aligned}\operatorname{rank} P = \max\{|I| \mid \operatorname{rank} Q_1[R_1, I] = |I| \\ = \operatorname{rank} T[I, J] = |J| = \operatorname{rank} Q_2[J, C_2]\}.\end{aligned}$$

We define an independent matching problem as follows. The vertex sets V^+ and V^- are the row set and the column set of T, respectively, and the arc set $A = \{(i, j) \mid T_{ij} \neq 0\}$. The matroids $\mathbf{M}^+$ and $\mathbf{M}^-$ attached to V^+ and V^- respectively are the linear matroids defined by Q_1 and the transpose of Q_2. Then we see from the above expression that $\operatorname{rank} P$ is equal to the maximum size of an independent matching. Namely,

$$\operatorname{rank}(Q_1 T Q_2) = \max\{|M| \mid M\text{: independent matching}\}. \tag{2.78}$$

If Q_i $(i = 1, 2)$ are identity matrices, this expression reduces to (2.10). □

The weighted version of the independent matching problem as well as the weighted matroid intersection problem will be treated in the framework of valuated matroids in §5.2.

2.3.6 Union

Given a bipartite graph $G = (V^+, V^-; A)$ and a matroid $\mathbf{M}^+ = (V^+, \mathcal{I}^+, \rho^+)$, we can induce another matroid through matchings. Define $\tilde{\mathcal{I}} \subseteq 2^{V^-}$ and $\tilde{\rho} : 2^{V^-} \to \mathbf{Z}$ by

$$\tilde{\mathcal{I}} = \{\partial^- M \mid M\text{: matching with } \partial^+ M \in \mathcal{I}^+\},$$
$$\tilde{\rho}(X) = \max\{|I| \mid I \in \tilde{\mathcal{I}}, I \subseteq X\}, \qquad X \subseteq V^-.$$

It follows from the Rado–Perfect theorem (2.75) that

$$\tilde{\rho}(X) = \min\{\rho^+(\Gamma(X')) + |X \setminus X'| \mid X' \subseteq X\}, \qquad X \subseteq V^-. \tag{2.79}$$

Theorem 2.3.38. $\tilde{\mathbf{M}} = (V^-, \tilde{\mathcal{I}}, \tilde{\rho})$ *is a matroid with the family of independent sets* $\tilde{\mathcal{I}}$ *and the rank function* $\tilde{\rho}$.

Proof. We show that $\tilde{\rho}$ in (2.79) satisfies the rank axiom of a matroid. Obviously, $0 \leq \tilde{\rho}(X) \leq |X|$ and $\tilde{\rho}(X) \leq \tilde{\rho}(Y)$ for $X \subseteq Y$. For the submodularity of $\tilde{\rho}$ we see

$$\tilde{\rho}(X) + \tilde{\rho}(Y) = \min_{X' \subseteq X, Y' \subseteq Y} \{\rho^+(\Gamma(X')) + \rho^+(\Gamma(Y')) + |X \setminus X'| + |Y \setminus Y'|\},$$

into which we substitute

$$\begin{aligned}\rho^+(\Gamma(X')) + \rho^+(\Gamma(Y')) &\geq \rho^+(\Gamma(X') \cup \Gamma(Y')) + \rho^+(\Gamma(X') \cap \Gamma(Y')) \\ &\geq \rho^+(\Gamma(X' \cup Y')) + \rho^+(\Gamma(X' \cap Y'))\end{aligned}$$

to obtain

$$\begin{aligned}\tilde{\rho}(X) + \tilde{\rho}(Y) \geq \min_{X' \subseteq X, Y' \subseteq Y} &\{\rho^+(\Gamma(X' \cup Y')) + |(X \cup Y) \setminus (X' \cup Y')| \\ &+ \rho^+(\Gamma(X' \cap Y')) + |(X \cap Y) \setminus (X' \cap Y')|\} \\ \geq \min_{Z' \subseteq X \cup Y} &\{\rho^+(\Gamma(Z')) + |(X \cup Y) \setminus Z'|\} \\ + \min_{Z'' \subseteq X \cap Y} &\{\rho^+(\Gamma(Z'')) + |(X \cap Y) \setminus Z''|\} \\ = \tilde{\rho}(X \cup Y) &+ \tilde{\rho}(X \cap Y).\end{aligned}$$

(It is also possible to show that $\tilde{\mathcal{I}}$ satisfies the axiom of independent sets of a matroid by a slight modification of the argument in the proof of Lemma 2.3.31. Remark 5.2.19 gives yet another alternative proof.) ■

Given two matroids $\mathbf{M}_1 = (V, \mathcal{I}_1, \rho_1)$ and $\mathbf{M}_2 = (V, \mathcal{I}_2, \rho_2)$ with the same ground set V, we can define another matroid called the *union* of $\mathbf{M}_1$ and $\mathbf{M}_2$, denoted as $\mathbf{M}_1 \vee \mathbf{M}_2 = (V, \mathcal{I}_1 \vee \mathcal{I}_2, \rho_1 \vee \rho_2)$. The family of independent sets $\mathcal{I}_1 \vee \mathcal{I}_2$ is defined by

$$\mathcal{I}_1 \vee \mathcal{I}_2 = \{I_1 \cup I_2 \mid I_1 \in \mathcal{I}_1,\ I_2 \in \mathcal{I}_2\} = \{I_1 \cup I_2 \mid I_1 \in \mathcal{I}_1,\ I_2 \in \mathcal{I}_2,\ I_1 \cap I_2 = \emptyset\}$$

and the rank function $\rho_1 \vee \rho_2 : 2^V \to \mathbf{Z}$ is given by

$$(\rho_1 \vee \rho_2)(X) = \min\{\rho_1(Y) + \rho_2(Y) + |X \setminus Y| \mid Y \subseteq X\}, \qquad X \subseteq V. \quad (2.80)$$

This construction is a special case of the induction of a matroid by a bipartite graph explained above. Let V_1 and V_2 be disjoint copies of V and put $V^+ = V_1 \cup V_2$ and $V^- = V$. Regarding $\mathbf{M}_i$ as being defined on V_i $(i = 1, 2)$, we consider $\mathbf{M}^+ = \mathbf{M}_1 \oplus \mathbf{M}_2$ (direct sum) defined on V^+. Define $A = \{(v_1, v), (v_2, v) \mid v \in V\}$, where $v_1 \in V_1$ and $v_2 \in V_2$ denote the copies of $v \in V$ (see Fig. 2.11). The matroid induced on V^- from $\mathbf{M}^+$ by the bipartite graph $(V^+, V^-; A)$ coincides with (is isomorphic to) $\mathbf{M}_1 \vee \mathbf{M}_2$. The expression (2.79) with $\rho^+(\Gamma(X')) = \rho_1(X') + \rho_2(X')$ yields (2.80).

The rank of the union is closely related to the maximum size of a common independent set of $\mathbf{M}_1$ and $\mathbf{M}_2^* = (V, \mathcal{I}_2^*, \rho_2^*)$ (the dual of $\mathbf{M}_2$). Namely,

$$\mathrm{rank}\,(\mathbf{M}_1 \vee \mathbf{M}_2) = \max\{|I| \mid I \in \mathcal{I}_1 \cap \mathcal{I}_2^*\} + \mathrm{rank}\,(\mathbf{M}_2). \qquad (2.81)$$

This relation follows from (2.80) combined with Theorem 2.3.28 and (2.64).

The union operation extends to a finite number of matroids $\mathbf{M}_i = (V, \mathcal{I}_i, \rho_i)$ $(i \in T)$ in an obvious way. The family of independent sets of their union is given by

$$\{\bigcup_{i \in T} I_i \mid I_i \in \mathcal{I}_i\ (i \in T)\} = \{\bigcup_{i \in T} I_i \mid I_i \in \mathcal{I}_i\ (i \in T), I_i \cap I_j = \emptyset\ (i \neq j)\}$$

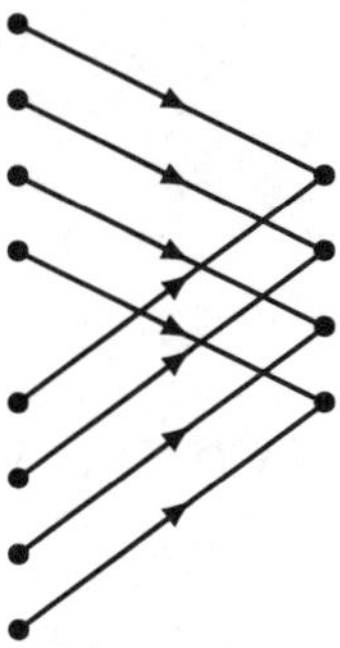

Fig. 2.11. Bipartite graph for union operation

and the rank function $\bigvee_{i \in T} \rho_i : 2^V \to \mathbf{Z}$ by

$$\left(\bigvee_{i \in T} \rho_i \right)(X) = \min\{\sum_{i \in T} \rho_i(Y) + |X \setminus Y| \mid Y \subseteq X\}, \qquad X \subseteq V. \tag{2.82}$$

In this connection we mention the following facts observed in Murota [206].

Proposition 2.3.39. *For a finite family of matroids $\mathbf{M}_i = (V, \rho_i)$ $(i \in T)$ on a common ground set V with rank functions ρ_i, define $\lambda(K, X)$ to be the rank of $X \subseteq V$ in the union of the partial family $\{\mathbf{M}_i \mid i \in K\}$, i.e.,*

$$\lambda(K, X) = \left(\bigvee_{i \in K} \rho_i \right)(X), \qquad K \subseteq T, X \subseteq V.$$

Then it holds that

$$\lambda(K,X)+\lambda(L,Y) \geq \lambda(K \cup L, X \cap Y)+\lambda(K \cap L, X \cup Y), \quad K, L \subseteq T; X, Y \subseteq V.$$

Proof. From (2.82) we have

$$\begin{aligned} & \lambda(K, X) + \lambda(L, Y) \\ & = \min_{X' \subseteq X, Y' \subseteq Y} \left\{ \sum_{i \in K} \rho_i(X') + \sum_{i \in L} \rho_i(Y') + |X \setminus X'| + |Y \setminus Y'| \right\}. \end{aligned}$$

Into this expression we substitute

$$\begin{aligned} & \sum_{i \in K} \rho_i(X') + \sum_{i \in L} \rho_i(Y') \\ & = \sum_{i \in K \setminus L} \rho_i(X') + \sum_{i \in K \cap L} [\rho_i(X') + \rho_i(Y')] + \sum_{i \in L \setminus K} \rho_i(Y') \end{aligned}$$

$$\geq \sum_{i \in K \setminus L} \rho_i(X' \cap Y') + \sum_{i \in K \cap L} [\rho_i(X' \cup Y') + \rho_i(X' \cap Y')] + \sum_{i \in L \setminus K} \rho_i(X' \cap Y')$$
$$= \sum_{i \in K \cup L} \rho_i(X' \cap Y') + \sum_{i \in K \cap L} \rho_i(X' \cup Y')$$

to obtain

$$\begin{aligned}\lambda(K, X) + \lambda(L, Y) \geq \min_{X' \subseteq X, Y' \subseteq Y} \Biggl\{ & \sum_{i \in K \cup L} \rho_i(X' \cap Y') + |(X \cap Y) \setminus (X' \cap Y')| \\ & + \sum_{i \in K \cap L} \rho_i(X' \cup Y') + |(X \cup Y) \setminus (X' \cup Y')| \Biggr\} \\ \geq \min_{Z' \subseteq X \cap Y} & \Biggl\{ \sum_{i \in K \cup L} \rho_i(Z') + |(X \cap Y) \setminus Z'| \Biggr\} \\ & + \min_{Z'' \subseteq X \cup Y} \Biggl\{ \sum_{i \in K \cap L} \rho_i(Z'') + |(X \cup Y) \setminus Z''| \Biggr\} \\ = \lambda(K \cup L, & X \cap Y) + \lambda(K \cap L, X \cup Y). \end{aligned}$$

■

Proposition 2.3.40. *For three matroids* $\mathbf{M}_i$ $(i = 1, 2, 3)$ *it holds that*

$$\text{rank}\,(\mathbf{M}_1 \vee \mathbf{M}_2 \vee \mathbf{M}_3) + \text{rank}\,(\mathbf{M}_2) \leq \text{rank}\,(\mathbf{M}_1 \vee \mathbf{M}_2) + \text{rank}\,(\mathbf{M}_2 \vee \mathbf{M}_3).$$

Proof. Take $X = V$ and $T = \{1, 2, 3\}$ in Proposition 2.3.39. An alternative proof is to make use of the fact that there exist disjoint $I_1, B_2, I_3 \subseteq V$ such that B_2 is a base of $\mathbf{M}_2$, $I_1 \cup B_2$ is a base of $\mathbf{M}_1 \vee \mathbf{M}_2$, and $I_1 \cup B_2 \cup I_3$ is a base of $\mathbf{M}_1 \vee \mathbf{M}_2 \vee \mathbf{M}_3$. Then we have $\text{rank}\,(\mathbf{M}_2 \vee \mathbf{M}_3) \geq |B_2| + |I_3|$. ■

As another application of Proposition 2.3.39 we mention the following observation of Kung [168].

Theorem 2.3.41. *For two matroids* $\mathbf{M}_1$ *and* $\mathbf{M}_2$ *it holds that* $\mathbf{M}_1 \vee \mathbf{M}_2 \to \mathbf{M}_1$, *where "$\to$" means a strong map.*

Proof. In Proposition 2.3.39, take $K = \{1, 2\}$ and $L = \{1\}$ to obtain

$$(\rho_1 \vee \rho_2)(X) + \rho_1(Y) \geq (\rho_1 \vee \rho_2)(X \cap Y) + \rho_1(X \cup Y).$$

If $X \supseteq Y$, this means $(\rho_1 \vee \rho_2)(X) - (\rho_1 \vee \rho_2)(Y) \geq \rho_1(X) - \rho_1(Y)$, the definition (2.65) of a strong map. ■

Remark 2.3.42. In Example 2.3.8 we have defined the matroid $\mathbf{M}\{U\}$ associated with a linear subspace U. For two subspaces $U_i = \ker A_i$ $(i = 1, 2)$, we have

$$U_1 \cap U_2 = \ker \begin{bmatrix} A_1 \\ A_2 \end{bmatrix}.$$

This implies

$$\mathbf{M}\{U_1 \cap U_2\} = \mathbf{M}\{U_1\} \vee \mathbf{M}\{U_2\}$$

when U_1 and U_2 are in the "general position" (in an appropriate sense). See §4.2 for more precise accounts. □

2.3.7 Bimatroid (Linking System)

The notion of a bimatroid was introduced first by Schrijver [290, 291] under the name of a linking system, and later by Kung [165] under the name bimatroid. Just as a matroid can be defined either by the basis family or by the rank function, a bimatroid can be defined either as a triple $\mathbf{L} = (S, T, \Lambda)$ of disjoint finite sets S, T and $\Lambda \subseteq 2^S \times 2^T$ (family of "linked pairs"), or equivalently as a triple $\mathbf{L} = (S, T, \lambda)$ of disjoint finite sets S, T and $\lambda : 2^S \times 2^T \to \mathbf{Z}$ ("birank function"). Unless otherwise indicated, the reader is referred to Schrijver [290, 291] for proofs not included here.

A canonical example of a bimatroid arises from a matrix. Let A be a matrix over a field $\boldsymbol{F}$, and put $S = \mathrm{Row}(A)$ and $T = \mathrm{Col}(A)$. Define Λ to be the family of all pairs (X, Y) such that $|X| = |Y|$ and the corresponding submatrix $A[X, Y]$ is nonsingular. It is an exercise in linear algebra to show that Λ has the following properties:

(L-1) If $(X, Y) \in \Lambda$ and $x \in X$, then $\exists\, y \in Y$ such that $(X \setminus \{x\}, Y \setminus \{y\}) \in \Lambda$;
(L-2) If $(X, Y) \in \Lambda$ and $y \in Y$, then $\exists\, x \in X$ such that $(X \setminus \{x\}, Y \setminus \{y\}) \in \Lambda$;
(L-3) If $(X_i, Y_i) \in \Lambda$ $(i = 1, 2)$, then $\exists\, X \subseteq S$, $\exists\, Y \subseteq T$ such that $(X, Y) \in \Lambda$, $X_1 \subseteq X \subseteq X_1 \cup X_2$, $Y_2 \subseteq Y \subseteq Y_1 \cup Y_2$.

The rank function for submatrices, defined by $\lambda(X, Y) = \mathrm{rank}\, A[X, Y]$ for $X \subseteq S$, $Y \subseteq T$, has the following properties (cf. Proposition 2.1.9 for (B-3)):

(B-1) $0 \le \lambda(X, Y) \le \min\{|X|, |Y|\}$ for $X \subseteq S$ and $Y \subseteq T$;
(B-2) $\lambda(X', Y') \le \lambda(X, Y)$ for $X' \subseteq X \subseteq S$ and $Y' \subseteq Y \subseteq T$;
(B-3) $\lambda(X, Y) + \lambda(X', Y') \ge \lambda(X \cup X', Y \cap Y') + \lambda(X \cap X', Y \cup Y')$ for $X, X' \subseteq S$ and $Y, Y' \subseteq T$.

The family Λ and the function λ determine each other by

$$\lambda(X, Y) = \max\{|X'| \mid (X', Y') \in \Lambda, X' \subseteq X, Y' \subseteq Y\}, \qquad X \subseteq S, Y \subseteq T, \quad (2.83)$$

$$\Lambda = \{(X, Y) \mid \lambda(X, Y) = |X| = |Y|,\ X \subseteq S, Y \subseteq T\}. \quad (2.84)$$

With this example in mind we start a formal description of bimatroids.

A *bimatroid* (or *linking system*) is a triple $\mathbf{L} = (S, T, \Lambda)$, where S and T are disjoint finite sets, and Λ is a nonempty subset of $2^S \times 2^T$ such that (L-1)–(L-3) above are satisfied. We call S the *row set* (or *exit set*) and T the *column set* (or *entrance set*) of $\mathbf{L}$, and write $S = \mathrm{Row}(\mathbf{L})$ and $T = \mathrm{Col}(\mathbf{L})$. A member (X, Y) of Λ is called a *linked pair*.

For a bimatroid $\mathbf{L} = (S, T, \Lambda)$ the *birank function* (or *linking function*) $\lambda : 2^S \times 2^T \to \mathbf{Z}$ is defined by (2.83). It can be proven that λ satisfies (B-1)–(B-3) above. Conversely, a function $\lambda : 2^S \times 2^T \to \mathbf{Z}$ satisfying (B-1)–(B-3) determines a bimatroid by (2.84). Namely, (L-1)–(L-3) for $\Lambda \subseteq 2^S \times 2^T$ are equivalent to (B-1)–(B-3) for $\lambda : 2^S \times 2^T \to \mathbf{Z}$. Thus, a bimatroid $\mathbf{L}$ is defined by a triple (S, T, Λ) with the properties (L-1)–(L-3) or equivalently by a triple (S, T, λ) with the properties (B-1)–(B-3).

It follows from (L-1) and (L-2) that $|X| = |Y|$ if $(X, Y) \in \Lambda$. A linked pair can be enlarged monotonically, i.e.,

$$\begin{aligned} &(X_1, Y_1) \in \Lambda, |X_1| \leq \lambda(X, Y), X_1 \subseteq X, Y_1 \subseteq Y \\ &\Longrightarrow \exists\, (X_2, Y_2) \in \Lambda, |X_2| = \lambda(X, Y), X_1 \subseteq X_2 \subseteq X, Y_1 \subseteq Y_2 \subseteq Y. \end{aligned} \tag{2.85}$$

The maximum size of a linked pair in $\mathbf{L}$ is referred to as the *rank* of $\mathbf{L}$, i.e., $\mathrm{rank}\, \mathbf{L} = \lambda(S, T)$. A bimatroid $\mathbf{L}$ is called *trivial* if $\mathrm{rank}\, \mathbf{L} = 0$, and *nonsingular* if $\mathrm{rank}\, \mathbf{L} = |S| = |T|$.

Example 2.3.43. Besides the canonical example from a matrix, another example of a bimatroid is obtained from linkings/matchings in a graph. Let $G = (V, A; S, T)$ be a directed graph with S and T being disjoint subsets of V. With reference to Menger-type linkings from S to T, define $\Lambda \subseteq 2^S \times 2^T$ as follows: $(X, Y) \in \Lambda$ if and only if there exists a Menger-type linking of size $|X| = |Y|$ from X to Y. Then $\mathbf{L} = (S, T, \Lambda)$ is a bimatroid, satisfying the conditions (L-1)–(L-3). □

As the name suggests, bimatroids are closely related to matroids. Given a bimatroid $\mathbf{L} = (S, T, \Lambda)$, define $\mathcal{B} \subseteq 2^{S \cup T}$ by

$$\mathcal{B} = \{(S \setminus X) \cup Y \mid (X, Y) \in \Lambda\}. \tag{2.86}$$

Then $\mathcal{B}$ is the basis family of a matroid on $S \cup T$ with $\mathcal{B} \ni S$. See Fig. 2.12 for this correspondence in the case of a matrix, where the left submatrix with column set S is an identity matrix. The rank function $\rho : 2^{S \cup T} \to \mathbf{Z}$ of the matroid is expressed as

$$\rho(X \cup Y) = \lambda(S \setminus X, Y) + |X|, \qquad X \subseteq S, Y \subseteq T$$

using the birank function λ. Conversely, if $(S \cup T, \mathcal{B})$ is a matroid with $\mathcal{B} \ni S$ and $S \cap T = \emptyset$, then

$$\Lambda = \{(X, Y) \mid X \subseteq S, Y \subseteq T, (S \setminus X) \cup Y \in \mathcal{B}\}$$

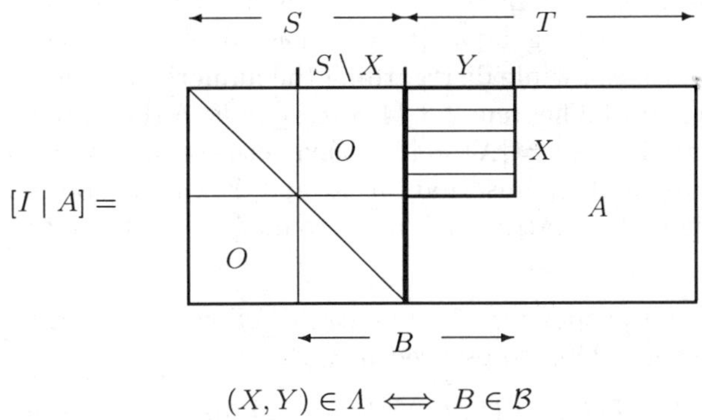

Fig. 2.12. Matroid associated with a bimatroid

is the family of linked pairs of a bimatroid. As such, the concept of bimatroids can be regarded as a variant of matroids. Matroids are more convenient in some cases and bimatroids are more natural in other cases.

The restriction of the associated matroid $(S \cup T, \mathcal{B})$ to $T = \mathrm{Col}(\mathbf{L})$ is called the *column matroid* of $\mathbf{L}$, denoted $\mathbf{CM}(\mathbf{L})$. By definition, $Y \subseteq T$ is independent in $\mathbf{CM}(\mathbf{L})$ if and only if $(X, Y) \in \Lambda$ for some $X \subseteq S$. Similarly, the *row matroid* $\mathbf{RM}(\mathbf{L})$ is the restriction to $S = \mathrm{Row}(\mathbf{L})$ of the dual of $(S \cup T, \mathcal{B})$. Namely, $X \subseteq S$ is independent in $\mathbf{RM}(\mathbf{L})$ if and only if $(X, Y) \in \Lambda$ for some $Y \subseteq T$.

The *underlying bipartite graph* of a bimatroid $\mathbf{L} = (S, T, \Lambda)$ is a bipartite graph $G(\mathbf{L}) = (S, T, \Delta)$ with vertex set $S \cup T$ and arc set $\Delta \subseteq S \times T$ such that

$$(x, y) \in \Delta \iff (\{x\}, \{y\}) \in \Lambda.$$

The information represented in the underlying bipartite graph is only partial in the sense that different bimatroids can have the same underlying bipartite graph. Still it carries some crucial portion of the combinatorial structure, as pointed out by Schrijver [290, 291].

Theorem 2.3.44. *Let* $\mathbf{L} = (S, T, \Lambda)$ *be a bimatroid and* $G(\mathbf{L}) = (S, T, \Delta)$ *be its underlying bipartite graph.*

(1) *If* $(X, Y) \in \Lambda$, *there exists a perfect matching between* X *and* Y *in* $G(\mathbf{L}) = (S, T, \Delta)$.

(2) *If there exists a unique perfect matching between* X *and* Y *in* $G(\mathbf{L}) = (S, T, \Delta)$, *then* $(X, Y) \in \Lambda$.

Proof. When translated to statements for the matroid associated with $\mathbf{L} = (S, T, \Lambda)$, these claims reduce respectively to the perfect-matching lemma (Lemma 2.3.16) and the unique matching lemma (Lemma 2.3.18). ∎

Remark 2.3.45. In case $\mathbf{L} = (S, T, \Lambda)$ is defined in terms of a matrix A, considering the underlying bipartite graph is to consider the zero/nonzero pattern of the matrix A while disregarding the numerical values of the entries. The first statement of Theorem 2.3.44 corresponds in this case to the obvious fact that, if $\text{rank}\, A[X,Y] = |X| = |Y|$, then $\text{term-rank}\, A[X,Y] = |X| = |Y|$. The second statement claims that, if X and Y can be permuted so that $A[X,Y]$ is a triangular matrix with nonzero diagonal entries, then $A[X,Y]$ is nonsingular. □

Two additional properties of a bimatroid follow. The first is an easy observation of Murota [220], to be used in §7.1.

Theorem 2.3.46. *Let* $\mathbf{L} = (S, T, \lambda)$ *be a bimatroid with birank function* λ. *For* $X_0 \subseteq S$, $Y_0 \subseteq T$, *and an integer* $k \geq \max(|X_0|, |Y_0|)$, *there exist* $X \subseteq S$ *and* $Y \subseteq T$ *such that* $X \supseteq X_0$, $Y \supseteq Y_0$, *and* $\lambda(X,Y) = |X| = |Y| = k$ *if and only if the following four conditions are satisfied:* (i) $\lambda(S,T) \geq k$, (ii) $\lambda(X_0, T) = |X_0|$, (iii) $\lambda(S, Y_0) = |Y_0|$, *and* (iv) $\lambda(X_0, Y_0) \geq |X_0| + |Y_0| - k$.

Proof. The conditions (i)–(iii) are obviously necessary. The necessity of (iv) can be shown as follows:

$$\begin{aligned} k = \lambda(X,Y) &\leq \lambda(X, Y_0) + \lambda(X, Y \setminus Y_0) \\ &\leq \lambda(X_0, Y_0) + \lambda(X \setminus X_0, Y_0) + \lambda(X, Y \setminus Y_0) \\ &\leq \lambda(X_0, Y_0) + |X \setminus X_0| + |Y \setminus Y_0| \\ &= \lambda(X_0, Y_0) - |X_0| - |Y_0| + 2k. \end{aligned}$$

For sufficiency, put $r_0 = \lambda(X_0, Y_0)$ and see: $\exists X_1 \subseteq X_0, \exists Y_1 \subseteq Y_0$ such that $\lambda(X_1, Y_1) = |X_1| = |Y_1| = r_0$. Hence $\lambda(X_0, Y_1) = |Y_1|$, whereas $\lambda(X_0, T) = |X_0|$ by (ii). Then, by (2.85), $\exists Y_2 \subseteq T \setminus Y_0$ such that $\lambda(X_0, Y_1 \cup Y_2) = |X_0| = |Y_1| + |Y_2|$. By (B-3) we have

$$\lambda(S, Y_0 \cup Y_2) + \lambda(X_0, Y_0) \geq \lambda(S, Y_0) + \lambda(X_0, Y_0 \cup Y_2) = |Y_0| + |Y_1| + |Y_2|,$$

where the (last) equality is due to (iii) and the above claim. Therefore $\lambda(S, Y_0 \cup Y_2) \geq |Y_0| + |Y_2|$, and hence $\lambda(S, Y_0 \cup Y_2) = |Y_0| + |Y_2|$. On the other hand,

$$|X_0| \geq \lambda(X_0, Y_0 \cup Y_2) \geq \lambda(X_0, Y_1 \cup Y_2) = |X_0|$$

implies $\lambda(X_0, Y_0 \cup Y_2) = |X_0|$. Hence, by (2.85), $\exists X_2 \subseteq S \setminus X_0$ such that

$$|X_0| + |X_2| = \lambda(X_0 \cup X_2, Y_0 \cup Y_2) = |Y_0| + |Y_2| = |X_0| + |Y_0| - r_0 \leq k,$$

where the last inequality is due to (iv). Hence, by (i) and (2.85), $\exists X \supseteq X_0 \cup X_2, \exists Y \supseteq Y_0 \cup Y_2$ such that $\lambda(X,Y) = |X| = |Y| = k$. This completes the proof of sufficiency. ■

The second is a matroid-theoretic abstraction of the König–Egerváry theorem (Theorem 2.2.15). This is due to Bapat [9] and will be used in §4.2 and §4.8.

Theorem 2.3.47 (König–Egerváry theorem for bimatroids). *Let* $\mathbf{L} = (S, T, \lambda)$ *be a bimatroid with birank function* λ. *Then there exists* $(X, Y) \in 2^S \times 2^T$ *such that*

(i) $|X| + |Y| - \lambda(X, Y) = |S| + |T| - \lambda(S, T)$,

(ii) $\lambda(X \setminus \{x\}, Y \setminus \{y\}) = \lambda(X, Y)$, $\forall x \in X$, $\forall y \in Y$.

If $\min(|S|, |T|) > \lambda(S, T)$, *then* $X \neq \emptyset$ *and* $Y \neq \emptyset$.

Proof. Obviously, there exists (X, Y) satisfying (i) (e.g., $X = S$, $Y = T$). Let (X, Y) be such a pair with $|X| + |Y|$ minimal. Put $\lambda(X, Y) = a$ and suppose that (ii) fails. Then $\lambda(X \setminus \{x\}, Y \setminus \{y\}) \leq a - 1$ for some $x \in X$, $y \in Y$. The inequality (B-3) shows

$$2a - 1 \geq \lambda(X, Y) + \lambda(X \setminus \{x\}, Y \setminus \{y\}) \geq \lambda(X \setminus \{x\}, Y) + \lambda(X, Y \setminus \{y\}).$$

This implies that either $\lambda(X \setminus \{x\}, Y) = a - 1$ or $\lambda(X, Y \setminus \{y\}) = a - 1$. In the former case, $(X \setminus \{x\}, Y)$ satisfies (i), contradicting the minimality of $|X|+|Y|$. Similarly for the latter case. Therefore, (X, Y) satisfies (ii). Suppose that $\min(|S|, |T|) > \lambda(S, T)$ and either $X = \emptyset$ or $Y = \emptyset$. Then we would have $|S| + |T| - \lambda(S, T) > \max(|S|, |T|) \geq \max(|X|, |Y|) = |X| + |Y| - \lambda(X, Y)$, a contradiction to (i). ■

A canonical choice of the pair (X, Y) in Theorem 2.3.47 can be made by way of the canonical partition of a bimatroid introduced by Geelen [92]. For a bimatroid $\mathbf{L} = (S, T, \lambda)$ and $Z \subseteq S \cup T$, we denote by $\mathbf{L} \setminus Z$ the bimatroid with Z deleted, i.e., $\mathbf{L} \setminus Z = (S \setminus Z, T \setminus Z, \lambda')$ with $\lambda'(X, Y) = \lambda(X, Y)$ for $X \subseteq S \setminus Z$ and $Y \subseteq T \setminus Z$. Define a partition of $S \cup T$ into three disjoint parts by

$$\begin{aligned} D(\mathbf{L}) &= \{z \in S \cup T \mid \operatorname{rank}(\mathbf{L} \setminus \{z\}) = \operatorname{rank} \mathbf{L}\}, \\ A(\mathbf{L}) &= \{z \in S \cup T \mid D(\mathbf{L} \setminus \{z\}) = D(\mathbf{L})\}, \\ C(\mathbf{L}) &= (S \cup T) \setminus (D(\mathbf{L}) \cup A(\mathbf{L})). \end{aligned}$$

The partition $(D(\mathbf{L}), A(\mathbf{L}), C(\mathbf{L}))$ is called the *canonical partition* of $\mathbf{L}$.

The canonical partition enjoys the following nice properties (Geelen [92]).

Proposition 2.3.48. *For* $x \in S \setminus D(\mathbf{L})$ *the following hold true.*

(1) $D(\mathbf{L} \setminus \{x\}) \cap S = D(\mathbf{L}) \cap S$.

(2) $D(\mathbf{L} \setminus \{x\}) \cap T \supseteq D(\mathbf{L}) \cap T$.

(3) *If* $y \in D(\mathbf{L} \setminus \{x\}) \setminus D(\mathbf{L})$, *then* $x \in D(\mathbf{L} \setminus \{y\})$.

(4) $A(\mathbf{L} \setminus \{x\}) \cap T \subseteq A(\mathbf{L}) \cap T$.

Proof. (1) Note the relation $S \setminus D(\mathbf{L}) = \{\text{coloops of } \mathbf{RM}(\mathbf{L})\}$ as well as the similar relation for $\mathbf{L} \setminus \{x\}$. Since x is a coloop of $\mathbf{RM}(\mathbf{L})$, we have $\{\text{coloops of } \mathbf{RM}(\mathbf{L})\} = \{x\} \cup \{\text{coloops of } \mathbf{RM}(\mathbf{L} \setminus \{x\})\}$.

(2) For $y \in D(\mathbf{L}) \cap T$ we have $\operatorname{rank}(\mathbf{L} \setminus \{x, y\}) = \operatorname{rank}(\mathbf{L} \setminus \{x\})$, since $\lambda(S \setminus \{x\}, T) \geq \lambda(S \setminus \{x\}, T \setminus \{y\}) \geq \lambda(S \setminus \{x\}, T) + \lambda(S, T \setminus \{y\}) - \lambda(S, T)$ by (B-2) and (B-3), and $\lambda(S, T) = \operatorname{rank} \mathbf{L} = \lambda(S, T \setminus \{y\})$.

(3) We have $\text{rank}\,(\mathbf{L} \setminus \{x,y\}) = \text{rank}\,(\mathbf{L} \setminus \{x\}) = \text{rank}\,\mathbf{L} - 1$ since $y \in D(\mathbf{L} \setminus \{x\})$, whereas $\text{rank}\,(\mathbf{L} \setminus \{y\}) = \text{rank}\,\mathbf{L} - 1$ since $y \notin D(\mathbf{L})$. Hence, $\text{rank}\,(\mathbf{L} \setminus \{x,y\}) = \text{rank}\,(\mathbf{L} \setminus \{y\})$.

(4) Let $y \in A(\mathbf{L} \setminus \{x\}) \cap T$. We have $x \notin D(\mathbf{L} \setminus \{y\})$ by (3) (with the roles of x and y interchanged), and hence $D(\mathbf{L} \setminus \{x,y\}) \cap S = D(\mathbf{L} \setminus \{y\}) \cap S$ by (1), whereas $D(\mathbf{L} \setminus \{x,y\}) \cap S = D(\mathbf{L} \setminus \{x\}) \cap S = D(\mathbf{L}) \cap S$ since $y \in A(\mathbf{L} \setminus \{x\})$ and $x \in S \setminus D(\mathbf{L})$. Therefore, $D(\mathbf{L} \setminus \{y\}) \cap S = D(\mathbf{L}) \cap S$. On the other hand, $D(\mathbf{L} \setminus \{y\}) \cap T = D(\mathbf{L}) \cap T$ since $y \in T \setminus D(\mathbf{L})$. Hence $D(\mathbf{L} \setminus \{y\}) = D(\mathbf{L})$, i.e., $y \in A(\mathbf{L})$. ∎

Proposition 2.3.49. *If $x \in A(\mathbf{L})$, then the canonical partition of $\mathbf{L} \setminus \{x\}$ is $(D(\mathbf{L}), A(\mathbf{L}) \setminus \{x\}, C(\mathbf{L}))$.*

Proof. We have $D(\mathbf{L} \setminus \{x\}) = D(\mathbf{L})$ by definition. For $y \in C(\mathbf{L})$ we see $D(\mathbf{L} \setminus \{x\}) = D(\mathbf{L}) \subsetneqq D(\mathbf{L} \setminus \{y\}) \subseteq D(\mathbf{L} \setminus \{x,y\})$ using Proposition 2.3.48, and hence $y \notin A(\mathbf{L} \setminus \{x\})$. Therefore, $C(\mathbf{L} \setminus \{x\}) \supseteq C(\mathbf{L})$. The proof is completed by showing $A(\mathbf{L} \setminus \{x\}) \supseteq A(\mathbf{L}) \setminus \{x\}$. Suppose that there exists $y \in A(\mathbf{L}) \setminus \{x\}$ with $y \notin A(\mathbf{L} \setminus \{x\})$, and take $z \in D(\mathbf{L} \setminus \{x,y\}) \setminus D(\mathbf{L} \setminus \{x\})$. It then follows that $\text{rank}\,(\mathbf{L} \setminus \{x\}) = \text{rank}\,(\mathbf{L} \setminus \{y\}) = \text{rank}\,(\mathbf{L} \setminus \{z\}) = \text{rank}\,\mathbf{L} - 1$ and $\text{rank}\,(\mathbf{L} \setminus \{x,y,z\}) = \text{rank}\,(\mathbf{L} \setminus \{x,y\}) = \text{rank}\,(\mathbf{L} \setminus \{y,z\}) = \text{rank}\,(\mathbf{L} \setminus \{z,x\}) = \text{rank}\,\mathbf{L} - 2$. This means $x \in D(\mathbf{L} \setminus \{y,z\}) \setminus D(\mathbf{L} \setminus \{y\})$, $y \in D(\mathbf{L} \setminus \{z,x\}) \setminus D(\mathbf{L} \setminus \{z\})$, and $z \in D(\mathbf{L} \setminus \{x,y\}) \setminus D(\mathbf{L} \setminus \{x\})$. The first of these implies, by Proposition 2.3.48(1), that x and z are on the opposite sides, i.e., $|\{z,x\} \cap S| = |\{z,x\} \cap T| = 1$. Similarly, $|\{x,y\} \cap S| = |\{x,y\} \cap T| = 1$ and $|\{y,z\} \cap S| = |\{y,z\} \cap T| = 1$. However, this is impossible. ∎

Proposition 2.3.50. *The conditions* (i) *and* (ii) *in Theorem 2.3.47 are satisfied by $(X,Y) = (D(\mathbf{L}) \cap S, (D(\mathbf{L}) \cup C(\mathbf{L})) \cap T)$, and symmetrically by $(X,Y) = ((D(\mathbf{L}) \cup C(\mathbf{L})) \cap S, D(\mathbf{L}) \cap T)$.*

Proof. We prove for the former. For $z \in A(\mathbf{L})$ we have $\text{rank}\,(\mathbf{L} \setminus \{z\}) = \text{rank}\,\mathbf{L} - 1$, while the canonical partition of $\mathbf{L} \setminus \{z\}$ remains the same as in Proposition 2.3.49. Hence $\mathbf{L}' = \mathbf{L} \setminus A(\mathbf{L})$ satisfies $\text{rank}\,\mathbf{L}' = \text{rank}\,\mathbf{L} - |A(\mathbf{L})|$, $D(\mathbf{L}') = D(\mathbf{L})$, $A(\mathbf{L}') = \emptyset$, and $C(\mathbf{L}') = C(\mathbf{L})$. For $x \in C(\mathbf{L}) \cap S$ we have $\text{rank}\,(\mathbf{L}' \setminus \{x\}) = \text{rank}\,\mathbf{L}' - 1$, $D(\mathbf{L}' \setminus \{x\}) \cap S = D(\mathbf{L}') \cap S$ by Proposition 2.3.48(1) and $A(\mathbf{L}' \setminus \{x\}) \cap T = \emptyset$ by Proposition 2.3.48(4). Hence $\mathbf{L}'' = \mathbf{L}' \setminus (C(\mathbf{L}) \cap S)$ satisfies $\text{rank}\,\mathbf{L}'' = \text{rank}\,\mathbf{L}' - |C(\mathbf{L}) \cap S|$, $D(\mathbf{L}'') \cap S = D(\mathbf{L}') \cap S = D(\mathbf{L}) \cap S = X$, and $A(\mathbf{L}'') \cap T = \emptyset$. Note that $\text{Row}(\mathbf{L}'') = X$ and $\text{Col}(\mathbf{L}'') = Y$. We claim $D(\mathbf{L}'') \cap Y = Y$. Suppose there exists $y \in Y \setminus D(\mathbf{L}'')$. Then $D(\mathbf{L}'' \setminus \{y\}) \cap Y = D(\mathbf{L}'') \cap Y$ and $D(\mathbf{L}'' \setminus \{y\}) \cap X \supseteq D(\mathbf{L}'') \cap X = X$, which together imply $D(\mathbf{L}'' \setminus \{y\}) = D(\mathbf{L}'')$, i.e., $y \in A(\mathbf{L}'')$, a contradiction to $A(\mathbf{L}'') \cap T = \emptyset$. Therefore, we have $D(\mathbf{L}'') = X \cup Y$, which is equivalent, by (B-3), to the condition (ii). As for the condition (i), we have $\lambda(X,Y) = \text{rank}\,\mathbf{L}'' = \text{rank}\,\mathbf{L} - |A(\mathbf{L})| - |C(\mathbf{L}) \cap S| = \lambda(S,T) - |S \setminus X| - |T \setminus Y|$. ∎

A number of natural operations can be defined for bimatroids, as introduced by Schrijver [290, 291]. Though all these operations can be transformed in principle to operations for the corresponding matroids, they are most naturally expressed for bimatroids. This is especially true for union and product operations.

For $X \subseteq S$ and $Y \subseteq T$, the *restriction* of $\mathbf{L} = (S, T, \Lambda)$ to (X, Y) is a bimatroid $\mathbf{L}[X, Y] = (X, Y, \Lambda')$ with

$$\Lambda' = \{(X', Y') \mid X' \subseteq X, Y' \subseteq Y, (X', Y') \in \Lambda\}.$$

We have $\mathbf{L}[X, Y] = \mathbf{L} \setminus Z$ for $Z = (S \setminus X) \cup (T \setminus Y)$.

The *dual* (or *transpose*) of $\mathbf{L} = (S, T, \Lambda)$ is a bimatroid $\mathbf{L}^* = (T, S, \Lambda^*)$ with $\Lambda^* = \{(Y, X) \mid (X, Y) \in \Lambda\}$.

For a nonsingular bimatroid $\mathbf{L} = (S, T, \Lambda)$, the *inverse* of $\mathbf{L}$ is a bimatroid $\mathbf{L}^{-1} = (T, S, \Lambda^{-1})$ with

$$\Lambda^{-1} = \{(Y, X) \mid (S \setminus X, T \setminus Y) \in \Lambda\}.$$

For two bimatroids $\mathbf{L}_i = (S_i, T_i, \Lambda_i)$ $(i = 1, 2)$, the *union* of $\mathbf{L}_1$ and $\mathbf{L}_2$ can be defined as a bimatroid $\mathbf{L}_1 \vee \mathbf{L}_2 = (S_1 \cup S_2, T_1 \cup T_2, \Lambda_1 \vee \Lambda_2)$ with

$$\begin{aligned} \Lambda_1 \vee \Lambda_2 = \{&(X_1 \cup X_2, Y_1 \cup Y_2) \mid \\ &X_1 \cap X_2 = \emptyset, Y_1 \cap Y_2 = \emptyset, (X_1, Y_1) \in \Lambda_1, (X_2, Y_2) \in \Lambda_2\}. \end{aligned}$$

It should be clear that $S_1 \cap S_2 \neq \emptyset$ and $T_1 \cap T_2 \neq \emptyset$ in general.

Theorem 2.3.51. *$\mathbf{L}_1 \vee \mathbf{L}_2 = (S_1 \cup S_2, T_1 \cup T_2, \Lambda_1 \vee \Lambda_2)$ is a bimatroid, and the birank function $\lambda_1 \vee \lambda_2$ of $\mathbf{L}_1 \vee \mathbf{L}_2$ is given by*

$$\begin{aligned} &(\lambda_1 \vee \lambda_2)(X, Y) \\ &= \min\{\lambda_1(X' \cap S_1, Y' \cap T_1) + \lambda_2(X' \cap S_2, Y' \cap T_2) + |X \setminus X'| + |Y \setminus Y'| \mid \\ &\qquad X' \subseteq X, Y' \subseteq Y\}, \qquad X \subseteq S_1 \cup S_2,\ Y \subseteq T_1 \cup T_2. \end{aligned}$$

□

Remark 2.3.52. The union operation of bimatroids corresponds roughly to the sum of matrices. See Theorem 4.2.9 for a precise statement. □

For two bimatroids $\mathbf{L}_i = (S_i, T_i, \Lambda_i)$ $(i = 1, 2)$ with $\mathrm{Col}(\mathbf{L}_1) = \mathrm{Row}(\mathbf{L}_2)$, the *product* of $\mathbf{L}_1$ and $\mathbf{L}_2$ can be defined as a bimatroid $\mathbf{L}_1 * \mathbf{L}_2 = (S_1, T_2, \Lambda_1 * \Lambda_2)$ with

$$\Lambda_1 * \Lambda_2 = \{(X, Z) \mid \exists Y \subseteq T_1 : (X, Y) \in \Lambda_1, (Y, Z) \in \Lambda_2\}.$$

Theorem 2.3.53. *$\mathbf{L}_1 * \mathbf{L}_2 = (S_1, T_2, \Lambda_1 * \Lambda_2)$ is a bimatroid, and the birank function $\lambda_1 * \lambda_2$ of $\mathbf{L}_1 * \mathbf{L}_2$ is given by*

$$(\lambda_1 * \lambda_2)(X, Z) = \min\{\lambda_1(X, T_1 \setminus Y) + \lambda_2(Y, Z) \mid Y \subseteq T_1\}, \quad X \subseteq S_1, Z \subseteq T_2.$$

□

Remark 2.3.54. The product operation for bimatroids is motivated by the Cauchy–Binet formula for the product of matrices (Proposition 2.1.6). Suppose that $\mathbf{L}_i = (S_i, T_i, \Lambda_i)$ $(i = 1, 2)$ are defined by matrices A_i $(i = 1, 2)$, and let $\mathbf{L}_{12} = (S_1, T_2, \Lambda_{12})$ denote the bimatroid defined by $A_1 A_2$. The Cauchy–Binet formula shows that, if $(A_1 A_2)[X, Z]$ is nonsingular, there exists Y such that both $A_1[X, Y]$ and $A_2[Y, Z]$ are nonsingular. When translated to bimatroids, this means that if $(X, Z) \in \Lambda_{12}$, then there exists Y such that $(X, Y) \in \Lambda_1$ and $(Y, Z) \in \Lambda_2$. The necessary condition here is adopted as the definition of the product of bimatroids. Therefore, if $(X, Z) \in \Lambda_{12}$, then $(X, Z) \in \Lambda_1 * \Lambda_2$.

The converse is not true because of possible numerical cancellations. Namely, $\mathbf{L}_1 * \mathbf{L}_2$ does not always agree with $\mathbf{L}_{12}$. Consider, for example, $A_1 = (\,1 \quad 1\,)$, $A_2 = \begin{pmatrix} 1 \\ -1 \end{pmatrix}$, for which $A_1 A_2 = O$ while $\mathrm{rank}\,(\mathbf{L}_1 * \mathbf{L}_2) = 1$. □

For three bimatroids $\mathbf{L}_i = (S_i, T_i, \Lambda_i)$ $(i = 1, 2, 3)$ with $\mathrm{Col}(\mathbf{L}_1) = \mathrm{Row}(\mathbf{L}_2)$ and $\mathrm{Col}(\mathbf{L}_2) = \mathrm{Row}(\mathbf{L}_3)$, we can define the triple product $\mathbf{L}_1 * \mathbf{L}_2 * \mathbf{L}_3$, which notation is justified since $(\mathbf{L}_1 * \mathbf{L}_2) * \mathbf{L}_3 = \mathbf{L}_1 * (\mathbf{L}_2 * \mathbf{L}_3)$. The following inequality is observed by Murota [211].

Theorem 2.3.55 (Frobenius inequality for bimatroids). *For three bimatroids $\mathbf{L}_i$ $(i = 1, 2, 3)$ such that $\mathbf{L}_1 * \mathbf{L}_2 * \mathbf{L}_3$ can be defined, it holds that*

$$\mathrm{rank}\,(\mathbf{L}_1 * \mathbf{L}_2 * \mathbf{L}_3) + \mathrm{rank}\,(\mathbf{L}_2) \geq \mathrm{rank}\,(\mathbf{L}_1 * \mathbf{L}_2) + \mathrm{rank}\,(\mathbf{L}_2 * \mathbf{L}_3).$$

Proof. Put $\mathbf{L}_i = (S_i, T_i, \lambda_i)$ $(i = 1, 2, 3)$, where $T_1 = S_2$ and $T_2 = S_3$. By Theorem 2.3.53 we have

$$\mathrm{rank}\,(\mathbf{L}_1 * \mathbf{L}_2) = \min\{\lambda_1(S_1, T_1 \setminus X_1) + \lambda_2(X_1, T_2) \mid X_1 \subseteq T_1\},$$
$$\mathrm{rank}\,(\mathbf{L}_2 * \mathbf{L}_3) = \min\{\lambda_2(S_2, T_2 \setminus X_2) + \lambda_3(X_2, T_3) \mid X_2 \subseteq T_2\}.$$

From these relations as well as

$$\lambda_2(X_1, T_2) + \lambda_2(S_2, T_2 \setminus X_2) \leq \lambda_2(X_1, T_2 \setminus X_2) + \lambda_2(S_2, T_2),$$

it follows that

$$\begin{aligned} &\mathrm{rank}\,(\mathbf{L}_1 * \mathbf{L}_2) + \mathrm{rank}\,(\mathbf{L}_2 * \mathbf{L}_3) \\ &\leq \min_{X_1, X_2} \{\lambda_1(S_1, T_1 \setminus X_1) + \lambda_2(X_1, T_2 \setminus X_2) + \lambda_3(X_2, T_3)\} + \lambda_2(S_2, T_2) \\ &= \mathrm{rank}\,(\mathbf{L}_1 * \mathbf{L}_2 * \mathbf{L}_3) + \mathrm{rank}\,(\mathbf{L}_2). \end{aligned}$$

■

Remark 2.3.56. The inequality in Theorem 2.3.55 may be compared with the similar inequality for matrix products:

$$\mathrm{rank}\,(A_1 \cdot A_2 \cdot A_3) + \mathrm{rank}\,(A_2) \geq \mathrm{rank}\,(A_1 \cdot A_2) + \mathrm{rank}\,(A_2 \cdot A_3),$$

which is sometimes referred to as the *Frobenius inequality*. It is emphasized that neither of these inequalities implies the other, because of the possible discrepancy (Remark 2.3.54) between the matrix multiplication and bimatroid multiplication. □

Suppose a matroid $\mathbf{M} = (T, \mathcal{I}, \mu)$ (with family $\mathcal{I}$ of independent sets and rank function μ) is defined on the column set $T = \mathrm{Col}(\mathbf{L})$ of a bimatroid $\mathbf{L} = (S, T, \lambda)$. Then another matroid, denoted by $\mathbf{L} * \mathbf{M}$, is induced on $S = \mathrm{Row}(\mathbf{L})$, as is noted by Schrijver [290, 291]. This generalizes the induction of a matroid through a bipartite graph in Theorem 2.3.38.

Theorem 2.3.57. *For a bimatroid* $\mathbf{L} = (S, T, \Lambda, \lambda)$ *and a matroid* $\mathbf{M} = (T, \mathcal{I}, \mu)$,

$$\tilde{\mathcal{I}} = \{X \subseteq S \mid \exists\, Y \subseteq T : \ (X, Y) \in \Lambda, Y \in \mathcal{I}\}$$

forms the family of independent sets of a matroid, denoted by $\mathbf{L} * \mathbf{M}$. *The rank function* $\lambda * \mu$ *of* $\mathbf{L} * \mathbf{M}$ *is given by*

$$(\lambda * \mu)(X) = \min\{\lambda(X, T \setminus Y) + \mu(Y) \mid Y \subseteq T\}, \qquad X \subseteq S.$$

□

Finally we mention the following facts concerning strong map relations, both due to Kung [165, 168].

Theorem 2.3.58. *For a bimatroid* $\mathbf{L} = (S, T, \lambda)$ *and a matroid* $\mathbf{M} = (T, \mu)$, $\mathbf{L} * \mathbf{M}$ *is a strong quotient of* $\mathbf{RM}(\mathbf{L})$, *i.e.*, $\mathbf{RM}(\mathbf{L}) \to \mathbf{L} * \mathbf{M}$. □

Theorem 2.3.59. *For two bimatroids* $\mathbf{L}_i$ $(i = 1, 2)$ *such that* $\mathbf{L}_1 * \mathbf{L}_2$ *can be defined,* $\mathbf{RM}(\mathbf{L}_1 * \mathbf{L}_2)$ *and* $\mathbf{CM}(\mathbf{L}_1 * \mathbf{L}_2)$ *are strong quotients of* $\mathbf{RM}(\mathbf{L}_1)$ *and* $\mathbf{CM}(\mathbf{L}_2)$, *respectively, namely,* $\mathbf{RM}(\mathbf{L}_1) \to \mathbf{RM}(\mathbf{L}_1 * \mathbf{L}_2)$ *and* $\mathbf{CM}(\mathbf{L}_2) \to \mathbf{CM}(\mathbf{L}_1 * \mathbf{L}_2)$.

Proof. This is a corollary of Theorem 2.3.58. Note that $\mathbf{RM}(\mathbf{L}_1 * \mathbf{L}_2) = \mathbf{L}_1 * \mathbf{RM}(\mathbf{L}_2)$. ■

Remark 2.3.60. The inequality (B-3) was first termed the bi-submodularity in Schrijver [290, 291]. Recently, however, bisubmodularity also denotes a similar but different inequality that appears in connection to delta-matroids and jump systems as in Bouchet–Cunningham [19]. In view of this situation we refrain from using the terms bi-submodularity and bisubmodularity to avoid possible confusions, though we still use the prefix "bi" in "bimatroid" and "birank function," admitting an inconsistent compromise. □

3. Physical Observations for Mixed Matrix Formulation

The dual viewpoint from structural analysis and dimensional analysis, as previewed in §1.2, is explained in more detail. Firstly, two different kinds, "accurate" and "inaccurate," are distinguished among numbers characterizing real-world systems, and secondly, algebraic implications of the principle of dimensional homogeneity are discussed. These observations lead to the concepts of "mixed matrices," "mixed polynomial matrices," and "physical matrices" as the mathematical models of matrices arising from real problems.

3.1 Mixed Matrix for Modeling Two Kinds of Numbers

3.1.1 Two Kinds of Numbers

A real-world physical/engineering system will be characterized by a set of relations among various kinds of numbers representing physical quantities, parameter values, incidence relations, etc., where it is important to recognize the difference in the nature of the quantities involved in the real-world problem and to establish a mathematical model that reflects the difference.

A primitive, yet fruitful, way of classifying numbers would be to distinguish nonvanishing elements from zeros. This dichotomy often leads to graph-theoretic methods for structural analysis, such as those described in §1.1.2 for the DAE-index problem, where the existence of nonvanishing numbers is represented by a set of arcs in a certain graph.

Closer investigation would reveal, however, that two different kinds can be distinguished among the nonvanishing numbers; that is, some of the nonvanishing numbers are accurate, and others are inaccurate but independent as a consequence of the fact that they are contaminated by random noises and errors. The purpose of this section[1] is to explain this statement by means of examples and to introduce the class of mixed matrices as a mathematical tool for handling those *two kinds of numbers*.

The distinction between accurate and inaccurate numbers, however, is not a matter in mathematics but in mathematical modeling, i.e., the way in

[1] This section deals with the same issue as previewed in §1.2.1, in more detail with different examples. Knowledge from §1.2.1 is not presupposed here.

which we recognize the problem, and therefore it is impossible in principle to give a mathematical definition to it. The following typical examples will help clarify what is meant by accurate and inaccurate numbers, and how numbers of different nature arise in mathematical descriptions of real systems.

Example 3.1.1. Consider a simple electrical network in Fig. 3.1 (taken from Iri [128]), which consists of five resistors of resistances r_i (branch i) ($i = 1, \cdots, 5$) and a voltage source of voltage e (branch 6). Then the current ξ^i in and the voltage η_i across branch i ($i = 1, \cdots, 6$) in the directions indicated in Fig. 3.1 are to satisfy the structural equations (Kirchhoff's laws) and the constitutive equations (Ohm's law), which altogether are expressed as

$$\left[\begin{array}{cccccc|cccccc} 1 & 0 & 0 & -1 & 0 & -1 & & & & & & \\ 0 & 1 & 0 & -1 & -1 & -1 & & & & & & \\ 0 & 0 & 1 & -1 & -1 & 0 & & & & & & \\ & & & & & & 1 & 1 & 1 & 1 & 0 & 0 \\ & & & & & & 0 & 1 & 1 & 0 & 1 & 0 \\ & & & & & & 1 & 1 & 0 & 0 & 0 & 1 \\ \hline r_1 & & & & & & -1 & & & & & \\ & r_2 & & & & & & -1 & & & & \\ & & r_3 & & & & & & -1 & & & \\ & & & r_4 & & & & & & -1 & & \\ & & & & r_5 & & & & & & -1 & \\ & & & & & 0 & & & & & & -1 \end{array}\right] \left[\begin{array}{c} \xi^1 \\ \xi^2 \\ \xi^3 \\ \xi^4 \\ \xi^5 \\ \xi^6 \\ \hline \eta_1 \\ \eta_2 \\ \eta_3 \\ \eta_4 \\ \eta_5 \\ \eta_6 \end{array}\right] = \left[\begin{array}{c} 0 \\ 0 \\ 0 \\ 0 \\ 0 \\ 0 \\ \hline 0 \\ 0 \\ 0 \\ 0 \\ 0 \\ e \end{array}\right]. \tag{3.1}$$

The upper six equations of (3.1) are the structural equations,[2] while the remaining six the constitutive equations.

The values of resistances r_i ($i = 1, \cdots, 5$), being subject to various kinds of noises, are expected to be inaccurate, or approximately equal to their nominal values to within an engineering tolerance. The nonvanishing coefficients appearing in the upper half of (3.1), on the other hand, are accurate and exactly equal to 1 or -1, since they stem from the incidence coefficients of the underlying graph.

The unique solvability of this electrical network reduces to the nonsingularity of the coefficient matrix of (3.1). By direct calculation, the determinant of (3.1) turns out to be $r_1 r_2 (r_3 + r_4) + (r_1 + r_2)(r_3 r_4 + r_4 r_5 + r_5 r_3)$, which is expected to be distinct from zero since r_i's ($i = 1, \cdots, 5$) are mutually independent, or uncorrelated, nonvanishing numbers (or, more directly, since $r_i > 0$). □

In general, the system of equations governing an electrical network is expressed in the following form:

[2] These equations express the Kirchhoff's laws with respect to a tree-cotree pair $(\{1, 2, 3\}, \{4, 5, 6\})$.

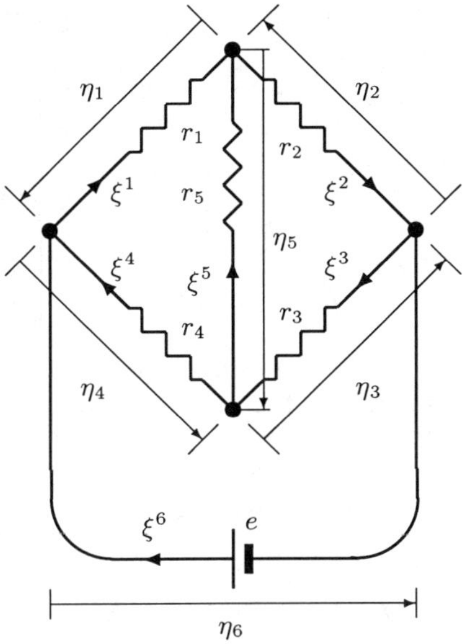

Fig. 3.1. An electrical network of Example 3.1.1

$$\begin{array}{|c|c|}\hline \text{K C L} & O \\ \hline O & \text{K V L} \\ \hline \multicolumn{2}{|c|}{\text{constitutive eqns}} \\ \hline \end{array} \begin{array}{|c|}\hline \boldsymbol{\xi} \\ \hline \boldsymbol{\eta} \\ \hline \end{array} = \begin{array}{|c|}\hline * \\ \hline * \\ \hline * \\ \hline \end{array}, \tag{3.2}$$

where for the submatrices labeled "KCL" and "KVL" the fundamental cutset matrix D and the fundamental circuit matrix R of the underlying graph may be taken (cf. Chen [34], Iri [123, 128], Recski [277]). The nonvanishing entries in "KCL" and "KVL" are accurate, being either 1 or -1, while some of the entries in "constitutive eqns" are inaccurate.

Another simple electrical network, with mutual couplings, is shown below.

Example 3.1.2. Consider the electrical network in Fig. 3.2, which consists of five elements: two resistors of resistances r_i (branch i) ($i = 1, 2$), a voltage source (branch 3) controlled by the voltage across branch 1, a current source (branch 4) controlled by the current in branch 2, and an independent voltage source of voltage e (branch 5). Namely,

$$\eta_1 = r_1 \xi^1, \quad \eta_2 = r_2 \xi^2, \quad \eta_3 = \alpha \eta_1, \quad \xi^4 = \beta \xi^2, \quad \eta_5 = e,$$

where ξ^i and η_i are the current in and the voltage across branch i ($i = 1, \cdots, 5$) in the directions indicated in Fig. 3.2. We then obtain the following system of equations:

$$
\left[\begin{array}{ccccc|ccccc}
0 & 0 & 1 & 1 & 1 & & & & & \\
1 & 0 & 0 & 0 & -1 & & & & & \\
0 & 1 & -1 & 0 & 0 & & & & & \\
 & & & & & 1 & 0 & 0 & -1 & 1 \\
 & & & & & 0 & 1 & 1 & -1 & 0 \\
\hline
r_1 & & & & & -1 & & & & \\
 & r_2 & & & & & -1 & & & \\
 & & 0 & & & \alpha & & -1 & & \\
 & \beta & & -1 & & & & & 0 & \\
 & & & & 0 & & & & & -1
\end{array}\right]
\left[\begin{array}{c}
\xi^1 \\ \xi^2 \\ \xi^3 \\ \xi^4 \\ \xi^5 \\ \hline \eta_1 \\ \eta_2 \\ \eta_3 \\ \eta_4 \\ \eta_5
\end{array}\right]
=
\left[\begin{array}{c}
0 \\ 0 \\ 0 \\ 0 \\ 0 \\ \hline 0 \\ 0 \\ 0 \\ 0 \\ e
\end{array}\right]. \tag{3.3}
$$

In accordance with (3.2), the upper five equations of (3.3) are the structural equations, while the remaining five the constitutive equations.

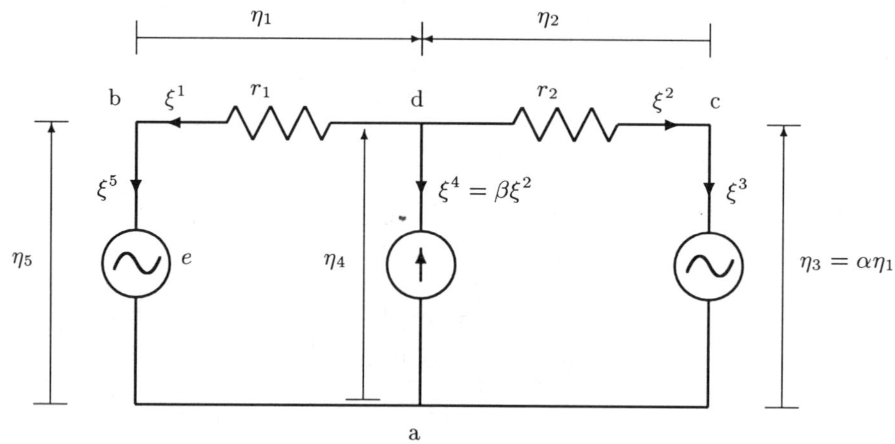

Fig. 3.2. An electrical network of Example 3.1.2

The values of the physical parameters r_1, r_2, α and β are inaccurate numbers which are only approximately equal to their nominal values on account of various kinds of noises and errors.

The unique solvability of this network amounts to the nonsingularity of the coefficient matrix of (3.3). If we calculate its determinant directly, we see it is equal to $-r_2-(1-\alpha)(1+\beta)r_1$, which is highly probably distinct from zero by the independence of the physical parameters $\{r_1, r_2, \alpha, \beta\}$. In this sense, we may say that the electrical network of this example is solvable in general, i.e., solvable generically with respect to the parameter set $\{r_1, r_2, \alpha, \beta\}$. The solvability of this system will be treated in §4.3.3 by a systematic combinatorial method (without a direct computation of the determinant). □

The third example is concerned with a chemical process simulation.

Example 3.1.3 (Ethylene dichloride production system). Consider a hypothetical system (Fig. 3.3) for the production of ethylene dichloride ($C_2H_4Cl_2$), which is slightly modified from an example used in "Users' Manual of Generalized Interrelated Flow Simulation" of "The Service Bureau Co."

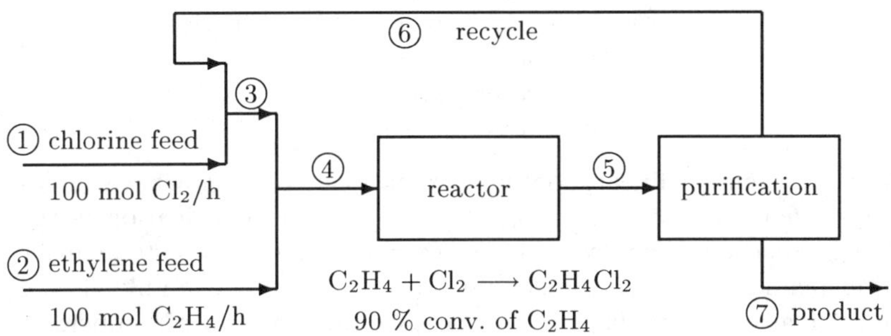

Fig. 3.3. Hypothetical ethylene dichloride production system of Example 3.1.3

Feeds to the system are 100 mol/h of pure chlorine (Cl_2) (stream 1), and 100 mol/h of pure ethylene (C_2H_4) (stream 2). In the reactor, 90% of the input ethylene is converted into ethylene dichloride according to the reaction formula

$$C_2H_4 + Cl_2 \rightarrow C_2H_4Cl_2. \tag{3.4}$$

At the purification stage, the product ethylene dichloride is recovered and the unreacted chlorine and ethylene are separated for recycle. The degree of purification is described in terms of component recovery ratios a_1, a_2 and a_3 of chlorine, ethylene and ethylene dichloride, respectively, which indicate the ratios of the amounts recovered in stream 6 of the respective components over those in stream 5.

We now consider the following problem.

> **[Problem]** Given the component recovery ratios a_1 and a_2 of chlorine and ethylene, determine the recovery ratio $x = a_3$ of ethylene dichloride with which a specified production rate y mol/h of ethylene dichloride is realized.

Let u_{i1}, u_{i2} and u_{i3} mol/h be the component flow rates of chlorine, ethylene and ethylene dichloride in stream i, respectively. The system of equations to be solved may be put in the following form, where u is an auxiliary variable in the reactor and r (= 0.90) is the conversion ratio of ethylene:

$$
\begin{array}{lll}
\text{str3=str1+str6:} & u_{31} = u_{61} + 100, & \\
 & u_{3j} = u_{6j} & (j = 2, 3); \\
\text{str4=str2+str3:} & u_{42} = u_{32} + 100, & \\
 & u_{4j} = u_{3j} & (j = 1, 3); \\
\text{reactor:} & u = r\ u_{42}, & \\
 & u_{5j} = u_{4j} - u & (j = 1, 2), \\
 & u_{53} = u_{43} + u, & \\
\text{purification:} & u_{6j} = a_j\ u_{5j} & (j = 1, 2), \\
 & u_{63} = x\ u_{53}, & \\
 & u_{7j} = u_{5j} - u_{6j} & (j = 1, 2), \\
 & y = u_{53} - u_{63}. &
\end{array}
\tag{3.5}
$$

This is a system of linear/nonlinear equations in unknown variables x, u and u_{ij}, where the equation "$u_{63} = x\ u_{53}$" in the purification is the only nonlinear equation. We may regard a_j $(j = 1, 2)$ and r $(= 0.90)$ as inaccurate and independent numbers. It should be noted in this example that, in the chemical reaction formula of (3.4), we encounter accurate numbers, ± 1, as the integer coefficients in the reaction formula, which are sometimes called the "stoichiometric coefficients." The Jacobian matrix J of (3.5) is shown in Fig. 3.4, and the solvability of (3.5) will be discussed in §4.3.3. □

	x	u_{31}	u_{32}	u_{33}	u_{41}	u_{42}	u_{43}	u_{51}	u_{52}	u_{53}	u_{61}	u_{62}	u_{63}	u_{71}	u_{72}	u
y										1			−1			
u_{31}		−1									1					
u_{32}			−1									1				
u_{33}				−1									1			
u_{41}		1			−1											
u_{42}			1			−1										
u_{43}				1			−1									
u_{51}					1			−1								−1
u_{52}						1			−1							−1
u_{53}							1			−1						1
u_{61}								a_1			−1					
u_{62}									a_2			−1				
u_{63}	u_{53}									x			−1			
u_{71}								1			−1			−1		
u_{72}									1			−1			−1	
u						r										−1

Fig. 3.4. Jacobian matrix of (3.5) (chemical process simulation in Example 3.1.3)

As illustrated by the examples above, the accurate numbers often appear in equations for conservation laws such as Kirchhoff's laws, the law of conservation of mass, energy or momentum, and the principle of action and reaction, where the nonvanishing coefficients are either 1 or −1, representing the underlying topological incidence relations. Another typical example is the integer coefficients, i.e., the *stoichiometric coefficients*, in chemical reactions.

If we consider a *gyrator* in electrical networks, which has the element characteristic represented by

$$\begin{pmatrix} \eta_1 \\ \eta_2 \end{pmatrix} = \begin{pmatrix} 0 & r_1 \\ r_2 & 0 \end{pmatrix} \begin{pmatrix} \xi^1 \\ \xi^2 \end{pmatrix},$$

the ratio r_1/r_2 is exactly equal to -1. Thus, accurate numbers arise also as ratios of inaccurate numbers, or in other words, as numbers representing mutual dependence among quantities which may be inaccurate by themselves. (Electrical networks containing gyrators are treated in §7.3.5.)

When we deal with dynamical systems, we encounter another example of accurate numbers which represent the defining relations such as those between velocity v and position x and between current ξ and charge Q:

$$v = 1 \cdot \frac{\mathrm{d}x}{\mathrm{d}t}, \qquad \xi = 1 \cdot \frac{\mathrm{d}Q}{\mathrm{d}t}.$$

Typical accurate numbers have been illustrated in Fig. 1.4.

To sum up, we can distinguish between accurate numbers and inaccurate numbers. We may alternatively refer to the numbers of the first kind as "fixed constants" and to those of the second kind as "system parameters." For easy reference we reiterate this distinction below:

Accurate numbers (fixed constants): Numbers accounting for various sorts of conservation laws such as Kirchhoff's laws which, stemming from topological incidence relations, are precise in value (often ± 1).

Inaccurate numbers (system parameters): Numbers representing independent physical parameters such as resistances in electrical networks and masses in mechanical systems which, being contaminated with noise and other errors, take values independent of one another.

In the above, we have explained informally what we mean by "two kinds of numbers." We now formulate this intuitive concept in more mathematical terms referring to a pair of nested fields.

Let us denote by $\mathcal{D}$ the (multi)set of finitely many numbers characterizing a system in question. Typically, for a linear system, the set of entries of the coefficient matrix may be taken for $\mathcal{D}$. As the basic assumption we postulate that the numbers in $\mathcal{D}$ are contained in a field $\boldsymbol{F}$, i.e.,

$$\text{Basic Assumption:} \quad \mathcal{D} \subseteq \boldsymbol{F}, \tag{3.6}$$

where it is assumed that $\boldsymbol{F}$ contains $\mathbf{Q}$ (the field of rational numbers).

In addition to the field $\boldsymbol{F}$ we consider a subfield $\boldsymbol{K}$ of $\boldsymbol{F}$:

$$\mathbf{Q} \subseteq \boldsymbol{K} \subseteq \boldsymbol{F} \tag{3.7}$$

with the intention that accurate numbers should belong to $\boldsymbol{K}$ and inaccurate ones to $\boldsymbol{F} \setminus \boldsymbol{K}$. Accordingly, the set $\mathcal{D}$ is divided into two disjoint subsets (multisets) as

$$\mathcal{D} = \mathcal{Q} \cup \mathcal{T} \tag{3.8}$$

with

$$\mathcal{Q} = \mathcal{D} \cap \boldsymbol{K}, \qquad \mathcal{T} = \mathcal{D} \setminus \boldsymbol{K}. \tag{3.9}$$

Our physical intuition that inaccurate numbers are independent of one another can be translated into a mathematical statement:

Generality Assumption: $\mathcal{T}$ is algebraically independent over $\boldsymbol{K}$. (3.10)

Assuming the algebraic independence of $\mathcal{T}$ is equivalent to regarding the members of $\mathcal{T}$ as independent parameters, and therefore to considering the family of systems parametrized by those parameters in $\mathcal{T}$.

We have so far assumed that the subfield $\boldsymbol{K}$ was given a priori. In practical situations, however, the choice of $\boldsymbol{K}$ is in some sense at our disposal and the statement (3.10) is adopted as a mathematical assumption in system modeling. That is, how to choose the subfield $\boldsymbol{K}$ in a real problem is not a matter of mathematics but is determined by how we model that problem. In contrast, the underlying field $\boldsymbol{F}$ is just a mathematical formality and it may be chosen to be sufficiently large.

For instance, in Example 3.1.1 above we may choose $\boldsymbol{K} = \mathbf{Q}$, $\boldsymbol{F} = \mathbf{Q}(r_1, r_2, r_3, r_4, r_5)$ and assume that $\mathcal{T} = \{r_1, r_2, r_3, r_4, r_5\}$ satisfies (3.10). In Example 3.1.2, we may take $\boldsymbol{K} = \mathbf{Q}$, $\boldsymbol{F} = \mathbf{Q}(g_1, g_2, \alpha, \beta)$ and $\mathcal{T} = \{g_1, g_2, \alpha, \beta\}$. A reasonable choice in Example 3.1.3 would be $\boldsymbol{K} = \mathbf{Q}$, $\boldsymbol{F} = \mathbf{Q}(a_1, a_2, r, x, u_{53})$ and $\mathcal{T} = \{a_1, a_2, r, x, u_{53}\}$.

Here are three generality assumptions that may possibly be adopted in system modeling. The first is

GA1: The nonvanishing elements of $\mathcal{D}$ are algebraically independent over $\mathbf{Q}$.

This requires that (3.10) should hold for $\boldsymbol{K} = \mathbf{Q}$ and $\mathcal{T} = \mathcal{D} \setminus \{0\}$. The generality assumption GA1 seems to be too stringent to be literally satisfied in practical situations, but is convenient for graph-theoretic methods for structural analysis such as those described in §1.1.2 and §4.3.2. The second is

GA2: Those elements of $\mathcal{D}$ which do not belong to the rational number field $\mathbf{Q}$ are algebraically independent over $\mathbf{Q}$.

This requires that (3.10) should hold for $\boldsymbol{K} = \mathbf{Q}$ and $\mathcal{T} = \mathcal{D} \setminus \mathbf{Q}$. The generality assumption GA2 is appropriate in many cases, including Examples 3.1.1 to 3.1.3 above, and will be adopted mostly in this book. The third is

GA3: Those elements of $\mathcal{D}$ which do not belong to the real number field $\mathbf{R}$ are algebraically independent over $\mathbf{R}$.

This requires that (3.10) should hold for $\boldsymbol{K} = \mathbf{R}$ and $\mathcal{T} = \mathcal{D} \setminus \mathbf{R}$. The generality assumption GA3 will be useful in dealing with a system of linear/nonlinear equations, where $\mathcal{D}$ denotes the set of partial derivatives (entries of the Jacobian matrix). Taking notice of the fact that the derivatives of linear functions are (real) constants, we classify the partial derivatives $\mathcal{D}$ into constants and nonconstants; the latter standing for nonlinearity. This classification conforms to the above choice of $\boldsymbol{K} = \mathbf{R}$. See also Remark 1.3.1.

It is important to recognize here that a generality assumption is concerned with the property of a mathematical description of a real system, and not of the system itself. The assumption GA2, for example, is often justified when the system in question is described by a collection of elementary relations among elementary variables rather than by a compact sophisticated representation. In Example 3.1.3, for instance, the auxiliary variable u in the reactor of (3.5) could have been eliminated, the reactor being then described more compactly by

$$u_{5j} = u_{4j} - ru_{42} \quad (j = 1, 2), \qquad u_{53} = u_{43} + ru_{42}.$$

If the system were so described, the assumption GA2 is no longer valid even if we may assume that r is independent of other quantities. In fact, the three occurrences of one and the same r themselves could never be independent of each other. The issue of mathematical description against generality assumption will be considered again for dynamical systems in §3.1.2.

Remark 3.1.4. In the above argument we have assumed that the subfield $\boldsymbol{K}$ is chosen from physical considerations in mathematical modeling. From the mathematical point of view, however, we may think of the following problem: Given $\mathcal{D} \subseteq \boldsymbol{F}$, find a subfield $\boldsymbol{K}$ and a bipartition $\mathcal{D} = \mathcal{Q} \cup \mathcal{T}$ such that $\mathcal{Q} \subseteq \boldsymbol{K}$ and $\mathcal{T}$ is algebraically independent over $\boldsymbol{K}$. It is not difficult to see that there exist a largest subset $\mathcal{T}$ and a smallest subfield $\boldsymbol{K}$ that satisfy these conditions, and they are given by

$$\mathcal{T} = \{t \in \mathcal{D} \mid t \text{ is transcendental over } \mathbf{Q}(\mathcal{D} \setminus \{t\})\}, \tag{3.11}$$
$$\boldsymbol{K} = \mathbf{Q}(\mathcal{D} \setminus \mathcal{T}). \tag{3.12}$$

The expressions (3.11) and (3.12) can be derived from a matroid-theoretic consideration as follows (see §2.3.2 for matroid-theoretic terms). Let $\mathbf{M}$ be the algebraic matroid (see Example 2.3.10) defined on $\mathcal{D}$ with respect to algebraic independence over $\mathbf{Q}$. For a given $\mathcal{T}$, (3.12) is an obvious choice of the smallest $\boldsymbol{K}$ to meet the condition that $\mathcal{Q} = \mathcal{D} \setminus \mathcal{T} \subseteq \boldsymbol{K}$. Then the condition (3.10) is equivalent to the statement that $\mathcal{T}$ is independent in the contraction of $\mathbf{M}$ to $\mathcal{T}$. This statement is tantamount to saying that $\mathcal{T}$ consists of coloops of $\mathbf{M}$. It follows, therefore, that the largest $\mathcal{T}$ is given by (3.11). □

Remark 3.1.5. Some comments would be in order here on the mutual relations among the generality assumptions GA1, GA2 and GA3 above. First of all, GA2 is weaker than GA1; that is, if $\mathcal{D}$ satisfies GA1, it satisfies GA2, too. No other implications exist, as exemplified below, where $\boldsymbol{F} = \mathbf{R}(x, \mathrm{e}^x)$ and $\mathcal{T}$ of (3.11) is also given. Note that the algebraic independence of $\{\mathrm{e}^x, x, \mathrm{e}^{\sqrt{2}}, \mathrm{e}^{\sqrt{3}}\}$ over $\mathbf{Q}$ follows from Theorem 3.1.6 below.

$\mathcal{D}$	GA1	GA2	GA3	$\mathcal{T}$
$\{\mathrm{e}^x, x, \mathrm{e}^{\sqrt{2}}, \mathrm{e}^{\sqrt{3}}\}$	yes	yes	yes	$\{\mathrm{e}^x, x, \mathrm{e}^{\sqrt{2}}, \mathrm{e}^{\sqrt{3}}\}$
$\{\mathrm{e}^x, x, \mathrm{e}^{\sqrt{2}}x, \mathrm{e}^{\sqrt{3}}\}$	yes	yes	no	$\{\mathrm{e}^x, x, \mathrm{e}^{\sqrt{2}}x, \mathrm{e}^{\sqrt{3}}\}$
$\{\mathrm{e}^x, x, \pi, 1\}$	no	yes	yes	$\{\mathrm{e}^x, x, \pi\}$
$\{x, \pi, \pi^2, \sqrt{2}\}$	no	no	yes	$\{x\}$
$\{\mathrm{e}^x, x, \pi x, 1\}$	no	yes	no	$\{\mathrm{e}^x, x, \pi x\}$
$\{\mathrm{e}^x, x, \pi x, \pi\}$	no	no	no	$\{\mathrm{e}^x\}$

This example is given for mathematical completeness, and not for physical significance. □

Theorem 3.1.6 (Lindemann–Weierstrass theorem). *Let $y_1, \cdots, y_q$ be algebraic numbers over* $\mathbf{Q}$ *that are linearly independent over* $\mathbf{Q}$. *Then the set* $\{\exp y_1, \cdots, \exp y_q\}$ *is algebraically independent over* $\mathbf{Q}$.

Proof. See, e.g., Jacobson [148, 149]. ■

3.1.2 Mixed Matrix and Mixed Polynomial Matrix

The distinction of two kinds of numbers can be embodied in the concept of mixed matrices. It is generalized to another concept of mixed polynomial matrices to deal with dynamical systems.

Consider a matrix A over a field $\boldsymbol{F}$ and denote by $\mathcal{D}$ the set of its entries. With reference to a subfield $\boldsymbol{K}$ of $\boldsymbol{F}$, the set $\mathcal{D}$ is divided into two parts, $\mathcal{D} = \mathcal{Q} \cup \mathcal{T}$ by (3.9), and accordingly, the matrix A is expressed as $A = Q + T$, where $\mathcal{T}$ is the set of the nonzero entries of T. The generality assumption (3.10) then amounts to an assumption of the algebraic independence over $\boldsymbol{K}$ of the nonzero entries of T. This leads to the following formal definition.

Let $\boldsymbol{K}$ be a subfield of a field $\boldsymbol{F}$. An $m \times n$ matrix A over $\boldsymbol{F}$ (i.e., $A_{ij} \in \boldsymbol{F}$) is called a *mixed matrix* with respect to $(\boldsymbol{K}, \boldsymbol{F})$ if

$$A = Q + T, \tag{3.13}$$

where

(M-Q) Q is an $m \times n$ matrix over $\boldsymbol{K}$ (i.e., $Q_{ij} \in \boldsymbol{K}$), and
(M-T) T is an $m \times n$ matrix over $\boldsymbol{F}$ (i.e., $T_{ij} \in \boldsymbol{F}$) such that the set $\mathcal{T}$ of its nonzero entries is algebraically independent over $\boldsymbol{K}$.

We usually assume

$$T_{ij} \neq 0 \quad \Rightarrow \quad Q_{ij} = 0$$

to make the decomposition (3.13) unique. The class of $m \times n$ mixed matrices with respect to $(\boldsymbol{K}, \boldsymbol{F})$ is denoted as $\mathrm{MM}(\boldsymbol{K}, \boldsymbol{F}; m, n)$ (or simply as $\mathrm{MM}(\boldsymbol{K}, \boldsymbol{F})$) and the subfield $\boldsymbol{K}$ will be called the *ground field*.

Mixed matrices are useful also in dealing with linear time-invariant dynamical systems. In this case, we encounter a field composed of, say, the

Laplace transforms, or a field consisting of operators such as Heaviside's and Mikusiński's.

Specifically, suppose that a dynamical system is written in the *descriptor form* (Katayama [155], Luenberger [182, 183]):

$$F \frac{\mathrm{d}\boldsymbol{x}}{\mathrm{d}t} = A\, \boldsymbol{x} + B\, \boldsymbol{u}, \tag{3.14}$$

where $\boldsymbol{x}$ is an n-dimensional vector called the descriptor-vector, $\boldsymbol{u}$ is an m-dimensional input-vector, and F, A and B are $n \times n$, $n \times n$, and $n \times m$ matrices, respectively. The Laplace transform of the equation (3.14) gives a frequency domain description:

$$s\, F\, \boldsymbol{x} = A\, \boldsymbol{x} + B\, \boldsymbol{u}, \qquad \text{or} \qquad \left[A - sF | B \right] \begin{bmatrix} \boldsymbol{x} \\ \boldsymbol{u} \end{bmatrix} = \mathbf{0}, \tag{3.15}$$

where $\boldsymbol{x}(0) = \mathbf{0}$, $\boldsymbol{u}(0) = \mathbf{0}$ is assumed (see Remark 1.1.1 for the Laplace transform).

Suppose further that the generality assumption GA2 is acceptable. Then the matrices F, A and B are mixed matrices with ground field $\mathbf{Q}$:

$$F = Q_F + T_F, \quad A = Q_A + T_A, \quad B = Q_B + T_B$$

such that the set of the nonvanishing entries of $[T_F \mid T_A \mid T_B]$ is algebraically independent over $\mathbf{Q}$.

The coefficient matrix $D(s) = [A - sF \mid B]$ in the frequency domain description is a polynomial matrix (a matrix pencil) with the expression

$$D(s) = D_0 + sD_1,$$

where the coefficient matrices, D_0 and D_1, are mixed matrices expressed as

$$\begin{aligned} D_0 &= [A \mid B] = [Q_A \mid Q_B] + [T_A \mid T_B], \\ D_1 &= [-F \mid O] = [-Q_F \mid O] + [-T_F \mid O]. \end{aligned}$$

Such matrix as $D(s)$ is called a mixed polynomial matrix (a formal definition is given later).

The matrix $D(s) = [A - sF \mid B]$ is also a mixed matrix with ground field $\boldsymbol{K} = \mathbf{Q}(s)$, since the expression

$$D(s) = Q(s) + T(s) \tag{3.16}$$

with

$$Q(s) = [Q_A - sQ_F \mid Q_B], \qquad T(s) = [T_A - sT_F \mid T_B]$$

satisfies the conditions (M-Q) and (M-T), in spite of the occurrences of the symbol s in both of the matrices $Q(s)$ and $T(s)$.

Example 3.1.7. Consider the mechanical system[3] in Fig. 3.5 consisting of two masses m_1 and m_2, two springs k_1 and k_2, and a damper $f; u$ is the force exerted from outside. We may choose $\boldsymbol{x} = (x_1, x_2, x_3, x_4, x_5, x_6)$ as the descriptor-vector, where x_1 (resp. x_2) is the vertical displacement (downwards) of mass m_1 (resp. m_2), x_3 (resp. x_4) is its velocity, x_5 is the force by the damper f, and x_6 is the relative velocity of the two masses.

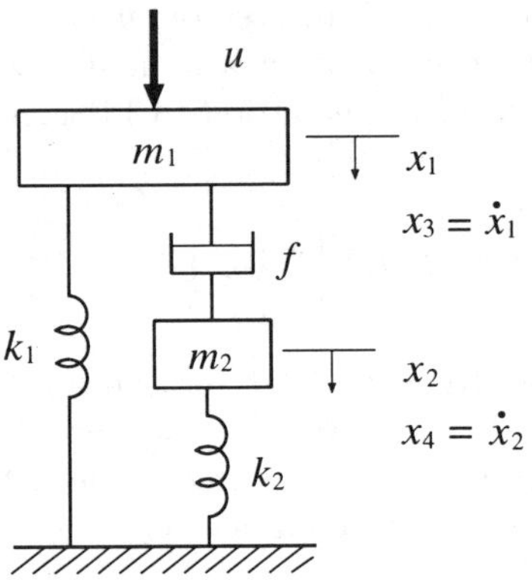

Fig. 3.5. A mechanical system of Example 3.1.7

Then the system can be expressed in the descriptor form (3.14) with

$$F = \begin{bmatrix} 1 & 0 & 0 & 0 & 0 & 0 \\ 0 & 1 & 0 & 0 & 0 & 0 \\ 0 & 0 & m_1 & 0 & 0 & 0 \\ 0 & 0 & 0 & m_2 & 0 & 0 \\ 0 & 0 & 0 & 0 & 0 & 0 \\ 1 & -1 & 0 & 0 & 0 & 0 \end{bmatrix}, \quad A = \begin{bmatrix} 0 & 0 & 1 & 0 & 0 & 0 \\ 0 & 0 & 0 & 1 & 0 & 0 \\ -k_1 & 0 & 0 & 0 & -1 & 0 \\ 0 & -k_2 & 0 & 0 & 1 & 0 \\ 0 & 0 & 0 & 0 & -1 & f \\ 0 & 0 & 0 & 0 & 0 & 1 \end{bmatrix}, \quad B = \begin{bmatrix} 0 \\ 0 \\ 1 \\ 0 \\ 0 \\ 0 \end{bmatrix}. \tag{3.17}$$

The matrix $D(s) = [A - sF \mid B]$ is given as

$$D(s) = \begin{array}{|cccccc|c|} \multicolumn{1}{c}{x_1} & x_2 & x_3 & x_4 & x_5 & \multicolumn{1}{c}{x_6} & \multicolumn{1}{c}{u} \\ \hline -s & 0 & 1 & 0 & 0 & 0 & 0 \\ 0 & -s & 0 & 1 & 0 & 0 & 0 \\ -k_1 & 0 & -sm_1 & 0 & -1 & 0 & 1 \\ 0 & -k_2 & 0 & -sm_2 & 1 & 0 & 0 \\ 0 & 0 & 0 & 0 & -1 & f & 0 \\ -s & s & 0 & 0 & 0 & 1 & 0 \\ \hline \end{array}.$$

If we regard $\{m_1, m_2, k_1, k_2, f\}$ as independent free parameters, i.e., as being algebraically independent, the additive decomposition $D(s) = Q(s) + T(s)$ in (3.16) is given by

[3] This mechanical system has been considered in §1.2.2.

$$
Q(s) = \begin{array}{c} \begin{array}{ccccccc} x_1 & x_2 & x_3 & x_4 & x_5 & x_6 & u \end{array} \\ \left[\begin{array}{cccccc|c} -s & 0 & 1 & 0 & 0 & 0 & 0 \\ 0 & -s & 0 & 1 & 0 & 0 & 0 \\ 0 & 0 & 0 & 0 & -1 & 0 & 1 \\ 0 & 0 & 0 & 0 & 1 & 0 & 0 \\ 0 & 0 & 0 & 0 & -1 & 0 & 0 \\ -s & s & 0 & 0 & 0 & 1 & 0 \end{array}\right] \end{array}, \tag{3.18}
$$

$$
T(s) = \begin{array}{c} \begin{array}{ccccccc} x_1 & x_2 & x_3 & x_4 & x_5 & x_6 & u \end{array} \\ \left[\begin{array}{cccccc|c} 0 & 0 & 0 & 0 & 0 & 0 & 0 \\ 0 & 0 & 0 & 0 & 0 & 0 & 0 \\ -k_1 & 0 & -sm_1 & 0 & 0 & 0 & 0 \\ 0 & -k_2 & 0 & -sm_2 & 0 & 0 & 0 \\ 0 & 0 & 0 & 0 & 0 & f & 0 \\ 0 & 0 & 0 & 0 & 0 & 0 & 0 \end{array}\right] \end{array}. \tag{3.19}
$$

In this way the mechanical system can be described by means of a mixed polynomial matrix with an appropriate choice of variables and equations.

The mechanical system could be described more compactly in the standard form $(\hat{A} - sI_4)\hat{\boldsymbol{x}} + \hat{B}\boldsymbol{u} = \mathbf{0}$ in terms of a four-dimensional state-vector $\hat{\boldsymbol{x}} = (x_1, x_2, x_3, x_4)$ and the input-vector $\boldsymbol{u} = (u)$. In such a compact representation, however, the generality assumptions will not be acceptable. In fact, the coefficient matrix

$$
[\hat{A} - sI_4 \mid \hat{B}] = \begin{array}{c} \begin{array}{ccccc} x_1 & x_2 & x_3 & x_4 & u \end{array} \\ \left[\begin{array}{cccc|c} -s & 0 & 1 & 0 & 0 \\ 0 & -s & 0 & 1 & 0 \\ -k_1/m_1 & 0 & -f/m_1 - s & f/m_1 & 1/m_1 \\ 0 & -k_2/m_2 & f/m_2 & -f/m_2 - s & 0 \end{array}\right] \end{array}
$$

may not be regarded as a mixed matrix.

This mechanical system will be considered again in Example 3.2.2. □

The above argument can be extended to a linear time-invariant dynamical system described by a polynomial matrix

$$
A(s) = \sum_{k=0}^{N} s^k A_k .
$$

The variable s here is primarily intended to mean the variable for the Laplace transform for continuous-time systems, though it could be interpreted as the variable for the z-transform for discrete-time systems (see Chen [33], Zadeh–Desoer [350] for the z-transform). In such reprepresentation, it is often justified to assume that the coefficient matrices A_k $(k = 0, 1, \cdots, N)$ are expressed in the form

$$
A_k = Q_k + T_k \qquad (k = 0, 1, \cdots, N)
$$

such that

(MP-Q1) Q_k $(k = 0, 1, \cdots, N)$ are matrices over $\boldsymbol{K}$, and
(MP-T) T_k $(k = 0, 1, \cdots, N)$ are matrices over $\boldsymbol{F}$ such that the set $\mathcal{T}$ of their nonzero entries is algebraically independent over $\boldsymbol{K}$.

Then $A(s)$ is split accordingly into two parts:

$$A(s) = Q(s) + T(s) \tag{3.20}$$

with

$$Q(s) = \sum_{k=0}^{N} s^k Q_k, \qquad T(s) = \sum_{k=0}^{N} s^k T_k.$$

Such a matrix $A(s)$ will be called a *mixed polynomial matrix* with respect to $(\boldsymbol{K}, \boldsymbol{F})$. Obviously, each A_k is a mixed matrix with respect to $(\boldsymbol{K}, \boldsymbol{F})$. Also note that $A(s)$ is a mixed matrix with respect to $(\boldsymbol{K}(s), \boldsymbol{F}(s))$ in spite of the occurrences of the symbol s in both of the matrices $Q(s)$ and $T(s)$.

In §3.2 we will discuss more on the mixed polynomial matrices from the viewpoint of the dimensional analysis to arrive at the concept of physical matrices.

Notes. This section is based on Murota–Iri [237, 238] as well as Murota [204].

3.2 Algebraic Implication of Dimensional Consistency

3.2.1 Introductory Comments

The concept of physical dimensions would be counted among the most fundamental in recognizing the nature of physical quantities. The principle of dimensional homogeneity claims that any equation describing a physical phenomenon must be consistent with respect to physical dimensions. This principle constitutes the basis of *dimensional analysis*, which has long been known to scientists and engineers, and has proved to be fruitful in various fields (de Jong [46], Huntley [116], Langhaar [169], Schouten [289, Chap. VI]). It is important to notice that we cannot talk of dimensional homogeneity before we recognize the difference in the nature of quantities from the viewpoint of physical dimensions.

Suppose a physical system is described by a system of equations, which may in turn be expressed by a matrix when linearized if necessary. With each entry of the matrix is associated a physical dimension in a physically consistent manner.

It is pointed out in the present section that, by virtue of the principle of dimensional homogeneity, the physically-dimensioned coefficient matrix describing a physical system enjoys a kind of total unimodularity in a certain ring defined appropriately with reference to physical dimensions. Several implications of this fact are discussed in §3.3 in connection to the mathematical framework for the structural analysis introduced in §3.1. To reflect the dual viewpoint from structural analysis and dimensional analysis, the notion of "physical matrix" is introduced as a mathematical model of a typical matrix that we encounter in real physical systems. The concept of physical matrix plays a central role, especially in the structural analysis of dynamical systems, to be treated in Chap. 6.

3.2.2 Dimensioned Matrix

A physical system is usually described by a set of relations among relevant physical quantities, to each of which is assigned a physical dimension. When a set of *fundamental dimensions*, or equivalently, a set of *fundamental quantities*, is chosen, the dimensions of the remaining physical quantities can be uniquely expressed by the so-called *dimensional formulas*. For example, a standard choice of fundamental quantities in mechanics consists of length L, mass M and time T, and the dimensional formula for force is then given by $[LMT^{-2}] = [L][M][T]^{-2}$ or simply by LMT^{-2}. In general, the exponents to the fundamental dimensions, namely the powers in the dimensional formula, may take on not only integers but also rational numbers.

Here we do not go into philosophical arguments such as those on what the physical dimensions are and which set of physical quantities are most fundamental. Instead we assume that the fundamental quantities with the associated fundamental dimensions are given along with the dimensional formulas for other quantities.

Let us consider a linear (or linearized) system represented by a system of linear equations:

$$A\boldsymbol{x} = \boldsymbol{b}, \tag{3.21}$$

where we assume that the entries of the $m \times n$ matrix $A = (A_{ij})$, as well as the components of $\boldsymbol{x} = (x_j)$ and $\boldsymbol{b} = (b_i)$, belong to some field $\boldsymbol{F}$, namely,

$$A_{ij}, x_j, b_i \in \boldsymbol{F} \quad (i = 1, \cdots, m; j = 1, \cdots, n).$$

It is also assumed that $\boldsymbol{F}$ is an extension of the field $\mathbf{Q}$ of rational numbers (i.e., $\boldsymbol{F} \supseteq \mathbf{Q}$).

Let $Z_1, \cdots, Z_d$ be a chosen set of fundamental quantities. Not only the components of $\boldsymbol{x}$ and $\boldsymbol{b}$ but also the entries of A have physical dimensions, expressed in the form of

$$[Z_1]^{p_1} \cdots [Z_d]^{p_d}$$

with exponents $p_k \in \mathbf{Q}$ $(k = 1, \cdots, d)$.

From the algebraic point of view, we may regard $Z_1, \cdots, Z_d$ as indeterminates over $\boldsymbol{F}$ and consider the extension field $\boldsymbol{E}$ of $\boldsymbol{F}$ generated over $\boldsymbol{F}$ by all the formal fractional powers of $Z_1, \cdots, Z_d$; i.e.,

$$\boldsymbol{E} = \boldsymbol{F}(\{Z_1{}^{p_1} \cdots Z_d{}^{p_d} \mid p_k \in \mathbf{Q}, k = 1, \cdots, d\}). \tag{3.22}$$

Accordingly, (3.21) may be replaced by the following system of equations in the extension field $\boldsymbol{E}$:

$$\tilde{A}\tilde{\boldsymbol{x}} = \tilde{\boldsymbol{b}}, \tag{3.23}$$

where

$$\tilde{A}_{ij} = A_{ij} \prod_{k=1}^{d} Z_k{}^{p_{ijk}}, \quad \tilde{x}_j = x_j \prod_{k=1}^{d} Z_k{}^{c_{jk}}, \quad \tilde{b}_i = b_i \prod_{k=1}^{d} Z_k{}^{r_{ik}} \tag{3.24}$$

with the exponents p_{ijk}, c_{jk}, r_{ik} of rational numbers representing the physical dimensions. The ith equation of (3.23) reads

$$\sum_{j:A_{ij} \neq 0} \left(A_{ij} x_j \cdot \prod_{k=1}^{d} Z_k{}^{p_{ijk}+c_{jk}} \right) = b_i \cdot \prod_{k=1}^{d} Z_k{}^{r_{ik}}. \tag{3.25}$$

The *principle of dimensional homogeneity* means the physical dimensional consistency of the ith equation in the sense that

[Dimension of (i, j) entry] × [Dimension of jth column]
= [Dimension of ith row]

for each (i, j) with $A_{ij} \neq 0$. It follows from (3.25) that the physical dimensional consistency is equivalent to the relations

$$p_{ijk} = r_{ik} - c_{jk} \qquad (i = 1, \cdots, m; j = 1, \cdots, n; k = 1, \cdots, d) \tag{3.26}$$

among the exponents p_{ijk}, c_{jk}, r_{ik}. Based on this observation we will define the notion of dimensioned matrix as follows.

Let $\tilde{A} = (\tilde{A}_{ij})$ be a matrix over $\boldsymbol{E}$ (defined by (3.22)) which is expressed as in (3.24) with exponents $p_{ijk} \in \mathbf{Q}$. We call $\tilde{A}$ a *dimensioned matrix* if (3.26) holds for some suitably chosen r_{ik} and c_{jk} ($\in \mathbf{Q}$). The set of $m \times n$ dimensioned matrices with ground field $\boldsymbol{F}$ and fundamental quantities (i.e., indeterminates) $Z_1, \cdots, Z_d$ will be denoted by $\mathcal{D}(\boldsymbol{F}; m, n; Z_1, \cdots, Z_d)$, or simply by $\mathcal{D}(\boldsymbol{F}; Z_1, \cdots, Z_d)$ if the size is not relevant.

The following proposition is a restatement of the definition, where

$$D_{\mathrm{r}} = \mathrm{diag}\left[\prod_{k=1}^{d} Z_k{}^{r_{1k}}, \cdots, \prod_{k=1}^{d} Z_k{}^{r_{mk}}\right], \tag{3.27}$$

$$D_{\mathrm{c}} = \mathrm{diag}\left[\prod_{k=1}^{d} Z_k{}^{c_{1k}}, \cdots, \prod_{k=1}^{d} Z_k{}^{c_{nk}}\right] \tag{3.28}$$

with $r_{ik} \in \mathbf{Q}$ and $c_{jk} \in \mathbf{Q}$ $(i = 1, \cdots, m; j = 1, \cdots, n; k = 1, \cdots, d)$.

Proposition 3.2.1. *A matrix $\tilde{A}$ over $\boldsymbol{E}$ belongs to $\mathcal{D}(\boldsymbol{F}; m, n; Z_1, \cdots, Z_d)$ if and only if it can be expressed as*

$$\tilde{A} = D_{\mathrm{r}}\ A\ D_{\mathrm{c}}{}^{-1},$$

where A is a matrix over $\boldsymbol{F}$, and D_{r} and D_{c} are nonsingular diagonal matrices of (3.27) *and* (3.28). □

Example 3.2.2. Recall the mechanical system (Fig. 3.5) of Example 3.1.7. As the fundamental quantities in dimensional analysis, we may choose time T, length L and mass M. Then the dimensions of velocity and force are given by $T^{-1}L$ and $T^{-2}LM$, respectively. The physical dimensions associated with the equations, i.e., with the rows of $D(s)$ are

row 1	row 2	row 3	row 4	row 5	row 6
velocity	velocity	force	force	force	velocity
$T^{-1}L$	$T^{-1}L$	$T^{-2}LM$	$T^{-2}LM$	$T^{-2}LM$	$T^{-1}L$

(3.29)

whereas those with the variables (x_i and u), i.e., with the columns of $D(s)$, are

col 1	col 2	col 3	col 4	col 5	col 6	col 7
length	length	velocity	velocity	force	velocity	force
L	L	$T^{-1}L$	$T^{-1}L$	$T^{-2}LM$	$T^{-1}L$	$T^{-2}LM$

(3.30)

Accordingly, the diagonal matrices D_{r} and D_{c} of (3.27) and (3.28) are given by

$$D_{\mathrm{r}} = \mathrm{diag}\,[T^{-1}L,\ T^{-1}L,\ T^{-2}LM,\ T^{-2}LM,\ T^{-2}LM,\ T^{-1}L],$$
$$D_{\mathrm{c}} = \mathrm{diag}\,[L,\ L,\ T^{-1}L,\ T^{-1}L,\ T^{-2}LM,\ T^{-1}L,\ T^{-2}LM],$$

where $T = Z_1$ =time, $L = Z_2$ =length and $M = Z_3$ =mass.

In this example, any minor of $Q(s) = [Q_A - sQ_F \mid Q_B]$ of (3.18) can easily be verified to be a monomial in s over $\mathbf{Q}$. In fact, this is a general phenomenon, to be discussed in §3.3; see Example 3.3.1. □

3.2.3 Total Unimodularity of a Dimensioned Matrix

Let R be an integral domain, i.e., a commutative ring without zero divisors and with a unit element. An element of R is called *invertible* if there exists another element of R such that their product equals the unit element. A matrix over R is said to be *totally unimodular* (over R) if every nonvanishing minor (=subdeterminant) is an invertible element of R. We will denote by $\mathcal{U}(R; m, n)$ the set of $m \times n$ totally unimodular matrices over R, or simply by $\mathcal{U}(R)$ if the size is not relevant. The significance of this concept[4] lies in

[4] In the canonical case of R being the ring of integers, the total unimodularity of incidence matrices of graphs is known to play substantial roles in combinatorial optimization (Lawler [171], Schrijver [292]).

the fact that, if a matrix is totally unimodular over R, not only its inverse but also all its pivotal transforms are matrices over R. The objective of this section is to point out that the dimensioned matrices can be characterized as totally unimodular matrices over a certain ring.

Consider a ring generated over $\boldsymbol{F}$ by all the formal fractional powers of $Z_1, \cdots, Z_d$:

$$\boldsymbol{F}\langle Z_1, \cdots, Z_d \rangle = \boldsymbol{F}[\{{Z_1}^{p_1} \cdots {Z_d}^{p_d} \mid p_k \in \mathbf{Q},\ k = 1, \cdots, d\}]. \qquad (3.31)$$

Obviously, $\boldsymbol{F}\langle Z_1, \cdots, Z_d \rangle$ is an integral domain whose quotient field is the field $\boldsymbol{E}$ defined in (3.22). An element of $\boldsymbol{F}\langle Z_1, \cdots, Z_d \rangle$ is invertible if and only if it is of the form:

$$\alpha \prod_{k=1}^{d} {Z_k}^{p_k} \qquad (\alpha \in \boldsymbol{F} \setminus \{0\},\ p_k \in \mathbf{Q} \text{ for } k = 1, \cdots, d).$$

As an immediate consequence of the definition, a dimensioned matrix is *totally unimodular* over $\boldsymbol{F}\langle Z_1, \cdots, Z_d \rangle$.

Proposition 3.2.3. $\mathcal{D}(\boldsymbol{F}; Z_1, \cdots, Z_d) \subseteq \mathcal{U}(\boldsymbol{F}\langle Z_1, \cdots, Z_d \rangle)$.
That is, for $\tilde{A} \in \mathcal{D}(\boldsymbol{F}; Z_1, \cdots, Z_d)$ *it holds that, for any* (I, J),

$$\det \tilde{A}[I, J] = \alpha \prod_{k=1}^{d} {Z_k}^{p_k}$$

for some $\alpha \in \boldsymbol{F}$ *and* $p_k \in \mathbf{Q}$ $(k = 1, \cdots, d)$.

Proof. The expression (3.24) of $\tilde{A}$ with $p_{ijk} = r_{ik} - c_{jk}$ implies

$$\det \tilde{A}[I, J] = \det A[I, J] \cdot \prod_{k=1}^{d} {Z_k}^{p_k}$$

with $p_k = \sum_{i \in I} r_{ik} - \sum_{j \in J} c_{jk} \in \mathbf{Q}$. ■

Moreover, these two classes of matrices coincide with each other, as stated in Theorem 3.2.4 below. This theorem, coupled with Proposition 3.2.1, provides us with a concrete representation of a totally unimodular matrix over $\boldsymbol{F}\langle Z_1, \cdots, Z_d \rangle$.

Theorem 3.2.4. $\mathcal{D}(\boldsymbol{F}; Z_1, \cdots, Z_d) = \mathcal{U}(\boldsymbol{F}\langle Z_1, \cdots, Z_d \rangle)$.

Proof. By Proposition 3.2.3, it suffices to show that every totally unimodular matrix over $\boldsymbol{F}\langle Z_1, \cdots, Z_d \rangle$ is a dimensioned matrix. Furthermore, the proof can be reduced to the case of $d = 1$ by induction on d, and the present theorem follows from Proposition 3.2.5 below. ■

Proposition 3.2.5. *Let Z be an indeterminate over $\boldsymbol{F}$, and let $\tilde{A} = (\tilde{A}_{ij})$ be an $m \times n$ totally unimodular matrix over $\boldsymbol{F}\langle Z \rangle$, i.e., $\tilde{A} \in \mathcal{U}(\boldsymbol{F}\langle Z \rangle; m, n)$. Then there exist $A_{ij} \in \boldsymbol{F}$, $r_i \in \mathbf{Q}$ and $c_j \in \mathbf{Q}$ $(i = 1, \cdots, m; j = 1, \cdots, n)$ such that*

$$\tilde{A}_{ij} = A_{ij} Z^{r_i - c_j}.$$

(A_{ij}'s are uniquely determined as $A_{ij} = \tilde{A}_{ij}|_{Z=1}$, while r_i's and c_j's are not.)

Proof. By definition, $\tilde{A}_{ij}$ can be expressed as

$$\tilde{A}_{ij} = A_{ij} Z^{p_{ij}} \qquad (A_{ij} \in \boldsymbol{F}, p_{ij} \in \mathbf{Q}).$$

Consider a bipartite (directed) graph $G = (V^+, V^-; E)$ associated with $\tilde{A}$, where V^+ corresponds to the row set of $\tilde{A}$ and V^- to the column set, and the arc set E is defined as $E = \{(i,j) \mid \tilde{A}_{ij} \neq 0\}$. By Theorem 2.2.35(2), p_{ij} can be expressed as $p_{ij} = r_i - c_j$ for some suitable r_i $(i = 1, \cdots, m)$ and c_j $(j = 1, \cdots, n)$ if the algebraic sum of p_{ij}'s along any circuit in G is equal to zero.

Suppose, to the contrary, that there exists a circuit in G along which the algebraic sum (cf. (2.59)) of p_{ij}'s is distinct from zero. Let C_0 be such a circuit with the minimal number of arcs, and let $i_1, j_1, i_2, j_2, \cdots, i_s, j_s (= j_0)$ be the sequence of vertices lying on C_0, where $I = \{i_1, \cdots, i_s\} \subseteq V^+$ and $J = \{j_1, \cdots, j_s\} \subseteq V^-$. Then, putting

$$p = \sum_{r=1}^{s} p_{i_r j_r}, \quad q = \sum_{r=1}^{s} p_{i_r j_{r-1}},$$

we have $p \neq q$.

The minimal circuit C_0 has no chord, that is, if $(i_{r'}, j_{r''})$ is an arc, then $r' - r'' \equiv 0$ or $1 \pmod{s}$. Therefore the determinant of the submatrix $\tilde{A}[I, J]$ is equal, up to a sign, to

$$\Delta = \prod_{r=1}^{s} \tilde{A}_{i_r j_r} + (-1)^{s-1} \prod_{r=1}^{s} \tilde{A}_{i_r j_{r-1}} = \alpha Z^p + \beta Z^q,$$

where

$$\alpha = \prod_{r=1}^{s} A_{i_r j_r} \neq 0, \qquad \beta = (-1)^{s-1} \prod_{r=1}^{s} A_{i_r j_{r-1}} \neq 0.$$

Since $p \neq q$, Δ is not invertible in $\boldsymbol{F}\langle Z \rangle$. This contradicts the total unimodularity of $\tilde{A}$. ∎

We conclude this section with a rather obvious remark on the use of the principle of dimensional consistency in structural analysis. When we are given a system of equations that is supposed to represent a physical system, we can sometimes detect errors in its description by verifying the condition (3.26):

$p_{ijk} = r_{ik} - c_{jk}$ for dimensional homogeneity. In case r_{ik} and c_{jk} are known along with p_{ijk}, the test for (3.26) is straightforward. Even in the case where only p_{ijk} are given, without information about r_{ik} and c_{jk}, we can efficiently decide whether (3.26) can be satisfied for some suitable r_{ik} and c_{jk}: Consider a tree in the bipartite graph associated with A, and for each k, set r_{ik} and c_{jk} so that the condition (3.26) may be satisfied for the tree arcs, and then check if (3.26) holds for all cotree arcs (see the proof of Proposition 3.2.5).

Notes. This section is based on Murota [200] as well as Murota [204].

3.3 Physical Matrix

3.3.1 Physical Matrix

We have already introduced two concepts of matrices that we encounter in the description of real systems, namely the mixed matrix and the dimensioned matrix. The former is motivated by structural analysis while the latter derives from dimensional analysis. In this section these two are combined to a third concept of "physical matrix" which models a matrix arising from real-world systems.

As has been discussed in §3.2.2, when we describe a physical system in the form of $A\boldsymbol{x} = \boldsymbol{b}$ with an $m \times n$ matrix A over $\boldsymbol{F}$, we usually know the physical dimensions associated with its rows and columns. Then we can determine the dimensioned matrix $\tilde{A}$ over $\boldsymbol{F}\langle Z_1, \cdots, Z_d \rangle$ that corresponds to A by (3.24) and (3.26) (see (3.31) for the definition of $\boldsymbol{F}\langle Z_1, \cdots, Z_d \rangle$). We call $\tilde{A}$ the dimensioned matrix corresponding to A (with the implicit understanding of the given physical dimensions). By Proposition 3.2.1, this correspondence between A and $\tilde{A} \in \mathcal{D}(\boldsymbol{F}; m, n; Z_1, \cdots, Z_d)$ is given by

$$\tilde{A} = D_{\mathrm{r}}\ A\ {D_{\mathrm{c}}}^{-1}, \tag{3.32}$$

where D_{r} and D_{c} are the known diagonal matrices of (3.27) and (3.28) representing the physical dimensions of the rows (equations) and the columns (variables).

When A is a mixed matrix of the form $A = Q + T$ with a ground field $\boldsymbol{K}$ ($\subseteq \boldsymbol{F}$), i.e., when $A \in \mathrm{MM}(\boldsymbol{K}, \boldsymbol{F}; m, n)$, we can express the corresponding dimensioned matrix $\tilde{A}$ of (3.32) as

$$\tilde{A} = \tilde{Q} + \tilde{T}$$

with

$$\tilde{Q} = D_{\mathrm{r}}\ Q\ {D_{\mathrm{c}}}^{-1}, \qquad \tilde{T} = D_{\mathrm{r}}\ T\ {D_{\mathrm{c}}}^{-1}.$$

This shows that $\tilde{A}$ is also a mixed matrix, but with the quotient field of $\boldsymbol{K}\langle Z_1 \cdots, Z_d \rangle$ as the ground field. In a slight abuse of notation we may write $\tilde{A} \in \mathrm{MM}(\boldsymbol{K}\langle Z_1, \cdots, Z_d \rangle, \boldsymbol{F}\langle Z_1, \cdots, Z_d \rangle; m, n)$. In particular, $\tilde{Q}$ is a matrix over $\boldsymbol{K}\langle Z_1, \cdots, Z_d\rangle$. Note also that the matrices $\tilde{Q}$ and $\tilde{T}$ constituting the mixed matrix $\tilde{A}$ coincide with the dimensioned matrices corresponding to Q and T, respectively.

The most important physical observation to be made here is concerned with the physical dimensions of the nonvanishing entries of Q. Usually, the matrix Q represents various kinds of conservation laws or structural equations, and consists of incidence coefficients such as those induced from the underlying topological/geometrical incidence relations in electrical networks and the stoichiometric coefficients in chemical reactions. Thus it is natural to expect that

$$\text{The nonvanishing entries of } Q \text{ are dimensionless.} \tag{3.33}$$

This statement is true of the examples considered above, provided that the generality assumption GA2 is accepted. In fact, the claim (3.33) can be verified easily for the coefficient matrix (3.3) of the electrical network of Example 3.1.2, for the matrices F, A and B in (3.17) of the mechanical system of Example 3.1.7, and also for the Jacobian matrix (Fig. 3.4) of the ethylene dichloride production system of Example 3.1.3.

The observation (3.33) above can be stated in algebraic terms as follows. Let $A = Q + T \in \mathrm{MM}(\boldsymbol{K}, \boldsymbol{F})$ be a mixed matrix and $\tilde{A}$ be the corresponding dimensioned matrix expressed as (3.24) with exponents p_{ijk} of dimensions. The condition that the nonvanishing entries of Q are dimensionless is equivalent to:

$$Q_{ij} \neq 0 \quad \text{implies} \quad p_{ijk} = 0 \quad \text{for} \quad k = 1, \cdots, d; \tag{3.34}$$

or alternatively,

$$D_{\mathrm{r}}\; Q\; {D_{\mathrm{c}}}^{-1} = Q \tag{3.35}$$

with reference to (3.32). The condition (3.34), or (3.35), does not exclude dimensionless nonvanishing entries from T; for instance, in Example 3.1.2, the parameters α and β contained in T are dimensionless.

We are now in the position to introduce the concept of physical matrices as a mathematical model of the matrices that we encounter in real-world systems. It reflects the dual viewpoint from structural analysis and dimensional analysis.

Suppose a matrix A over $\boldsymbol{F}$ is given along with a subfield $\boldsymbol{K}$ of $\boldsymbol{F}$ and a pair $(D_{\mathrm{r}}, D_{\mathrm{c}})$ of diagonal matrices of the forms (3.27) and (3.28), respectively. We say that a matrix A over $\boldsymbol{F}$ is a *physical matrix* with respect to $(\boldsymbol{K}, \boldsymbol{F}; D_{\mathrm{r}}, D_{\mathrm{c}})$, if

(i) $A = Q + T$ is a mixed matrix with respect to $(\boldsymbol{K}, \boldsymbol{F})$, and
(ii) $D_{\mathrm{r}}\; Q\; {D_{\mathrm{c}}}^{-1} = Q$.

When $(D_{\mathrm{r}}, D_{\mathrm{c}})$ is understood, we say simply that A is a physical matrix with respect to $(\boldsymbol{K}, \boldsymbol{F})$.

3.3.2 Physical Matrices in a Dynamical System

In the previous section we considered the physical dimensional consistency for mixed matrices to arrive at the concept of physical matrix. We now extend this approach to mixed polynomial matrices that describe linear time-invariant dynamical systems.

Recalling the notation from §3.1.2 let

$$A(s) = \sum_{k=0}^{N} s^k A_k = \sum_{k=0}^{N} s^k Q_k + \sum_{k=0}^{N} s^k T_k = Q(s) + T(s)$$

be an $m \times n$ mixed polynomial matrix with $A_k = Q_k + T_k$ $(k = 0, 1, \cdots, N)$. Let $(D_\mathrm{r}, D_\mathrm{c})$ be the pair of matrices of (3.27) and (3.28) representing the physical dimensions of $A = A(s)$, where we assume that time is chosen as one of the fundamental dimensions, say Z_1.

The most important fact to note here is:

> The symbol s should have the dimension of ${Z_1}^{-1}$ (the inverse of time) since it represents "d/dt" (the differentiation with respect to time).

It then follows that, for each $k = 0, 1, \cdots, N$, the dimensions associated with the coefficient matrix A_k is given by $(D_\mathrm{r}, {Z_1}^{-k} D_\mathrm{c})$, since $(D_\mathrm{r}, D_\mathrm{c})$ is associated with $s^k A_k$.

Let us now assume that the coefficient matrix $A_k = Q_k + T_k$ is a physical matrix, namely, that

$$D_\mathrm{r}\ Q_k\ ({Z_1}^{-k} D_\mathrm{c})^{-1} = Q_k.$$

This implies

$$D_\mathrm{r}\ Q(s)\ {D_\mathrm{c}}^{-1} = \sum_{k=0}^{N} s^k D_\mathrm{r} Q_k {D_\mathrm{c}}^{-1} = \sum_{k=0}^{N} (s{Z_1}^{-1})^k Q_k = Q(s{Z_1}^{-1}).$$

Substitution of $Z_1 = s$ (and $Z_k = 1$ for $k \geq 2$) into this expression reveals a remarkable identity: $Q(s) = (D_\mathrm{r}|_{Z_1=s})^{-1}\, Q(1)\, (D_\mathrm{c}|_{Z_1=s})$, that is,

$$Q(s) = \mathrm{diag}\,[s^{-r_{11}}, \cdots, s^{-r_{m1}}] \cdot Q(1) \cdot \mathrm{diag}\,[s^{c_{11}}, \cdots, s^{c_{n1}}]$$

using the exponents $r_{i1} \in \mathbf{Q}$ and $c_{j1} \in \mathbf{Q}$ in (3.27) and (3.28). This relation shows the existence of $r_i \in \mathbf{Q}$ $(i = 1, \cdots, m)$ and $c_j \in \mathbf{Q}$ $(j = 1, \cdots, n)$ such that $r_i - c_j = \deg_s Q_{ij} \in \mathbf{Z}$ for all (i, j) with $Q_{ij} \neq 0$. This implies the existence of such $r_i \in \mathbf{Z}$ $(i = 1, \cdots, m)$ and $c_j \in \mathbf{Z}$ $(j = 1, \cdots, n)$ by Theorem 2.2.35(2). That is,

$$Q(s) = \mathrm{diag}\,[s^{r_1}, \cdots, s^{r_m}] \cdot Q(1) \cdot \mathrm{diag}\,[s^{-c_1}, \cdots, s^{-c_n}] \tag{3.36}$$

for some $r_i \in \mathbf{Z}$ $(i = 1, \cdots, m)$ and $c_j \in \mathbf{Z}$ $(j = 1, \cdots, n)$. Note that, if $r_{i1} \in \mathbf{Z}$ and $c_{j1} \in \mathbf{Z}$, we may take $r_i = -r_{i1}$ and $c_j = -c_{j1}$, and the exponents r_i and c_j have a natural physical meaning.

We have thus revealed an important property (3.36) of the Q-part of a mixed polynomial matrix that is subject to dimensional consistency. In other words, we have identified a subclass of mixed polynomial matrices on the basis of dimensional analysis.

Example 3.3.1. For the mechanical system (Fig. 3.5) treated in Examples 3.1.7 and 3.2.2, the matrix $Q(s) = [Q_A - sQ_F \mid Q_B]$ of (3.18) admits an expression of the form (3.36):

$$\left[\begin{array}{cccccc|c} -s & 0 & 1 & 0 & 0 & 0 & 0 \\ 0 & -s & 0 & 1 & 0 & 0 & 0 \\ 0 & 0 & 0 & 0 & -1 & 0 & 1 \\ 0 & 0 & 0 & 0 & 1 & 0 & 0 \\ 0 & 0 & 0 & 0 & -1 & 0 & 0 \\ -s & s & 0 & 0 & 0 & 1 & 0 \end{array}\right]$$

$$= \left[\begin{array}{cccccc} s & 0 & 0 & 0 & 0 & 0 \\ 0 & s & 0 & 0 & 0 & 0 \\ 0 & 0 & s^2 & 0 & 0 & 0 \\ 0 & 0 & 0 & s^2 & 0 & 0 \\ 0 & 0 & 0 & 0 & s^2 & 0 \\ 0 & 0 & 0 & 0 & 0 & s \end{array}\right] \cdot \left[\begin{array}{cccccc|c} -1 & 0 & 1 & 0 & 0 & 0 & 0 \\ 0 & -1 & 0 & 1 & 0 & 0 & 0 \\ 0 & 0 & 0 & 0 & -1 & 0 & 1 \\ 0 & 0 & 0 & 0 & 1 & 0 & 0 \\ 0 & 0 & 0 & 0 & -1 & 0 & 0 \\ -1 & 1 & 0 & 0 & 0 & 1 & 0 \end{array}\right] \cdot \left[\begin{array}{cccccc|c} 1 & 0 & 0 & 0 & 0 & 0 & 0 \\ 0 & 1 & 0 & 0 & 0 & 0 & 0 \\ 0 & 0 & s^{-1} & 0 & 0 & 0 & 0 \\ 0 & 0 & 0 & s^{-1} & 0 & 0 & 0 \\ 0 & 0 & 0 & 0 & s^{-2} & 0 & 0 \\ 0 & 0 & 0 & 0 & 0 & s^{-1} & 0 \\ \hline 0 & 0 & 0 & 0 & 0 & 0 & s^{-2} \end{array}\right].$$

Note that the diagonal entries s^{r_i} and s^{-c_j} are determined from the negative of the exponents to T (time) in (3.29) and (3.30) as

$$(r_1, \cdots, r_6) = (1, 1, 2, 2, 2, 1), \qquad (c_1, \cdots, c_7) = (0, 0, 1, 1, 2, 1, 2).$$

□

The decomposition property (3.36) is equivalent to the total unimodularity in $\boldsymbol{K}[s, 1/s]$, the ring of Laurent polynomials, as follows. Recall that

$$\boldsymbol{K}[s, 1/s] = \{ \sum_{k=-N_1}^{N_2} \alpha_k s^k \mid 0 \le N_1, N_2 \in \mathbf{Z}, \alpha_k \in \boldsymbol{K} \ (-N_1 \le k \le N_2) \}$$

and that an element of $\boldsymbol{K}[s, 1/s]$ is invertible if and only if it is of the form αs^p for some $\alpha \in \boldsymbol{K} \setminus \{0\}$ and $p \in \mathbf{Z}$. If an $m \times n$ matrix $Q(s)$ admits a decomposition (3.36) with some $r_i \in \mathbf{Z}$ $(i = 1, \cdots, m)$ and $c_j \in \mathbf{Z}$ $(j = 1, \cdots, n)$, then every subdeterminant of $Q(s)$ is of the form αs^p with $\alpha \in \boldsymbol{K}$ and $p \in \mathbf{Z}$. That is, $Q(s)$ is a total unimodular matrix over $\boldsymbol{K}[s, 1/s]$. The following theorem reveals that the converse is also true.

Theorem 3.3.2. *Let $Q(s)$ be an $m \times n$ matrix over $\boldsymbol{K}[s, 1/s]$. Then $Q(s)$ is totally unimodular over $\boldsymbol{K}[s, 1/s]$ if and only if*

$$Q(s) = \mathrm{diag}\,[s^{r_1}, \cdots, s^{r_m}] \cdot Q(1) \cdot \mathrm{diag}\,[s^{-c_1}, \cdots, s^{-c_n}] \tag{3.37}$$

for some integers r_i $(i = 1, \cdots, m)$ *and* c_j $(j = 1, \cdots, n)$. *For a polynomial matrix* $Q(s)$, *in particular, every (nonvanishing) subdeterminant of* $Q(s)$ *is a monomial in* s *over* $\boldsymbol{K}$ *if and only if* (3.37) *is true for some integers* r_i $(i = 1, \cdots, m)$ *and* c_j $(j = 1, \cdots, n)$.

Proof. This is a corollary of Theorem 3.2.4 (or rather Proposition 3.2.5). ∎

Theorem 3.3.2 allows us to characterize the mixed polynomial matrices having property (3.36) as those polynomial matrices $A(s) = Q(s) + T(s)$ which satisfy

(MP-Q2) Every nonvanishing subdeterminant of $Q(s)$ is a monomial over $\boldsymbol{K}$, i.e., of the form αs^p with $\alpha \in \boldsymbol{K}$ and an integer p, and
(MP-T) The collection of nonzero coefficients in $T(s)$ is algebraically independent over $\boldsymbol{K}$.

Namely, (MP-Q2) and (MP-T) characterize the physically meaningful subclass of mixed polynomial matrices subject to dimensional consistency.

The extra property (MP-Q2), or equivalently (3.36), has significant implications with respect to computational complexity in applications of mixed polynomial matrices.

Proposition 3.3.3. *Let* $\mathbf{M}(Q(s))$ *denote the matroid defined on the column set of* $Q(s)$ *by the linear independence of the column vectors over* $\boldsymbol{K}(s)$. *If* $Q(s)$ *is a total unimodular matrix over* $\boldsymbol{K}[s, 1/s]$, *then* $\mathbf{M}(Q(s)) = \mathbf{M}(Q(1))$, *and therefore* $\mathbf{M}(Q(s))$ *is representable over* $\boldsymbol{K}$. □

Proposition 3.3.4. *If* $Q(s)$ *admits the decomposition* (3.36), *then*

$$\deg_s \det Q[I,J] = \begin{cases} \sum_{i\in I} r_i - \sum_{j \in J} c_j & (\textit{if } \det Q(1)[I,J] \neq 0) \\ -\infty & (\textit{otherwise}). \end{cases}$$

Here $Q(1)$ *is a matrix over* $\boldsymbol{K}$ *and therefore* $\det Q(1)[I, J]$ *can be computed by means of arithmetic operations over* $\boldsymbol{K}$. □

In Chap. 6 we shall investigate some problems such as dynamical degree and controllability for dynamical systems described by mixed polynomial matrices $A(s) = Q(s) + T(s)$ having the additional property (MP-Q2) for $\boldsymbol{K} = \mathbf{Q}$.

Notes. This section is based on Murota [200] as well as Murota [204].

4. Theory and Application of Mixed Matrices

This chapter is devoted to a study on mixed matrices and layered mixed matrices using matroid-theoretic methods. Particular emphasis is laid on the combinatorial canonical form (CCF) of layered mixed matrices and related decompositions, which generalize the Dulmage–Mendelsohn decomposition. Applications to the structural solvability of systems of equations are also discussed.

4.1 Mixed Matrix and Layered Mixed Matrix

In the previous section we have introduced the concept of mixed matrices as a possible mathematical tool for systems analysis by means of matroid-theoretic combinatorial methods. A mixed matrix is a matrix A expressed as $A = Q + T$, where Q is a "constant" matrix and T is a "generic" matrix in the sense that the nonzero entries of T are algebraically independent parameters. A layered mixed (or LM-) matrix is defined as a mixed matrix such that Q and T have disjoint nonzero rows, i.e., no row of $A = Q + T$ has both a nonzero entry from Q and a nonzero entry from T.

The concept of a mixed matrix has been motivated by the physical observation that, when we describe a physical system in terms of elementary variables, we can often distinguish two kinds of numbers, accurate numbers and inaccurate numbers, together characterizing the physical system. The "accurate numbers" constitute the matrix Q whereas the "inaccurate numbers" the matrix T. We may also refer to the numbers of the first kind as "fixed constants" and to those of the second kind as "system parameters."

In this chapter we shall investigate the mathematical properties of a mixed matrix and an LM-matrix. Here is a preview of some nice properties enjoyed by a mixed matrix or an LM-matrix.

- The rank is expressed as the minimum of a submodular function and can be computed efficiently by a matroid-theoretic algorithm (§4.2).
- A concept of irreducibility is defined with respect to a natural transformation of physical significance. Irreducibility for an LM-matrix is an extension of the well-known concept of full indecomposability for a generic matrix. The irreducibility of an LM-matrix can be characterized, e.g., in terms

of the irreducibility of determinant, which is an extension of Frobenius's characterization of a fully indecomposable generic matrix (§4.5).

- There exists a unique canonical block-triangular decomposition, called the combinatorial canonical form (CCF for short), into irreducible components. This is a generalization of the Dulmage–Mendelsohn decomposition. The CCF can be computed by an efficient algorithm (§4.4).

We now give the precise mathematical definitions of mixed matrices and layered mixed matrices.

Let $\boldsymbol{K}$ be a subfield of a field $\boldsymbol{F}$. An $m \times n$ matrix $A = (A_{ij})$ over $\boldsymbol{F}$ (i.e., $A_{ij} \in \boldsymbol{F}$) is called a *mixed matrix* with respect to $(\boldsymbol{K}, \boldsymbol{F})$ if

$$A = Q + T, \tag{4.1}$$

where

(M-Q) Q is an $m \times n$ matrix over $\boldsymbol{K}$ (i.e., $Q_{ij} \in \boldsymbol{K}$), and
(M-T) T is an $m \times n$ matrix over $\boldsymbol{F}$ (i.e., $T_{ij} \in \boldsymbol{F}$) such that the set $\mathcal{T}$ of its nonzero entries is algebraically independent over $\boldsymbol{K}$.

The class of $m \times n$ mixed matrices is denoted as $\mathrm{MM}(\boldsymbol{K}, \boldsymbol{F}; m, n)$ (or simply as $\mathrm{MM}(\boldsymbol{K}, \boldsymbol{F})$ without reference to the size (m, n)) and the subfield $\boldsymbol{K}$ will be called the *ground field.*

A mixed matrix A of (4.1) is called a *layered mixed matrix* (or an *LM-matrix*) with respect to $(\boldsymbol{K}, \boldsymbol{F})$ if the nonzero rows of Q and T are disjoint. In other words, A is an LM-matrix if it can be put into the following form with a permutation of rows:

$$A = \begin{pmatrix} Q \\ T \end{pmatrix} = \begin{pmatrix} Q \\ O \end{pmatrix} + \begin{pmatrix} O \\ T \end{pmatrix}, \tag{4.2}$$

where

(L-Q) Q is an $m_Q \times n$ matrix over $\boldsymbol{K}$ (i.e., $Q_{ij} \in \boldsymbol{K}$), and
(L-T) T is an $m_T \times n$ matrix over $\boldsymbol{F}$ (i.e., $T_{ij} \in \boldsymbol{F}$) such that the set $\mathcal{T}$ of its nonzero entries is algebraically independent over $\boldsymbol{K}$.

The class of such LM-matrices is denoted as $\mathrm{LM}(\boldsymbol{K}, \boldsymbol{F}; m_Q, m_T, n)$ (or simply as $\mathrm{LM}(\boldsymbol{K}, \boldsymbol{F})$). Obviously we have

$$\mathrm{LM}(\boldsymbol{K}, \boldsymbol{F}; m_Q, m_T, n) \subseteq \mathrm{MM}(\boldsymbol{K}, \boldsymbol{F}; m_Q + m_T, n).$$

Though an LM-matrix is a special case of mixed matrix, the following argument would indicate that the class of LM-matrices is as general as the class of mixed matrices both in theory and in application. Consider a system of equations $A\boldsymbol{x} = \boldsymbol{b}$ in $\boldsymbol{x} \in \boldsymbol{F}^n$ described with an $m \times n$ mixed matrix $A = Q + T$. By introducing an auxiliary variable $\boldsymbol{w} \in \boldsymbol{F}^m$ we can equivalently rewrite the equation as

$$\begin{pmatrix} I_m & Q \\ -I_m & T \end{pmatrix} \begin{pmatrix} \boldsymbol{w} \\ \boldsymbol{x} \end{pmatrix} = \begin{pmatrix} \boldsymbol{b} \\ \boldsymbol{0} \end{pmatrix}.$$

It may be assumed that $\boldsymbol{F}$ is so large that we can choose m numbers in $\boldsymbol{F}$, say $t_1, \cdots, t_m$, which are algebraically independent over the field $\boldsymbol{K}(\mathcal{T})$, where $\mathcal{T}$ denotes the set of the nonzero entries of T. Then, multiplying each of the last m equations by $t_1, \cdots, t_m$, we obtain a system of equations

$$\begin{pmatrix} I_m & Q \\ -\mathrm{diag}\,[t_1, \cdots, t_m] & T' \end{pmatrix} \begin{pmatrix} \boldsymbol{w} \\ \boldsymbol{x} \end{pmatrix} = \begin{pmatrix} \boldsymbol{b} \\ \boldsymbol{0} \end{pmatrix}, \tag{4.3}$$

where $\mathrm{diag}\,[t_1, \cdots, t_m]$ is a diagonal matrix with "new" parameters $t_1, \cdots, t_m$, and $T'_{ij} = t_i T_{ij}$. The coefficient matrix of (4.3) is an LM-matrix with respect to $(\boldsymbol{K}, \boldsymbol{F})$ since the nonvanishing entries of $[-\mathrm{diag}\,[t_1, \cdots, t_m] \mid T']$ are algebraically independent over $\boldsymbol{K}$. In this way any system of equations with a mixed matrix as its coefficient can be equivalently rewritten into an augmented system using an LM-matrix. Hence we may restrict ourselves to LM-matrices when we deal with the unique solvability of a system of equations having a mixed matrix as its coefficient matrix.

In general, with a mixed matrix $A = Q + T \in \mathrm{MM}(\boldsymbol{K}, \boldsymbol{F}; m, n)$ we will associate a $(2m) \times (m+n)$ LM-matrix $\tilde{A} \in \mathrm{LM}(\boldsymbol{K}, \boldsymbol{F}; m, m, m+n)$ defined by

$$\tilde{A} = \begin{pmatrix} \tilde{Q} \\ \tilde{T} \end{pmatrix} = \begin{pmatrix} I_m & Q \\ -\mathrm{diag}\,[t_1, \cdots, t_m] & T' \end{pmatrix}. \tag{4.4}$$

Note that the column set of $\tilde{A}$ has a natural one-to-one correspondence with the union of the column set and the row set of A. Evidently, $\mathrm{rank}\,\tilde{A} = \mathrm{rank}\,A + m$.

Example 4.1.1. Let $\{\alpha, \beta, \gamma, t_1, t_2\}$ be algebraically independent over $\mathbf{Q}$, and put $\boldsymbol{K} = \mathbf{Q}$ and $\boldsymbol{F} = \mathbf{Q}(\alpha, \beta, \gamma, t_1, t_2)$. An equation

$$\begin{pmatrix} 2+\alpha & 3 \\ \beta & 4+\gamma \end{pmatrix} \begin{pmatrix} x_1 \\ x_2 \end{pmatrix} = \begin{pmatrix} b_1 \\ b_2 \end{pmatrix}$$

described with a 2×2 mixed matrix with respect to $(\boldsymbol{K}, \boldsymbol{F})$ can be rewritten as

$$\left(\begin{array}{cccc} 1 & 0 & 2 & 3 \\ 0 & 1 & 0 & 4 \\ \hline -t_1 & 0 & t_1\alpha & 0 \\ 0 & -t_2 & t_2\beta & t_2\gamma \end{array}\right) \begin{pmatrix} w_1 \\ w_2 \\ x_1 \\ x_2 \end{pmatrix} = \left(\begin{array}{c} b_1 \\ b_2 \\ \hline 0 \\ 0 \end{array}\right)$$

by means of a 4×4 LM-matrix with respect to $(\boldsymbol{K}, \boldsymbol{F})$. Note that $\tilde{\mathcal{T}} = \{-t_1, -t_2, t_1\alpha, t_2\beta, t_2\gamma\}$ is algebraically independent over $\boldsymbol{K}$. □

A more size-efficient transformation can be obtained following the same principle as above but by distinguishing "mixed" rows from "pure" rows that consist either solely of constants or solely of independent nonzero entries.

Suppose the coefficient matrix $A \in \mathrm{MM}(\boldsymbol{K}, \boldsymbol{F}; m, n)$ has m_1 ($\leq m$) "mixed" rows, say,

$$A = Q + T = \begin{pmatrix} Q_1 \\ \hline Q_2 \\ \hline O \end{pmatrix} + \begin{pmatrix} T_1 \\ \hline O \\ \hline T_3 \end{pmatrix} \begin{matrix} \updownarrow R_1 \\ \updownarrow R_2 \\ \updownarrow R_3 \end{matrix} \left. \vphantom{\begin{matrix} R_1 \\ R_2 \\ R_3 \end{matrix}} \right\} R, \tag{4.5}$$

where R_1, R_2, and R_3 are disjoint row subsets of the row set R of A such that $R_1 \cup R_2 \cup R_3 = R$ and $|R_i| = m_i$ $(i = 1, 2, 3)$, Q_1 and Q_2 are matrices over $\boldsymbol{K}$, and T_1 and T_3 are matrices over $\boldsymbol{F}$ with independent nonzero entries. Then by introducing an auxiliary m_1-dimensional vector $\boldsymbol{w}$, we obtain a similar augmented system as (4.3) but now with an $(m + m_1) \times (n + m_1)$ LM-matrix

$$\tilde{A} = \begin{pmatrix} \tilde{Q} \\ \hline \tilde{T} \end{pmatrix} = \begin{pmatrix} I_{m_1} & Q_1 \\ O & Q_2 \\ \hline -\mathrm{diag}[t_1, \cdots, t_{m_1}] & T_1' \\ O & T_3 \end{pmatrix}, \tag{4.6}$$

where diag $[t_1, \cdots, t_{m_1}]$ is a diagonal matrix with "new" parameters $t_1, \cdots, t_{m_1} \in \boldsymbol{F}$ and $(T_1')_{ij} = t_i (T_1)_{ij}$. We have $\tilde{A} \in \mathrm{LM}(\boldsymbol{K}, \boldsymbol{F}; m_1 + m_2, m_1 + m_3, m_1 + n)$ and $\mathrm{rank}\, \tilde{A} = \mathrm{rank}\, A + m_1$.

When $m_1 = m$, this transformation is equivalent to the above transformation (4.4). In the other extreme case of $m_1 = 0$, i.e., when A is already an LM-matrix, this transformation does not change A (i.e., $\tilde{A} = A$). The transformation (4.6) will obviously be more attractive than transformation (4.4) in practical situations.

Notes. The concept of mixed matrices was introduced in Murota–Iri [237, 238], whereas that of LM-matrices was subsequently introduced in Murota [201] and Murota–Iri–Nakamura [239]. As survey papers on mixed matrices, we may mention Murota [218] for mathematical properties and Murota [208, 214, 215] for applications to systems analysis.

4.2 Rank of Mixed Matrices

In this section we consider combinatorial characterizations of the rank of a mixed matrix A with respect to $(\boldsymbol{K}, \boldsymbol{F})$. The rank of A is defined with reference to the field $\boldsymbol{F}$, and not to the ground field $\boldsymbol{K}$. Hence the rank of A is equal to (i) the maximum number of linearly independent column vectors of A with coefficients taken from $\boldsymbol{F}$, (ii) the maximum number of linearly independent row vectors of A with coefficients taken from $\boldsymbol{F}$, and (iii) the maximum size of a submatrix of A for which the determinant does not vanish in $\boldsymbol{F}$.

4.2.1 Rank Identities for LM-matrices

We will first concentrate on an LM-matrix $A = \binom{Q}{T} \in \mathrm{LM}(\boldsymbol{K}, \boldsymbol{F}; m_Q, m_T, n)$. Then all the results for an LM-matrix will be translated to those for a general mixed matrix through the trick of (4.4). We put $C = \mathrm{Col}(A)$, $R = \mathrm{Row}(A)$, $R_Q = \mathrm{Row}(Q)$ and $R_T = \mathrm{Row}(T)$; then $\mathrm{Col}(Q) = \mathrm{Col}(T) = C$, and $R = R_Q \cup R_T$.

Before dealing with a general LM-matrix, let us review what is known for the matrix T, all the nonzero entries of which are algebraically independent over $\boldsymbol{K}$. Note that a generic matrix T can be regarded as a special case of an LM-matrix with $m_Q = 0$.

The structure of T is represented by the functions τ, $\gamma : 2^C \to \mathbf{Z}$ and $\Gamma : 2^C \to 2^R$ defined by

$$\tau(J) = \text{term-rank}\, T[R_T, J], \qquad J \subseteq C, \tag{4.7}$$

$$\Gamma(J) = \{i \in R_T \mid \exists j \in J : T_{ij} \neq 0\}, \qquad J \subseteq C, \tag{4.8}$$

$$\gamma(J) = |\Gamma(J)|, \qquad J \subseteq C. \tag{4.9}$$

It should be clear that $\Gamma(J)$ stands for the set of nonzero rows of the submatrix $T[R_T, J]$, and $\gamma(J)$ for the number of nonzero rows of $T[R_T, J]$. The functions τ and γ both enjoy submodularity (cf. Propositions 2.1.9 and 2.1.12, Lemma 2.2.16), that is,

$$\tau(J_1) + \tau(J_2) \geq \tau(J_1 \cup J_2) + \tau(J_1 \cap J_2), \qquad J_1, J_2 \subseteq C,$$

$$\gamma(J_1) + \gamma(J_2) \geq \gamma(J_1 \cup J_2) + \gamma(J_1 \cap J_2), \qquad J_1, J_2 \subseteq C.$$

These two functions are related by

$$\tau(J) = \min\{\gamma(J') - |J'| \mid J' \subseteq J\} + |J|, \qquad J \subseteq C. \tag{4.10}$$

This is a version of the fundamental minimax relation (cf. Theorem 2.2.17) concerning the maximum matchings and the minimum covers of a bipartite graph, which is called the Hall–Ore theorem in §2.2.3. Recall also that the function $\gamma(J) - |J|$ is called the surplus function.

Combining Proposition 2.1.12 and (4.10) we obtain a rank formula for T:

$$\mathrm{rank}\, T = \text{term-rank}\, T = \min\{\gamma(J) - |J| \mid J \subseteq C\} + |C|. \tag{4.11}$$

It is emphasized that the first equality, connecting the algebraic quantity (rank) to a combinatorial quantity (term-rank), is a consequence of the algebraic independence of the set $\mathcal{T}$ of the nonzero entries of T, whereas the second equality is due to a purely combinatorial min-max duality theorem. We will use an argument of this type to derive a rank formula for a general LM-matrix.

We are now in the position to consider a general LM-matrix A. Note that a submatrix of T is nonsingular if and only if it is term-nonsingular.

Lemma 4.2.1. *A square LM-matrix* $A = \binom{Q}{T} \in \mathrm{LM}(\boldsymbol{K}, \boldsymbol{F})$ *is nonsingular if and only if both* $Q[R_Q, J]$ *and* $T[R_T, C \setminus J]$ *are nonsingular for some* $J \subseteq C$.

Proof. Consider the generalized Laplace expansion (cf. Proposition 2.1.3):

$$\det A = \sum_{J \subseteq C, |J| = m_Q} \pm \det Q[R_Q, J] \cdot \det T[R_T, C \setminus J].$$

If $\det A \neq 0$, the summation contains at least one nonvanishing term, namely, $\det Q[R_Q, J] \neq 0$ and $\det T[R_T, C \setminus J] \neq 0$ for some J. Conversely, suppose that both $Q[R_Q, J_0]$ and $T[R_T, C \setminus J_0]$ are nonsingular for some J_0. Let $t_1 t_2 \cdots t_{m_T}$ be a term contained in $\det T[R_T, C \setminus J_0]$. The algebraic independence of $\mathcal{T}$ ensures that no similar terms arise from different J's. Hence $t_1 t_2 \cdots t_{m_T}$ appears in $\det A$ with a nonzero coefficient, which is equal to $\det Q[R_Q, J_0]$. Hence $\det A \neq 0$. ■

The following fact is a basic rank identity for an LM-matrix.

Theorem 4.2.2. *For an LM-matrix* $A = \binom{Q}{T} \in \mathrm{LM}(\boldsymbol{K}, \boldsymbol{F})$,

$$\operatorname{rank} A = \max\{\operatorname{rank} Q[R_Q, J] + \text{term-rank}\, T[R_T, C \setminus J] \mid J \subseteq C\}. \quad (4.12)$$

Proof. Lemma 4.2.1 applied to submatrices of A establishes the claim. ■

As the second step of the derivation of the rank formula for A, the right-hand side of the basic identity (4.12) in Theorem 4.2.2 should be rewritten using a combinatorial min-max duality result. To this end, we will first translate (4.12) into a matroid-theoretic expression.

Let $\mathbf{M}(A)$, $\mathbf{M}(Q)$ and $\mathbf{M}(T)$ be the matroids defined on C by matrices A, Q and T, respectively, with respect to the linear independence among column vectors. The rank function of $\mathbf{M}(Q)$ is given by

$$\rho(J) = \operatorname{rank} Q[R_Q, J], \qquad J \subseteq C, \quad (4.13)$$

while that of $\mathbf{M}(T)$ is τ of (4.7). Then (4.12) in Theorem 4.2.2 is rewritten as

$$\operatorname{rank} A = \max\{\rho(J) + \tau(C \setminus J) \mid J \subseteq C\}, \quad (4.14)$$

and Lemma 4.2.1 is rephrased as follows, where $\vee$ means the union of matroids.

Theorem 4.2.3. *For an LM-matrix* $A = \binom{Q}{T} \in \mathrm{LM}(\boldsymbol{K}, \boldsymbol{F})$, *it holds that* $\mathbf{M}(A) = \mathbf{M}(Q) \vee \mathbf{M}(T)$ *and that*

$$\operatorname{rank} A = \min\{\rho(J) + \tau(J) - |J| \mid J \subseteq C\} + |C|. \quad (4.15)$$

Proof. It follows from Lemma 4.2.1, applied to submatrices of A, that $\operatorname{rank} A[R, J] = |J|$ if and only if $\operatorname{rank} Q[R_Q, J'] = |J'|$ and $\operatorname{rank} T[R_T, J \setminus J'] = |J \setminus J'|$ for some $J' \subseteq J$. That is, J is independent in $\mathbf{M}(A)$ if and only if it can be partitioned into two disjoint subsets that are independent in $\mathbf{M}(Q)$ and $\mathbf{M}(T)$, respectively. Namely, $\mathbf{M}(A) = \mathbf{M}(Q) \vee \mathbf{M}(T)$. Then we have $\operatorname{rank} A = \operatorname{rank} \mathbf{M}(A) = \operatorname{rank} (\mathbf{M}(Q) \vee \mathbf{M}(T))$, in which $\operatorname{rank} (\mathbf{M}(Q) \vee \mathbf{M}(T))$ equals the right-hand side of (4.15) by the rank formula (2.80) for matroid union. ■

Remark 4.2.4. A combination of (4.14) and (4.15) yields

$$\rho(J_1) + \tau(C \setminus J_1) \leq \operatorname{rank} A \leq \rho(J_2) + \tau(J_2) + |C \setminus J_2| \qquad (J_1, J_2 \subseteq C).$$

This makes it possible to estimate $\operatorname{rank} A$ using any $J_1, J_2 \subseteq C$. Furthermore, by Theorem 4.2.2 and Theorem 4.2.3, there exist J_1, J_2 for which this estimate is tight. In particular, Theorem 4.2.2 guarantees the existence of a "certificate" (namely, J attaining the maximum) for the nonsingularity of A, whereas Theorem 4.2.3 the existence of a "certificate" (namely, J attaining the minimum) for the singularity of A. □

The rank formula (4.15) is certainly a nontrivial and meaningful expression, giving a combinatorial characterization of $\operatorname{rank} A$ in terms of the minimum value of a submodular function. But it is not satisfactory enough in that it does not extend the rank formula (4.11) for T. In fact, in this special case (with $A = T$ and $\rho = 0$) the expression (4.15) reduces to

$$\operatorname{rank} T = \min\{\tau(J) - |J| \mid J \subseteq C\} + |C|,$$

which is almost a triviality, since the minimum on the right-hand side is obviously attained by $J = C$ and therefore the formula claims that $\operatorname{rank} T = \tau(C)$, i.e., $\operatorname{rank} T = \text{term-rank}\, T$.

In order to obtain a more useful rank formula for A, we will introduce a set function $p : 2^C \to \mathbf{Z}$ defined by

$$p(J) = \rho(J) + \gamma(J) - |J|, \qquad J \subseteq C, \tag{4.16}$$

as an extension of the surplus function $\gamma(J) - |J|$ in the rank formula (4.11) for T. We name p the *LM-surplus function*. This function p expresses a combination of the combinatorial structures of the constituent matrices Q and T, with ρ standing for Q and γ for T. The LM-surplus function p is submodular, namely,

$$p(J_1) + p(J_2) \geq p(J_1 \cup J_2) + p(J_1 \cap J_2), \qquad J_1, J_2 \subseteq C, \tag{4.17}$$

since both ρ and γ are submodular. In the special case where $A = T$ (i.e., $m_Q = 0$ and $\rho = 0$), the LM-surplus function $p(J)$ reduces indeed to $\gamma(J) - |J|$, which is the surplus function appearing in (4.11).

The following theorem gives another minimax expression for the rank of an LM-matrix, due to Murota [201, 204] and Murota–Iri–Nakamura [239].

Theorem 4.2.5. *For an LM-matrix* $A = \binom{Q}{T} \in \mathrm{LM}(\boldsymbol{K}, \boldsymbol{F})$, *it holds that*

$$\operatorname{rank} A = \min\{\rho(J) + \gamma(J) - |J| \mid J \subseteq C\} + |C|. \tag{4.18}$$

Using the notation p this formula can be written as

$$\operatorname{rank} A = \min\{p(J) \mid J \subseteq C\} + |C|. \tag{4.19}$$

Proof. The right-hand sides of (4.15) and (4.18) are equal, since

$$\begin{aligned}
&\min_J\{\rho(J) + \tau(J) - |J| \mid J \subseteq C\} \\
&= \min_J\{\rho(J) + \min_{J'}\{\gamma(J') - |J'| \mid J' \subseteq J\} \mid J \subseteq C\} \\
&= \min_{J'}\{\min_J\{\rho(J) \mid J \supseteq J'\} + \gamma(J') - |J'| \mid J' \subseteq C\} \\
&= \min_{J'}\{\rho(J') + \gamma(J') - |J'| \mid J' \subseteq C\},
\end{aligned}$$

where the first equality is by (4.10) and the last equality is due to the monotonicity of $\rho(J)$ with respect to J. ■

The two expressions, (4.15) in Theorem 4.2.3 and (4.18) in Theorem 4.2.5, look very similar, with τ in (4.15) replaced by γ in (4.18). Moreover, in both formulas, the functions to be minimized are submodular in J. However, the second expression (4.18) is superior to the first (4.15) for two reasons:

1. It contains the rank formula (4.11) for T (=the Hall–Ore theorem for bipartite matchings) as a special case;
2. It leads to a canonical block-triangular decomposition, to be explained at length in §4.4.

Example 4.2.6. Consider a 4×5 LM-matrix

$$A = \begin{pmatrix} Q \\ T \end{pmatrix} = \begin{array}{c|ccccc|} & x_1 & x_2 & x_3 & x_4 & x_5 \\ \hline & 1 & 1 & 1 & 1 & 0 \\ & 0 & 2 & 1 & 1 & 0 \\ \hline f_1 & t_1 & 0 & 0 & 0 & t_2 \\ f_2 & 0 & t_3 & 0 & 0 & t_4 \\ \hline \end{array}$$

with $\mathcal{T} = \{t_1, t_2, t_3, t_4\}$ being algebraically independent over $\mathbf{Q}$. The columns and the rows are indexed as $\mathrm{Col}(A) = C = \{x_1, x_2, x_3, x_4, x_5\}$ and $\mathrm{Row}(T) = R_T = \{f_1, f_2\}$. It turns out that $J = \{x_1, x_3\}$ attains the maximum on the right-hand side of (4.12) with $\operatorname{rank} Q[R_Q, J] = \text{term-rank}\, T[R_T, C \setminus J] = 2$. Hence $\operatorname{rank} A = 2 + 2 = 4$. This can be obtained also from the rank formulas (4.15) in Theorem 4.2.3 and (4.18) in Theorem 4.2.5. It can be verified that, in either formula, $J = \{x_3, x_4\}$ attains the minimum value -1 with $\rho(J) = 1$, $\tau(J) = \gamma(J) = 0$ and $|J| = 2$. □

4.2.2 Rank Identities for Mixed Matrices

In this subsection we provide some results on the rank of a general mixed matrix A. In principle, these results are straightforward translations of the above results for an LM-matrix applied to the associated LM-matrix $\tilde{A}$ of (4.4), for which we have $\mathrm{rank}\,\tilde{A} = \mathrm{rank}\,A + m$. We use the notations $R = \mathrm{Row}(A)$ and $C = \mathrm{Col}(A)$.

Lemma 4.2.1 yields the following counterpart. Note that a submatrix of T is nonsingular if and only if it is term-nonsingular.

Lemma 4.2.7. *A square mixed matrix $A = Q+T$ is nonsingular if and only if both $Q[I,J]$ and $T[R\setminus I, C\setminus J]$ are nonsingular for some $I \subseteq R$ and $J \subseteq C$.* □

The following identity is obtained from Theorem 4.2.2.

Theorem 4.2.8. *For a mixed matrix $A = Q+T$,*

$$\mathrm{rank}\,A = \max\{\mathrm{rank}\,Q[I,J] + \text{term-rank}\,T[R\setminus I, C\setminus J] \mid I \subseteq R, J \subseteq C\}. \tag{4.20}$$

□

The content of Lemma 4.2.7 can be expressed in matroid-theoretic terms as follows. We denote by $\mathbf{L}(A)$, $\mathbf{L}(Q)$, and $\mathbf{L}(T)$ the bimatroids (cf. §2.3.7) defined respectively by matrices A, Q, and T; for example, (I,J) is a linked pair in $\mathbf{L}(A)$ if and only if $A[I,J]$ is nonsingular.

Theorem 4.2.9. *For a mixed matrix $A = Q+T$, it holds that $\mathbf{L}(A) = \mathbf{L}(Q) \vee \mathbf{L}(T)$, where $\vee$ means the union of bimatroids.* □

Theorem 4.2.10. *For a mixed matrix $A = Q+T \in \mathrm{MM}(\boldsymbol{K},\boldsymbol{F};m,n)$, it holds that*

$$\begin{aligned}\mathrm{rank}\,A &= \mathrm{rank}\,[\ \mathbf{M}([I_m \mid Q]) \vee \mathbf{M}([I_m \mid T])\ ] - m\\ &= \text{maximum size of a common independent set}\\ &\quad\ \text{of } \mathbf{M}([I_m \mid Q])^* \text{ and } \mathbf{M}([I_m \mid T]),\end{aligned}$$

where $\mathbf{M}([I_m \mid Q])^*$ ($\simeq \mathbf{M}([-Q^{\mathrm{T}} \mid I_n])$) is the dual of $\mathbf{M}([I_m \mid Q])$.

Proof. Apply Theorem 4.2.3 to the LM-matrix $\tilde{A}$ of (4.4) and use the relation $\mathrm{rank}\,\tilde{A} = \mathrm{rank}\,A + |R|$. The second equality is due to the general relation (2.81) between union and intersection of two matroids. ■

Define $\hat{\rho}, \hat{\tau}, \hat{\gamma} : 2^R \times 2^C \to \mathbf{Z}$ and $\hat{\Gamma} : 2^R \times 2^C \to 2^R$ by

$$\begin{aligned}\hat{\rho}(I,J) &= \mathrm{rank}\,Q[I,J], \qquad I \subseteq R, J \subseteq C,\\ \hat{\tau}(I,J) &= \text{term-rank}\,T[I,J], \qquad I \subseteq R, J \subseteq C,\\ \hat{\Gamma}(I,J) &= \{i \in I \mid \exists j \in J : T_{ij} \neq 0\}, \qquad I \subseteq R, J \subseteq C,\\ \hat{\gamma}(I,J) &= |\hat{\Gamma}(I,J)|, \qquad I \subseteq R, J \subseteq C.\end{aligned}$$

Note that $\hat{\Gamma}(I,J) = I \cap \hat{\Gamma}(R,J)$.

Theorem 4.2.11. *For a mixed matrix $A = Q + T$, it holds that*

$$\operatorname{rank} A = \min_{I \subseteq R, J \subseteq C} \{\hat{\rho}(I,J) + \hat{\tau}(I,J) - |I| - |J|\} + |R| + |C|, \quad (4.21)$$

$$\operatorname{rank} A = \min_{I \subseteq R, J \subseteq C} \{\hat{\rho}(I,J) + \hat{\gamma}(I,J) - |I| - |J|\} + |R| + |C|. \quad (4.22)$$

Proof. Consider the LM-matrix $\tilde{A}$ of (4.4) and let $\tilde{\rho}, \tilde{\tau}, \tilde{\gamma} : 2^{R \cup C} \to \mathbf{Z}$ and $\tilde{\Gamma} : 2^{R \cup C} \to 2^R$ be the functions associated with $\tilde{A}$ by (4.13), (4.7), (4.9) and (4.8). For $I \subseteq R$ and $J \subseteq C$ we have

$$\begin{aligned}
\tilde{\rho}(I \cup J) &= \hat{\rho}(R \setminus I, J) + |I|, \\
\tilde{\tau}(I \cup J) &= \hat{\tau}(R \setminus I, J) + |I|, \\
\tilde{\Gamma}(I \cup J) &= \hat{\Gamma}(R, J) \cup I = \hat{\Gamma}(R \setminus I, J) \cup I, \\
\tilde{\gamma}(I \cup J) &= \hat{\gamma}(R \setminus I, J) + |I|.
\end{aligned}$$

Substituting these expressions into the rank formulas (4.15) and (4.18) for $\tilde{A}$ and noting the relation $\operatorname{rank} \tilde{A} = \operatorname{rank} A + |R|$, we obtain the claims. ∎

From the second formula in Theorem 4.2.11 we can derive some variants.

Corollary 4.2.12. *For a mixed matrix $A = Q + T$, it holds that*

$$\operatorname{rank} A = \min_{I \subseteq R, J \subseteq C} \{\operatorname{rank} Q[I,J] - |I| - |J| \mid \operatorname{rank} T[I,J] = 0\} + |R| + |C|, \quad (4.23)$$

$$\operatorname{rank} A = \min_{J \subseteq C} \{\operatorname{rank} Q[R \setminus \hat{\Gamma}(R,J), J] + |\hat{\Gamma}(R,J)| - |J|\} + |C|. \quad (4.24)$$

Proof. The second rank formula (4.22) in Theorem 4.2.11 can be rewritten as

$$\operatorname{rank} A = \min_{I,J} \{\operatorname{rank} Q[I,J] - |I \setminus \hat{\Gamma}(R,J)| - |J|\} + |R| + |C|.$$

Since the function to be minimized does not increase when I is replaced by $I \setminus \hat{\Gamma}(R,J)$, we may assume $I \cap \hat{\Gamma}(R,J) = \emptyset$, i.e., $\operatorname{rank} T[I,J] = 0$. Hence follows the first expression. Furthermore, we may choose I as large as possible under this condition, i.e., $I = R \setminus \hat{\Gamma}(R,J)$, which results in the second formula. ∎

If A is an LM-matrix, in particular, the expression (4.24) specializes directly to the rank formula (4.18) in Theorem 4.2.5, from which (4.24) itself has been derived by way of (4.22) and (4.23). Note that $\hat{\Gamma}(R,J) = \Gamma(J) \subseteq R_T$ and $\operatorname{rank} Q[R \setminus \hat{\Gamma}(R,J), J] = \operatorname{rank} Q[R_Q, J]$ for an LM-matrix. This demonstrates the equivalence of these formulas.

Just as the duality concerning bipartite matchings can be expressed equivalently by the Hall–Ore theorem and by the König–Egerváry theorem, the rank formula above, which is a generalization of the Hall–Ore theorem, admits a reformulation of the König–Egerváry type found by Bapat [9].

Theorem 4.2.13 (König–Egerváry theorem for mixed matrix). *Let $A = Q + T$ be a mixed matrix. Then there exist $I \subseteq R$ and $J \subseteq C$ such that*
(i) $|I| + |J| - \mathrm{rank}\, Q[I, J] = |R| + |C| - \mathrm{rank}\, A$,
(ii) $\mathrm{rank}\, T[I, J] = 0$.

Proof. Take (I, J) that attains the minimum on the right-hand side of (4.23). ■

Remark 4.2.14. Theorem 4.2.13 is a refinement of the previous result of Hartfiel–Loewy [102] for a square mixed matrix, named the "determinantal version of the Frobenius–König theorem," and its extension to a general rectangular mixed matrix by Murota [218]. The original proof of Hartfiel–Loewy (for the square case) was quite involved, based on factorizations of determinants. □

Remark 4.2.15. As is naturally expected, the König–Egerváry-type result (Theorem 4.2.13) is equivalent to the rank formula (4.23), under an obvious inequality

$$\begin{aligned} \mathrm{rank}\, A &\leq \mathrm{rank}\, A[R, J] + \mathrm{rank}\, A[R, C \setminus J] \\ &\leq \mathrm{rank}\, A[I, J] + \mathrm{rank}\, A[R \setminus I, J] + \mathrm{rank}\, A[R, C \setminus J] \\ &\leq \mathrm{rank}\, Q[I, J] + |R \setminus I| + |C \setminus J| \end{aligned}$$

valid for (I, J) with $T[I, J] = O$ (i.e., with $I \cap \hat{\Gamma}(R, J) = \emptyset$).

Furthermore, it is pointed out by Bapat [9] that Theorem 4.2.13 can be proved independently of the rank formula using a fundamental property of a general bimatroid. By Theorem 2.3.47 (applied to the bimatroid associated with A), there exist $I \subseteq R$ and $J \subseteq C$ such that
(i) $|I| + |J| - \mathrm{rank}\, A[I, J] = |R| + |C| - \mathrm{rank}\, A$,
(ii) $\mathrm{rank}\, A[I \setminus \{i\}, J \setminus \{j\}] = \mathrm{rank}\, A[I, J]$, $\forall i \in I$, $\forall j \in J$.
We claim that (ii) implies $T[I, J] = O$. Suppose, to the contrary, that $T_{ij} \neq 0$ for some $i \in I$ and $j \in J$. Since $\mathrm{rank}\, A[I \setminus \{i\}, J \setminus \{j\}] = \mathrm{rank}\, A[I, J]$ $(=: r)$, there exist $I' \subseteq I \setminus \{i\}$ and $J' \subseteq J \setminus \{j\}$ such that $\mathrm{rank}\, A[I', J'] = |I'| = |J'| = r$. Consider the Laplace expansion of $\det A[I' \cup \{i\}, J' \cup \{j\}]$. It contains a nonvanishing term $T_{ij} \cdot \det A[I', J']$, which is not cancelled out by virtue of the algebraic independence of the nonzero entries of T. This implies a contradiction that $r = \mathrm{rank}\, A[I, J] \geq \mathrm{rank}\, A[I' \cup \{i\}, J' \cup \{j\}] = |I'| + 1 = r + 1$. □

Remark 4.2.16. When numerical values are substituted into the nonzero entries of the T-part of a mixed matrix, the rank of the resulting numerical matrix can possibly decrease. On the basis of Theorem 4.2.13 a systematic procedure has been given by Geelen [92] that assigns numerical values so that the rank remains the same. □

Finally we mention the following fact in connection with the basic rank identity (4.20).

Theorem 4.2.17. *For a maximizer* (I, J) *in* (4.20) *that is minimal with respect to set inclusion, we have* $|I| = |J|$ *and* $\det A[I, J] \in \boldsymbol{K}^*$, *where* $\boldsymbol{K}^* = \boldsymbol{K} \setminus \{0\}$.

Proof. Suppose $|I| > \operatorname{rank} Q[I, J]$. Then there exists $i \in I$ such that $\operatorname{rank} Q[I, J] = \operatorname{rank} Q[I \setminus \{i\}, J]$. This implies that $(I \setminus \{i\}, J)$ is also a maximizer in (4.20), which contradicts the minimality of (I, J). Similarly for $|J|$. Hence $|I| = |J| = \operatorname{rank} Q[I, J]$, that is, $Q[I, J]$ is nonsingular, and a fortiori $A[I, J]$ is nonsingular, i.e., $\det A[I, J] \neq 0$.

Suppose $\det A[I, J] \notin \boldsymbol{K}$. Then there exist nonempty $I' \subseteq I$ and nonempty $J' \subseteq J$ such that both $T[I', J']$ and $Q[I \setminus I', J \setminus J']$ are nonsingular, which implies that $(I \setminus I', J \setminus J')$ is also a maximizer in (4.20), a contradiction. ∎

A mixed matrix A with the property $\det A \in \boldsymbol{K}^*$ will be investigated in §4.6.1.

Notes. The rank formulas for mixed matrices (Lemma 4.2.7, Theorem 4.2.8, Theorem 4.2.9, Theorem 4.2.10) are due to Murota–Iri [237, 238]. Theorem 4.2.17 is by Murota [198].

4.2.3 Reduction to Independent Matching Problems

We explain how the computation of rank A for an LM-matrix $A = \binom{Q}{T}$ can be reduced to solving an independent matching problem. This leads to an efficient algorithm, to be described in §4.2.4, for computing the rank of an LM-matrix with arithmetic operations in the ground field $\boldsymbol{K}$.

Here and henceforth $C_Q = \{j_Q \mid j \in C\}$ denotes a disjoint copy of $C = \operatorname{Col}(A)$ (with $j_Q \in C_Q$ denoting the copy of $j \in C$), whereas $R_Q = \operatorname{Row}(Q)$, $R_T = \operatorname{Row}(T)$, $|R_Q| = m_Q$, $|R_T| = m_T$ and $|C| = n$. Denote by $\mathbf{M}_Q = (C_Q, \mathcal{I}_Q)$ the matroid associated with Q, where $C_Q = \operatorname{Col}(Q)$ and $\mathcal{I}_Q$ is the family of independent sets, namely,

$$\mathcal{I}_Q = \{J_Q \subseteq C_Q \mid \operatorname{rank} Q[R_Q, J_Q] = |J_Q|\}.$$

We consider an independent matching problem defined on a bipartite graph $G = (V^+, V^-; E)$ with $V^+ = R_T \cup C_Q$, $V^- = C$ and $E = E_T \cup E_Q$, where

$$E_T = \{(i, j) \mid i \in R_T, j \in C, T_{ij} \neq 0\}, \quad E_Q = \{(j_Q, j) \mid j \in C\}.$$

The matroid $\mathbf{M}^+ = (V^+, \mathcal{I}^+)$ attached to V^+ is the direct sum of the free matroid on R_T and $\mathbf{M}_Q$ on C_Q, i.e.,

$$\mathcal{I}^+ = \{I^+ \subseteq V^+ \mid I^+ \cap C_Q \in \mathcal{I}_Q\},$$

whereas the free matroid $\mathbf{M}^- = (V^-, \mathcal{I}^-)$ (with $\mathcal{I}^- = 2^{V^-}$) is attached to V^-. The set of the end-vertices of a matching M will be designated as ∂M $(\subseteq V)$.

We then have the following characterization of rank A in terms of the maximum size of an independent matching.

Theorem 4.2.18. *For an LM-matrix A,* rank A *coincides with the maximum size of an independent matching in the independent matching problem defined above. That is,*

$$\mathrm{rank}\, A = \max\{|M| \mid M\text{: independent matching}\}.$$

Proof. The proof is based on the basic rank identity of Theorem 4.2.2. Consider $J \subseteq C$ that attains the maximum on the right-hand side of (4.12). We may assume that $\mathrm{rank}\, Q[R_Q, J] = |J|$. Then there exists an independent matching M such that $\partial M \cap C_Q = J_Q$ and $|\partial M \cap R_T| = \text{term-rank}\, T[R_T, C \setminus J]$, where $J_Q \subseteq C_Q$ is the copy of J. Hence follows $\mathrm{rank}\, A \leq |M|$. Conversely, given an independent matching M, we put $J_Q = \partial(M \cap C_Q)$ to obtain $\mathrm{rank}\, Q[R_Q, J] = |J|$ and $\text{term-rank}\, T[R_T, C \setminus J] \geq |\partial M \cap R_T|$. This shows $\mathrm{rank}\, A \geq \mathrm{rank}\, Q[R_Q, J] + \text{term-rank}\, T[R_T, C \setminus J] \geq |M|$. ■

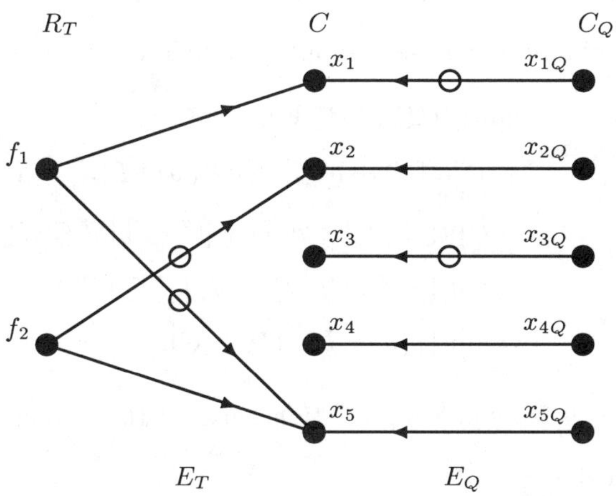

Fig. 4.1. Graph G (○: arc in a maximum independent matching M)

Example 4.2.19. The independent matching problem associated with the 4×5 LM-matrix in Example 4.2.6:

$$A = \begin{pmatrix} Q \\ T \end{pmatrix} = \begin{array}{c|ccccc|} & x_1 & x_2 & x_3 & x_4 & x_5 \\ \hline & 1 & 1 & 1 & 1 & 0 \\ & 0 & 2 & 1 & 1 & 0 \\ \hline f_1 & t_1 & 0 & 0 & 0 & t_2 \\ f_2 & 0 & t_3 & 0 & 0 & t_4 \\ \hline \end{array}$$

is illustrated in Fig. 4.1. The columns and the rows are indexed as $\mathrm{Col}(A) = C = \{x_1, x_2, x_3, x_4, x_5\}$ and $\mathrm{Row}(T) = R_T = \{f_1, f_2\}$ and accordingly $C_Q = \{x_{1Q}, x_{2Q}, x_{3Q}, x_{4Q}, x_{5Q}\}$. A maximum independent matching $M = \{(f_1, x_5), (f_2, x_2), (x_{1Q}, x_1), (x_{3Q}, x_3)\}$ is marked by ○. We have $\mathrm{rank}\, A = |M| = 4$. Note also that $J = \{x_1, x_3\}$, corresponding to $J_Q = \partial(M \cap C_Q) = \{x_{1Q}, x_{3Q}\}$, attains the maximum on the right-hand side of (4.12). □

The LM-surplus function p characterizing the rank of A in the identity (4.19) of Theorem 4.2.5 is closely related to the cut function of the independent matching problem. The rank functions ρ^+ and ρ^- of $\mathbf{M}^+$ and $\mathbf{M}^-$ are given by

$$\rho^+(X) = |X \cap R_T| + \rho(X \cap C_Q), \qquad X \subseteq V^+,$$
$$\rho^-(Y) = |Y|, \qquad Y \subseteq V^-.$$

For $U \subseteq V^+ \cup V^-$, there is no arc going out of U if and only if $\Gamma(J) \subseteq I$ and $J \subseteq K$, where $I = R_T \setminus U$, $J = C \setminus U$, and K ($\subseteq C$) is the copy of $K_Q = C_Q \setminus U$ ($\subseteq C_Q$). Then the cut function $\kappa(U)$ (cf. (2.71)) is given by

$$\kappa(U) = \rho^+(V^+ \setminus U) + \rho^-(V^- \cap U) = |I| + \rho(K) + |C \setminus J|, \tag{4.25}$$

and its minimum can be computed as follows:

$$\begin{aligned}
&\min_U \{\kappa(U) \mid U \subseteq V^+ \cup V^-\} \\
&= \min_{I,J,K} \{|I| + \rho(K) + |C \setminus J| \mid \Gamma(J) \subseteq I, J \subseteq K\} \\
&= \min_J \{|\Gamma(J)| + \rho(J) + |C \setminus J| \mid J \subseteq C\} \\
&= \min_J \{\gamma(J) + \rho(J) - |J| \mid J \subseteq C\} + |C| \\
&= \min_J \{p(J) \mid J \subseteq C\} + |C|.
\end{aligned} \tag{4.26}$$

This reveals the following relation between the minimizers of p and κ.

Lemma 4.2.20.

$$\mathcal{L}_{\min}(p) = \{J \subseteq C \mid J = C \setminus U,\ U \in \mathcal{L}_{\min}(\kappa)\},$$

where $\mathcal{L}_{\min}(p)$ and $\mathcal{L}_{\min}(\kappa)$ denote the families of the minimizers of $p : 2^C \to \mathbf{Z}$ and $\kappa : 2^{V^+ \cup V^-} \to \mathbf{Z}$, respectively. □

Remark 4.2.21. By combining Theorem 4.2.18, the above relation (4.26), and the general min-max theorem for the independent matching problem (Theorem 2.3.27 or (2.72)), we obtain

$$\mathrm{rank}\, A = \max_{\substack{M:\ \text{indep.}\\ \text{matching}}} |M| = \min_U \kappa(U) = \min_J p(J) + |C|.$$

This argument affords an alternative proof of the rank formula of Theorem 4.2.5, though the min-max theorem for the independent matching problem is almost equivalent to the matroid union/partition theorem of Edmonds used in deriving Theorem 4.2.5. Note that the relation (4.26) has been derived independently of Theorem 4.2.5. □

4.2.4 Algorithms for the Rank

Algorithm for LM-matrices. An efficient (polynomial time) algorithm is described here which computes the rank of an LM-matrix $A = \binom{Q}{T} \in \mathrm{LM}(\boldsymbol{K}, \boldsymbol{F}; m_Q, m_T, n)$. On the basis of Theorem 4.2.18 the algorithm solves the associated independent matching problem by specializing the general algorithmic scheme described in §2.3.5.

Recall that the associated independent matching problem is defined on the bipartite graph $G = (V^+, V^-; E) = (R_T \cup C_Q, C; E_T \cup E_Q)$, where $R_T = \mathrm{Row}(T)$, $C = \mathrm{Col}(A)$, C_Q is a disjoint copy of C (with $j_Q \in C_Q$ denoting the copy of $j \in C$), and

$$E_T = \{(i,j) \mid i \in R_T, j \in C, T_{ij} \neq 0\}, \quad E_Q = \{(j_Q, j) \mid j \in C\}.$$

The algorithm works with a directed graph $\tilde{G} = \tilde{G}_M = (\tilde{V}, \tilde{E})$ with vertex set $\tilde{V} = R_T \cup C_Q \cup C$ and arc set $\tilde{E} = E_T \cup E_Q \cup E^+ \cup M^\circ$, where E^+ and M° are defined and updated in the algorithm; E^+ represents the structure of the matroid $\mathbf{M}(Q)$ and M° expresses an independent matching $M \subseteq E_T \cup E_Q$ as

$$M^\circ = \{\overline{a} \mid a \in M\} \qquad (\overline{a}\text{: reorientation of } a).$$

It is noted that the arcs in E^+ have both ends in C_Q and the arcs in M° are directed from C to $R_T \cup C_Q$, i.e., $\partial^+ M^\circ \subseteq C$ and $\partial^- M^\circ \subseteq R_T \cup C_Q$.

Since M is an independent matching, $I = \{i \in C \mid i_Q \in \partial^- M^\circ \cap C_Q\}$ is an independent set of $\mathbf{M}(Q)$, whereas we denote by J the set of elements of $C \setminus I$ which are dependent on I in $\mathbf{M}(Q)$. Namely,

$$\mathrm{rank}\, Q[R_Q, I] = |I|, \qquad J = \{j \in C \setminus I \mid \mathrm{rank}\, Q[R_Q, I \cup \{j\}] = |I|\}.$$

Besides the graph $\tilde{G}$ we use a matrix (or two-dimensional array) P and a vector (or one-dimensional array) *base* to implement the structure of the matroid $\mathbf{M}(Q)$. The array P represents a matrix over $\boldsymbol{K}$, of size $m_Q \times n$, which is obtained from Q by row-transformations; we have $P = Q$ at the beginning of the algorithm (Step 1 below). The variable *base* is a vector of size m_Q, which represents a mapping (correspondence): $R_Q \to C \cup \{0\}$. The sets I and J are represented as

$$I = \{i \in C \mid i = \mathit{base}[h] \neq 0, h \in R_Q\},$$
$$J = \{j \in C \setminus I \mid \forall h:\ \mathit{base}[h] = 0 \Rightarrow P[h,j] = 0\}.$$

For $i \in I$ and $j \in J$, $I-i+j$ is independent in $\mathbf{M}(Q)$ if and only if $P[h,j] \neq 0$ for the $h \in R_Q$ such that $i = base[h]$. Optionally, it computes an $m_Q \times m_Q$ matrix S over $\boldsymbol{K}$ such that $SQ = P$. If such information is not needed, the matrix S may simply be eliminated from the computation without any side effect.

The entrance $S^+ \subseteq \tilde{V}$ and the exit $S^- \subseteq \tilde{V}$ are defined by

$$S^+ = (R_T \setminus \partial^- M^\circ) \cup \{j_Q \in C_Q \mid j \in C \setminus (I \cup J)\}, \quad S^- = C \setminus \partial^+ M^\circ.$$

The algorithm looks for a shortest path from the entrance S^+ to the exit S^- to augment the matching M.

Algorithm for computing the rank of an LM-matrix A

Step 1:
  $M^\circ := \emptyset$; $base[i] := 0 \ (i \in R_Q)$; $P[i,j] := Q_{ij} \ (i \in R_Q, j \in C)$;
  $S :=$ unit matrix of order m_Q.
Step 2:
  $I := \{i \in C \mid i_Q \in \partial^- M^\circ \cap C_Q\}$;
  $J := \{j \in C \setminus I \mid \forall h : \ base[h] = 0 \Rightarrow P[h,j] = 0\}$;
  $S_T^+ := R_T \setminus \partial^- M^\circ$; $S_Q^+ := \{j_Q \in C_Q \mid j \in C \setminus (I \cup J)\}$;
  $S^+ := S_T^+ \cup S_Q^+$; $S^- := C \setminus \partial^+ M^\circ$;
  $E^+ := \{(i_Q, j_Q) \mid h \in R_Q, j \in J, P[h,j] \neq 0, i = base[h] \neq 0\}$;
  [$\tilde{E}$ is updated accordingly]
  If there exists in $\tilde{G} = (\tilde{V}, \tilde{E})$ a directed path from S^+ to S^- then go to Step 3; otherwise (including the case where $S^+ = \emptyset$ or $S^- = \emptyset$) stop with the conclusion that $\mathrm{rank}\, A = |M^\circ|$.
Step 3:
  Let L ($\subseteq \tilde{E}$) be (the set of arcs on) a shortest path from S^+ to S^- ("shortest" in the number of arcs);
  $M^\circ := (M^\circ \setminus L) \cup \{(j,i) \mid (i,j) \in L \cap E_T\} \cup \{(j, j_Q) \mid (j_Q, j) \in L \cap E_Q\}$;
  If the initial vertex ($\in S^+$) of the path L belongs to S_Q^+, then do the following:
    {Let j_Q ($\in S_Q^+ \subseteq C_Q$) be the initial vertex;
    Find h such that $base[h] = 0$ and $P[h,j] \neq 0$;
    [$j \in C$ corresponds to $j_Q \in C_Q$]
    $base[h] := j$; $w := 1/P[h,j]$;
    $P[k,l] := P[k,l] - w \times P[k,j] \times P[h,l] \quad (k \in R_Q \setminus \{h\}, l \in C \setminus \{j\})$;
    $S[k,l] := S[k,l] - w \times P[k,j] \times S[h,l] \quad (k \in R_Q \setminus \{h\}, l \in R_Q)$;
    $P[k,j] := 0 \quad (k \in R_Q \setminus \{h\})$ };
  For all $(i_Q, j_Q) \in L \cap E^+$ (in the order from S^+ to S^- along L) do the following:
    {Find h such that $i = base[h]$;   [$j \in C$ corresponds to $j_Q \in C_Q$]
    $base[h] := j$; $w := 1/P[h,j]$;
    $P[k,l] := P[k,l] - w \times P[k,j] \times P[h,l] \quad (k \in R_Q \setminus \{h\}, l \in C \setminus \{j\})$;
    $S[k,l] := S[k,l] - w \times P[k,j] \times S[h,l] \quad (k \in R_Q \setminus \{h\}, l \in R_Q)$;

$P[k, j] := 0 \quad (k \in R_Q \setminus \{h\})$ };
Go to Step 2. □

In the above algorithm, $\partial^+ M^\circ \cap C$ is an independent set in $\mathbf{M}(Q) \vee \mathbf{M}(T)$, since I is independent in $\mathbf{M}(Q)$ and $(\partial^+ M^\circ \cap C) \setminus I$ is independent in $\mathbf{M}(T)$. Since $\mathbf{M}(Q) \vee \mathbf{M}(T) = \mathbf{M}(A)$ by Theorem 4.2.3, we have $\mathrm{rank}\, A[R, \partial^+ M^\circ \cap C] = |M^\circ|$. At each execution of Step 3 the size of M° increases by one, and at the termination of the algorithm we have the relation: $\mathrm{rank}\, A = |M^\circ|$.

The updates of P in Step 3 are the standard pivoting operations on P, which is a matrix over the subfield $\boldsymbol{K}$. The sparsity of P should be taken into account in actual implementations of the algorithm. Computational techniques developed for solving sparse linear programs can be utilized here. As indicated in Step 3, pivoting operations are required for each arc $(i_Q, j_Q) \in L \cap E^+$. It is important to traverse the path L from S^+ to S^-, not from S^- to S^+, to avoid unnecessary fill-ins. See Murota–Scharbrodt [241] for other implementation issues.

The above algorithm uses arithmetic operations in the subfield $\boldsymbol{K}$ only, and, according to the result of Cunningham [43], runs in $\mathrm{O}(n^3 \log n)$, where $m = m_Q + m_T = \mathrm{O}(n)$ is assumed for simplicity in this complexity bound. The algorithm will be efficient enough for practical applications (see §4.4.6). Theoretically (but not practically) the rank of an LM-matrix can be computed in $(n^{2.62})$ time, according to Gabow and Xu [84].

Example 4.2.22. The algorithm above is illustrated here for the 4×5 LM-matrix used in Example 4.2.19:

$$A = \begin{pmatrix} Q \\ T \end{pmatrix} = \begin{array}{c|ccccc|} & x_1 & x_2 & x_3 & x_4 & x_5 \\ \hline & 1 & 1 & 1 & 1 & 0 \\ & 0 & 2 & 1 & 1 & 0 \\ \hline f_1 & t_1 & 0 & 0 & 0 & t_2 \\ f_2 & 0 & t_3 & 0 & 0 & t_4 \\ \hline \end{array}$$

where $\mathrm{Col}(A) = C = \{x_1, x_2, x_3, x_4, x_5\}$ and $\mathrm{Row}(T) = R_T = \{f_1, f_2\}$. We work with a 2×5 matrix P, a 2×2 matrix S, and a vector *base* of size 2. The copy of C is denoted as $C_Q = \{x_{1Q}, x_{2Q}, x_{3Q}, x_{4Q}, x_{5Q}\}$.

The flow of computation is traced below.

Step 1: $M^\circ := \emptyset$;

$$\mathit{base} := \begin{array}{c|c|} r_1 & 0 \\ r_2 & 0 \end{array}, \quad P := \begin{array}{c|ccccc|} & x_1 & x_2 & x_3 & x_4 & x_5 \\ \hline r_1 & 1 & 1 & 1 & 1 & 0 \\ r_2 & 0 & 2 & 1 & 1 & 0 \\ \hline \end{array}, \quad S := \begin{array}{|cc|} 1 & 0 \\ 0 & 1 \end{array}.$$

Step 2: $I := \emptyset$; $J := \{x_5\}$;
$S_T^+ := \{f_1, f_2\}$; $S_Q^+ := \{x_{1Q}, x_{2Q}, x_{3Q}, x_{4Q}\}$;
$S^+ := \{f_1, f_2, x_{1Q}, x_{2Q}, x_{3Q}, x_{4Q}\}$; $S^- := \{x_1, x_2, x_3, x_4, x_5\}$;
$E^+ := \emptyset$;
There exists a path from S^+ to S^-. [See $\tilde{G}^{(0)}$ in Fig .4.2]

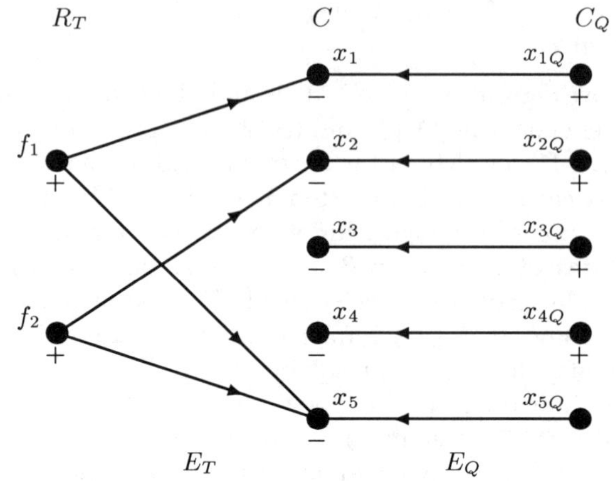

Fig. 4.2. Graph $\tilde{G}^{(0)}$ (+: vertex in S^+; −: vertex in S^-)

Step 3: $L := \{(x_{1Q}, x_1)\}$; $M^\circ := \{(x_1, x_{1Q})\}$;
The initial vertex x_{1Q} of L is in S_Q^+, and the matrices are updated (with $h = r_1$) to

$$base := \begin{array}{c|c|}\cline{2-2} r_1 & x_1 \\ r_2 & 0 \\ \cline{2-2}\end{array}, \quad P := \begin{array}{c|ccccc|} & x_1 & x_2 & x_3 & x_4 & x_5 \\ \cline{2-6} r_1 & 1 & 1 & 1 & 1 & 0 \\ r_2 & 0 & 2 & 1 & 1 & 0 \\ \cline{2-6}\end{array}, \quad S := \begin{array}{|cc|}\hline 1 & 0 \\ 0 & 1 \\ \hline\end{array}.$$

Noting $L \cap E^+ = \emptyset$ we return to Step 2.
Step 2: $I := \{x_1\}$; $J := \{x_5\}$;
$S_T^+ := \{f_1, f_2\}$; $S_Q^+ := \{x_{2Q}, x_{3Q}, x_{4Q}\}$; $S^+ := \{f_1, f_2, x_{2Q}, x_{3Q}, x_{4Q}\}$;
$S^- := \{x_2, x_3, x_4, x_5\}$;
$E^+ := \emptyset$;
There exists a path from S^+ to S^-. [See $\tilde{G}^{(1)}$ in Fig. 4.3]
Step 3: $L := \{(x_{2Q}, x_2)\}$; $M^\circ := \{(x_1, x_{1Q}), (x_2, x_{2Q})\}$;
The initial vertex x_{2Q} of L is in S_Q^+, and the matrices are updated (with $h = r_2$) to

$$base := \begin{array}{c|c|}\cline{2-2} r_1 & x_1 \\ r_2 & x_2 \\ \cline{2-2}\end{array}, \quad P := \begin{array}{c|ccccc|} & x_1 & x_2 & x_3 & x_4 & x_5 \\ \cline{2-6} r_1 & 1 & 0 & 1/2 & 1/2 & 0 \\ r_2 & 0 & 2 & 1 & 1 & 0 \\ \cline{2-6}\end{array}, \quad S := \begin{array}{|cc|}\hline 1 & -1/2 \\ 0 & 1 \\ \hline\end{array}.$$

Noting $L \cap E^+ = \emptyset$ we return to Step 2.
Step 2: $I := \{x_1, x_2\}$; $J := \{x_3, x_4, x_5\}$;
$S_T^+ := \{f_1, f_2\}$; $S_Q^+ := \emptyset$; $S^+ := \{f_1, f_2\}$; $S^- := \{x_3, x_4, x_5\}$;
$E^+ := \{(x_{1Q}, x_{3Q}), (x_{1Q}, x_{4Q}), (x_{2Q}, x_{3Q}), (x_{2Q}, x_{4Q})\}$;
There exists a path from S^+ to S^-. [See $\tilde{G}^{(2)}$ in Fig. 4.4]

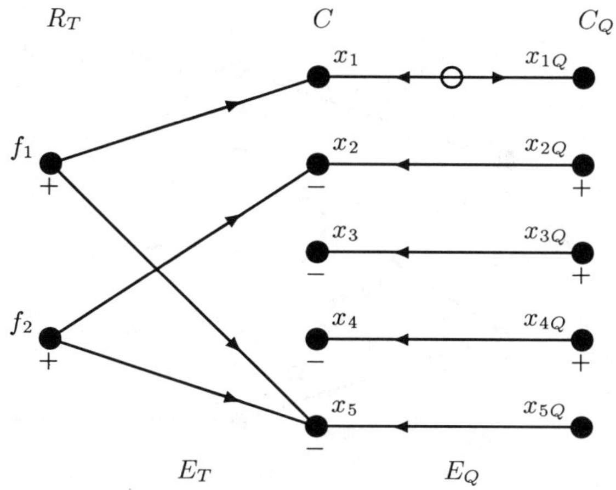

Fig. 4.3. Graph $\tilde{G}^{(1)}$ (○: arc in M; +: vertex in S^+; −: vertex in S^-)

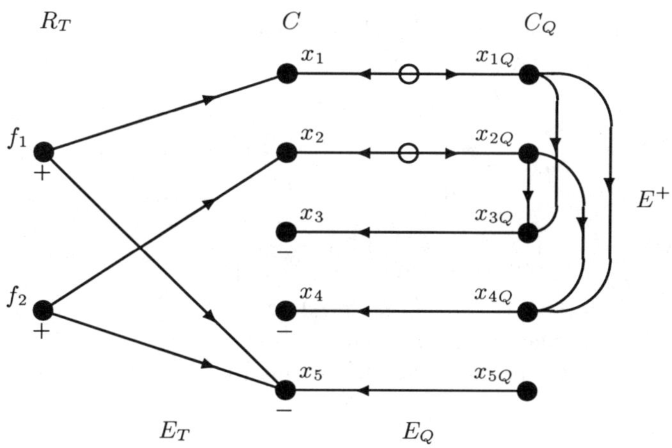

Fig. 4.4. Graph $\tilde{G}^{(2)}$ (○: arc in M; +: vertex in S^+; −: vertex in S^-)

Step 3: $L := \{(f_1, x_5)\}$; $M^\circ := \{(x_1, x_{1Q}), (x_2, x_{2Q}), (x_5, f_1)\}$;
The initial vertex $f_1 \notin S_Q^+$ and $L \cap E^+ = \emptyset$, and therefore the matrices remain unchanged and we return to Step 2.

Step 2: $I := \{x_1, x_2\}$; $J := \{x_3, x_4, x_5\}$;
$S_T^+ := \{f_2\}$; $S_Q^+ := \emptyset$; $S^+ := \{f_2\}$; $S^- := \{x_3, x_4\}$;
$E^+ := \{(x_{1Q}, x_{3Q}), (x_{1Q}, x_{4Q}), (x_{2Q}, x_{3Q}), (x_{2Q}, x_{4Q})\}$;
There exists a path from S^+ to S^-. [See $\tilde{G}^{(3)}$ in Fig. 4.5]

Step 3: $L := \{(f_2, x_2), (x_2, x_{2Q}), (x_{2Q}, x_{3Q}), (x_{3Q}, x_3)\}$;
$M^\circ := \{(x_1, x_{1Q}), (x_3, x_{3Q}), (x_5, f_1), (x_2, f_2)\}$;

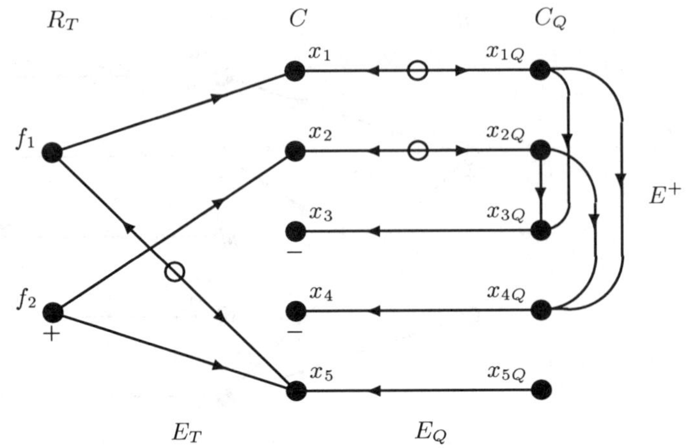

Fig. 4.5. Graph $\tilde{G}^{(3)}$ (○: arc in M; +: vertex in S^+; −: vertex in S^-)

The initial vertex $f_2 \notin S_Q^+$ and $L \cap E^+ = \{(x_{2Q}, x_{3Q})\}$, and the matrices are updated (with $h = r_2$) to

$$\mathit{base} := \begin{array}{c|c|} & \\ \cline{2-2} r_1 & x_1 \\ r_2 & x_3 \\ \cline{2-2} \end{array}, \quad P := \begin{array}{c|ccccc|} & x_1 & x_2 & x_3 & x_4 & x_5 \\ \cline{2-6} r_1 & 1 & -1 & 0 & 0 & 0 \\ r_2 & 0 & 2 & 1 & 1 & 0 \\ \cline{2-6} \end{array}, \quad S := \begin{array}{|cc|} \hline 1 & -1 \\ 0 & 1 \\ \hline \end{array}.$$

Step 2: $I := \{x_1, x_3\}$; $J := \{x_2, x_4, x_5\}$;
$S_T^+ := \emptyset$; $S_Q^+ := \emptyset$; $S^+ := \emptyset$; $S^- := \{x_4\}$;
$E^+ := \{(x_{1Q}, x_{2Q}), (x_{3Q}, x_{2Q}), (x_{3Q}, x_{4Q})\}$;
There exists no path from S^+ $(= \emptyset)$ to S^-;
We stop with the conclusion that rank $A = |M^\circ| = 4$.

[See $\tilde{G}^{(4)}$ in Fig. 4.6]

□

Remark 4.2.23. It is easy to observe that the arcs in M, directed from $R_T \cup C_Q$ to C, are never used in the shortest path from S^+ to S^-, whereas the reoriented arcs, implemented as M°, are indispensable. This means that the arcs of M could have been eliminated in the above algorithm for computing the rank. They are included, however, for the consistency with the algorithm for computing the CCF, to be presented in §4.4.4, in which the arcs of M are necessary. □

Algorithm for Mixed Matrices. The rank of a mixed matrix $A = Q + T$ can be computed by applying the above algorithm to the associated LM-matrix $\tilde{A} = \binom{\tilde{Q}}{\tilde{T}}$ of (4.4). Adaptation to the special form of $\tilde{A}$ results in some simplifications in the algorithm. Put $R = \mathrm{Row}(A)$ and $C = \mathrm{Col}(A)$.

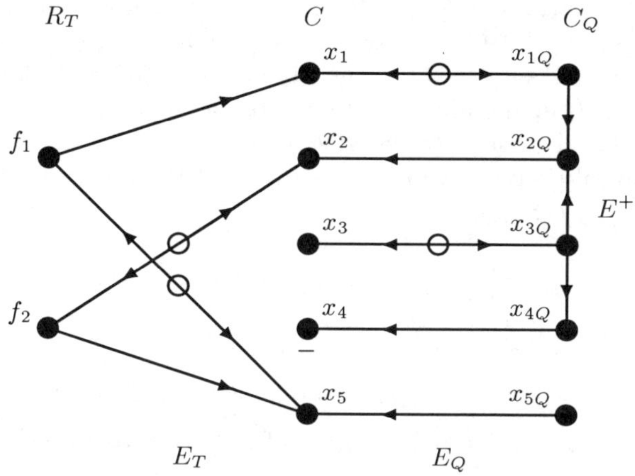

Fig. 4.6. Graph $\tilde{G}^{(4)}$ (○: arc in M; $S^+ = \emptyset$, −: vertex in S^-)

The vertex set in the general algorithm for $\tilde{A}$ would consist of three disjoint parts, say $\tilde{R}_T \cup \tilde{C}_Q \cup \tilde{C}$, where $\tilde{R}_T$ corresponds to $R \simeq \mathrm{Row}(\tilde{T})$, and $\tilde{C}$ and $\tilde{C}_Q$ are copies of $R \cup C \simeq \mathrm{Col}(\tilde{A})$.

First we exploit the structure of the matrix $\tilde{Q}$. In the matroid $\mathbf{M}(\tilde{Q})$, the column set corresponding to R is a basis because of the identity submatrix. Let $\tilde{M}_0$ be the set of arcs, from $\tilde{C}_Q$ to $\tilde{C}$, connecting the corresponding copies of R. Then we may take $\tilde{M}_0$ as the initial independent matching.

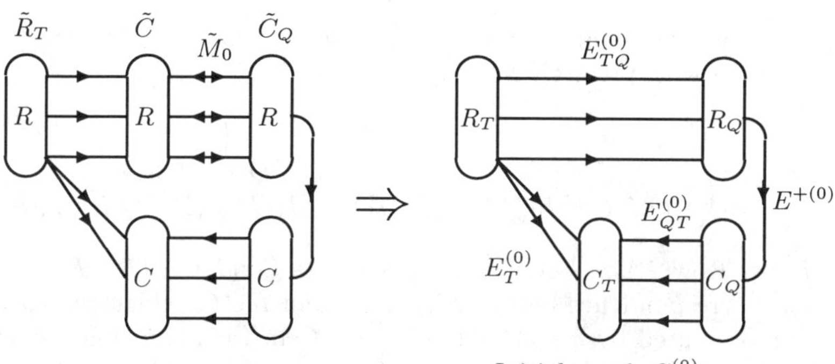

Fig. 4.7. Auxiliary graph for a mixed matrix

Next, by virtue of the diagonal matrix contained in the matrix $\tilde{T}$, the underlying graph can be simplified: the arcs in $\tilde{M}_0$ may be contracted (see Fig. 4.7). Namely, we may work on a graph with the vertex set consisting of four disjoint parts, $R_Q \cup C_Q \cup R_T \cup C_T$, where $R_Q = \mathrm{Row}(Q)$, $C_Q = \mathrm{Col}(Q)$, $R_T = \mathrm{Row}(T)$, and $C_T = \mathrm{Col}(T)$. We denote by $\varphi_Q : R \cup C \to R_Q \cup C_Q$ and $\varphi_T : R \cup C \to R_T \cup C_T$ the obvious one-to-one correspondences. The copies of $i \in R$ in R_Q and R_T are written as i_Q and i_T, respectively, that is, $i_Q = \varphi_Q(i)$ and $i_T = \varphi_T(i)$. Similarly, we write $j_Q = \varphi_Q(j)$ and $j_T = \varphi_T(j)$.

The initial graph, say $G^{(0)}$, has the arc set $E^{*(0)} \cup E^{+(0)}$, where $E^{*(0)} = E_{TQ}^{(0)} \cup E_{QT}^{(0)} \cup E_T^{(0)}$ and

$$\begin{aligned} E_{TQ}^{(0)} &= \{(i_T, i_Q) \mid i \in R\}, \\ E_{QT}^{(0)} &= \{(j_Q, j_T) \mid j \in C\}, \\ E_T^{(0)} &= \{(i_T, j_T) \mid T_{ij} \neq 0, i \in R, j \in C\}, \\ E^{+(0)} &= \{(i_Q, j_Q) \mid Q_{ij} \neq 0, i \in R, j \in C\}. \end{aligned}$$

The initial matching $\tilde{M}_0$ turns into an empty matching in the graph $G^{(0)}$, and a shortest path L is sought from $S^+ = R_T$ to $S^- = C_T$ in $G^{(0)}$ at the first stage of the algorithm.

At a general stage, we maintain $I \subseteq R$, $J \subseteq C$ and a matching $M \subseteq E_T^{(0)}$ in $(R_T, C_T; E_T^{(0)})$ such that $\partial^+ M \subseteq \varphi_T(I)$, $\partial^- M \subseteq \varphi_T(J)$, and $\hat{Q} \equiv Q[R \setminus I, C \setminus J]$ is nonsingular. This means that $I \cup (C \setminus J)$ is independent in $\mathbf{M}(\tilde{Q})$ and $(R \setminus I) \cup \varphi_T^{-1}(\partial^- M)$ is independent in $\mathbf{M}(\tilde{T})$, and therefore, $R \cup (C \setminus J) \cup \varphi_T^{-1}(\partial^- M)$ is independent in $\mathbf{M}(\tilde{A})$. Noting

$$\tilde{Q} = \begin{array}{c} \\ R \setminus I \\ I \end{array} \!\!\begin{array}{c} \begin{array}{cccc} R\setminus I & I & C \setminus J & J \end{array} \\ \begin{pmatrix} I_* & O & \hat{Q} & Q[R \setminus I, J] \\ O & I_* & Q[I, C \setminus J] & Q[I, J] \end{pmatrix} \end{array},$$

where I_* denotes an identity matrix of appropriate size, let P be the pivotal transform of Q with pivot $\hat{Q}$. Namely,

$$P = \begin{array}{c} \\ C \setminus J \\ I \end{array} \!\!\begin{array}{c} \begin{array}{cc} R\setminus I & J \end{array} \\ \begin{pmatrix} \hat{Q}^{-1} & \hat{Q}^{-1} Q[R \setminus I, J] \\ -Q[I, C \setminus J]\hat{Q}^{-1} & Q[I, J] - Q[I, C \setminus J]\hat{Q}^{-1} Q[R \setminus I, J] \end{pmatrix} \end{array},$$

where $R_P \equiv \mathrm{Row}(P) \simeq I \cup (C \setminus J)$ and $C_P \equiv \mathrm{Col}(P) \simeq (R \setminus I) \cup J$. The one-to-one correspondence between $R_P \cup C_P$ and $R \cup C$, which changes with (I, J), is represented by $\sigma : R_P \cup C_P \to R \cup C$ in the algorithm below. We start the algorithm with $I = R$, $J = C$, $M = \emptyset$, and $P = Q$.

The matching M, the sets I and J and the structure of P are represented by a graph $G = (V, E)$ with vertex set $V = R_Q \cup C_Q \cup R_T \cup C_T$ and arc set $E = E^* \cup E^+$, where $E^* = E_{TQ} \cup E_{QT} \cup E_T \cup M^\circ$ and

$$\begin{aligned} E_{TQ} &= \{(i_T, i_Q) \mid i \in I\} \cup \{(j_T, j_Q) \mid j \in C \setminus J\}, \\ E_{QT} &= \{(i_Q, i_T) \mid i \in R \setminus I\} \cup \{(j_Q, j_T) \mid j \in J\}, \end{aligned}$$

$$E_T = E_T^{(0)} \setminus M,$$
$$M^\circ = \{\overline{a} \mid a \in M\} \qquad (\overline{a}\text{: reorientation of } a),$$
$$E^+ = \{(i_Q, j_Q) \mid P_{ij} \neq 0, i \in (C \setminus J) \cup I, j \in (R \setminus I) \cup J\}.$$

Note that $I = \{i \in R \mid (i_T, i_Q) \in E\}$, $J = \{j \in C \mid (j_Q, j_T) \in E\}$, and $M = \{(i_T, j_T) \mid (j_T, i_T) \in E\}$. The entrance S^+ and the exit S^- are defined by

$$S^+ = \{i_T \in R_T \mid i \in I\} \setminus \partial^- M^\circ, \quad S^- = \{j_T \in C_T \mid j \in J\} \setminus \partial^+ M^\circ,$$

and a shortest path L is sought from S^+ to S^- in G.

Algorithm for computing the rank of a mixed matrix $A = Q + T$

Step 1:

$E^* := \{(i_T, i_Q) \mid i \in R\} \cup \{(j_Q, j_T) \mid j \in C\}$
$\qquad \cup \{(i_T, j_T) \mid T_{ij} \neq 0, i \in R, j \in C\}$;
$P[i,j] := Q_{ij}\ (i \in R, j \in C)$; $\quad \sigma(i) := i\ (i \in R \cup C)$;

Step 2:

$I := \{i \in R \mid (i_T, i_Q) \in E^*\}$; $\quad J := \{j \in C \mid (j_Q, j_T) \in E^*\}$;
$M^\circ := \{(j_T, i_T) \in E^*\}$;
$S^+ := \{i_T \in R_T \mid i \in I\} \setminus \partial^- M^\circ$; $\quad S^- := \{j_T \in C_T \mid j \in J\} \setminus \partial^+ M^\circ$;
$E^+ := \{(i_Q, j_Q) \mid P[\sigma^{-1}(i), \sigma^{-1}(j)] \neq 0, i \in I \cup (C \setminus J), j \in (R \setminus I) \cup J\}$;
$E := E^* \cup E^+$;
If there exists in $G = (V, E)$ a directed path from S^+ to S^- then go to Step 3; otherwise (including the case where $S^+ = \emptyset$ or $S^- = \emptyset$) stop with the conclusion that $\operatorname{rank} A = |M^\circ| + |C \setminus J|$.

Step 3:

Let $L\ (\subseteq E)$ be (the set of arcs on) a shortest path from S^+ to S^- ("shortest" in the number of arcs);
$E^* := (E^* \setminus L) \cup \{\overline{a} \mid a \in L \cap E^*\}$; [Reverse arcs in $L \cap E^*$]
For all $(i_Q, j_Q) \in L \cap E^+$ (in the order from S^+ to S^- along L) do the following:
$\{h := \sigma^{-1}(i)$; $\quad g := \sigma^{-1}(j)$; $\quad \sigma(h) := j$; $\quad \sigma(g) := i$;
$w := 1/P[h,g]$; $\quad P[h,g] := w$;
$P[k,g] := -w \times P[k,g] \quad (k \in R_P \setminus \{h\})$;
$P[h,l] := w \times P[h,l] \quad (l \in C_P \setminus \{g\})$;
$P[k,l] := P[k,l] - w \times P[k,g] \times P[h,l] \quad (k \in R_P \setminus \{h\}, l \in C_P \setminus \{g\})$
$\}$;
Go to Step 2. □

4.3 Structural Solvability of Systems of Equations

4.3.1 Formulation of Structural Solvability

The unique solvability of a system of linear equations is obviously equivalent to the nonsingularity of the coefficient matrix. In this section we consider the solvability of a system of linear/nonlinear equations from a combinatorial structural point of view. A mathematical formalism is given to the intuitive idea that a system of linear/nonlinear equations has a structure that admits a unique solution in general. The notion of "structural solvability" in its crude form seems to have been proposed first by Iri–Tsunekawa–Yajima [135] along with a graph-theoretic criterion for checking it. The present formulation is due to Iri–Tsunekawa–Murota [134] and Murota–Iri [237, 238].

We consider a system of equations in the following "standard form" with unknowns x_j $(j = 1, \cdots, N)$ and u_k $(k = 1, \cdots, K)$, and parameters y_i $(i = 1, \cdots, M)$:

$$\begin{aligned} y_i &= f_i(\boldsymbol{x}, \boldsymbol{u}) \qquad (i = 1, \cdots, M), \\ u_k &= g_k(\boldsymbol{x}, \boldsymbol{u}) \qquad (k = 1, \cdots, K), \end{aligned} \tag{4.27}$$

where f_i $(i = 1, \cdots, M)$ and g_k $(k = 1, \cdots, K)$ are assumed to be sufficiently smooth real-valued functions. This form is most natural and convenient when treating a physical/engineering system represented by a set of functional relations among elementary state variables, where, for arbitrarily given values of y-variables, the values of x- and u-variables are adjusted so that all the equations may be satisfied.

We are concerned with whether the system (4.27) of equations has a structure which admits a unique solution. In the following we assume that $M = N$, since the number of equations must usually be equal to the number of unknowns in order for (4.27) to have a unique solution. We denote the Jacobian matrix of (4.27) with respect to $\boldsymbol{x}$ and $\boldsymbol{u}$ by

$$J(\boldsymbol{x}, \boldsymbol{u}) = \begin{pmatrix} J[f, x] & J[f, u] \\ J[g, x] & J[g, u] - I_K \end{pmatrix}, \tag{4.28}$$

where

$$J[f, x] = \left(\frac{\partial f_i}{\partial x_j} \right), \quad J[f, u] = \left(\frac{\partial f_i}{\partial u_l} \right),$$
$$J[g, x] = \left(\frac{\partial g_k}{\partial x_j} \right), \quad J[g, u] = \left(\frac{\partial g_k}{\partial u_l} \right).$$

Suppose that (4.27) has a solution $(\boldsymbol{x}, \boldsymbol{u}) = (\hat{\boldsymbol{x}}, \hat{\boldsymbol{u}})$ for some $\boldsymbol{y} = \hat{\boldsymbol{y}}$. It follows from the implicit-function theorem (cf., e.g., Spivak [302]) that, if

$$\det J(\hat{\boldsymbol{x}}, \hat{\boldsymbol{u}}) \neq 0, \tag{4.29}$$

(4.27) has a unique solution $(\boldsymbol{x}, \boldsymbol{u})$ around $(\hat{\boldsymbol{x}}, \hat{\boldsymbol{u}})$ in accordance with an arbitrary perturbation of $\boldsymbol{y}$ in a neighborhood of $\hat{\boldsymbol{y}}$. It should be noted also that, from the computational point of view, the condition (4.29) guarantees the feasibility of a Newton-like iterative method for the numerical solution of (4.27) with x_j $(j = 1, \cdots, N)$ and u_k $(k = 1, \cdots, K)$ as unknowns.

The above condition (4.29), however, depends not only on the functional forms of f_i and g_k but also on particular values of $(\hat{\boldsymbol{x}}, \hat{\boldsymbol{u}})$, which are usually not known before we start numerical computation. Furthermore, it is difficult to distinguish numerically the "exact zero" from a "very small" number due to the existence of rounding errors. Hence, we will consider an alternative condition that the Jacobian, as a function in x_j ($j = 1, \cdots, N$) and u_k ($k = 1, \cdots, K$), does not vanish identically:

$$\det J(\boldsymbol{x}, \boldsymbol{u}) \neq 0.$$

More precisely, we shall regard the partial derivatives of functions f_i and g_k as elements of a field $\boldsymbol{F}$ which is an extension of the rational number field $\mathbf{Q}$. That is, denoting by

$$\mathcal{D} = \{\partial f_i/\partial x_j, \partial f_i/\partial u_l, \partial g_k/\partial x_j, \partial g_k/\partial u_l \mid \\ i = 1, \cdots M; j = 1, \cdots, N; k, l = 1, \cdots, K\}$$

the collection of partial derivatives of f_i and g_k, we adopt

$$\text{Basic Assumption:} \quad \mathcal{D} \subseteq \boldsymbol{F}.$$

This assumption is literally valid, for example, if f_i and g_k are rational functions of x_j ($j = 1, \cdots, N$) and u_l ($l = 1, \cdots, K$), in which case the field of all rational functions in x_j ($j = 1, \cdots, N$) and u_l ($l = 1, \cdots, K$) may be taken as the field $\boldsymbol{F}$. We say that the system (4.27) of equations is *structurally solvable* if the Jacobian matrix $J(\boldsymbol{x}, \boldsymbol{u})$ of (4.28), as a matrix over $\boldsymbol{F}$, is nonsingular, i.e., if

$$\det J(\boldsymbol{x}, \boldsymbol{u}) \neq 0 \quad \text{in } \boldsymbol{F}. \tag{4.30}$$

The structural solvability condition (4.30) implies a one-to-one correspondence between x_j ($j = 1, \cdots, N$) and y_i ($i = 1, \cdots, M$) in the following (structural) sense. The submatrix $J[g, u] - I$ is term-nonsingular under a plausible assumption that the diagonal entries of $J[g, u]$ are distinct from one (see §2.1.3 for the terminology of "term-nonsingular"). This means further that $J[g, u] - I$ is likely to be nonsingular. Then, by the implicit-function theorem, the second subsystem of (4.27):

$$u_k = g_k(\boldsymbol{x}, \boldsymbol{u}) \qquad (k = 1, \cdots, K) \tag{4.31}$$

can be solved for u_k as

$$u_k = u_k(\boldsymbol{x}) \qquad (k = 1, \cdots, K). \tag{4.32}$$

Substitution of (4.32) into the first subsystem of (4.27) yields a system of equations

$$y_i = f_i(\boldsymbol{x}, \boldsymbol{u}(\boldsymbol{x})) \qquad (i = 1, \cdots, M) \tag{4.33}$$

in unknowns x_j ($j = 1, \cdots, N$). The Jacobian matrix $J[y, x]$ of (4.33) is given by

$$J[y, x] = J[f, x] - J[f, u](J[g, u] - I)^{-1} J[g, x] \tag{4.34}$$

as long as $J[g,u] - I$ is nonsingular. In fact, this expression is derived from the relations among the differentials of (4.27):

$$\begin{aligned} \mathrm{d}\boldsymbol{y} &= J[f,x]\mathrm{d}\boldsymbol{x} + J[f,u]\mathrm{d}\boldsymbol{u}, \\ \mathrm{d}\boldsymbol{u} &= J[g,x]\mathrm{d}\boldsymbol{x} + J[g,u]\mathrm{d}\boldsymbol{u} \end{aligned}$$

through elimination of $\mathrm{d}\boldsymbol{u}$. If (4.27) is structurally solvable (4.30), the Jacobian matrix $J[y,x]$ above is nonsingular by a formula in matrix algebra (cf. Proposition 2.1.7):

$$\det \begin{bmatrix} A & B \\ C & D \end{bmatrix} = \det D \cdot \det[A - BD^{-1}C]$$

(where A and D are square matrices and $\det D \neq 0$), and hence x_j ($j = 1,\cdots,N$) and y_i ($i = 1,\cdots,M$) are in one-to-one correspondence, at least locally.

In the following we consider combinatorial characterizations of the structural solvability under two different generality assumptions, GA1 and GA2 introduced in §3.1.1. Generality assumption GA1 leads to a graph-theoretic method in §4.3.2, whereas GA2 to a matroid-theoretic method in §4.3.3.

4.3.2 Graphical Conditions for Structural Solvability

The structure of a system of equations in the standard form (4.27) can be expressed in terms of a graph with vertices corresponding to variables (i.e., unknowns $(\boldsymbol{x},\boldsymbol{u})$ and parameters $\boldsymbol{y}$) and arcs representing the existence of the explicit direct functional dependence. To be concrete, we consider the vertex set $X \cup U \cup Y$, where $X = \{x_1,\cdots,x_N\}$, $U = \{u_1,\cdots,u_K\}$ and $Y = \{y_1,\cdots,y_M\}$. The functional dependence $y_i = f_i(\boldsymbol{x},\boldsymbol{u})$ is expressed by a set of arcs coming into y_i from those x_j and u_l which effectively appear in f_i. In a similar manner, the functional dependence $u_k = g_k(\boldsymbol{x},\boldsymbol{u})$ is expressed by a set of arcs coming into u_k from x_j and u_l appearing effectively in g_k.

The graph thus obtained may be regarded as a kind of signal-flow graph representing the causal relation among variables, or the flow of information in the system. This graph is called the *representation graph* (Iri–Tsunekawa–Murota [134]) of the system of equations. When it is acyclic, it is also called the *computational graph* (Bauer [10]), in which emphasis is laid on the aspect that it represents the order of successive function evaluations according to which the values of y_i are computed from those of x_j.

Example 4.3.1. For a system of equations:

$$\begin{aligned} y_1 &= f_1(x_1,u_1,u_3), & u_1 &= g_1(x_1,u_2), \\ y_2 &= f_2(u_1,u_3), & u_2 &= g_2(x_1,x_2,u_3), \\ y_3 &= f_3(x_2,u_3,u_4), & u_3 &= g_3(u_1), \\ & & u_4 &= g_4(x_2,x_3,u_3), \end{aligned}$$

the representation graph G is shown in Fig. 4.8. □

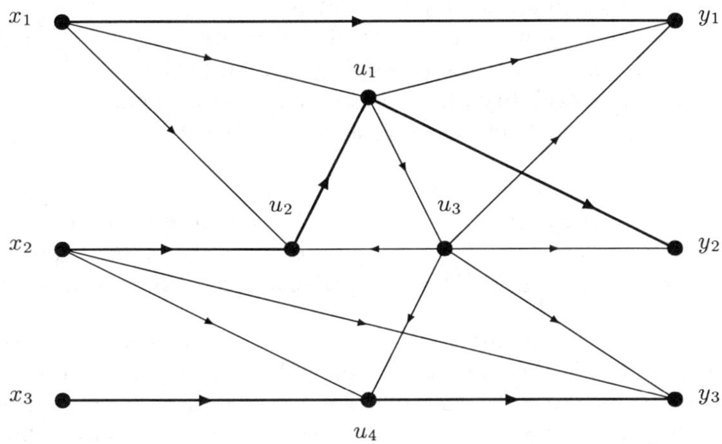

Fig. 4.8. Representation graph of Example 4.3.1 (linking arcs in thick lines)

By the definition of the sets X, Y, and U, the representation graph of a system of equations in the standard form satisfies the following properties:

i) Each vertex $x_j \in X$ has no in-coming arcs, and vice versa.
ii) Each vertex $y_i \in Y$ has no out-going arcs. (Some of the vertices of U may possibly have no out-going arcs.)

Note that the representation graph expresses nothing more than the existence of functional dependence among variables, concrete functional forms being disregarded.

The objective of this section is to translate the structural solvability condition (4.30) into a condition on the representation graph under the generality assumption

GA1: The nonvanishing elements of $\mathcal{D}$ are algebraically independent over $\mathbf{Q}$

about the collection $\mathcal{D}$ of the partial derivatives. Structural solvability under GA1 is equivalent to generic solvability when the nonvanishing elements of $\mathcal{D}$ are regarded as independent parameters. It is also noted that the structural solvability under GA1 is equivalent to the nonsingularity of $J[y,x]$ of (4.34) since GA1 guarantees the nonsingularity of $J[g,u]-I$.

The generality assumption GA1 can be partly justified as follows. When the system of equations describes a physical system, the functions f_i and g_k represent element characteristics which cannot be free from noises and/or errors. Hence the nonvanishing partial derivatives of f_i and g_k, even when they are constant (i.e., when the functions are linear), are so "general" that they do not satisfy any polynomial relation with integer coefficients. Thus we are led to GA1. It is admitted at the same time that concrete numerical data stored in a computer with a finite number of digits cannot satisfy this assumption in the rigorous mathematical sense, and that the assumption is sometimes too stringent to be satisfied in practical problems.

The structural solvability (4.30) under GA1 is equivalent to the existence of a Menger-type perfect linking in G as follows (see §2.2.4 for Menger-type linkings).

Theorem 4.3.2. *A system of equations in the standard form* (4.27) *is structurally solvable under* GA1 *if and only if there exists on the representation graph* $G = (X \cup U \cup Y, A; X, Y)$ *a Menger-type perfect linking from* X *to* Y.

Proof. First, the diagonal entries of $J[g,u] - I$ are distinct from zero by GA1. Next, $J(\boldsymbol{x}, \boldsymbol{u})$ of (4.28) is nonsingular if and only if it is term-nonsingular. This follows from Proposition 2.1.12 when combined with a simple observation that multiplication of the last K rows of $J(\boldsymbol{x}, \boldsymbol{u})$ by algebraically independent numbers yields a matrix with algebraically independent nonvanishing entries.

It remains to show the equivalence of the term-nonsingularity of $J(\boldsymbol{x}, \boldsymbol{u})$ and the existence of a Menger-type perfect linking.

Suppose $J(\boldsymbol{x}, \boldsymbol{u})$ is term-nonsingular. Fix a bijection $\pi : X \cup U \to Y \cup U$ such that $J_{\pi(v)v} \neq 0$ ($\forall\, v \in X \cup U$). Obviously $M = N$. For each $x_j \in X$ ($1 \leq j \leq N$) determine a sequence $u_{k_1}, u_{k_2}, \cdots, u_{k_{m(j)}} \in U$ and $y_{\sigma(j)} \in Y$ by $\pi(x_j) = u_{k_1}$, $\pi(u_{k_1}) = u_{k_2}$, $\cdots$, $\pi(u_{k_{m(j)-1}}) = u_{k_{m(j)}}$, and $\pi(u_{k_{m(j)}}) = y_{\sigma(j)}$. Such sequences for different j have no vertex in common, and the collection of such sequences gives a Menger-type perfect linking in G.

Conversely, suppose that there exists a Menger-type perfect linking in G, and let U' ($\subseteq U$) denote the set of u-vertices lying on the linking. Then $M = N$ and the linking gives a bijection $\pi : X \cup U' \to Y \cup U'$ such that $J_{\pi(v)v} \neq 0$ ($\forall\, v \in X \cup U'$). The bijection π can be extended to $\pi : X \cup U \to Y \cup U$ such that $J_{\pi(v)v} \neq 0$ ($\forall\, v \in X \cup U$) by defining $\pi(v) = v$ for $v \in U \setminus U'$. This shows the term-nonsingularity of $J(\boldsymbol{x}, \boldsymbol{u})$. ■

The above criterion for structural solvability was put to practical use in a chemical process simulator developed in Japan in the seventies (IJUSE [121], ITPA [118], ITPA–IJUSE [119, 120], and Sebastian–Noble–Thambynayagam–Wood [293]). See Murota [204, §10] for an account of other graph-theoretic techniques employed there.

Example 4.3.3. Recall the system of equations in Example 4.3.1. As shown in Fig. 4.8, there exists a Menger-type perfect linking: $x_1 \to y_1$, $x_2 \to u_2 \to u_1 \to y_2$, $x_3 \to u_4 \to y_3$, in the representation graph G. Hence, by Theorem 4.3.2, this system of equations is structurally solvable under GA1. □

Next we turn to another example which is not structurally solvable. This motivates us to look at minimum separators as the reason for the failure of structural solvability.

Example 4.3.4. A system of equations:

$$\begin{aligned} y_1 &= f_1(u_1, u_3), & u_1 &= g_1(x_1, x_2, u_2), \\ y_2 &= f_2(u_2, u_3), & u_2 &= g_2(x_2, x_3, u_3), \\ y_3 &= f_3(u_2), & u_3 &= g_3(u_1) \end{aligned}$$

has the representation graph G shown in Fig. 4.9, in which there is a path from any x_j $(j = 1, 2, 3)$ to any y_i $(i = 1, 2, 3)$. However, since a maximum linking from $X = \{x_1, x_2, x_3\}$ to $Y = \{y_1, y_2, y_3\}$ is of size 2, not a perfect linking, Theorem 4.3.2 reveals that this system is not structurally solvable.

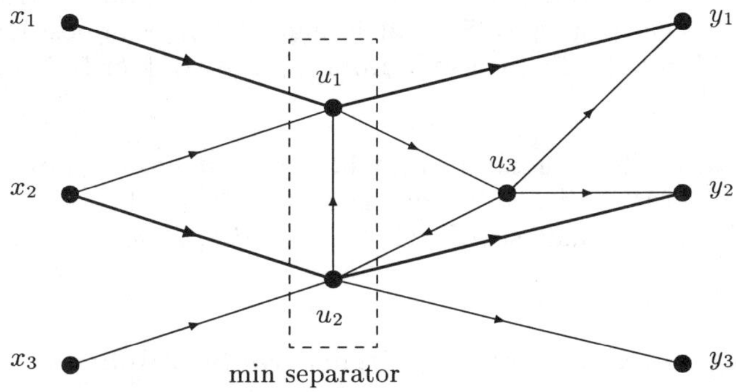

Fig. 4.9. Representation graph of Example 4.3.4 (linking arcs in thick lines)

An intuitive interpretation of this fact would be that three degrees of freedom at the entrance X are reduced to two of the intermediate variables, u_1 and u_2, and, as a result, it is not possible in general to adjust the values of x_1, x_2 and x_3 so as to make y_1, y_2 and y_3 equal to arbitrarily prescribed values.

More precisely, we may say the following, referring to Menger's theorem (Theorem 2.2.31). In the representation graph G of Fig. 4.9, $\{u_1, u_2\}$ is a minimum separator of (X, Y). The cardinality of a minimum separator in the representation graph of the system (4.27) of equations may be interpreted as the effective degrees of freedom of the system. Thus, for a system of equations not necessarily structurally solvable, a minimum separator in its representation graph can reveal where the inconsistency comes from. □

Remark 4.3.5. Theorem 4.3.2, as well as the argument in Example 4.3.4, suggests that some meaningful decomposition of a system of equations should be obtained through a decomposition of its representation graph based on maximum linkings and minimum separators. In fact, this idea has been worked out by Murota [196, 205] and the obtained decomposition is named "Menger-decomposition" (or "M-decomposition" for short). The M-decomposition is constructed as follows. The linking problem can be formulated as a network flow problem (see §2.2.4), maximum linkings corresponding to maximum flows and minimum separators to minimum cuts. On the other hand, the submodularity (2.52) of the cut capacity function of a network leads to a canonical decomposition with respect to minimum cuts, according to the Jordan–Hölder-type theorem for submodular functions explained in §2.2.2. The essence of the M-decomposition is a straightforward combination of these two results. See Murota [196, 197, 205] as well as Murota [204, §8, §11] for details about M-decomposition and its application to systems of equations, and van der Woude [327] for its application to control theoretic problems. Another decomposition of the representation graph is also proposed by Iri–Tsunekawa–Yajima [135], and is named "L-decomposition" by Iri–Tsunekawa–Murota [134]; see also Murota [204, §8, §11] for L-decomposition. □

Remark 4.3.6. The structural solvability for systems of equations with degrees of freedom is discussed by Sugihara [304, 305]. This is closely related to the combinatorial analysis of rigidity in statics, as expounded in Recski [277]. □

4.3.3 Matroidal Conditions for Structural Solvability

With the aid of the combinatorial characterizations of the rank of a mixed matrix, we can deal with the structural solvability of a system of equations (4.27) under more realistic generality assumptions such as

GA2: Those elements of $\mathcal{D}$ which do not belong to the rational number field $\mathbf{Q}$ are algebraically independent over $\mathbf{Q}$, and

GA3: Those elements of $\mathcal{D}$ which do not belong to the real number field $\mathbf{R}$ are algebraically independent over $\mathbf{R}$,

where $\mathcal{D}$ denotes the collection of the partial derivatives of the equations. Recall that the generality assumptions GA2 and GA3 have been introduced in §3.1.1 on the basis of the physical observation on the two kinds of numbers.

To be specific, we assume GA2 (and GA3 can be treated similarly). The set $\mathcal{D}$ is divided into two parts, $\mathcal{D} = \mathcal{Q} \cup \mathcal{T}$ with $\mathcal{Q} = \mathcal{D} \cap \mathbf{Q}$ and $\mathcal{T} = \mathcal{D} \setminus \mathbf{Q}$. Accordingly, the Jacobian matrix $A = J(\boldsymbol{x}, \boldsymbol{u})$ is expressed as $A = Q + T$, which is, by GA2, a mixed matrix with respect to $(\mathbf{Q}, \boldsymbol{F})$.

The following theorem gives a matroid-theoretic criterion for the structural solvability of the system (4.27) of equations under the realistic assumption GA2.

Theorem 4.3.7. *Let the Jacobian matrix* $A = J(\boldsymbol{x}, \boldsymbol{u})$ *of* (4.28) *be decomposed into two parts,* $A = Q + T$*, such that* Q *is a matrix over* $\mathbf{Q}$ *and the nonzero entries of* T *do not belong to* $\mathbf{Q}$*. Then the system* (4.27) *of equations* *is structurally solvable under* GA2 *if and only if* $M = N$ *and the maximum size of a common independent set of* $\mathbf{M}([I_{M+K} \mid Q])^*$ *and* $\mathbf{M}([I_{M+K} \mid T])$ *is equal to* $N + K$.

Proof. Apply the rank formula of Theorem 4.2.10 to $A = Q + T$, which is a mixed matrix under GA2. ■

Theorem 4.3.7 implies that we can test for the structural solvability under GA2 by the efficient matroid-theoretic algorithm of §4.2.4 using arithmetic operations on rational numbers. It is important in practice that the entries of Q are often simple integers and it seems, empirically, that no serious numerical difficulty arises from the round-off errors in handling those "rational" numbers.

Example 4.3.8. By way of the hypothetical ethylene dichloride production system described in Example 3.1.3, we will demonstrate the effectiveness of Theorem 4.3.7 as compared to the graph-theoretic criterion (Theorem 4.3.2) for the structural solvability under the assumption GA1.

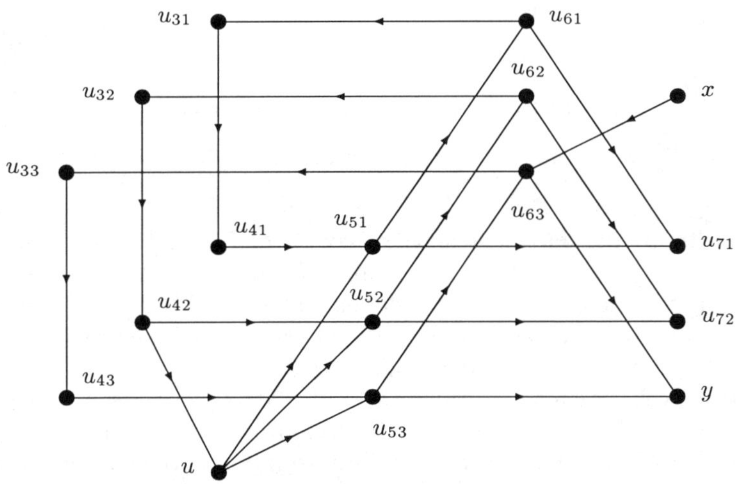

Fig. 4.10. Representation graph of the system (3.5) (cf. Example 3.1.3)

The system (3.5) of equations is in the form (4.27) with $M = N = 1$ and $K = 15$. The representation graph of this system, as defined in §4.3.2, is depicted in Fig. 4.10, on which a Menger-type perfect linking (e.g., $x \to u_{63} \to y$) exists from the x-vertex $\{x\}$ to the y-vertex $\{y\}$. Therefore the graph-theoretic method (Theorem 4.3.2), assuming GA1, would conclude that this system was structurally solvable, in contradiction to the fact that the Jacobian of this system (Fig. 3.4) vanishes for any value of a_1, a_2, r, x, and u_{53}.

This contradiction stems from the assumption GA1, which obviously fails to hold in this case. In fact, in the DM-decomposition of the Jacobian matrix, shown in Fig. 4.11, we can detect the rank deficiency in the 4×4 block corresponding to variables $\{u_{43}, u_{33}, u_{63}, u_{53}\}$. A more adequate assumption for this problem would be the GA2, which implies the choice of $\boldsymbol{K} = \mathbf{Q}$ and $\mathcal{T} = \{a_1, a_2, r, x, u_{53}\}$.

	u_{71}	u_{61}	u_{51}	u_{41}	u_{31}	x	u_{43}	u_{33}	u_{63}	u_{53}	u_{72}	u	u_{62}	u_{52}	u_{42}	u_{32}
u_{71}	-1	-1	1													
u_{61}		-1	a_1	0	0											
u_{51}		0	-1	1	0							-1				
u_{41}		0	0	-1	1											
u_{31}		1	0	0	-1											
u_{63}						u_{53}			-1	x						
u_{53}							1	0	0	-1		1				
u_{43}							-1	1	0	0						
u_{33}							0	-1	1	0						
y							0	0	-1	1						
u_{72}											-1		-1	1		
u												-1	0	0	r	0
u_{62}												0	-1	a_2	0	0
u_{52}												-1	0	-1	1	0
u_{42}												0	0	0	-1	1
u_{32}												0	1	0	0	-1

Fig. 4.11. DM-decomposition of the Jacobian matrix of (3.5) (cf. Fig. 3.4)

Accordingly, the Jacobian matrix, say A, of Fig. 3.4 is recognized as a mixed matrix with respect to $\mathbf{Q}$, to which the algorithm of §4.2.4 is applied. The maximum size of a common independent set of $\mathbf{M}([I \mid Q])^*$ and $\mathbf{M}([I \mid T])$ is equal to 15, whereas $N + K = 16$. Theorem 4.3.7 then reveals that the system (3.5) is not structurally solvable.

Alternatively, we may consider the LM-matrix (4.6) associated with A, which is given in Fig. 4.12. Since A contains four mixed rows, namely, the rows indexed by u_{61}, u_{62}, u_{63} and u, it suffices to increase the size of the matrix by four. As a result the associated LM-matrix is 20×20. An implementation of the algorithm of §4.2.4 found the rank to be 19, with deficiency 1. Hence the system (3.5) is not structurally solvable. □

Example 4.3.9. Recall the electrical network of Example 3.1.2 containing mutual couplings. If we regard the set of the physical parameters $\{r_1, r_2, \alpha, \beta\}$ as being algebraically independent over $\mathbf{Q}$, assuming GA2, the coefficient matrix, say A, of (3.3) is a mixed matrix with respect to $(\mathbf{Q}, \boldsymbol{F})$ for $\boldsymbol{F} = \mathbf{Q}(r_1, r_2, \alpha, \beta)$, i.e., $A \in \mathrm{MM}(\mathbf{Q}, \boldsymbol{F}, 10, 10)$. It is expressed as $A = Q + T$ with

	w_1	w_2	w_3	w_4	x	u_{31}	u_{32}	u_{33}	u_{41}	u_{42}	u_{43}	u_{51}	u_{52}	u_{53}	u_{61}	u_{62}	u_{63}	u_{71}	u_{72}	u
u_{61}	1														-1					
u_{62}		1														-1				
u_{63}			1														-1			
u				1																-1
y														1			-1			
u_{31}						-1									1					
u_{32}							-1									1				
u_{33}								-1									1			
u_{41}						1			-1											
u_{42}							1			-1										
u_{43}								1			-1									
u_{51}									1			-1								-1
u_{52}										1			-1							-1
u_{53}											1			-1						1
u_{71}												1			-1			-1		
u_{72}													1			-1			-1	
u_{61}	$-t_1$											a_1								
u_{62}		$-t_2$											a_2							
u_{63}			$-t_3$		u_{53}									x						
u				$-t_4$						r										

Fig. 4.12. LM-matrix associated with Jacobian matrix of (3.5) (chemical process simulation in Example 3.1.3)

$$Q = \left[\begin{array}{ccccc|ccccc}
0 & 0 & 1 & 1 & 1 & & & & & \\
1 & 0 & 0 & 0 & -1 & & & & & \\
0 & 1 & -1 & 0 & 0 & & & & & \\
 & & & & & 1 & 0 & 0 & -1 & 1 \\
 & & & & & 0 & 1 & 1 & -1 & 0 \\
0 & & & & & -1 & & & & \\
 & 0 & & & & & -1 & & & \\
 & & 0 & & & & & -1 & & \\
 & & & -1 & & & & & 0 & \\
 & & & & 0 & & & & & -1
\end{array}\right],$$

$$T = \left[\begin{array}{ccccc|ccccc}
0 & 0 & 0 & 0 & 0 & & & & & \\
0 & 0 & 0 & 0 & 0 & & & & & \\
0 & 0 & 0 & 0 & 0 & & & & & \\
 & & & & & 0 & 0 & 0 & 0 & 0 \\
 & & & & & 0 & 0 & 0 & 0 & 0 \\
r_1 & & & & & 0 & & & & \\
 & r_2 & & & & & 0 & & & \\
 & & 0 & & & \alpha & & 0 & & \\
 & \beta & & 0 & & & & & 0 & \\
 & & & & 0 & & & & & 0
\end{array}\right].$$

Then we can apply Theorem 4.2.10 to check for the solvability of this electrical network.

Or alternatively, we may treat A as if it were a layered mixed matrix, $A \in \mathrm{LM}(\mathbf{Q}, \boldsymbol{F}; 5, 5, 10)$, as follows. On expressing A as

$$A = \begin{pmatrix} Q \\ T \end{pmatrix}$$

with

$$Q = \left[\begin{array}{cccccccccc} 0 & 0 & 1 & 1 & 1 & & & & & \\ 1 & 0 & 0 & 0 & -1 & & & & & \\ 0 & 1 & -1 & 0 & 0 & & & & & \\ & & & & & 1 & 0 & 0 & -1 & 1 \\ & & & & & 0 & 1 & 1 & -1 & 0 \end{array}\right],$$

$$T = \left[\begin{array}{cccccccccc} r_1 & & & & & -1 & & & & \\ & r_2 & & & & & -1 & & & \\ & & 0 & & & \alpha & & -1 & & \\ & \beta & & -1 & & & & & 0 & \\ & & & & 0 & & & & & -1 \end{array}\right]$$

and conceptually multiplying the rows of T by algebraically independent transcendentals, we can apply Theorem 4.2.2 or Theorem 4.2.3. We may take $J = \{\xi^3, \xi^4, \xi^5, \eta_3, \eta_4\}$ for the subset that attains the maximum (=10) on the right-hand side of (4.12). Therefore A is nonsingular.

The latter approach agrees with the established method for testing the unique solvability of an electrical network (Iri [127, 128], Iri–Tomizawa [131, 132], Petersen [267], Recski [275, 276, 277]). It is remarkable in the case of an electrical network that the matrix Q above, expressing the incidence relations in the underlying graph, is totally unimodular over $\mathbf{Z}$, and hence totally free from rounding errors in the pivoting operations. □

The structural solvability of two realistic problems in chemical engineering is investigated below by the matroid-theoretic method under the realistic assumption GA2.

Example 4.3.10 (Reactor-separator model). This example is taken from the reactor-separator model (EV-6) of Yajima–Tsunekawa–Kobayashi [344]. The original problem, involving 218 variables, is modified to the standard form (4.27) with 120 unknowns and as many equations; $N = M = 18$ and $K = 102$ in the notation of (4.27). The Jacobian matrix in Fig. 4.13 is sparse, containing 351 nonvanishing entries.

Before the matroid-theoretic method is considered, it is confirmed by the graph-theoretic method (by Theorem 4.3.2) that the whole system of equations is structurally solvable under the generality assumption GA1.

Of the 351 nonvanishing entries of the Jacobian matrix of size 120, 172 entries are rational constants (1 or -1) and the remaining 179 entries are

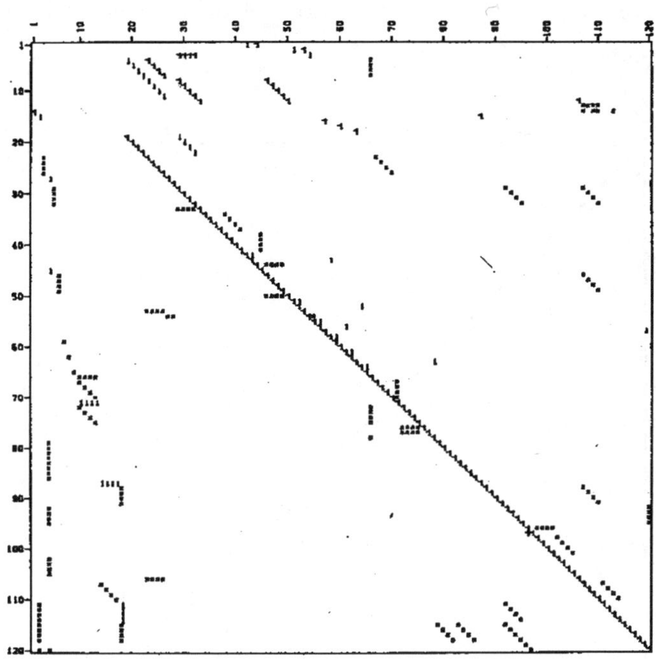

Fig. 4.13. Jacobian matrix of the reactor-separator model (Example 4.3.10)

regarded here as being algebraically independent, by assuming GA2. Accordingly, the Jacobian matrix A belongs to $\mathrm{MM}(\mathbf{Q}, \boldsymbol{F}; 120, 120)$. Then the maximum size of a common independent set $I \cup J$ ($I \subseteq \mathrm{Row}(A)$, $J \subseteq \mathrm{Col}(A)$) of $\mathbf{M}([I_{120} \mid Q])^*$ and $\mathbf{M}([I_{120} \mid T])$ is found to be 120 with $|I| = 91$ and $|J| = 29$. Therefore this system of equations remains to be structurally solvable under the more realistic assumption GA2.

In passing we mention the M- and L-decompositions (cf. Remark 4.3.5). This system is decomposed by the M-decomposition into 71 structurally solvable subproblems, of which only four components have more than one unknown variable; more precisely, the four components have 25, 10, 9 and 9 unknowns, respectively. The L-decomposition, on the other hand, leads to 47 structurally solvable subproblems, the largest being of size 48. See Murota [204, §11] for detailed data about these decompositions. □

Example 4.3.11 (Hydrogen production system). This example arises from an analysis of an industrial hydrogen production system. The standard form (4.27) of equations with $N = M = 13$ and $K = 531$ is obtained; it involves $N + K = 544$ unknowns and as many equations. Fig. 4.14 demonstrates the sparsity of the Jacobian matrix A, which has 1464 nonvanishing entries.

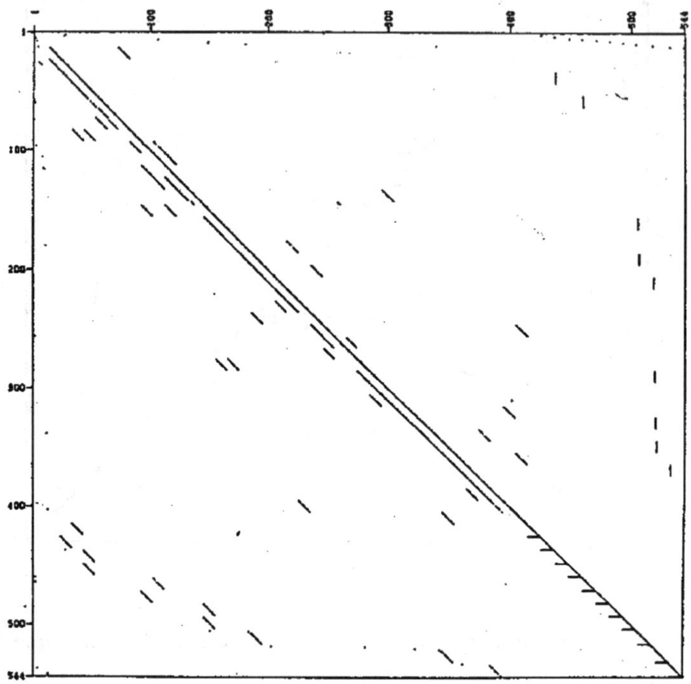

Fig. 4.14. Jacobian matrix of the hydrogen production system (Example 4.3.11)

The whole system is structurally solvable under GA1, as verified by the graph-theoretic method of Theorem 4.3.2.

Under the generality assumption GA2, the 1464 nonvanishing entries of the Jacobian matrix A are divided into 1142 rational constants (1 or -1) and 322 algebraically independent transcendentals. For a common independent set $I \cup J$ ($I \subseteq \mathrm{Row}(A), J \subseteq \mathrm{Col}(A)$) of $\mathbf{M}([I_{544} \mid Q])^*$ and $\mathbf{M}([I_{544} \mid T])$ such that $|J|$ is maximal, we have $|I| = 455$ and $|J| = 89$. It is noteworthy that the maximum size of J is much smaller than term-rank $T = 178$. It may also be remarked that no fractions are involved in the course of pivotal transformations of Q-matrix, although Q has not been proved to be totally unimodular.

We mention again the M- and L-decompositions (cf. Remark 4.3.5). The M-decomposition yields 268 structurally solvable subproblems, while the L-decomposition 234 subproblems. The size of an M-component varies from 1 to 104, whereas that of an L-component from 1 to 120. There is no substantial difference between the two decompositions in this example. See Murota [204, §11] for detailed data about these decompositions. □

We will return to the above problems in §4.4.6 to illustrate the application of a matroid-theoretic decomposition technique for systems of equations.

Notes. In Examples 4.3.8, 4.3.10, and 4.3.11, the problem data was provided by J. Tsunekawa and S. Kobayashi of the Institute of Japanese Union of Scientists and Engineers and the computation was done by M. Ichikawa [117].

4.4 Combinatorial Canonical Form of LM-matrices

4.4.1 LM-equivalence

For an LM-matrix $A = \binom{Q}{T} \in \mathrm{LM}(\boldsymbol{K}, \boldsymbol{F}; m_Q, m_T, n)$ we define an *LM-admissible transformation* to be a transformation of the form:

$$P_{\mathrm{r}} \begin{pmatrix} S & O \\ O & I \end{pmatrix} \begin{pmatrix} Q \\ T \end{pmatrix} P_{\mathrm{c}}, \tag{4.35}$$

where P_{r} and P_{c} are permutation matrices, and S is a nonsingular matrix over the ground field $\boldsymbol{K}$ (i.e., $S \in \mathrm{GL}(m_Q, \boldsymbol{K})$). An LM-admissible transformation brings an LM-matrix into another LM-matrix, since $\begin{pmatrix} S & O \\ O & I \end{pmatrix} \begin{pmatrix} Q \\ T \end{pmatrix} = \begin{pmatrix} SQ \\ T \end{pmatrix}$ and SQ is again a matrix over $\boldsymbol{K}$. Two LM-matrices are said to be *LM-equivalent* if they are connected by an LM-admissible transformation. If A' is LM-equivalent to A, then $\mathrm{Col}(A')$ may be identified with $\mathrm{Col}(A)$ through the permutation P_{c}.

The objective of this section is to consider a block-triangular decomposition of LM-matrices under the LM-admissible transformation (4.35). It will be shown that there exists a canonical proper block-triangular form ("proper" in the sense of §2.1.4) among the matrices LM-equivalent to a given LM-matrix. The canonical form is called the *combinatorial canonical form* or *CCF* for short.

In the special case where $m_Q = 0$, the LM-admissible transformation (4.35) reduces to $P_{\mathrm{r}} T P_{\mathrm{c}}$, involving permutations only. Accordingly the decomposition by means of the LM-admissible transformation reduces to the Dulmage–Mendelsohn decomposition. In the other extreme case where $m_T = 0$, the transformation (4.35) reduces to $P_{\mathrm{r}} S Q P_{\mathrm{c}}$, and the decomposition by means of (4.35) agrees with the ordinary Gauss–Jordan elimination in matrix computation. Hence, the theory of CCF to be developed here amounts to a natural amalgamation of the results on the DM-decomposition and the LU-decomposition.

Example 4.4.1. Consider a 3×3 LM-matrix

$$A = \begin{pmatrix} Q \\ \hline T \end{pmatrix} = \begin{pmatrix} 1 & 1 & 0 \\ 1 & 2 & 3 \\ \hline 0 & t_1 & t_2 \end{pmatrix},$$

where $\mathcal{T} = \{t_1, t_2\}$ is the set of algebraically independent parameters. This matrix cannot be decomposed into smaller blocks by means of permutations of rows and columns (DM-irreducible). However, by choosing $S = \begin{pmatrix} 1 & 0 \\ -1 & 1 \end{pmatrix}$ and $P_{\rm r} = P_{\rm c} = I$ in the LM-admissible transformation (4.35), we can obtain a block-triangular decomposition:

$$\bar{A} = \begin{pmatrix} SQ \\ T \end{pmatrix} = \left(\begin{array}{c|cc} 1 & 1 & 0 \\ \hline & 1 & 3 \\ & t_1 & t_2 \end{array}\right).$$

Thus the LM-admissible transformation is more powerful than mere permutations. □

Example 4.4.2. Here is an example containing a "tail" (nonsquare diagonal block as in the DM-decomposition). Recall the 4×5 LM-matrix

$$A = \begin{pmatrix} Q \\ \hline T \end{pmatrix} = \begin{array}{c} \\ \\ \\ f_1 \\ f_2 \end{array} \begin{array}{|ccccc|} \multicolumn{1}{c}{x_1} & x_2 & x_3 & x_4 & \multicolumn{1}{c}{x_5} \\ \hline 1 & 1 & 1 & 1 & 0 \\ 0 & 2 & 1 & 1 & 0 \\ \hline t_1 & 0 & 0 & 0 & t_2 \\ 0 & t_3 & 0 & 0 & t_4 \\ \hline \end{array}$$

used in Examples 4.2.19 and 4.2.22. By choosing

$$S = \begin{bmatrix} 1 & -1 \\ 0 & 1 \end{bmatrix}, \quad P_{\rm r} = \begin{bmatrix} 0&1&0&0 \\ 1&0&0&0 \\ 0&0&1&0 \\ 0&0&0&1 \end{bmatrix}, \quad P_{\rm c} = \begin{bmatrix} 0&0&1&0&0 \\ 0&0&0&1&0 \\ 1&0&0&0&0 \\ 0&1&0&0&0 \\ 0&0&0&0&1 \end{bmatrix}$$

in the LM-admissible transformation (4.35), we obtain the CCF:

$$\bar{A} = \begin{array}{c} \\ \\ \\ f_1 \\ f_2 \end{array} \begin{array}{|cc|ccc|} \multicolumn{1}{c}{x_3} & \multicolumn{1}{c}{x_4} & x_1 & x_2 & \multicolumn{1}{c}{x_5} \\ \hline 1 & 1 & 0 & 2 & 0 \\ \hline & & 1 & -1 & 0 \\ & & t_1 & 0 & t_2 \\ & & 0 & t_3 & t_4 \\ \hline \end{array}$$

with a nonempty horizontal tail $C_0 = \{x_3, x_4\}$, a single square block $C_1 = \{x_1, x_2, x_5\}$, and an empty vertical tail $C_\infty = \emptyset$. Note the rank deficiency is localized to the tail, and accordingly, this is a proper block-triangularization in the sense of §2.1.4. □

Example 4.4.3. Let us discuss a physical meaning of the LM-equivalence with reference to the electrical network of Example 3.1.2. Consider an LM-matrix

$$A = \begin{pmatrix} Q \\ \hline T \end{pmatrix} = \begin{array}{c} \begin{matrix} \xi^1 & \xi^2 & \xi^3 & \xi^4 & \xi^5 & \eta_1 & \eta_2 & \eta_3 & \eta_4 & \eta_5 \end{matrix} \\ \left(\begin{array}{cccccccccc} 0 & 0 & 1 & 1 & 1 & 0 & 0 & 0 & 0 & 0 \\ 1 & 0 & 0 & 0 & -1 & 0 & 0 & 0 & 0 & 0 \\ 0 & 1 & -1 & 0 & 0 & 0 & 0 & 0 & 0 & 0 \\ 0 & 0 & 0 & 0 & 0 & 1 & 0 & 0 & -1 & 1 \\ 0 & 0 & 0 & 0 & 0 & 0 & 1 & 1 & -1 & 0 \\ \hline r_1 & 0 & 0 & 0 & 0 & t_1 & 0 & 0 & 0 & 0 \\ 0 & r_2 & 0 & 0 & 0 & 0 & t_2 & 0 & 0 & 0 \\ 0 & 0 & 0 & 0 & 0 & \alpha & 0 & t_3 & 0 & 0 \\ 0 & \beta & 0 & t_4 & 0 & 0 & 0 & 0 & 0 & 0 \\ 0 & 0 & 0 & 0 & 0 & 0 & 0 & 0 & 0 & t_5 \end{array}\right) \end{array},$$

where $\mathcal{T} = \{r_1, r_2, \alpha, \beta; t_1, \cdots, t_5\}$ is the set of algebraically independent parameters. This matrix is essentially the same as the coefficient matrix of (3.3).

A block-triangular form under the LM-admissible transformation (4.35) is obtained as follows. Choosing

$$S = \begin{pmatrix} 0 & -1 & 0 & 0 & 0 \\ 0 & 0 & -1 & 0 & 0 \\ 1 & 1 & 1 & 0 & 0 \\ 0 & 0 & 0 & -1 & 0 \\ 0 & 0 & 0 & -1 & 1 \end{pmatrix}$$

in (4.35) we first transform Q to

$$Q' = SQ = \begin{array}{c} \begin{matrix} \xi^1 & \xi^2 & \xi^3 & \xi^4 & \xi^5 & \eta_1 & \eta_2 & \eta_3 & \eta_4 & \eta_5 \end{matrix} \\ \left(\begin{array}{cccccccccc} -1 & 0 & 0 & 0 & 1 & 0 & 0 & 0 & 0 & 0 \\ 0 & -1 & 1 & 0 & 0 & 0 & 0 & 0 & 0 & 0 \\ 1 & 1 & 0 & 1 & 0 & 0 & 0 & 0 & 0 & 0 \\ 0 & 0 & 0 & 0 & 0 & -1 & 0 & 0 & 1 & -1 \\ 0 & 0 & 0 & 0 & 0 & -1 & 1 & 1 & 0 & -1 \end{array}\right) \end{array}, \qquad (4.36)$$

and then permute the rows and the columns of $\begin{pmatrix} Q' \\ T \end{pmatrix}$ with permutation matrices P_r and P_c defined respectively by

$$
\begin{pmatrix}
0 & 1 & 0 & 0 & 0 & 0 & 0 & 0 & 0 & 0 \\
1 & 0 & 0 & 0 & 0 & 0 & 0 & 0 & 0 & 0 \\
0 & 0 & 0 & 1 & 0 & 0 & 0 & 0 & 0 & 0 \\
0 & 0 & 1 & 0 & 0 & 0 & 0 & 0 & 0 & 0 \\
0 & 0 & 0 & 0 & 1 & 0 & 0 & 0 & 0 & 0 \\
0 & 0 & 0 & 0 & 0 & 1 & 0 & 0 & 0 & 0 \\
0 & 0 & 0 & 0 & 0 & 0 & 1 & 0 & 0 & 0 \\
0 & 0 & 0 & 0 & 0 & 0 & 0 & 1 & 0 & 0 \\
0 & 0 & 0 & 0 & 0 & 0 & 0 & 0 & 1 & 0 \\
0 & 0 & 0 & 0 & 0 & 0 & 0 & 0 & 0 & 1
\end{pmatrix},
\quad
\begin{pmatrix}
0 & 0 & 0 & 1 & 0 & 0 & 0 & 0 & 0 & 0 \\
0 & 0 & 0 & 0 & 1 & 0 & 0 & 0 & 0 & 0 \\
1 & 0 & 0 & 0 & 0 & 0 & 0 & 0 & 0 & 0 \\
0 & 0 & 0 & 0 & 0 & 1 & 0 & 0 & 0 & 0 \\
0 & 1 & 0 & 0 & 0 & 0 & 0 & 0 & 0 & 0 \\
0 & 0 & 0 & 0 & 0 & 0 & 1 & 0 & 0 & 0 \\
0 & 0 & 0 & 0 & 0 & 0 & 0 & 1 & 0 & 0 \\
0 & 0 & 0 & 0 & 0 & 0 & 0 & 0 & 1 & 0 \\
0 & 0 & 1 & 0 & 0 & 0 & 0 & 0 & 0 & 0 \\
0 & 0 & 0 & 0 & 0 & 0 & 0 & 0 & 0 & 1
\end{pmatrix}
$$

to obtain an explicit block-triangular LM-matrix

$$
\bar{A} = P_{\mathrm{r}} \begin{pmatrix} Q' \\ T \end{pmatrix} P_{\mathrm{c}} =
\begin{array}{cccccccccc}
\xi^3 & \xi^5 & \eta_4 & \xi^1 & \xi^2 & \xi^4 & \eta_1 & \eta_2 & \eta_3 & \eta_5 \\
\hline
1 & & & & -1 & & & & & \\
 & 1 & & -1 & & & & & & \\
 & & 1 & & & & -1 & & & -1 \\
 & & & 1 & 1 & 1 & 0 & 0 & 0 & \\
 & & & 0 & 0 & 0 & -1 & 1 & 1 & -1 \\
 & & & r_1 & 0 & 0 & t_1 & 0 & 0 & \\
 & & & 0 & r_2 & 0 & 0 & t_2 & 0 & \\
 & & & 0 & 0 & 0 & \alpha & 0 & t_3 & \\
 & & & 0 & \beta & t_4 & 0 & 0 & 0 & \\
 & & & & & & & & & t_5
\end{array} .
$$

It turns out that this is the finest block-triangular matrix which is LM-equivalent to A. Namely, $\bar{A}$ is the CCF of A.

The column set of $\bar{A}$ is partitioned into five blocks:

$$
C_1 = \{\xi^3\},\ C_2 = \{\xi^5\},\ C_3 = \{\eta_4\},\ C_4 = \{\xi^1, \xi^2, \xi^4, \eta_1, \eta_2, \eta_3\},\ C_5 = \{\eta_5\}.
$$

A partial order among the blocks:

$$
\begin{array}{ccc}
 & C_5 & \\
 & \uparrow & \\
 & C_4 & \\
\nearrow & \uparrow & \nwarrow \\
C_1 & C_2 & C_3
\end{array}
\tag{4.37}
$$

is defined by the zero/nonzero structure of $\bar{A}$. This partial order indicates, for example, that the blocks C_1 and C_2, having no order relation, could be exchanged in position without destroying the block-triangular form provided the corresponding rows are exchanged in position accordingly. This corresponds to the fact that the entry in the first row of the column ξ^5 is equal to 0.

The matrix Q has been obtained from the Kirchhoff's conservation laws. In Example 3.1.3 we have chosen three nodes a, b, c in Fig. 3.2 for the KCL (Kirchhoff's current law) to obtain

$$\xi^3 + \xi^4 + \xi^5 = 0, \quad \xi^1 - \xi^5 = 0, \quad \xi^2 - \xi^3 = 0,$$

and two loops consisting of branches 1–5–4 and 2–3–4 for the KVL (Kirchhoff's voltage law) to obtain

$$\eta_1 + \eta_5 - \eta_4 = 0, \quad \eta_2 + \eta_3 - \eta_4 = 0.$$

These conservation equations have been written in a matrix form:

$$Q \begin{pmatrix} \boldsymbol{\xi} \\ \boldsymbol{\eta} \end{pmatrix} = \mathbf{0}. \tag{4.38}$$

The Kirchhoff's conservation laws could have been represented equally well in a different way. For example, another set of three nodes b, c, d for the KCL yields

$$-\xi^1 + \xi^5 = 0, \quad -\xi^2 + \xi^3 = 0, \quad \xi^1 + \xi^2 + \xi^4 = 0,$$

and another choice of two independent loops 1–5–4 and 1–2–3–5 for the KVL leads to

$$-\eta_1 - \eta_5 + \eta_4 = 0, \quad -\eta_1 + \eta_2 + \eta_3 - \eta_5 = 0.$$

Then we would obtain

$$Q' \begin{pmatrix} \boldsymbol{\xi} \\ \boldsymbol{\eta} \end{pmatrix} = \mathbf{0} \tag{4.39}$$

with another coefficient matrix Q', which is identical with Q' of (4.36).

There seems to be nothing with which to choose between the two expressions (4.38) and (4.39) in themselves. The conservation laws claim that $(\boldsymbol{\xi}, \boldsymbol{\eta})$ should belong to a linear subspace, but does not prescribe how the linear space should be described. Both (4.38) and (4.39) are legitimate descriptions of one and the same subspace, and as a consequence, the coefficient matrices Q and Q' are related as $Q' = SQ$. The transformation matrix S in the LM-admissible transformation (4.35) serves to yield a hierarchical decomposition independent of an arbitrary choice in the description of the conservation laws. Such invariance or stability of the CCF should be compared favorably with the susceptibility of the DM-decomposition. In fact, the DM-decomposition of A yields a coarser decomposition

$$\begin{array}{c} C_5 \\ \uparrow \\ C_1 \cup C_2 \cup C_3 \cup C_4 \end{array}$$

whereas that of $\bar{A}$ is given by (4.37). □

4.4.2 Theorem of CCF

This section is to establish the combinatorial canonical form (CCF) of LM-matrices, which has already been introduced informally by means of examples in §4.4.1. We shall prove the existence and uniqueness of the finest proper block-triangular decomposition of an LM-matrix ("proper" in the sense of §2.1.4) under the LM-admissible transformation (4.35).

The LM-surplus function $p : 2^C \to \mathbf{Z}$ defined as $p(J) = \rho(J) + \gamma(J) - |J|$ in (4.16) plays a crucial role. Key facts about p are the following:

1. The rank of an LM-matrix is characterized by the minimum of p (cf. Theorem 4.2.5).
2. The function p is submodular (cf. (4.17)).
3. The function p is invariant under LM-equivalence. Namely, if $\hat{A}$ is LM-equivalent to A, the LM-surplus function $\hat{p}$ associated with $\hat{A}$ is the same as p.

Seeing that the rank of A is expressed by the minimum of p, it is natural to look at the family of minimizers:

$$\mathcal{L}_{\min}(p) = \{J \subseteq C \mid p(J) \leq p(J'), \forall\, J' \subseteq C\}, \tag{4.40}$$

which forms a sublattice of 2^C by virtue of the submodularity (4.17) of p (cf. Theorem 2.2.5).

We are going to make use of a general decomposition principle, the Jordan–Hölder-type theorem for submodular functions, explained in §2.2.2. According to this, the sublattice $\mathcal{L}_{\min}(p)$ determines a pair of a partition of $C = \mathrm{Col}(A)$ and a partial order $\preceq$:

$$\mathcal{P}(\mathcal{L}_{\min}(p)) = (\{C_0; C_1, \cdots, C_b; C_\infty\}, \preceq). \tag{4.41}$$

Here $b \geq 0$ and $C_k \neq \emptyset$ for $k = 1, \cdots, b$, whereas C_0 and C_∞ are distinguished blocks that can be empty. It is assumed that the blocks are indexed consistently with the partial order in the sense that

$$C_k \preceq C_l \quad \Rightarrow \quad k \leq l. \tag{4.42}$$

The following theorem claims the existence of the CCF, a proper block-triangular decomposition of an LM-matrix under LM-equivalence. It was established first in an unpublished report of Murota [201] in 1985 and published by Murota [204] and Murota–Iri–Nakamura [239].

Theorem 4.4.4 (Combinatorial Canonical Form). *For an LM-matrix $A \in \mathrm{LM}(\boldsymbol{K}, \boldsymbol{F})$ there exists another LM-matrix $\bar{A}$ which is LM-equivalent to A and satisfies the following properties.*

(B1) [Nonzero structure and partial order $\preceq$] *$\bar{A}$ is block-triangularized, i.e.,*

$$\bar{A}[R_k, C_l] = O \qquad \text{if} \quad 0 \leq l < k \leq \infty, \tag{4.43}$$

with respect to partitions $(R_0; R_1, \cdots, R_b; R_\infty)$ *and* $(C_0; C_1, \cdots, C_b; C_\infty)$ *of the row set* $\mathrm{Row}(\bar{A})$ *and the column set* $\mathrm{Col}(\bar{A})$ *respectively, where* $b \geq 0$, $R_k \neq \emptyset$ *and* $C_k \neq \emptyset$ *for* $k = 1, \cdots, b$, *and* R_0, R_∞, C_0 *and* C_∞ *can be empty.*

Moreover, when $\mathrm{Col}(\bar{A})$ *is identified with* $\mathrm{Col}(A)$, *the above partition* $(C_0; C_1, \cdots, C_b; C_\infty)$ *agrees with that defined by the lattice* $\mathcal{L}_{\min}(p)$ *and the partial order on* $\{C_1, \cdots, C_b\}$ *induced by the zero/nonzero structure of* $\bar{A}$ *agrees with the partial order* $\preceq$ *defined by* $\mathcal{L}_{\min}(p)$; *i.e.,*

$$\bar{A}[R_k, C_l] = O \quad \text{unless} \quad C_k \preceq C_l \quad (1 \leq k, l \leq b); \tag{4.44}$$

$$\bar{A}[R_k, C_l] \neq O \quad \text{if} \quad C_k \prec\cdot\, C_l \quad (1 \leq k, l \leq b). \tag{4.45}$$

(B2) [Size of diagonal blocks]

$$\begin{array}{lll} |R_0| < |C_0| & \text{or} & |R_0| = |C_0| = 0, \\ |R_k| = |C_k| > 0 & \text{for} & k = 1, \cdots, b, \\ |R_\infty| > |C_\infty| & \text{or} & |R_\infty| = |C_\infty| = 0. \end{array}$$

(B3) [Rank of diagonal blocks]

$$\begin{array}{l} \operatorname{rank} \bar{A}[R_0, C_0] = |R_0|, \\ \operatorname{rank} \bar{A}[R_k, C_k] = |R_k| = |C_k| \quad \text{for} \quad k = 1, \cdots, b, \\ \operatorname{rank} \bar{A}[R_\infty, C_\infty] = |C_\infty|. \end{array}$$

(B4) [Rank of submatrices of diagonal blocks]

$$\begin{array}{l} \operatorname{rank} \bar{A}[R_0, C_0 \setminus \{j\}] = |R_0| \qquad (j \in C_0), \\ \operatorname{rank} \bar{A}[R_k \setminus \{i\}, C_k \setminus \{j\}] = |R_k| - 1 = |C_k| - 1 \quad (i \in R_k, j \in C_k) \\ \qquad\qquad\qquad\qquad\qquad\qquad\qquad\qquad\qquad \text{for} \quad k = 1, \cdots, b, \\ \operatorname{rank} \bar{A}[R_\infty \setminus \{i\}, C_\infty] = |C_\infty| \qquad (i \in R_\infty). \end{array}$$

(B5) [Uniqueness] $\bar{A}$ *is the finest proper block-triangular matrix that is LM-equivalent to* A. *Namely, if* $\hat{A}$ *is LM-equivalent to* A *and is block-triangularized with respect to certain partitions* $(\hat{R}_0; \hat{R}_1, \cdots, \hat{R}_q; \hat{R}_\infty)$ *and* $(\hat{C}_0; \hat{C}_1, \cdots, \hat{C}_q; \hat{C}_\infty)$ *of* $\mathrm{Row}(\hat{A})$ *and* $\mathrm{Col}(\hat{A})$ $(= \mathrm{Col}(A))$ *with the diagonal blocks satisfying the conditions* (B2) *and* (B3), *then each* $\hat{C}_k$ *is a union of some blocks in* $(C_0; C_1, \cdots, C_b; C_\infty)$.

Proof. A constructive proof is given in §4.4.3. ∎

The matrix $\bar{A}$ in the theorem is called the CCF of A. The CCF is uniquely determined so far as the partitions of the row and column sets as well as the partial order among the blocks are concerned, whereas there remains some indeterminacy, or degree of freedom, in the numerical values of the entries in the Q-part. For example, elementary row transformations within a block change numerical values without affecting the block structure. When the numerical indeterminacy is to be emphasized, such $\bar{A}$ will be called *a* CCF, instead of *the* CCF. In Example 4.4.1, for instance, both of

$$\bar{A} = \left[\begin{array}{c|cc} 1 & 1 & 0 \\ \hline & 1 & 3 \\ & t_1 & t_2 \end{array}\right], \qquad \bar{A}' = \left[\begin{array}{c|cc} -1 & 0 & -1 \\ \hline & 3 & 1 \\ & t_2 & t_1 \end{array}\right]$$

are qualified as the CCF of A.

The submatrices $\bar{A}[R_0, C_0]$ and $\bar{A}[R_\infty, C_\infty]$ are called the *horizontal tail* and the *vertical tail*, respectively. The tails are nonsquare if they are not empty, and the "discrepancies from squareness":

$$\delta_0 = |C_0| - |R_0|, \qquad \delta_\infty = |R_\infty| - |C_\infty|$$

indicate the rank deficiencies of the whole matrix A, since

$$\delta_0 = |C| - \operatorname{rank} \bar{A} = |C| - \operatorname{rank} A, \qquad \delta_\infty = |R| - \operatorname{rank} \bar{A} = |R| - \operatorname{rank} A$$

by (B1) and (B3). Thus the rank deficiency is localized to the tails. In particular, A is nonsingular if and only if both tails are empty (i.e., $C_0 = R_\infty = \emptyset$).

For a nonsingular A, the CCF gives a finer decomposition than the DM-decomposition, as is expected from the comparison of the admissible transformations: $P_{\mathrm{r}} \begin{pmatrix} S & O \\ O & I \end{pmatrix} A P_{\mathrm{c}}$ for the CCF and $P_{\mathrm{r}} A P_{\mathrm{c}}$ for the DM-decomposition. This can be explained also in terms of the LM-surplus function p of (4.16) and the (original) surplus function p_0 of (2.39) defined as

$$p_0(J) = \gamma_A(J) - |J|, \qquad J \subseteq C,$$

with

$$\gamma_A(J) = |\{i \in \operatorname{Row}(A) \mid \exists j \in J: \ A_{ij} \neq 0\}|, \qquad J \subseteq C. \tag{4.46}$$

Proposition 4.4.5. *If $A \in \mathrm{LM}(\boldsymbol{K}, \boldsymbol{F})$ is nonsingular, then* $\min p = \min p_0 = 0$ *and* $\mathcal{L}_{\min}(p) \supseteq \mathcal{L}_{\min}(p_0)$. *Hence the decomposition of* $\operatorname{Col}(A)$ *in the CCF of A is a refinement of the one in the DM-decomposition.*

Proof. First note that $p(J) \leq p_0(J)$ for $J \subseteq C$. By Theorem 4.2.5, A is nonsingular if and only if $\min p = 0$. This implies $\min p_0 = 0$. The inclusion $\mathcal{L}_{\min}(p) \supseteq \mathcal{L}_{\min}(p_0)$ is then evident. ∎

Example 4.4.6. For a singular matrix the CCF is not necessarily a refinement of the DM-decomposition. Consider, e.g., a matrix $A \in \mathrm{LM}(\mathbf{Q}, \boldsymbol{F}; 4, 0, 4)$ (for any $\boldsymbol{F} \supseteq \mathbf{Q}$) and its CCF $\bar{A}$:

$$A = \begin{array}{|cccc|} \hline 1 & 1 & 1 & 1 \\ \hline 1 & 1 & 1 & 1 \\ \hline 0 & 0 & 1 & 1 \\ \hline 0 & 0 & 1 & 1 \\ \hline \end{array}\,, \qquad \bar{A} = \begin{array}{|cccc||} \hline 1 & 1 & 1 & 1 \\ \hline 0 & 0 & 1 & 1 \\ \hline & & & \\ & & & \\ \hline \end{array}\,.$$

The CCF consists of tails only with $C_0 = \operatorname{Col}(A)$, $|R_0| = 2$, $C_\infty = \emptyset$ and $|R_\infty| = 2$. On the other hand, the DM-decomposition evidently decomposes A into two square blocks. □

4.4.3 Construction of CCF

This subsection gives a constructive proof of Theorem 4.4.4. The following mathematical construction of the CCF will be polished up to a practically efficient algorithm in §4.4.4.

As already mentioned, we consider the submodular function p, the sublattice $\mathcal{L}_{\min}(p)$ of its minimizers, and the associated partition

$$\mathcal{P}(\mathcal{L}_{\min}(p)) = (\{C_0; C_1, \cdots, C_b; C_\infty\}, \preceq)$$

of $C = \mathrm{Col}(A) = \mathrm{Col}(Q) = \mathrm{Col}(T)$ (see (4.16), (4.40), (4.41) for the definitions of p, $\mathcal{L}_{\min}(p)$, and $\mathcal{P}(\mathcal{L}_{\min}(p))$). In accordance with (2.23) we define $X_k = \bigcup_{l=0}^{k} C_l$ for $k = 0, 1, \cdots, b$ to obtain

$$X_0 \ (= \min \mathcal{L}_{\min}(p)) \subsetneqq X_1 \subsetneqq X_2 \subsetneqq \cdots \subsetneqq X_b \ (= \max \mathcal{L}_{\min}(p)), \tag{4.47}$$

which is a maximal chain of $\mathcal{L}_{\min}(p)$ by (4.42).

Note that the LM-admissible transformation (4.35) is equivalent to

$$P_{\mathrm{r}} \begin{pmatrix} S & O \\ O & P_T \end{pmatrix} \begin{pmatrix} Q \\ T \end{pmatrix} P_{\mathrm{c}}, \tag{4.48}$$

which contains another permutation matrix P_T. In what follows we will find these four matrices $P_{\mathrm{c}}, S, P_T, P_{\mathrm{r}}$ that bring about the CCF.

[Matrix P_{c}]: The permutation matrix P_{c} is such that the column set C is reordered as $C_0, C_1, \cdots, C_b, C_\infty$, where the ordering of columns within each block is arbitrary.

[Matrix S]: Recall the notation $\rho(J) = \mathrm{rank}\, Q[R_Q, J]$, $J \subseteq C$, and note that $\mathrm{Col}(QP_{\mathrm{c}})$ can be identified with C through permutation P_{c}. By the usual row elimination operations, we can bring QP_{c} into a block-triangular matrix (in the sense of §2.1.4) with column partition $(C_0; C_1, \cdots, C_b; C_\infty)$. More precisely, we can find a nonsingular matrix $S \in \mathrm{GL}(m_Q, \boldsymbol{K})$ and a partition

$$(R_{Q0}; R_{Q1}, \cdots, R_{Qb}; R_{Q\infty}) \tag{4.49}$$

of $\mathrm{Row}(SQP_{\mathrm{c}})$ such that $\bar{Q} = SQP_{\mathrm{c}}$ satisfies

$$\bar{Q}[R_{Qk}, C_l] = O \qquad (0 \le l < k \le \infty) \tag{4.50}$$

and

$$\begin{aligned} \mathrm{rank}\, \bar{Q}[R_{Q0}, C_0] &= |R_{Q0}| = \rho(X_0), \\ \mathrm{rank}\, \bar{Q}[R_{Qk}, C_k] &= |R_{Qk}| = \rho(X_k) - \rho(X_{k-1}) \quad (k = 1, \cdots, b), \\ |R_{Q\infty}| &= m_Q - \rho(X_b). \end{aligned} \tag{4.51}$$

We may further impose that

For $0 \leq k < l \leq \infty$, the nonzero row vectors of $\bar{Q}[R_{Qk}, C_l]$ are linearly independent of the row vectors of $\bar{Q}[R_{Ql}, C_l]$. (4.52)

[Matrix P_T]: Define a partition of $R_T = \mathrm{Row}(T)$:

$$(R_{T0}; R_{T1}, \cdots, R_{Tb}; R_{T\infty}) \tag{4.53}$$

by

$$\begin{aligned} R_{T0} &= \Gamma(X_0), \\ R_{Tk} &= \Gamma(X_k) \setminus \Gamma(X_{k-1}) \qquad (k = 1, \cdots, b), \\ R_{T\infty} &= R_T \setminus \Gamma(X_b) \end{aligned} \tag{4.54}$$

using the notation $\Gamma(J) = \{i \in R_T \mid \exists j \in J : T_{ij} \neq 0\}$ of (4.8). Let P_T be a permutation matrix which permutes R_T compatibly with (4.53), so that $\bar{T} = P_T T P_\mathrm{c}$ is in an explicit block-triangular form:

$$T[R_{Tk}, C_l] = \bar{T}[R_{Tk}, C_l] = O \qquad (0 \leq l < k \leq \infty), \tag{4.55}$$

where it is understood that $\mathrm{Row}(\bar{T}) = \mathrm{Row}(T)$ and $\mathrm{Col}(\bar{T}) = \mathrm{Col}(T) = C$ through the permutations P_T and P_c. Note that

$$\begin{aligned} |R_{T0}| &= \gamma(X_0), \\ |R_{Tk}| &= \gamma(X_k) - \gamma(X_{k-1}) \qquad (k = 1, \cdots, b), \\ |R_{T\infty}| &= m_T - \gamma(X_b) \end{aligned} \tag{4.56}$$

with $\gamma(J) = |\Gamma(J)|$.

[Matrix P_r]: We have constructed two block-triangular matrices $\bar{Q}$ and $\bar{T}$, the former being block-triangularized with respect to the partitions (4.41) and (4.49) and the latter with respect to (4.41) and (4.53). Put these two matrices together:

$$\bar{A} = \begin{pmatrix} \bar{Q} \\ \bar{T} \end{pmatrix},$$

and define a partition of $\mathrm{Row}(\bar{A})$:

$$(R_0; R_1, \cdots, R_b; R_\infty) \tag{4.57}$$

by $R_k = R_{Qk} \cup R_{Tk}$ for $k = 0, 1, \cdots, b, \infty$. By (4.50) and (4.55), $\bar{A}$ is (essentially) block-triangularized with respect to the partitions (4.41) and (4.57), namely,

$$\bar{A}[R_k, C_l] = O \qquad (0 \leq l < k \leq \infty).$$

To obtain an explicit block-triangular form, we use a matrix P_r that reorders $\mathrm{Row}(\bar{A})$ compatibly with (4.57), and redefine $\bar{A}$ to be $P_\mathrm{r}\bar{A}$.

The matrix $\bar{A}$ constructed above is LM-equivalent to A. Obviously, it is block-triangularized, satisfying (4.43) in (B1). We go on to prove (B2), (B3), (B4), and (B5), while deferring the proof of the claims (4.44) and (4.45) concerning partial order.

[Proof of (B2)]: Since $C_0 \in \mathcal{L}_{\min}(p)$, $\rho(C_0) = |R_{Q0}|$, and $\gamma(C_0) = |R_{T0}|$, we have

$$0 = p(\emptyset) \geq \min p = p(C_0) = \rho(C_0) + \gamma(C_0) - |C_0| = |R_0| - |C_0|.$$

Hence $|R_0| \leq |C_0|$. If the equality holds here, then $p(\emptyset) = \min p$, i.e., $\emptyset \in \mathcal{L}_{\min}(p)$. Since $C_0 = \min \mathcal{L}_{\min}(p)$, this implies $C_0 = \emptyset$ and therefore $R_0 = \emptyset$.

For $k = 1, \cdots, b$, we have $p(X_{k-1}) = p(X_k)$ $(= \min p)$, i.e.,

$$\rho(X_{k-1}) + \gamma(X_{k-1}) - |X_{k-1}| = \rho(X_k) + \gamma(X_k) - |X_k|.$$

This reduces to $|R_k| = |C_k|$ by (4.51) and (4.56).

If $C_\infty \neq \emptyset$, then $p(C) > \min p = p(X_b)$, i.e.,

$$\rho(C) + \gamma(C) - |C| > \rho(X_b) + \gamma(X_b) - |X_b|.$$

Combination of this with

$$|R| \geq \rho(C) + \gamma(C), \quad |R_\infty| = |R| - \rho(X_b) - \gamma(X_b), \quad |C_\infty| = |C| - |X_b|$$

yields $|R_\infty| > |C_\infty|$.

[Proof of (B3)]: Put $Y_k = \bigcup_{l=0}^{k} R_l$ for $k = 0, 1, \cdots, b$, and note

$$\min p = p(X_k) = \rho(X_k) + \gamma(X_k) - |X_k| = |Y_k| - |X_k| \qquad (k = 0, 1, \cdots, b). \tag{4.58}$$

For $k = 0, 1, \cdots, b$, it follows from (4.43), Theorem 4.2.5, and (4.58) that

$$\begin{aligned} \operatorname{rank} \bar{A}[Y_k, X_k] &= \operatorname{rank} \bar{A}[\operatorname{Row}(\bar{A}), X_k] = \operatorname{rank} A[R, X_k] \\ &= \min\{p(X) \mid X \subseteq X_k\} + |X_k| = |Y_k|, \end{aligned}$$

which implies

$$\operatorname{rank} \bar{A}[R_k, C_k] = |R_k| \qquad (k = 0, 1, \cdots, b).$$

Since $\bar{A}[R_\infty, X_b] = O$ and $\operatorname{rank} \bar{A}[Y_b, X_b] = |Y_b|$, we have

$$\operatorname{rank} \bar{A}[R_\infty, C_\infty] = \operatorname{rank} \bar{A} - |Y_b| = \min p + |C| - |Y_b| = |C| - |X_b| = |C_\infty|.$$

[Proof of (B4)]: By (4.43) and Theorem 4.2.5,

$$\begin{aligned} \operatorname{rank} \bar{A}[R_0, C_0 \setminus \{j\}] &= \min\{p(X) \mid X \subseteq C_0 \setminus \{j\}\} + |C_0| - 1 \\ &\geq \min p + |C_0| = |R_0|. \end{aligned}$$

For $k = 1, \cdots, b$, put $C' = X_k \setminus \{j\}$. Similarly we have

$$\begin{aligned} \operatorname{rank} \bar{A}[R_k \setminus \{i\}, C_k \setminus \{j\}] &= \operatorname{rank} \bar{A}[R \setminus \{i\}, C'] - |Y_{k-1}| \\ &= \min\{p'(X) \mid X \subseteq C'\} + |C'| - |Y_{k-1}|, \end{aligned} \tag{4.59}$$

where $p' : 2^{C'} \to \mathbf{Z}$, the LM-surplus function associated with $\bar{A}[R \setminus \{i\}, C']$, is given by

$$p'(X) = \operatorname{rank} \bar{Q}[\operatorname{Row}(\bar{Q}) \setminus \{i\}, X] + |\Gamma(X) \setminus \{i\}| - |X|, \qquad X \subseteq C'.$$

We see $p(X) - 1 \le p'(X) \le p(X)$ for $X \subseteq C'$, and $p'(X) = p(X)$ for $X \subseteq X_{k-1}$. We further claim that $\min p' = \min p$, since otherwise there exists $X \subseteq C'$ with $X \in \mathcal{L}_{\min}(p)$ and $X \not\subseteq X_{k-1}$, which, combined with $X_{k-1} \in \mathcal{L}_{\min}(p)$, implies (cf. Theorem 2.2.5) that $X \cup X_{k-1} \in \mathcal{L}_{\min}(p)$ and $X_{k-1} \subsetneqq X \cup X_{k-1} \subsetneqq X_k$, a contradiction to our assumption that (4.47) is a maximal chain. Hence (4.59) is equal to $\min p + |C'| - |Y_{k-1}| = |C'| - |X_{k-1}| = |C_k| - 1$. The final case, $\operatorname{rank} \bar{A}[R_\infty \setminus \{i\}, C_\infty]$, can be treated mutatis mutandis (by replacing C' with C, the index k with ∞, and $k-1$ with b).

[Proof of (B5)]: Since $\hat{A}$ is a block-triangular matrix with the rank conditions in (B3), it holds that

$$\operatorname{rank} \hat{A} = |C| - |\hat{X}_k| + |\hat{Y}_k| \qquad (k = 0, 1, \cdots, q),$$

where

$$\hat{Y}_k = \bigcup_{l=0}^{k} \hat{R}_l, \quad \hat{X}_k = \bigcup_{l=0}^{k} \hat{C}_l \qquad (k = 0, 1, \cdots, q).$$

On the other hand, since $\hat{A}$ and A are LM-equivalent, they share the same LM-surplus function p (i.e., the same ρ and γ). By virtue of this invariance of p as well as Theorem 4.2.5, the rank of $\hat{A}$ can be expressed as

$$\operatorname{rank} \hat{A} = \min p + |C|$$

in terms of the original LM-surplus function p. Hence follows

$$\min p = |\hat{Y}_k| - |\hat{X}_k| = \rho(\hat{X}_k) + \gamma(\hat{X}_k) - |\hat{X}_k| = p(\hat{X}_k),$$

where the second equality is due to the assumed rank condition of $\hat{A}$. That is, $\hat{X}_k \in \mathcal{L}_{\min}(p)$ for $k = 0, 1, \cdots, q$. This implies the claim of (B5).

[Proof of (4.44)]: First we note that $\mathcal{L}_{\min}(p)$ is both ρ- and γ-skeleton, since ρ and γ are submodular while $p = \rho + \gamma - |\cdot|$ is modular on $\mathcal{L}_{\min}(p)$ (see §2.2.2 for the definition of skeleton). It then follows from Theorem 2.2.13 that

$$|R_{Qk}| = \rho(\langle C_k \rangle \cup C_k) - \rho(\langle C_k \rangle) \qquad (k = 0, 1, \cdots, b), \tag{4.60}$$

$$|R_{Tk}| = \gamma(\langle C_k \rangle \cup C_k) - \gamma(\langle C_k \rangle) \qquad (k = 0, 1, \cdots, b), \tag{4.61}$$

where

$$\langle C_k \rangle = \bigcup \{C_j \mid C_j \prec C_k\} = \bigcup \{C_j \mid C_j \preceq C_k, C_j \neq C_k\}$$

for $k = 0, 1, \cdots, b, \infty$. It is emphasized that $C_0 \subseteq \langle C_k \rangle$ for $k = 1, \cdots, b, \infty$ according to the convention of (2.26), whereas $\langle C_0 \rangle = \emptyset$.

Lemma 4.4.7. *For $1 \leq k, l \leq b$, it holds that $\bar{Q}[R_{Qk}, C_l] \neq O \Rightarrow C_k \preceq C_l$.*

Proof. Put $\langle R_{Ql} \rangle = \bigcup \{R_{Qk} \mid C_k \prec C_l\}$ for $l = 0, 1, \cdots, b$. It suffices to show

$$\bar{Q}[\mathrm{Row}(\bar{Q}) \setminus (\langle R_{Ql} \rangle \cup R_{Ql}), C_l] = O \tag{4.62}$$

for $l = 0, 1, \cdots, b$. We prove (4.62) by induction on l. First, (4.62) for $l = 0$ holds true by (4.50). Next we consider a general $l \geq 1$. If $C_j \subseteq \langle C_l \rangle$, then $j < l$ by (4.42) and therefore, by the induction hypothesis,

$$\bar{Q}[\mathrm{Row}(\bar{Q}) \setminus (\langle R_{Qj} \rangle \cup R_{Qj}), C_j] = O.$$

Since $\langle R_{Qj} \rangle \cup R_{Qj} \subseteq \langle R_{Ql} \rangle$, this yields

$$\bar{Q}[\mathrm{Row}(\bar{Q}) \setminus \langle R_{Ql} \rangle, C_j] = O$$

for all j with $C_j \subseteq \langle C_l \rangle$. Therefore,

$$\bar{Q}[\mathrm{Row}(\bar{Q}) \setminus \langle R_{Ql} \rangle, \langle C_l \rangle] = O. \tag{4.63}$$

From this and (4.51) follows

$$|\langle R_{Ql} \rangle| = \mathrm{rank}\, \bar{Q}[\langle R_{Ql} \rangle, \langle C_l \rangle] = \mathrm{rank}\, \bar{Q}[\mathrm{Row}(\bar{Q}), \langle C_l \rangle] = \rho(\langle C_l \rangle).$$

Hence, by (4.63) and (4.60),

$$\begin{aligned} &\mathrm{rank}\, \bar{Q}[\mathrm{Row}(\bar{Q}) \setminus \langle R_{Ql} \rangle, C_l] = \mathrm{rank}\, \bar{Q}[\mathrm{Row}(\bar{Q}), \langle C_l \rangle \cup C_l] - |\langle R_{Ql} \rangle| \\ &= \rho(\langle C_l \rangle \cup C_l) - \rho(\langle C_l \rangle) = |R_{Ql}|. \end{aligned}$$

This means (4.62) by the condition (4.52). ∎

Lemma 4.4.8. $\quad R_{Tk} = \Gamma(C_k) \setminus \Gamma(\langle C_k \rangle) \qquad (k = 1, \cdots, b).$

Proof. Since $R_{Tk} = \Gamma(X_{k-1} \cup C_k) \setminus \Gamma(X_{k-1}) = \Gamma(C_k) \setminus \Gamma(X_{k-1})$ (cf. (4.54)) and $\Gamma(\langle C_k \rangle \cup C_k) \setminus \Gamma(\langle C_k \rangle) = \Gamma(C_k) \setminus \Gamma(\langle C_k \rangle)$, it follows from (4.61) that

$$|R_{Tk}| = |\Gamma(C_k) \setminus \Gamma(X_{k-1})| = |\Gamma(C_k) \setminus \Gamma(\langle C_k \rangle)|,$$

in which $\Gamma(X_{k-1}) \supseteq \Gamma(\langle C_k \rangle)$. Therefore, $R_{Tk} = \Gamma(C_k) \setminus \Gamma(\langle C_k \rangle)$. ∎

Lemma 4.4.9. *For $1 \leq k, l \leq b$, it holds that $\bar{T}[R_{Tk}, C_l] \neq O \Rightarrow C_k \preceq C_l$.*

Proof. Suppose that $\bar{T}[R_{Tk}, C_l] \neq O$, where we may assume $1 \leq k < l \leq b$. Then, $\exists\, i \in R_{Tk}$, $\exists\, j \in C_l : T_{ij} \neq 0$. This implies $i \notin R_{Tl}$ and $i \in \Gamma(C_l)$. With the expression $R_{Tl} = \Gamma(C_l) \setminus \Gamma(\langle C_l \rangle)$ (cf. Lemma 4.4.8) we see that $i \in \Gamma(\langle C_l \rangle)$, i.e.,

$$\exists\, l_1 : \quad k \leq l_1 < l_0,\ C_{l_1} \prec C_{l_0},\ \bar{T}[R_{Tk}, C_{l_1}] \neq O,$$

where $l_0 = l$. Repeating this, we see that $\exists\, l_1, \exists\, l_2, \cdots, \exists\, l_s$: $C_k = C_{l_s} \prec C_{l_{s-1}} \prec \cdots \prec C_{l_1} \prec C_{l_0} = C_l$. ■

The claim (4.44) is established by Lemmas 4.4.7 and 4.4.9, since $\bar{A}[R_k, C_l] \neq O$ if and only if $\bar{Q}[R_{Qk}, C_l] \neq O$ or $\bar{T}[R_{Tk}, C_l] \neq O$.

[Proof of (4.45)]: Suppose that $C_k \prec\cdot\, C_l$. We may assume $1 \leq k < l \leq b$ by (4.42). Put

$$I = \{i \mid k < i < l, C_k \prec C_i\}, \qquad I^* = I \cup \{k\},$$
$$J = \{j \mid k < j < l\} \setminus I, \qquad J^* = J \cup \{l\}.$$

We have (i) $i \in I^*, j \in J \;\Rightarrow\; C_i \not\preceq C_j$, and (ii) $i \in I \;\Rightarrow\; C_i \not\preceq C_l$. The statement (i) is due to the transitivity of the partial order and the statement (ii) is by the assumption $C_k \prec\cdot\, C_l$. It then follows that

$$i \in I^*, j \in J^*, (i,j) \neq (k,l) \;\Rightarrow\; C_i \not\preceq C_j \;\Rightarrow\; \bar{A}[R_i, C_j] = O,$$

where (4.44) is used.

If $\bar{A}[R_k, C_l] = O$ were the case, we would have $\bar{A}[\bigcup_{i \in I^*} R_i, \bigcup_{j \in J^*} C_j] = O$. Then $\rho(X) + \gamma(X) = |Y_{k-1}| + \sum_{j \in J^*} |R_j|$ for $X = X_{k-1} \cup \left(\bigcup_{j \in J^*} C_j\right)$ and $Y_{k-1} = \bigcup_{l=0}^{k-1} R_l$, and therefore X belongs to $\mathcal{L}_{\min}(p)$, since

$$\begin{aligned} p(X) &= (\rho(X) + \gamma(X)) - |X| \\ &= \left(|Y_{k-1}| + \sum_{j \in J^*} |R_j| \right) - \left(|X_{k-1}| + \sum_{j \in J^*} |C_j| \right) \\ &= |Y_{k-1}| - |X_{k-1}| = \min p, \end{aligned}$$

where (B2) and (4.58) are used. Hence we have $X \in \mathcal{L}_{\min}(p)$, $C_k \not\subseteq X$, and $C_l \subseteq X$. This contradicts the definition (2.24) of $C_k \preceq C_l$. Hence we must have $\bar{A}[R_k, C_l] \neq O$, completing the proof of (4.45).

The proof of Theorem 4.4.4 is completed.

Remark 4.4.10. The construction of the CCF is a natural generalization of that of the DM-decomposition. Note in particular that both rely on the same decomposition principle, the Jordan–Hölder-type theorem for submodular functions, which is applied to the surplus functions, p_0 and p, respectively. The admissible transformations are not explicit in this. The properties of the resulting decompositions are established in relation to the admissible transformations by virtue of the rank formulas expressed in terms of the surplus functions. The corresponding items are compared in Table 4.1. □

Table 4.1. DM-decomposition and CCF

	DM-decomposition	CCF
matrix	generic matrix T	LM-matrix $\binom{Q}{T}$
surplus function	$p_0 = \gamma - \vert \cdot \vert$	$p = \rho + \gamma - \vert \cdot \vert$
rank formula	Theorem 2.2.17 (Hall–Ore):	Theorem 4.2.5:
	$\mathrm{rank}\, T = \min p_0 + \vert C \vert$	$\mathrm{rank} \binom{Q}{T} = \min p + \vert C \vert$
transformation	$P_\mathrm{r} T P_\mathrm{c}$	$P_\mathrm{r} \begin{pmatrix} S & O \\ O & I \end{pmatrix} \begin{pmatrix} Q \\ T \end{pmatrix} P_\mathrm{c}$

4.4.4 Algorithm for CCF

We describe here an efficient (polynomial time) algorithm for computing the CCF of an LM-matrix $A = \binom{Q}{T} \in \mathrm{LM}(\boldsymbol{K}, \boldsymbol{F}; m_Q, m_T, n)$. The fundamental idea of the algorithm is to combine the following facts that have already been established.

1. The rank of an LM-matrix is characterized by the associated LM-surplus function p (cf. Theorem 4.2.5), and this characterization can be reformulated in terms of an independent matching problem (cf. Theorem 4.2.18).
2. The CCF is constructed from $\mathcal{L}_{\min}(p)$, the family of the minimizers of the submodular function p (cf. Theorem 4.4.4), and moreover $\mathcal{L}_{\min}(p)$ is closely related to $\mathcal{L}_{\min}(\kappa)$, the family of the minimizers of the cut capacity function κ of the associated independent matching problem (cf. Lemma 4.2.20).
3. In an independent matching problem, in general, $\mathcal{L}_{\min}(\kappa)$ induces a decomposition of the vertex set (the min-cut decomposition) and moreover the decomposition can be computed by an efficient algorithm (cf. Lemma 2.3.35).

Namely, the fundamental idea of the algorithm is to compute the CCF of an LM-matrix by finding the min-cut decomposition of the independent matching problem associated with the LM-matrix.

Before going on to a detailed description of the concrete procedure, it would be worth while demonstrating a connection between the CCF and the DM-decomposition in the case of an LM-matrix A with Q of full-row rank. The validity of this procedure follows from that of the general case.

Algorithm for the CCF of A with Q of full-row rank

Step 1: Find $J \subseteq C$ such that $Q[R_Q, J]$ is nonsingular and $\mathrm{rank}\, A = |J| + \text{term-rank}\, T[R_T, C \setminus J]$ (such J exists by Lemma 4.2.1 or Theorem 4.2.3).

Step 2: Put

$$S := Q[R_Q, J]^{-1}, \qquad A' := \begin{pmatrix} S & O \\ O & I \end{pmatrix} A. \tag{4.64}$$

Step 3: Find the DM-decomposition $\bar{A}$ of A', namely, $\bar{A} := P_{\mathrm{r}}\ A'\ P_{\mathrm{c}}$ with suitable permutation matrices P_{r} and P_{c}. ($\bar{A}$ is the CCF of A.) □

Example 4.4.11. For the LM-matrix of Example 4.4.3, which is nonsingular, we can take $J = \{\xi^5, \xi^3, \xi^4, \eta_4, \eta_3\}$ in Step 1. The inverse of

$$Q[R_Q, J] = \begin{array}{c} \begin{array}{ccccc} \xi^5 & \xi^3 & \xi^4 & \eta_4 & \eta_3 \end{array} \\ \boxed{\begin{array}{ccccc} 1 & 1 & 1 & 0 & 0 \\ -1 & 0 & 0 & 0 & 0 \\ 0 & -1 & 0 & 0 & 0 \\ 0 & 0 & 0 & -1 & 0 \\ 0 & 0 & 0 & -1 & 1 \end{array}} \end{array}$$

coincides with the transformation matrix S in Example 4.4.3. □

The following is an algorithm for finding the CCF of a general LM-matrix. Steps 1–3 are identical with the algorithm for computing the rank given in §4.2.4, except that at the end of Step 2 here we go on to Step 4 for decomposition. The algorithm works with the same directed graph $\tilde{G} = \tilde{G}_M = (\tilde{V}, \tilde{E})$ that has vertex set $\tilde{V} = R_T \cup C_Q \cup C$ and arc set $\tilde{E} = E_T \cup E_Q \cup E^+ \cup M^\circ$, where $R_T = \mathrm{Row}(T)$, $C_Q = \{j_Q \mid j \in C\}$ is a disjoint copy of $C = \mathrm{Col}(A)$,

$$E_T = \{(i,j) \mid i \in R_T, j \in C, T_{ij} \neq 0\}, \quad E_Q = \{(j_Q, j) \mid j \in C\},$$

and E^+ and M° are defined and updated in the algorithm; E^+ represents the structure of the matroid $\mathbf{M}(Q)$ and M° is the set of reoriented arcs in an independent matching $M \subseteq E_T \cup E_Q$. Recall also that E^+ and M° consist of arcs, respectively, from C_Q to C_Q and from C to $R_T \cup C_Q$. The array S, at the termination of the algorithm, gives the matrix S in the LM-admissible transformation (4.35). When the transformation matrix is not needed, it may simply be eliminated from the computation without any side effect.

Algorithm for the CCF of an LM-matrix A

Step 1:
$M^\circ := \emptyset$; $\mathit{base}[i] := 0$ $(i \in R_Q)$; $P[i,j] := Q_{ij}$ $(i \in R_Q, j \in C)$;
$S :=$ unit matrix of order m_Q.

Step 2:
$I := \{i \in C \mid i_Q \in \partial^- M^\circ \cap C_Q\}$;
$J := \{j \in C \setminus I \mid \forall h : \mathit{base}[h] = 0 \Rightarrow P[h,j] = 0\}$;
$S_T^+ := R_T \setminus \partial^- M^\circ$; $S_Q^+ := \{j_Q \in C_Q \mid j \in C \setminus (I \cup J)\}$;
$S^+ := S_T^+ \cup S_Q^+$; $S^- := C \setminus \partial^+ M^\circ$;
$E^+ := \{(i_Q, j_Q) \mid h \in R_Q, j \in J, P[h,j] \neq 0, i = \mathit{base}[h] \neq 0\}$;
[$\tilde{E}$ is updated accordingly]
If there exists in $\tilde{G} = (\tilde{V}, \tilde{E})$ a directed path from S^+ to S^- then go to Step 3; otherwise (including the case where $S^+ = \emptyset$ or $S^- = \emptyset$) go to Step 4.

Step 3:

Let L ($\subseteq \tilde{E}$) be (the set of arcs on) a shortest path from S^+ to S^- ("shortest" in the number of arcs);
$M^\circ := (M^\circ \setminus L) \cup \{(j,i) \mid (i,j) \in L \cap E_T\} \cup \{(j,j_Q) \mid (j_Q,j) \in L \cap E_Q\}$;
If the initial vertex ($\in S^+$) of the path L belongs to S_Q^+, then do the following:

{Let j_Q ($\in S_Q^+ \subseteq C_Q$) be the initial vertex;
Find h such that $base[h] = 0$ and $P[h,j] \neq 0$;
[$j \in C$ corresponds to $j_Q \in C_Q$]
$base[h] := j$; $w := 1/P[h,j]$;
$P[k,l] := P[k,l] - w \times P[k,j] \times P[h,l]$ $(k \in R_Q \setminus \{h\}, l \in C \setminus \{j\})$;
$S[k,l] := S[k,l] - w \times P[k,j] \times S[h,l]$ $(k \in R_Q \setminus \{h\}, l \in R_Q)$;
$P[k,j] := 0$ $(k \in R_Q \setminus \{h\})$ };

For all $(i_Q, j_Q) \in L \cap E^+$ (in the order from S^+ to S^- along L) do the following:

{Find h such that $i = base[h]$; [$j \in C$ corresponds to $j_Q \in C_Q$]
$base[h] := j$; $w := 1/P[h,j]$;
$P[k,l] := P[k,l] - w \times P[k,j] \times P[h,l]$ $(k \in R_Q \setminus \{h\}, l \in C \setminus \{j\})$;
$S[k,l] := S[k,l] - w \times P[k,j] \times S[h,l]$ $(k \in R_Q \setminus \{h\}, l \in R_Q)$;
$P[k,j] := 0$ $(k \in R_Q \setminus \{h\})$ };

Go to Step 2.

Step 4:

Let V_∞ ($\subseteq \tilde{V}$) be the set of vertices reachable from S^+ by a directed path in $\tilde{G}$;
Let V_0 ($\subseteq \tilde{V}$) be the set of vertices reachable to S^- by a directed path in $\tilde{G}$;
$C_0 := C \cap V_0$; $C_\infty := C \cap V_\infty$;
Let $\tilde{G}'$ denote the graph obtained from $\tilde{G}$ by deleting the vertices $V_0 \cup V_\infty$ (and arcs incident thereto);
Decompose $\tilde{G}'$ into strongly connected components $\{V_\lambda \mid \lambda \in \Lambda\}$ ($V_\lambda \subseteq \tilde{V}$), where $V_\lambda \preceq V_{\lambda'}$ if and only if there is a directed path from V_λ to $V_{\lambda'}$;
Let $\{C_k \mid k = 1, \cdots, b\}$ be the subcollection of $\{C \cap V_\lambda \mid \lambda \in \Lambda\}$ consisting of all the nonempty sets $C \cap V_\lambda$, where C_k's are indexed in such a way that for $l < k$ there does not exist a directed path in $\tilde{G}'$ from C_k to C_l;
$R_0 := (R_T \cap V_0) \cup \{h \in R_Q \mid base[h] \in C_0\}$;
$R_\infty := (R_T \cap V_\infty) \cup \{h \in R_Q \mid base[h] \in C_\infty \cup \{0\}\}$;
$R_k := (R_T \cap V_k) \cup \{h \in R_Q \mid base[h] \in C_k\}$ $(k = 1, \cdots, b)$;
$\bar{A} := P_{\mathrm{r}} \begin{pmatrix} P \\ T \end{pmatrix} P_{\mathrm{c}}$, where the permutation matrices P_{r} and P_{c} are determined so that the rows and the columns of $\bar{A}$ are ordered as $(R_0; R_1, \cdots, R_b; R_\infty)$ and $(C_0; C_1, \cdots, C_b; C_\infty)$, respectively. □

The matrix $\bar{A}$ obtained in Step 4 is the CCF of the input matrix A, where $(R_0; R_1, \cdots, R_b; R_\infty)$ and $(C_0; C_1, \cdots, C_b; C_\infty)$ give the partitions of the row set and the column set, respectively. The partial order among the blocks is induced from the partial order among the strongly connected components $\{V_\lambda \mid \lambda \in \Lambda\}$. The strong component decomposition $\{V_\lambda \mid \lambda \in \Lambda\}$ is essentially the same as the min-cut decomposition (cf. §2.3.5) of the associated independent matching problem, except that the partial order is reversed here.

The above algorithm runs in $\mathrm{O}(n^3 \log n)$ time with arithmetic operations in the subfield $\boldsymbol{K}$ only, where $m = m_Q + m_T = \mathrm{O}(n)$ is assumed, for simplicity, in this complexity bound. Note that Step 4 runs in $\mathrm{O}(n^2)$ time, whereas Steps 1–3 in $\mathrm{O}(n^3 \log n)$ time (cf. §4.2.4). The algorithm will be efficient enough also for practical applications. It can be made more efficient if we first compute the DM-decomposition by purely graph-theoretic algorithm and then apply the above algorithm to each of the DM-irreducible components; such two-stage procedure works for a nonsingular A, since the CCF is a refinement of the DM-decomposition. See Murota–Scharbrodt [241] for improvements in implementation and Gabow and Xu [84] for a theoretical complexity bound of $(n^{2.62})$ for the CCF computation.

Example 4.4.12. The algorithm above is illustrated here for the 4×5 LM-matrix used in Examples 4.2.6, 4.2.19 and 4.2.22. After repeating Step 1 to Step 3 (cf. Example 4.2.22) the algorithm reaches

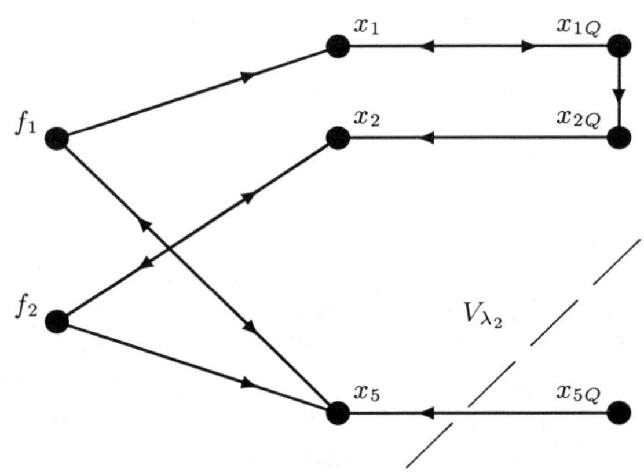

Fig. 4.15. Graph $\tilde{G}'$ (V_{λ_1}, V_{λ_2}: strongly connected components)

Step 4: $V_\infty := \emptyset$; $V_0 := \{x_3, x_4, x_{3Q}, x_{4Q}\}$; $C_0 := \{x_3, x_4\}$; $C_\infty := \emptyset$;
 The graph $\tilde{G}'$ of Fig. 4.15 is obtained;
 Strongly connected components of $\tilde{G}'$ are given by $\{V_{\lambda_1}, V_{\lambda_2}\}$, where $V_{\lambda_1} = \{x_{5Q}\}$, $V_{\lambda_2} = \{x_1, x_2, x_5, x_{1Q}, x_{2Q}, f_1, f_2\}$ and $V_{\lambda_1} \preceq V_{\lambda_2}$;
 Since $C \cap V_{\lambda_1} = \emptyset$, we have $b := 1$ and $C_1 := C \cap V_{\lambda_2} = \{x_1, x_2, x_5\}$;
 $R_0 := \{r_2\}$; $R_\infty := \emptyset$; $R_1 := \{r_1, f_1, f_2\}$;

$$\bar{A} := P_{\mathrm{r}} \begin{pmatrix} P \\ T \end{pmatrix} P_{\mathrm{c}} = \begin{array}{c|cc|ccc|} & x_3 & x_4 & x_1 & x_2 & x_5 \\ \cline{2-6} r_2 & 1 & 1 & 0 & 2 & 0 \\ \cline{4-6} r_1 & & & 1 & -1 & 0 \\ f_1 & & & t_1 & 0 & t_2 \\ f_2 & & & 0 & t_3 & t_4 \\ \cline{2-6} \end{array}$$

is the CCF. □

Example 4.4.13. This is an example with a nonempty vertical tail. Consider

$$A = \begin{pmatrix} Q \\ T \end{pmatrix} = \begin{array}{c|ccccccc|} & x_1 & x_2 & x_3 & x_4 & x_5 & x_6 & x_7 \\ \cline{2-8} & 1 & 0 & 0 & 1 & 0 & 1 & -1 \\ & -2 & 0 & 1 & -2 & 0 & 0 & 2 \\ & 1 & 0 & 0 & 1 & 1 & 1 & -1 \\ \cline{2-8} f_1 & t_1 & & & & & t_2 & \\ f_2 & t_3 & & & & t_4 & & \\ f_3 & & t_5 & & t_6 & t_7 & & \\ f_4 & & t_8 & & t_9 & t_{10} & & t_{11} \\ f_5 & & & & & & t_{12} & \\ f_6 & & & & & & t_{13} & \\ \cline{2-8} \end{array},$$

where $\mathcal{T} = \{t_i \mid i = 1, \cdots, 13\}$ is the set of algebraically independent parameters. The bipartite graph $G = (V^+, V^-; E)$ with $V^+ = R_T \cup C_Q$, $V^- = C$ for the independent matching problem is depicted in Fig. 4.16, where $R_T = \{f_1, \cdots, f_6\}$, $C = \{x_1, \cdots, x_7\}$ and $C_Q = \{x_{1Q}, \cdots, x_{7Q}\}$. A maximum independent matching M of size 7 is found. The auxiliary graph $\tilde{G}$ is shown in Fig. 4.17, from which we obtain the partition $\{V_0; V_{\lambda_1}, \cdots, V_{\lambda_5}; V_\infty\}$ of $\tilde{V} = R_T \cup C_Q \cup C$, where

$$\begin{aligned} &V_0 = \emptyset, \quad V_\infty = \{f_1, f_2, f_5, f_6, x_{5Q}, x_1, x_5, x_6\}, \\ &V_{\lambda_1} = \{x_{2Q}\}, \quad V_{\lambda_2} = \{f_3, f_4, x_{4Q}, x_{7Q}, x_2, x_4, x_7\}, \quad V_{\lambda_3} = \{x_{3Q}, x_3\} \\ &V_{\lambda_4} = \{x_{1Q}\}, \quad V_{\lambda_5} = \{x_{6Q}\}, \end{aligned}$$

with the partial order shown in Fig. 4.18. For the partition of the column set we have $(C_0; C_1, C_2; C_\infty)$ with

$$\begin{aligned} &C_0 = \emptyset, \quad C_1 = C \cap V_{\lambda_2} = \{x_2, x_4, x_7\}, \quad C_2 = C \cap V_{\lambda_3} = \{x_3\}, \\ &C_\infty = C \cap V_\infty = \{x_1, x_5, x_6\}. \end{aligned}$$

Notice that C_1 and C_2 have no order relation with each other. The CCF of A is given by

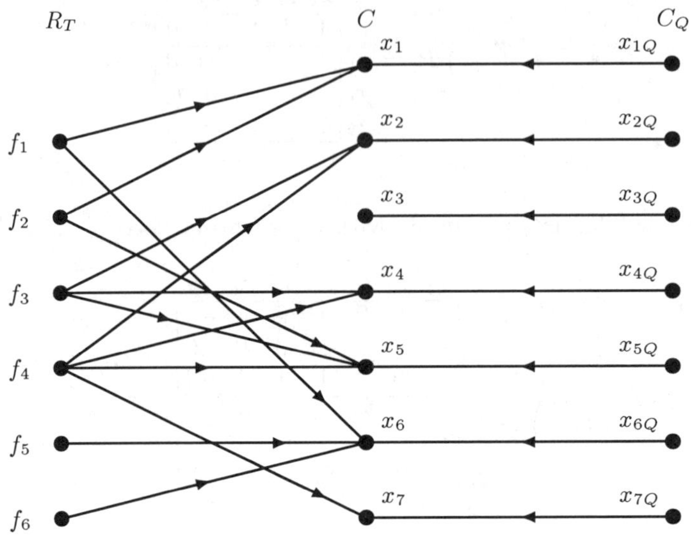

Fig. 4.16. Independent matching problem of Example 4.4.13

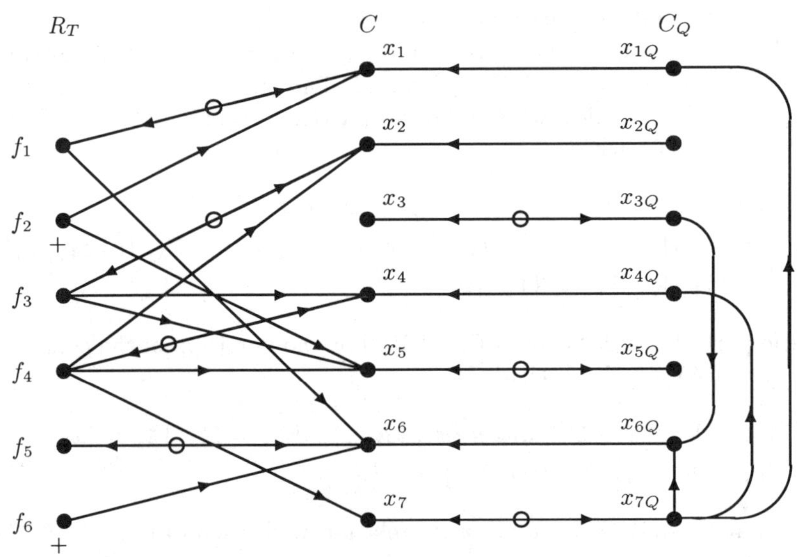

Fig. 4.17. Auxiliary graph $\tilde{G}$ of Example 4.4.13 (○: arc in a maximum independent matching M; +: vertex in S^+; $S^- = \emptyset$)

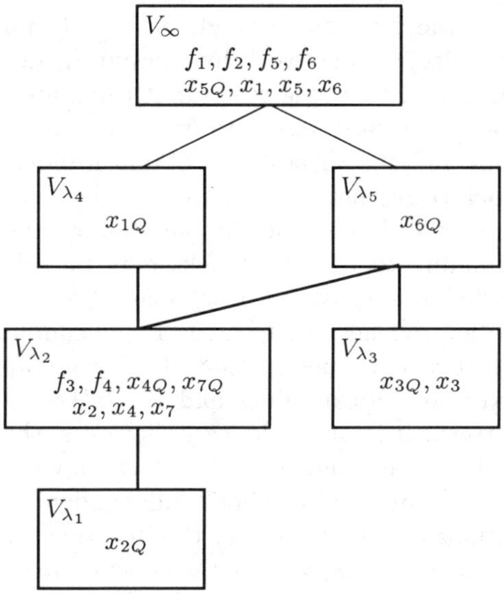

Fig. 4.18. Partial order of Example 4.4.13 $(V_0 = \emptyset)$

$$
\bar{A} = \begin{array}{c|ccc|c|ccc|}
 & \multicolumn{3}{c}{C_1} & \multicolumn{1}{c}{C_2} & \multicolumn{3}{c}{C_\infty} \\
 & x_2 & x_4 & x_7 & x_3 & x_1 & x_5 & x_6 \\
\hline
 & 0 & 1 & -1 & & 1 & & 1 \\
f_3 & t_5 & t_6 & 0 & & & t_7 & \\
f_4 & t_8 & t_9 & t_{11} & & & t_{10} & \\
\cline{2-5}
 & & & & 1 & & & 2 \\
\cline{5-8}
 & & & & & 0 & 1 & 0 \\
f_1 & & & & & t_1 & 0 & t_2 \\
f_2 & & & & & t_3 & t_4 & 0 \\
f_5 & & & & & 0 & 0 & t_{12} \\
f_6 & & & & & 0 & 0 & t_{13} \\
\hline
\end{array} .
$$

□

4.4.5 Decomposition of Systems of Equations by CCF

When solving a system of linear equations $A\boldsymbol{x} = \boldsymbol{b}$ repeatedly for right-hand side vectors $\boldsymbol{b} = \boldsymbol{b}(\theta)$ with varying parameters θ but with a fixed coefficient matrix A, it is standard to first decompose A (possibly with permutations of rows and columns) into LU-factors as $A = LU$, and then solve the triangular systems $L\boldsymbol{y} = \boldsymbol{b}$, $U\boldsymbol{x} = \boldsymbol{y}$ for different values of $\boldsymbol{b} = \boldsymbol{b}(\theta)$. It is important here that the LU-factors of A can be determined independently of the parameters θ.

No less of interest are the cases where the coefficient A, as well as $\boldsymbol{b}$, changes with parameters, but with its zero/nonzero pattern kept fixed. Such situations often arise in practice, for example, in solving a system of nonlinear equations by the Newton method, or in determining the frequency characteristic of an electrical network by computing its responses to inputs of various frequencies. In this case we cannot calculate the LU-factors of A in advance, so that we usually resort to graph-theoretic methods and rearrange the equations and the variables to obtain a block-triangular form. In particular, the block-triangularization based on the DM-decomposition is of fundamental importance. Each time the parameter values are specified, the equations corresponding to the DM-blocks may be solved either by direct inversion through *LU-decomposition* or by some iterative method.

The above two approaches, namely, the LU-decomposition and the DM-decomposition, are two extremes in that the former applies to a constant matrix A with fixed numerical values and the latter to a symbolic matrix A with a fixed pattern. It is often the case, however, that the matrix A is a mixture of constant numbers and symbols, which may be modeled as a mixed matrix under the assumption of algebraic independence of the symbols.

As a typical situation, let us consider the iterative solution of a system of linear/nonlinear equations $\boldsymbol{f}(\boldsymbol{x}) = \boldsymbol{0}$ by the *Newton method.* This amounts to solving $J(\boldsymbol{x})\Delta\boldsymbol{x} = -\boldsymbol{f}(\boldsymbol{x})$ for a correction $\Delta\boldsymbol{x}$ through the LU-decomposition of $J(\boldsymbol{x})$, where $J(\boldsymbol{x})$ is the Jacobian matrix of $\boldsymbol{f}(\boldsymbol{x})$. The equations may be divided into linear and nonlinear parts as $\boldsymbol{f}(\boldsymbol{x}) = Q\boldsymbol{x} + \boldsymbol{g}(\boldsymbol{x})$, where Q is a constant matrix. Accordingly, we have $J(\boldsymbol{x}) = Q + T(\boldsymbol{x})$, where $T(\boldsymbol{x})$ is the Jacobian matrix of the nonlinear part $\boldsymbol{g}(\boldsymbol{x})$. Then we may treat $J(\boldsymbol{x}) = Q+T(\boldsymbol{x})$ as a mixed matrix, regarding (or modeling) the nonvanishing entries of $T(\boldsymbol{x})$ as independent symbols, even when the nonvanishing entries of $T(\boldsymbol{x})$ are subject to algebraic relations.

We will describe here how the CCF can be utilized to generate an efficient solution of a system of equations

$$A(\theta)\boldsymbol{x} = \boldsymbol{b}(\theta) \tag{4.65}$$

for varying values of parameters θ, where $\boldsymbol{x}, \boldsymbol{b} \in \mathbf{R}^n$. We express the coefficient matrix as

$$A(\theta) = Q + T(\theta)$$

and regard it as a mixed matrix with ground field $\mathbf{Q}$ or $\mathbf{R}$ treating the nonvanishing entries of $T(\theta)$ as if they were algebraically independent. As discussed in §4.1, we may introduce an auxiliary vector $\boldsymbol{w}$ to obtain the augmented system of equations with the LM-matrix $\tilde{A}$ of (4.4) or (4.6) as the coefficient matrix (where we may put $t_i = 1$ for all i).

The CCF of $\tilde{A}$, being a block-triangular matrix, determines a hierarchical decomposition of the whole augmented system into smaller subsystems. Since the LM-admissible transformation (4.35) is more general than permutations, the problem decomposition by the CCF is finer than the one by the DM-decomposition. The crucial point is that the transformation (4.35) needed in the CCF decomposition is determined independently of the particular values of θ and hence this procedure is feasible in practice. That is, we can use one and the same decomposition for varying values of θ and then we may repeatedly solve the subproblems with the diagonal CCF-blocks as the coefficients whenever the parameter values are specified.

For the subproblems to be solved uniquely, the diagonal blocks of the CCF of $\tilde{A}$ must be nonsingular. If the assumption of the algebraic independence of the nonvanishing entries of $T(\theta)$ is literally met, the nonsingularity of the diagonal blocks is guaranteed by Theorem 4.4.4. Even if the assumption is not satisfied, the diagonal blocks must be nonsingular if the original coefficient matrix A is nonsingular at all, which fact is obvious from the block-triangular structure of the matrix. Therefore the decomposition procedure above can be carried out successfully if the original system is uniquely solvable at all.

Let $\bar{A}_k$ be the kth diagonal block of the CCF of $\tilde{A}$ in (4.4) (with $t_i = 1$), which is the coefficient matrix of the kth subproblem. The row set of $\bar{A}_k$ is divided into R_{Qk} and R_{Tk}. The column set, say C_k, is also partitioned as $C_k = C_{wk} \cup C_{xk}$, where C_{wk} and C_{xk} correspond to part of the variables $\boldsymbol{w}$ and $\boldsymbol{x}$, respectively. In what follows we show that

$$\min(|C_{xk}|, |R_{Tk}|) \tag{4.66}$$

can be adopted as a rough measure for the substantial size of the subproblem.

The kth subproblem may be solved as follows. Since $\bar{A}_k$ is LM-irreducible, the T-part of $\bar{A}_k$ does not have zero columns. Hence the subproblem can be expressed as

$$\begin{array}{c} \\ R_{Qk} \\ R_{Tk} \\ \downarrow \end{array} \begin{array}{c} \begin{array}{cc} C_{wk} & C_{xk} \end{array} \\ \begin{pmatrix} Q_1 & Q_2 \\ -I & T_1 \\ O & T_2 \end{pmatrix} \end{array} \begin{pmatrix} \boldsymbol{w}_k \\ \boldsymbol{x}_k \end{pmatrix} = \begin{pmatrix} \bar{\boldsymbol{b}}_k \\ \boldsymbol{0} \\ \boldsymbol{0} \end{pmatrix},$$

where $\bar{\boldsymbol{b}}_k = \bar{\boldsymbol{b}}_k(\theta)$ is to be computed from $\boldsymbol{b}(\theta)$ each time θ is given. On eliminating the auxiliary variable $\boldsymbol{w}_k$ we obtain a system of equations

$$\begin{pmatrix} Q_1 T_1 + Q_2 \\ T_2 \end{pmatrix} \boldsymbol{x}_k = \begin{pmatrix} \bar{\boldsymbol{b}}_k \\ \boldsymbol{0} \end{pmatrix}$$

in $|C_{xk}|$ variables. The amount of computation needed to determine $\boldsymbol{x}_k$ in this way may be estimated roughly by

$$2\left(|R_{Qk}||C_{wk}||C_{xk}| + |C_{xk}|^3/3\right).$$

Another approach to the subproblem may be conceivable that makes no distinction between $\boldsymbol{w}_k$ and $\boldsymbol{x}_k$. Since $\bar{A}_k[R_{Qk}, C_k]$ is of full-row rank, we can make the subsystem into the form

$$\begin{array}{c} R_{Qk} \\ R_{Tk} \end{array} \begin{pmatrix} I & Q_1 \\ T_1 & T_2 \end{pmatrix} \begin{pmatrix} z_1 \\ z_2 \end{pmatrix} = \begin{pmatrix} \bar{\boldsymbol{b}}_k \\ \boldsymbol{0} \end{pmatrix} \tag{4.67}$$

by a nonsingular transformation independent of θ, where $(\boldsymbol{z}_1, \boldsymbol{z}_2)$ is a rearrangement of $(\boldsymbol{w}_k, \boldsymbol{x}_k)$. The Gaussian elimination procedure applied to (4.67), possibly with permutations of rows in R_{Tk}, can be done with

$$2\left(|R_{Tk}|^2|R_{Qk}| + |R_{Tk}|^3/3\right)$$

arithmetic operations.

In practical applications, the following procedure would be recommended for the solution of (4.65).

[Problem decomposition by the CCF]

1. Introduce auxiliary variables to separate the equations that depend on the parameters. Denote by $\tilde{A}$ the coefficient matrix of the augmented system, which is now in the form:

$$\tilde{A} = \left(\frac{\tilde{Q}}{\tilde{T}(\theta)}\right).$$

To be more precise, express the ith equation of (4.65) as

$$\sum_{j \in J_i} a_{ij} x_j + \sum_{j \in K_i} a_{ij}(\theta) x_j = b_i(\theta).$$

In case $|J_i| \geq 1$ and $|K_i| \geq 1$, we introduce an auxiliary variable, say w_i, to obtain

$$\sum_{j \in J_i} a_{ij} x_j + w_i = b_i(\theta),$$

$$\sum_{j \in K_i} a_{ij}(\theta) x_j - w_i = 0.$$

Denoting by m_1 the number of auxiliary variables thus introduced, we see that $m_1 \leq n$ and $\tilde{A}$ is an $(m_1 + n) \times (m_1 + n)$ matrix.

2. Find the DM-decomposition of $\tilde{A}$ into blocks $(\tilde{A}_{kl} \mid 1 \leq k, l \leq D)$ to obtain the block-triangularization:

$$\begin{pmatrix} \tilde{A}_{11} & \tilde{A}_{12} & \tilde{A}_{13} & \cdots & \tilde{A}_{1D} \\ O & \tilde{A}_{22} & \tilde{A}_{23} & \cdots & \tilde{A}_{2D} \\ O & O & \ddots & & \vdots \\ O & O & \ddots & \ddots & \vdots \\ O & O & \cdots & O & \tilde{A}_{DD} \end{pmatrix} \begin{pmatrix} \boldsymbol{z}_1 \\ \boldsymbol{z}_2 \\ \vdots \\ \vdots \\ \boldsymbol{z}_D \end{pmatrix} = \begin{pmatrix} \tilde{\boldsymbol{b}}_1(\theta) \\ \tilde{\boldsymbol{b}}_2(\theta) \\ \vdots \\ \vdots \\ \tilde{\boldsymbol{b}}_D(\theta) \end{pmatrix},$$

where $\boldsymbol{z} = (\boldsymbol{z}_1, \cdots, \boldsymbol{z}_D)$ is a rearrangement of the variables $(\boldsymbol{x}, \boldsymbol{w})$.

3. For each DM-component $\tilde{A}_{kk}$, which is an LM-matrix of a smaller size:

$$\tilde{A}_{kk} = \left(\frac{\tilde{Q}_k}{\tilde{T}_k(\theta)} \right),$$

find its CCF:

$$S_k \tilde{A}_{kk} = (\bar{A}_{k;ij} \mid 1 \le i, j \le D_k)$$

where $\bar{A}_{k;ij} = O$ for $i > j$, and S_k is a constant matrix representing the row transformation of (4.35) and the column permutation is suppressed for simplicity. Accordingly put $z_k = (z_{k;1}, \cdots, z_{k;D_k})$.

4. Each time the value of θ is given, solve the subproblems as follows:
   for $k := D$ **downto** 1 **do**
   Put $\bar{b}_k := S_k\ [\bar{b}_k - (\tilde{A}_{k,k+1} z_{k+1} + \cdots + \tilde{A}_{kD} z_D)]$.
   for $i := D_k$ **downto** 1 **do**
   Solve

$$\bar{A}_{k;ii} z_{k;i} = \bar{b}_{k;i} - (\bar{A}_{k;i,i+1} z_{k;i+1} + \cdots + \bar{A}_{k;iD_k} z_{D_k}) \qquad (4.68)$$

   for $z_{k;i}$, where $(\bar{b}_{k;1}, \cdots, \bar{b}_{k;D_k}) = \bar{b}_k$. □

It should be noted that there is no need to keep S_k explicitly. In solving (4.68), the LU-decomposition of $\bar{A}_{k;ii}$ is to be determined each time θ is given.

4.4.6 Application of CCF

The decomposition technique based on the CCF, as described in §4.4.5, is applied to a series of example problems: an electrical network, the hypothetical ethylene dichloride production system of Example 3.1.3, the reactor-separator model of Example 4.3.10, the hydrogen production system of Example 4.3.11, and a collection of test matrices taken from the Harwell–Boeing database.

Example 4.4.14. The decomposition by the CCF is applied to a simple electrical network of Fig. 4.19, which is taken from Nakamura [243, Example 4.1.3]. It consists of six resistors of resistances r_i (branch i) ($i = 1, \cdots, 6$), and three voltage-controlled current sources (branch i) with mutual conductances g_i ($i = 7, 8, 9$); the current sources of branches 7, 8, and 9 are controlled respectively by the voltages across branches 2, 4, and 5. Then the current ξ^i in and the voltage η_i across branch i ($i = 1, \cdots, 9$) are to satisfy a system of equations of the form (3.2) with the coefficient matrix

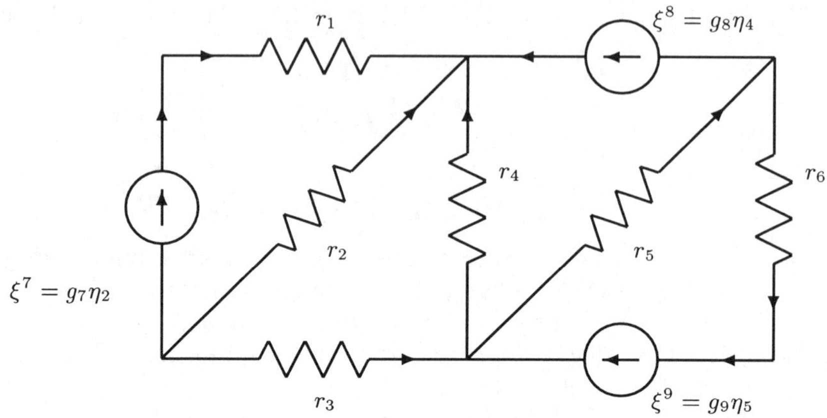

Fig. 4.19. An electrical network of Example 4.4.14

$$
A = \begin{array}{|ccccccccc|ccccccccc|}
\multicolumn{1}{c}{\xi^1} & \xi^2 & \xi^3 & \xi^4 & \xi^5 & \xi^6 & \xi^7 & \xi^8 & \multicolumn{1}{c}{\xi^9} & \eta_1 & \eta_2 & \eta_3 & \eta_4 & \eta_5 & \eta_6 & \eta_7 & \eta_8 & \multicolumn{1}{c}{\eta_9} \\
\hline
1 & 0 & 0 & 0 & 0 & 0 & -1 & 0 & 0 & & & & & & & & & \\
0 & 1 & 1 & 0 & 0 & 0 & 1 & 0 & 0 & & & & & & & & & \\
0 & 1 & 0 & 1 & 1 & 0 & 1 & 0 & -1 & & & & & & & & & \\
0 & 0 & 0 & 0 & 0 & 1 & 0 & 0 & -1 & & & & & & & & & \\
0 & 1 & 0 & 1 & 0 & 0 & 1 & 1 & 0 & & & & & & & & & \\
 & & & & & & & & & 0 & -1 & 1 & 1 & 0 & 0 & 0 & 0 & 0 \\
 & & & & & & & & & -1 & 0 & 1 & 1 & 0 & 0 & -1 & 0 & 0 \\
 & & & & & & & & & 0 & 0 & 0 & 1 & -1 & 0 & 0 & -1 & 0 \\
 & & & & & & & & & 0 & 0 & 0 & 0 & -1 & -1 & 0 & 0 & -1 \\
\hline
r_1 & & & & & & & & & -1 & & & & & & & & \\
 & r_2 & & & & & & & & & -1 & & & & & & & \\
 & & r_3 & & & & & & & & & -1 & & & & & & \\
 & & & r_4 & & & & & & & & & -1 & & & & & \\
 & & & & r_5 & & & & & & & & & -1 & & & & \\
 & & & & & r_6 & & & & & & & & & -1 & & & \\
 & & & & & & -1 & & & & g_7 & & & & & 0 & & \\
 & & & & & & & -1 & & & & & g_8 & & & & 0 & \\
 & & & & & & & & -1 & & & & & g_9 & & & & 0 \\
\hline
\end{array} .
$$

The unique solvability of the network reduces to the nonsingularity of the matrix A.

We will regard r_i $(i = 1, \cdots, 6)$ and g_i $(i = 7, 8, 9)$ as real numbers which are algebraically independent over the field of rationals. Then we have $A \in \mathrm{MM}(\mathbf{Q}, \mathbf{R}; 18, 18)$. Here we would rather treat A as an LM-matrix, just as we did in Example 4.3.9, by multiplying the last 9 rows by independent transcendentals. That is, we multiply the last 9 equations by transcendental numbers and express the modified coefficient, which we denote also as A, in the form of $A = \binom{Q}{T}$ with Q being the first 9 rows and T being the last 9 rows: $A \in \mathrm{LM}(\mathbf{Q}, \mathbf{R}; 9, 9, 18)$.

Then the CCF of A is found to be

$$
\begin{array}{c|c|c|c|c|c|c|ccc|cccccccc}
C_1 & C_2 & C_3 & C_4 & C_5 & C_6 & C_7 & & C_8 & & & & & C_9 & & & & \\
\eta_7 & \eta_1 & \xi^1 & \eta_8 & \eta_9 & \eta_6 & \xi^6 & \eta_5 & \xi^5 & \xi^9 & \xi^2 & \eta_2 & \xi^3 & \eta_3 & \xi^4 & \eta_4 & \xi^7 & \xi^8 \\
\hline
-1 & -1 & & & & & & & & & & & & 1 & & 1 & & \\
 & -1 & r_1 & & & & & & & & & & & & & & & \\
 & & 1 & & & & & & & & & & & & & & -1 & \\
 & & & -1 & & & & -1 & & & & & & & & 1 & & \\
 & & & & -1 & -1 & & -1 & & & & & & & & & & \\
 & & & & & -1 & r_6 & & & & & & & & & & & \\
 & & & & & & 1 & & & -1 & & & & & & & & \\
 & & & & & & & 0 & 1 & -1 & 1 & & & & 1 & & 1 & \\
 & & & & & & & -1 & r_5 & 0 & & & & & & & & \\
 & & & & & & & g_9 & 0 & -1 & & & & & & & & \\
\hline
 & & & & & & & & & & 1 & 0 & 1 & 0 & 0 & 0 & 1 & 0 \\
 & & & & & & & & & & 1 & 0 & 0 & 0 & 1 & 0 & 1 & 1 \\
 & & & & & & & & & & 0 & -1 & 0 & 1 & 0 & 1 & 0 & 0 \\
 & & & & & & & & & & r_2 & -1 & 0 & 0 & 0 & 0 & 0 & 0 \\
 & & & & & & & & & & 0 & 0 & r_3 & -1 & 0 & 0 & 0 & 0 \\
 & & & & & & & & & & 0 & 0 & 0 & 0 & r_4 & -1 & 0 & 0 \\
 & & & & & & & & & & 0 & g_7 & 0 & 0 & 0 & 0 & -1 & 0 \\
 & & & & & & & & & & 0 & 0 & 0 & 0 & 0 & g_8 & 0 & -1 \\
\hline
\end{array}
.
$$

It has empty tails ($C_0 = \emptyset$, $R_\infty = \emptyset$) and nine square diagonal blocks; the partial order among the column sets are shown in Fig. 4.20. □

Example 4.4.15. In Example 4.3.8 we have seen that the graph-theoretic method is not sufficient for the analysis of the hypothetical ethylene dichloride production system of Example 3.1.3. Though the DM-decomposition (Fig. 4.11) can be useful to localize the source of singularity, it fails to fully identify the rank structure of the Jacobian matrix. Here we apply the CCF-decomposition technique to the associated LM-matrix given in Fig. 4.12. The CCF, shown in Fig. 4.21, contains a 5×6 horizontal tail $C_0 = \{w_3, x, u_{33}, u_{43}, u_{53}, u_{63}\}$, a 5×4 vertical tail $C_\infty = \{w_2, w_4, u_{52}, u_{42}\}$, and nine nonsingular blocks $C_1 = \{u_{32}\}$, $C_2 = \{u_{71}\}$, $C_3 = \{u_{72}\}$, $C_4 = \{u\}$, $C_5 = \{u_{61}\}$, $C_6 = \{u_{31}\}$, $C_7 = \{u_{41}\}$, $C_8 = \{w_1, u_{51}\}$, $C_9 = \{u_{62}\}$. The partial order among the nonsingular blocks is given by $C_5 \prec C_8$, $C_6 \prec C_8$, $C_7 \prec C_8$. The existence of the nonempty tails shows the rank deficiency of the matrix. □

Example 4.4.16. The decomposition technique described in §4.4.5 is applied to the reactor-separator model used in Example 4.3.10. The Jacobian matrix, say A, is regarded as a mixed matrix, i.e., $A \in \mathrm{MM}(\mathbf{Q}, \boldsymbol{F}; 120, 120)$.

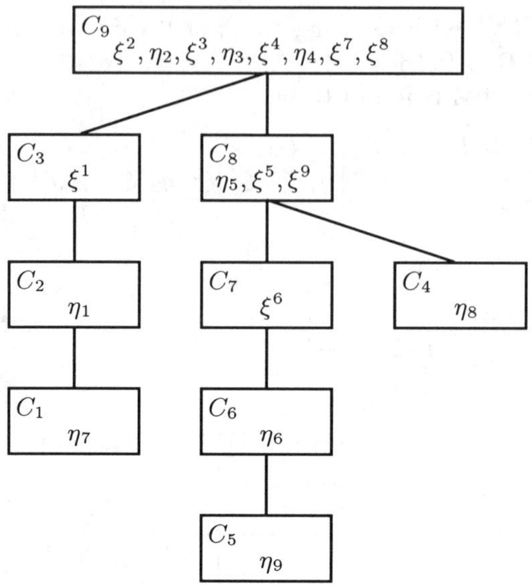

Fig. 4.20. Partial order of Example 4.4.14

	C_0						C_1	C_2	C_3	C_4	C_5	C_6	C_7	C_8		C_9	C_∞			
	w_3	x	u_{33}	u_{43}	u_{53}	u_{63}	u_{32}	u_{71}	u_{72}	u	u_{61}	u_{31}	u_{41}	w_1	u_{51}	u_{62}	w_2	w_4	u_{42}	u_{52}
u_{63}	1					-1														
y					1	-1														
u_{33}			-1			1														
u_{43}			1	-1																
u_{63}	$-t_3$	u_{53}			x															
u_{42}			⇑				1												-1	
u_{71}			horizontal tail					-1												
u_{72}									-1										-1	1
u_{53}										1										
u_{61}											-1			1						
u_{31}												-1		1						
u_{41}													-1	1						
u_{51}														1	-1					
u_{61}														$-t_1$	a_1					
u_{32}																1			-1	
u_{62}																	1		-1	
u																		1		
u_{52}														vertical tail ⇒					1	-1
u_{62}																	$-t_2$			a_2
u																		$-t_4$	r	

Fig. 4.21. CCF of the LM-matrix associated with Jacobian matrix of (3.5) (chemical process simulation in Example 3.1.3)

The DM-decomposition yields four nontrivial blocks involving more than one unknown variable. The maximum size of the blocks is 25. See Table 4.2.

The CCF of the corresponding LM-matrix $\tilde{A} \in \mathrm{LM}(\mathbf{Q}, \boldsymbol{F}; 120, 120, 240)$ in the sense of (4.4) provides a decomposition of the augmented system of equations with 120 auxiliary variables. The CCF of $\tilde{A}$ has empty tails and five nontrivial blocks, the maximum size of which being equal to 17. In Table 4.2, the DM-decomposition of A and the CCF of $\tilde{A}$ are compared, where $|R_{Tk}|$ (the number of rows of the T-part of each block) is indicated in brackets. Recall that the substantial size of a subproblem can be measured by $\min(|C_{xk}|, |R_{Tk}|)$ in (4.66).

The more compact transformation (4.6) to another LM-matrix, say $\tilde{A}_{\mathrm{cpt}}$, is also applied to A, for which m_1 (number of mixed rows) $= 85$, m_2 (number of purely constant rows) $= 34$, m_3 (number of purely symbolic rows) $= 1$ in the notation of (4.5). Hence we obtain $\tilde{A}_{\mathrm{cpt}} \in \mathrm{LM}(\mathbf{Q}, \boldsymbol{F}; 119, 86, 205)$. As expected, the CCF of $\tilde{A}_{\mathrm{cpt}}$ agrees with that of $\tilde{A}$ up to singleton blocks. The matrix $\tilde{A}_{\mathrm{cpt}}$, the DM-decomposition and the CCF of $\tilde{A}_{\mathrm{cpt}}$ are depicted in Fig. 4.22. □

Table 4.2. Decompositions for the reactor-separator model (Example 4.4.16)

DM-decomposition of $A \in \mathrm{MM}(\mathbf{Q}, \boldsymbol{F}; 120, 120)$		CCF of $\tilde{A} \in \mathrm{LM}(\mathbf{Q}, \boldsymbol{F}; 120, 120, 240)$		
size	# blocks	size		# blocks
C_{xk}		$C_k = C_{wk} + C_{xk}$	$[R_{Tk}]$	
25	1	$17 = 8 + 9$	[9]	1
10	1	$15 = 6 + 9$	[6]	1
9	2	$14 = 4 + 10$	[9]	1
		$8 = 0 + 8$	[4]	1
		$5 = 0 + 5$	[5]	1
1	67	1		181

Example 4.4.17. The decomposition technique is applied to the problem of the industrial hydrogen production system described in Example 4.3.11. The Jacobian matrix A is thought of as a mixed matrix: $A \in \mathrm{MM}(\mathbf{Q}, \boldsymbol{F}; 544, 544)$. The CCF of the corresponding LM-matrix $\tilde{A} \in \mathrm{LM}(\mathbf{Q}, \boldsymbol{F}; 544, 544, 1088)$ in the sense of (4.4) has empty tails and contains 23 nontrivial blocks with more than one variable. The DM-decomposition of A and the CCF of $\tilde{A}$ are summarized in Table 4.3. Note that the substantial sizes of the subproblems in terms of $\min(|C_{xk}|, |R_{Tk}|)$ are much smaller than those obtained by the DM-decomposition. □

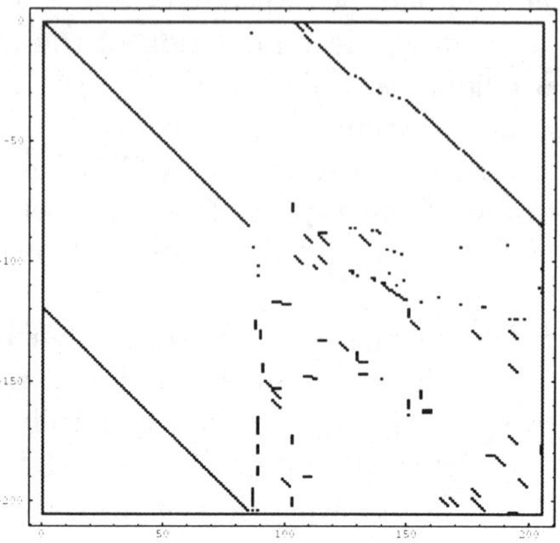

LM-matrix $\tilde{A}_{\text{cpt}} \in \text{LM}(\mathbf{Q}, \boldsymbol{F}; 119, 86, 205)$

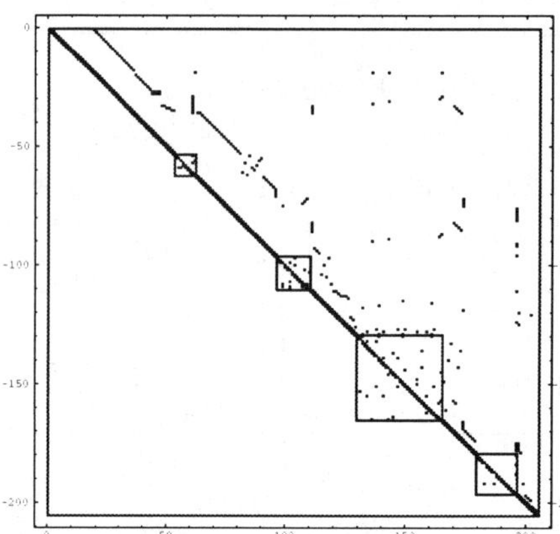

DM-decomposition of $\tilde{A}_{\text{cpt}}$

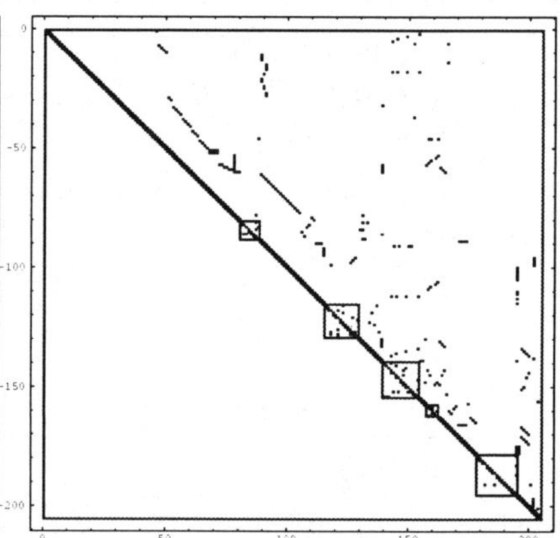

CCF of $\tilde{A}_{\text{cpt}}$

Fig. 4.22. LM-matrix $\tilde{A}_{\text{cpt}}$ and its decompositions in Example 4.4.16 (reactor-separator model)

Table 4.3. Decompositions for the hydrogen production system (Example 4.4.17)

DM-decomposition of $A \in \mathrm{MM}(\mathbf{Q}, \boldsymbol{F}; 544, 544)$ size	# blocks	CCF of $\tilde{A} \in \mathrm{LM}(\mathbf{Q}, \boldsymbol{F}; 544, 544, 1088)$ size		# blocks
C_{xk}		$C_k = C_{wk} + C_{xk}$	$[R_{Tk}]$	
104	1	$114 = 75 + 39$	[75]	1
28	1	$24 = 15 + 9$	[15]	1
23	1	$18 = 10 + 8$	[10]	1
14	1	$14 = 8 + 6$	[8]	1
10	5	$6 = 4 + 2$	[4]	1
8	1	$4 = 2 + 2$	[2]	15
6	7	$2 = 1 + 1$	[1]	3
4	2			
3	9			
1	240	1		846

Example 4.4.18. A collection of matrices are taken from the Harwell–Boeing database (Duff–Grimes–Lewis [61, 62]), Problems IMPCOL and WEST in particular, for test LM-matrices. Each matrix is thought of as a mixed matrix, where integer entries of absolute value ≤ 10 are included in the Q-part and the remaining entries are put into the T-part. The resulting mixed matrix is then converted into an LM-matrix according to the compact transformation (4.6). Table 4.4 summarizes properties of the test LM-matrices. All the matrices are square.

Table 4.4. LM-matrices made from Harwell–Boeing matrices (Example 4.4.18)

Problem	# Cols (n)	# Q-Rows (m_Q)	# T-Rows (m_T)	# Entries in Q	# Entries in T
IMPCOL A	228	171	57	338	276
IMPCOL B	89	58	31	137	194
IMPCOL C	154	136	18	399	35
IMPCOL D	483	425	58	1255	116
IMPCOL E	364	223	141	566	1015
WEST0067	86	31	55	94	238
WEST0132	211	93	118	203	368
WEST0156	229	135	94	264	244
WEST0167	262	115	147	244	452

Table 4.5 describes the block structures of the DM-decompositions and the CCF of those matrices. It turned out, in particular, that all the LM-matrices are nonsingular.

Table 4.5. Decompositions of the LM-matrices in Example 4.4.18

Problem	Rank	DM-decomp.		CCF	
		size	# blocks	size	# blocks
IMPCOL A	228	30	1	27	1
		12	1	10	1
		2	9	2	9
		1	168	1	173
IMPCOL B	89	66	1	45	1
		1	23	1	44
IMPCOL C	154	8	1	4	1
		7	3		
		1	125	1	150
IMPCOL D	483	115	1	5	1
				4	1
				2	1
		1	368	1	472
IMPCOL E	364	73	1	70	1
		36	1	27	1
		19	1	10	1
		2	1	2	1
		1	234	1	255
WEST0067	86	85	1	85	1
		1	1	1	1
WEST0132	211	127	1	115	1
		1	84	1	96
WEST0156	229	35	1	32	1
		1	194	1	197
WEST0167	262	129	1	117	1
		1	133	1	145

Table 4.6. Behavior of the CCF algorithm

Problem	# Pivots	# Base exchanges	Change of # Entries
IMPCOL A	787	6	−60
IMPCOL B	146	0	−8
IMPCOL C	6426	1	−144
IMPCOL D	76096	3	+24
IMPCOL E	2015	16	+157
WEST0067	21	0	+9
WEST0132	332	4	−14
WEST0156	227	6	−21
WEST0167	334	5	−22

Table 4.6 shows some data about the behavior of the CCF algorithm. The first column counts the total number of pivoting operations and the second the total number of pairs $(i_Q, j_Q) \in L \cap E^+$ in Step 3. The third column designates difference of the number of nonzero Q-entries in the CCF and in the input LM-matrix. The data is based on an implementation of a minor variant of the CCF algorithm (called "new algorithm" in [241]). Though such data are implementation dependent, they would serve to convey a rough idea about the behavior of the CCF algorithm. □

Notes. The examples in this section has been computed using a sightly modified version of the FORTRAN program originally coded by M. Ichikawa [117] and the Mathematica program coded by M. Scharbrodt [241].

4.4.7 CCF over Rings

We consider an extension of the concept of LM-matrix and its CCF when the ground field is replaced by a ring. Let $\boldsymbol{D}$ be an integral domain, and $\boldsymbol{K}$ the field of quotients of $\boldsymbol{D}$; it is still assumed that $\boldsymbol{K}$ is a subfield of $\boldsymbol{F}$. To be more concrete, we are mainly interested in the cases where $\boldsymbol{D}$ is the ring of integers $\mathbf{Z}$ or the ring of univariate polynomials over a field.

We say that a matrix $A = \binom{Q}{T}$ is an LM-matrix with respect to $(\boldsymbol{D}, \boldsymbol{F})$, denoted as $A \in \mathrm{LM}(\boldsymbol{D}, \boldsymbol{F})$, if $A \in \mathrm{LM}(\boldsymbol{K}, \boldsymbol{F})$ and furthermore, Q is a matrix over $\boldsymbol{D}$. Accordingly, the admissible transformation over $\boldsymbol{D}$ is defined to be an invertible transformation of the form (cf. (4.35))

$$P_{\mathrm{r}} \begin{pmatrix} S & O \\ O & I \end{pmatrix} \begin{pmatrix} Q \\ T \end{pmatrix} P_{\mathrm{c}} \tag{4.69}$$

with a matrix S over $\boldsymbol{D}$. For the invertibility of the transformation it is imposed that S is invertible over $\boldsymbol{D}$, i.e., that S has an inverse S^{-1} over $\boldsymbol{K}$ and furthermore each entry of S^{-1} belongs to $\boldsymbol{D}$. As is well known, matrix S has its inverse S^{-1} over $\boldsymbol{D}$ if and only if $\det S$ is an invertible element of $\boldsymbol{D}$, in which case S is called *unimodular* over $\boldsymbol{D}$. With this terminology we can say that an admissible transformation over $\boldsymbol{D}$ is defined to be a transformation of the form (4.69) with S unimodular over $\boldsymbol{D}$. It is obvious from the definition that such an admissible transformation is a transformation in the class $\mathrm{LM}(\boldsymbol{D}, \boldsymbol{F})$.

Given $A \in \mathrm{LM}(\boldsymbol{D}, \boldsymbol{F})$, we can regard it as a member of $\mathrm{LM}(\boldsymbol{K}, \boldsymbol{F})$ and construct its CCF, say $\bar{A}$, using an LM-admissible transformation with a nonsingular matrix S over $\boldsymbol{K}$. Here we can assume that S is a matrix over $\boldsymbol{D}$, since we may multiply S with any nonvanishing number in $\boldsymbol{D}$. This means that $\bar{A} \in \mathrm{LM}(\boldsymbol{D}, \boldsymbol{F})$ (see the matrix $\bar{A}_2$ in Example 4.4.20 below). Note, however, the transformation that brings A to $\bar{A}$ is not necessarily admissible for $\mathrm{LM}(\boldsymbol{D}, \boldsymbol{F})$, since S may not be unimodular over $\boldsymbol{D}$.

The following theorem claims that, if $\boldsymbol{D}$ is a well-behaved ring called *principal ideal domain* (*PID*), there exists an admissible transformation over

$\boldsymbol{D}$ such that the resulting matrix agrees with a CCF in its diagonal blocks. The ring of integers $\mathbf{Z}$ and the ring of univariate polynomials over a field are typical examples of PID, where the reader is referred to van der Waerden [325] for the definition of PID.

In the statement of the theorem below, a linear extension of a partial order means a linear order (=total order) that is compatible with the partial order, also called a topological sorting in computer science. Our indexing convention (4.42) for the blocks $\{C_k\}$ in the CCF of A represents a linear extension of the partial order $\preceq$ in the CCF.

Theorem 4.4.19. *Let A be an LM-matrix with respect to $(\boldsymbol{D}, \boldsymbol{F})$, where $\boldsymbol{D}$ is a PID. Let $(C_0; C_1, \cdots, C_b; C_\infty)$ denote the partition of C in the CCF of A and $\preceq$ the partial order among the blocks (using the notation of Theorem 4.4.4). For any linear extension of $\preceq$, which is represented by the linear order of the index k of the blocks, there exist permutation matrices P_r and P_c, a unimodular matrix S over $\boldsymbol{D}$, and a CCF $\bar{A}$ of A (as an LM-matrix with respect to $(\boldsymbol{K}, \boldsymbol{F})$) such that*

$$\hat{A} = P_\mathrm{r} \begin{pmatrix} S & O \\ O & I \end{pmatrix} \begin{pmatrix} Q \\ T \end{pmatrix} P_\mathrm{c}$$

is in the same block-triangular form as $\bar{A}$, having the same diagonal blocks, i.e., $\hat{A}[R_k, C_l] = \bar{A}[R_k, C_l] = O$ for $k > l$ and $\hat{A}[R_k, C_k] = \bar{A}[R_k, C_k]$ for $k = 0, 1, \cdots, b, \infty$. (It is not claimed that $\hat{A}[R_k, C_l]$ coincides with $\bar{A}[R_k, C_l]$ for $k < l$.)

Proof. In the proof of Theorem 4.4.4, the transformation to the Hermite normal form (see Newman [252], Schrijver [292]) under a unimodular transformation guarantees the existence of a unimodular matrix S such that $\bar{Q} = SQP_\mathrm{c}$ satisfies (4.50) and (4.51). However, we cannot impose the further condition (4.52), which fact causes the discrepancy in the upper off-diagonal blocks of $\hat{A}$ and $\bar{A}$. ■

Example 4.4.20. Let $\boldsymbol{D} = \mathbf{Z}$, $\boldsymbol{K} = \mathbf{Q}$, and $\boldsymbol{F} = \mathbf{Q}(t_1, t_2)$, where t_1 and t_2 are indeterminates. Consider a 3×3 LM-matrix with respect to $(\boldsymbol{D}, \boldsymbol{F}) = (\mathbf{Z}, \mathbf{Q}(t_1, t_2))$:

$$A = \begin{pmatrix} Q \\ T \end{pmatrix} = \begin{array}{c} \begin{array}{ccc} x_1 & x_2 & x_3 \end{array} \\ \begin{array}{|ccc|} \hline 2 & -2 & -4 \\ 3 & 1 & 2 \\ \hline 0 & t_1 & t_2 \\ \hline \end{array} \end{array} .$$

First regard A as a member of $\mathrm{LM}(\mathbf{Q}, \mathbf{Q}(t_1, t_2))$. By choosing $S = S_1$ below (with $\det S_1 = 1$) in the LM-admissible transformation (4.35) we obtain a CCF $\bar{A}_1$, where

$$\bar{A}_1 = \begin{array}{c} \begin{array}{ccc} x_1 & x_2 & x_3 \end{array} \\ \begin{array}{|c|cc|} \hline 2 & & \\ \hline & 4 & 8 \\ & t_1 & t_2 \\ \hline \end{array} \end{array}, \qquad S_1 = \begin{array}{|cc|} \hline 1/4 & 1/2 \\ -3/2 & 1 \\ \hline \end{array} .$$

The CCF $\bar{A}_1$ has two square blocks $C_1 = \{x_1\}$ and $C_2 = \{x_2, x_3\}$ with no order relation between them.

The transformation matrix $S = S_1$ is not a matrix over $\mathbf{Z}$. However, a transformation over $\mathbf{Z}$ can be constructed easily by putting $S = S_2 = 4 \cdot S_1$, which yields another CCF:

$$\bar{A}_2 = \begin{array}{c} \begin{array}{ccc} x_1 & x_2 & x_3 \end{array} \\ \begin{array}{|c|cc|} \hline 8 & & \\ \hline & 16 & 32 \\ & t_1 & t_2 \\ \hline \end{array} \end{array} .$$

It is noted, however, that the transformation with $S = S_2$ is not invertible over $\mathbf{Z}$ since S_2, with $\det S_2 = 16$, is not unimodular.

Restricting S to a unimodular matrix over $\mathbf{Z}$, we may take $S = \hat{S}$ below (with $\det \hat{S} = 1$) to transform A to a block-triangular matrix $\hat{A}$, where

$$\hat{A} = \begin{array}{c} \begin{array}{ccc} x_1 & x_2 & x_3 \end{array} \\ \begin{array}{|c|cc|} \hline 1 & 3 & 6 \\ \hline & 8 & 16 \\ & t_1 & t_2 \\ \hline \end{array} \end{array}, \qquad \hat{S} = \begin{array}{|cc|} \hline -1 & 1 \\ -3 & 2 \\ \hline \end{array} .$$

The order relation of the blocks in $\hat{A}$ is given by: $C_1 = \{x_1\} \preceq C_2 = \{x_2, x_3\}$. This matrix $\hat{A}$ has the same diagonal blocks as yet another CCF of A, $\bar{A}_3$ below, that is obtained with $S = S_3$, where

$$\bar{A}_3 = \begin{array}{c} \begin{array}{ccc} x_1 & x_2 & x_3 \end{array} \\ \begin{array}{|c|cc|} \hline 1 & & \\ \hline & 8 & 16 \\ & t_1 & t_2 \\ \hline \end{array} \end{array}, \qquad S_3 = \begin{array}{|cc|} \hline 1/8 & 1/4 \\ -3 & 2 \\ \hline \end{array} .$$

□

A concrete instance of Theorem 4.4.19 with $\boldsymbol{D}$ being a ring of univariate polynomials will be given in Example 6.3.10.

Notes. The content of Theorem 4.4.19 was observed by Murota [216] in the case where $\boldsymbol{D}$ is a ring of univariate polynomials, whereas the present form is found in Murota [218]. We consider some variants and extensions of the CCF later in §4.6, §4.7, §4.8, and §4.9. It is mentioned in this connection that a hierarchical decomposition of discrete systems possessing group symmetry has been investigated in Murota [217] by combining the combinatorial method for the CCF and the conventional group representation theory.

4.5 Irreducibility of LM-matrices

In this section we investigate into a concept of irreducibility for LM-matrices. Most of the results below are natural extensions of those concerning DM-irreducibility (or full indecomposability) for generic matrices treated in §2.2.3.

4.5.1 Theorems on LM-irreducibility

With respect to the LM-admissible transformation (4.35) we can define a concept of irreducibility for LM-matrices. An LM-matrix $A \in \mathrm{LM}(\boldsymbol{K}, \boldsymbol{F})$ is defined to be *LM-reducible* if it can be decomposed into smaller submatrices by means of the LM-admissible transformation (4.35). Otherwise, A is called *LM-irreducible*. In other words, A is LM-reducible if there exists a DM-reducible matrix A' that is LM-equivalent to A, and A is LM-irreducible if any LM-matrix A' that is LM-equivalent to A is DM-irreducible. An LM-irreducible matrix is DM-irreducible, and not conversely. Note that an LM-matrix A is LM-irreducible if $\mathrm{Row}(A) = \emptyset$ or $\mathrm{Col}(A) = \emptyset$, since the whole matrix A is a (horizontal or vertical) tail. On the other hand, a zero matrix $A = O$ with $\mathrm{Row}(A) \neq \emptyset$ and $\mathrm{Col}(A) \neq \emptyset$ is LM-reducible, since it can be decomposed into the horizontal tail with $(R_0, C_0) = (\emptyset, \mathrm{Col}(A))$ and the vertical tail with $(R_\infty, C_\infty) = (\mathrm{Row}(A), \emptyset)$.

From Theorem 4.4.4 we obtain the following characterization of LM-irreducibility in terms of the lattice $\mathcal{L}_{\min}(p)$ of the minimizers of the LM-surplus function p. This is a kind of "dual" characterization of the LM-irreducibility in contrast to the "primal" characterization (definition) in terms of the indecomposability with respect to the LM-admissible transformation.

Theorem 4.5.1. *Let* $A \in \mathrm{LM}(\boldsymbol{K}, \boldsymbol{F})$ *be an LM-matrix.*
(a) *In case* $|R| < |C|$: *A is LM-irreducible* $\iff \mathcal{L}_{\min}(p) = \{C\}$;
(b) *In case* $|R| = |C|$: *A is LM-irreducible* $\iff \mathcal{L}_{\min}(p) = \{\emptyset, C\}$;
(c) *In case* $|R| > |C|$: *A is LM-irreducible* $\iff \mathcal{L}_{\min}(p) = \{\emptyset\}$. □

The validity of the algorithm for CCF in §4.4.4 yields a characterization of the LM-irreducibility in terms of the graph $\tilde{G}$ used in the algorithm. In particular, we mention the case of square LM-matrices.

Theorem 4.5.2. *A square LM-matrix A is LM-irreducible (and hence nonsingular) if and only if in Step 4 of the algorithm of §4.4.4 both* V_0 *and* V_∞ *are empty and* $R_T \cup C$ *is contained in a single strong component of* $\tilde{G}$. □

The following theorem refers to the rank of submatrices of an LM-irreducible matrix. This is an extension of Theorem 2.2.24 for a generic matrix.

Theorem 4.5.3. *Let* $A \in \mathrm{LM}(\boldsymbol{K}, \boldsymbol{F})$ *be LM-irreducible.*
(a) *In case* $|R| < |C|$: $\mathrm{rank}\, A[R, C \setminus \{j\}] = |R| \quad (\forall j \in C)$;
(b) *In case* $|R| = |C|$: $\mathrm{rank}\, A[R \setminus \{i\}, C \setminus \{j\}] = |R| - 1 \quad (\forall i \in R, \forall j \in C)$;
(c) *In case* $|R| > |C|$: $\mathrm{rank}\, A[R \setminus \{i\}, C] = |C| \quad (\forall i \in R)$.

Proof. See the proof of (B4) of Theorem 4.4.4. ■

As immediate corollaries we obtain the following properties of a nonsingular irreducible LM-matrix. We regard the determinant of $A \in \mathrm{LM}(\boldsymbol{K}, \boldsymbol{F})$ as a polynomial in $\mathcal{T}$ (=set of nonzero entries of T) with coefficients from the ground field $\boldsymbol{K}$, that is, $\det A \in \boldsymbol{K}[\mathcal{T}]$.

Theorem 4.5.4. *Let $A \in \mathrm{LM}(\boldsymbol{K}, \boldsymbol{F})$ be nonsingular and LM-irreducible.*
(1) *A^{-1} is completely dense, i.e., $(A^{-1})_{ji} \neq 0$ for all (j, i).*
(2) *Each element of $\mathcal{T}$ appears in* $\det A$.

Proof. (1) This follows from Theorem 4.5.3(b), since $(A^{-1})_{ji} = \det A[R \setminus \{i\}, C \setminus \{j\}] / \det A$.

(2) If $t \in \mathcal{T}$ is the (i, j) entry of A, $\det A$ contains $t \cdot \det A[R \setminus \{i\}, C \setminus \{j\}] \neq 0$, which is not cancelled out. ■

The converse of Theorem 4.5.4(1) does not hold true. That is, a nonsingular LM-matrix A may possibly be LM-reducible even if $(A^{-1})_{ji} \neq 0$ for all (j, i).

Example 4.5.5. Consider an LM-matrix A and its CCF $\bar{A}$:

$$A = \begin{pmatrix} Q \\ T \end{pmatrix} = \begin{array}{|ccc|} \hline 1 & -1 & 0 \\ 1 & 0 & 1 \\ \hline 0 & t_1 & t_2 \\ \hline \end{array}\,, \qquad \bar{A} = \begin{array}{|c|cc|} \hline 1 & -1 & 0 \\ \hline 0 & 1 & 1 \\ 0 & t_1 & t_2 \\ \hline \end{array}\,,$$

where t_1 and t_2 are indeterminates over $\boldsymbol{K} = \mathbf{Q}$. Every minor of order two of A is nonsingular and hence $(A^{-1})_{ji} \neq 0$ for all (j, i). But A is LM-reducible with its CCF splitting into two blocks. □

The following theorem (Murota [207]) states that the combinatorial irreducibility (namely LM-irreducibility) is essentially equivalent to the algebraic irreducibility of the determinant as a polynomial in $\mathcal{T}$ over $\boldsymbol{K}$. This is an extension of Theorem 2.2.28 for a generic matrix.

Theorem 4.5.6. *Let $A \in \mathrm{LM}(\boldsymbol{K}, \boldsymbol{F})$ be a nonsingular LM-matrix. If A is LM-irreducible,* $\det A$ *is irreducible as a polynomial in $\mathcal{T}$ over $\boldsymbol{K}$. Conversely, if* $\det A$ *is an irreducible polynomial, then in the CCF of A, there is at most one diagonal block that contains elements of $\mathcal{T}$ and all the other diagonal blocks are* 1×1 *matrices over $\boldsymbol{K}$.*

Proof. The proof of the first half is given later in §4.5.2. The second half follows easily from Theorem 4.4.4 and Theorem 4.5.4(2). ■

Remark 4.5.7. If two square LM-matrices $A^{(k)} \in \mathrm{LM}(\boldsymbol{K}, \boldsymbol{F})$ $(k = 1, 2)$ are LM-equivalent, being connected by an LM-admissible transformation (4.35), they have an identical determinant up to a constant factor: $\det A^{(1)} = c \cdot$

$\det A^{(2)}$ with $c \in \boldsymbol{K}^* = \boldsymbol{K} \setminus \{0\}$. The converse of this statement, however, is not true. For example, the LM-matrices

$$A^{(1)} = \left[\begin{array}{cccccc} 1 & 0 & 0 & 1 & 1 & 0 \\ 0 & 1 & 0 & 0 & 1 & 1 \\ 0 & 0 & 1 & 1 & 0 & 1 \\ \hline t_1 & 0 & 0 & t_4 & 0 & 0 \\ 0 & t_2 & 0 & 0 & t_5 & 0 \\ 0 & 0 & t_3 & 0 & 0 & t_6 \end{array}\right], \quad A^{(2)} = \left[\begin{array}{cccccc} 1 & 0 & 0 & 1 & 0 & 1 \\ 0 & 1 & 0 & 1 & 1 & 0 \\ 0 & 0 & 1 & 0 & 1 & 1 \\ \hline t_1 & 0 & 0 & t_4 & 0 & 0 \\ 0 & t_2 & 0 & 0 & t_5 & 0 \\ 0 & 0 & t_3 & 0 & 0 & t_6 \end{array}\right]$$

have an identical determinant (both being LM-irreducible). However, these matrices are not LM-equivalent. Thus, the determinant does not characterize the LM-equivalence.

In this connection we mention an observation of Ryser [285] that a fully indecomposable (DM-irreducible) matrix is determined, up to scaling, by the determinant of its formal incidence matrix. To be precise, let $B^{(k)}$ $(k = 1, 2)$ be two fully indecomposable square matrices over $\boldsymbol{K}$, and $M^{(k)}$ $(k = 1, 2)$ be the *formal incidence matrices* defined by $M^{(k)}_{ij} = B^{(k)}_{ij} T_{ij}$ with indeterminates T_{ij}. Then, $B^{(1)} = D_{\rm r}\, B^{(2)}\, D_{\rm c}$ for some nonsingular diagonal matrices $D_{\rm r}$ and $D_{\rm c}$ over $\boldsymbol{K}$ if and only if $\det M^{(1)} = c \cdot \det M^{(2)}$ for some $c \in \boldsymbol{K}^*$. □

A minor (=subdeterminant) of $A \in \mathrm{LM}(\boldsymbol{K}, \boldsymbol{F})$ is also a polynomial in $\mathcal{T}$ over $\boldsymbol{K}$. Let $d_k(\mathcal{T}) \in \boldsymbol{K}[\mathcal{T}]$ denote the kth *determinantal divisor* of A, i.e., the greatest common divisor of all minors of order k as polynomials in $\mathcal{T}$ over $\boldsymbol{K}$. Note that $d_k(\mathcal{T}) \in \boldsymbol{K}^*$ means $d_k(\mathcal{T})$ is a nonzero "constant" free from any variables in $\mathcal{T}$.

Theorem 4.5.8. *Let $A \in \mathrm{LM}(\boldsymbol{K}, \boldsymbol{F})$ be LM-irreducible.*
(a) *In case $|R| < |C|$: $d_k(\mathcal{T}) \in \boldsymbol{K}^*$ for $k = 1, \cdots, |R|$;*
(b) *In case $|R| = |C|$: $d_k(\mathcal{T}) \in \boldsymbol{K}^*$ for $k = 1, \cdots, |R| - 1$;*
(c) *In case $|R| > |C|$: $d_k(\mathcal{T}) \in \boldsymbol{K}^*$ for $k = 1, \cdots, |C|$.*

Proof. We prove (b), while (a) and (c) can be proven similarly. By the Laplace expansion (Proposition 2.1.2), $d_{k-1}(\mathcal{T})$ divides $d_k(\mathcal{T})$ for $k = 2, \cdots, |R| - 1$, and hence it suffices to show that $d_k(\mathcal{T})$ is free from any $t \in \mathcal{T}$ for $k = |R| - 1$. Suppose t appears at position (i, j). It follows from Theorem 4.5.3(b) that $\delta \equiv \det A[R \setminus \{i\}, C \setminus \{j\}] \neq 0$. Obviously δ does not contain t, and, a fortiori, $d_k(\mathcal{T})$ does not contain t, since $d_k(\mathcal{T})$ is a divisor of δ. ■

The determinantal divisors of a general (possibly reducible) LM-matrix can be expressed in terms of the CCF as follows.

Theorem 4.5.9. *Let $d_k(\mathcal{T})$ denote the kth determinantal divisor of $A \in \mathrm{LM}(\boldsymbol{K}, \boldsymbol{F})$ for $k = 1, \cdots, r$, where $r = \operatorname{rank} A$. Then $d_k(\mathcal{T}) \in \boldsymbol{K}^*$ for $k = 1, \cdots, r - 1$, and $d_r(\mathcal{T})$ is decomposed into irreducible factors (in the ring $\boldsymbol{K}[\mathcal{T}]$) as*

$$d_r(\mathcal{T}) = \alpha \cdot \prod_{l \in I} \det \bar{A}[R_l, C_l],$$

where $\alpha \in \boldsymbol{K}^*$, *and* $\bar{A}[R_l, C_l]$ $(l \in I)$ *are the LM-irreducible square blocks in the CCF of* A *that contain an element of* $\mathcal{T}$.

Proof. The claim follows from Theorems 4.4.4, 4.5.6 and 4.5.8. ∎

4.5.2 Proof of the Irreducibility of Determinant

The proof of the first half of Theorem 4.5.6 is given below. Assume that A is nonsingular and LM-irreducible, and suppose that $\det A$ is decomposed as

$$\det A = f_1 \cdot f_2 \tag{4.70}$$

with $f_k \in \boldsymbol{K}[\mathcal{T}] \setminus \boldsymbol{K}$ $(k = 1, 2)$. For $k = 1, 2$, we denote by $\mathcal{T}_k$ $(\neq \emptyset)$ the subset of $\mathcal{T}$ consisting of the indeterminates appearing in f_k. Since $\det A$ is linear in each element of $\mathcal{T}$ and since Theorem 4.5.4(2) holds, $\{\mathcal{T}_1, \mathcal{T}_2\}$ is a nontrivial partition of $\mathcal{T}$, i.e., $\mathcal{T}_1 \cap \mathcal{T}_2 = \emptyset$ and $\mathcal{T}_1 \cup \mathcal{T}_2 = \mathcal{T}$. Put

$$R_{Tk} = \{i \in R_T \mid T_{ij} \in \mathcal{T}_k\}, \quad C_k = \{j \in C \mid T_{ij} \in \mathcal{T}_k\} \qquad (k = 1, 2). \tag{4.71}$$

Then we have $R_{Tk} \neq \emptyset$, $C_k \neq \emptyset$ $(k = 1, 2)$ and

$$R_{T1} \cap R_{T2} = \emptyset, \quad R_{T1} \cup R_{T2} = R_T, \quad C_1 \cap C_2 = \emptyset, \quad C_1 \cup C_2 = C,$$

where the first and the third relation are obvious, the second relation $R_{T1} \cup R_{T2} = R_T$ is due to the fact that the matrix T has no zero row because of the nonsingularity of A, and the last relation $C_1 \cup C_2 = C$ follows from the fact that the matrix T has no zero column because of the LM-irreducibility of A.

We claim that

$$T[R_{T1}, C_2] = O, \qquad T[R_{T2}, C_1] = O. \tag{4.72}$$

For, if there exist $i \in R_{T1}$, $j \in C_2$ such that $T_{ij} \neq 0$, the indeterminate T_{ij} must appear in $\det A$ by Theorem 4.5.4(2), which contradicts the definitions (4.71). Thus we have the first assertion in (4.72). Similarly for the second.

By Lemma 4.2.1 there exists $\hat{J} \subseteq C$ such that

$$Q[R_Q, \hat{J}] \text{ and } T[R_T, C \setminus \hat{J}] \text{ are nonsingular.} \tag{4.73}$$

Fixing arbitrarily a one-to-one correspondence $\varphi : R_Q \to \hat{J}$ and choosing $S = Q[R_Q, \hat{J}]^{-1}$ in the LM-admissible transformation (4.35), we may assume $Q[R_Q, \hat{J}] = I$. It should be noted here that the LM-admissible transformation changes the determinant only by a factor in $\boldsymbol{K} \setminus \{0\}$. Put

$$\begin{aligned} &C_{Tk} = C_k \setminus \hat{J} \quad (k=1,2), && C_T = C_{T1} \cup C_{T2} \ (= C \setminus \hat{J}), \\ &C_{Qk} = C_k \cap \hat{J} \quad (k=1,2), && C_Q = C_{Q1} \cup C_{Q2} \ (= \hat{J}), \\ &R_{Qk} = \varphi^{-1}(C_{Qk}) \quad (k=1,2), && R_k = R_{Qk} \cup R_{Tk} \quad (k=1,2). \end{aligned}$$

Then we see that $T[R_{Tk}, C_{Tk}]$ is nonsingular for $k = 1,2$ by (4.72) and (4.73), and that $|R_k| = |C_k|$ $(k = 1,2)$ and $R_{Q1} \cup R_{Q2} = R_Q$. Hence we can extend $\varphi : R_Q \to \hat{J}$ to $\varphi : R \to C$ in such a way that

$$Q_{i,\varphi(i)} = 1 \quad (i \in R_Q), \qquad T_{i,\varphi(i)} \neq 0 \quad (i \in R_T). \tag{4.74}$$

Hence we have the following picture of the matrix A:

$$\begin{array}{c|cc|cc|} & C_{Q1} & C_{T1} & C_{Q2} & C_{T2} \\ \hline R_{Q1} & I & Q[R_{Q1}, C_{T1}] & O & Q[R_{Q1}, C_{T2}] \\ R_{T1} & T[R_{T1}, C_{Q1}] & T[R_{T1}, C_{T1}] & O & O \\ \hline R_{Q2} & O & Q[R_{Q2}, C_{T1}] & I & Q[R_{Q2}, C_{T2}] \\ R_{T2} & O & O & T[R_{T2}, C_{Q2}] & T[R_{T2}, C_{T2}] \\ \hline \end{array}, \tag{4.75}$$

where the diagonal submatrices are all nonsingular.

By the Laplace expansion of (4.75) with respect to the rows of R_{T2} we obtain

$$\det A = \sum_{J_2 \subseteq C_2} \pm \det A[R \setminus R_{T2}, C \setminus J_2] \cdot \det T[R_{T2}, J_2],$$

in which no similar terms appear among the nonvanishing terms for distinct J_2. Consider the particular term for $J_2 = C_{T2}$. By (4.74), $\det T[R_{T2}, C_{T2}]$ contains a nonvanishing term $\tau = \prod_{i \in R_{T2}} T_{i,\varphi(i)}$, which, multiplied by $\det A[R \setminus R_{T2}, C \setminus C_{T2}] = \det A[R_1, C_1]$, appears in $\det A$. In other words, if we think of $\det A$ as a polynomial in $\mathcal{T}_2$ over $\boldsymbol{K}[\mathcal{T}_1]$, the coefficient of τ is equal to $\det A[R_1, C_1]$. On the other hand, since the term τ is contained in f_2 of (4.70) with a nonzero coefficient, say c', in $\boldsymbol{K}$, the coefficient of τ in $f_1 \cdot f_2$ is equal to $c' \cdot f_1 \in \boldsymbol{K}[\mathcal{T}_1]$. Therefore, we see that

$$f_1 = c_1 \cdot \det A[R_1, C_1], \quad c_1 \in \boldsymbol{K}.$$

Similarly we have

$$f_2 = c_2 \cdot \det A[R_2, C_2], \quad c_2 \in \boldsymbol{K}.$$

That is,

$$\det A = c_1 c_2 \cdot \det A[R_1, C_1] \cdot \det A[R_2, C_2] = \det A[R_1, C_1] \cdot \det A[R_2, C_2], \tag{4.76}$$

where we see $c_1 c_2 = 1$ from (4.75).

Remark 4.5.10. Before proceeding further, we explain our intuition behind the rigorous arguments that follow, though this remark is not necessary from the mathematical logical point of view. Since A is LM-irreducible, $Q[R_{Q2}, C_{T1}]$ is not a zero matrix (see (4.75)), that is, there exist $i_0 \in R_{Q2}$, $j_0 \in C_{T1}$ such that $Q_{i_0 j_0} \neq 0$. Since $A[R \setminus \{i_0\}, C \setminus \{j_0\}]$ is nonsingular by Theorem 4.5.3, we have a nonzero term :

$$\prod_{i \in R} A_{i,\sigma(i)} = \prod_{i \in R_Q} Q_{i,\sigma(i)} \cdot \prod_{i \in R_T} T_{i,\sigma(i)}$$

involving $Q_{i_0 j_0}$ in the usual expansion of $\det A$ into the sum of products, where $\sigma : R \to C$ is a permutation, or a one-to-one correspondence such that $\sigma(i_0) = j_0$.

One might be tempted to claim that this contradicts (4.76) based on the observation that this term cannot appear on the right hand side of (4.76) since $i_0 \in R_{Q2} \subseteq R_2$ and $\sigma(i_0) = j_0 \in C_{T1} \subseteq C_1$. This reasoning, however, is not rigorous since this term may or may not be cancelled out by similar terms. In fact, the following example of an LM-irreducible matrix demonstrates that such cancellation does occur:

$$A = \begin{array}{ll|cc|c|cc|c|}
 & & \multicolumn{2}{c}{C_{Q1}} & \multicolumn{1}{c}{C_{T1}} & \multicolumn{2}{c}{C_{Q2}} & \multicolumn{1}{c}{C_{T2}} \\
 & & c_1 & c_2 & c_3 & c_4 & c_5 & c_6 \\ \hline
R_{Q1} & r_1 & 1 & 0 & 1 & 0 & 0 & 1 \\
 & r_2 & 0 & 1 & 1 & 0 & 0 & 1 \\ \hline
R_{T1} & r_3 & x_1 & x_2 & x_3 & 0 & 0 & 0 \\ \hline
R_{Q2} & r_4 & 0 & 0 & 1^{(*)} & 1 & 0 & 1 \\
 & r_5 & 0 & 0 & 1 & 0 & 1 & 1 \\ \hline
R_{T2} & r_6 & 0 & 0 & 0 & y_1 & y_2 & y_3 \\ \hline
\end{array}$$

where $R_{Q1} = \{r_1, r_2\}, R_{T1} = \{r_3\}, R_{Q2} = \{r_4, r_5\}, R_{T2} = \{r_6\}; C_{Q1} = \{c_1, c_2\}, C_{T1} = \{c_3\}, C_{Q2} = \{c_4, c_5\}, C_{T2} = \{c_6\}$. If we choose $i_0 = r_4, j_0 = c_3$ (at the position $(*)$), we have

$$\det A[R \setminus \{i_0\}, C \setminus \{j_0\}] = x_1 y_1 + x_2 y_1.$$

These terms, however, are cancelled out and do not appear in

$$\det A = -x_1 y_3 - x_2 y_3 - x_3 y_1 - x_3 y_2 + x_3 y_3.$$

A rather sophisticated argument below is to overcome this complication in deriving a contradiction. It is noted in passing that $\det A$ above is an irreducible polynomial in $\mathbf{Q}[x_1, x_2, x_3, y_1, y_2, y_3]$. □

Associated with the matrix A of (4.75), we define, with reference to φ, a graph $G = (V, E)$ with vertex set $V = R_T \cup C$ and arc set $E = E_Q \cup E_T \cup E_M$, where

$$\begin{aligned} E_Q &= \{(\varphi(i), j) \mid i \in R_Q, j \in C_T, Q_{ij} \neq 0\}, \\ E_T &= \{(i, j) \mid i \in R_T, j \in C, T_{ij} \neq 0\}, \\ E_M &= \{(\varphi(i), i) \mid i \in R_T\}. \end{aligned}$$

Note that by (4.72) there exist no arcs between R_{T1} and C_2 nor between R_{T2} and C_1. Also note that (i) a vertex of R_T has exactly one in-coming arc, which is in E_M, (ii) a vertex of C_T has exactly one out-going arc, which is in E_M, (iii) the arcs coming into a vertex of C_Q belong to E_T, whereas the arcs going out of a vertex of C_Q belong to E_Q.

The graph G is strongly connected by the LM-irreducibility of A and Theorem 4.5.2, since G is obtained from the graph $\tilde{G}$ in the CCF-algorithm of §4.4.4 by identifying the corresponding copies of C. Therefore, there exists in G a directed simple cycle which contains both a vertex of C_1 and a vertex of C_2. Choose such a directed cycle H having the minimum number of arcs, and let E_H ($\subseteq E$) denote the set of arcs in H. We index the arcs of $E_H \cap E_Q$ in such a way that

$$e_1, e_2, \cdots, e_m \ (= e_0)$$

appear in this order along the cycle H, $\partial^+ e_m \in C_2$, and $\partial^- e_m \in C_1$, where, for $e \in E$ in general, $\partial^+ e$ and $\partial^- e$ designate the initial and terminal vertices. Note here that $\{\partial^+ e_r, \partial^- e_r\} \subseteq C \ (= C_1 \cup C_2)$ for $r = 1, \cdots, m$ since $e_r \in E_Q$.

From among the arcs $e_1, e_2, \cdots, e_m$, we pick up those which connect from C_1 to C_2 or from C_2 to C_1, and denote them as

$$\hat{e}_1, \hat{e}_2, \cdots, \hat{e}_{\hat{m}} \ (= \hat{e}_0),$$

where they are indexed again along H and $\hat{e}_{\hat{m}} = e_m$; put $\hat{E} = \{\hat{e}_1, \cdots, \hat{e}_{\hat{m}}\}$. Note that $\hat{m}$ is even and that

$$\begin{cases} \partial^+ \hat{e}_s \in C_1, \ \partial^- \hat{e}_s \in C_2 \ (s\text{: odd}) \\ \partial^+ \hat{e}_s \in C_2, \ \partial^- \hat{e}_s \in C_1 \ (s\text{: even}) \end{cases}$$

for $s = 1, \cdots, \hat{m}$.

With reference to φ we define

$$\begin{aligned} i_r &= \varphi^{-1}(\partial^+ e_r) \in R_Q, \quad j_r = \partial^- e_{r-1} \in C_T \qquad (r = 1, \cdots, m), \\ \hat{\imath}_s &= \varphi^{-1}(\partial^+ \hat{e}_s) \in R_Q, \quad \hat{\jmath}_s = \partial^- \hat{e}_{s-1} \in C_T \qquad (s = 1, \cdots, \hat{m}). \end{aligned}$$

This means

$$e_r = (\varphi(i_r), j_{r+1}) \quad (r = 1, \cdots, m), \qquad \hat{e}_s = (\varphi(\hat{\imath}_s), \hat{\jmath}_{s+1}) \quad (s = 1, \cdots, \hat{m}).$$

Here and below the indices r and s are to be understood with modulo m and modulo $\hat{m}$, respectively. We further define $r(s)$ ($s = 1, \cdots, \hat{m}$) by $j_{r(s)} = \hat{\jmath}_s$ ($s = 1, \cdots, \hat{m}$), and put

$$I_s^- = \{i_r \mid r(s) \le r \le r(s+1)-1\} \qquad (s = 1,\cdots,\hat{m}),$$
$$J_s^- = \varphi(I_s^-) = \{\varphi(i_r) \mid r(s) \le r \le r(s+1)-1\} \qquad (s = 1,\cdots,\hat{m}),$$
$$J_s^+ = \{j_r \mid r(s) \le r \le r(s+1)-1\} \qquad (s = 1,\cdots,\hat{m});$$
$$\hat{I}_1^- = \bigcup_{s:\text{odd}} I_s^-,\ \hat{I}_2^- = \bigcup_{s:\text{even}} I_s^-,\ \hat{I}^- = \hat{I}_1^- \cup \hat{I}_2^- = \{i_r \mid r = 1,\cdots,m\},$$
$$\hat{J}_1^- = \bigcup_{s:\text{odd}} J_s^-,\ \hat{J}_2^- = \bigcup_{s:\text{even}} J_s^-,\ \hat{J}^- = \hat{J}_1^- \cup \hat{J}_2^- = \{\varphi(i_r) \mid r = 1,\cdots,m\},$$
$$\hat{J}_1^+ = \bigcup_{s:\text{odd}} J_s^+,\ \hat{J}_2^+ = \bigcup_{s:\text{even}} J_s^+,\ \hat{J}^+ = \hat{J}_1^+ \cup \hat{J}_2^+ = \{j_r \mid r = 1,\cdots,m\}.$$

Then the end vertices of the arcs of $E_H \cap E_Q$ are partitioned into $\hat{m}\,(\ge 2)$ disjoint "blocks" $J_s^+ \cup J_s^-$ $(s = 1,\cdots,\hat{m})$, the consecutive blocks being connected by arcs of $\hat{E}$. Also note that

$$\begin{cases} J_s^+ \subseteq C_{T1},\ J_s^- \subseteq C_{Q1}\ (s\text{: odd}) \\ J_s^+ \subseteq C_{T2},\ J_s^- \subseteq C_{Q2}\ (s\text{: even}) \end{cases}$$

for $s = 1,\cdots,\hat{m}$.

With the notation above we now make key observations which are consequences of the minimality of the chosen cycle H. The first observation is that

$$e \in E_Q,\ \varphi(i_r) = \partial^+ e \in J_s^-,\ j_{r'} = \partial^- e \in J_s^+ \quad \Rightarrow \quad r' \le r+1. \tag{4.77}$$

This states that there are no "forward" arcs in E_Q within each block. The second observation is that

$$e \in E_Q,\ \partial^+ e \in J_s^-,\ \partial^- e \in J_{s'}^+,\ s \ne s' \quad \Rightarrow \quad e \in \hat{E}. \tag{4.78}$$

This says that the arcs of $\hat{E}$ are the only arcs of E_Q connecting vertices in different blocks.

In terms of matrix Q, (4.77) and (4.78) are rephrased as follows:

$$Q_{i_r j_{r'}} \ne 0,\ i_r \in I_s^-,\ j_{r'} \in J_s^+ \quad \Rightarrow \quad r' \le r+1, \tag{4.79}$$

that is, $Q[I_s^-, J_s^+]$ is in a lower-left Hessenberg form for each s; and

$$Q_{i_r j_{r'}} \ne 0,\ i_r \in I_s^-,\ j_{r'} \in J_{s'}^+,\ s \ne s' \quad \Rightarrow \begin{cases} s' = s+1 \pmod{\hat{m}} \\ r' = r+1 = r(s') \pmod{m}. \end{cases} \tag{4.80}$$

In addition, we have

$$Q_{i_r j_{r+1}} \ne 0 \qquad (r = 1,\cdots,m) \tag{4.81}$$

that correspond to arcs of $E_H \cap E_Q$.

In the case of $\hat{m} = 4$, for example, $Q[\hat{I}^-, \hat{J}^+]$ looks as follows

$$Q[\hat{I}^-, \hat{J}^+] = \begin{array}{c} \\ I_1^- \\ I_2^- \\ I_3^- \\ I_4^- \end{array} \!\!\begin{array}{c} \begin{array}{cccc} J_1^+ & J_2^+ & J_3^+ & J_4^+ \end{array} \\ \boxed{\begin{array}{cccc} H_{11} & D_{12} & O & O \\ O & H_{22} & D_{23} & O \\ O & O & H_{33} & D_{34} \\ D_{41} & O & O & H_{44} \end{array}} \end{array}, \tag{4.82}$$

where the rows are indexed by $i_1, i_2, \cdots, i_m$ in this order, and the columns by $j_1, j_2, \cdots, j_m$ in this order. For each $s = 1, \cdots, \hat{m}$, $H_{ss} = Q[I_s^-, J_s^+]$ is a lower-left Hessenberg matrix with nonzero upper subdiagonal entries that correspond to arcs in $(E_H \cap E_Q) \setminus \hat{E}$, and $D_{s,s+1} = Q[I_s^-, J_{s+1}^+]$ contains the only one nonzero entry in the lower-left corner that corresponds to an arc in $\hat{E}$.

It follows from (4.79), (4.80), and (4.81) (also (4.82)) that

$$\det Q[\hat{I}^-, \hat{J}^+] = \det Q[\hat{I}_1^-, \hat{J}_1^+] \cdot \det Q[\hat{I}_2^-, \hat{J}_2^+] + \alpha \tag{4.83}$$

with $\alpha = \pm \prod_{r=1}^{m} Q_{i_r j_{r+1}} \neq 0$.

By introducing $\hat{I}_k^-$, $\hat{J}_k^-$, and $\hat{J}_k^+$ ($k = 1, 2$) into (4.75), we obtain the following more detailed picture:

$$A = \begin{array}{cc} & \\ & \\ R_{Q1} & \hat{I}_1^Q \\ & \hat{I}_1^- \\ R_{T1} & \\ R_{Q2} & \hat{I}_2^Q \\ & \hat{I}_2^- \\ R_{T2} & \end{array} \begin{array}{cc|cc|cc|cc} \multicolumn{2}{c}{C_{Q1}} & \multicolumn{2}{c}{C_{T1}} & \multicolumn{2}{c}{C_{Q2}} & \multicolumn{2}{c}{C_{T2}} \\ \hat{J}_1^Q & \hat{J}_1^- & \hat{J}_1^T & \hat{J}_1^+ & \hat{J}_2^Q & \hat{J}_2^- & \hat{J}_2^T & \hat{J}_2^+ \\ \hline I & O & Q[*] & Q[*] & O & O & Q[*] & Q[*] \\ O & I & Q[*] & Q[1,1] & O & O & Q[*] & Q[1,2] \\ \hline T[*] & T[\#] & T[\#] & T[*] & O & O & O & O \\ \hline O & O & Q[*] & Q[*] & I & O & Q[*] & Q[*] \\ O & O & Q[*] & Q[2,1] & O & I & Q[*] & Q[2,2] \\ \hline O & O & O & O & T[*] & T[\#] & T[\#] & T[*] \\ \hline \multicolumn{4}{c}{\leftarrow C_1 \setminus \hat{J}_1^* \rightarrow} & \multicolumn{4}{c}{\leftarrow C_2 \setminus \hat{J}_2^* \rightarrow} \end{array}. \tag{4.84}$$

Here

$$\hat{I}_k^Q = R_{Qk} \setminus \hat{I}_k^-, \quad \hat{J}_k^Q = C_{Qk} \setminus \hat{J}_k^-, \quad \hat{J}_k^T = C_{Tk} \setminus \hat{J}_k^+ \quad (k = 1, 2);$$

$T[*]$ and $T[\#]$ are each a submatrix of T; $Q[*]$ denotes a submatrix of Q, and $Q[k, l] = Q[\hat{I}_k^-, \hat{J}_l^+]$ ($k, l = 1, 2$) are submatrices of $Q[\hat{I}^-, \hat{J}^+]$ of (4.82), i.e.,

$$Q[\hat{I}^-, \hat{J}^+] = \begin{array}{c} \\ \hat{I}_1^- \\ \hat{I}_2^- \end{array} \!\!\begin{array}{c} \begin{array}{cc} \hat{J}_1^+ & \hat{J}_2^+ \end{array} \\ \boxed{\begin{array}{cc} Q[1,1] & Q[1,2] \\ Q[2,1] & Q[2,2] \end{array}} \end{array}. \tag{4.85}$$

Since H is a directed cycle and E_M is a matching in G,

$$E_M^* = (E_M \setminus E_H) \cup \{(j, i) \mid i \in R_T, j \in C, (i, j) \in E_T \cap E_H\}$$

is again a matching of size $|R_T|$. This implies that $T[R_{Tk}, C_k \setminus \hat{J}_k^*]$ is nonsingular for $k = 1, 2$, where

$$\hat{J}_k^* = (C_{Qk} \setminus \hat{J}_k^-) \cup \hat{J}_k^+ = \hat{J}_k^Q \cup \hat{J}_k^+ \qquad (k = 1, 2).$$

To be specific, $\varphi^* : R_T \to C$ defined by $(\varphi^*(i), i) \in E_M^*$ gives a one-to-one correspondence between R_T and $C \setminus (\hat{J}_1^* \cup \hat{J}_2^*)$ such that $\varphi^*(R_{Tk}) = C_k \setminus \hat{J}_k^*$ $(k = 1, 2)$, and that

$$\tau_k \equiv \prod_{i \in R_{Tk}} T_{i,\varphi^*(i)} \neq 0 \qquad (k = 1, 2).$$

We consider the term $\tau_1 \cdot \tau_2$ in $\det A$. By the Laplace expansion of $\det A$ with A of (4.84) we see that the coefficient $c^* \in \boldsymbol{K}$ of $\tau_1 \cdot \tau_2$ in $\det A$ is given by

$$c^* = \det Q[R_Q, \hat{J}_1^* \cup \hat{J}_2^*].$$

For the determinant on the right-hand side we have

$$\det Q[R_Q, \hat{J}_1^* \cup \hat{J}_2^*] = \det Q[\hat{I}^-, \hat{J}^+] = \det Q[\hat{I}_1^-, \hat{J}_1^+] \cdot \det Q[\hat{I}_2^-, \hat{J}_2^+] + \alpha,$$

where the first equality follows from

$$Q[R_Q, \hat{J}_1^* \cup \hat{J}_2^*] = \begin{array}{c|cc|cc|} & \hat{J}_1^Q & \hat{J}_1^+ & \hat{J}_2^Q & \hat{J}_2^+ \\ \hline \hat{I}_1^Q & I & Q[*] & O & Q[*] \\ \hat{I}_1^- & O & Q[1,1] & O & Q[1,2] \\ \hline \hat{I}_2^Q & O & Q[*] & I & Q[*] \\ \hat{I}_2^- & O & Q[2,1] & O & Q[2,2] \\ \hline \end{array}$$

and (4.85), and the second equality from (4.83). Therefore,

$$c^* = \det Q[\hat{I}_1^-, \hat{J}_1^+] \cdot \det Q[\hat{I}_2^-, \hat{J}_2^+] + \alpha. \tag{4.86}$$

On the other hand, since τ_k is contained in $\det A[R_k, C_k]$ with the coefficient equal to $Q[\hat{I}_k^-, \hat{J}_k^+]$ for $k = 1, 2$, the expression (4.76) requires that

$$c^* = \det Q[\hat{I}_1^-, \hat{J}_1^+] \cdot \det Q[\hat{I}_2^-, \hat{J}_2^+]. \tag{4.87}$$

Thus we are led to two contradictory expressions, (4.86) and (4.87), for c^*. This completes the proof of Theorem 4.5.6.

Notes. This section is based on Murota [207] and Murota [218].

4.6 Decomposition of Mixed Matrices

The decomposition of LM-matrices has been established by the CCF in §4.4. The decomposition of general mixed matrices is considered in this section.

4.6.1 LU-decomposition of Invertible Mixed Matrices

We investigate here the invertibility of a mixed matrix $A = Q + T$ in $\boldsymbol{K}[\mathcal{T}]$ (=the ring of polynomials in the nonvanishing entries $\mathcal{T}$ of T over $\boldsymbol{K}$). More specifically, we are interested in whether we can compute A^{-1} by means of pivot operations in $\boldsymbol{K}[\mathcal{T}]$ and also in how simple we can make the LU-factors of A by applying suitable permutations to the rows and columns.

Let $A = Q+T$ be a square mixed matrix, $A \in \mathrm{MM}(\boldsymbol{K}, \boldsymbol{F}; n, n)$, which we regard as a matrix over $\boldsymbol{K}[\mathcal{T}]$. Recall a well-known fact that A is invertible in $\boldsymbol{K}[\mathcal{T}]$, i.e., $A^{-1} \in \boldsymbol{K}[\mathcal{T}]$, if and only if $\det A \in \boldsymbol{K}^*(= \boldsymbol{K} \setminus \{0\})$. By way of illustration of our problem we start with an example.

Example 4.6.1. A matrix

$$A = \begin{array}{c|ccccc|} & 1 & 2 & 3 & 4 & 5 \\ \hline 1 & -1 & 1 & 1 & 0 & 1 \\ 2 & 1 & 0 & x & 1 & 0 \\ 3 & 0 & 1 & 1 & 0 & 1 \\ 4 & y & -1 & 1 & 0 & -1 \\ 5 & 1 & 1 & 0 & z & 0 \\ \hline \end{array}$$

is a mixed matrix, $A = Q + T \in \mathrm{MM}(\boldsymbol{K}, \boldsymbol{F}; 5, 5)$ for $\boldsymbol{K} = \mathbf{Q}$ and $\boldsymbol{F} = \mathbf{Q}(x, y, z)$ with

$$Q = \begin{array}{c|ccccc|} & 1 & 2 & 3 & 4 & 5 \\ \hline 1 & -1 & 1 & 1 & 0 & 1 \\ 2 & 1 & 0 & 0 & 1 & 0 \\ 3 & 0 & 1 & 1 & 0 & 1 \\ 4 & 0 & -1 & 1 & 0 & -1 \\ 5 & 1 & 1 & 0 & 0 & 0 \\ \hline \end{array}, \qquad T = \begin{array}{c|ccccc|} & 1 & 2 & 3 & 4 & 5 \\ \hline 1 & 0 & 0 & 0 & 0 & 0 \\ 2 & 0 & 0 & x & 0 & 0 \\ 3 & 0 & 0 & 0 & 0 & 0 \\ 4 & y & 0 & 0 & 0 & 0 \\ 5 & 0 & 0 & 0 & z & 0 \\ \hline \end{array},$$

if $\mathcal{T} = \{x, y, z\}$ is algebraically independent over $\mathbf{Q}$. Note that $\det A = 2$ and hence A is invertible in $\mathbf{Q}[x, y, z]$. The matrix A is decomposed into LU-factors in $\mathbf{Q}(x, y, z)$ as $A = L\,U$ with

$$L = \begin{bmatrix} 1 & 0 & 0 & 0 & 0 \\ -1 & 1 & 0 & 0 & 0 \\ 0 & 1 & 1 & 0 & 0 \\ -y & y-1 & y-1-2/x & 1 & 0 \\ -1 & 2 & 2+1/x & -(xz+1)/2 & 1 \end{bmatrix}, \quad U = \begin{bmatrix} -1 & 1 & 1 & 0 & 1 \\ 0 & 1 & x+1 & 1 & 1 \\ 0 & 0 & -x & -1 & 0 \\ 0 & 0 & 0 & -2/x & 0 \\ 0 & 0 & 0 & 0 & -1 \end{bmatrix}.$$

Observe that some of the entries of L and U do not belong to $\mathbf{Q}[x, y, z]$. However, after rearranging the rows and the columns of A as

$$P_{\mathrm{r}} A P_{\mathrm{c}} = \begin{array}{c|ccccc|} & 5 & 2 & 4 & 3 & 1 \\ \hline 1 & 1 & 1 & 0 & 1 & -1 \\ 5 & 0 & 1 & z & 0 & 1 \\ 2 & 0 & 0 & 1 & x & 1 \\ 4 & -1 & -1 & 0 & 1 & y \\ 3 & 1 & 1 & 0 & 1 & 0 \\ \hline \end{array},$$

we have the LU-decomposition $P_r A P_c = LU$ with

$$L = \begin{bmatrix} 1 & 0 & 0 & 0 & 0 \\ 0 & 1 & 0 & 0 & 0 \\ 0 & 0 & 1 & 0 & 0 \\ -1 & 0 & 0 & 1 & 0 \\ 1 & 0 & 0 & 0 & 1 \end{bmatrix}, \qquad U = \begin{bmatrix} 1 & 1 & 0 & 1 & -1 \\ 0 & 1 & z & 0 & 1 \\ 0 & 0 & 1 & x & 1 \\ 0 & 0 & 0 & 2 & y-1 \\ 0 & 0 & 0 & 0 & 1 \end{bmatrix}.$$

These LU-factors are much simpler in the sense that all the entries of L are numbers in $\mathbf{Q}$ and consequently the entries of U are polynomials in x, y, and z over $\mathbf{Q}$ of degree at most one. □

In the following, we establish a theorem (Theorem 4.6.4) stating that it is always possible to find permutations of rows and columns through which an invertible matrix A can be brought to a form decomposable into LU-factors with the L-factor being a matrix over $\boldsymbol{K}$. Furthermore, we show how to find suitable permutations.

First, a necessary and sufficient condition for the invertibility of a mixed matrix is given. A matrix is said to be *strictly upper triangular* if it is an upper triangular matrix with zero diagonals.

Theorem 4.6.2. *A square mixed matrix* $A = Q + T \in \mathrm{MM}(\boldsymbol{K}, \boldsymbol{F}; n, n)$ *is invertible in* $\boldsymbol{K}[T]$, *if and only if* $\det Q \neq 0$ *and* ${P_c}^{\mathrm{T}}(Q^{-1}T)P_c$ *is strictly upper triangular for some permutation matrix* P_c.

Proof. Firstly suppose that ${P_c}^{\mathrm{T}}(Q^{-1}T)P_c$ is strictly upper triangular for some permutation matrix P_c. Then, since $\det Q \neq 0$ and $A = Q + T$, we have

$$\det A = \det[Q(I + Q^{-1}T)] = \det Q \cdot \det[I + {P_c}^{\mathrm{T}}(Q^{-1}T)P_c] = \det Q \in \boldsymbol{K}^*.$$

Conversely, if $\det A \in \boldsymbol{K}^*$, then $\det Q = \det A \neq 0$. Put $S = Q^{-1}$. Suppose that ${P_c}^{\mathrm{T}}(Q^{-1}T)P_c = {P_c}^{\mathrm{T}}(ST)P_c$ is not strictly upper triangular for any permutation matrix P_c. Then ST has a cycle of nonzero entries, that is, there exist an integer $M \geq 1$ and a sequence of indices $i(m)$ and $j(m)$ ($m = 1, \cdots, M$) such that $S_{j(m),i(m)} \neq 0$ and $T_{i(m),j(m+1)} \neq 0$ for $m = 1, \cdots, M$, where $j(M+1) = j(1)$. Choose M to be the minimum of such integers. We write $S_{j(m),i(m)} = s_m$ and $T_{i(m),j(m+1)} = t_m$.

For $k = 0, 1, \cdots$, consider the expression of the $(j(1), i(1))$ entry of $(ST)^{kM} S$ in the form of the sum of products of S_{ji}'s and T_{ij}'s. Corresponding to the above cycle, it contains a term

$$s_1 (s_1 s_2 \cdots s_M)^k \cdot (t_1 t_2 \cdots t_M)^k,$$

since no other similar terms of $(t_1 t_2 \cdots t_M)^k$ exist due to the minimality of M and since it cannot be canceled out by nonsimilar terms by virtue of the algebraic independence of T.

Next we formally expand A^{-1} as[1]

$$A^{-1} = [Q(I + Q^{-1}T)]^{-1} = S - STS + STSTS - \cdots .$$

Each entry of A^{-1} on the left-hand side is a polynomial in $\mathcal{T}$ over $\boldsymbol{K}$ since A is invertible. On the right-hand side, the $(j(1), i(1))$ entry contains a term of arbitrarily high degree, since the nonzero term $(t_1 t_2 \cdots t_M)^k$ of degree kM, stemming from $(ST)^{kM} S$, as above, cannot be canceled out for $k = 0, 1, \cdots$. This is a contradiction. ■

Example 4.6.3. For the matrix $A = Q + T$ in Example 4.6.1, we have $\det Q = 2$ and

$$P_{\rm c}{}^{\rm T}(Q^{-1}T)P_{\rm c} = \begin{array}{c|ccccc|} & 5 & 2 & 4 & 3 & 1 \\ \hline 5 & 0 & 0 & -z & 0 & -y/2 \\ 2 & 0 & 0 & z & 0 & 0 \\ 4 & 0 & 0 & 0 & x & 0 \\ 3 & 0 & 0 & 0 & 0 & y/2 \\ 1 & 0 & 0 & 0 & 0 & 0 \\ \hline \end{array} ,$$

which is strictly upper triangular. □

We now state the theorem of LU-decomposition of mixed matrices due to Murota [198].

Theorem 4.6.4. *A square mixed matrix* $A = Q + T \in \mathrm{MM}(\boldsymbol{K}, \boldsymbol{F}; n, n)$ *is invertible in* $\boldsymbol{K}[\mathcal{T}]$, *if and only if there exist permutation matrices* $P_{\rm r}$ *and* $P_{\rm c}$, *an* $n \times n$ *matrix* $L = (L_{ij})$ *over* $\boldsymbol{K}$ *and an* $n \times n$ *matrix* $U = (U_{ij})$ *over* $\boldsymbol{F}$ *such that* (i) $P_{\rm r} A P_{\rm c} = L\, U$, (ii) $L_{ij} = 0$ *for* $i < j$ *and* $L_{ii} = 1$ *for* $i = 1, \cdots, n$, *and* (iii) $U_{ij} = 0$ *for* $i > j$, $U_{ii} \in \boldsymbol{K}^*$, *and* U_{ij} *is a polynomial of degree at most one in* $\mathcal{T}$ *over* $\boldsymbol{K}$.

Proof. It suffices to prove the "only if" part. Let $P_{\rm c}$ be the permutation matrix in Theorem 4.6.2 for which $P_{\rm c}{}^{\rm T}(Q^{-1}T)P_{\rm c}$ is strictly upper triangular. Since $\det Q \neq 0$, a standard result on the LU-decomposition or the Gaussian elimination (cf., e.g., Gantmacher [87], Golub–Van Loan [97]) shows that there exist a permutation matrix $P_{\rm r}$, a lower triangular matrix with unit diagonals $L \in \mathrm{GL}(n, \boldsymbol{K})$, and a nonsingular upper triangular matrix $V \in \mathrm{GL}(n, \boldsymbol{K})$ such that $P_{\rm r} Q P_{\rm c} = L V$. Then we obtain

$$\begin{aligned} P_{\rm r}\, A\, P_{\rm c} &= P_{\rm r}(Q + T)P_{\rm c} = (P_{\rm r} Q P_{\rm c})[I + P_{\rm c}{}^{\rm T}(Q^{-1}T)P_{\rm c}] \\ &= (L V)[I + P_{\rm c}{}^{\rm T}(Q^{-1}T)P_{\rm c}] = L U \end{aligned}$$

with $U = V[I + P_{\rm c}{}^{\rm T}(Q^{-1}T)P_{\rm c}]$, which is an upper triangular matrix. Obviously L is a matrix over $\boldsymbol{K}$, and consequently the entries of $U = L^{-1} P_{\rm r} A P_{\rm c}$ are polynomials in $\mathcal{T}$ of degree at most one. ■

[1] This expansion converges for sufficiently small absolute values of $\mathcal{T}$.

Remark 4.6.5. In Theorem 4.6.4 the assumption of algebraic independence of $\mathcal{T}$ as a whole cannot be weakened to element-wise transcendency of the members of $\mathcal{T}$. Consider, e.g., $A = \begin{bmatrix} 1+x & x \\ -x & 1-x \end{bmatrix}$, which can be expressed as $A = Q + T$ with $Q = \begin{bmatrix} 1 & 0 \\ 0 & 1 \end{bmatrix}$ and $T = \begin{bmatrix} x & x \\ -x & -x \end{bmatrix}$. Although $\det A = 1$ and each entry of T is transcendental over $\mathbf{Q}$, there exists no LU-decomposition with the L-factor over $\mathbf{Q}$. □

Given a mixed matrix $A \in \mathrm{MM}(\boldsymbol{K}, \boldsymbol{F}; n, n)$, we can test for its invertibility with $\mathrm{O}(n^3)$ arithmetic operations in $\boldsymbol{K}$ on the basis of Theorem 4.6.2: first compute Q^{-1} by elimination operations in $\boldsymbol{K}$, then determine the zero/nonzero pattern of $Q^{-1}T$ by boolean operations and finally check for the acyclicity of the graph associated with $Q^{-1}T$ as defined in §2.2.1. This procedure simultaneously provides the permutation matrix P_{c}. In this connection we may recall Theorem 4.2.17, which shows how an invertible submatrix can be extracted.

Theorem 4.6.4 reads that if $A \in \mathrm{MM}(\boldsymbol{K}, \boldsymbol{F}; n, n)$ is invertible, it can be brought to an upper triangular form U over $\boldsymbol{F}$ by a transformation $(L^{-1}P_{\mathrm{r}})\, A\, P_{\mathrm{c}} = U$ with $L \in \mathrm{GL}(n, \boldsymbol{K})$ and permutation matrices P_{r} and P_{c}. In the next subsection we will consider the problem of reducing a general mixed matrix $A \in \mathrm{MM}(\boldsymbol{K}, \boldsymbol{F}; m, n)$ to an upper block-triangular form by a transformation $S\,A\,P$ with $S \in \mathrm{GL}(m, \boldsymbol{K})$ and a permutation matrix P.

4.6.2 Block-triangularization of General Mixed Matrices

We consider a block-triangularization of a mixed matrix $A = Q + T \in \mathrm{MM}(\boldsymbol{K}, \boldsymbol{F}; m, n)$ under a transformation of the form

$$\hat{A} = S\,A\,P, \tag{4.88}$$

where $S \in \mathrm{GL}(m, \boldsymbol{K})$ and P is a permutation matrix. For a proper block-triangularization (in the sense of §2.1.4) the following conditions are required of $\hat{A}$ and of partitions $(\hat{R}_0; \hat{R}_1, \cdots, \hat{R}_{\hat{b}}; \hat{R}_\infty)$ and $(\hat{C}_0; \hat{C}_1, \cdots, \hat{C}_{\hat{b}}; \hat{C}_\infty)$ of $\mathrm{Row}(\hat{A})$ and $\mathrm{Col}(\hat{A})$:

$$\begin{aligned}
&\hat{R}_k \neq \emptyset,\ \hat{C}_k \neq \emptyset\ (k = 1, \cdots, \hat{b});\ \hat{R}_0, \hat{R}_\infty, \hat{C}_0, \hat{C}_\infty \text{ can be empty},\\
&\hat{A}[\hat{R}_k, \hat{C}_l] = O \qquad \text{if} \quad 0 \le l < k \le \infty,\\
&\mathrm{rank}\,\hat{A}[\hat{R}_0, \hat{C}_0] = |\hat{R}_0| \quad (< |\hat{C}_0| \ \text{ if } \hat{R}_0 \neq \emptyset),\\
&\mathrm{rank}\,\hat{A}[\hat{R}_k, \hat{C}_k] = |\hat{R}_k| = |\hat{C}_k| > 0 \quad \text{for} \quad k = 1, \cdots, \hat{b},\\
&\mathrm{rank}\,\hat{A}[\hat{R}_\infty, \hat{C}_\infty] = |\hat{C}_\infty| \quad (< |\hat{R}_\infty| \ \text{ if } \hat{C}_\infty \neq \emptyset).
\end{aligned}$$

Note that the existence of such $\hat{A}$ is by no means obvious and that the transformed matrix $\hat{A} = (SQP) + (STP)$ no longer belongs to $\mathrm{MM}(\boldsymbol{K}, \boldsymbol{F}; m, n)$

in general. Roughly speaking, such $\hat{A}$ can be constructed as an aggregation of the CCF of the associated LM-matrix.

Let $\tilde{A} \in \mathrm{LM}(\boldsymbol{K}, \boldsymbol{F}; m, m, m+n)$ be the LM-matrix associated with A in the sense of (4.4). Putting $R = \mathrm{Row}(A)$ and $C = \mathrm{Col}(A)$, we identify $\tilde{C} = \mathrm{Col}(\tilde{A})$ with $R \cup C$ through a one-to-one correspondence $\psi : R \cup C \to \tilde{C}$. Let $\tilde{\rho}, \tilde{\gamma}, \tilde{p} : 2^{\tilde{C}} \to \mathbf{Z}$ be the functions associated with $\tilde{A}$ by (4.13), (4.9), and (4.16).

Recalling that the CCF of $\tilde{A}$ is obtained (cf. §4.4.3) from the lattice $\mathcal{L}_{\min}(\tilde{p})$ ($\subseteq 2^{\tilde{C}}$) of the minimizers of the LM-surplus function $\tilde{p}$, we consider here a subfamily of $\mathcal{L}_{\min}(\tilde{p})$ defined by

$$\hat{\mathcal{L}} = \{X \in \mathcal{L}_{\min}(\tilde{p}) \mid I \supseteq \hat{\Gamma}(R, J) \text{ for } I = \psi^{-1}(X) \cap R,\ J = \psi^{-1}(X) \cap C\}, \tag{4.89}$$

where

$$\hat{\Gamma}(R, J) = \{i \in R \mid \exists j \in J : T_{ij} \neq 0\}, \qquad J \subseteq C.$$

Lemma 4.6.6. *$\hat{\mathcal{L}} \neq \emptyset$ and $\hat{\mathcal{L}}$ is a sublattice of $\mathcal{L}_{\min}(\tilde{p})$.*

Proof. For $X \in \mathcal{L}_{\min}(\tilde{p})$, put $I = \psi^{-1}(X) \cap R$ and $J = \psi^{-1}(X) \cap C$, and define $I' = I \cup \hat{\Gamma}(R, J)$ and $X' = \psi(I' \cup J) \supseteq X$. Since $\tilde{\gamma}(X') = \tilde{\gamma}(X)$, $\tilde{\rho}(X') \leq \tilde{\rho}(X) + |I' \setminus I|$, and $|X'| = |X| + |I' \setminus I|$, we have $\tilde{p}(X') \leq \tilde{p}(X)$, which shows $X' \in \mathcal{L}_{\min}(\tilde{p})$. If X is the maximum element of $\mathcal{L}_{\min}(\tilde{p})$, we must have $X' = X$, i.e., $I \supseteq \hat{\Gamma}(R, J)$, which means $X \in \hat{\mathcal{L}}$, and therefore $\hat{\mathcal{L}} \neq \emptyset$. It follows from Lemma 2.2.16 that $\mathcal{L}_0 = \{X \subseteq \tilde{C} \mid I \supseteq \hat{\Gamma}(R, J) \text{ for } I = \psi^{-1}(X) \cap R, J = \psi^{-1}(X) \cap C\}$ forms a sublattice of $2^{\tilde{C}}$. Hence $\hat{\mathcal{L}} = \mathcal{L}_0 \cap \mathcal{L}_{\min}(\tilde{p})$ is a sublattice of $\mathcal{L}_{\min}(\tilde{p})$. ∎

By Lemma 4.6.6 above and Birkhoff's representation theorem (Theorem 2.2.10), $\hat{\mathcal{L}}$ determines a partition of $\tilde{C}$ which is an aggregation of the one induced by $\mathcal{L}_{\min}(\tilde{p})$. Accordingly (cf. §4.4.3), $\hat{\mathcal{L}}$ induces a block-triangularization, coarser than the CCF, of the LM-matrix $\tilde{A}$ under a transformation (4.35) with $S \in \mathrm{GL}(m, \boldsymbol{K})$ and permutation matrices P_{r} and P_{c}. We shall see that this matrix S gives the desired transformation in $\hat{A} = SAP$.

Let

$$\bar{A} = P_{\mathrm{r}} \begin{pmatrix} S & \\ & I_m \end{pmatrix} \begin{pmatrix} I_m & Q \\ -I_m & T \end{pmatrix} P_{\mathrm{c}}$$

be the block-triangular matrix induced by $\hat{\mathcal{L}}$, and let $(\bar{R}_0; \bar{R}_1, \cdots, \bar{R}_b; \bar{R}_\infty)$ and $(\bar{C}_0; \bar{C}_1, \cdots, \bar{C}_b; \bar{C}_\infty)$ be the associated partitions of $\mathrm{Row}(\bar{A})$ and $\mathrm{Col}(\bar{A})$. Then we have

$$\bar{A}[\bar{R}_k, \bar{C}_l] = O \qquad \text{for} \quad l < k. \tag{4.90}$$

Define

$$\bar{Q} = S[\, I_m \mid Q\,], \qquad \bar{T} = [\, -I_m \mid T\,]$$

so that

$$\bar{A} = P_{\mathrm{r}} \begin{pmatrix} \bar{Q} \\ \bar{T} \end{pmatrix} P_{\mathrm{c}}.$$

Note that

$$S = \bar{Q}[\mathrm{Row}(\bar{Q}), \psi(R)], \tag{4.91}$$

where ψ is regarded as $\psi : R \cup C \to \mathrm{Col}(\bar{Q})$ through the natural identification of $\mathrm{Col}(\bar{Q})$ with $\tilde{C}$; similarly for $\mathrm{Col}(\bar{T})$. From the identity

$$\begin{pmatrix} O & SA \\ -I_m & T \end{pmatrix} = \begin{pmatrix} I_m & S \\ O & I_m \end{pmatrix} \begin{pmatrix} S & \\ & I_m \end{pmatrix} \begin{pmatrix} I_m & Q \\ -I_m & T \end{pmatrix} = \begin{pmatrix} \bar{Q} + S\bar{T} \\ \bar{T} \end{pmatrix}$$

we see that

$$SA = (\bar{Q} + S\bar{T})[\mathrm{Row}(\bar{Q}), \psi(C)]. \tag{4.92}$$

For $k = 0, 1, \cdots, b, \infty$, put $\bar{R}_{Qk} = \mathrm{Row}(\bar{Q}) \cap \bar{R}_k$ and $\bar{R}_{Tk} = \mathrm{Row}(\bar{T}) \cap \bar{R}_k$, where $\mathrm{Row}(\bar{A})$ is identified with $\mathrm{Row}(\bar{Q}) \cup \mathrm{Row}(\bar{T})$ through the permutation P_{r}. Similarly, $\mathrm{Col}(\bar{A})$ is identified with $\mathrm{Col}(\bar{Q})$ $(= \mathrm{Col}(\bar{T}))$ through the permutation P_{c}. By the condition $I \supseteq \hat{\Gamma}(R, J)$ in the definition of $\hat{\mathcal{L}}$ and the construction of the CCF (cf. (4.54) in particular), it holds that

$$\psi(\bar{R}_{Tk}) = \bar{C}_k \cap \psi(R). \tag{4.93}$$

Hence the diagonal submatrix $\bar{A}[\bar{R}_k, \bar{C}_k]$ is of the following form:

$$\bar{A}[\bar{R}_k, \bar{C}_k] = \begin{array}{c} \\ \bar{R}_{Qk} \\ \bar{R}_{Tk} \end{array} \begin{array}{c} \begin{array}{cc} \bar{C}_k \cap \psi(R) & \bar{C}_k \cap \psi(C) \end{array} \\ \begin{pmatrix} Q_{1k} & \qquad Q_{2k} \\ -I & \qquad T_k \end{pmatrix} \end{array},$$

that is, the submatrix $\bar{A}[\bar{R}_{Tk}, \bar{C}_k \cap \psi(R)]$ is equal to $-I$ (the negative of an identity matrix).

We now claim

$$(\bar{Q} + S\bar{T})[\bar{R}_{Qk}, \bar{C}_l \cap \psi(C)] = O \qquad \text{for} \quad l < k. \tag{4.94}$$

To show this, first note from (4.90) that

$$\bar{Q}[\bar{R}_{Qk}, \bar{C}_l \cap \psi(C)] = O, \quad \bar{T}[\bar{R}_{Tk}, \bar{C}_l \cap \psi(C)] = O \qquad \text{for} \quad l < k.$$

Then we have

$$(S\bar{T})[\bar{R}_{Qk}, \bar{C}_l \cap \psi(C)] = \sum_j S[\bar{R}_{Qk}, \bar{R}_{Tj}] \cdot \bar{T}[\bar{R}_{Tj}, \bar{C}_l \cap \psi(C)]$$

$$= \sum_j \bar{Q}[\bar{R}_{Qk}, \bar{C}_j \cap \psi(R)] \cdot \bar{T}[\bar{R}_{Tj}, \bar{C}_l \cap \psi(C)] = O$$

with the aid of (4.91) and (4.93). Thus the claim (4.94) is proven.

Noting that $|\bar{R}_{Qk}| = |\bar{C}_k \cap \psi(C)|$ for $k = 1, \cdots, b$, define $\{\hat{R}_l \mid l = 1, \cdots, \hat{b}\}$ to be the family of nonempty blocks among $\{\bar{R}_{Qk} \mid k = 1, \cdots, b\}$ and likewise

$\{\hat{C}_l \mid l = 1, \cdots, \hat{b}\}$ to be the family of nonempty blocks among $\{\bar{C}_k \cap \psi(C) \mid k = 1, \cdots, b\}$. Also define

$$\hat{R}_0 = \bar{R}_{Q0}, \quad \hat{R}_\infty = \bar{R}_{Q\infty}, \quad \hat{C}_0 = \bar{C}_0 \cap \psi(C), \quad \hat{C}_\infty = \bar{C}_\infty \cap \psi(C).$$

The expressions (4.92) and (4.94) show that

$$(SA)[\hat{R}_k, \hat{C}_l] = O \qquad \text{if} \quad 0 \le l < k \le \infty, \tag{4.95}$$

namely, the matrix SA is block-triangularized with respect to the partitions $(\hat{R}_0; \hat{R}_1, \cdots, \hat{R}_{\hat{b}}; \hat{R}_\infty)$ and $(\hat{C}_0; \hat{C}_1, \cdots, \hat{C}_{\hat{b}}; \hat{C}_\infty)$. Hence, $\hat{A} = S\,A\,P$ with some permutation matrix P is explicitly block-triangularized. Denote by $\preceq$ the partial order on $\{\hat{C}_0; \hat{C}_1, \cdots, \hat{C}_{\hat{b}}; \hat{C}_\infty\}$ that is induced from the partial order defined by $\hat{\mathcal{L}}$ on $\{\bar{C}_0; \bar{C}_1, \cdots, \bar{C}_b; \bar{C}_\infty\}$.

Example 4.6.7. The construction explained above is illustrated for a mixed matrix $A \in \mathrm{MM}(\mathbf{Q}, \boldsymbol{F}; 5, 5)$:

$$A = \begin{array}{c|ccccc|} & x_1 & x_2 & x_3 & x_4 & x_5 \\ \hline w_1 & 1 & 1 & t_1 & 1 & t_2 \\ w_2 & -1 & -1 & 1 & t_3 & 0 \\ w_3 & 0 & 0 & t_4 & t_5 & t_6 \\ w_4 & 0 & 0 & 0 & 0 & 1 \\ w_5 & t_7 & t_8 & 0 & 0 & 0 \\ \hline \end{array},$$

where t_i $(i = 1, \cdots, 8)$ are indeterminates over $\mathbf{Q}$ and $\boldsymbol{F} = \mathbf{Q}(t_1, \cdots, t_8)$. By the CCF of the associated LM-matrix $\tilde{A} \in \mathrm{LM}(\mathbf{Q}, \boldsymbol{F}; 5, 5, 10)$ we see that

$$\tilde{A} = P_{\mathrm{r}} \begin{pmatrix} S & O \\ O & I_5 \end{pmatrix} \begin{pmatrix} I_5 & Q \\ -I_5 & T \end{pmatrix} P_{\mathrm{c}}$$

$$\begin{array}{c|cc|c|cccc|c|c|c|} & \multicolumn{2}{c}{\longleftarrow \bar{C}_1 \longrightarrow} & \multicolumn{5}{c}{\longleftarrow \quad \bar{C}_2 \quad \longrightarrow} & \bar{C}_3 & \bar{C}_4 \\ & \multicolumn{2}{c}{\leftarrow C_1 \rightarrow} & C_2 & \multicolumn{4}{c}{\longleftarrow \quad C_3 \quad \longrightarrow} & C_4 & C_5 & C_6 \\ & x_1 & x_2 & w_5 & w_1 & w_2 & x_3 & x_4 & w_3 & x_5 & w_4 \\ \hline r_1 & 1 & 1 & & 1 & & & 1 & & & \\ w_5 & t_7 & t_8 & -1 & & & & & & & \\ r_2 & & & 1 & & & & & & & \\ r_3 & & & & 1 & 1 & 1 & 1 & & & \\ w_1 & & & & -1 & 0 & t_1 & 0 & & t_2 & \\ w_2 & & & & 0 & -1 & 0 & t_3 & & & \\ w_3 & & & & 0 & 0 & t_4 & t_5 & -1 & t_6 & \\ r_4 & & & & & & & & 1 & & \\ r_5 & & & & & & & & & 1 & 1 \\ w_4 & & & & & & & & & & -1 \\ \hline \end{array},$$

where

$$S = \begin{bmatrix} 1\,0\,0\,0\,0 \\ 0\,0\,0\,0\,1 \\ 1\,1\,0\,0\,0 \\ 0\,0\,1\,0\,0 \\ 0\,0\,0\,1\,0 \end{bmatrix}.$$

In the CCF of $\tilde{A}$, the column set $\mathrm{Col}(\tilde{A})$, identified with $\{w_1, \cdots, w_5\} \cup \{x_1, \cdots, x_5\}$, is divided into six (nonempty) blocks:

$$\begin{array}{l} C_1 = \{x_1, x_2\}, \quad C_2 = \{w_5\}, \quad C_3 = \{w_1, w_2, x_3, x_4\}, \quad C_4 = \{w_3\}, \\ C_5 = \{x_5\}, \quad C_6 = \{w_4\} \qquad (C_0 = C_\infty = \emptyset) \end{array}$$

with the partial order being the transitive closure of the relations:

$$C_1 \preceq C_2; \quad C_3 \preceq C_4; \quad C_1 \preceq C_3 \preceq C_5 \preceq C_6.$$

This corresponds to the lattice $\mathcal{L}_{\min}(\tilde{p})$.

The sublattice $\hat{\mathcal{L}}$ of (4.89), on the other hand, yields a coarser partition consisting of four blocks:

$$\bar{C}_1 = \{x_1, x_2, w_5\}, \quad \bar{C}_2 = \{x_3, x_4, w_1, w_2, w_3\}, \quad \bar{C}_3 = \{x_5\}, \quad \bar{C}_4 = \{w_4\}$$

with $\bar{C}_1 \preceq \bar{C}_2 \preceq \bar{C}_3 \preceq \bar{C}_4$, where $\bar{C}_0 = \bar{C}_\infty = \emptyset$. Namely, $\hat{\mathcal{L}} = \{\emptyset, \bar{C}_1, \bar{C}_1 \cup \bar{C}_2, \bar{C}_1 \cup \bar{C}_2 \cup \bar{C}_3, \bar{C}_1 \cup \bar{C}_2 \cup \bar{C}_3 \cup \bar{C}_4\}$. Note, for example, that $\bar{C}_1 = C_1 \cup C_2 \in \mathcal{L}_{\min}(\tilde{p})$, and $I = \psi^{-1}(\bar{C}_1) \cap R = \{w_5\}$, and $J = \psi^{-1}(\bar{C}_1) \cap C = \{x_1, x_2\}$ satisfy the condition $I \supseteq \Gamma(R, J)$.

Finally, for the partition of $\mathrm{Col}(A)$, we obtain

$$\hat{C}_1 = \{x_1, x_2\}, \quad \hat{C}_2 = \{x_3, x_4\}, \quad \hat{C}_3 = \{x_5\}$$

with the partial order $\hat{C}_1 \preceq \hat{C}_2 \preceq \hat{C}_3$, where $\hat{C}_0 = \hat{C}_\infty = \emptyset$. Accordingly, the following block-triangular form is obtained:

$$SAP = \begin{array}{c|cc|cc|c|} & \multicolumn{2}{c}{\hat{C}_1} & \multicolumn{2}{c}{\hat{C}_2} & \multicolumn{1}{c}{\hat{C}_3} \\ & x_1 & x_2 & x_3 & x_4 & \multicolumn{1}{c}{x_5} \\ \cline{2-6} r_1 & 1 & 1 & t_1 & 1 & t_2 \\ r_2 & t_7 & t_8 & & & \\ \cline{2-6} r_3 & & & t_1+1 & t_3+1 & t_2 \\ r_4 & & & t_4 & t_5 & t_6 \\ \cline{4-6} r_5 & & & & & 1 \\ \cline{2-6} \end{array},$$

where $P = I_5$. □

The theorem on the block-triangularization of a general mixed matrix is now stated.

Theorem 4.6.8. *The matrix $\hat{A}$ as well as partitions $(\hat{R}_0; \hat{R}_1, \cdots, \hat{R}_{\hat{b}}; \hat{R}_\infty)$ and $(\hat{C}_0; \hat{C}_1, \cdots, \hat{C}_{\hat{b}}; \hat{C}_\infty)$ constructed above gives a proper block-triangular form, having the following properties.*

(1) *$\hat{A}$ is block-triangularized, i.e.,*

$$\hat{A}[\hat{R}_k, \hat{C}_l] = O \quad \text{if} \quad 0 \le l < k \le \infty. \tag{4.96}$$

Moreover, the partial order on $\{\hat{C}_1, \cdots, \hat{C}_{\hat{b}}\}$ induced by the zero/nonzero structure of $\hat{A}$ agrees with the partial order $\preceq$ defined from $\hat{\mathcal{L}}$; i.e.,

$$\begin{aligned} \hat{A}[\hat{R}_k, \hat{C}_l] = O \ \ &\text{unless} \ \ \hat{C}_k \preceq \hat{C}_l \quad (1 \le k, l \le \hat{b}); \\ \hat{A}[\hat{R}_k, \hat{C}_l] \ne O \ \ &\text{if} \ \ \hat{C}_k \prec\cdot\, \hat{C}_l \quad (1 \le k, l \le \hat{b}). \end{aligned}$$

(2)

$$\begin{aligned} &\operatorname{rank} \hat{A}[\hat{R}_0, \hat{C}_0] = |\hat{R}_0| \quad (< |\hat{C}_0| \ \text{ if } \hat{R}_0 \ne \emptyset), \\ &\operatorname{rank} \hat{A}[\hat{R}_k, \hat{C}_k] = |\hat{R}_k| = |\hat{C}_k| > 0 \quad \text{for} \quad k = 1, \cdots, \hat{b}, \\ &\operatorname{rank} \hat{A}[\hat{R}_\infty, \hat{C}_\infty] = |\hat{C}_\infty| \quad (< |\hat{R}_\infty| \ \text{ if } \hat{C}_\infty \ne \emptyset). \end{aligned}$$

(3) *$\hat{A}$ is the finest proper block-triangular matrix ("proper" in the sense of §2.1.4) among those obtained by a transformation of the form* (4.88).

Proof. (1)–(2) The claim (4.96) has been shown in (4.95). The other claims in (1) and (2) can be proven similarly to the corresponding claims in Theorem 4.4.4.

(3) Suppose that there exist $S \in \mathrm{GL}(m, \boldsymbol{K})$, $W \subseteq R_S \equiv \mathrm{Row}(S)$, and $J \subseteq C$ such that

$$\begin{aligned} &(SA)[R_S \setminus W, J] = O, \\ &\operatorname{rank}(SA)[W, J] = |W|, \\ &\operatorname{rank}(SA)[R_S \setminus W, C \setminus J] = |C \setminus J|. \end{aligned} \tag{4.97}$$

This means

$$\operatorname{rank} A = \operatorname{rank} SA = n - |J| + |W|,$$

which implies by Theorem 4.2.5 that

$$\min \tilde{p} = \operatorname{rank} \tilde{A} - (m+n) = \operatorname{rank} A - n = |W| - |J|. \tag{4.98}$$

To show that $\hat{A}$ is the finest proper block-triangularization, it suffices to prove that $X = \psi(I \cup J) \in \mathcal{L}_{\min}(\tilde{p})$ for $I = \hat{\Gamma}(R, J)$, which implies $X \in \hat{\mathcal{L}}$. By the algebraic independence of $\mathcal{T}$, (4.97) is equivalent to

$$(SQ)[R_S \setminus W, J] = O, \qquad (ST)[R_S \setminus W, J] = O. \tag{4.99}$$

Moreover, the latter condition is further equivalent, again by the algebraic independence of $\mathcal{T}$, to

$$S[R_S \setminus W, I] = O. \tag{4.100}$$

From the first of (4.99) and (4.100), we see that

$$\tilde{p}(X) = \operatorname{rank}(I_m \mid Q)[R, X] = \operatorname{rank}(S(I_m \mid Q))[R_S, X] \le |W|. \tag{4.101}$$

On the other hand, the definition of I implies

$$\tilde{\gamma}(X) = |I \cup \hat{\Gamma}(R, J)| = |I|. \tag{4.102}$$

Combining (4.101) and (4.102), we obtain

$$\tilde{p}(X) = \tilde{\rho}(X) + \tilde{\gamma}(X) - |X| \le |W| - |J|,$$

which shows $X \in \mathcal{L}_{\min}(\tilde{p})$ by (4.98). Hence $\hat{A}$ is the finest proper block-triangular form under the admissible transformation (4.88). ∎

Notes. Section 4.6.1 is based on Murota [198]. Theorem 4.6.8 was first stated in §24 of Murota [204], whereas the proof is improved here.

4.7 Related Decompositions

In the literature of electrical network theory, it has been known that a system of equations describing an electrical network can be put in a block-triangular form if one chooses appropriate bases (tree-cotree pairs) for Kirchhoff's laws and rearranges the variables and the equations (for both Kirchhoff's laws and element characteristics). A decomposition method, called *2-graph method*, referring to a pair of current-graph and voltage-graph is investigated in Ozawa [260, 261, 262] for networks involving controlled sources. Based on the result of Tomizawa–Iri [317], a decomposition of networks with admittance expressions is considered by Iri [127] in relation to the independent-matching problem. An attempt has been made in Nakamura–Iri [246] and Nakamura [243, 244] to define a block-triangularization for a system of equations describing the most general class of networks with arbitrary mutual couplings (such as those containing controlled sources, nullators, and norators) as an application of the principal partition for a pair of matroids. This section is devoted to clarifying the relationship of the CCF-based decomposition to some of those decomposition techniques and to extending the concept of LM-equivalence to multilayered matrices.

4.7.1 Decomposition as Matroid Union

For an LM-matrix $A = \binom{Q}{T} \in \mathrm{LM}(\boldsymbol{K}, \boldsymbol{F})$ the CCF of A has been constructed on the basis of the LM-surplus function p which is submodular and characterizes the rank of A (cf. Theorem 4.2.5). In Theorem 4.2.3, on the other hand, we have encountered another submodular function that characterizes the rank of A, namely, $p_\tau : 2^C \to \mathbf{Z}$ defined by

$$p_\tau(J) = \rho(J) + \tau(J) - |J|, \qquad J \subseteq C, \tag{4.103}$$

which is only slightly different from the LM-surplus function

$$p(J) = \rho(J) + \gamma(J) - |J|, \qquad J \subseteq C.$$

It is quite natural to be tempted to apply to p_τ the Jordan–Hölder-type theorem for submodular functions with a vague hope that some meaningful decomposition of an LM-matrix might be obtained. Indeed it was claimed in Nakamura–Iri [246] and Nakamura [243, 244] (before the CCF is established) that this approach yielded a block-triangularization of a system of equations (3.2) for an electrical network. It is certainly true that, for an LM-matrix in general, the Jordan–Hölder-type theorem applied to p_τ yields a partition of the column set (currents and voltages of branches in the case of electrical networks) into partially ordered blocks. Let us call this decomposition the *principal partition with respect to the matroid union* $\mathbf{M}(Q) \vee \mathbf{M}(T)$, as ρ and τ are the rank functions of $\mathbf{M}(Q)$ and $\mathbf{M}(T)$, respectively.

The objective of this subsection is to compare the CCF and the principal partition with respect to $\mathbf{M}(Q) \vee \mathbf{M}(T)$ and to discuss the irrelevance of the latter by identifying the corresponding admissible transformation, which is different from the LM-admissible transformation (4.35) for LM-equivalence. Remember that partition of C in the CCF corresponds to $\mathcal{L}_{\min}(p)$ (the family of the minimizers of p), while that in the principal partition with respect to $\mathbf{M}(Q) \vee \mathbf{M}(T)$ is given by $\mathcal{L}_{\min}(p_\tau)$.

Let us begin with two simple examples which demonstrate that the principal partition with respect to $\mathbf{M}(Q) \vee \mathbf{M}(T)$ provides a finer partition of C than the CCF does, and that it is too fine for a useful block-triangularization.

Example 4.7.1. Consider an LM-matrix

$$A = \left[\frac{Q}{T}\right] = \begin{array}{|cccc|} \multicolumn{1}{c}{\xi^1} & \xi^2 & \eta_1 & \multicolumn{1}{c}{\eta_2} \\ \hline 1 & & & \\ & 1 & & \\ \hline -t_1 & & y^{11} & y^{12} \\ & -t_2 & y^{21} & y^{22} \\ \hline \end{array} .$$

It is easy to verify that

$$\mathcal{L}_{\min}(p_\tau) = \{\emptyset, \{\eta_1\}, \{\eta_2\}, \{\eta_1, \eta_2\}, \{\xi^1, \eta_1, \eta_2\}, \{\xi^2, \eta_1, \eta_2\}, \{\xi^1, \xi^2, \eta_1, \eta_2\}\}$$

and therefore the principal partition with respect to $\mathbf{M}(Q) \vee \mathbf{M}(T)$ based on p_τ yields a partition of $C = \{\xi^1, \xi^2, \eta_1, \eta_2\}$ into four singletons with partial order given by $\{\eta_i\} \prec \{\xi^j\}$ $(i, j = 1, 2)$. However, it is clear by inspection that $\{\eta_1, \eta_2\}$ cannot be split in solving the system of equations. On the other hand, the CCF is based on

$$\mathcal{L}_{\min}(p) = \{\emptyset, \{\eta_1, \eta_2\}, \{\xi^1, \eta_1, \eta_2\}, \{\xi^2, \eta_1, \eta_2\}, \{\xi^1, \xi^2, \eta_1, \eta_2\}\}$$

and gives a more natural partition $C = \{\xi^1\} \cup \{\xi^2\} \cup \{\eta_1, \eta_2\}$ with partial order $\{\eta_1, \eta_2\} \prec \{\xi^i\}$ $(i = 1, 2)$.

It is mentioned that the above matrix A (with $t_1 = t_2 = 1$) appears as the coefficient matrix of a system of equations (3.2) that describes a free electrical network consisting of two branches connected in series with complete mutual couplings given in terms of admittances. As the notation shows, ξ^i and η_i are the current in and the voltage across branch i ($i = 1, 2$). □

Example 4.7.2. Consider an electrical network consisting of two branches connected in parallel, where branch 1 is a current source controlled by the voltage across branch 2, i.e., $\xi^1 = g\eta_2$, and the branch 2 is an ohmic resistor, i.e., $\eta_2 = r\xi^2$. These equations, together with Kirchhoff's laws $\eta_1 - \eta_2 = 0$ and $\xi^1 + \xi^2 = 0$, are put into the form (3.2) with

$$A = \left[\frac{Q}{T}\right] = \begin{array}{c} \begin{array}{cccc} \xi^1 & \xi^2 & \eta_1 & \eta_2 \end{array} \\ \left[\begin{array}{cccc} 1 & 1 & & \\ & & 1 & -1 \\ \hline -1 & & & g \\ & r & & -1 \end{array}\right] \end{array},$$

where ξ^i and η_i are, as usual, the current in and the voltage across branch i ($i = 1, 2$). We have

$$\mathcal{L}_{\min}(p_\mathcal{T}) = \{\emptyset, \{\eta_1\}, \{\eta_1, \eta_2\}, \{\xi^1, \xi^2, \eta_1, \eta_2\}\},$$
$$\mathcal{L}_{\min}(p) = \{\emptyset, \{\eta_1\}, \{\xi^1, \xi^2, \eta_1, \eta_2\}\}.$$

The former yields the partition $\{\eta_1\} \cup \{\eta_2\} \cup \{\xi^1, \xi^2\}$ with partial order $\{\eta_1\} \prec \{\eta_2\} \prec \{\xi^1, \xi^2\}$, whereas the latter gives $\{\eta_1\} \cup \{\eta_2, \xi^1, \xi^2\}$ with partial order $\{\eta_1\} \prec \{\eta_2, \xi^1, \xi^2\}$. It is obvious from the CCF of A:

$$\begin{array}{c} \begin{array}{cccc} \eta_1 & \eta_2 & \xi^1 & \xi^2 \end{array} \\ \left[\begin{array}{c|ccc} 1 & -1 & & \\ \hline & & 1 & 1 \\ & g & -1 & \\ & -1 & & r \end{array}\right] \end{array}$$

that the variables $\{\xi^1, \xi^2\}$ cannot be determined independently of η_2. □

The following proposition gives a precise comparison of the two decompositions.

Proposition 4.7.3. *For an LM-matrix $A \in \mathrm{LM}(\boldsymbol{K}, \boldsymbol{F})$, the following hold.*

(1) $p_\mathcal{T}(J) \le p(J)$ *for* $J \subseteq C$.

(2) $\min p_\mathcal{T} = \min p$.

(3) $\mathcal{L}_{\min}(p_\mathcal{T}) \supseteq \mathcal{L}_{\min}(p)$.

(4) $\min \mathcal{L}_{\min}(p_\mathcal{T}) = \min \mathcal{L}_{\min}(p)$.

Proof. (1) This is obvious from (4.10).

(2) This has been shown in the proof of Theorem 4.2.5.

(3) Immediate from (1) and (2) above.

(4) For $J = \min \mathcal{L}_{\min}(p_\tau)$ let J' ($\subseteq J$) be a minimizer in (4.10), i.e., such that $\tau(J) = \gamma(J') - |J'| + |J|$. From (2), we have

$$\min p = \min p_\tau = \rho(J) + \gamma(J') - |J'| \geq \rho(J') + \gamma(J') - |J'| = p(J'),$$

i.e., $J' \in \mathcal{L}_{\min}(p)$. This implies $J' = J = \min \mathcal{L}_{\min}(p)$ by (3). ■

In view of the correspondence between the distributive sublattices and the partition into partially ordered blocks (§2.2.2), the inclusion $\mathcal{L}_{\min}(p_\tau) \supseteq \mathcal{L}_{\min}(p)$ shows that the hierarchical decomposition of the column set C by the principal partition with respect to $\mathbf{M}(Q) \vee \mathbf{M}(T)$ is finer than that of the CCF. In other words, the column set of each block of the CCF is an aggregation of some blocks of the principal partition with respect to $\mathbf{M}(Q) \vee \mathbf{M}(T)$.

In Theorem 4.4.4 we have seen that the decomposition of C based on p provides the finest block-triangular form under the equivalence transformation of the form (4.35). By a similar argument it can be shown on the basis of Theorem 4.2.3 that the principal partition of C with respect to $\mathbf{M}(Q) \vee \mathbf{M}(T)$ corresponds to a block-triangularization under a wider class of transformations of the following form:

$$P_{\mathrm{r}} \begin{pmatrix} S_Q & 0 \\ 0 & S_T \end{pmatrix} \begin{pmatrix} Q \\ T \end{pmatrix} P_{\mathrm{c}}, \tag{4.104}$$

where $S_Q \in \mathrm{GL}(m_Q, \boldsymbol{K})$, $S_T \in \mathrm{GL}(m_T, \boldsymbol{F})$, and P_{r} and P_{c} are permutation matrices of orders m ($= m_Q + m_T$) and n, respectively. That is, we have the following.

Theorem 4.7.4. *For $A \in \mathrm{LM}(\boldsymbol{K}, \boldsymbol{F}; m_Q, m_T, n)$, the partition of* $\mathrm{Col}(A)$ *by the principal partition with respect to* $\mathbf{M}(Q) \vee \mathbf{M}(T)$ *yields the finest proper block-triangularization ("proper" in the sense of §2.1.4) under the transformation* (4.104). □

The transformation (4.104), however, does not seem natural and would be different from what is intended in considering a hierarchical decomposition of a system into subsystems. Recall, for instance, the matrix A of Example 4.7.1. Since its column set is decomposed into singletons by $\mathcal{L}_{\min}(p_\tau)$, it can be put in a triangular form by the transformation of the form (4.104) with $S_T = (y^{ij})^{-1}$, which can be determined only after the parameter values y^{ij} are fixed. This simple example demonstrates that the transformation (4.35) for LM-equivalence is more suitable in practical situations than (4.104), and hence p is more appropriate than p_τ. Note also that the transformed matrix in (4.104) no longer belongs to $\mathrm{LM}(\boldsymbol{K}, \boldsymbol{F}; m_Q, m_T, n)$. We shall come back to this issue in §4.7.2.

4.7.2 Multilayered Matrix

In §4.7.1 (Theorem 4.7.4 in particular) we have seen that, for an LM-matrix $A = \binom{Q}{T} \in \mathrm{LM}(\boldsymbol{K}, \boldsymbol{F})$, the principal partition with respect to $\mathbf{M}(Q) \vee \mathbf{M}(T)$ yields the finest proper block-triangularization under a transformation of the form

$$P_{\mathrm{r}} \begin{pmatrix} S_Q & 0 \\ 0 & S_T \end{pmatrix} \begin{pmatrix} Q \\ T \end{pmatrix} P_{\mathrm{c}} \tag{4.105}$$

with $S_Q \in \mathrm{GL}(m_Q, \boldsymbol{K})$ and $S_T \in \mathrm{GL}(m_T, \boldsymbol{F})$. It has been mentioned at the same time that the transformed matrix no longer belongs to the class of LM-matrices. This suggests that the block-triangularization under a transformation of the form (4.105) should be considered in a wider class of matrices.

Let $\boldsymbol{F}_0$ be an intermediate field of $\boldsymbol{K} \subseteq \boldsymbol{F}$, i.e., $\boldsymbol{K} \subseteq \boldsymbol{F}_0 \subseteq \boldsymbol{F}$, and consider an $(m_Q + m_T) \times n$ matrix A over $\boldsymbol{F}$ of the form $A = \binom{Q}{T}$ such that

(i) Q is an $m_Q \times n$ matrix over $\boldsymbol{K}$,
(ii) $T = Q_1 T_1$ is an $m_T \times n$ matrix over $\boldsymbol{F}$, where Q_1 is an $m_T \times n$ matrix over $\boldsymbol{F}_0$, and T_1 is a diagonal matrix of order n with its diagonal entries being algebraically independent numbers in $\boldsymbol{F}$ over $\boldsymbol{F}_0$.

The set of such matrices A will be denoted by $\mathrm{LC}(\boldsymbol{K}, \boldsymbol{F}_0, \boldsymbol{F}; m_Q, m_T, n)$. This class of matrices is closed under the transformation (4.105) with $S_T \in \mathrm{GL}(m_T, \boldsymbol{F}_0)$. Moreover, Theorem 4.2.3 and Theorem 4.7.4 can be extended to this class.

Theorem 4.7.5. *For* $A = \binom{Q}{T} \in \mathrm{LC}(\boldsymbol{K}, \boldsymbol{F}_0, \boldsymbol{F}; m_Q, m_T, n)$ *it holds that* $\mathbf{M}(A) = \mathbf{M}(Q) \vee \mathbf{M}(T)$ *and that*

$$\operatorname{rank} A = \min\{\rho_Q(J) + \rho_T(J) - |J| \mid J \subseteq C\} + |C|,$$

where $\rho_Q(J) = \operatorname{rank} Q[\mathrm{Row}(Q), J]$ *and* $\rho_T(J) = \operatorname{rank} T[\mathrm{Row}(T), J]$ *for* $J \subseteq C = \mathrm{Col}(A)$. *The partition of* C *by the principal partition with respect to* $\mathbf{M}(Q) \vee \mathbf{M}(T)$ *yields the finest proper block-triangularization ("proper" in the sense of §2.1.4) under the transformation* (4.105) *with* $S_Q \in \mathrm{GL}(m_Q, \boldsymbol{K})$ *and* $S_T \in \mathrm{GL}(m_T, \boldsymbol{F}_0)$.

Proof. Lemma 4.2.1 and Theorem 4.2.2 are still valid for $A \in \mathrm{LC}(\boldsymbol{K}, \boldsymbol{F}_0, \boldsymbol{F})$. Then the subsequent arguments carry over to this case. ∎

The considerations above naturally suggest an extension to a *multilayered matrix*, which, by definition, is a matrix of the form

$$A = \begin{bmatrix} A_0 \\ A_1 \\ \vdots \\ A_\mu \end{bmatrix} \tag{4.106}$$

such that A_0 is an $m_0 \times n$ matrix over $\boldsymbol{K}$, and $A_\alpha = Q_\alpha T_\alpha$ is an $m_\alpha \times n$ matrix over $\boldsymbol{F}_\alpha$, where $\boldsymbol{K} \subseteq \boldsymbol{F}_0 \subseteq \boldsymbol{F}_1 \subseteq \cdots \subseteq \boldsymbol{F}_\mu$ is a sequence of field extensions, Q_α is an $m_\alpha \times n$ matrix over $\boldsymbol{F}_{\alpha-1}$, and T_α is a diagonal matrix of order n with its diagonal entries being algebraically independent numbers in $\boldsymbol{F}_\alpha$ over $\boldsymbol{F}_{\alpha-1}$ ($\alpha = 1, \cdots, \mu$). Putting $C = \mathrm{Col}(A)$ we define $\tilde{p} : 2^C \to \mathbf{Z}$ by

$$\tilde{p}(J) = \rho_0(J) + \rho_1(J) + \cdots + \rho_\mu(J) - |J|, \qquad J \subseteq C, \tag{4.107}$$

where $\rho_\alpha(J) = \mathrm{rank}\, A_\alpha[\mathrm{Row}(A_\alpha), J]$, $J \subseteq C$, for $\alpha = 0, 1, \cdots, \mu$.

Theorem 4.7.6. *For a multilayered matrix A of* (4.106) *it holds that*

$$\mathrm{rank}\, A = \min\{\tilde{p}(J) \mid J \subseteq C\} + |C|. \tag{4.108}$$

Furthermore, the family $\mathcal{L}_{\min}(\tilde{p})$ of the minimizers of $\tilde{p}$ yields the finest proper block-triangularization ("proper" in the sense of §2.1.4) under the transformation

$$P_\mathrm{r} \begin{bmatrix} S_0 & & & \\ & S_1 & & \\ & & \ddots & \\ & & & S_\mu \end{bmatrix} \begin{bmatrix} A_0 \\ A_1 \\ \vdots \\ A_\mu \end{bmatrix} P_\mathrm{c}, \tag{4.109}$$

where $S_0 \in \mathrm{GL}(m_0, \boldsymbol{K})$; $S_\alpha \in \mathrm{GL}(m_\alpha, \boldsymbol{F}_{\alpha-1})$ ($\alpha = 1, \cdots, \mu$); and P_r and P_c are permutation matrices. □

Remark 4.7.7. An LM-matrix $A = \binom{Q}{T} \in \mathrm{LM}(\boldsymbol{K}, \boldsymbol{F}; m_Q, m_T, n)$ can be regarded as a multilayered matrix (4.106) in a number of different ways. A canonical way is to take $A_0 = Q$ and A_α to be the αth row of T for $\alpha = 1, \cdots, m_T$ (with $\mu = m_T$). Then the function $\tilde{p}$ of (4.107) agrees with the LM-surplus function, and the transformation (4.109) is equivalent to the LM-admissible transformation (4.35) for LM-matrices, so far as the block-triangular decomposition is concerned. □

Remark 4.7.8. The canonical form of multilayered matrices introduced above seems to have a natural meaning for electrical networks involving *multiports* (see Recski [277, §8.1] for an exposition on multiports from the viewpoint of matroid theory). To be specific, consider an electrical network consisting of μ multiports, each of which is described by a set of equations with coefficient matrix A_α ($\alpha = 1, \cdots, \mu$). Let A_0 denote the matrix (over $\mathbf{Q}$) for Kirchhoff's laws. Then the coefficient matrix for the whole system is written as (4.106), and the admissible transformation (4.109) reflects the locality in the sense that we can choose an appropriate description for each device. Furthermore, the assumption of the algebraic independence among different devices would be fairly realistic. □

As an application of Theorem 4.7.5 we derive here the maximum-rank minimum-term rank theorem for the pivotal transforms of a matrix, due to Iri [122, 124]. For an $m \times n$ matrix N over $\boldsymbol{K}$, a *pivotal transform* means

$$N = \begin{bmatrix} N_{11} & N_{12} \\ N_{21} & N_{22} \end{bmatrix} \quad \mapsto \quad N' = \begin{bmatrix} {N_{11}}^{-1} & {N_{11}}^{-1} N_{12} \\ -N_{21} {N_{11}}^{-1} & N_{22} - N_{21} {N_{11}}^{-1} N_{12} \end{bmatrix}$$

for a nonsingular submatrix N_{11}. This transformation is invertible, and hence defines an equivalence relation $\underset{\text{pv}}{\sim}$ among matrices (of the same size).

Theorem 4.7.9 (Maximum-rank minimum-term rank theorem). *For an $m \times n$ matrix N over $\boldsymbol{K}$ it holds that*

$$\max_{N' \underset{\text{pv}}{\sim} N} \operatorname{rank} N' = \min_{N' \underset{\text{pv}}{\sim} N} \text{term-rank} \, N'.$$

Moreover, there exists an $m \times n$ matrix N° over $\boldsymbol{K}$ such that $N^\circ \underset{\text{pv}}{\sim} N$ and

$$\operatorname{rank} N^\circ = \text{term-rank} \, N^\circ. \tag{4.110}$$

Proof. Put $Q = [I_m \; N]$ and $T = [I_m \; N] \, D$, where $D = \operatorname{diag}(t_1, \cdots, t_{m+n})$ with t_i $(i = 1, \cdots, m+n)$ being indeterminates over $\boldsymbol{K}$, and consider $A = \binom{Q}{T} \in \mathrm{LC}(\boldsymbol{K}, \boldsymbol{K}, \boldsymbol{F}; m, m, m+n)$ where $\boldsymbol{F} = \boldsymbol{K}(t_1, \cdots, t_{m+n})$. The column set C of A is given by $C = \mathrm{Col}(Q) \simeq \mathrm{Row}(N) \cup \mathrm{Col}(N)$. By Theorem 4.7.5 there exists $B \subseteq C$ such that $\operatorname{rank} Q[\mathrm{Row}(Q), B] = |B| = m$ and $\operatorname{rank} A = m + \operatorname{rank} T[\mathrm{Row}(T), C \setminus B]$. Put $S = Q[\mathrm{Row}(Q), B]^{-1}$, $\bar{Q} = SQ$, and $\bar{T} = ST$, to obtain $\bar{A} = \binom{\bar{Q}}{\bar{T}}$. This is the canonical block-triangular form of A (cf. (4.64)). Since $\bar{A}$ (with columns permuted) is of the form

$$\bar{A} = \begin{pmatrix} I_m & N^\circ \\ I_m \cdot D_B & N^\circ \cdot D_{C \setminus B} \end{pmatrix},$$

with column blocks labelled B and $C \setminus B$, where $N^\circ = \bar{Q}[\mathrm{Row}(\bar{Q}), C \setminus B]$, $D_B = \operatorname{diag}(t_i \mid i \in B)$, $D_{C\setminus B} = \operatorname{diag}(t_i \mid i \in C \setminus B)$, it can be seen that

$$\text{term-rank} \, \bar{A} = \text{term-rank} \begin{pmatrix} I_m & N^\circ \\ I_m & N^\circ \end{pmatrix} = m + \text{term-rank} \, N^\circ.$$

On the other hand, $\operatorname{rank} \bar{A} = \operatorname{rank} A = m + \operatorname{rank} T[\mathrm{Row}(T), C \setminus B] = m + \operatorname{rank} \bar{T}[\mathrm{Row}(\bar{T}), C \setminus B] = m + \operatorname{rank} N^\circ$ by the choice of B. Finally, we have $\operatorname{rank} \bar{A} = \text{term-rank} \, \bar{A}$, since $\bar{A}$ is in a proper block-triangular form. Hence follows (4.110). Note that N° is a pivotal transform of N with respect to $N[\mathrm{Row}(N) \setminus B, \mathrm{Col}(N) \cap B]$. ■

Remark 4.7.10. The matrix N° constructed in the proof of Theorem 4.7.9 coincides with the combinatorial canonical form of N with respect to its pivotal transforms introduced by Iri [125]. Note that N° can be put into a proper block-triangular form by (4.110) and Proposition 2.1.17. See Iri [125] for its relationship to the principal partition of graphs and to the topological degrees of freedom of electrical networks considered by Iri [122], Kishi–Kajitani [157, 158, 159], Ohtsuki–Ishizaki–Watanabe [254]. Also see Maurer [189] for a matroid theoretic generalization of Theorem 4.7.9. □

4.7.3 Electrical Network with Admittance Expression

A decomposition method has been proposed by Iri [127] for electrical networks with admittance expressions (see Iri [128] for an explicit illustration). We discuss here its relationship to the CCF.

When the branch characteristics of an electrical network are given in terms of *self-* and *mutual admittances* Y, the coefficient matrix A of the system (3.2) of equations in $(\boldsymbol{\xi}, \boldsymbol{\eta})$ takes the form:

$$A = \begin{array}{c} \quad B_\xi \; B_\eta \\ \begin{array}{|cc|} \hline D & O \\ O & R \\ \hline -I & Y \\ \hline \end{array} \end{array}, \tag{4.111}$$

where D is a fundamental cutset matrix and R is a fundamental circuit matrix of the underlying graph. The column set $C = \mathrm{Col}(A)$ is the disjoint union of two copies, say B_ξ and B_η, of the set B of branches; i.e., $C = B_\xi \cup B_\eta$. Note that $\mathrm{Row}(Y)$ is identified with B_ξ and $\mathrm{Col}(Y)$ with B_η. It is mentioned that the above system of equations represents the "free" network that is obtained after the branches of voltage sources are contracted and those of current sources are deleted (see Recski [277] for more about this).

The unique solvability of the network is equivalent to the nonsingularity of DYD^{T} by the following lemma.

Lemma 4.7.11. *Suppose D and R in* (4.111) *are of full-row rank and* $\ker D = (\ker R)^\perp$. *Then* $\det A = c \cdot \det(DYD^{\mathrm{T}})$ *for some* $c \neq 0$.

Proof. By the assumption there exist a matrix N, nonsingular matrices S_D and S_R, and permutation matrix P such that $D = S_D[I \mid N]P$ and $R = S_R[-N^{\mathrm{T}} \mid I]P$. We assume $P = I$ for notational simplicity, and partition Y as $Y = (Y_{ij} \mid i,j = 1,2)$ accordingly. We observe

$$\begin{bmatrix} I & O & D \\ O & I & O \\ O & O & I \end{bmatrix} \begin{bmatrix} D & O \\ O & R \\ -I & Y \end{bmatrix} = \begin{bmatrix} O & DY \\ O & R \\ -I & Y \end{bmatrix},$$

$$\begin{bmatrix} {S_D}^{-1} & O \\ O & {S_R}^{-1} \end{bmatrix} \begin{bmatrix} DY \\ R \end{bmatrix} = \begin{bmatrix} {S_D}^{-1}(DYD^{\mathrm{T}}){S_D}^{-\mathrm{T}} & Y_{12} + NY_{22} \\ O & I \end{bmatrix} \begin{bmatrix} I & O \\ -N^{\mathrm{T}} & I \end{bmatrix}$$

to prove the claim with $c = \pm(\det S_R / \det S_D) \neq 0$. ∎

The decomposition proposed by Iri [127, 128] may be described as follows. Under the assumption that the nonvanishing entries of Y are algebraically independent over $\mathbf{Q}$, the nonsingularity of DYD^{T} can be expressed in terms of an independent matching problem, as has been explained in Remark 2.3.37. Namely, we consider an independent matching problem on the bipartite graph $G = (V^+, V^-; \tilde{A})$ with vertex sets $V^+ = \mathrm{Row}(Y)$ $(= B_\xi)$, $V^- = \mathrm{Col}(Y)$ $(=$

B_η), and arc set $\tilde{A} = \{(i,j) \mid Y_{ij} \neq 0\}$. The matroids $\mathbf{M}^+$ and $\mathbf{M}^-$ attached to V^+ and V^- are both isomorphic to the linear matroid $\mathbf{M}(D) = (B, \mu)$ defined by the matrix D. It should be clear that $\mu(I) = \operatorname{rank} D[\mathrm{Row}(D), I]$ $(I \subseteq B)$, which is equal to the rank of I (= maximum size of a circuit-free subset of I) in the underlying graph. Then $\operatorname{rank}(DYD^{\mathrm{T}})$ is equal to the maximum size of an independent matching (cf. (2.78)). Though not explicit in Iri [127, 128], Iri's decomposition can be identified as the min-cut decomposition (§2.3.5) for this independent matching problem.

To be specific, define

$$\begin{aligned}
\Gamma_Y(J) &= \{i \in B_\xi \mid \exists j \in J : Y_{ij} \neq 0\}, \qquad J \subseteq B_\eta, \\
\mathcal{H} &= \{(I,J) \in 2^{B_\xi} \times 2^{B_\eta} \mid I \supseteq \Gamma_Y(J)\}, \\
p_\mu(I,J) &= \begin{cases} \mu(I) + \mu(B_\eta \setminus J) - \mu(B) & ((I,J) \in \mathcal{H}) \\ +\infty & ((I,J) \notin \mathcal{H}), \end{cases} \\
\mathcal{L}_{\min}(p_\mu) &= \{I \cup J \subseteq B_\xi \cup B_\eta \mid p_\mu(I,J) = \min p_\mu\}.
\end{aligned}$$

Then $(I,J) \in \mathcal{H}$ if and only if $(I, B_\eta \setminus J)$ is a cover in the independent matching problem, and the cut capacity function $\kappa : B_\xi \cup B_\eta \to \mathbf{Z}$, as defined in (2.71), is given by

$$\kappa(I \cup J) = p_\mu(B_\xi \setminus I, B_\eta \setminus J) + \mu(B).$$

The min-cut decomposition is induced from the lattice $\mathcal{L}_{\min}(\kappa)$ of the minimizers of κ. However, it is more convenient here to consider $\mathcal{L}_{\min}(p_\mu)$ in place of $\mathcal{L}_{\min}(\kappa)$; obviously, $I \cup J \in \mathcal{L}_{\min}(\kappa) \iff (B_\xi \setminus I) \cup (B_\eta \setminus J) \in \mathcal{L}_{\min}(p_\mu)$. The decomposition of $C = B_\xi \cup B_\eta$ proposed by Iri [127] is the one induced from $\mathcal{L}_{\min}(p_\mu)$ according to the general principle of §2.2.2.

On the other hand, we may regard A as a member of $\mathrm{LM}(\mathbf{Q}, \mathbf{R})$ with

$$A = \left[\frac{Q}{T}\right], \qquad Q = \begin{bmatrix} D & O \\ O & R \end{bmatrix}, \qquad T = \begin{bmatrix} -I & Y \end{bmatrix},$$

where a trivial scaling of the constitutive equations (matrix T) using transcendental numbers is assumed to bring A into the class of $\mathrm{LM}(\mathbf{Q}, \mathbf{R})$ (as in Example 4.3.9). We shall show that the decomposition through the CCF of A agrees essentially with Iri's decomposition.

The LM-surplus function p associated with A is given by

$$p(I \cup J) = \mu(I) + \nu(J) + |I \cup \Gamma_Y(J)| - |I \cup J|, \qquad I \subseteq B_\xi, J \subseteq B_\eta,$$

where $\nu(J) = \operatorname{rank} R[\mathrm{Row}(R), J]$ $(J \subseteq B_\eta)$, which is equal to the nullity of J (= maximum size of a cutset-free subset of J) in the underlying graph. Since

$$\nu(J) = \mu(B_\eta \setminus J) + |J| - \mu(B), \qquad J \subseteq B_\eta,$$

we have

$$p(I \cup J) = p_\mu(I,J), \qquad (I,J) \in \mathcal{H}. \tag{4.112}$$

Proposition 4.7.12.
(1) $\min\{p(I \cup J) \mid I \subseteq B_\xi, J \subseteq B_\eta\} = \min\{p_\mu(I, J) \mid I \subseteq B_\xi, J \subseteq B_\eta\}$.
(2) $\mathcal{L}_{\min}(p) \supseteq \mathcal{L}_{\min}(p_\mu)$.
(3) $\{J \subseteq B_\eta \mid \exists\, I \subseteq B_\xi : I \cup J \in \mathcal{L}_{\min}(p)\} = \{J \subseteq B_\eta \mid \exists\, I \subseteq B_\xi : I \cup J \in \mathcal{L}_{\min}(p_\mu)\}$.

Proof. (1) We have

$$\begin{aligned}\min p &= \min\{\min\{\mu(I) + |\Gamma_Y(J) \setminus I| \mid I \subseteq B_\xi\} + \nu(J) - |J| \mid J \subseteq B_\eta\} \\ &= \min\{\mu(\Gamma_Y(J)) + \nu(J) - |J| \mid J \subseteq B_\eta\},\end{aligned}$$

which shows that the minimum of p is attained by an (I, J) in $\mathcal{H}$.

(2) This is immediate from (4.112) and (1) above.

(3) Both sides of (3) agrees with the minimizers J ($\subseteq B_\eta$) of $\mu(\Gamma_Y(J)) + \nu(J) - |J|$. ■

Proposition 4.7.12(2) shows that the decomposition by the CCF applied to (4.111) yields a finer partition of the variables $\{\boldsymbol{\xi}, \boldsymbol{\eta}\}$. However, the difference is not substantial, since, as indicated by Proposition 4.7.12(3), they provide an identical partition for the voltage variables $\boldsymbol{\eta}$ which play the primary role in (4.111); the current variables $\boldsymbol{\xi}$ are only secondary as they are readily obtained from $\boldsymbol{\eta}$ as $\boldsymbol{\xi} = Y\boldsymbol{\eta}$. In this way, we may say that they give essentially the same decomposition. However, the inclusion in Proposition 4.7.12(2) is proper in general, as is exemplified below.

Example 4.7.13. For a matrix

$$A = \begin{array}{c} \begin{array}{cccc} \xi^1 & \xi^2 & \eta_1 & \eta_2 \end{array} \\ \left[\begin{array}{cccc} 1 & & & \\ & 1 & & \\ \hline -1 & & y^{11} & 0 \\ & -1 & y^{21} & y^{22} \end{array}\right] \end{array},$$

the CCF based on $\mathcal{L}_{\min}(p)$ decomposes $\{\xi^1, \xi^2, \eta_1, \eta_2\}$ into four singletons with partial order: $\{\eta_2\} \prec \{\eta_1\} \prec \{\xi^1\}$, $\{\eta_2\} \prec \{\xi^2\}$. The decomposition by $\mathcal{L}_{\min}(p_\mu)$, on the other hand, gives a partition into two blocks with $\{\xi^2, \eta_2\} \prec \{\xi^1, \eta_1\}$. □

4.8 Partitioned Matrix

"Partitioned matrix," investigated in Ito–Iwata–Murota [138], offers a general framework in which we can gain a deeper understanding of proper block-triangularizations of matrices with respect to existence, uniqueness, and algorithmic construction. Some of the nice properties enjoyed by the DM-decomposition of generic matrices and the CCF of LM-matrices carry over to this general setting, whereas the construction by combinatorial algorithms does not.

4.8.1 Definitions

We consider an $m \times n$ matrix over a field $\boldsymbol{F}$ whose row set and column set are independently divided into groups:

$$A = \begin{bmatrix} A_{11} & A_{12} & \cdots & A_{1\nu} \\ A_{21} & A_{22} & \cdots & A_{2\nu} \\ \vdots & \vdots & \ddots & \vdots \\ A_{\mu 1} & A_{\mu 2} & \cdots & A_{\mu\nu} \end{bmatrix}, \tag{4.113}$$

which we call a *partitioned matrix*, where $A_{\alpha\beta}$ is an $m_\alpha \times n_\beta$ matrix called the (α, β)-submatrix of A for $\alpha = 1, \cdots, \mu$ and $\beta = 1, \cdots, \nu$. We are concerned with a proper block-triangularization of A by means of an equivalence transformation of the form

$$S_{\mathrm{r}}^{-1} A S_{\mathrm{c}} = \begin{bmatrix} S_{\mathrm{r}1} & O & \cdots & O \\ O & S_{\mathrm{r}2} & \ddots & \vdots \\ \vdots & \ddots & \ddots & O \\ O & \cdots & O & S_{\mathrm{r}\mu} \end{bmatrix}^{-1} \begin{bmatrix} A_{11} & A_{12} & \cdots & A_{1\nu} \\ A_{21} & A_{22} & \cdots & A_{2\nu} \\ \vdots & \vdots & \ddots & \vdots \\ A_{\mu 1} & A_{\mu 2} & \cdots & A_{\mu\nu} \end{bmatrix} \begin{bmatrix} S_{\mathrm{c}1} & O & \cdots & O \\ O & S_{\mathrm{c}2} & \ddots & O \\ \vdots & \ddots & \ddots & O \\ O & \cdots & O & S_{\mathrm{c}\nu} \end{bmatrix} \tag{4.114}$$

with

$$S_{\mathrm{r}} = \bigoplus_{\alpha=1}^{\mu} S_{\mathrm{r}\alpha} = \begin{bmatrix} S_{\mathrm{r}1} & O & \cdots & O \\ O & S_{\mathrm{r}2} & \ddots & \vdots \\ \vdots & \ddots & \ddots & O \\ O & \cdots & O & S_{\mathrm{r}\mu} \end{bmatrix}, \quad S_{\mathrm{c}} = \bigoplus_{\beta=1}^{\nu} S_{\mathrm{c}\beta} = \begin{bmatrix} S_{\mathrm{c}1} & O & \cdots & O \\ O & S_{\mathrm{c}2} & \ddots & O \\ \vdots & \ddots & \ddots & O \\ O & \cdots & O & S_{\mathrm{c}\nu} \end{bmatrix} \tag{4.115}$$

being nonsingular matrices over $\boldsymbol{F}$. Such an equivalence transformation, preserving the given partition structure, is called a *partition-respecting equivalence transformation* (or *PE-transformation* for short). Then our problem is to bring a partitioned matrix A into a proper block-triangular form by means of a PE-transformation.

The block-diagonal structure imposed on the transformation matrices S_{r} and S_{c} can be expressed in terms of two families of projection matrices, $\Pi = \{\Pi_\alpha\}_{\alpha=1}^{\mu}$ and $\Gamma = \{\Gamma_\beta\}_{\beta=1}^{\nu}$. The matrix Π_α is an $m \times m$ projection matrix such that the (α, α)-submatrix of Π_α is the unit matrix I_{m_α} of dimension m_α and the other submatrices are zeroes. Similarly, Γ_β is an $n \times n$ projection matrix such that the (β, β)-submatrix of Γ_β is the unit matrix I_{n_β} of dimension n_β and all the other submatrices are zeroes. Then a transformation $S_{\mathrm{r}}^{-1} A S_{\mathrm{c}}$ with nonsingular S_{r} and S_{c} is a PE-transformation if and only if

$$\Pi_\alpha S_{\mathrm{r}} = S_{\mathrm{r}} \Pi_\alpha \quad (\alpha = 1, \cdots, \mu), \qquad \Gamma_\beta S_{\mathrm{c}} = S_{\mathrm{c}} \Gamma_\beta \quad (\beta = 1, \cdots, \nu). \tag{4.116}$$

Note that this condition is equivalent to the off-diagonal submatrices of $S_{\rm r}$ and $S_{\rm c}$ being zeroes. Sometimes we denote a partitioned matrix by a triple (A, Π, Γ) of matrix A and two families of projection matrices $\Pi = \{\Pi_\alpha\}_{\alpha=1}^{\mu}$ and $\Gamma = \{\Gamma_\beta\}_{\beta=1}^{\nu}$.

The concepts of partitioned matrices and PE-transformations may be described in terms of linear maps, as follows. Let $U \cong \boldsymbol{F}^m$ and $V \cong \boldsymbol{F}^n$ be $\boldsymbol{F}$-linear spaces which are, respectively, expressed as direct sums of lower dimensional component spaces:

$$U = \bigoplus_{\alpha=1}^{\mu} U_\alpha, \qquad \dim_{\boldsymbol{F}} U_\alpha = m_\alpha \quad (\alpha = 1, \cdots, \mu), \tag{4.117}$$

$$V = \bigoplus_{\beta=1}^{\nu} V_\beta, \qquad \dim_{\boldsymbol{F}} V_\beta = n_\beta \quad (\beta = 1, \cdots, \nu). \tag{4.118}$$

A linear transformation $f : V \to U$ is defined by a family of linear transformations

$$f_{\alpha\beta} : V_\beta \to U_\alpha, \qquad \alpha = 1, \cdots, \mu; \quad \beta = 1, \cdots, \nu.$$

When a family of bases for $\{U_\alpha\}_{\alpha=1}^{\mu}$ and one for $\{V_\beta\}_{\beta=1}^{\nu}$ are chosen, f is represented by a partitioned matrix A, where $A_{\alpha\beta}$ corresponds to $f_{\alpha\beta}$. A PE-transformation corresponds to a change of "local basis families" for $\{U_\alpha\}_{\alpha=1}^{\mu}$ and $\{V_\beta\}_{\beta=1}^{\nu}$. The block-triangularization of a partitioned matrix by a PE-transformation amounts to finding a global hierarchical decomposition by means of a local basis change.

Example 4.8.1. The block-triangularization of a partitioned matrix by a PE-transformation is illustrated for a 4×5 partitioned matrix over $\boldsymbol{F} = \mathbf{Q}$:

$$A = \begin{array}{c} \\ \star \\ \star \\ \diamond \\ \diamond \end{array} \begin{array}{c} \begin{array}{ccccc} \circ & \circ & \circ & \bullet & \bullet \end{array} \\ \left[\begin{array}{ccc|cc} 1 & 1 & 1 & 1 & 0 \\ 0 & 2 & 1 & 1 & 1 \\ \hline 2 & -2 & 0 & 0 & 2 \\ 0 & 3 & 0 & 0 & 4 \end{array}\right] \end{array},$$

where $\mu = 2$, $\nu = 2$, $m_1 = 2$, $m_2 = 2$, $n_1 = 3$, $n_2 = 2$. With the choice of

$$S_{\rm r} = \left[\begin{array}{cc|cc} 1 & 1 & \multicolumn{2}{c}{O} \\ 1 & 0 & & \\ \hline \multicolumn{2}{c|}{O} & 0 & 1 \\ & & 1 & 0 \end{array}\right], \qquad S_{\rm c} = \left[\begin{array}{ccc|cc} 0 & 1 & 1 & & \\ 0 & 1 & 0 & \multicolumn{2}{c}{O} \\ 1 & 0 & 0 & & \\ \hline \multicolumn{3}{c|}{O} & 1 & 0 \\ & & & 0 & 1 \end{array}\right]$$

we have

$$\tilde{A} = S_{\rm r}^{-1} A S_{\rm c} = \begin{array}{c} \\ \star \\ \star \\ \diamond \\ \diamond \end{array} \begin{array}{c} \begin{array}{ccccc} \circ & \circ & \circ & \bullet & \bullet \end{array} \\ \left[\begin{array}{ccc|cc} 1 & 2 & 0 & 1 & 1 \\ 0 & 0 & 1 & 0 & -1 \\ \hline 0 & 3 & 0 & 0 & 4 \\ 0 & 0 & 2 & 0 & 2 \end{array}\right] \end{array}.$$

Using permutation matrices

$$P_{\rm r} = \begin{bmatrix} 1&0&0&0 \\ 0&0&1&0 \\ 0&1&0&0 \\ 0&0&0&1 \end{bmatrix}, \qquad P_{\rm c} = \begin{bmatrix} 1&0&0&0&0 \\ 0&0&1&0&0 \\ 0&0&0&1&0 \\ 0&1&0&0&0 \\ 0&0&0&0&1 \end{bmatrix},$$

we can transform $\tilde{A}$ into an explicit upper block-triangular form:

$$\bar{A} = P_{\rm r}\tilde{A}P_{\rm c} = \begin{array}{c} \\ \star \\ \diamond \\ \star \\ \diamond \end{array} \begin{array}{c} \begin{array}{ccccc} \circ & \bullet & \circ & \circ & \bullet \end{array} \\ \left[\begin{array}{c|c|c|c|c} 1 & 1 & 2 & 0 & 1 \\ \hline & & 3 & 0 & 4 \\ \hline & O & & 1 & -1 \\ & & & 2 & 2 \end{array}\right] \end{array}$$

with two square blocks, a nonempty horizontal tail ($|R_0| = 1$, $|C_0| = 2$) and an empty vertical tail. Note that this is a proper block-triangular matrix. □

We now introduce a submodular function $p_{\rm PE}$ for a partitioned matrix, which is a generalization of the (LM-)surplus function for a generic (or LM-) matrix. We denote by $\mathcal{W}$ the family of all the subspaces of V that can be represented as a direct sum of subspaces of V_β's, i.e.,

$$\mathcal{W} = \{W \mid W\text{: subspace of } V,\ \Gamma_\beta W \subseteq W\ (\beta = 1, \cdots, \nu)\}. \tag{4.119}$$

For $W_1 \in \mathcal{W}$ and $W_2 \in \mathcal{W}$, we have $W_1 + W_2 \in \mathcal{W}$ and $W_1 \cap W_2 \in \mathcal{W}$, which means that $\mathcal{W}$ forms a lattice. Furthermore, we have

$$\dim W_1 + \dim W_2 = \dim(W_1 + W_2) + \dim(W_1 \cap W_2),$$

which means $\mathcal{W}$ is a *modular lattice* (not distributive in general). Regarding $A_\alpha = \Pi_\alpha A$ as a linear map, we define $p_{\rm PE} : \mathcal{W} \to \mathbf{Z}$ by

$$p_{\rm PE}(W) = \sum_{\alpha=1}^{\mu} \dim(A_\alpha W) - \dim W, \qquad W \in \mathcal{W}. \tag{4.120}$$

We call $p_{\rm PE} : \mathcal{W} \to \mathbf{Z}$ the *PE-surplus function* associated with a partitioned matrix A.

Lemma 4.8.2. *The PE-surplus function $p_{\rm PE} : \mathcal{W} \to \mathbf{Z}$ is submodular, i.e.,*

$$p_{\rm PE}(W_1) + p_{\rm PE}(W_2) \geq p_{\rm PE}(W_1 + W_2) + p_{\rm PE}(W_1 \cap W_2), \qquad W_1, W_2 \in \mathcal{W}.$$

Proof. It suffices to show that $\dim(A_\alpha W)$ is submodular for each α. This follows from $A_\alpha(W_1 + W_2) = A_\alpha W_1 + A_\alpha W_2$ and $A_\alpha(W_1 \cap W_2) \subseteq A_\alpha W_1 \cap A_\alpha W_2$. ■

The following proposition shows that the PE-surplus function $p_{\rm PE}$ is relevant in dealing with the rank of partitioned matrices.

Proposition 4.8.3. *For an $m \times n$ partitioned matrix A,*

$$\operatorname{rank} A \leq \min\{p_{\rm PE}(W) \mid W \in \mathcal{W}\} + n \leq \text{term-rank}\, A.$$

Proof. For any $W \in \mathcal{W}$ there exists a matrix $S_{\rm c}$ of the form of (4.115) such that a subfamily of the column vectors of $S_{\rm c}$ is a basis of W. Let

$$\tilde{A} = \begin{array}{c} \begin{array}{cc} J \simeq W & C \setminus J \end{array} \\ \begin{pmatrix} \tilde{A}_1[R_1, J] & \tilde{A}_1[R_1, C \setminus J] \\ \tilde{A}_2[R_2, J] & \tilde{A}_2[R_2, C \setminus J] \\ \vdots & \vdots \\ \tilde{A}_\mu[R_\mu, J] & \tilde{A}_\mu[R_\mu, C \setminus J] \end{pmatrix} \end{array} \tag{4.121}$$

be obtained from A by a PE-transformation using such $S_{\rm c}$, where $C = \operatorname{Col}(\tilde{A})$, $R_\alpha = \operatorname{Row}(\tilde{A}_\alpha)$, and the subset of C indicated by "$J \simeq W$" corresponds to the subspace W (hence $|J| = \dim W$). Then, $\operatorname{rank} \tilde{A}_\alpha[R_\alpha, J] = \dim(A_\alpha W)$ for each α, and

$$\begin{aligned} \operatorname{rank} A = \operatorname{rank} \tilde{A} &\leq \sum_{\alpha=1}^{\mu} \operatorname{rank} \tilde{A}_\alpha[R_\alpha, J] + \operatorname{rank} \tilde{A}[R, C \setminus J] \\ &\leq \sum_{\alpha=1}^{\mu} \dim(A_\alpha W) + n - \dim W = p_{\rm PE}(W) + n. \end{aligned} \tag{4.122}$$

This establishes the first inequality.

For the second inequality, let $p_0 : \operatorname{Col}(A) \to \mathbf{Z}$ be the surplus function (2.39) associated with A, and let $J \subseteq \operatorname{Col}(A)$ be a minimizer of p_0. Then we have $p_0(J) + n = \text{term-rank}\, A$ by the Hall–Ore theorem (Theorem 2.2.17). Define W to be the subspace of V spanned by the unit vectors corresponding to J. Then $W \in \mathcal{W}$, $\dim W = |J|$ and $\sum_{\alpha=1}^{\mu} \dim(A_\alpha W) \leq \gamma_A(J)$ (cf. (4.46)). Therefore $p_{\rm PE}(W) + n \leq p_0(J) + n = \text{term-rank}\, A$. ■

It may be said that the relation

$$\operatorname{rank} A \leq \min\{p_{\rm PE}(W) \mid W \in \mathcal{W}\} + n \tag{4.123}$$

states a weak duality of the same kind as the easier part of min-max relations. Though the equality is not always guaranteed in (4.123), this weak duality turns out to be most fundamental in that the strong duality (the equality in (4.123)) is equivalent to the existence of a proper block-triangularization, as will be stated in Theorem 4.8.6. Note in this connection that both rank A and $\min p_{\rm PE}$ are invariant under PE-transformations, whereas term-rank A is not.

Example 4.8.4. A simplest partitioned matrix that cannot be brought into a proper block-triangular form is $A = \left[\begin{array}{c|c} 1 & 1 \\ \hline 1 & 1 \end{array}\right]$ with $\mu = \nu = 2$. We have $\operatorname{rank} A = 1$ while $\min p_{\rm PE} + 2 = 2 = \text{term-rank}\, A$. □

Remark 4.8.5. An LM-matrix $A = \binom{Q}{T} \in \mathrm{LM}(\boldsymbol{K}, \boldsymbol{F}; m_Q, m_T, n)$ can be regarded as a partitioned matrix such that the column set is partitioned into singletons ($\nu = n$, $n_\beta = 1$ for $\beta = 1, \cdots, \nu$) and the row set is partitioned into $\mathrm{Row}(Q)$ and singletons from $\mathrm{Row}(T)$ ($\mu = 1 + m_T$; $m_1 = m_Q$, $m_\alpha = 1$ for $\alpha = 2, \cdots, \mu$). Then, $\mathcal{W} \cong 2^{\mathrm{Col}(A)}$, and the associated PE-surplus function p_{PE} agrees with the LM-surplus function p defined in (4.16). □

4.8.2 Existence of Proper Block-triangularization

In Proposition 4.8.3 we have seen a weak duality between the rank and the PE-surplus function p_{PE} for a partitioned matrix. We shall show that the strong duality (the validity of the minimax formula) is equivalent to the existence of a proper block-triangularization. The block-triangularization can be constructed on the basis of the family of the minimizers of p_{PE}:

$$\mathcal{L}_{\min}(p_{\mathrm{PE}}) = \{W \in \mathcal{W} \mid p_{\mathrm{PE}}(W) = \min_{W' \in \mathcal{W}} p_{\mathrm{PE}}(W')\}, \tag{4.124}$$

which forms a modular lattice (cf. Lemma 4.8.2 and Theorem 2.2.5).

Theorem 4.8.6. *For an $m \times n$ partitioned matrix A, a proper block-triangular matrix can be obtained by a PE-transformation* (4.114) *if and only if*

$$\mathrm{rank}\, A = \min\{p_{\mathrm{PE}}(W) \mid W \in \mathcal{W}\} + n. \tag{4.125}$$

Proof. ["only if" part] Suppose $\tilde{A}$ is a proper block-triangular matrix obtained from A by a PE-transformation. Since $\mathrm{rank}\, A$ and $\min p_{\mathrm{PE}}$ are invariant under PE-transformations, Proposition 4.8.3 implies

$$\mathrm{rank}\, A = \mathrm{rank}\, \tilde{A} \le \min\{p_{\mathrm{PE}}(W) \mid W \in \mathcal{W}\} + n \le \text{term-rank}\, \tilde{A},$$

in which $\mathrm{rank}\, \tilde{A} = \text{term-rank}\, \tilde{A}$ by Proposition 2.1.17.

["if" part] Let $\mathcal{C}$ be a maximal chain of $\mathcal{L}_{\min}(p_{\mathrm{PE}})$:

$$\mathcal{C} : W_0 \subsetneqq W_1 \subsetneqq \cdots \subsetneqq W_b.$$

Denoting $W_k \cap V_\beta$ by $W_{k\beta}$, we obtain from $\mathcal{C}$ a family of increasing chains

$$\mathcal{C}_\beta : W_{0\beta} \subseteq W_{1\beta} \subseteq \cdots \subseteq W_{b\beta}$$

for $\beta = 1, \cdots, \nu$. Let $\Psi_{k\beta}$ be a set of linearly independent column vectors spanning $W_{k\beta}$ for $k = 0, 1, \cdots, b$ and $\Psi_{\infty\beta}$ spanning V_β such that

$$\Psi_{0\beta} \subseteq \Psi_{1\beta} \subseteq \cdots \subseteq \Psi_{b\beta} \subseteq \Psi_{\infty\beta}.$$

Then $\Psi_k = \bigcup_{\beta=1}^{\nu} \Psi_{k\beta}$ spans W_k for $k = 0, 1, \cdots, b$, and $\Psi = \bigcup_{\beta=1}^{\nu} \Psi_{\infty\beta}$ becomes a basis of V. Order the n column vectors of Ψ as $[\Psi_{\infty 1}, \Psi_{\infty 2}, \cdots, \Psi_{\infty\nu}]$ to get a nonsingular matrix $S_{\mathrm{c}} = \bigoplus_{\beta=1}^{\nu} S_{\mathrm{c}\beta}$.

Similarly, we obtain from $\mathcal{C}$ another family of increasing chains

$$A_\alpha \mathcal{C} : A_\alpha W_0 \subseteq A_\alpha W_1 \subseteq \cdots \subseteq A_\alpha W_b$$

for $\alpha = 1, \cdots, \mu$. Let $\Phi_{k\alpha}$ be a set of linearly independent column vectors spanning $A_\alpha W_k$ for $k = 0, 1, \cdots, b$ and $\Phi_{\infty\alpha}$ spanning U_α such that

$$\Phi_{0\alpha} \subseteq \Phi_{1\alpha} \subseteq \cdots \subseteq \Phi_{b\alpha} \subseteq \Phi_{\infty\alpha}.$$

Then $\Phi_k = \bigcup_{\alpha=1}^{\mu} \Phi_{k\alpha}$ spans AW_k for $k = 0, 1, \cdots, b$, and $\Phi = \bigcup_{\alpha=1}^{\mu} \Phi_{\infty\alpha}$ becomes a basis of U. Order the m column vectors of Φ as $[\Phi_{\infty 1}, \Phi_{\infty 2}, \cdots, \Phi_{\infty\mu}]$ to get a nonsingular matrix $S_\mathrm{r} = \bigoplus_{\alpha=1}^{\mu} S_{\mathrm{r}\alpha}$.

Put $\tilde{A} = {S_\mathrm{r}}^{-1} A S_\mathrm{c}$. Let $C_k \subseteq \mathrm{Col}(\tilde{A})$ be the column subset corresponding to $\hat{\Psi}_k$, and $R_k \subseteq \mathrm{Row}(\tilde{A})$ the row subset corresponding to $\hat{\Phi}_k$, where

$$\begin{array}{lll} \hat{\Psi}_0 = \Psi_0, & \hat{\Phi}_0 = \Phi_0, & \\ \hat{\Psi}_k = \Psi_k \setminus \Psi_{k-1}, & \hat{\Phi}_k = \Phi_k \setminus \Phi_{k-1}, & \text{for} \quad k = 1, \cdots, b, \\ \hat{\Psi}_\infty = \Psi \setminus \Psi_b, & \hat{\Phi}_\infty = \Phi \setminus \Phi_b. & \end{array}$$

Then the consistency of the basis vectors Ψ and Φ with the chains $\mathcal{C}$ and $A_\alpha \mathcal{C}$ implies that

$$\tilde{A}[R_k, C_l] = O \qquad \text{if} \quad 0 \le l < k \le \infty.$$

Since

$$p_\mathrm{PE}(W_k) = \sum_{l=0}^{k} |R_l| - \sum_{l=0}^{k} |C_l|$$

and $p_\mathrm{PE}(W_{k-1}) = p_\mathrm{PE}(W_k)$ $(= \min p_\mathrm{PE})$ for $k = 1, \cdots, b$, it holds that

$$|R_k| = |C_k| \quad \text{for } k = 1, \cdots, b.$$

Furthermore it can be shown from (4.125) (see the proof of Theorem 4.4.4) that

$$\mathrm{rank}\, \tilde{A}[R_k, C_k] = \min(|R_k|, |C_k|) \quad \text{for} \quad k = 0, 1, \cdots, b, \infty.$$

That is to say, $\tilde{A}$ is in a proper block-triangular form, where the number of square blocks b is given by the length of $\mathcal{C}$. This completes the proof of Theorem 4.8.6. ∎

When the strong duality (or the rank formula) (4.125) holds true, the lattice $\mathcal{L}_{\min}(p_\mathrm{PE})$ admits an alternative expression. Define $\mathcal{L}(A, \Pi, \Gamma)$ to be the family of subspaces W of V such that

$$\Gamma_\beta W \subseteq W \ (\beta = 1, \cdots, \nu), \quad \Pi_\alpha A W \subseteq AW \ (\alpha = 1, \cdots, \mu), \quad \ker A \subseteq W. \tag{4.126}$$

Theorem 4.8.7. *For an $m \times n$ partitioned matrix A, the rank formula (4.125) holds true if and only if $\mathcal{L}(A, \Pi, \Gamma) \neq \emptyset$. If $\mathcal{L}(A, \Pi, \Gamma) \neq \emptyset$, then $\mathcal{L}_{\min}(p_\mathrm{PE}) = \mathcal{L}(A, \Pi, \Gamma)$.*

Proof. For $W \in \mathcal{W}$ we have (cf. (4.122))

$$\begin{aligned} p_{\mathrm{PE}}(W) &= \sum_{\alpha=1}^{\mu} \dim(\Pi_\alpha AW) - \dim W \\ &\geq \dim(AW) - \dim W \geq -\dim(\ker A) = \operatorname{rank} A - n. \end{aligned}$$

It then follows that $p_{\mathrm{PE}}(W) + n = \operatorname{rank} A$ if and only if

$$\sum_{\alpha=1}^{\mu} \dim(\Pi_\alpha AW) = \dim(AW) = \dim W - \dim(\ker A).$$

The first equality here is equivalent to $\Pi_\alpha AW \subseteq AW$ ($\alpha = 1, \cdots, \mu$), and the second to $\ker A \subseteq W$. Therefore, the rank formula (4.125) holds true if and only if $\mathcal{L}(A, \Pi, \Gamma) \neq \emptyset$. The final claim is obvious from the above argument. ■

Remark 4.8.8. Theorems 4.8.6 and 4.8.7 imply that the nonemptyness of $\mathcal{L}(A, \Pi, \Gamma)$ is necessary and sufficient for the existence of a proper block-triangular form under PE-transformations. It may be said that $\mathcal{L}(A, \Pi, \Gamma)$ captures the geometric aspect of proper block-triangularization more directly, while $\mathcal{L}_{\min}(p_{\mathrm{PE}})$ gives a combinatorial representation of the same lattice. □

A partitioned matrix A of full rank (i.e., $\operatorname{rank} A = \min(m, n)$) has a proper block-triangular form by Theorem 4.8.6 and the following fact.

Proposition 4.8.9. *The rank formula* (4.125) *holds true for a partitioned matrix of full rank.*

Proof. This follows from $p_{\mathrm{PE}}(0) = 0$, $p_{\mathrm{PE}}(V) \leq m - n$ and the first inequality in Proposition 4.8.3. ■

A partitioned matrix A admitting a proper block-triangularization is called *PE-reducible* if it can be transformed into a proper block-triangular form with two or more nonempty blocks by a PE-transformation; otherwise it is called *PE-irreducible.* Note that PE-reducibility or PE-irreducibility is defined only if A possesses a proper block-triangular form.

Proposition 4.8.10.

(1) *A PE-irreducible partitioned matrix is of full rank.*

(2) *A partitioned matrix of full rank is PE-irreducible if and only if*

$$\mathcal{L}_{\min}(p_{\mathrm{PE}}) = \begin{cases} \{V\} & (\text{if } m < n) \\ \{0, V\} & (\text{if } m = n) \\ \{0\} & (\text{if } m > n) \end{cases}.$$

Proof. (1) and the "only if" part of (2) follow from the proof of Theorem 4.8.6. For the "if" part of (2), suppose that A can be brought to a proper block-triangular matrix $\tilde{A}$ with two or more nonempty blocks by a PE-transformation. Then there exists $\emptyset \neq I \subseteq \mathrm{Row}(\tilde{A})$ and $\emptyset \neq J \subseteq \mathrm{Col}(\tilde{A})$ such that $\mathrm{rank}\,\tilde{A}[I,J] = 0$, $\mathrm{rank}\,\tilde{A}[\mathrm{Row}(\tilde{A}) \setminus I, J] = |\mathrm{Row}(\tilde{A}) \setminus I|$, and $\mathrm{rank}\,\tilde{A}[I, \mathrm{Col}(\tilde{A}) \setminus J] = |\mathrm{Col}(\tilde{A}) \setminus J|$. The subspace $W \in \mathcal{W}$ that corresponds to J (as in (4.121)) satisfies $p_{\mathrm{PE}}(W) = \mathrm{rank}\,A - n$. Furthermore, $W \neq V$ if $m \leq n$ and $W \neq 0$ if $m \geq n$. ∎

If $\tilde{A}$ is a proper block-triangular matrix obtained from A by a PE-transformation and if, in addition, all the diagonal blocks $\tilde{A}[R_k, C_k]$ for $k = 0, 1, \cdots, b, \infty$ are PE-irreducible, we say that $\tilde{A}$ is a *PE-irreducible decomposition* of A, whereas the diagonal blocks $\tilde{A}[R_k, C_k]$ $(k = 0, 1, \cdots, b, \infty)$ are called the *PE-irreducible components* of A. The matrix $\tilde{A}$ constructed in the proof of Theorem 4.8.6 is a PE-irreducible decomposition due to the maximality of the chain $\mathcal{C}$. The PE-irreducible components of a partitioned matrix are uniquely determined up to PE-transformations, as follows.

Theorem 4.8.11. *The set of PE-irreducible components of a partitioned matrix is unique to within PE-transformations of each component.*

Proof. The proof relies on a module-theoretic argument, in particular, on the Jordan–Hölder theorem for modules. See Ito–Iwata–Murota [138] for details. ∎

4.8.3 Partial Order Among Blocks

For a block-triangular matrix in general a partial order is defined among the blocks by the zero/nonzero structure of the off-diagonal blocks. Unlike the CCF of LM-matrices, the partial order among the blocks is not uniquely determined for partitioned matrices. Recall, by contrast, that the CCF gives a unique decomposition of an LM-matrix that is finest not only with respect to the partition into blocks but also with respect to the partial order among the blocks. Mathematically, the nonuniqueness of the partial order for partitioned matrices is ascribed to the nondistributivity of the lattice $\mathcal{L}_{\min}(p_{\mathrm{PE}})$ of the minimizers of the PE-surplus function p_{PE}, whereas the uniqueness for LM-matrices is due to the distributivity of the lattice $\mathcal{L}_{\min}(p)$ of the minimizers of the LM-surplus function p.

Let A be a partitioned matrix, as in §4.8.1, and $\tilde{A} = {S_{\mathrm{r}}}^{-1} A S_{\mathrm{c}}$ be a proper block-triangular matrix obtained from A by a PE-transformation. Denote by $(R_0; R_1, \cdots, R_b; R_\infty)$ and $(C_0; C_1, \cdots, C_b; C_\infty)$ the partitions of $R = \mathrm{Row}(\tilde{A})$ and $C = \mathrm{Col}(\tilde{A})$, respectively. The partial order $\preceq$ defined among the blocks is the reflexive and transitive closure of the relation given by: $C_k \preceq C_l$ if $\tilde{A}[R_k, C_l] \neq O$ with the convention (2.15). We denote this partially ordered set $(\{C_0; C_1, \cdots, C_b; C_\infty\}, \preceq)$ by $\mathcal{P}(\tilde{A})$. The order ideals of $\mathcal{P}(\tilde{A})$ constitute a distributive sublattice of 2^C, which we denote by $\mathcal{D}(\tilde{A})$; namely, $\mathcal{D}(\tilde{A}) = \mathcal{L}(\mathcal{P})$ for $\mathcal{P} = \mathcal{P}(\tilde{A})$ in the notation of (2.27).

A subset J of $\mathrm{Col}(\tilde{A})$ can be naturally identified with a subspace of V, on which the given A acts. We denote this subspace by $\psi(J, S_{\mathrm{c}})$, i.e.,

$$\psi(J, S_{\mathrm{c}}) = \mathrm{span}\{S_{\mathrm{c}}(0, \cdots, 0, \overset{\overset{j}{\vee}}{1}, 0, \cdots, 0)^{\mathrm{T}} \mid j \in J\}. \tag{4.127}$$

Then

$$\psi(J_1 \cup J_2, S_{\mathrm{c}}) = \psi(J_1, S_{\mathrm{c}}) + \psi(J_2, S_{\mathrm{c}}), \quad \psi(J_1 \cap J_2, S_{\mathrm{c}}) = \psi(J_1, S_{\mathrm{c}}) \cap \psi(J_2, S_{\mathrm{c}}),$$

and hence

$$\psi(\mathcal{D}(\tilde{A}), S_{\mathrm{c}}) = \{\psi(J, S_{\mathrm{c}}) \mid J \in \mathcal{D}(\tilde{A})\}$$

is a distributive sublattice of the modular lattice formed by the subspaces of V.

Proposition 4.8.12. *If $\tilde{A} = S_{\mathrm{r}}^{-1} A S_{\mathrm{c}}$ is a proper block-triangular matrix obtained from a partitioned matrix A by a PE-transformation, then $\psi(\mathcal{D}(\tilde{A}), S_{\mathrm{c}})$ is a sublattice of $\mathcal{L}_{\min}(p_{\mathrm{PE}}) = \mathcal{L}(A, \Pi, \Gamma)$.*

Proof. For $J \in \mathcal{D}(\tilde{A})$ put $W = \psi(J, S_{\mathrm{c}})$. Since $\tilde{A} = S_{\mathrm{r}}^{-1} A S_{\mathrm{c}}$ is a PE-transformation, we have $\Gamma_\beta W \subseteq W$ for $\beta = 1, \cdots, \nu$. It follows from the definition of a proper block-triangular form that $\ker A \subseteq W$ and from the definition of the partial order that $\Pi_\alpha A W \subseteq A W$ for $\alpha = 1, \cdots, \mu$. Hence $W \in \mathcal{L}(A, \Pi, \Gamma)$. ■

Suppose we have two proper block-triangular matrices, $\tilde{A} = S_{\mathrm{r}}^{-1} A S_{\mathrm{c}}$ and $\tilde{A}' = S_{\mathrm{r}}'^{-1} A S_{\mathrm{c}}'$, obtained from A by PE-transformations. We say that $\tilde{A}$ is a *finer decomposition* than $\tilde{A}'$ if $\psi(\mathcal{D}(\tilde{A}'), S_{\mathrm{c}}')$ is a proper sublattice of $\psi(\mathcal{D}(\tilde{A}), S_{\mathrm{c}})$. Furthermore, we say that $\tilde{A}$ is a *finest-possible decomposition* of A if there exists no proper block-triangular matrix $\tilde{A}'$ which is obtained from A by a PE-transformation and is finer than $\tilde{A}$.

Theorem 4.8.13. *Suppose that a partitioned matrix A has a proper block-triangular form under PE-transformations. Then $\tilde{A} = S_{\mathrm{r}}^{-1} A S_{\mathrm{c}}$ is a finest-possible decomposition of A if and only if $\psi(\mathcal{D}(\tilde{A}), S_{\mathrm{c}})$ is a maximal distributive sublattice of $\mathcal{L}_{\min}(p_{\mathrm{PE}}) = \mathcal{L}(A, \Pi, \Gamma)$.*

Proof. This follows from Proposition 4.8.12 and Lemma 4.8.14 below. ■

Lemma 4.8.14. *For any distributive sublattice $\mathcal{D}'$ of $\mathcal{L}(A, \Pi, \Gamma)$ there exists a PE-transformation $\tilde{A} = S_{\mathrm{r}}^{-1} A S_{\mathrm{c}}$ such that $\psi(\mathcal{D}(\tilde{A}), S_{\mathrm{c}}) \supseteq \mathcal{D}'$.*

Proof. According to Birkhoff's representation theorem, the distributive lattice $\mathcal{D}'$ can be represented by a partially ordered set $\mathcal{P}' = (\{Z_1, \cdots, Z_b\}, \subseteq)$

consisting of the join-irreducible elements of $\mathcal{D}'$ except the minimum element of $\mathcal{D}'$ (see Remark 2.2.6). We assume that $k \le l$ if $Z_k \subseteq Z_l$, and put $Z_0 = \min \mathcal{D}'$. For each $\beta = 1, \cdots, \nu$, let $\hat{\Psi}_{0\beta}$ be a set of basis vectors of $Z_0 \cap V_\beta$. For $k = 1, \cdots, b$ inductively define $\hat{\Psi}_{k\beta}$ to be a set of vectors such that $\hat{\Psi}_{k\beta} \cup \left(\bigcup_{Z_l \subset Z_k} \hat{\Psi}_{l\beta} \right)$ is a basis of $Z_k \cap V_\beta$, and finally let $\hat{\Psi}_{\infty\beta}$ be such that $\Psi_\beta = \hat{\Psi}_{\infty\beta} \cup \left(\bigcup_{k=0}^{b} \hat{\Psi}_{k\beta} \right)$ is a basis of V_β. Then define the matrix $S_{\mathrm{c}\beta}$ from the column vectors of Ψ_β for $\beta = 1, \cdots, \nu$, and put $S_\mathrm{c} = \bigoplus_{\beta=1}^{\nu} S_{\mathrm{c}\beta}$. The other transformation matrix $S_\mathrm{r} = \bigoplus_{\alpha=1}^{\mu} S_{\mathrm{r}\alpha}$ should be constructed similarly from the basis vectors $\{\hat{\Phi}_{k\alpha} \mid k = 0, 1, \cdots, b, \infty; \alpha = 1, \cdots, \mu\}$ compatible with $\Pi_\alpha A Z_k$ for $k = 0, 1, \cdots, b$ and $\alpha = 1, \cdots, \mu$. We claim that

$$\tilde{A}[R_k, C_l] = O \quad \text{unless} \quad Z_k \subseteq Z_l.$$

This is because $Z_l \in \mathcal{D}' \subseteq \mathcal{L}(A, \Pi, \Gamma)$ implies $\Pi_\alpha A Z_l \subseteq A Z_l$ $(\alpha = 1, \cdots, \mu)$, where $\Pi_\alpha A Z_l$ is spanned by $\bigcup_{Z_k \subseteq Z_l} \hat{\Phi}_{k\alpha}$. The claim in turn implies that $C_k \preceq C_l \Rightarrow Z_k \subseteq Z_l$. Hence $\psi(\mathcal{D}(\tilde{A}), S_\mathrm{c}) \supseteq \mathcal{D}'$. ■

An instance of nonunique partial order will be given in Example 4.8.24.

4.8.4 Generic Partitioned Matrix

We have seen that not every partitioned matrix admits a proper block-triangularization. In this section we introduce a certain genericity assumption on the nonzero entries with a view to identifying a subclass of partitioned matrices for which the proper block-triangularization exists.

We consider a partitioned matrix A that is generic in the following sense. Let $\boldsymbol{K}$ be a subfield of a field $\boldsymbol{F}$, $A^\natural_{\alpha\beta}$ be an $m_\alpha \times n_\beta$ matrix over $\boldsymbol{K}$ for $\alpha = 1, \cdots, \mu$ and $\beta = 1, \cdots, \nu$, and $\mathcal{T} = \{t_{\alpha\beta} \in \boldsymbol{F} \mid \alpha = 1, \cdots, \mu;\ \beta = 1, \cdots, \nu\}$ be algebraically independent over $\boldsymbol{K}$. Then

$$A = \begin{bmatrix} A_{11} & A_{12} & \cdots & A_{1\nu} \\ A_{21} & A_{22} & \cdots & A_{2\nu} \\ \vdots & \vdots & \ddots & \vdots \\ A_{\mu 1} & A_{\mu 2} & \cdots & A_{\mu\nu} \end{bmatrix} = \begin{bmatrix} t_{11}A^\natural_{11} & t_{12}A^\natural_{12} & \cdots & t_{1\nu}A^\natural_{1\nu} \\ t_{21}A^\natural_{21} & t_{22}A^\natural_{22} & \cdots & t_{2\nu}A^\natural_{2\nu} \\ \vdots & \vdots & \ddots & \vdots \\ t_{\mu 1}A^\natural_{\mu 1} & t_{\mu 2}A^\natural_{\mu 2} & \cdots & t_{\mu\nu}A^\natural_{\mu\nu} \end{bmatrix}$$

is a partitioned matrix over the field $\boldsymbol{F}$, where $A_{\alpha\beta} = t_{\alpha\beta}A^\natural_{\alpha\beta}$ for $\alpha = 1, \cdots, \mu$ and $\beta = 1, \cdots, \nu$. Such a matrix A is named a *generic partitioned matrix* (or *GP-matrix* for short) of type $(m_1, \cdots, m_\mu; n_1, \cdots, n_\nu)$ with ground field $\boldsymbol{K}$. A GP-matrix of type $(1, \cdots, 1; 1, \cdots, 1)$ is nothing but a generic matrix. A GP-matrix is called a *GP(2)-matrix* if $m_\alpha \le 2$ for $\alpha = 1, \cdots, \mu$ and $n_\beta \le 2$ for $\beta = 1, \cdots, \nu$.

For a generic partitioned matrix it is natural to assume that the matrices S_r and S_c in the PE-transformation (4.114) are nonsingular matrices over the ground field $\boldsymbol{K}$. We call such a transformation a *GP-transformation.* It is emphasized that a GP-transformation preserves not only the partition structure but the genericity in the above sense, and therefore the resulting matrix remains a generic partitioned matrix.

A generic partitioned matrix A admitting a proper block-triangularization under GP-transformations is called *GP-reducible* if it can be transformed into a proper block-triangular form with two or more nonempty blocks by a GP-transformation; otherwise it is called *GP-irreducible.* Note that GP-reducibility or GP-irreducibility is defined only if A possesses a proper block-triangular form.

The main objective of this section is to show that the proper block-triangularization is possible for GP(2)-matrices, whereas this is not the case with generic partitioned matrices of general type.

Example 4.8.15. The basic concepts introduced above are illustrated here. Consider a 6×6 matrix

$$A = \left[\begin{array}{cc|cc|cc} 2t_{11} & t_{11} & t_{12} & t_{12} & t_{13} & t_{13} \\ & & & & t_{13} & \\ \hline t_{21} & t_{21} & t_{22} & & t_{23} & \\ & & & & t_{23} & \\ \hline t_{31} & t_{31} & t_{32} & -t_{32} & & \\ & & & t_{32} & t_{33} & \end{array}\right]$$

which is a GP(2)-matrix of type $(2,2,2;2,2,2)$ with $\boldsymbol{K} = \mathbf{Q}$. Using admissible transformation matrices:

$$S_{\mathrm{r}} = \left[\begin{array}{cc|cc|cc} 1 & 0 & & & & \\ 0 & 1 & & & & \\ \hline & & 1 & 0 & & \\ & & 0 & 1 & & \\ \hline & & & & 1 & -1 \\ & & & & 0 & 1 \end{array}\right], \qquad S_{\mathrm{c}} = \left[\begin{array}{cc|cc|cc} 1 & 0 & & & & \\ -1 & 1 & & & & \\ \hline & & 1 & 0 & & \\ & & 0 & 1 & & \\ \hline & & & & 1 & 0 \\ & & & & 0 & 1 \end{array}\right]$$

we obtain

$$\tilde{A} = {S_{\mathrm{r}}}^{-1} A S_{\mathrm{c}} = \left[\begin{array}{cc|cc|cc} t_{11} & t_{11} & t_{12} & t_{12} & t_{13} & t_{13} \\ & & & & t_{13} & \\ \hline & t_{21} & t_{22} & & t_{23} & \\ & & & & t_{23} & \\ \hline & t_{31} & t_{32} & & t_{33} & \\ & & & t_{32} & t_{33} & \end{array}\right],$$

which can be put into an explicit upper block-triangular form:

$$\bar{A} = P_{\mathrm{r}} \tilde{A} P_{\mathrm{c}} = \left[\begin{array}{cc|cccc} t_{13} & t_{11} & t_{11} & t_{12} & t_{12} & t_{13} \\ \hline & & t_{21} & t_{22} & & t_{23} \\ & & t_{31} & t_{32} & & t_{33} \\ & & & & t_{32} & t_{33} \\ & & & & & t_{13} \\ & & & & & t_{23} \end{array}\right]$$

with permutation matrices

$$P_{\mathrm{r}} = \begin{bmatrix} 1&0&0&0&0&0 \\ 0&0&1&0&0&0 \\ 0&0&0&0&1&0 \\ 0&0&0&0&0&1 \\ 0&1&0&0&0&0 \\ 0&0&0&1&0&0 \end{bmatrix}, \qquad P_{\mathrm{c}} = \begin{bmatrix} 0&1&0&0&0&0 \\ 0&0&1&0&0&0 \\ 0&0&0&1&0&0 \\ 0&0&0&0&1&0 \\ 0&0&0&0&0&1 \\ 1&0&0&0&0&0 \end{bmatrix}.$$

The matrix $\bar{A}$ is a proper block-triangular matrix (in an explicit block-triangular form), giving a GP-irreducible decomposition with the horizontal tail $\bar{A}[R_0, C_0] = [\, t_{13}\ t_{11} \,]$, the vertical tail $\bar{A}[R_\infty, C_\infty] = \begin{bmatrix} t_{13} \\ t_{23} \end{bmatrix}$, and two square diagonal blocks $\bar{A}[R_1, C_1] = \begin{bmatrix} t_{21} & t_{22} \\ t_{31} & t_{32} \end{bmatrix}$ and $\bar{A}[R_2, C_2] = [t_{32}]$. □

Compatibly with the restriction of PE-transformations to GP-transformations, namely, the restriction to transformations over $\boldsymbol{K}$, we introduce a variant of the PE-surplus function p_{PE} as follows. Let $U^\circ \cong \boldsymbol{K}^m$ and $V^\circ \cong \boldsymbol{K}^n$ be $\boldsymbol{K}$-linear spaces which are, respectively, expressed as direct sums of lower dimensional component spaces:

$$U^\circ = \bigoplus_{\alpha=1}^{\mu} U^\circ_\alpha, \qquad \dim_{\boldsymbol{K}} U^\circ_\alpha = m_\alpha \quad (\alpha = 1, \cdots, \mu),$$

$$V^\circ = \bigoplus_{\beta=1}^{\nu} V^\circ_\beta, \qquad \dim_{\boldsymbol{K}} V^\circ_\beta = n_\beta \quad (\beta = 1, \cdots, \nu).$$

Then we have the relations (4.117) and (4.118) for $U = U^\circ \otimes_{\boldsymbol{K}} \boldsymbol{F}$, $V = V^\circ \otimes_{\boldsymbol{K}} \boldsymbol{F}$, $U_\alpha = U^\circ_\alpha \otimes_{\boldsymbol{K}} \boldsymbol{F}$ and $V_\beta = V^\circ_\beta \otimes_{\boldsymbol{K}} \boldsymbol{F}$, where it should be clear that $U^\circ \otimes_{\boldsymbol{K}} \boldsymbol{F}$, for example, denotes the linear space obtained from U° ($\cong \boldsymbol{K}^m$) by extending the base field to $\boldsymbol{F}$, and hence $U^\circ \otimes_{\boldsymbol{K}} \boldsymbol{F} \cong \boldsymbol{F}^m$. We denote by $\mathcal{W}^\circ$ the family of all the subspaces of V that can be generated by a direct sum of subspaces of V°_β's, i.e.,

$$\mathcal{W}^\circ = \{ W^\circ \otimes_{\boldsymbol{K}} \boldsymbol{F} \mid W^\circ\text{: subspace of } V^\circ,\ \Gamma_\beta W^\circ \subseteq W^\circ \ (\beta = 1, \cdots, \nu) \}. \tag{4.128}$$

Similarly, we denote by $\mathcal{Y}^\circ$ the family of all the subspaces of U that can be generated by a direct sum of subspaces of U°_α's. Then both $\mathcal{W}^\circ$ and $\mathcal{Y}^\circ$ form a modular lattice. We define $p_{\mathrm{GP}} : \mathcal{W}^\circ \to \mathbf{Z}$ and $\lambda : \mathcal{Y}^\circ \times \mathcal{W}^\circ \to \mathbf{Z}$ by

$$p_{\mathrm{GP}}(W) = \sum_{\alpha=1}^{\mu} \dim(A_\alpha W) - \dim W, \qquad W \in \mathcal{W}^\circ, \tag{4.129}$$

$$\lambda(Y, W) = \dim(AW/Y) = \dim(AW) - \dim(AW \cap Y), \qquad Y \in \mathcal{Y}^\circ, W \in \mathcal{W}^\circ, \tag{4.130}$$

and call them the *GP-surplus function* and the *GP-birank function*, respectively. Note that p_{GP} is the restriction of $p_{\mathrm{PE}} : \mathcal{W} \to \mathbf{Z}$ to $\mathcal{W}^\circ$, and hence, by Lemma 4.8.2, p_{GP} is submodular on $\mathcal{W}^\circ$.

A combination of Proposition 4.8.3 and Theorem 4.8.6 can be adapted as follows.

Proposition 4.8.16. *For an $m \times n$ generic partitioned matrix A,*

$$\mathrm{rank}\, A \le \min\{p_{\mathrm{GP}}(W) \mid W \in \mathcal{W}^\circ\} + n \le \text{term-rank}\, A,$$

and a proper block-triangular matrix can be obtained by a GP-transformation if and only if

$$\mathrm{rank}\, A = \min\{p_{\mathrm{GP}}(W) \mid W \in \mathcal{W}^\circ\} + n. \tag{4.131}$$

Proof. The proof is similar to those of Proposition 4.8.3 and Theorem 4.8.6. ■

Next, we have the following lemmas, the latter of which should be compared with Theorem 2.3.47.

Lemma 4.8.17. *For a generic partitioned matrix A the function $\lambda : \mathcal{Y}^\circ \times \mathcal{W}^\circ \to \mathbf{Z}$ is submodular, i.e.,*

$$\lambda(Y_1, W_1) + \lambda(Y_2, W_2) \ge \lambda(Y_1 + Y_2, W_1 + W_2) + \lambda(Y_1 \cap Y_2, W_1 \cap W_2)$$

for $Y_i \in \mathcal{Y}^\circ$, $W_i \in \mathcal{W}^\circ$ $(i = 1, 2)$.

Proof. It is clear from the following calculation:

$$\begin{aligned}
&\lambda(Y_1, W_1) + \lambda(Y_2, W_2)\\
&= \dim(AW_1) + \dim(AW_2) - \dim(AW_1 \cap Y_1) - \dim(AW_2 \cap Y_2)\\
&= \dim(A(W_1 + W_2)) + \dim((AW_1 \cap AW_2)/(Y_1 \cap Y_2))\\
&\quad - \dim((AW_1 \cap Y_1) + (AW_2 \cap Y_2))\\
&\ge \dim(A(W_1 + W_2)) + \dim(A(W_1 \cap W_2)/(Y_1 \cap Y_2))\\
&\quad - \dim(A(W_1 + W_2) \cap (Y_1 + Y_2))\\
&= \lambda(Y_1 + Y_2, W_1 + W_2) + \lambda(Y_1 \cap Y_2, W_1 \cap W_2).
\end{aligned}$$

■

Lemma 4.8.18. *For an $m \times n$ generic partitioned matrix A, there exists a pair $Y^* \in \mathcal{Y}^\circ$ and $W^* \in \mathcal{W}^\circ$ such that*

(i) $\dim W^* - \dim Y^* - \lambda(Y^*, W^*) = n - \mathrm{rank}\, A$, *and*

(ii) $\lambda(Y', W') = \lambda(Y^*, W^*)$ *for any* $Y' \supset Y^*$ *and* $W' \subset W^*$ *such that* $\dim Y' = \dim Y^* + 1$ *and* $\dim W' = \dim W^* - 1$.

Proof. Consider a pair (Y^*, W^*) which minimizes $\dim W^* - \dim Y^*$ subject to (i). Such (Y^*, W^*) certainly exists since (i) is satisfied by $(\{\mathbf{0}\}, V)$. Then for any $Y' \supset Y^*$ and $W' \subset W^*$ such that $\dim Y' = \dim Y^* + 1$ and $\dim W' = \dim W^* - 1$, it follows from Lemma 4.8.17 that

$$\lambda(Y^*, W^*) + \lambda(Y', W') \geq \lambda(Y', W^*) + \lambda(Y^*, W').$$

Because of the minimality of $\dim W^* - \dim Y^*$ we have $\lambda(Y', W^*) = \lambda(Y^*, W^*)$ since otherwise (Y', W^*) would satisfy (i) with $\dim W^* - \dim Y' < \dim W^* - \dim Y^*$. Likewise we have $\lambda(Y^*, W') = \lambda(Y^*, W^*)$. Therefore $\lambda(Y', W') \geq \lambda(Y^*, W^*)$. On the other hand, it is clear that $\lambda(Y', W') \leq \lambda(Y^*, W^*)$ since $Y' \supset Y^*$ and $W' \subset W^*$. Hence (Y^*, W^*) satisfies (ii). ∎

Whereas the above two lemmas are valid for generic partitioned matrices in general (even the genericity is irrelevant), the following theorem states a key property valid for GP(2)-matrices (and not for generic partitioned matrices of general type). We call this the *König–Egerváry theorem for GP(2)-matrices*, since for a generic matrix it reduces to the König–Egerváry theorem for bipartite graphs.

Theorem 4.8.19. *For an $m \times n$ GP(2)-matrix A, there exists a pair $Y^* \in \mathcal{Y}^\circ$ and $W^* \in \mathcal{W}^\circ$ such that*

(i) $\dim W^* - \dim Y^* = n - \operatorname{rank} A$, *and*

(ii) $\lambda(Y^*, W^*) = 0$.

In other words, there exists a GP-transformation $\tilde{A} = S_{\mathrm{r}}{}^{-1} A S_{\mathrm{c}}$ and subsets $R^* \subseteq \mathrm{Row}(\tilde{A})$ and $C^* \subseteq \mathrm{Col}(\tilde{A})$ such that

(i′) $|R^*| + |C^*| = m + n - \operatorname{rank} A$, *and*

(ii″) $\operatorname{rank} \tilde{A}[R^*, C^*] = 0$.

Proof. Given a pair (Y^*, W^*) of Lemma 4.8.18, consider a GP-transformation $\tilde{A} = S_{\mathrm{r}}{}^{-1} A S_{\mathrm{c}}$ such that a subset of the column vectors of S_{r} spans Y^* and a subset of the column vectors of S_{c} spans W^*. We denote by R^* the complement of the subset of $\mathrm{Row}(\tilde{A})$ corresponding to Y^* and by C^* the subset of $\mathrm{Col}(\tilde{A})$ corresponding to W^*. Note that $\mathrm{Row}(\tilde{A})$ and $\mathrm{Col}(\tilde{A})$ have natural one-to-one correspondences with $\mathrm{Col}(S_{\mathrm{r}})$ and $\mathrm{Col}(S_{\mathrm{c}})$, respectively, and that $|R^*| = m - \dim Y^*$, $|C^*| = \dim W^*$ and $\lambda(Y^*, W^*) = \operatorname{rank} \tilde{A}[R^*, C^*]$.

We claim that $\operatorname{rank} \tilde{A}_{\alpha\beta}[R^*, C^*] \neq 1$ for each (α, β), where $\tilde{A}_{\alpha\beta}[R^*, C^*]$ is a short-hand notation for $\tilde{A}_{\alpha\beta}[R^* \cap \mathrm{Row}(\tilde{A}_{\alpha\beta}), C^* \cap \mathrm{Col}(\tilde{A}_{\alpha\beta})]$. Assume, to the contrary, that $\tilde{A}_{\alpha\beta}[R^*, C^*]$ has rank 1 for some (α, β). We may further assume that $\tilde{A}_{\alpha\beta}[R^*, C^*]$ is in the rank normal form, i.e.,

$$\begin{pmatrix} t_{\alpha\beta} & 0 \\ 0 & 0 \end{pmatrix}, \quad \begin{pmatrix} t_{\alpha\beta} \\ 0 \end{pmatrix}, \quad \begin{pmatrix} t_{\alpha\beta} & 0 \end{pmatrix}, \quad \text{or} \quad \begin{pmatrix} t_{\alpha\beta} \end{pmatrix},$$

and that $\tilde{A}_{\alpha\beta}[R^*, C^*]$ has the only nonzero element at (i, j)-entry of $\tilde{A}$. Let $\tilde{A}[I^*, J^*]$ be a maximum-sized nonsingular submatrix of $\tilde{A}[R^* \setminus \{i\}, C^* \setminus \{j\}]$, and then $\tilde{A}[I^* \cup \{i\}, J^* \cup \{j\}]$ is nonsingular since the nonzero terms arising from $t_{\alpha\beta} \det \tilde{A}[I^*, J^*]$ would not vanish in the determinant expansion of $\tilde{A}[I^* \cup \{i\}, J^* \cup \{j\}]$ because of the genericity. Therefore we have

$$\operatorname{rank} \tilde{A}[R^*, C^*] > \operatorname{rank} \tilde{A}[R^* \setminus \{i\}, C^* \setminus \{j\}],$$

which contradicts the condition (ii) of Lemma 4.8.18. Hence $\operatorname{rank} \tilde{A}_{\alpha\beta}[R^*, C^*]$ is 0 or 2.

Consider a generic matrix $B = (b_{\alpha\beta})$ with $\operatorname{Row}(B) = \{1, \cdots, \mu\}$ and $\operatorname{Col}(B) = \{1, \cdots, \nu\}$ defined by

$$b_{\alpha\beta} = \begin{cases} t_{\alpha\beta} & \text{if} \quad \operatorname{rank} \tilde{A}_{\alpha\beta}[R^*, C^*] = 2, \\ 0 & \text{if} \quad \operatorname{rank} \tilde{A}_{\alpha\beta}[R^*, C^*] = 0. \end{cases}$$

Note the correspondence between the entry $b_{\alpha\beta}$ of B and the submatrix $A_{\alpha\beta}$ of A. The DM-decomposition of B splits $\overline{R} = \operatorname{Row}(B)$ and $\overline{C} = \operatorname{Col}(B)$ into blocks $(\overline{R}_0; \overline{R}_1, \cdots, \overline{R}_b; \overline{R}_\infty)$ and $(\overline{C}_0; \overline{C}_1, \cdots, \overline{C}_b; \overline{C}_\infty)$, respectively. Accordingly, R^* and C^* are split into blocks $(R^*_0; R^*_1, \cdots, R^*_b; R^*_\infty)$ and $(C^*_0; C^*_1, \cdots, C^*_b; C^*_\infty)$, respectively. Since $\operatorname{rank} \tilde{A}_{\alpha\beta}[R^*, C^*]$ is either 2 or 0, it follows from the genericity that

$$\operatorname{rank} \tilde{A}[R^*, C^*] = 2 \operatorname{rank} B.$$

Moreover, $\tilde{A}[R^*, C^*]$ is in a proper block-triangular form with respect to the blocks $(R^*_0; R^*_1, \cdots, R^*_b; R^*_\infty)$ and $(C^*_0; C^*_1, \cdots, C^*_b; C^*_\infty)$. For any $i \in R^* \setminus R^*_\infty$, we would have

$$\operatorname{rank} \tilde{A}[R^* \setminus \{i\}, C^*] < \operatorname{rank} \tilde{A}[R^*, C^*],$$

which contradicts the condition (ii) in Lemma 4.8.18. Similarly, for any $j \in C^* \setminus C^*_0$, we would have

$$\operatorname{rank} \tilde{A}[R^*, C^* \setminus \{j\}] < \operatorname{rank} \tilde{A}[R^*, C^*],$$

which also contradicts the condition (ii) in Lemma 4.8.18. Therefore $R^* = R^*_\infty$ and $C^* = C^*_0$. That is to say, $\tilde{A}[R^*, C^*] = O$, i.e., $\operatorname{rank} \tilde{A}[R^*, C^*] = 0$. ■

We now state the main result of this section, namely the rank identity for GP(2)-matrices, due to Iwata–Murota [144]. This is an extension of the Hall–Ore theorem for generic matrices.

Theorem 4.8.20. *For an $m \times n$ GP(2)-matrix A,*

$$\operatorname{rank} A = \min\{p_{\mathrm{GP}}(W) \mid W \in \mathcal{W}^\circ\} + n. \tag{4.132}$$

Hence a proper block-triangular form can be obtained by a GP-transformation.

Proof. Let (Y^*, W^*) be the pair of Theorem 4.8.19. From (ii) it follows that $A_\alpha W^* \subseteq Y^* \cap U_\alpha$. Using this and (i) we obtain

$$\begin{aligned}\operatorname{rank} A &= \dim Y^* - \dim W^* + n \\ &\geq \sum_{\alpha=1}^{\mu} \dim(A_\alpha W^*) - \dim W^* + n = p_{\mathrm{GP}}(W^*) + n.\end{aligned}$$

The other direction of the inequality follows from Proposition 4.8.16. ■

Remark 4.8.21. A compilation of Theorem 4.8.20 and the previous results show that the rank identity (4.132) holds for the following types of generic partitioned matrices:

- Generic matrix: $m_\alpha = 1$ for $\alpha = 1, \cdots, \mu$ and $n_\beta = 1$ for $\beta = 1, \cdots, \nu$.
- Multilayered matrix: $n_\beta = 1$ for $\beta = 1, \cdots, \nu$.
- Transposed multilayered matrix: $m_\alpha = 1$ for $\alpha = 1, \cdots, \mu$.
- GP(2)-matrix: $m_\alpha \leq 2$ for $\alpha = 1, \cdots, \mu$ and $n_\beta \leq 2$ for $\beta = 1, \cdots, \nu$.

The second case above is easily seen from Theorem 4.7.6, whereas the third follows from this with Theorem 4.8.6 and an observation that A has a proper block-triangular form if and only if the transpose of A does also. The identity (4.132) is not valid for generic partitioned matrices in general, as illustrated in the following example. □

Example 4.8.22. The identity (4.132) is not valid for generic partitioned matrices in general. Consider a 6×6 generic partitioned matrix of type $(3, 3; 2, 2, 2)$:

$$A = \left[\begin{array}{cc|cc|cc} t_{11} & & t_{12} & & & \\ & t_{11} & & & t_{13} & \\ & & & t_{12} & & t_{13} \\ \hline & & t_{22} & & t_{23} & \\ t_{21} & & & & & t_{23} \\ & t_{21} & & t_{22} & & \end{array}\right].$$

It can be easily verified that $\operatorname{rank} A = 5 < \min p_{\mathrm{GP}} + n = 6$. □

Example 4.8.23. The block-triangular decomposition of a GP(2)-matrix depends on the ground field $\boldsymbol{K}$. Consider, for example, a 4×4 GP(2)-matrix

$$A = \left[\begin{array}{cc|cc} t_{11} & & & t_{12} \\ & 2\,t_{11} & t_{12} & \\ \hline & t_{21} & & t_{22} \\ t_{21} & & t_{22} & \end{array}\right].$$

If regarded as a GP(2)-matrix with the ground field $\boldsymbol{K} = \mathbf{Q}$, A is GP-irreducible. If $\mathbf{R}$ is the ground field $\boldsymbol{K}$, on the other hand, the following block-triangularization of A is obtained:

$$\tilde{A} = S_{\mathrm{r}}^{-1} A S_{\mathrm{c}} = \left[\begin{array}{cc|cc} t_{11} & & -t_{12} & \\ & t_{11} & & t_{12} \\ \hline t_{21} & & \sqrt{2}\, t_{22} & \\ & t_{21} & & \sqrt{2}\, t_{22} \end{array}\right]$$

with

$$S_{\mathrm{r}} = \left[\begin{array}{cc|cc} \sqrt{2} & \sqrt{2} & & \\ -2 & 2 & & \\ \hline & & -1 & 1 \\ & & \sqrt{2} & \sqrt{2} \end{array}\right], \qquad S_{\mathrm{c}} = \left[\begin{array}{cc|cc} \sqrt{2} & \sqrt{2} & & \\ -1 & 1 & & \\ \hline & & 2 & 2 \\ & & -\sqrt{2} & \sqrt{2} \end{array}\right].$$

By using permutation matrices

$$P_{\mathrm{r}} = \begin{bmatrix} 1 & 0 & 0 & 0 \\ 0 & 0 & 1 & 0 \\ 0 & 1 & 0 & 0 \\ 0 & 0 & 0 & 1 \end{bmatrix}, \qquad P_{\mathrm{c}} = \begin{bmatrix} 1 & 0 & 0 & 0 \\ 0 & 0 & 1 & 0 \\ 0 & 1 & 0 & 0 \\ 0 & 0 & 0 & 1 \end{bmatrix},$$

we obtain an explicit upper block-triangular (block-diagonal) matrix

$$\bar{A} = P_{\mathrm{r}} \tilde{A} P_{\mathrm{c}} = \left[\begin{array}{cc|cc} t_{11} & -t_{12} & & \\ t_{21} & \sqrt{2}\, t_{22} & & \\ \hline & & t_{11} & t_{12} \\ & & t_{21} & \sqrt{2}\, t_{22} \end{array}\right].$$

Thus, when $\boldsymbol{K} = \mathbf{R}$, the matrix A is decomposed into a block-triangular (block-diagonal) form with two square blocks and empty tails. □

Example 4.8.24. Unlike the CCF of LM-matrices, the partial order among the blocks is not uniquely determined in the block-triangularization of GP(2)-matrices. The analysis in §4.8.3 for partitioned matrices carries over, mutatis mutandis, to generic partitioned matrices. Here is an illustration by means of an 8×8 GP(2)-matrix

$$A = \left[\begin{array}{cc|cc|cc|cc} t_{11} & & & & & & & t_{14} \\ & 2t_{11} & & & & & t_{14} & \\ \hline & & t_{22} & & t_{23} & -2t_{23} & & \\ & & & t_{22} & & t_{23} & & \\ \hline & & t_{32} & & t_{33} & -2t_{33} & & \\ & & & t_{32} & & t_{33} & & \\ \hline & t_{41} & t_{42} & 2t_{42} & t_{43} & & & t_{44} \\ t_{41} & & 2t_{42} & 4t_{42} & t_{43} & & t_{44} & \end{array}\right]$$

with the ground field $\mathbf{Q}$. For nonsingular matrices

$$
S_{\mathrm{r}} = \left[\begin{array}{cc|cc|cc|cc}
1 & 0 & & & & & & \\
0 & 1 & & & & & & \\ \hline
 & & 1 & 0 & & & & \\
 & & 0 & 1 & & & & \\ \hline
 & & & & 1 & 0 & & \\
 & & & & 0 & 1 & & \\ \hline
 & & & & & & 0 & 1 \\
 & & & & & & 1 & 0
\end{array}\right], \qquad
S_{\mathrm{c}} = \left[\begin{array}{cc|cc|cc|cc}
1 & 0 & & & & & & \\
0 & 1 & & & & & & \\ \hline
 & & 1 & 0 & & & & \\
 & & 0 & 1 & & & & \\ \hline
 & & & & 1 & 2 & & \\
 & & & & 0 & 1 & & \\ \hline
 & & & & & & 0 & 1 \\
 & & & & & & 1 & 0
\end{array}\right]
$$

we have

$$
\tilde{A} = {S_{\mathrm{r}}}^{-1} A S_{\mathrm{c}} = \left[\begin{array}{cc|cc|cc|cc}
t_{11} & & & & & & t_{14} & \\
 & 2t_{11} & & & & & & t_{14} \\ \hline
 & & t_{22} & & t_{23} & & & \\
 & & & t_{22} & & t_{23} & & \\ \hline
 & & t_{32} & & t_{33} & & & \\
 & & & t_{32} & & t_{33} & & \\ \hline
t_{41} & & 2t_{42} & 4t_{42} & t_{43} & 2t_{43} & & t_{44} \\
 & t_{41} & t_{42} & 2t_{42} & t_{43} & 2t_{43} & t_{44} &
\end{array}\right].
$$

With suitable permutation matrices P_{r} and P_{c}, we obtain an explicit upper block-triangular form

$$
\bar{A} = P_{\mathrm{r}} \tilde{A} P_{\mathrm{c}} = \left[\begin{array}{cccc|cc|cc}
t_{11} & & t_{14} & & & & & \\
 & 2t_{11} & & t_{14} & & & & \\
t_{41} & & & t_{44} & 2t_{42} & t_{43} & 4t_{42} & 2t_{43} \\
 & t_{41} & t_{44} & & t_{42} & t_{43} & 2t_{42} & 2t_{43} \\ \hline
 & & & & t_{22} & t_{23} & & \\
 & & & & t_{32} & t_{33} & & \\ \hline
 & & & & & & t_{22} & t_{23} \\
 & & & & & & t_{32} & t_{33}
\end{array}\right].
$$

Thus $\bar{A}$ is a GP-irreducible decomposition of A with empty tails and square diagonal blocks $\bar{A}[R_1, C_1] = \left[\begin{array}{cc|cc} t_{11} & & t_{14} & \\ & 2t_{11} & & t_{14} \\ \hline t_{41} & & & t_{44} \\ & t_{41} & t_{44} & \end{array}\right]$, $\bar{A}[R_2, C_2] = \left[\begin{array}{c|c} t_{22} & t_{23} \\ \hline t_{32} & t_{33} \end{array}\right]$, and $\bar{A}[R_3, C_3] = \left[\begin{array}{c|c} t_{22} & t_{23} \\ \hline t_{32} & t_{33} \end{array}\right]$. For another pair of transformation matrices

$$
S'_{\mathrm{r}} = \left[\begin{array}{cc|cc|cc|cc}
1 & 0 & & & & & & \\
0 & 1 & & & & & & \\ \hline
 & & 1 & -2 & & & & \\
 & & 0 & 1 & & & & \\ \hline
 & & & & 1 & -2 & & \\
 & & & & 0 & 1 & & \\ \hline
 & & & & & & 1 & 0 \\
 & & & & & & 0 & 1
\end{array}\right], \qquad
S'_{\mathrm{c}} = \left[\begin{array}{cc|cc|cc|cc}
1 & 0 & & & & & & \\
0 & 1 & & & & & & \\ \hline
 & & 1 & -2 & & & & \\
 & & 0 & 1 & & & & \\ \hline
 & & & & 1 & 0 & & \\
 & & & & 0 & 1 & & \\ \hline
 & & & & & & 1 & 0 \\
 & & & & & & 0 & 1
\end{array}\right]
$$

we have

$$
\tilde{A}' = S_{\mathrm{r}}'^{-1} A S_{\mathrm{c}}' = \left[\begin{array}{cc|cc|cc|cc}
t_{11} & & & & & & & t_{14} \\
 & 2t_{11} & & & & & t_{14} & \\ \hline
 & & t_{22} & & t_{23} & & & \\
 & & & t_{22} & & t_{23} & & \\ \hline
 & & t_{32} & & t_{33} & & & \\
 & & & t_{32} & & t_{33} & & \\ \hline
 & t_{41} & t_{42} & & t_{43} & & & t_{44} \\
t_{41} & & 2t_{42} & & t_{43} & & t_{44} &
\end{array}\right].
$$

With suitable permutation matrices P_{r}' and P_{c}', we obtain another explicit upper block-triangular form

$$
\bar{A}' = P_{\mathrm{r}}' \tilde{A}' P_{\mathrm{c}}' = \left[\begin{array}{cccc|cc|cc}
t_{11} & & & t_{14} & & & & \\
 & 2t_{11} & t_{14} & & & & & \\
 & t_{41} & & t_{44} & t_{42} & t_{43} & & \\
t_{41} & & t_{44} & & 2t_{42} & t_{43} & & \\ \hline
 & & & & t_{22} & t_{23} & & \\
 & & & & t_{32} & t_{33} & & \\ \hline
 & & & & & & t_{22} & t_{23} \\
 & & & & & & t_{32} & t_{33}
\end{array}\right].
$$

Thus $\bar{A}'$ is another GP-irreducible decomposition of A, which has empty tails and square diagonal blocks $\bar{A}'[R_1', C_1'] = \left[\begin{array}{cc|cc} t_{11} & & & t_{14} \\ & 2t_{11} & t_{14} & \\ \hline & t_{41} & & t_{44} \\ t_{41} & & t_{44} & \end{array}\right]$, $\bar{A}'[R_2', C_2'] = \left[\begin{array}{c|c} t_{22} & t_{23} \\ \hline t_{32} & t_{33} \end{array}\right]$, and $\bar{A}'[R_3', C_3'] = \left[\begin{array}{c|c} t_{22} & t_{23} \\ \hline t_{32} & t_{33} \end{array}\right]$. The partial orders among the block in the two block-triangular forms are given by

$$
\bar{A}: \quad \begin{array}{ccc} C_2 & & C_3 \\ & \diagdown\!\diagup & \\ & C_1 & \end{array} \qquad\qquad \bar{A}': \quad \begin{array}{cc} C_2' & \\ | & C_3' \\ C_1' & \end{array} .
$$

$\square$

Notes. This section is a reorganization of the results from Ito–Iwata–Murota [138] and Iwata–Murota [144]. In particular, Theorems 4.8.6, 4.8.7, 4.8.11, and 4.8.13, are from Ito–Iwata–Murota [138], and Theorems 4.8.19 and 4.8.20 are from Iwata–Murota [144]. Ito–Iwata–Murota [138] deals also with the block-triangularization under partition-respecting similarity transformations with a motivation from the hidden Markov information sources investigated in Ito [136] and Ito–Amari–Kobayashi [137]. Partition matrices are investigated also in the context of "Matrix Problem," though from a different viewpoint (Gabriel–Roiter [85] and Simson [301]).

4.9 Principal Structures of LM-matrices

4.9.1 Motivations

Suppose an engineering system is described by a system of nonlinear equations

$$f_i(\boldsymbol{x}) = 0, \qquad i \in R, \tag{4.133}$$

in a set of variables $\boldsymbol{x} = (x_j \mid j \in C)$. Then a physical state of the engineering system is specified by a point $\boldsymbol{x}$ of the manifold (in a loose sense) described by (4.133). Let $A = A(\boldsymbol{x})$ denote the Jacobian matrix of $\boldsymbol{f}(\boldsymbol{x})$ and assume that $\mathrm{rank}\, A = |R| < |C|$ (for all $\boldsymbol{x}$ in an open set).

If $|C| - |R|$ variables $\{x_j \mid j \in C \setminus J\}$, where $J \subseteq C$ and $|J| = |R|$, are chosen in such a way that the submatrix $A[R, J]$ is nonsingular, the remaining variables $\{x_j \mid j \in J\}$ can be determined uniquely in general by (4.133). Such variables $\{x_j \mid j \in C \setminus J\}$ are sometimes called *design variables* in the engineering literature. Design variables can be regarded as local coordinates of the manifold described by (4.133) (see Takamatsu–Hashimoto–Tomita [308]).

The choice of design variables is, to some extent, at our disposal. From a computational point of view, it is advantageous to select a set of design variables so that the system (4.133) of equations in the remaining dependent variables may be hierarchically decomposable as fine as possible. Even though we may not expect an optimal one in this respect, we would like to ask: "How fine can we decompose the system with a clever choice of design variables?" Assuming that the Jacobian matrix is an LM-matrix, we shall give a combinatorial answer to this question in terms of the horizontal principal structure of LM-matrices in §4.9.5.

As a second motivation for the same question, suppose we are given a linear program:

$$\text{Maximize } \boldsymbol{c}^{\mathrm{T}}\boldsymbol{x} \quad \text{subject to} \quad A\boldsymbol{x} = \boldsymbol{b},\ \boldsymbol{x} \geq \boldsymbol{0}$$

with an $m \times n$ LM-matrix of rank m as the coefficient matrix A. A basic solution, corresponding to an $m \times m$ nonsingular submatrix $A[R, J]$ for some $J \subseteq C$, is computed by solving $A[R, J]\boldsymbol{x}[J] = \boldsymbol{b}$ and putting $\boldsymbol{x}[C \setminus J] = \boldsymbol{0}$. The submatrix $A[R, J]$ is again an LM-matrix, for which a canonical decomposition is obtained by the CCF. Furthermore we may be interested in the family of the decompositions of the submatrices $A[R, J]$ for all possible basic solutions. Mathematically this leads to the same question as the above on design variable selection, and a combinatorial answer is given as the horizontal principal structure of LM-matrices in §4.9.5.

Next, consider a pair of primal and dual linear programs:

$$\begin{aligned} &\text{(P) Maximize } \boldsymbol{c}^{\mathrm{T}}\boldsymbol{x} \quad \text{subject to} \quad A\boldsymbol{x} \leq \boldsymbol{b}, \\ &\text{(D) Minimize } \boldsymbol{y}^{\mathrm{T}}\boldsymbol{b} \quad \text{subject to} \quad \boldsymbol{y}^{\mathrm{T}}A = \boldsymbol{c}^{\mathrm{T}},\ \boldsymbol{y} \geq \boldsymbol{0}, \end{aligned}$$

where A is assumed to be an $m \times n$ LM-matrix of rank n. A basic solution to the dual problem (D) corresponds to an $n \times n$ nonsingular submatrix $A[I,C]$ for some $I \subseteq R$, and is computed by solving $\boldsymbol{y}[I]^{\mathrm{T}} A[I,C] = \boldsymbol{c}[I]^{\mathrm{T}}$ and putting $\boldsymbol{y}[R \setminus I] = \mathbf{0}$. In this case we may be interested in the family of the decompositions obtained by the CCF of the submatrices $A[I,C]$ for all $I \subseteq R$ such that $A[I,C]$ is nonsingular. This question is not the same as the previous one, since the transpose of an LM-matrix is not an LM-matrix. We shall give a combinatorial answer to this second question by the vertical principal structure of LM-matrices in §4.9.4.

We may ask a more general question: For an LM-matrix A of rank r, what is the coarsest decomposition of the row and the column sides which is finer than any decomposition induced by the CCF of $r \times r$ nonsingular submatrices of A? We address this problem in Remark 4.9.20.

Example 4.9.1. The idea of the vertical principal structure is illustrated here in an informal manner. Consider a 5×3 LM-matrix

$$A = \begin{array}{c|ccc|} & x_1 & x_2 & x_3 \\ \hline r_1 & 1 & 2 & 1 \\ r_2 & 1 & 1 & -1 \\ r_3 & 0 & t_1 & t_2 \\ r_4 & 0 & t_3 & t_4 \\ r_5 & t_5 & t_6 & 0 \\ \hline \end{array}$$

with ground field $\mathbf{Q}$, where $C = \{x_1, x_2, x_3\}$, $R = \{r_1, r_2, r_3, r_4, r_5\}$, and t_i $(i = 1, \cdots, 6)$ are indeterminates. This matrix is LM-irreducible, the whole matrix being a vertical tail.

For a nonsingular submatrix $A[I,C]$ we denote by $\mathcal{P}_{\mathrm{CCF}}(I,C)$ the partition of C (together with the partial order) in the CCF of the submatrix $A[I,C]$. For $I = \{r_1, r_2, r_3\}$, for instance, the CCF of $A[I,C] = A[\{r_1, r_2, r_3\}, C]$ is given by

$$\begin{array}{c|c|cc|} & x_1 & x_2 & x_3 \\ \hline r_1 & 1 & 2 & 1 \\ \cline{3-4} r_2 & & -1 & -2 \\ r_3 & & t_1 & t_2 \\ \hline \end{array},$$

which is obtained from $A[\{r_1, r_2, r_3\}, C]$ by subtracting row r_1 from row r_2. Hence, $\mathcal{P}_{\mathrm{CCF}}(\{r_1, r_2, r_3\}, C)$ is given by $\{x_1\} \prec \{x_2, x_3\}$.

By inspection we see that $A[I,C]$ is a nonsingular submatrix for any $I \subseteq R$ with $|I| = 3$, and $\mathcal{P}_{\mathrm{CCF}}(I,C)$ for all I are given as follows:

$\mathcal{P}_{\mathrm{CCF}}(I,C)$	I
$\{x_3\} \prec \{x_1, x_2\}$	$\{r_1, r_2, r_5\}$
$\{x_1, x_2, x_3\}$	$\{r_i, r_j, r_5\} (i = 1, 2; j = 3, 4)$
$\{x_1\} \prec \{x_2, x_3\}$	otherwise

This shows that the decomposition of C defined by $\{x_1\} \prec \{x_2\}$, $\{x_3\} \prec \{x_2\}$ gives the coarsest common refinement of $\mathcal{P}_{\mathrm{CCF}}(I, C)$ for all I, which we denote by $\bigwedge_I \mathcal{P}_{\mathrm{CCF}}(I, C)$ (see Theorem 2.2.10 for this notation). The vertical principal structure of A will give a succinct description of $\bigwedge_I \mathcal{P}_{\mathrm{CCF}}(I, C)$ in Example 4.9.5. □

4.9.2 Principal Structure of Submodular Systems

The concept of the principal structure of submodular systems was introduced first by Fujishige [81], and subsequently generalized for submodular functions on arbitrary lattices by Tomizawa–Fujishige [315]. This section is devoted to a description of this concept.

Let $\mathcal{L}$ be a lattice with finite length and f a *submodular function* on it:

$$f(X) + f(Y) \geq f(X \vee Y) + f(X \wedge Y), \qquad X, Y \in \mathcal{L},$$

where $\vee$ and $\wedge$ are the join and the meet in $\mathcal{L}$. The partial order $\preceq$ in $\mathcal{L}$ is defined by:

$$X \preceq Y \iff X \vee Y = Y \quad (\iff X \wedge Y = X).$$

For each $X \in \mathcal{L}$,

$$\mathcal{L}_{\min}(f; X) = \{Y \in \mathcal{L} \mid X \preceq Y,\ f(Y) = \min\{f(Y') \mid X \preceq Y' \in \mathcal{L}\}\} \tag{4.134}$$

forms a sublattice of $\mathcal{L}$ by the submodularity of f (the proof is similar to that of Theorem 2.2.5). We denote by $D(f; X)$ the minimum element of this sublattice, i.e.,

$$D(f; X) = \min \mathcal{L}_{\min}(f; X). \tag{4.135}$$

A mapping $\phi : \mathcal{L} \to \mathcal{L}$ is said to be a *closure function* if it satisfies the following three conditions:

(CL0) $\forall X \in \mathcal{L} : X \preceq \phi(X)$,
(CL1) $\forall X, Y \in \mathcal{L} : X \preceq Y \Rightarrow \phi(X) \preceq \phi(Y)$,
(CL2) $\forall X \in \mathcal{L} : \phi(\phi(X)) = \phi(X)$.

With this terminology we have the following lemma.

Lemma 4.9.2. *The mapping $D(f; \cdot) : \mathcal{L} \to \mathcal{L}$ is a closure function on $\mathcal{L}$.*

Proof. The conditions (CL0) and (CL2) are immediate from the definition. The condition (CL1) is proved as follows. Because of the definition of $D(f; \cdot)$, we have

$$f(D(f; Y)) \leq f(D(f; X) \vee D(f; Y)).$$

It follows from the submodularity of f and the above inequality that

$$f(D(f; X)) \geq f(D(f; X) \wedge D(f; Y)).$$

On the other hand, if $X \preceq Y$, it holds that $X \preceq D(f;X) \wedge D(f;Y)$. Hence, from the minimality of $D(f;X)$ we have $D(f;X) = D(f;X) \wedge D(f;Y)$, which implies $D(f;X) \preceq D(f;Y)$. ■

For a closure function ϕ, it can be easily shown that $\phi(X \wedge Y) = X \wedge Y$ if $\phi(X) = X$ and $\phi(Y) = Y$. That is to say, the family $\{X \in \mathcal{L} \mid \phi(X) = X\}$ of "closed sets" is a lower semilattice. Therefore the subset $\mathcal{K}_{\rm PS}(f)$ defined by

$$\mathcal{K}_{\rm PS}(f) = \{X \in \mathcal{L} \mid D(f;X) = X\} \tag{4.136}$$

is a lower semilattice containing the maximum element of $\mathcal{L}$. We say that $\mathcal{K}_{\rm PS}(f)$ is the *principal structure* of $(\mathcal{L}, f)$. Denoting the minimum sublattice which contains $\mathcal{K}_{\rm PS}(f)$ by $\mathcal{L}_{\rm PS}(f)$, we will call $\mathcal{L}_{\rm PS}(f)$ the *principal sublattice* of $(\mathcal{L}, f)$.

The principal structure $\mathcal{K}_{\rm PS}(f)$, which is derived from (4.134), is closely related to the family of the (global) minimizers of f:

$$\mathcal{L}_{\min}(f) = \{Y \in \mathcal{L} \mid f(Y) = \min\{f(Y') \mid Y' \in \mathcal{L}\}\ \}, \tag{4.137}$$

which forms a sublattice of $\mathcal{L}$. Denote the maximum elements of $\mathcal{L}$ and $\mathcal{L}_{\min}(f)$ by $\max \mathcal{L}$ and $\max \mathcal{L}_{\min}(f)$, respectively.

Lemma 4.9.3. *For* $X \preceq \max \mathcal{L}_{\min}(f)$, *it holds that*

$$X \in \mathcal{K}_{\rm PS}(f) \iff X \in \mathcal{L}_{\min}(f).$$

Therefore, if $\max \mathcal{L} \in \mathcal{L}_{\min}(f)$, *then* $\mathcal{K}_{\rm PS}(f) = \mathcal{L}_{\rm PS}(f) = \mathcal{L}_{\min}(f)$.

Proof. If $X \preceq \max \mathcal{L}_{\min}(f)$, the lattice (4.134) is a sublattice of $\mathcal{L}_{\min}(f)$. ■

Originally, the principal structure is defined in the case of $\mathcal{L} = 2^V$ for a finite set V, as follows. Let $f : 2^V \to \mathbf{R}$ be a submodular function:

$$f(X) + f(Y) \geq f(X \cup Y) + f(X \cap Y), \qquad X, Y \subseteq V,$$

with $f(\emptyset) = 0$. Such a pair $(2^V, f)$ is called a *submodular system.* Given an element $v \in V$, we denote by $D(f;v)$ the minimum element of the distributive lattice

$$\mathcal{L}_{\min}(f;v) = \{X \subseteq V \mid v \in X,\ f(X) = \min\{f(Y) \mid v \in Y \subseteq V\}\ \}. \tag{4.138}$$

Since the relation $\sqsubseteq$ defined by

$$v \sqsubseteq v' \iff v \in D(f;v')$$

is reflexive and transitive, the relation $\sim$ defined by

$$v \sim v' \iff v \sqsubseteq v',\ v' \sqsubseteq v$$

is an equivalence relation, and determines a partition of V into equivalence classes $\{V_1, \cdots, V_s\}$. A partial order $\preceq$ is induced among the equivalence classes in such a way that $V_k \preceq V_l$ if and only if $v \sqsubseteq v'$ for $v \in V_k$ and $v' \in V_l$. This decomposition, together with the partial order $\preceq$ among the blocks, is called the *principal structure of the submodular system* $(2^V, f)$. We denote this by $\mathcal{P}_{\rm PS}(f)$. According to Birkhoff's representation theorem (Theorem 2.2.10), the principal structure $\mathcal{P}_{\rm PS}(f)$ corresponds to a sublattice of 2^V, which we denote by $\mathcal{L}(\mathcal{P}_{\rm PS}(f))$. This sublattice coincides with the principal sublattice $\mathcal{L}_{\rm PS}(f)$ for $\mathcal{L} = 2^V$, as stated below.

Lemma 4.9.4. $\mathcal{L}(\mathcal{P}_{\rm PS}(f)) = \mathcal{L}_{\rm PS}(f)$ *for a submodular function* $f : 2^V \to \mathbf{R}$.

Proof. $\mathcal{L}(\mathcal{P}_{\rm PS}(f))$ is the sublattice of $\mathcal{L} = 2^V$ generated by $\{D(f;v) \mid v \in V\}$, whereas $\mathcal{L}_{\rm PS}(f)$ is by $\mathcal{K}_{\rm PS}(f) = \{D(f;X) \mid X \subseteq V\}$. Since $\{D(f;v) \mid v \in V\} \subseteq \mathcal{K}_{\rm PS}(f)$, we have $\mathcal{L}(\mathcal{P}_{\rm PS}(f)) \subseteq \mathcal{L}_{\rm PS}(f)$. Conversely, for $X = D(f;X) \in \mathcal{K}_{\rm PS}(f)$, we have $X = \bigcup_{v\in X} D(f;v) \in \mathcal{L}(\mathcal{P}_{\rm PS}(f))$ since $D(f;X) \supseteq D(f;v) \supseteq \{v\}$ for $v \in X$. This implies $\mathcal{L}_{\rm PS}(f) \subseteq \mathcal{L}(\mathcal{P}_{\rm PS}(f))$. ■

Example 4.9.5. As an example of a submodular function we consider the LM-surplus function $p : 2^C \to \mathbf{Z}$ associated with the LM-matrix A of Example 4.9.1, where $C = \{x_1, x_2, x_3\}$ and $p(X) = \rho(X) + \gamma(X) - |X|$ as defined in (4.16). From the values of p shown in Fig. 4.23, we see that $D(p;x_1) = \{x_1\}$, $D(p;x_2) = \{x_1, x_2, x_3\}$, and $D(p;x_3) = \{x_3\}$. Hence $\mathcal{P}_{\rm PS}(p)$ is given by: $\{x_1\} \prec \{x_2\}$, $\{x_3\} \prec \{x_2\}$. We have $\mathcal{K}_{\rm PS}(p) = \{\emptyset, \{x_1\}, \{x_3\}, \{x_1, x_2, x_3\}\}$ and $\mathcal{L}_{\rm PS}(p) = \mathcal{K}_{\rm PS}(p) \cup \{\{x_1, x_3\}\}$. We may observe here that $\mathcal{P}_{\rm PS}(p)$ agrees with $\bigwedge_{I\in\mathcal{B}_{\rm row}} \mathcal{P}_{\rm CCF}(I, C)$, the coarsest common refinement of $\{\mathcal{P}_{\rm CCF}(I, C) \mid I \in \mathcal{B}_{\rm row}\}$ that we considered in Example 4.9.1. Corollary 4.9.11 will reveal that this is always the case. □

4.9.3 Principal Structure of Generic Matrices

Before entering into the general case of LM-matrices we consider here the principal structure of generic matrices. This special case deserves a separate consideration not only because it is the origin of the main idea, but also because it has an interesting application to the problem of making matrices sparser.

Let A be a generic matrix with $R = \mathrm{Row}(A)$ and $C = \mathrm{Col}(A)$, and

$$\mathcal{B}_{\rm row} = \{I \subseteq R \mid \mathrm{rank}\, A = \mathrm{rank}\, A[I, C] = |I|\} \tag{4.139}$$

be the family of row-bases of A. We assume in this subsection that $\mathrm{rank}\, A = |C|$.

For each $I \in \mathcal{B}_{\rm row}$ the submatrix $A[I, C]$ is nonsingular, and the DM-decomposition of $A[I, C]$ determines a block-triangularization with nonsingular diagonal blocks. Denote by $\mathcal{P}_{\rm DM}(I, C)$ the pair of the partition of C

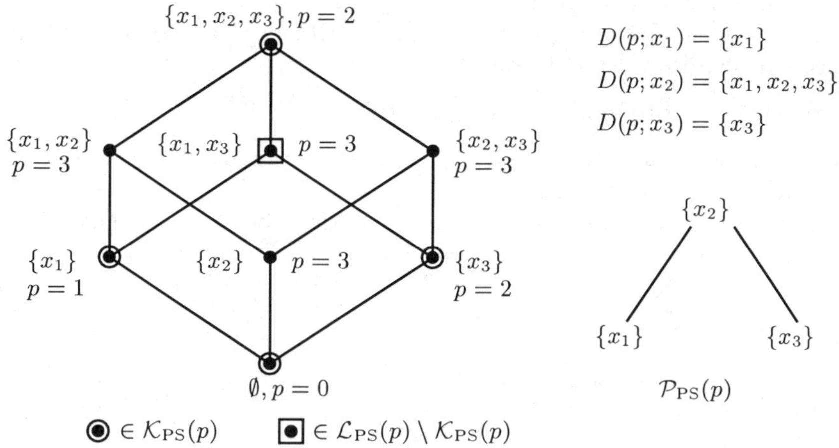

Fig. 4.23. The principal structure of the LM-surplus function p of the LM-matrix in Example 4.9.1

and the partial order among the blocks in the DM-decomposition of the submatrix $A[I, C]$, and by $\bigwedge_{I \in \mathcal{B}_{\rm row}} \mathcal{P}_{\rm DM}(I, C)$ the coarsest partition of C which is finer than all $\mathcal{P}_{\rm DM}(I, C)$ with $I \in \mathcal{B}_{\rm row}$.

The surplus function $p_0 : 2^C \to \mathbf{Z}$ defined as $p_0(X) = \gamma(X) - |X|$ for $X \subseteq C$ in (2.39) is submodular, and hence we may think of the principal structure $\mathcal{P}_{\rm PS}(p_0)$ of p_0, which we call the *principal structure of the generic matrix* A. It is observed by Murota [210] that the principal structure $\mathcal{P}_{\rm PS}(p_0)$ of the surplus function p_0 is identical with the *SP-decomposition* introduced by McCormick [190]. With this observation a result of McCormick [190] can be formulated as follows.

Theorem 4.9.6. *For a generic matrix A of full-column rank and its surplus function $p_0 : 2^C \to \mathbf{Z}$, we have*

$$\mathcal{P}_{\rm PS}(p_0) = \bigwedge_{I \in \mathcal{B}_{\rm row}} \mathcal{P}_{\rm DM}(I, C).$$

Proof. This is proven later as a special case of Corollary 4.9.11. ∎

The principal structure $\mathcal{P}_{\rm PS}(p_0)$, or rather the family $\{D(p_0; j) \mid j \in C\}$ defining $\mathcal{P}_{\rm PS}(p_0)$, plays a key role in the algorithm of Hoffman–McCormick [112] for making matrices optimally sparse.

Suppose we are given a matrix $A = (A_{ij})$ of full-column rank and we want to find a nonsingular matrix $S = (S_{jk})$ such that $\bar{A} = AS$ has the minimum number of nonzero entries. For a generic matrix A this problem has a nice combinatorial answer with an efficient algorithm, while for a general numerical matrix it is NP-hard due to "unexpected" numerical cancellations.

The algorithm of Hoffman–McCormick [112] may be described as follows, where $C = \mathrm{Col}(A) \cong \mathrm{Row}(S)$ and we fix an arbitrary one-to-one correspondence between $\mathrm{Row}(S)$ and $\mathrm{Col}(S)$.

Algorithm for optimally sparse matrix $\bar{A} = AS$

1. Fix a one-to-one mapping $\sigma : C \to R$ such that $A_{\sigma(j),j} \neq 0$ for $j \in C$ (such σ exists since $\mathrm{rank}\, A = |C|$).
2. For each $k \in C$, solve the system of equations in $\{S_{jk} \mid j \in D(p_0; k)\}$:

$$\sum_{j \in D(p_0;k)} A_{ij} S_{jk} = \begin{cases} 1 & (i = \sigma(k)) \\ 0 & (i \in \sigma(D(p_0;k) \setminus \{k\})) \end{cases} \tag{4.140}$$

   and put $S_{jk} = 0$ for $j \in C \setminus D(p_0; k)$.
3. Put $\bar{A} = AS$. □

For the validity of the algorithm we have the following two lemmas.

Lemma 4.9.7. *The matrix S constructed by the algorithm is nonsingular.*

Proof. Let $D_1, D_2, \cdots,$ be the distinct elements among $\{D(p_0; j) \mid j \in C\}$, and put $X_t = \bigcup\{D_s \mid D_s \subset D_t, D_s \neq D_t\}$ and $C_t = D_t \setminus X_t$ for $t = 1, 2, \cdots$. By construction, S is block-triangular with respect to the partition $\{C_t\}$ of C ($\cong \mathrm{Row}(S) \cong \mathrm{Col}(S)$), and therefore it suffices to show that $S[C_t, C_t]$ is nonsingular for each t. From (4.140) we have an identity: $A[\sigma(D_t), D_t] \cdot S[D_t, D_t] = \bar{A}[\sigma(D_t), D_t]$, which can be rewritten as

$$\begin{array}{c} \\ \sigma(X_t) \\ \sigma(C_t) \end{array} \begin{array}{c} \begin{array}{cc} X_t & C_t \end{array} \\ \begin{pmatrix} A_{11} & A_{12} \\ A_{21} & A_{22} \end{pmatrix} \end{array} \begin{array}{c} \begin{array}{cc} X_t & C_t \end{array} \\ \begin{pmatrix} S_{11} & S_{12} \\ O & S_{22} \end{pmatrix} \end{array} = \begin{array}{c} \begin{array}{cc} X_t & C_t \end{array} \\ \begin{pmatrix} \bar{A}_{11} & O \\ \bar{A}_{21} & I \end{pmatrix} \end{array},$$

where $\bar{A}[\sigma(X_t), C_t] = O$ and $\bar{A}[\sigma(C_t), C_t] = I$. In the above expression, $A_{11} = A[\sigma(X_t), X_t]$ is nonsingular since $A_{\sigma(j),j} \neq 0$ for $j \in X_t$, and then it follows that $(A_{22} - A_{21}{A_{11}}^{-1}A_{12})S_{22} = I$. This implies that $S_{22} = S[C_t, C_t]$ is nonsingular. ■

Lemma 4.9.8. *The matrix $\bar{A}$ constructed by the algorithm contains the minimum number of nonzero entries.*

Proof. Consider $\bar{A} = AS$ for a nonsingular matrix S in general, and put $J_k = \{j \mid S_{jk} \neq 0\}$ for $k \in C$. By the nonsingularity of S, we may assume $k \in J_k$ for all $k \in C$ with the understanding of the above-mentioned one-to-one correspondence between $\mathrm{Row}(S)$ and $\mathrm{Col}(S)$. The assumed genericity of the nonzero entries of A implies that the number of the nonzero entries in the column k of $\bar{A}$ is not smaller than $|\{i \in R \mid \exists j \in J_k : A_{ij} \neq 0\}| - |J_k| + 1 = p_0(J_k) + 1$, whereas this lower bound is attained in the algorithm with $J_k = D(p_0; k)$. ■

Example 4.9.9. The above argument is illustrated here for a generic matrix

$$A = \begin{array}{c} \begin{array}{ccc} x_1 & x_2 & x_3 \end{array} \\ \begin{array}{|ccc|} \hline a_{11} & a_{12} & 0 \\ \underline{a_{21}} & a_{22} & 0 \\ a_{31} & \underline{a_{32}} & a_{33} \\ 0 & a_{42} & \underline{a_{43}} \\ 0 & a_{52} & a_{53} \\ \hline \end{array} \end{array},$$

where the underlined entries indicate $\sigma : C \to R$. We have $D(p_0; x_1) = \{x_1\}$, $D(p_0; x_2) = \{x_1, x_2, x_3\}$, and $D(p_0; x_3) = \{x_3\}$, since $p_0(\emptyset) = 0$, $p_0(\{x_1\}) = 2$, $p_0(\{x_2\}) = 4$, $p_0(\{x_3\}) = 2$, $p_0(\{x_1, x_2\}) = 3$, $p_0(\{x_1, x_3\}) = 3$, $p_0(\{x_2, x_3\}) = 3$, $p_0(\{x_1, x_2, x_3\}) = 2$. By the algorithm above, the matrix A is transformed to a sparser matrix $\bar{A} = SA$, where

$$S = \begin{array}{|ccc|} \hline s_{11} & s_{12} & 0 \\ 0 & s_{22} & 0 \\ 0 & s_{32} & s_{33} \\ \hline \end{array}, \qquad \bar{A} = \begin{array}{|ccc|} \hline \bar{a}_{11} & \bar{a}_{12} & 0 \\ 1 & \mathbf{0} & 0 \\ \bar{a}_{31} & 1 & \bar{a}_{33} \\ 0 & \mathbf{0} & 1 \\ 0 & \bar{a}_{52} & \bar{a}_{53} \\ \hline \end{array}.$$

The two boldface zeros in $\bar{A}$ are created. The entries of S are determined from

$$a_{21} s_{11} = 1, \qquad \begin{bmatrix} a_{21} & a_{22} & 0 \\ a_{31} & a_{32} & a_{33} \\ 0 & a_{42} & a_{43} \end{bmatrix} \begin{bmatrix} s_{12} \\ s_{22} \\ s_{32} \end{bmatrix} = \begin{bmatrix} 0 \\ 1 \\ 0 \end{bmatrix}, \qquad a_{43} s_{33} = 1.$$

□

Computational aspects of this algorithm and its variants are reported in Chang–McCormick [31, 32], McCormick [191], and McCormick–Chang [192].

4.9.4 Vertical Principal Structure of LM-matrices

Let $A = \binom{Q}{T} \in \mathrm{LM}(\boldsymbol{K}, \boldsymbol{F})$ be an LM-matrix with $R = \mathrm{Row}(A)$ and $C = \mathrm{Col}(A)$, and $\mathcal{B}_{\mathrm{row}} \subseteq 2^R$ be the family of row-bases of A as in (4.139). For $I \in \mathcal{B}_{\mathrm{row}}$ the submatrix $A[I, C]$ is an LM-matrix of full-row rank, and the CCF of $A[I, C]$ determines a block-triangularization with an empty vertical tail. Denote by $\mathcal{P}_{\mathrm{CCF}}(I, C)$ the pair of the partition of C and the partial order among the blocks in the CCF of the submatrix $A[I, C]$, and by $\bigwedge_{I \in \mathcal{B}_{\mathrm{row}}} \mathcal{P}_{\mathrm{CCF}}(I, C)$ the coarsest partition of C which is finer than all $\mathcal{P}_{\mathrm{CCF}}(I, C)$ with $I \in \mathcal{B}_{\mathrm{row}}$. We also denote by $\mathcal{L}_{\mathrm{CCF}}(I, C)$ the sublattice of 2^C that corresponds to $\mathcal{P}_{\mathrm{CCF}}(I, C)$, and by $\bigvee_{I \in \mathcal{B}_{\mathrm{row}}} \mathcal{L}_{\mathrm{CCF}}(I, C)$ the sublattice generated by $\{\mathcal{L}_{\mathrm{CCF}}(I, C) \mid I \in \mathcal{B}_{\mathrm{row}}\}$. Note that $\mathcal{L}(\mathcal{P}_{\mathrm{CCF}}(I, C)) = \mathcal{L}_{\mathrm{CCF}}(I, C)$ and

$$\mathcal{L}\Big(\bigwedge_{I \in \mathcal{B}_{\mathrm{row}}} \mathcal{P}_{\mathrm{CCF}}(I, C)\Big) = \bigvee_{I \in \mathcal{B}_{\mathrm{row}}} \mathcal{L}_{\mathrm{CCF}}(I, C)$$

with the notation of Theorem 2.2.10.

The LM-surplus function $p : 2^C \to \mathbf{Z}$ defined as $p(X) = \rho(X)+\gamma(X)-|X|$ for $X \subseteq C$ in (4.16) is submodular, and hence we may think of the principal structure of p. We name this the *vertical principal structure of an LM-matrix* A. The following theorem of Murota [210] connects this with the family of the CCF.

Theorem 4.9.10. *For an LM-matrix A and its LM-surplus function p : $2^C \to \mathbf{Z}$, we have*

$$\mathcal{K}_{\mathrm{PS}}(p) = \bigcup_{I \in \mathcal{B}_{\mathrm{row}}} \mathcal{L}_{\mathrm{CCF}}(I, C).$$

Proof. The proof is given later. ■

Theorem 4.9.10 can be reformulated in terms of the principal sublattice and the corresponding partitions, as follows.

Corollary 4.9.11. *For an LM-matrix A and its LM-surplus function p : $2^C \to \mathbf{Z}$, we have*

$$\mathcal{L}_{\mathrm{PS}}(p) = \bigvee_{I \in \mathcal{B}_{\mathrm{row}}} \mathcal{L}_{\mathrm{CCF}}(I, C), \qquad \mathcal{P}_{\mathrm{PS}}(p) = \bigwedge_{I \in \mathcal{B}_{\mathrm{row}}} \mathcal{P}_{\mathrm{CCF}}(I, C). \tag{4.141}$$

□

This corollary shows that the above result is a generalization of Theorem 4.9.6. We mention that the vertical principal structure $\mathcal{P}_{\mathrm{PS}}(p)$ of an LM-matrix $A \in \mathrm{LM}(\boldsymbol{K}, \boldsymbol{F})$ can be found by an efficient algorithm using arithmetic operations in $\boldsymbol{K}$.

Example 4.9.12. In Examples 4.9.1 and 4.9.5 we have seen an instance of the identity (4.141). Note that $\mathcal{P}_{\mathrm{PS}}(p) \neq \mathcal{P}_{\mathrm{CCF}}(I, C)$ for each $I \in \mathcal{B}_{\mathrm{row}}$. □

Remark 4.9.13. The LM-surplus function p remains invariant against LM-admissible transformations, whereas $\mathcal{B}_{\mathrm{row}}$ does not. For example, consider a pair of LM-equivalent LM-matrices

$$A^{(1)} = \begin{array}{c} r_1 \\ r_2 \\ r_3 \end{array}\!\!\begin{array}{|ccc|} \hline 1 & 1 & 0 \\ 0 & 0 & 1 \\ 0 & 0 & t \\ \hline \end{array}, \qquad A^{(2)} = \begin{array}{c} r_1 \\ r_2 \\ r_3 \end{array}\!\!\begin{array}{|ccc|} \hline 1 & 1 & 1 \\ 1 & 1 & 2 \\ 0 & 0 & t \\ \hline \end{array}.$$

We have $\mathcal{B}_{\mathrm{row}}^{(1)} = \{\{r_1, r_2\}, \{r_1, r_3\}\}$ and $\mathcal{B}_{\mathrm{row}}^{(2)} = \mathcal{B}_{\mathrm{row}}^{(1)} \cup \{\{r_2, r_3\}\}$. One of the consequences of Theorem 4.9.10 is that $\bigcup_{I \in \mathcal{B}_{\mathrm{row}}} \mathcal{L}_{\mathrm{CCF}}(I, C)$ remains invariant under LM-equivalence in spite of its apparent dependence on $\mathcal{B}_{\mathrm{row}}$. □

Proof of Theorem 4.9.10. We need to introduce the LM-surplus function for submatrices of $A = \binom{Q}{T}$. Putting $R_Q = \mathrm{Row}(Q)$ and $R_T = \mathrm{Row}(T)$, define $\Gamma : 2^{R_T} \times 2^C \to 2^{R_T}$, $\gamma : 2^{R_T} \times 2^C \to \mathbf{Z}$, and $\rho : 2^{R_Q} \times 2^C \to \mathbf{Z}$ by

$$\begin{aligned}
\Gamma(I,J) &= \{i \in I \mid \exists j \in J : T_{ij} \neq 0\}, \quad I \subseteq R_T, J \subseteq C, \\
\gamma(I,J) &= |\Gamma(I,J)|, \quad I \subseteq R_T, J \subseteq C, \\
\rho(I,J) &= \mathrm{rank}\, Q[I,J], \quad I \subseteq R_Q, J \subseteq C.
\end{aligned}$$

Then for $I \subseteq R$ the LM-surplus function $p_I : 2^C \to \mathbf{Z}$ of the submatrix $A[I,C]$ is given by

$$p_I(J) = \rho(I \cap R_Q, J) + \gamma(I \cap R_T, J) - |J|, \quad J \subseteq C.$$

We have

$$\mathcal{L}_{\mathrm{CCF}}(I,C) = \mathcal{L}_{\min}(p_I) = \mathcal{K}_{\mathrm{PS}}(p_I), \qquad I \in \mathcal{B}_{\mathrm{row}}, \tag{4.142}$$

by the construction of the CCF (cf. §4.4.3) and Lemma 4.9.3. Hence, in order to prove Theorem 4.9.10, we shall reveal the relation between $\mathcal{K}_{\mathrm{PS}}(p_R)$ and $\mathcal{K}_{\mathrm{PS}}(p_I)$, i.e., the relation between $D(p_R;X)$ and $D(p_I;X)$, defined in (4.135).

Lemma 4.9.14. *For $X \subseteq C$ and $I \subseteq R$ we have $D(p_R;X) \subseteq D(p_I;X)$.*

Proof. Put $D_R = D(p_R;X)$ and $D_I = D(p_I;X)$. By Proposition 2.1.9(2) we have

$$\rho(R_Q,J) - \rho(R_Q, J \cap J') \geq \rho(I'', J \cup J') - \rho(I'', J'), \quad I'' \subseteq R_Q, \ J, J' \subseteq C.$$

Similarly, it can be shown that

$$\gamma(R_T,J) - \gamma(R_T, J \cap J') \geq \gamma(I', J \cup J') - \gamma(I', J'), \quad I' \subseteq R_T, \ J, J' \subseteq C.$$

These inequalities imply

$$p_R(D_R) - p_R(D_R \cap D_I) \geq p_I(D_R \cup D_I) - p_I(D_I).$$

The right-hand side of this is nonnegative since $X \subseteq D_R \cup D_I$ and $D_I \in \mathcal{L}_{\min}(p_I;X)$. Hence follows $p_R(D_R) \geq p_R(D_R \cap D_I)$. This implies $D_R \cap D_I \in \mathcal{L}_{\min}(p_R;X)$, from which follows $D_R = D_R \cap D_I$ by the minimality of D_R. Therefore, $D_R \subseteq D_I$. ■

Lemma 4.9.15. *For $X \subseteq C$, there exists $I \in \mathcal{B}_{\mathrm{row}}$ such that $D(p_R;X) = D(p_I;X)$.*

Proof. Put $D_R = D(p_R;X)$, $C' = C \setminus D_R$, and $\Gamma_T = \Gamma(R_T, D_R)$.

(i) First choose $I_1 \subseteq R_Q$ such that

$$\mathrm{rank}\, Q[R_Q, D_R] = \mathrm{rank}\, Q[I_1, D_R] = |I_1|.$$

The row vectors of $Q[R_Q \setminus I_1, D_R]$ can be expressed as linear combinations of those of $Q[I_1, D_R]$, i.e., $Q[R_Q \setminus I_1, D_R] = SQ[I_1, D_R]$ for some matrix S over $\boldsymbol{K}$. If we put

$$\bar{Q} = \begin{array}{c} \\ I_1 \\ R_Q \setminus I_1 \end{array} \!\!\begin{array}{c} \begin{array}{cc} I_1 & R_Q \setminus I_1 \end{array} \\ \left(\begin{array}{cc} I & O \\ -S & I \end{array} \right) \end{array} Q, \tag{4.143}$$

we have $\bar{Q}[R_Q \setminus I_1, D_R] = O$. Furthermore, put

$$\bar{A} = \begin{pmatrix} \bar{Q} \\ T \end{pmatrix} = \begin{array}{r} \\ I_1 \\ R_Q \setminus I_1 \\ \Gamma_T \\ R_T \setminus \Gamma_T \end{array} \!\!\begin{array}{c} \begin{array}{cc} D_R & C' = C \setminus D_R \end{array} \\ \left(\begin{array}{cc} \bar{Q}[I_1, D_R] & \bar{Q}[I_1, C'] \\ O & \bar{Q}[R_Q \setminus I_1, C'] \\ T[\Gamma_T, D_R] & T[\Gamma_T, C'] \\ O & T[R_T \setminus \Gamma_T, C'] \end{array} \right) \end{array}, \tag{4.144}$$

which is an LM-matrix. Denoting by $\bar{p}_I$ the LM-surplus function associated with $\bar{A}[I, C]$ we have $\bar{p}_I(J) = p_I(J)$ if $I_1 \subseteq I \subseteq R$ and $J \subseteq C$.

(ii) Next choose $I_2 \subseteq \Gamma_T \subseteq R_T$ such that

$$\operatorname{rank} A[R, D_R] = \operatorname{rank} A[I_1 \cup I_2, D_R] = |I_1| + |I_2|.$$

This is equivalent, by (4.143), to

$$\operatorname{rank} \bar{A}[R, D_R] = \operatorname{rank} \bar{A}[I_1 \cup I_2, D_R] = |I_1| + |I_2|. \tag{4.145}$$

(iii) Put

$$R' = R \setminus (I_1 \cup \Gamma_T) = (R_Q \setminus I_1) \cup (R_T \setminus \Gamma_T),$$

and note that $\bar{A}[R', D_R] = O$ as seen in (4.144). We claim

$$\operatorname{rank} \bar{A}[R', C'] = |C'|. \tag{4.146}$$

To compute the rank of $\bar{A}[R', C']$ by Theorem 4.2.5, we consider $\bar{p}_{R'}(J)$, $J \subseteq C'$. Since

$$\bar{p}_{R'}(J) = \operatorname{rank} \bar{Q}[R_Q \setminus I_1, J] + \gamma(R_T \setminus \Gamma_T, J) - |J|,$$

by the definition, and furthermore

$$\begin{aligned} \operatorname{rank} \bar{Q}[R_Q \setminus I_1, J] &= \operatorname{rank} \bar{Q}[R_Q, D_R \cup J] - |I_1| \\ &= \operatorname{rank} Q[R_Q, D_R \cup J] - \operatorname{rank} Q[R_Q, D_R], \\ \gamma(R_T \setminus \Gamma_T, J) &= \gamma(R_T, D_R \cup J) - \gamma(R_T, D_R), \end{aligned}$$

by the choice of I_1 and the definition of Γ_T, we have

$$\bar{p}_{R'}(J) = p_R(D_R \cup J) - p_R(D_R).$$

This is nonnegative for all $J \subseteq C'$ since $X \subseteq D_R \cup J$ and $D_R \in \mathcal{L}_{\min}(p_R; X)$. Hence follows (4.146) from Theorem 4.2.5. Therefore there exists $I_3 \subseteq R'$ such that

$$\operatorname{rank} \bar{A}[R', C'] = \operatorname{rank} \bar{A}[I_3, C'] = |I_3| = |C'|. \tag{4.147}$$

(iv) We claim that $I = I_1 \cup I_2 \cup I_3$ belongs to $\mathcal{B}_{\text{row}}$. By (4.145), (4.147), and $\bar{A}[I_3, D_R] = O$, we see

$$\operatorname{rank} A[I, C] = \operatorname{rank} \bar{A}[I, C] = |I|. \tag{4.148}$$

On the other hand, since $\bar{A}[R', D_R] = O$ and $\bar{A}[R', C']$ is of full-column rank by (4.147), we see

$$\operatorname{rank} A = \operatorname{rank} \bar{A} = \operatorname{rank} \bar{A}[R \setminus R', D_R] + |C'| = |I_1| + |I_2| + |I_3| = |I|. \tag{4.149}$$

Combination of this and (4.148) shows that $I \in \mathcal{B}_{\text{row}}$.

(v) Furthermore we claim that

$$p_I(D_R) = \min\{p_I(J) \mid J \subseteq C\}. \tag{4.150}$$

By the definitions of I_1 and I_2, we have

$$p_I(D_R) = p_{I_1 \cup I_2}(D_R) = |I_1| + |I_2| - |D_R|.$$

We also have

$$\min\{p_I(J) \mid J \subseteq C\} = |I| - |C|$$

from (4.148) and Theorem 4.2.5. Noting the relation $|I_1| + |I_2| - |D_R| = |I| - |C|$ due to (4.147), we establish (4.150).

(vi) Since $X \subseteq D_R$, (4.150) means $D_R \in \mathcal{L}_{\min}(p_I; X)$. The minimality of $D_I = D(p_I; X)$ then implies that $D_R \supseteq D_I$. The other direction $D_R \subseteq D_I$ is already shown in Lemma 4.9.14. ∎

With the above lemmas we can now prove Theorem 4.9.10. It follows from Lemma 4.9.14 that $D(p_I; X) = X$ implies $D(p_R; X) = X$. Hence, $\mathcal{K}_{\text{PS}}(p_I) \subseteq \mathcal{K}_{\text{PS}}(p_R)$ holds for any $I \subseteq R$. On the other hand, Lemma 4.9.15 implies $\mathcal{K}_{\text{PS}}(p_R) \subseteq \bigcup_{I \in \mathcal{B}_{\text{row}}} \mathcal{K}_{\text{PS}}(p_I)$. Therefore, we have

$$\mathcal{K}_{\text{PS}}(p_R) = \bigcup_{I \in \mathcal{B}_{\text{row}}} \mathcal{K}_{\text{PS}}(p_I),$$

which establishes Theorem 4.9.10 when combined with the relation (4.142).

4.9.5 Horizontal Principal Structure of LM-matrices

Let $A = \binom{Q}{T} \in \text{LM}(\boldsymbol{K}, \boldsymbol{F}; m_Q, m_T, n)$ be an LM-matrix with $R = \text{Row}(A)$ and $C = \text{Col}(A)$. In the previous subsection we have characterized the common refinement of the CCF of all $A[I, C]$ with $I \in \mathcal{B}_{\text{row}}$ by means of the principal structure of the LM-surplus function $p : 2^C \to \mathbf{Z}$. Here we are concerned with the "transpose version" of the problem by considering the family of submatrices $A[R, J]$ for $J \in \mathcal{B}_{\mathrm{col}}$, where

$$\mathcal{B}_{\mathrm{col}} = \{J \subseteq C \mid \mathrm{rank}\, A = \mathrm{rank}\, A[R, J] = |J|\}. \tag{4.151}$$

The present problem is not reduced to the previous one, since the transpose of an LM-matrix is no longer an LM-matrix. The transpose of an LM-matrix, however, may be regarded as a partitioned matrix, for which we have developed a general framework in §4.8. In particular, the function q to be introduced in place of p is essentially the same as the PE-surplus function associated with the transpose of A.

We regard the $m_Q \times n$ matrix Q as a representation of a linear transformation from $\boldsymbol{K}^n$ to the dual space ${V_Q}^* \cong \boldsymbol{K}^{m_Q}$ of $V_Q \cong \boldsymbol{K}^{m_Q}$. Let $\mathcal{L}$ be the set of the pairs of a subspace W of V_Q and a subset I of $R_T = \mathrm{Row}(T)$, i.e.,

$$\mathcal{L} = \{(W, I) \mid W\text{: subspace of } V_Q,\ I \subseteq R_T\},$$

which forms a modular lattice with the operations $\wedge$ and $\vee$ defined by

$$(W_1, I_1) \wedge (W_2, I_2) = (W_1 \cap W_2, I_1 \cap I_2)$$
$$(W_1, I_1) \vee (W_2, I_2) = (W_1 + W_2, I_1 \cup I_2)$$

for $W_h \subseteq V_Q$ and $I_h \subseteq R_T$ $(h = 1, 2)$. We define $\kappa, q : \mathcal{L} \to \mathbf{Z}$ by

$$\kappa(W, I) = |\{j \in C \mid Q_j \notin W^\perp \text{ or } \exists i \in I : T_{ij} \neq 0\}|,\ (W, I) \in \mathcal{L}, \tag{4.152}$$
$$q(W, I) = \kappa(W, I) - \dim W - |I|, \quad (W, I) \in \mathcal{L}, \tag{4.153}$$

where Q_j denotes the column vector of Q indexed by $j \in C$, T_{ij} is the (i, j) entry of T, and $W^\perp$ is the subspace of ${V_Q}^*$ annihilating W, i.e.,

$$W^\perp = \{v \in {V_Q}^* \mid \langle w, v\rangle = 0,\ \forall w \in W\},$$

in which $\langle \cdot\, , \cdot \rangle$ means the inner product (pairing). Note that $Q_j \in {V_Q}^*$.

Remark 4.9.16. For $A = \binom{Q}{T} \in \mathrm{LM}(\boldsymbol{K}, \boldsymbol{F}; m_Q, m_T, n)$ we regard A^{T} as a partitioned matrix (4.113) with the parameters

$$\mu = n, \quad \nu = 1 + m_T, \quad m_\alpha = 1\ (\alpha = 1, \cdots, \mu), \quad n_\beta = \begin{cases} m_Q & (\beta = 1) \\ 1 & (\beta = 2, \cdots, \nu). \end{cases}$$

The function q defined above is essentially the same as the PE-surplus function p_{PE} defined by (4.120), in which "$W \in \mathcal{W}$" is replaced by "$(W, I) \in \mathcal{L}$", "$\sum_{\alpha=1}^{\mu} \dim(A_\alpha W)$" by "$\kappa(W, I)$", and "$\dim W$" by "$\dim W + |I|$". □

The following identity holds true.

Lemma 4.9.17. *For an $m \times n$ LM-matrix A,*

$$\operatorname{rank} A = \min\{q(W,I) \mid (W,I) \in \mathcal{L}\} + m.$$

Proof. The CCF of A gives a proper block-triangularization under an LM-admissible transformation, which is a PE-transformation for A^{T}. Then Theorem 4.8.6 shows the validity of the claimed identity. ∎

For $J \in \mathcal{B}_{\mathrm{col}}$ the submatrix $A[R,J]$ is an LM-matrix of full-column rank. Let $\mathcal{L}_{\mathrm{CCF}}(R,J)$ denote the sublattice of $\mathcal{L}$ that corresponds to the CCF of $A[R,J]$ in the sense of Proposition 4.8.12 (applied to the transpose of $A[R,J]$). Note that $\mathcal{L}_{\mathrm{CCF}}(R,J)$ contains $\max \mathcal{L} = (V_Q, R_T)$, since the CCF of $A[R,J]$ has an empty horizontal tail.

The function $q : \mathcal{L} \to \mathbf{Z}$ is submodular, and hence we may think of the principal structure $\mathcal{K}_{\mathrm{PS}}(q)$ as well as the principal sublattice $\mathcal{L}_{\mathrm{PS}}(q)$. We name this the *horizontal principal structure of an LM-matrix A*. The following theorem of Iwata–Murota [145] connects this with the family of CCF.

Theorem 4.9.18. *For an LM-matrix A and the function $q : \mathcal{L} \to \mathbf{Z}$ of* (4.153), *we have*

$$\mathcal{K}_{\mathrm{PS}}(q) = \bigcup_{J \in \mathcal{B}_{\mathrm{col}}} \mathcal{L}_{\mathrm{CCF}}(R,J), \qquad \mathcal{L}_{\mathrm{PS}}(q) = \bigvee_{J \in \mathcal{B}_{\mathrm{col}}} \mathcal{L}_{\mathrm{CCF}}(R,J).$$

Proof. The proof is given later. ∎

Example 4.9.19. Let us illustrate the above result for an LM-matrix

$$A = \begin{array}{c|ccccc|} & x_1 & x_2 & x_3 & x_4 & x_5 \\ \hline y_1 & 0 & 1 & 1 & 1 & 1 \\ y_2 & 0 & 2 & 0 & 2 & 0 \\ \hline z_1 & t_1 & 0 & 0 & 0 & t_2 \\ z_2 & t_3 & 0 & t_4 & t_5 & 0 \\ \hline \end{array},$$

where $C = \{x_1, x_2, x_3, x_4, x_5\}$, $R_Q = \{y_1, y_2\}$ and $R_T = \{z_1, z_2\}$. The whole matrix A is the horizontal tail of its CCF.

As is easily verified, we have

$$\mathcal{K}_{\mathrm{PS}}(q) = \{(\mathbf{0},\emptyset), (V_1,\emptyset), (V_2,\emptyset), (\mathbf{0},Z_1), (V_2,Z_1), (V_2,R_T), (V_Q,R_T)\},$$

where $Z_1 = \{z_1\}$, $Z_2 = \{z_2\}$, and V_1 and V_2 are the 1-dimensional vector spaces spanned by $(0,1) \in V_Q$ and $(2,-1) \in V_Q$, respectively. Note, for example, that $q(V_1,Z_1) = q(V_Q,Z_1) = 2$ and $q(V_Q,R_T) = 1$. The principal structure $\mathcal{K}_{\mathrm{PS}}(q)$ and the principal sublattice $\mathcal{L}_{\mathrm{PS}}(q)$ are illustrated in Fig. 4.24.

On the other hand, we have $\mathcal{B}_{\mathrm{col}} = \{J_h \mid h = 1,\cdots,5\}$ with $J_h = C \setminus \{x_h\}$ for $h = 1,\cdots,5$. In view of

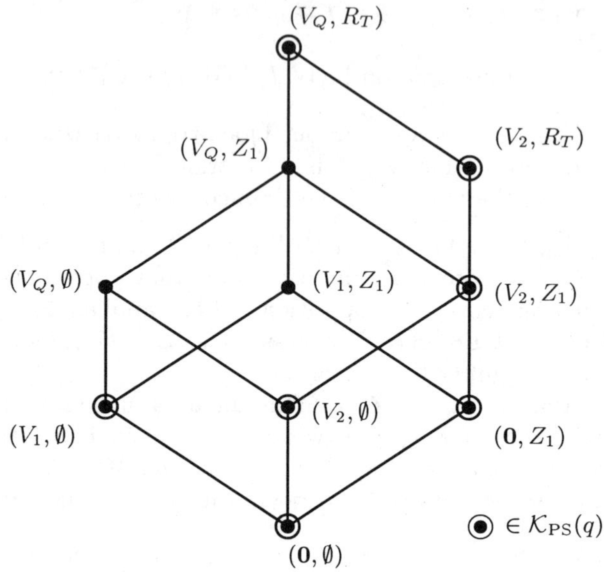

Fig. 4.24. Principal sublattice $\mathcal{L}_{\rm PS}(q)$ of the horizontal principal structure of the LM-matrix in Example 4.9.19

$$\tilde{A} = \begin{array}{c} \\ V_1 \\ V_2 \\ z_1 \\ z_2 \end{array} \left[\begin{array}{cc|cc|ccccc} & & & & x_1 & x_2 & x_3 & x_4 & x_5 \\ 0 & 1 & & & 0 & 1 & 1 & 1 & 1 \\ 2 & -1 & & & 0 & 2 & 0 & 2 & 0 \\ \hline & & 1 & 0 & t_1 & 0 & 0 & 0 & t_2 \\ & & 0 & 1 & t_3 & 0 & t_4 & t_5 & 0 \end{array}\right] = \left[\begin{array}{ccccc} x_1 & x_2 & x_3 & x_4 & x_5 \\ 0 & 2 & 0 & 2 & 0 \\ 0 & 0 & 2 & 0 & 2 \\ \hline t_1 & 0 & 0 & 0 & t_2 \\ t_3 & 0 & t_4 & t_5 & 0 \end{array}\right],$$

we see that the CCF of $A[R, J_h]$, denoted by $\tilde{A}_h$, are given as follows:

$$\tilde{A}_1 = \begin{array}{c|cccc} & x_2 & x_4 & x_3 & x_5 \\ \hline V_1 & 2 & 2 & & \\ z_2 & & t_5 & t_4 & \\ V_2 & & & 2 & 2 \\ z_1 & & & & t_2 \end{array}, \quad \tilde{A}_2 = \begin{array}{c|cccc} & x_1 & x_3 & x_5 & x_4 \\ \hline V_2 & 0 & 2 & 2 & \\ z_1 & t_1 & 0 & t_2 & \\ z_2 & t_3 & t_4 & 0 & t_5 \\ V_1 & & & & 2 \end{array}, \quad \tilde{A}_3 = \begin{array}{c|cccc} & x_2 & x_4 & x_1 & x_5 \\ \hline V_1 & 2 & 2 & & \\ z_2 & & t_5 & t_3 & \\ z_1 & & & t_1 & t_2 \\ V_2 & & & & 2 \end{array},$$

$$\tilde{A}_4 = \begin{array}{c|cccc} & x_1 & x_3 & x_5 & x_2 \\ \hline V_2 & 0 & 2 & 2 & \\ z_1 & t_1 & 0 & t_2 & \\ z_2 & t_3 & t_4 & 0 & \\ V_1 & & & & 2 \end{array}, \quad \tilde{A}_5 = \begin{array}{c|cccc} & x_2 & x_4 & x_1 & x_3 \\ \hline V_1 & 2 & 2 & & \\ z_2 & & t_5 & t_3 & t_4 \\ z_1 & & & t_1 & \\ V_2 & & & & 2 \end{array}.$$

Figure 4.25 illustrates the sublattices $\mathcal{L}_{\rm CCF}(R, J_h)$ for $h = 1, \cdots, 5$, which correspond to $\tilde{A}_h$ above. For example, the height of $\mathcal{L}_{\rm CCF}(R, J_5)$ is four, since $\tilde{A}_5$ is in a block-triangular form with four diagonal blocks, and a parallelepiped appears in $\mathcal{L}_{\rm CCF}(R, J_5)$, since the $(3,4)$ entry of $\tilde{A}_5$ is zero. We

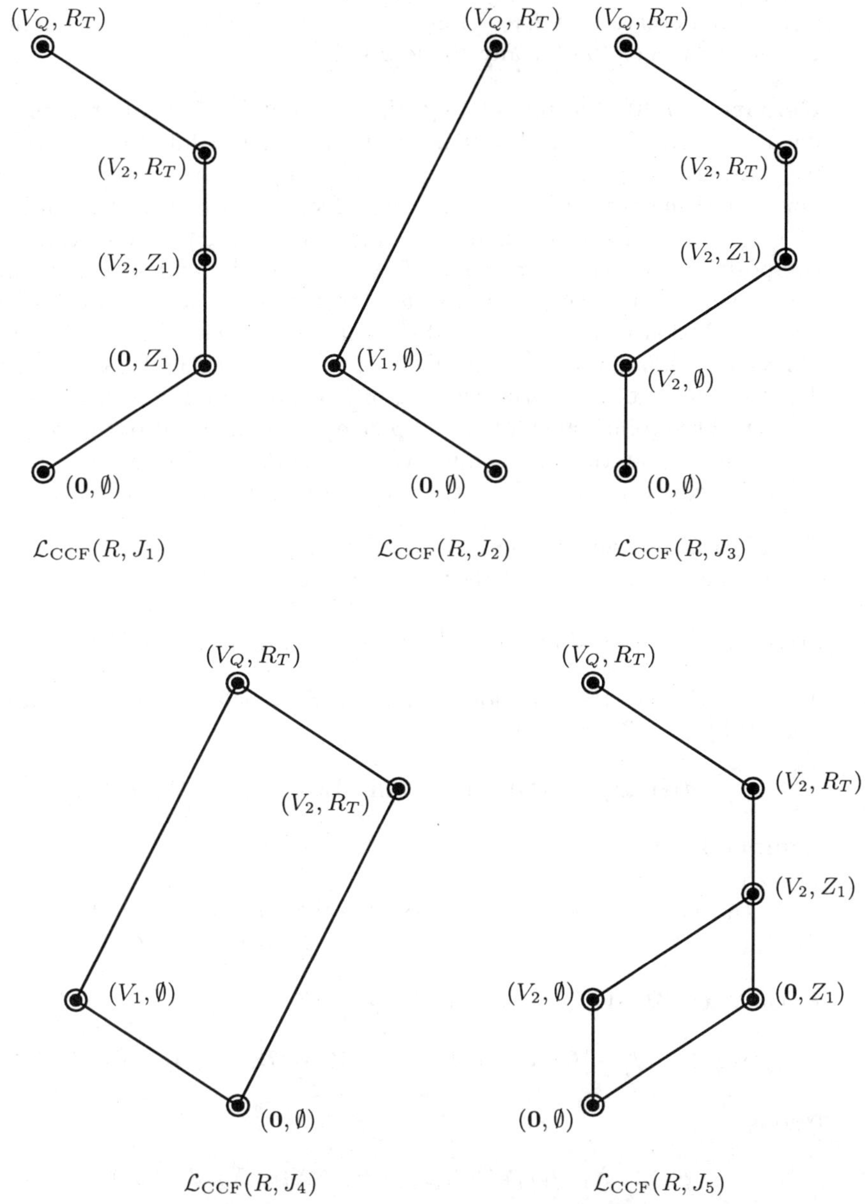

Fig. 4.25. The Hasse diagrams for $\mathcal{L}_{\mathrm{CCF}}(R, J_h)$'s in Example 4.9.19.

can easily observe that $\mathcal{K}_{\rm PS}(q)$ agrees with $\bigcup_{h=1}^{5} \mathcal{L}_{\rm CCF}(R, J_h)$. Furthermore $\mathcal{L}_{\rm PS}(q) \neq \mathcal{L}_{\rm CCF}(R, J)$ for any single $J \in \mathcal{B}_{\rm col}$. □

Remark 4.9.20. We intend that the horizontal principal structure is defined primarily for an LM-matrix of full-row rank, while the vertical principal structure for an LM-matrix of full-column rank. It will be natural to ask: For an LM-matrix A of rank r in general what is the coarsest simultaneous decomposition of the row and the column sides which is finer than any decomposition induced by the CCF of an $r \times r$ nonsingular submatrix of A? A simple combination of the results for the horizontal and vertical structures gives a solution to this question if A is already in the CCF. All the diagonal blocks of A should remain in the CCF of an $r \times r$ nonsingular submatrix of A. The refinement of the horizontal tail of A is given by the principal structure of q while the principal structure of p gives the refinement of the vertical tail. This explains, at the same time, why we name the former the "horizontal principal structure" and the latter the "vertical principal structure." □

Proof of Theorem 4.9.18. We need to define the function "q" of (4.153) for submatrices of $A = \binom{Q}{T}$. Define $\kappa : \mathcal{L} \times 2^C \to \mathbf{Z}$ by

$$\kappa((W,I),J) = |\{j \in J \mid Q_j \notin W^\perp \text{ or } \exists i \in I : T_{ij} \neq 0\}|, \quad (W,I) \in \mathcal{L},\ J \subseteq C.$$

Then for $J \subseteq C$ the function $q_J : \mathcal{L} \to \mathbf{Z}$ associated with the submatrix $A[R,J]$ by (4.153) is given by

$$q_J(W,I) = \kappa((W,I),J) - \dim W - |I|, \qquad (W,I) \in \mathcal{L}.$$

Lemma 4.9.21.

$$q_{J_1}(X_1) + q_{J_2}(X_2) \geq q_{J_1 \cap J_2}(X_1 \vee X_2) + q_{J_1 \cup J_2}(X_1 \wedge X_2), \\ X_i \in \mathcal{L},\ J_i \subseteq C\ (i = 1,2).$$

Proof. With $\Omega((W,I),J) = \{j \in J \mid Q_j \in W^\perp,\ T[I,j] = \mathbf{0}\}$ we have

$$q_J(W,I) = |J| - |\Omega((W,I),J)| - \dim W - |I|, \qquad (W,I) \in \mathcal{L},\ J \subseteq C.$$

Noting

$$\Omega(X_1,J_1) \cap \Omega(X_2,J_2) = \Omega(X_1 \vee X_2, J_1 \cap J_2), \\ \Omega(X_1,J_1) \cup \Omega(X_2,J_2) \subseteq \Omega(X_1 \wedge X_2, J_1 \cup J_2),$$

we obtain

$$|\Omega(X_1,J_1)| + |\Omega(X_2,J_2)| \\ = |\Omega(X_1,J_1) \cap \Omega(X_2,J_2)| + |\Omega(X_1,J_1) \cup \Omega(X_2,J_2)| \\ \leq |\Omega(X_1 \vee X_2, J_1 \cap J_2)| + |\Omega(X_1 \wedge X_2, J_1 \cup J_2)|.$$

This implies the desired inequality. ■

Consider a submatrix $A[R, J]$ for $J \in \mathcal{B}_{\rm col}$. Since $A[R, J]$ is of full-column rank, having no horizontal tail in its CCF, we have

$$\mathcal{K}_{\rm PS}(q_J) = \mathcal{L}_{\min}(q_J) = \mathcal{L}_{\rm CCF}(R, J) \tag{4.154}$$

by Lemma 4.9.3 and the arguments in §4.8.

Lemma 4.9.22. *For $X \in \mathcal{L}$ and $J \subseteq C$ we have $D(q_C; X) \preceq D(q_J; X)$.*

Proof. Put $D_C = D(q_C; X)$ and $D_J = D(q_J; X)$. By Lemma 4.9.21 we have

$$q_C(D_C) - q_C(D_C \wedge D_J) \geq q_J(D_C \vee D_J) - q_J(D_J).$$

The right-hand side is nonnegative since $X \preceq D_C \vee D_J$ and $D_J \in \mathcal{L}_{\min}(q_J; X)$. This implies $D_C \preceq D_J$. See the proof of Lemma 4.9.14. ■

Lemma 4.9.23. *For $X \in \mathcal{L}$, there exists $J \in \mathcal{B}_{\rm col}$ such that $D(q_C; X) = D(q_J; X)$.*

Proof. Put $D_C = (W_C, I_C) = D(q_C; X)$. By an LM-admissible transformation that takes a basis in V_Q compatible with W_C, the matrix A can be transformed to the following form:

$$\hat{A} = \begin{bmatrix} \hat{Q} \\ T \end{bmatrix} = \begin{array}{c} \\ R'_Q \\ H \\ R'_T \\ I_C \end{array} \!\!\begin{array}{c} \begin{array}{cc} \quad C' \quad & \quad K \quad \end{array} \\ \begin{pmatrix} \hat{Q}[R'_Q, C'] & \hat{Q}[R'_Q, K] \\ O & \hat{Q}[H, K] \\ T[R'_T, C'] & T[R'_T, K] \\ O & T[I_C, K] \end{pmatrix} \end{array},$$

where $|H| = \dim W_C$, $R'_Q = \mathrm{Row}(\hat{Q}) \setminus H$, $R'_T = R_T \setminus I_C$,

$$K = \{j \in C \mid Q_j \notin {W_C}^{\perp} \text{ or } \exists\, i \in I_C : T_{ij} \neq 0\}$$

and $C' = C \setminus K$. We put $\hat{R} = \mathrm{Row}(\hat{A})$.

Putting $R' = R'_Q \cup R'_T$ we claim that

$$\mathrm{rank}\, \hat{A}[R', C'] = |R'|. \tag{4.155}$$

This can be shown as follows. Since $\hat{A}[H \cup I_C, C'] = O$, it holds that

$$\mathrm{rank}\, \hat{A}[R', C'] = \mathrm{rank}\, \hat{A}[\hat{R}, C'] = \mathrm{rank}\, A[R, C']. \tag{4.156}$$

Applying Lemma 4.9.17 to $A[R, C']$ and noting that $Q_j \in {W_C}^{\perp}$ and $T[I_C, j] = \mathbf{0}$ for all $j \in C'$, we obtain

$$\begin{aligned}\operatorname{rank} A[R, C'] &= \min\{q_{C'}(W,I) \mid (W,I) \in \mathcal{L}\} + m \\ &= \min\{q_{C'}(W,I) \mid W \supseteq W_C,\ I \supseteq I_C\} + m. \qquad (4.157)\end{aligned}$$

For $W \supseteq W_C$ and $I \supseteq I_C$, we have

$$\kappa((W,I),C') = \kappa((W,I),C) - |K| = \kappa((W,I),C) - \kappa((W_C,I_C),C).$$

Hence it holds that

$$\begin{aligned}q_{C'}(W,I) &= \kappa((W,I),C') - |I| - \dim W \\ &= q_C(W,I) - q_C(W_C,I_C) - |I_C| - \dim W_C, \qquad (4.158)\end{aligned}$$

in which

$$q_C(W,I) - q_C(W_C,I_C) \geq 0 \qquad (4.159)$$

by the definition of $D_C = (W_C, I_C)$ and $(W,I) \succeq X$. Combining (4.156), (4.157), (4.158), (4.159), and $m - |I_C| - \dim W_C = |R'|$, we obtain (4.155).

Therefore there exists $J' \subseteq C'$ such that

$$\operatorname{rank} \hat{A}[R', J'] = |R'| = |J'|.$$

At the same time, there exists $J_K \subseteq K$ such that

$$\operatorname{rank} \hat{A}[H \cup I_C, K] = \operatorname{rank} \hat{A}[H \cup I_C, J_K] = |J_K|.$$

Put $J = J' \cup J_K$. We have

$$\operatorname{rank} \hat{A}[\hat{R}, J] = |J|, \qquad (4.160)$$

since both $\hat{A}[R', J']$ and $\hat{A}[H \cup I_C, J_K]$ are of full-column rank, and $\hat{A}[H \cup I_C, J'] = O$. On the other hand, since $\hat{A}[R', C']$ is of full-row rank,

$$\operatorname{rank} \hat{A} = \operatorname{rank} \hat{A}[R', C'] + \operatorname{rank} \hat{A}[H \cup I_C, K] = |J'| + |J_K| = |J|.$$

Thus we obtain $J \in \mathcal{B}_{\mathrm{col}}$.

Applying Lemma 4.9.17 to $A[R, J]$ and using $\operatorname{rank} A[R, J] = \operatorname{rank} \hat{A}[\hat{R}, J]$ and (4.160), we obtain

$$\min\{q_J(Y) \mid Y \in \mathcal{L}\} = |J| - m,$$

which together with $q_J(D_C) = |J_K| - |I_C| - \dim W_C = |J| - m$ and $X \preceq D_C$ implies

$$q_J(D_C) = \min\{q_J(Y) \mid X \preceq Y \in \mathcal{L}\}.$$

Thus we obtain $D_C \succeq D(q_J; X)$, which completes the proof since we have already shown $D_C \preceq D(q_J; X)$ in Lemma 4.9.22. ■

We are now ready to complete the proof of Theorem 4.9.18. It follows from Lemma 4.9.22 that $D(q_J; X) = X$ implies $D(q_C; X) = X$. Hence, $\mathcal{K}_{\mathrm{PS}}(q_J) \subseteq \mathcal{K}_{\mathrm{PS}}(q_C)$ holds for any $J \subseteq C$. On the other hand, from Lemma 4.9.23, $X = D(q_C; X)$ implies the existence of $J \in \mathcal{B}_{\mathrm{col}}$ such that $X = D(q_J; X)$. Hence

$$\mathcal{K}_{\mathrm{PS}}(q_C) = \bigcup_{J \in \mathcal{B}_{\mathrm{col}}} \mathcal{K}_{\mathrm{PS}}(q_J),$$

which establishes Theorem 4.9.18 when combined with (4.154).

Notes. A series of theorems described in this section, due to McCormick [190], Murota [210], and Iwata–Murota [145], clarified a relationship between the two general decomposition principles for submodular functions, i.e., between the Jordan–Hölder-type theorem of §2.2.2 and the principal structure of §4.9.2, in the context of block-triangularization of matrices. The combinatorial essence of those theorems has been extracted by Iwata–Murota [143] and Iwata [139] without reference to matrices.

5. Polynomial Matrix and Valuated Matroid

Matrices consisting of polynomials or rational functions play fundamental roles in various branches in engineering. Combinatorial properties of polynomial matrices are abstracted in the language of valuated matroids. This chapter is mostly devoted to an exposition of the theory of valuated matroids, whereas the first section describes a number of canonical forms of polynomial/rational matrices that are amenable to combinatorial methods of analysis to be explained in Chap. 6.

5.1 Polynomial/Rational Matrix

Matrices consisting of polynomials or rational functions play fundamental roles in various branches in engineering (Gohberg–Lancaster–Rodman [95]). In dynamical system theory, for example, a linear time-invariant system is described by a polynomial matrix called the system matrix (the Laplace transform of the state-space equations), or by a rational function matrix called the transfer function matrix (Rosenbrock [284], Vidyasagar [331]).

In this section three canonical forms of polynomial/rational matrices are described: the Smith form of polynomial matrices, the Smith–McMillan form at infinity of rational matrices, and the Kronecker form of matrix pencils.

5.1.1 Polynomial Matrix and Smith Form

Let $A(s) = (A_{ij}(s))$ be an $m \times n$ polynomial matrix with $A_{ij}(s)$ being a polynomial in s with coefficients from a certain field $\boldsymbol{F}$ (i.e., $A_{ij}(s) \in \boldsymbol{F}[s]$). Typically $\boldsymbol{F}$ is the real number field $\mathbf{R}$.

The kth *determinantal divisor*, denoted by $d_k(s)$, is defined to be the greatest common divisor of all the minors of order k:

$$d_k(s) = \gcd\{\det A[I,J] \mid |I| = |J| = k\} \qquad (k = 0, 1, \cdots, r), \tag{5.1}$$

where $r = \operatorname{rank} A$ and $d_0(s) = 1$ by convention. The kth *invariant factor* (or *invariant polynomial*), denoted by $e_k(s)$, is defined by

$$e_k(s) = \frac{d_k(s)}{d_{k-1}(s)} \qquad (k = 1, \cdots, r). \tag{5.2}$$

Note that $d_{k-1}(s)$ divides $d_k(s)$ by the Laplace expansion (Proposition 2.1.2). Furthermore, it is known that $e_{k-1}(s)$ divides $e_k(s)$ for $k = 1, \cdots, r$ (see Example 5.2.16 for a combinatorial proof).

We call a polynomial matrix $U(s)$ *unimodular* if it is square and its determinant is a nonvanishing constant (in $\boldsymbol{F}$). A square polynomial matrix $U(s)$ is invertible (i.e., $\exists\, U^{-1}$ with $(U^{-1})_{ji} \in \boldsymbol{F}[s]$) if and only if it is unimodular. The following fundamental result claims the existence of a canonical diagonal form under the unimodular equivalence.

Theorem 5.1.1 (Smith normal form). *For a polynomial matrix $A(s)$, there exist unimodular matrices $U(s)$ and $V(s)$ such that*

$$U(s)A(s)V(s) = \mathrm{diag}\,(e_1(s), \cdots, e_r(s), 0, \cdots, 0),$$

where $r = \mathrm{rank}\, A(s)$, and $e_k(s)$ $(k = 1, \cdots, r)$ are polynomials such that $e_{k-1}(s)$ divides $e_k(s)$ for $k = 2, \cdots, r$. Furthermore, $e_k(s)$ coincides with the kth invariant factor given by (5.2). □

The Smith normal form is uniquely determined by (5.2) and is invariant under unimodular equivalence transformations. A significance of the Smith normal form is indicated by the following fact.

Lemma 5.1.2. *For a polynomial matrix $A(s)$ and a polynomial vector $\boldsymbol{b}(s)$, there exists a polynomial vector $\boldsymbol{x}(s)$ such that $A(s)\boldsymbol{x}(s) = \boldsymbol{b}(s)$ if and only if $A(s)$ and $[A(s) \mid \boldsymbol{b}(s)]$ have the same invariant factors.* □

Remark 5.1.3. The invariant factors $e_k(s)$ are also called the *elementary divisors*, though some books (e.g., Gantmacher [87]) distinguish between the two, defining elementary divisors to be the prime powers appearing in the factorization of the invariant factors. □

Remark 5.1.4. The theorem on the Smith normal form holds true more generally for a matrix over a PID (principal ideal domain), of which a Euclidean domain (e.g., the ring of polynomials in a single variable) is a special case. A square matrix U over a PID, say R, is invertible (i.e., $\exists\, U^{-1}$ with $(U^{-1})_{ji} \in R$) if and only if its determinant is an invertible element in R. Such a matrix U is said to be *unimodular* over R. Theorem 5.1.1 can be generalized as follows: For a matrix A over R, there exist unimodular matrices U and V such that $UAV = \mathrm{diag}\,(e_1, \cdots, e_r, 0, \cdots, 0)$, where $r = \mathrm{rank}\, A$ and e_{k-1} divides e_k for $k = 1, \cdots, r$. Furthermore, $e_k = d_k / d_{k-1}$ with $d_k = \gcd\{\det A[I, J] \mid |I| = |J| = k\}$ $(k = 1, \cdots, r)$. See Newman [252] for the proof. □

5.1.2 Rational Matrix and Smith–McMillan Form at Infinity

Let $A(s) = (A_{ij}(s))$ be an $m \times n$ rational function matrix with $A_{ij}(s)$ being a rational function in s with coefficients from a certain field $\boldsymbol{F}$ (i.e., $A_{ij}(s) \in$

$\boldsymbol{F}(s)$). Typically $\boldsymbol{F}$ is the real number field $\mathbf{R}$. In this book, we are often concerned with the highest degree of a minor (subdeterminant) of order k of $A(s)$:

$$\delta_k = \delta_k(A) = \max\{\deg_s \det A[I,J] \mid |I| = |J| = k\} \quad (k = 0, 1, 2, \cdots). \tag{5.3}$$

By convention $\delta_0(A) = 0$, and $\delta_k(A) = -\infty$ for $k > r$.

A rational function $f(s)$ is called *proper* if $\deg_s f(s) \leq 0$, where the degree of a rational function $f(s) = p(s)/q(s)$ (with $p(s)$ and $q(s)$ being polynomials) is defined by

$$\deg_s f(s) = \deg_s p(s) - \deg_s q(s), \qquad f(s) = p(s)/q(s).$$

By convention we put $\deg_s(0) = -\infty$.

We call a rational matrix $A(s) = (A_{ij}(s))$ *proper* if its entries are proper rational functions, i.e., $\deg_s A_{ij}(s) \leq 0$ for all (i,j). A square proper rational matrix is called *biproper* if it is invertible and its inverse is a proper rational matrix. A proper rational matrix $A(s)$ is biproper if and only if $\deg_s \det A(s) = 0$.

Since the proper rational functions form a Euclidean ring, any proper rational matrix can be brought into a canonical form (the Smith form) according to the general principle explained in Remark 5.1.4. This is sometimes referred to as the *structure at infinity* in the control literature. From this we see further that any rational matrix can be brought into the Smith–McMillan form at infinity, as stated below (Verghese–Kailath [329]).

Theorem 5.1.5 (Smith–McMillan form at infinity). *For a rational function matrix $A(s)$, there exist biproper matrices $U(s)$ and $V(s)$ such that*

$$U(s)A(s)V(s) = \begin{pmatrix} \Gamma(s) & O \\ O & O \end{pmatrix},$$

where

$$\Gamma(s) = \mathrm{diag}\,(s^{t_1}, \cdots, s^{t_r}),$$

$r = \mathrm{rank}\, A(s)$, *and* $t_k = t_k(A)$ $(k = 1, \cdots, r)$ *are integers with* $t_1 \geq \cdots \geq t_r$. *Furthermore,* t_k *can be expressed in terms of the minors of* A *as*

$$t_k(A) = \delta_k(A) - \delta_{k-1}(A) \qquad (k = 1, \cdots, r) \tag{5.4}$$

using $\delta_k(A)$ *in* (5.3). □

The integers t_k $(k = 1, \cdots, r)$, uniquely determined by (5.4), are referred to as the *contents at infinity*. If they are positive, t_k $(k = 1, \cdots, r)$ are the *orders of the poles at infinity*, and if negative, $-t_k$ $(k = 1, \cdots, r)$ are the *orders of the zeroes at infinity*. It should be noted in (5.4) that $\delta_k(A)$'s are invariant under biproper equivalence transformations, that is,

$$\delta_k(A) = \delta_k(A') \qquad (k = 1, \cdots, r) \tag{5.5}$$

if $A'(s) = U(s)A(s)V(s)$ with biproper matrices $U(s)$ and $V(s)$.

A significance of the Smith–McMillan normal form at infinity is indicated by the following fact.

Lemma 5.1.6. *For a rational function matrix $A(s)$ and a rational function vector $\boldsymbol{b}(s)$, there exists a vector $\boldsymbol{x}(s)$ of proper rational functions such that $A(s)\boldsymbol{x}(s) = \boldsymbol{b}(s)$ if and only if $A(s)$ and $[A(s) \mid \boldsymbol{b}(s)]$ have the same contents at infinity.* □

Remark 5.1.7. A (proper) rational function matrix typically appears as the transfer function matrix of a linear time-invariant dynamical system, and the Smith–McMillan form at infinity of the transfer function matrix has control-theoretic significances (decoupling, disturbance rejection, etc.). See Bhattacharyya [11], Commault–Dion [37], Descusse–Dion [47], Hautus [103], Hautus–Heymann [104, 105], Svaricek [307], and Verghese–Kailath [329].

The *transfer function matrix* of a system in the descriptor form

$$F\dot{\boldsymbol{x}}(t) = A\boldsymbol{x}(t) + B\boldsymbol{u}(t), \quad \boldsymbol{y}(t) = C\boldsymbol{x}(t)$$

with "state" $\boldsymbol{x}(t) \in \mathbf{R}^N$, input $\boldsymbol{u}(t)$, and output $\boldsymbol{y}(t)$, is given by

$$P(s) = C(sF - A)^{-1}B,$$

provided that $\det(sF - A) \neq 0$ (while F can be singular). In such a case it is desirable to express the Smith–McMillan form at infinity of $P(s)$ directly from the matrices F, A, B and C, without referring to the entries of $P(s)$ explicitly. From the formula (cf. Proposition 2.1.7)

$$\det \begin{pmatrix} A - sF & B' \\ C' & O \end{pmatrix} = \det(A - sF) \cdot \det[-C'(A - sF)^{-1}B'],$$

where C' denotes a submatrix of C with k rows and B' is a submatrix of B with k columns, it follows that

$$\delta_k(P) = \delta_{N+k}(D; I_0, J_0) - \delta_N(A - sF),$$

where

$$D(s) = \begin{pmatrix} A - sF & B \\ C & O \end{pmatrix},$$

I_0 and J_0 are respectively the row and column sets corresponding to the $N \times N$ nonsingular submatrix $A - sF$, and

$$\delta_{N+k}(D; I_0, J_0) = \max\{\deg_s \det D[I, J] \mid I \supseteq I_0,\ J \supseteq J_0, |I| = |J| = N + k\}$$

means the highest degree of a minor of order $N + k$ that contains row set I_0 and column set J_0. Note that $\delta_{N+k}(D; I_0, J_0) = \delta_{N+k}(\tilde{D}) - 2Nd$ for a sufficiently large integer d and

$$\tilde{D}(s) = \begin{pmatrix} \mathrm{diag}\,(s^d, \cdots, s^d) & O \\ O & I \end{pmatrix} \begin{pmatrix} A - sF & B \\ C & O \end{pmatrix} \begin{pmatrix} \mathrm{diag}\,(s^d, \cdots, s^d) & O \\ O & I \end{pmatrix}.$$

□

5.1.3 Matrix Pencil and Kronecker Form

A polynomial matrix $A(s) = (A_{ij}(s))$ with $\deg_s A_{ij}(s) \leq 1$ for all (i,j) is called a *pencil.* Obviously, a pencil $A(s)$ can be represented as $A(s) = sX + Y$ in terms of a pair of constant matrices X and Y. A pencil $A(s)$ is said to be *regular* if it is square and $\det A(s)$ is a nonvanishing polynomial. A pencil is called *singular* if it is not regular.

A pencil can be brought into a canonical block-diagonal matrix by means of *strict equivalence* $PA(s)Q$ using constant nonsingular matrices P and Q. The block-diagonal matrix is known as the Kronecker form (Gantmacher [87, Chap. XII]). For $m \geq 1$ and $\varepsilon \geq 0$, we define an $m \times m$ bidiagonal matrix $N_m(s)$ and an $\varepsilon \times (\varepsilon+1)$ bidiagonal matrix $L_\varepsilon(s)$ by

$$N_m(s) = \begin{pmatrix} 1 & s & & & \\ & 1 & s & & \\ & & \ddots & \ddots & \\ & & & \ddots & s \\ & & & & 1 \end{pmatrix}, \quad L_\varepsilon(s) = \begin{pmatrix} 1 & s & & & & \\ & 1 & s & & & \\ & & \ddots & \ddots & & \\ & & & \ddots & s & \\ & & & & 1 & s \end{pmatrix}.$$

For $\eta \geq 0$ we define $U_\eta(s)$ to be the transpose of $L_\eta(s)$.

Theorem 5.1.8 (Kronecker form). *For a pencil $A(s)$ over a field $\boldsymbol{F}$, there exist nonsingular matrices P and Q over $\boldsymbol{F}$ such that*

$$\begin{aligned} PA(s)Q = \text{block-diag}\,(sI_{m_0} + B; N_{m_1}(s), \cdots, N_{m_b}(s); \\ L_{\varepsilon_1}(s), \cdots, L_{\varepsilon_c}(s); U_{\eta_1}(s), \cdots, U_{\eta_d}(s)), \end{aligned} \quad (5.6)$$

where

$$m_1 \geq \cdots \geq m_b \geq 1, \quad \varepsilon_1 \geq \cdots \geq \varepsilon_c \geq 0, \quad \eta_1 \geq \cdots \geq \eta_d \geq 0,$$

and B is an $m_0 \times m_0$ matrix over $\boldsymbol{F}$. The indices, m_0; b, $m_1, \cdots, m_b$; c, $\varepsilon_1, \cdots, \varepsilon_c$; d, $\eta_1, \cdots, \eta_d$, are uniquely determined. Denoting $r = \operatorname{rank} A$ and using $\delta_k(A)$ $(k = 0, 1, 2, \cdots)$ in (5.3), we have

$$b = r - \max_{k \geq 0} \delta_k(A), \quad c = |\mathrm{Col}(A)| - r, \quad d = |\mathrm{Row}(A)| - r, \quad (5.7)$$

$$m_0 = \delta_r(A) - \sum_{i=1}^{c} \varepsilon_i - \sum_{j=1}^{d} \eta_j, \quad (5.8)$$

$$m_k = \delta_{r-k}(A) - \delta_{r-k+1}(A) + 1 \qquad (k = 1, \cdots, b). \quad (5.9)$$

For a regular pencil, in particular, the indices, m_0; b, $m_1, \cdots, m_b$, are determined by $\delta_k(A)$ $(k = 0, 1, 2, \cdots)$.

Proof.[1] The formulas (5.7)–(5.9) can be derived from the block diagonal structure (5.6). The uniqueness of the indices $\varepsilon_1, \cdots, \varepsilon_c$ and $\eta_1, \cdots, \eta_d$ is not difficult to establish.

The existence of block diagonal form (5.6) is proven here. Put $A(s) = sX + Y$, where X and Y are matrices over $\boldsymbol{F}$.

Claim: There exist nonsingular matrices $\bar{P}$ and $\bar{Q}$ over $\boldsymbol{F}$ and partitions $(R_1, \cdots, R_\mu; R_\infty)$ and $(C_1, \cdots, C_\mu; C_\infty)$ of the row set R and the column set C of $\bar{A}(s) = s\bar{X} + \bar{Y} = \bar{P}(sX + Y)\bar{Q}$ such that

$$\begin{aligned}
&\operatorname{rank} \bar{X}[R_i, C_j] = 0 \qquad (1 \le j \le \mu, j \le i \le \infty),\\
&\operatorname{rank} \bar{Y}[R_i, C_j] = 0 \qquad (1 \le j \le \mu, j+1 \le i \le \infty),\\
&\operatorname{rank} \bar{X}[R_{j-1}, C_j] = |C_j| \qquad (2 \le j \le \mu),\\
&\operatorname{rank} \bar{X}[R_\infty, C_\infty] = |C_\infty|,\\
&\operatorname{rank} \bar{Y}[R_i, C_i] = |R_i| \qquad (1 \le i \le \mu).
\end{aligned}$$

Here R_μ, R_∞, and C_∞ can be empty, whereas other blocks are nonempty. Note that $\bar{A}(s)$ is an upper block-triangular matrix and that the rank conditions imply

$$|C_1| \ge |R_1| \ge |C_2| \ge |R_2| \ge \cdots \ge |C_\mu| \ge |R_\mu|, \qquad |R_\infty| \ge |C_\infty|.$$

The upper block-triangular form in the claim can be constructed as follows. The column set C_1 is determined by a column-transformation for X, since $\bar{X}[R, C_1] = O$ and $\bar{X}[R, C \setminus C_1]$ is of full-column rank by the rank conditions. Then the row set R_1 is determined by a row-transformation for the submatrix $Y[R, C_1]$, since $\bar{Y}[R \setminus R_1, C_1] = O$ and $\bar{Y}[R_1, C_1]$ is of full-row rank. Next, C_2 is determined by a column-transformation for the submatrix $X[R \setminus R_1, C \setminus C_1]$ (with X denoting the modified X), since $\bar{X}[R \setminus R_1, C_2] = O$ and $\bar{X}[R \setminus R_1, (C \setminus C_1) \setminus C_2)]$ is of full-column rank. Then the row set R_2 is determined from the submatrix $Y[R \setminus R_1, C_2]$. Continuing this way, we eventually arrive at $C_{\mu+1} = \emptyset$ for some $\mu \ge 0$; then we terminate by defining R_∞ and C_∞ to be the complements of $\bigcup_{i=1}^{\mu} R_i$ and $\bigcup_{j=1}^{\mu} C_j$, respectively.

In the above claim we may further assume that

$$\begin{aligned}
&\bar{X}[R_i, C_j] = O \qquad \text{unless } 2 \le j = i+1 \le \mu \text{ or } i = j = \infty,\\
&\bar{Y}[R_i, C_j] = O \qquad \text{unless } 1 \le i = j \le \infty,\\
&\bar{X}[R_{j-1}, C_j] = \begin{bmatrix} I_{|C_j|} \\ O \end{bmatrix} \qquad (2 \le j \le \mu),\\
&\bar{X}[R_\infty, C_\infty] = \begin{bmatrix} I_{|C_\infty|} \\ O \end{bmatrix},\\
&\bar{Y}[R_i, C_i] = \begin{bmatrix} I_{|R_i|} & O \end{bmatrix} \qquad (1 \le i \le \mu),
\end{aligned}$$

[1] The present proof, valid for an arbitrary $\boldsymbol{F}$, is communicated by S. Iwata. See Gantmacher [87, Chap. XII, §2] for an alternative proof in the case of $\boldsymbol{F} = \mathbf{C}$.

where I_N denotes the identity matrix of order N. Then the matrix $\bar{A}(s)$ takes the form depicted in Fig. 5.1 for $\mu = 4$.

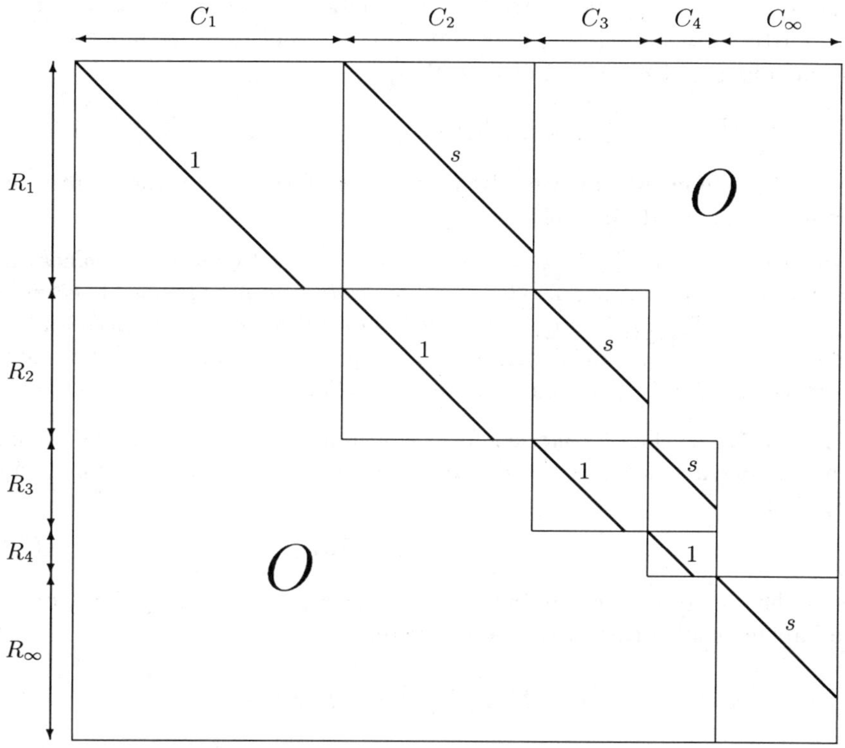

Fig. 5.1. Matrix $\bar{A}(s)$ in the proof for the Kronecker form ($\mu = 4$)

Consider the submatrix $\bar{A}[\bigcup_{i=1}^{\mu} R_i, \bigcup_{j=1}^{\mu} C_j]$. With suitable permutations of rows and columns, it can be put into a block-diagonal form with each diagonal block being equal to $N_m(s)$ with $1 \le m \le \mu$ or $L_\varepsilon(s)$ with $0 \le \varepsilon \le \mu - 1$, where $N_m(s)$ appears with multiplicity $|R_m| - |C_{m+1}|$ (note: $|C_{\mu+1}| = 0$) and $L_\varepsilon(s)$ with multiplicity $|C_{\varepsilon+1}| - |R_{\varepsilon+1}|$. Namely,

$$\bar{A}[\bigcup_{i=1}^{\mu} R_i, \bigcup_{j=1}^{\mu} C_j] = \text{block-diag}\,(N_{m_1}(s), \cdots, N_{m_b}(s); L_{\varepsilon_1}(s), \cdots, L_{\varepsilon_c}(s))$$

with

$$b = \sum_{i=1}^{\mu} |R_i| - \sum_{j=2}^{\mu} |C_j|, \qquad c = \sum_{j=1}^{\mu} |C_j| - \sum_{i=1}^{\mu} |R_i|,$$

$$|\{k \mid m_k = m\}| = |R_m| - |C_{m+1}| \qquad (1 \le m \le \mu),$$

$$|\{k \mid \varepsilon_k = \varepsilon\}| = |C_{\varepsilon+1}| - |R_{\varepsilon+1}| \qquad (0 \le \varepsilon \le \mu - 1).$$

For the submatrix $\bar{A}[R_\infty, C_\infty]$, we apply the above argument to its transpose. Let $(\hat{R}_1, \cdots, \hat{R}_{\hat{\mu}}; \hat{R}_\infty)$ and $(\hat{C}_1, \cdots, \hat{C}_{\hat{\mu}}; \hat{C}_\infty)$ be the resulting partitions of R_∞ and C_∞, respectively. Since $\mathrm{rank}\, \bar{X}[R_\infty, C_\infty] = |C_\infty|$, we have $|\hat{C}_i| = |\hat{R}_{i+1}|$ for $i = 1, \cdots, \hat{\mu} - 1$ and $|\hat{C}_{\hat{\mu}}| = 0$. This means that there exist nonsingular matrices $\hat{P}$ and $\hat{Q}$ such that

$$\hat{P}\bar{A}[R_\infty, C_\infty]\hat{Q} = \text{block-diag}\,(U_{\eta_1}(s), \cdots, U_{\eta_d}(s); s\hat{X}_\infty + \hat{Y}_\infty),$$

where $\hat{X}_\infty$ is nonsingular. Finally, we can transform $\hat{X}_\infty$ to the identity matrix to obtain the desired block-diagonal form (5.6). ■

The matrices $N_{m_k}(s)$ $(k = 1, \cdots, b)$ are called the *nilpotent blocks* and the number $m_1 = \max_{1 \le k \le b} m_k$ is the *index of nilpotency.* The indices $\{\varepsilon_1, \cdots, \varepsilon_c\}$ and $\{\eta_1, \cdots, \eta_d\}$ are called *Kronecker column indices* and *row indices*, respectively. For algorithms to compute the Kronecker form, see the references in Golub–Van Loan [97, pp. 389–390].

Remark 5.1.9. The Kronecker form is a fundamental tool for the analysis of a dynamical system in the *descriptor form* (Katayama [155], Luenberger [182, 183]):

$$F\frac{\mathrm{d}\boldsymbol{x}}{\mathrm{d}t} = A\boldsymbol{x} + B\boldsymbol{u}, \tag{5.10}$$

where the matrix F is square $(n \times n)$ but not necessarily nonsingular. Consider the Laplace transform[2] of the above system:

$$\left(A - sF \middle| B \right) \begin{pmatrix} \bar{\boldsymbol{x}} \\ \bar{\boldsymbol{u}} \end{pmatrix} = -F\, \boldsymbol{x}(0-),$$

where s is the symbol (or indeterminate) standing for the differentiation with respect to time, and $\bar{\boldsymbol{x}}$ and $\bar{\boldsymbol{u}}$ are the Laplace transforms of $\boldsymbol{x}(t)$ and $\boldsymbol{u}(t)$, respectively. For the unique solvability, as a system of differential equations in $\boldsymbol{x}$, the matrix $A - sF$ is usually assumed to be a regular pencil.

Then, by Theorem 5.1.8, there exist two real-constant nonsingular matrices P and Q such that

$$P(A - sF)Q = \text{block-diag}\,(A_0 - sI_{m_0}; I_{m_1} - sJ_{m_1}, \cdots, I_{m_b} - sJ_{m_b}),$$

where J_m is an $m \times m$ matrix defined by $N_m(s) = I_m + sJ_m$. Using this we can rewrite the descriptor form (5.10) into

$$\frac{\mathrm{d}\boldsymbol{x}_0}{\mathrm{d}t} = A_0\boldsymbol{x}_0 + B_0\boldsymbol{u}, \tag{5.11}$$

$$J_{m_k}\frac{\mathrm{d}\boldsymbol{x}_k}{\mathrm{d}t} = \boldsymbol{x}_k + B_k\boldsymbol{u} \qquad (k = 1, \cdots, b), \tag{5.12}$$

[2] To be more precise, $\mathcal{L}_-$ transform (Kailath [152, §1.2]) defined by $\bar{\boldsymbol{x}}(s) = \int_{0-}^{\infty} \boldsymbol{x}(t)\mathrm{e}^{-st}\mathrm{d}t$.

where $\boldsymbol{x}_k \in \mathbf{R}^{m_k}$ for $k = 0, 1, \cdots, b$ and

$$\begin{pmatrix} \boldsymbol{x}_0 \\ \boldsymbol{x}_1 \\ \vdots \\ \boldsymbol{x}_b \end{pmatrix} = Q^{-1}\boldsymbol{x}, \qquad \begin{pmatrix} B_0 \\ B_1 \\ \vdots \\ B_b \end{pmatrix} = PB.$$

The subsystems in (5.11) and (5.12) admit explicit solutions:

$$\boldsymbol{x}_0(t) = \exp(A_0 t)\boldsymbol{x}_0(0) + \int_0^t \exp[A_0(t-\tau)]B_0\boldsymbol{u}(\tau)\mathrm{d}\tau,$$

$$\boldsymbol{x}_k(t) = -\left(\sum_{p=0}^{m_k-2} \delta^{(p)}(t) {J_{m_k}}^{p+1}\right)\boldsymbol{x}_k(0-) - \sum_{p=0}^{m_k-1} {J_{m_k}}^{p} B_k \boldsymbol{u}^{(p)}(t)$$
$$(k = 1, \cdots, b),$$

where $\delta^{(p)}(t)$ and $\boldsymbol{u}^{(p)}(t)$ are the pth derivatives of the Dirac delta function (the unit impulse function) and the input-vector $\boldsymbol{u}(t)$, respectively. Thus the first subsystem (5.11), in the standard form, expresses the *exponential modes*, while the second (5.12) accounts for the *impulse modes*. In this context, $m_0 = \deg_s \det(A - sF)$ is sometimes called the *dynamical degree* (Hayakawa–Hosoe–Ito [108]), which stands for the number of exponential modes, whereas the number of impulse modes is represented by $\sum_{k=1}^{b}(m_k - 1) = \operatorname{rank} F - \deg_s \det(A - sF)$. See Suda [303] for more about the role of the Kronecker form in control theory. □

Remark 5.1.10. The index of nilpotency has an important significance in numerical analysis of a system of equations consisting of a mixture of differential and algebraic relations, which is often abbreviated to DAE in the literature of numerical analysis. For a linear time-invariant DAE in general, say $A\boldsymbol{x} = \boldsymbol{b}$ with $A = A(s)$ being an $n \times n$ nonsingular polynomial matrix in s, the index is defined by

$$\nu(A) = \max_{i,j} \deg_s (A^{-1})_{ji} + 1.$$

Here it should be clear that each entry $(A^{-1})_{ji}$ of A^{-1} is a rational function in s. An alternative expression for $\nu(A)$ is

$$\nu(A) = \delta_{n-1}(A) - \delta_n(A) + 1.$$

When $\deg_s A_{ij}(s) \leq 1$ for all (i, j), the index $\nu(A)$ agrees with the index of nilpotency of A as a matrix pencil; namely, we have $\nu(A) = m_1$.

The solution $\boldsymbol{x}$ to $A\boldsymbol{x} = \boldsymbol{b}$ is of course given by $\boldsymbol{x} = A^{-1}\boldsymbol{b}$, and therefore $\nu(A) - 1$ equals the highest order of the derivatives of the input $\boldsymbol{b}$ that can possibly appear in the solution $\boldsymbol{x}$. As such, a high index indicates the difficulty in numerical solution of the DAE, and sometimes even the inadequacy in mathematical modeling. The structural approach to the DAE index has been expounded in Chap. 1. See Brenan–Campbell–Petzold [21], Gear [88, 89], Hairer–Wanner [101], and Ungar–Kröner–Marquardt [324] for more about the index of DAE. □

Remark 5.1.11. It turns out that the Smith form of a polynomial matrix is closely related, at least in the generic case, to the DM-decomposition and the CCF of LM-matrices (to be explained in §6.3). The combinatorial properties of the degree of subdeterminants, on the other hand, will be investigated in the next section in a more abstract framework of "valuated matroids." □

5.2 Valuated Matroid

5.2.1 Introduction

While matroids are a combinatorial abstraction of matrices over a field with respect to linear independence, valuated matroids originate from a combinatorial structure of polynomial/rational matrices with respect to the degree of determinants. The axiomatic development will be motivated and illustrated by the special case of polynomial/rational matrices. The concept of valuated matroids was introduced by Dress–Wenzel [54, 57].

A *valuated matroid* is a pair $\mathbf{M} = (V, \omega)$ of a finite set V and a function $\omega : 2^V \to \mathbf{R} \cup \{-\infty\}$ such that

$$\mathcal{B} = \{B \subseteq V \mid \omega(B) \neq -\infty\} \tag{5.13}$$

is nonempty and that the following exchange property holds:

(VM) For $B, B' \in \mathcal{B}$ and $u \in B \setminus B'$, there exists $v \in B' \setminus B$ such that $B - u + v \in \mathcal{B}$, $B' + u - v \in \mathcal{B}$, and

$$\omega(B) + \omega(B') \leq \omega(B - u + v) + \omega(B' + u - v). \tag{5.14}$$

If this is the case, $\mathcal{B}$ satisfies the simultaneous exchange property:

(BM$_\pm$) For $B, B' \in \mathcal{B}$ and for $u \in B \setminus B'$, there exists $v \in B' \setminus B$ such that $B - u + v \in \mathcal{B}$ and $B' + u - v \in \mathcal{B}$

introduced in §2.3.4, and accordingly $\mathcal{B}$ forms the basis family of a matroid. Therefore, we can alternatively say that a valuated matroid is a triple $\mathbf{M} = (V, \mathcal{B}, \omega)$, where $(V, \mathcal{B})$ is a matroid (defined in terms of the basis family) and $\omega : \mathcal{B} \to \mathbf{R}$ is a function satisfying (VM). It is also said that ω is a *valuation* of the matroid $(V, \mathcal{B})$. We denote by r the rank of the underlying matroid $(V, \mathcal{B})$.

A valuated matroid $\mathbf{M} = (V, \mathcal{B}, \omega)$ such that $\omega(B) = 0$ for all $B \in \mathcal{B}$ can be identified with the underlying matroid $(V, \mathcal{B})$. In fact, (VM) for $\omega : 2^V \to \{0, -\infty\}$ reduces to (BM$_\pm$). This ω is called the *trivial valuation.*

Remark 5.2.1. As we have seen in §2.3, the theory of matroids offers deep and useful results for a pair of matroids (independent matchings as well as intersection in §2.3.5 and union in §2.3.6). The most interesting part of the theory of valuated matroids lies in a generalization of these results, to be described in §5.2.9. Specifically, for two valuated matroids $\mathbf{M}_1 = (V, \mathcal{B}_1, \omega_1)$ and $\mathbf{M}_2 = (V, \mathcal{B}_2, \omega_2)$, the "sum" $\omega_1 + \omega_2$ turns out to be a nice combinatorial object, though it is not a valuated matroid in general. Note that $(\omega_1 + \omega_2)(B) > -\infty$ if and only if B is a common base (i.e., $B \in \mathcal{B}_1 \cap \mathcal{B}_2$). □

5.2.2 Examples

Examples of valuated matroids are shown.

Example 5.2.2. A linear weighting on a matroid $(V, \mathcal{B})$ is a valuation. That is, for $p : V \to \mathbf{R}$ and $\alpha \in \mathbf{R}$, the function $\omega : \mathcal{B} \to \mathbf{R}$ defined by

$$\omega(B) = \alpha + \sum \{p(u) \mid u \in B\} \qquad (B \in \mathcal{B})$$

is a matroid valuation, satisfying (VM) with equality in (5.14). Such ω is called a *separable valuation.* □

Example 5.2.3. Let $A(s)$ be an $m \times n$ matrix of rank m with each entry being a rational function in a variable s, and let $(C, \mathcal{B})$ denote the (linear) matroid defined on the column set C of $A(s)$ in terms of the linear independence of the column vectors (cf. Example 2.3.8). Namely, $\mathcal{B} = \{B \subseteq C \mid \det A[R, B] \neq 0\}$, where R denotes the row set of A. Then $\omega : \mathcal{B} \to \mathbf{Z}$ defined by

$$\omega(B) = \deg_s \det A[R, B] \qquad (B \in \mathcal{B}) \tag{5.15}$$

is a valuation of $(C, \mathcal{B})$. In fact, by considering the degree of the terms in the Grassmann–Plücker identity (Proposition 2.1.4):

$$\det A[R, B] \cdot \det A[R, B'] = \sum_{j \in B' \setminus B} \det A[R, B - i + j] \cdot \det A[R, B' + i - j]$$

for $i \in B \setminus B'$, we obtain

$$\omega(B) + \omega(B') \leq \max_{j \in B' \setminus B} [\omega(B - i + j) + \omega(B' + i - j)],$$

the exchange axiom (VM). This observation by Dress–Wenzel [57] is the origin of the concept of valuated matroids.

A concrete instance of (nonseparable) valuation of this kind is provided by

$$A(s) = \begin{array}{c} \begin{matrix} x_1 & x_2 & x_3 & x_4 \end{matrix} \\ \boxed{\begin{matrix} s+1 & s & 1 & 0 \\ 1 & 1 & 1 & 1 \end{matrix}} \end{array},$$

where $C = \{x_1, x_2, x_3, x_4\}$. For $B = \{x_1, x_2\}$ and $B' = \{x_3, x_4\}$ we have $\omega(B) = \omega(B') = 0$ and $\omega(B - x_i + x_j) = \omega(B' + x_i - x_j) = 1$ for $i = 1, 2$ and $j = 3, 4$. □

Example 5.2.4. This is an example from combinatorial optimization due to Murota [224]. Let $G = (V, A)$ be a directed graph, and S and T be disjoint subsets of the vertex set V. By L ($\subseteq A$) we denote (the arc set of) a Menger-type vertex-disjoint linking from S to T, and by $\partial^+ L$ ($\subseteq S$) the set of its initial vertices. Put $\mathcal{B} = \{\partial^+ L \mid L \in \mathcal{L}\}$, where $\mathcal{L}$ ($\subseteq 2^A$) denotes the family of maximum linkings. It is well known (see also the augmenting-path argument below) that $\mathcal{B}$ forms the basis family of a matroid $(S, \mathcal{B})$. Given a cost function $\gamma : A \to \mathbf{R}$, define a function $\omega : \mathcal{B} \to \mathbf{R}$ by

$$\omega(B) = -\min_{L}\{\sum_{a \in L} \gamma(a) \mid \partial^+ L = B, L \in \mathcal{L}\} \qquad (B \in \mathcal{B}).$$

By definition, $-\omega(B)$ means the minimum cost of a maximum linking L with initial vertex set B.

The function ω is a valuation of $(S, \mathcal{B})$, satisfying the exchange axiom (VM). To see this, let $L, L' \in \mathcal{L}$ be such that $\partial^+ L = B$, $\partial^+ L' = B'$, $\omega(B) = -\sum_{a \in L} \gamma(a)$, and $\omega(B') = -\sum_{a \in L'} \gamma(a)$. For $u \in B \setminus B'$, a standard augmenting-path argument shows that there exists $P \subseteq (L \setminus L') \cup (L' \setminus L)$ such that $(P \cap L) \cup (\overline{P \cap L'})$ forms a directed path from u to some $v \in B' \setminus B$, where $\overline{P \cap L'}$ means the set of arcs in $P \cap L'$ reoriented. Note the maximality of L and L' is used here. For $\tilde{L} = (L \setminus (P \cap L)) \cup (P \cap L')$ and $\tilde{L}' = (L' \setminus (P \cap L')) \cup (P \cap L)$, we have $\tilde{L}, \tilde{L}' \in \mathcal{L}$, $\partial^+ \tilde{L} = B - u + v$, $\partial^+ \tilde{L}' = B' + u - v$, and therefore,

$$\begin{aligned} \omega(B) + \omega(B') &= -\sum_{a \in L} \gamma(a) - \sum_{a \in L'} \gamma(a) = -\sum_{a \in \tilde{L}} \gamma(a) - \sum_{a \in \tilde{L}'} \gamma(a) \\ &\le \omega(B - u + v) + \omega(B' + u - v). \end{aligned}$$

An example of this kind is provided by $G = (V, A)$ with $V = S \cup T$, $S = \{s_1, s_2, s_3, s_4\}$, $T = \{t_1, t_2\}$,

$$A = \{(s_1, t_1), (s_2, t_2), (s_3, t_1), (s_3, t_2), (s_4, t_1), (s_4, t_2)\},$$

$\gamma(a) = 0$ except for $\gamma(s_3, t_2) = 1$ and $\gamma(s_4, t_2) = 2$. We have, for example, $\omega(\{s_1, s_2\}) = 0$, $\omega(\{s_1, s_3\}) = -1$, $\omega(\{s_1, s_4\}) = -2$. This valuation is not separable, since the system of equations $\alpha + p_i + p_j = \omega(\{s_i, s_j\})$ $(i \ne j)$ in $(\alpha, p_1, p_2, p_3, p_4) \in \mathbf{R}^5$ has no solution. □

5.2.3 Basic Operations

Basic operations for a valuated matroid, such as dual, restriction, contraction, truncation, and elongation, are explained here, whereas other more sophisticated operations such as union are treated in §5.2.6.

Let $\mathbf{M} = (V, \mathcal{B}, \omega)$ be a valuated matroid. For $\alpha \in \mathbf{R}$ and $0 \leq \beta \in \mathbf{R}$, the function $\omega_{\alpha,\beta} : \mathcal{B} \to \mathbf{R}$ defined by

$$\omega_{\alpha,\beta}(B) = \alpha + \beta\omega(B) \qquad (B \in \mathcal{B})$$

is a matroid valuation. For $p : V \to \mathbf{R}$ we define $\omega[p] : \mathcal{B} \to \mathbf{R}$ (or $\omega[p] : 2^V \to \mathbf{R} \cup \{-\infty\}$) by

$$\omega[p](B) = \omega(B) + \sum\{p(u) \mid u \in B\}. \tag{5.16}$$

$\mathbf{M}[p] = (V, \mathcal{B}, \omega[p])$ is again a valuated matroid, called a *similarity transformation* of $\mathbf{M}$ by p. A linear weighting is a similarity transformation of the trivial valuation.

The *dual* of $\mathbf{M} = (V, \mathcal{B}, \omega)$ is a valuated matroid $\mathbf{M}^* = (V, \mathcal{B}^*, \omega^*)$ defined by

$$\mathcal{B}^* = \{B \subseteq V \mid V \setminus B \in \mathcal{B}\}, \qquad \omega^*(B) = \omega(V \setminus B) \ \text{ for } \ B \in \mathcal{B}^*.$$

The *restriction* and the *contraction* of $\mathbf{M} = (V, \mathcal{B}, \omega)$ to U ($\subseteq V$) are defined as follows (Dress–Wenzel [57]). Let $(U, \mathcal{B}^U)$ and $(U, \mathcal{B}_U)$ be the restriction and the contraction of the underlying matroid $(V, \mathcal{B})$ to U. Similarly for $(V \setminus U, \mathcal{B}^{V\setminus U})$ and $(V \setminus U, \mathcal{B}_{V\setminus U})$. Fix a base I of $(V \setminus U, \mathcal{B}_{V\setminus U})$ and a base J of $(V \setminus U, \mathcal{B}^{V\setminus U})$, and define $\omega_I^U : \mathcal{B}^U \to \mathbf{R}$ and $\omega_U^J : \mathcal{B}_U \to \mathbf{R}$ by

$$\omega_I^U(X) = \omega(I \cup X), \quad X \in \mathcal{B}^U; \qquad \omega_U^J(X) = \omega(J \cup X), \quad X \in \mathcal{B}_U.$$

Theorem 5.2.5.
(1) $\mathbf{M}_I^U = (U, \mathcal{B}^U, \omega_I^U)$ *is a valuated matroid, and for* $I, I' \in \mathcal{B}_{V\setminus U}$ *there exists* $\alpha_{I,I'}^U \in \mathbf{R}$, *independent of* $X \in \mathcal{B}^U$, *such that*

$$\omega_{I'}^U(X) = \omega_I^U(X) + \alpha_{I,I'}^U, \qquad X \in \mathcal{B}^U.$$

(2) $\mathbf{M}_U^J = (U, \mathcal{B}_U, \omega_U^J)$ *is a valuated matroid, and for* $J, J' \in \mathcal{B}^{V\setminus U}$ *there exists* $\beta_U^{J,J'} \in \mathbf{R}$, *independent of* $X \in \mathcal{B}_U$, *such that*

$$\omega_U^{J'}(X) = \omega_U^J(X) + \beta_U^{J,J'}, \qquad X \in \mathcal{B}_U.$$

Proof. (1) It is obvious that ω_I^U satisfies (VM). We may assume $I \setminus I' = \{u\}$ and $I' \setminus I = \{u'\}$. For $X, Y \in \mathcal{B}^U$ we apply (VM) to $(I \cup X, I' \cup Y)$ and $u \in (I \cup X) \setminus (I' \cup Y)$ to obtain

$$\omega(I \cup X) + \omega(I' \cup Y) \leq \omega(I' \cup X) + \omega(I \cup Y),$$

since u' is the only exchangeable element of $(I' \cup Y) \setminus (I \cup X)$. The reverse inequality can be shown similarly, and therefore

$$\omega(I' \cup X) - \omega(I \cup X) = \omega(I' \cup Y) - \omega(I \cup Y) = \alpha_{I,I'}^U.$$

(2) This can be proven similarly. ∎

Let $(V, \mathcal{B}_k)$ and $(V, \mathcal{B}^l)$ denote the truncation to k and the elongation to l, respectively, of the underlying matroid $(V, \mathcal{B})$ of a valuated matroid $\mathbf{M} = (V, \mathcal{B}, \omega)$, where $k \le r \le l$. By definition,

$$\mathcal{B}_k = \{I \subseteq V \mid |I| = k,\ \exists B : I \subseteq B \in \mathcal{B}\}, \tag{5.17}$$
$$\mathcal{B}^l = \{S \subseteq V \mid |S| = l,\ \exists B : S \supseteq B \in \mathcal{B}\}. \tag{5.18}$$

For a spanning set S_0 of $(V, \mathcal{B})$ and an independent set I_0 of $(V, \mathcal{B})$, define $\omega_{k,S_0} : \mathcal{B}_k \to \mathbf{R}$ and $\omega^{l,I_0} : \mathcal{B}^l \to \mathbf{R}$ by

$$\omega_{k,S_0}(I) = \max\{\omega(B) \mid I \cup S_0 \supseteq B \supseteq I, B \in \mathcal{B}\} \qquad (I \in \mathcal{B}_k), \tag{5.19}$$

$$\omega^{l,I_0}(S) = \max\{\omega(B) \mid S \cap I_0 \subseteq B \subseteq S, B \in \mathcal{B}\} \qquad (S \in \mathcal{B}^l). \tag{5.20}$$

The following theorem (Murota [229]) states that these constructions yield valuated matroids. We call $\mathbf{M}_{k,S_0} = (V, \mathcal{B}_k, \omega_{k,S_0})$ the *truncation* of $\mathbf{M}$ to rank k relative to a spanning set S_0, and $\mathbf{M}^{l,I_0} = (V, \mathcal{B}^l, \omega^{l,I_0})$ the *elongation* of $\mathbf{M}$ to rank l relative to an independent set I_0.

Theorem 5.2.6. *For a valuated matroid $\mathbf{M} = (V, \mathcal{B}, \omega)$ of rank r, let ω_{k,S_0} and ω^{l,I_0} be defined by* (5.19) *and* (5.20) *for a spanning set S_0 and an independent set I_0, where $0 \le k \le r \le l \le |V|$.*

(1) *$\mathbf{M}_{k,S_0} = (V, \mathcal{B}_k, \omega_{k,S_0})$ is a valuated matroid (of rank k).*

(2) *$\mathbf{M}^{l,I_0} = (V, \mathcal{B}^l, \omega^{l,I_0})$ is a valuated matroid (of rank l).*

(3) *$(\omega^{r+s,I_0})^* = (\omega^*)_{(r+s)^*,I_0^*}$ for s with $0 \le s \le |V| - r$, where $(r+s)^* = |V| - (r+s)$ and $I_0^* = V \setminus I_0$.*

Proof. First note that (2) follows from (1) and (3), and that (3) can be proven by a direct calculation:

$$\begin{aligned}(\omega^*)_{(r+s)^*,I_0^*}(I) &= \max\{\omega^*(B') \mid I \cup I_0^* \supseteq B' \supseteq I\} \\ &= \max\{\omega(B) \mid (V \setminus I) \cap I_0 \subseteq B \subseteq V \setminus I\} \\ &= \omega^{r+s,I_0}(V \setminus I) = (\omega^{r+s,I_0})^*(I),\end{aligned}$$

where $I \subseteq V$ and $|I| = (r+s)^*$.

Next, we show that the proof of (1) reduces to the case of $S_0 = V$. Define $p : V \to \mathbf{R}$ by

$$p(u) = \begin{cases} \alpha & (u \in S_0) \\ 0 & (u \in V \setminus S_0) \end{cases}$$

with a sufficiently large number α, and consider $\omega[p]$ of (5.16), which is also a valuation of $(V, \mathcal{B})$. Then we have

$$\omega_{k,S_0}(I) = (\omega[p])_{k,V}(I) - \sum\{p(u) \mid u \in I\} - \alpha(r-k).$$

Hence it suffices to prove that $(\omega[p])_{k,V}$ is a valuation, which is done later in Lemma 5.2.21. ■

For two valuated matroids $\mathbf{M}_1 = (V_1, \mathcal{B}_1, \omega_1)$ and $\mathbf{M}_2 = (V_2, \mathcal{B}_2, \omega_2)$ with disjoint ground sets ($V_1 \cap V_2 = \emptyset$), the *direct sum* is a valuated matroid $\mathbf{M} = (V_1 \cup V_2, \mathcal{B}, \omega)$ defined by $\mathcal{B} = \{B_1 \cup B_2 \mid B_1 \in \mathcal{B}_1, B_2 \in \mathcal{B}_2\}$ and $\omega(B) = \omega_1(B \cap V_1) + \omega_2(B \cap V_2)$. It is noted that this construction does not necessarily yield a valuated matroid if $V_1 \cap V_2 \neq \emptyset$.

5.2.4 Greedy Algorithms

A greedy algorithm works for valuated matroids and this property in turn characterizes valuated matroids.

Let $\mathbf{M} = (V, \mathcal{B}, \omega)$ be a valuated matroid. For $B \in \mathcal{B}$, $u \in B$, and $v \in V \backslash B$, we define

$$\omega(B, u, v) = \omega(B - u + v) - \omega(B), \tag{5.21}$$

which we refer to as the *exchange gain* of ω at B for the pair (u, v). Note that $\omega(B, u, v) = -\infty$ if $B - u + v \notin \mathcal{B}$.

The following fact is most fundamental, showing the local optimality implies the global optimality. This is due to Dress–Wenzel [57].

Theorem 5.2.7. *Let $B \in \mathcal{B}$. Then $\omega(B) \geq \omega(B')$ for any $B' \subseteq V$ if and only if*

$$\omega(B, u, v) \leq 0 \quad \text{for any } u \in B \text{ and } v \in V \setminus B. \tag{5.22}$$

Proof. We prove $\omega(B) \geq \omega(B')$ by induction on $d = |B' \setminus B|$. Obviously, this is true for $d = 0$. For $d \geq 1$, there exist $u \in B \setminus B'$ and $v \in B' \setminus B$ such that

$$\omega(B) + \omega(B') \leq \omega(B - u + v) + \omega(B' + u - v),$$

in which $\omega(B - u + v) \leq \omega(B)$ by (5.22) and $\omega(B' + u - v) \leq \omega(B)$ by the induction hypothesis. Hence follows $\omega(B') \leq \omega(B)$. ■

For the maximization of ω the *greedy algorithm* of Dress–Wenzel [54] starts with an arbitrary base $B_0 = \{u_1, u_2, \cdots, u_r\} \in \mathcal{B}$ with an arbitrary ordering of the elements, and repeats the following for $k = 1, 2, \cdots, r$ ($= \operatorname{rank} \mathbf{M}$):

> Find $v_k \in (V \setminus B_{k-1}) \cup \{u_k\} = V \setminus \{v_1, \cdots, v_{k-1}, u_{k+1}, \cdots, u_r\}$ such that
>
> $$\omega(B_{k-1} - u_k + v_k) \geq \omega(B_{k-1} - u_k + v) \quad (\forall v \in (V \setminus B_{k-1}) \cup \{u_k\})$$
>
> and put $B_k = B_{k-1} - u_k + v_k$.

In this way an optimal base (maximizing ω) can be found with $r(|V| - r) + 1$ function evaluations of ω. Moreover, the success of this algorithm characterizes valuations, a result of Dress–Wenzel [54].

Theorem 5.2.8. *Let $(V, \mathcal{B})$ be a matroid, and $\omega : \mathcal{B} \to \mathbf{R}$. If ω is a valuation, the above algorithm yields an optimal base. Conversely, if for any $p \in \mathbf{R}^V$ the above algorithm applied to $\omega[p]$ yields an optimal base with respect to $\omega[p]$, then ω is a valuation.*

Proof. The algorithm works for a valuation due to Lemma 5.2.9 below, applied to a sequence of the contractions of $(V, \mathcal{B}, \omega)$ to $V \setminus \{v_1, \cdots, v_{k-1}\}$ ($k = 1, 2, \cdots$). For the converse, suppose that (VM) fails for $B, B' \in \mathcal{B}$ and $u_* \in B \setminus B'$. Note that $|B \setminus B'| \geq 2$. Define $p : V \to \mathbf{R}$ by

$$p(v) = \begin{cases} 0 & (v = u_*) \\ -\omega(B, u_*, v) & (v \in B' \setminus B, B - u_* + v \in \mathcal{B}) \\ +M_1 & (v \in B' \setminus B, B - u_* + v \notin \mathcal{B}) \\ +M_2 & (v \in B \cap B') \\ -M_2 & (v \in V \setminus (B' \cup \{u_*\})) \end{cases}$$

with sufficiently large positive numbers M_1 and M_2 such that $M_1 \ll M_2$. Then B' is the unique maximizer of $\omega[p]$ and $\omega[p](B) \geq \omega[p](B - u_* + v)$ for $v \in V \setminus B$. The latter means that the algorithm applied to $B_0 = B$ with $u_1 = u_*$ fails by choosing $v_1 = u_*$. ∎

The proof above relies on the following fundamental fact due to Shioura [298].

Lemma 5.2.9. *Let $(V, \mathcal{B}, \omega)$ be a valuated matroid, $\hat{u} \in B \in \mathcal{B}$, and $\hat{v} \in (V \setminus B) \cup \{\hat{u}\}$. If $\omega(B - \hat{u} + \hat{v}) = \max_{v \in (V \setminus B) \cup \{\hat{u}\}} \omega(B - \hat{u} + v)$, there exists $\hat{B} \in \mathcal{B}$ such that $\hat{v} \in \hat{B}$ and $\omega(\hat{B}) = \max \omega$.*

Proof. Take any $B' \in \mathcal{B}$ with $\omega(B') = \max \omega$. If $\hat{v} \in B'$, put $\hat{B} = B'$. Otherwise, $\hat{v} \in (B - \hat{u} + \hat{v}) \setminus B'$, and therefore, there exists $u \in B' \setminus (B - \hat{u} + \hat{v})$ such that

$$\omega(B - \hat{u} + \hat{v}) + \omega(B') \leq \omega(B - \hat{u} + u) + \omega(B' + \hat{v} - u),$$

which must be satisfied with equality since $\omega(B - \hat{u} + u) \leq \omega(B - \hat{u} + \hat{v})$ and $\omega(B' + \hat{v} - u) \leq \max \omega = \omega(B')$. Hence $\hat{B} = B' + \hat{v} - u$ is a valid choice. ∎

Example 5.2.10. In case ω is defined by an $m \times n$ rational matrix $A(s)$ of rank m (cf. Example 5.2.3), the exchange gain (5.21) can be represented as

$$\omega(B, i, j) = \deg_s (A[R, B]^{-1} A[R, C \setminus B])_{ij} \qquad (i \in B, j \in C \setminus B), \quad (5.23)$$

where the right-hand side designates the degree of the (i, j) entry of the $m \times (n - m)$ rational matrix $A[R, B]^{-1} A[R, C \setminus B]$. Hence, by Theorem 5.2.7, we see

$$\begin{aligned} & B \text{ maximizes } \deg_s \det A[R, B] \\ \iff & A[R, B]^{-1} A[R, C \setminus B] \text{ is a proper rational matrix.} \end{aligned} \quad (5.24)$$

□

Remark 5.2.11. Greedy algorithms are fundamental in combinatorial optimization. Investigation of variants of greedy algorithms leads to many interesting combinatorial structures. See, for example, Edmonds [68, 69], Faigle [74], Lawler [171], and Welsh [333] (matroids and polymatroids); Korte–Lovász–Schrader [163] (greedoids); Bouchet [15] and Chandrasekaran–Kabadi [30] (delta matroids); Dress–Terhalle [51, 52, 53] (variants of valuated matroids); and Dress–Wenzel [55] and Murota [222] (valuated delta matroids). □

5.2.5 Valuated Bimatroid

In §2.3.7 we explained about bimatroids, which can be regarded as a variant of matroids (see (2.86) in particular). Following Murota [221] we introduce here a variant of valuated matroids in a similar vein.

Let S and T be disjoint finite sets and $\delta : 2^S \times 2^T \to \mathbf{R} \cup \{-\infty\}$ be a map such that

$$\delta(X, Y) = \omega((S \setminus X) \cup Y) \qquad (X \subseteq S, Y \subseteq T) \tag{5.25}$$

for some valuated matroid $(S \cup T, \omega)$ with $\omega(S) \neq -\infty$. Then (S, T, Λ) with

$$\Lambda = \{(X, Y) \mid \delta(X, Y) \neq -\infty,\ X \subseteq S,\ Y \subseteq T\}$$

is a bimatroid due to (2.86). Note that $(\emptyset, \emptyset) \in \Lambda$ and that $|X| = |Y|$ for $(X, Y) \in \Lambda$.

The axiom (VM) for ω can be translated into a condition on δ that (VB-1) and (VB-2) below hold true for any $(X, Y) \in \Lambda$ and $(X', Y') \in \Lambda$:

(VB-1) For any $x' \in X' \setminus X$, either (a1) or (b1) (or both) holds, where
(a1) $\exists y' \in Y' \setminus Y$:
$\delta(X, Y) + \delta(X', Y') \leq \delta(X + x', Y + y') + \delta(X' - x', Y' - y')$,
(b1) $\exists x \in X \setminus X'$:
$\delta(X, Y) + \delta(X', Y') \leq \delta(X - x + x', Y) + \delta(X' - x' + x, Y')$.

(VB-2) For any $y \in Y \setminus Y'$, either (a2) or (b2) (or both) holds, where
(a2) $\exists x \in X \setminus X'$:
$\delta(X, Y) + \delta(X', Y') \leq \delta(X - x, Y - y) + \delta(X' + x, Y' + y)$,
(b2) $\exists y' \in Y' \setminus Y$:
$\delta(X, Y) + \delta(X', Y') \leq \delta(X, Y - y + y') + \delta(X', Y' - y' + y)$.

A triple (S, T, δ) (or quadruple (S, T, Λ, δ)) satisfying (VB-1) and (VB-2) is named a *valuated bimatroid.*

We are now interested in optimal linked pairs $(X, Y) \in \Lambda$ of a specified size k with respect to the value of δ. Define

$$\begin{aligned} r &= \max\{|X| \mid \exists (X, Y) \in \Lambda\}, \\ \Lambda_k &= \{(X, Y) \in \Lambda \mid |X| = |Y| = k\} \qquad (0 \leq k \leq r), \\ \delta_k &= \max\{\delta(X, Y) \mid (X, Y) \in \Lambda_k\} \qquad (0 \leq k \leq r), \\ \mathcal{M}_k &= \{(X, Y) \in \Lambda_k \mid \delta(X, Y) = \delta_k\} \qquad (0 \leq k \leq r). \end{aligned}$$

The optimal linked pairs can be chosen to be nested (Murota [221]).

Theorem 5.2.12. *For any* $(X_k, Y_k) \in \mathcal{M}_k$ *with* $1 \leq k \leq r-1$, *there exist* $(X_l, Y_l) \in \mathcal{M}_l$ $(0 \leq l \leq r,\ l \neq k)$ *such that*

$$(\emptyset =)\ X_0 \subset X_1 \subset \cdots \subset X_{k-1} \subset X_k \subset X_{k+1} \subset \cdots \subset X_r,$$
$$(\emptyset =)\ Y_0 \subset Y_1 \subset \cdots \subset Y_{k-1} \subset Y_k \subset Y_{k+1} \subset \cdots \subset Y_r.$$

Proof. Take any $(X_-, Y_-) \in \mathcal{M}_{k-1}$ and apply (VB-1) to $(X, Y) = (X_-, Y_-)$, $(X', Y') = (X_k, Y_k)$ and any $x' \in X_k \setminus X_-$ $(\neq \emptyset)$. In case (a1),

$$\begin{aligned}\delta_{k-1} + \delta_k &= \delta(X_-, Y_-) + \delta(X_k, Y_k)\\ &\leq \delta(X_- + x', Y_- + y') + \delta(X_k - x', Y_k - y') \leq \delta_k + \delta_{k-1},\end{aligned}$$

and therefore $(X_{k-1}, Y_{k-1}) = (X_k - x', Y_k - y')$ is eligible. In case (b1),

$$\begin{aligned}\delta_{k-1} + \delta_k &= \delta(X_-, Y_-) + \delta(X_k, Y_k)\\ &\leq \delta(X_- - x + x', Y_-) + \delta(X_k - x' + x, Y_k) \leq \delta_{k-1} + \delta_k,\end{aligned}$$

which shows $(\tilde{X}_-, \tilde{Y}_-) = (X_- - x + x', Y_-) \in \mathcal{M}_{k-1}$ with $|\tilde{X}_- \setminus X_k| = |X_- \setminus X_k| - 1$. We now apply the same argument to $(\tilde{X}_-, \tilde{Y}_-)$. This process eventually reaches the case (a1). ■

The nesting property implies the following fact (Murota [221]).

Theorem 5.2.13. *The sequence* $(\delta_0, \delta_1, \cdots, \delta_r)$ *is concave, i.e.,*

$$\delta_{k-1} + \delta_{k+1} \leq 2\delta_k \qquad (1 \leq k \leq r-1).$$

Proof. Take $(X_{k-1}, Y_{k-1}) \in \mathcal{M}_{k-1}$ and $(X_{k+1}, Y_{k+1}) \in \mathcal{M}_{k+1}$ with $X_{k-1} \subset X_{k+1}$ and $Y_{k-1} \subset Y_{k+1}$ (cf. Theorem 5.2.12). By (VB-1) there exist $x \in X_{k+1} \setminus X_{k-1}$ and $y \in Y_{k+1} \setminus Y_{k-1}$ such that

$$\delta_{k-1} + \delta_{k+1} \leq \delta(X_{k-1} + x, Y_{k-1} + y) + \delta(X_{k+1} - x, Y_{k+1} - y) \leq 2\delta_k.$$

■

The nesting property revealed in Theorem 5.2.12 justifies the following incremental greedy algorithm for computing δ_k for $k = 0, 1, \cdots, r$.

Greedy algorithm for δ_k $(k = 1, 2, \cdots)$
$X_0 := \emptyset$; $Y_0 := \emptyset$;
for $k := 1, 2, \cdots$ **do**
  Find $x \in S \setminus X_{k-1}$, $y \in T \setminus Y_{k-1}$ maximizing $\delta(X_{k-1} + x, Y_{k-1} + y)$
  and put $X_k := X_{k-1} + x$, $Y_k := Y_{k-1} + y$, and $\delta_k := \delta(X_k, Y_k)$.

The iteration stops when $\delta(X_k, Y_k) = -\infty$, and then we see $r = k - 1$. This algorithm involves $\mathrm{O}(r|S|\,|T|)$ evaluations of δ to compute the whole sequence $(\delta_0, \delta_1, \cdots, \delta_r)$.

For a valuated bimatroid $\delta : 2^S \times 2^T \to \mathbf{R} \cup \{-\infty\}$ we define a function $\tilde{\delta} : 2^S \times 2^T \to \mathbf{R}$ by

$$\tilde{\delta}(X, Y) = \max\{\delta(X', Y') \mid X' \subseteq X, Y' \subseteq Y\}.$$

Note that $\tilde{\delta}(X, Y)$ is finite for all (X, Y) since $\delta(\emptyset, \emptyset) > -\infty$.

Theorem 5.2.14. *The function $\tilde{\delta} : 2^S \times 2^T \to \mathbf{R}$ derived from a valuated bimatroid $\delta : 2^S \times 2^T \to \mathbf{R} \cup \{-\infty\}$ satisfies*

$$\tilde{\delta}(X, Y) + \tilde{\delta}(X', Y') \geq \tilde{\delta}(X \cup X', Y \cap Y') + \tilde{\delta}(X \cap X', Y \cup Y') \qquad (X, X' \subseteq S; Y, Y' \subseteq T).$$

Proof. It suffices to show

$$\tilde{\delta}(X, Y) + \tilde{\delta}(X + x, Y + y) \geq \tilde{\delta}(X + x, Y) + \tilde{\delta}(X, Y + y), \tag{5.26}$$
$$\tilde{\delta}(X + x_1, Y) + \tilde{\delta}(X + x_2, Y) \geq \tilde{\delta}(X, Y) + \tilde{\delta}(X + \{x_1, x_2\}, Y), \tag{5.27}$$
$$\tilde{\delta}(X, Y + y_1) + \tilde{\delta}(X, Y + y_2) \geq \tilde{\delta}(X, Y) + \tilde{\delta}(X, Y + \{y_1, y_2\}), \tag{5.28}$$

where $x, x_1, x_2 \in S \setminus X$ and $y, y_1, y_2 \in T \setminus Y$.

To show (5.26) take $X_1 \subseteq X + x$, $Y_1 \subseteq Y$, $X_2 \subseteq X$, and $Y_2 \subseteq Y + y$ such that $\tilde{\delta}(X + x, Y) = \delta(X_1, Y_1)$ and $\tilde{\delta}(X, Y + y) = \delta(X_2, Y_2)$. If $x \notin X_1$, we are done, since $\tilde{\delta}(X, Y) \geq \delta(X_1, Y_1)$ and $\tilde{\delta}(X + x, Y + y) \geq \delta(X_2, Y_2)$. Otherwise, apply (VB-1) for $x \in X_1 \setminus X_2$ to obtain either (a) $\delta(X_1, Y_1) + \delta(X_2, Y_2) \leq \delta(X_1 - x, Y_1 - y_1) + \delta(X_2 + x, Y_2 + y_1) \leq \tilde{\delta}(X, Y) + \tilde{\delta}(X + x, Y + y)$ for some $y_1 \in Y_1 \setminus Y_2$ or (b) $\delta(X_1, Y_1) + \delta(X_2, Y_2) \leq \delta(X_1 - x + x_2, Y_1) + \delta(X_2 - x_2 + x, Y_2) \leq \tilde{\delta}(X, Y) + \tilde{\delta}(X + x, Y + y)$ for some $x_2 \in X_2 \setminus X_1$.

To show (5.27) take $X_1 \subseteq X$, $X_2 \subseteq X + \{x_1, x_2\}$, and $Y_1, Y_2 \subseteq Y$ such that $\tilde{\delta}(X, Y) = \delta(X_1, Y_1)$ and $\tilde{\delta}(X + \{x_1, x_2\}, Y) = \delta(X_2, Y_2)$. If $x_2 \notin X_2$, we are done, since $\tilde{\delta}(X + x_1, Y) \geq \delta(X_2, Y_2)$ and $\tilde{\delta}(X + x_2, Y) \geq \delta(X_1, Y_1)$. Otherwise, apply (VB-1) for $x_2 \in X_2 \setminus X_1$ to obtain either (a) $\delta(X_1, Y_1) + \delta(X_2, Y_2) \leq \delta(X_1 + x_2, Y_1 + y_2) + \delta(X_2 - x_2, Y_2 - y_2) \leq \tilde{\delta}(X + x_2, Y) + \tilde{\delta}(X + x_1, Y)$ for some $y_2 \in Y_2 \setminus Y_1$ or (b) $\delta(X_1, Y_1) + \delta(X_2, Y_2) \leq \delta(X_1 - x + x_2, Y_1) + \delta(X_2 - x_2 + x, Y_2) \leq \tilde{\delta}(X + x_2, Y) + \tilde{\delta}(X + x_1, Y)$ for some $x \in X_1 \setminus X_2$.

The final case (5.28) can be shown similarly. ■

Example 5.2.15. From an $m \times n$ rational matrix $A(s)$ with $R = \mathrm{Row}(A)$ and $C = \mathrm{Col}(A)$, a valuated bimatroid (R, C, δ) is defined by

$$\delta(I, J) = \deg_s \det A[I, J] \qquad (I \subseteq R, J \subseteq C). \tag{5.29}$$

The associated valuated matroid in (5.25) is the one defined by an $m \times (m+n)$ matrix $[I_m \; A]$ according to (5.15) (see also Fig. 2.12). A weaker form of the nesting property of Theorem 5.2.12 in this special case has been observed by Svaricek [306], [307, Satz 6.23]. The concavity of the sequence δ_k in Theorem 5.2.13 is equivalent to the monotone decrease of the sequence $t_k = \delta_k - \delta_{k-1}$, which is called the sequence of contents at infinity in connection to the Smith–McMillan form at infinity of $A(s)$ (cf. Theorem 5.1.5). □

Example 5.2.16. Another valuated bimatroid arises from a polynomial matrix $A(s)$. For a number α and a polynomial $f(s)$ we denote by $\mathrm{ord}_s^{(\alpha)} f(s)$ the maximum p such that $(s-\alpha)^p$ divides $f(s)$. Then

$$\delta^{(\alpha)}(I,J) = -\mathrm{ord}_s^{(\alpha)} \det A[I,J] \qquad (I \subseteq R, J \subseteq C) \tag{5.30}$$

defines a valuated bimatroid $(R, C, \delta^{(\alpha)})$, where $R = \mathrm{Row}(A)$ and $C = \mathrm{Col}(A)$. This is in fact a variant of the valuated bimatroid in Example 5.2.15, since $\delta^{(\alpha)}(I,J)$ for $A(s)$ coincides with $\delta(I,J)$ for $A(\alpha + 1/s)$. The exponent to $(s-\alpha)$ in the factorization of the kth determinantal divisor $d_k(s)$ of $A(s)$ is given by $\min\{\mathrm{ord}_s^{(\alpha)} \det A[I,J] \mid |I| = |J| = k\}$, which is equal to $-\max\{\delta^{(\alpha)}(I,J) \mid |I| = |J| = k\}$. Then Theorem 5.2.13 implies that $(k-1)$st invariant factor $e_{k-1}(s) = d_{k-1}(s)/d_{k-2}(s)$ divides the kth invariant factor $e_k(s) = d_k(s)/d_{k-1}(s)$ for $k = 2, \cdots, r$ (cf. Theorem 5.1.1). □

Example 5.2.17. Let $G = (V^+, V^-; A)$ be a bipartite graph and $w : A \to \mathbf{R}$ be a weight function. A valuated bimatroid (V^+, V^-, δ) is defined by

$$\delta(I,J) = \max\{w(M) \mid M\text{: matching}, \partial^+ M = I, \partial^- M = J\} \qquad (I \subseteq V^+, J \subseteq V^-).$$

Then Theorem 5.2.13 shows the concavity of the sequence of

$$\delta_k = \max\{w(M) \mid M\text{: } k\text{-matching}\}.$$

□

5.2.6 Induction Through Bipartite Graphs

A valuated matroid can be transformed into another valuated matroid through matchings in a bipartite graph. This is an extension of the well-known fact (cf. §2.3.6) that a matroid can be transformed into another matroid through a bipartite graph.

Let $G = (V^+, V^-; A)$ be a bipartite graph, $w : A \to \mathbf{R}$ a weight function, and $\mathbf{M} = (V^+, \mathcal{B}, \omega)$ a valuated matroid. Then, by Theorem 2.3.38,

$$\tilde{\mathcal{B}} = \{\partial^- M \mid M : \text{matching}, \ \partial^+ M \in \mathcal{B}\}$$

forms the basis family of a matroid, provided $\tilde{\mathcal{B}} \neq \emptyset$. Here $\partial^+ M \subseteq V^+$ and $\partial^- M \subseteq V^-$ denote the sets of vertices incident to M. Define $\tilde{\omega} : \tilde{\mathcal{B}} \to \mathbf{R}$ by

$$\tilde{\omega}(X) = \max_M \{w(M) + \omega(\partial^+ M) \mid M : \text{matching}, \ \partial^+ M \in \mathcal{B}, \partial^- M = X\}, \qquad X \in \tilde{\mathcal{B}}. \tag{5.31}$$

This construction yields a valuated matroid as follows (Murota [227]).

Theorem 5.2.18. *$\tilde{\mathbf{M}} = (V^-, \tilde{\mathcal{B}}, \tilde{\omega})$ is a valuated matroid, provided $\tilde{\mathcal{B}} \neq \emptyset$.*

Proof. The induction through $G = (V^+, V^-; A)$ with $w : A \to \mathbf{R}$ can be decomposed into two stages. Let $G^+ = (V^+, A; A^+)$ and $G^- = (A, V^-; A^-)$ be bipartite graphs with

$$A^+ = \{(v, a) \mid v \in V^+, a \in A, v = \partial^+ a\},$$
$$A^- = \{(a, v) \mid v \in V^-, a \in A, v = \partial^- a\},$$

and define $w^+ : A^+ \to \mathbf{R}$ and $w^- : A^- \to \mathbf{R}$ by

$$w^+(v, a) = w(a), \quad w^-(a, v) = 0 \qquad (a \in A).$$

It is not difficult to see that the transformation of $\mathbf{M} = (V^+, \mathcal{B}, \omega)$ by (G, w) is equivalent to the transformation of $\mathbf{M}$ to $\mathbf{M}^\circ = (A, \mathcal{B}^\circ, \omega^\circ)$ by (G^+, w^+) followed by the transformation of $\mathbf{M}^\circ$ by (G^-, w^-). Hence it suffices to consider the following two special cases of $G = (V^+, V^-; A)$ and $w : A \to \mathbf{R}$:

Case 1: $\deg v = 1$ for $\forall\, v \in V^-$,
Case 2: $\deg v = 1$ for $\forall\, v \in V^+$, and $w(a) = 0$ for $\forall\, a \in A$.

In either case we shall show (VM) for $\tilde{\omega}$, i.e., for $B, B' \in \tilde{\mathcal{B}}$ and $u \in B \setminus B'$, there exists $v \in B' \setminus B$ such that

$$\tilde{\omega}(B) + \tilde{\omega}(B') \le \tilde{\omega}(B - u + v) + \tilde{\omega}(B' + u - v). \tag{5.32}$$

Fix $u \in B \setminus B'$.

Case 1: For $v \in V^-$ let $\varphi(v)$ denote the unique element of V^+ such that $(\varphi(v), v) \in A$. Then $|\varphi^{-1}(x) \cap B| \le 1$, $|\varphi^{-1}(x) \cap B'| \le 1$ for all $x \in V^+$, and $\varphi(B), \varphi(B') \in \mathcal{B}$, where $\varphi(B) = \{\varphi(v) \mid v \in B\}$. By definition, we have

$$\tilde{\omega}(B) = \omega(\varphi(B)) + \bar{w}(B), \quad \tilde{\omega}(B') = \omega(\varphi(B')) + \bar{w}(B')$$

with $\bar{w}(B) = \sum_{v \in B} w(\varphi(v), v)$.

If $\varphi(u) \in \varphi(B')$, then $\varphi(u) = \varphi(v)$ for some $v \in B' \setminus B$, and

$$\omega(\varphi(B - u + v)) + \omega(\varphi(B' + u - v)) = \omega(\varphi(B)) + \omega(\varphi(B')),$$
$$\bar{w}(B - u + v) + \bar{w}(B' + u - v) = \bar{w}(B) + \bar{w}(B'). \tag{5.33}$$

Consequently, (5.32) holds true with equality. If $x = \varphi(u) \notin \varphi(B')$, then by (VM), there exists $y \in \varphi(B') \setminus \varphi(B)$ with

$$\begin{aligned}\omega(\varphi(B)) + \omega(\varphi(B')) &\le \omega(\varphi(B) - x + y) + \omega(\varphi(B') + x - y)\\ &= \omega(\varphi(B - u + v)) + \omega(\varphi(B' + u - v)),\end{aligned}$$

where $v \in B' \setminus B$, $\varphi(v) = y$. Combination of this and (5.33) yields (5.32).

Case 2: We may restrict ourselves to a further special case where

$$\exists t \in V^- : \quad \deg t = 2, \quad \deg v = 1 \ (\forall\, v \in V^- \setminus \{t\}),$$

in addition to $\deg v = 1$ for all $v \in V^+$, since the general case can be obtained as a composition of a series of such special cases. Let $t_1, t_2 \in V^+$ denote the elements of V^+ such that $(t_1, t), (t_2, t) \in A$, and for $v \in V^- \setminus \{t\}$ let $\varphi(v)$ denote the unique element of V^+ such that $(\varphi(v), v) \in A$. Put $x = \varphi(u)$.

We consider the case where $t \in B \cap B'$, since otherwise the proof is easier. We have

$$\tilde{\omega}(B) = \omega(\bar{B} + t_k), \qquad \tilde{\omega}(B') = \omega(\bar{B}' + t_l)$$

with $\bar{B} = \varphi(B - t)$ and $\bar{B}' = \varphi(B' - t)$ for some $k, l \in \{1, 2\}$.

Case 2(a) $[k = l]$: Since $x = \varphi(u) \in (\bar{B} + t_k) \setminus (\bar{B}' + t_k)$, there exist $y \in \bar{B}' \setminus \bar{B}$ such that

$$\omega(\bar{B} + t_k) + \omega(\bar{B}' + t_k) \leq \omega(\bar{B} + t_k - x + y) + \omega(\bar{B}' + t_k + x - y).$$

We can take $v = \varphi^{-1}(y) \in B' \setminus B$ in (5.32).

Case 2(b) $[k \neq l]$: We may assume $k = 1$, $l = 2$ and

$$\omega(\bar{B} + t_1) > \omega(\bar{B} + t_2), \qquad \omega(\bar{B}' + t_2) > \omega(\bar{B}' + t_1). \tag{5.34}$$

Since $x = \varphi(u) \in (\bar{B} + t_1) \setminus (\bar{B}' + t_2)$, there exists $y \in (\bar{B}' + t_2) \setminus (\bar{B} + t_1)$ with

$$\omega(\bar{B} + t_1) + \omega(\bar{B}' + t_2) \leq \omega(\bar{B} + t_1 - x + y) + \omega(\bar{B}' + t_2 + x - y). \tag{5.35}$$

If $y \neq t_2$, RHS of (5.35) $\leq$ RHS of (5.32) with $v = \varphi^{-1}(y) \in B' \setminus B$ (RHS=right hand side). If $y = t_2$,

$$\begin{aligned}\text{RHS of (5.35)} &= \omega(\bar{B} - x + t_1 + t_2) + \omega(\bar{B}' + x) \\ &\leq \omega(\bar{B} - x + t_2 + z) + \omega(\bar{B}' + x + t_1 - z)\end{aligned}$$

for some $z \in (\bar{B}' + x) \setminus (\bar{B} - x + t_1 + t_2)$. By (5.34) we must have $z \neq x$, and therefore $z \in \bar{B}' \setminus \bar{B}$. Then (5.32) is satisfied with $v = \varphi^{-1}(z)$. ■

Remark 5.2.19. The present proof of Theorem 5.2.18 is elementary, as compared with the original proof by Murota [227] and an alternative proof by Shioura [297]. The present proof, if specialized to $\omega = 0$ and $w = 0$, serves also as an alternative proof of Theorem 2.3.38, showing that $\tilde{\mathcal{B}}$ satisfies (BM$_\pm$). □

Theorem 5.2.18 has important consequences. Let $\mathbf{M}_1 = (V, \mathcal{B}_1, \omega_1)$ and $\mathbf{M}_2 = (V, \mathcal{B}_2, \omega_2)$ be valuated matroids. Denote by $(V, \mathcal{B}_1 \vee \mathcal{B}_2)$ the union of the underlying matroids $(V, \mathcal{B}_1)$ and $(V, \mathcal{B}_2)$, where, by definition, $\mathcal{B}_1 \vee \mathcal{B}_2$ is the family of the maximal elements of $\{X_1 \cup X_2 \mid X_1 \in \mathcal{B}_1, X_2 \in \mathcal{B}_2\}$ (cf. §2.3.6). Define $\omega_1 \vee \omega_2 : \mathcal{B}_1 \vee \mathcal{B}_2 \to \mathbf{R}$ by

$$(\omega_1 \vee \omega_2)(X) = \max\{\omega_1(X_1) + \omega_2(X_2) \mid X_1 \cup X_2 = X, X_1 \in \mathcal{B}_1, X_2 \in \mathcal{B}_2\}, \qquad X \in \mathcal{B}_1 \vee \mathcal{B}_2.$$

We call $\mathbf{M}_1 \vee \mathbf{M}_2 = (V, \mathcal{B}_1 \vee \mathcal{B}_2, \omega_1 \vee \omega_2)$ the *union* of $\mathbf{M}_1$ and $\mathbf{M}_2$ on the basis of the following fact (Murota [227]).

Theorem 5.2.20. $\mathbf{M}_1 \vee \mathbf{M}_2 = (V, \mathcal{B}_1 \vee \mathcal{B}_2, \omega_1 \vee \omega_2)$ *is a valuated matroid.*

Proof. Let V_1 and V_2 be disjoint copies of V, and U be a set with $|U| = r_1 + r_2 - r_{12}$, where r_1, r_2 and r_{12} denote the ranks of $(V, \mathcal{B}_1)$, $(V, \mathcal{B}_2)$ and $(V, \mathcal{B}_1 \vee \mathcal{B}_2)$. Consider a bipartite graph $G = (V^+, V^-; A)$ with $V^+ = V_1 \cup V_2$, $V^- = V \cup U$, and

$$A = \{(v_1, v) \mid v \in V\} \cup \{(v_2, v) \mid v \in V\} \cup \{(v_2, u) \mid v \in V, u \in U\},$$

where $v_i \in V_i$ is the copy of $v \in V$ $(i = 1, 2)$. Let $\tilde{\omega}$ be the valuation induced on V^- from the valuation ω on V^+ defined by $\omega(X_1 \cup X_2) = \omega_1(X_1) + \omega_2(X_2)$ $(X_i \in \mathcal{B}_i\ (i = 1, 2))$ and weight function $w = 0$ on A. Then $(\omega_1 \vee \omega_2)(X) = \tilde{\omega}(X \cup U)$ for $X \subseteq V$. ■

By showing the following lemma using Theorem 5.2.18 we can complete the proof of Theorem 5.2.6 concerning truncation. For a valuated matroid $\mathbf{M} = (V, \mathcal{B}, \omega)$ and $k \leq r = \operatorname{rank} \mathbf{M}$, define $\mathcal{B}_k \subseteq 2^V$ by (5.17) and $\omega_k : \mathcal{B}_k \to \mathbf{R}$ by

$$\omega_k(I) = \max\{\omega(B) \mid I \subseteq B \in \mathcal{B}\}, \qquad I \in \mathcal{B}_k.$$

Lemma 5.2.21. $\mathbf{M}_k = (V, \mathcal{B}_k, \omega_k)$ *is a valuated matroid, where* $k \leq r$.

Proof. Let V' be a copy of V, and U be a set with $|U| = r - k$. Consider a bipartite graph $G = (V^+, V^-; A)$ with $V^+ = V'$, $V^- = V \cup U$, and

$$A = \{(v', v) \mid v \in V\} \cup \{(v', u) \mid v \in V, u \in U\},$$

where $v' \in V'$ is the copy of $v \in V$. Let $\tilde{\omega}$ be the valuation induced on V^- from ω on V^+ $(\simeq V)$ and $w = 0$ on A. Then $\omega_k(I) = \tilde{\omega}(I \cup U)$ for $I \subseteq V$. See also Murota [229] for the original proof that does not rely on Theorem 5.2.18. ■

Just as a matroid can be induced by a bimatroid (Theorem 2.3.57), a valuated matroid can be induced by a valuated bimatroid, as follows. This construction appears more general than the induction of a valuated matroid by a bipartite graph, though the proof shows that it is just a variant thereof.

Theorem 5.2.22. *For a valuated bimatroid* (S, T, Λ, δ) *and a valuated matroid* $(T, \mathcal{B}, \omega)$*, assume that*

$$\Lambda * \mathcal{B} = \{X \subseteq S \mid \exists Y \subseteq T : (X, Y) \in \Lambda, Y \in \mathcal{B}\}$$

is nonempty. Then $(S, \Lambda * \mathcal{B}, \delta * \omega)$ *with*

$$(\delta * \omega)(X) = \max_{Y}\{\delta(X, Y) + \omega(Y) \mid (X, Y) \in \Lambda, Y \in \mathcal{B}\}, \qquad X \in \Lambda * \mathcal{B}$$

is a valuated matroid.

Proof. Let S' be a copy of S, and T' and T'' be copies of T. Consider a bipartite graph $G = (V^+, V^-; A)$ with $V^+ = S' \cup T' \cup T''$, $V^- = S \cup T$, and

$$A = \{(x', x) \mid x \in S\} \cup \{(y', y) \mid y \in T\} \cup \{(y'', y) \mid y \in T\},$$

where $x' \in S'$ is the copy of $x \in S$, and $y' \in T'$ [resp. $y'' \in T''$] is the copy of $y \in T$. Define $\omega' : 2^{S' \cup T' \cup T''} \to \mathbf{R} \cup \{-\infty\}$ by $\omega'(X' \cup Y_1' \cup Y_2'') = \delta(X, T \setminus Y_1) + \omega(Y_2)$ for $X' \subseteq S'$, $Y_1' \subseteq T'$, and $Y_2'' \subseteq T''$. Let $\tilde{\omega}'$ be the valuation induced on V^- from ω' with $w = 0$ on A. Then $(\delta * \omega)(X) = \tilde{\omega}'(X \cup T)$ for $X \subseteq S$. ■

Using a similar proof technique we can show that a *product* operation can be defined for valuated bimatroids in a manner compatible with the bimatroid product in Theorem 2.3.53.

Theorem 5.2.23. *For two valuated bimatroids* (S_i, T_i, δ_i) $(i = 1, 2)$ *with* $T_1 = S_2$, *define* $\delta_1 * \delta_2 : 2^{S_1} \times 2^{T_2} \to \mathbf{R} \cup \{-\infty\}$ *by*

$$(\delta_1 * \delta_2)(X, Z) = \max\{\delta_1(X, Y) + \delta_2(Y, Z) \mid Y \subseteq T_1\}, \quad X \subseteq S_1, Z \subseteq T_2.$$

Then $(S_1, T_2, \delta_1 * \delta_2)$ *is a valuated bimatroid.*

Proof. Let S_i' be a copy of S_i, and T_i' be a copy of T_i $(i = 1, 2)$, where $T_1' \cap S_2' = \emptyset$. Consider a bipartite graph $G = (V^+, V^-; A)$ with $V^+ = S_1' \cup T_1' \cup S_2' \cup T_2'$, $V^- = S_1 \cup T_1 \cup T_2$, and

$$A = \{(x', x) \mid x \in S_1\} \cup \{(y', y) \mid y \in T_1\} \cup \{(y'', y) \mid y \in T_1\} \cup \{(z', z) \mid z \in T_2\},$$

where $x' \in S_1'$ is the copy of $x \in S_1$, $y' \in T_1'$ [resp. $y'' \in S_2'$] is the copy of $y \in T_1$, and $z' \in T_2'$ is the copy of $z \in T_2$. Define $\omega' : 2^{S_1' \cup T_1' \cup S_2' \cup T_2'} \to \mathbf{R} \cup \{-\infty\}$ by $\omega'(X' \cup Y_1' \cup Y_2'' \cup Z') = \delta_1(S_1 \setminus X, Y_1) + \delta_2(S_2 \setminus Y_2, Z)$ for $X' \subseteq S_1'$, $Y_1' \subseteq T_1'$, $Y_2'' \subseteq S_2'$, and $Z' \subseteq T_2'$. Let $\tilde{\omega}'$ be the valuation induced on V^- from ω' with $w = 0$ on A. Then $(\delta_1 * \delta_2)(X, Z) = \tilde{\omega}'((S_1 \setminus X) \cup T_1 \cup Z)$ for $X \subseteq S_1$ and $Z \subseteq T_2$. ■

Theorem 5.2.23 above implies as an immediate corollary that a *union* operation can be defined for valuated bimatroids compatibly with the bimatroid union in Theorem 2.3.51. It should be clear that $S_1 \cap S_2 \neq \emptyset$ and $T_1 \cap T_2 \neq \emptyset$ in general.

Theorem 5.2.24. *For two valuated bimatroids* (S_i, T_i, δ_i) $(i = 1, 2)$, *define* $\delta_1 \vee \delta_2 : 2^{S_1 \cup S_2} \times 2^{T_1 \cup T_2} \to \mathbf{R} \cup \{-\infty\}$ *by*

$$\begin{aligned}(\delta_1 \vee \delta_2)&(X, Y) \\ = \max\{&\delta_1(X_1, Y_1) + \delta_2(X_2, Y_2) \mid X_1 \cup X_2 = X, Y_1 \cup Y_2 = Y, \\ &X_1 \cap X_2 = \emptyset, Y_1 \cap Y_2 = \emptyset\}, \quad X \subseteq S_1 \cup S_2, Y \subseteq T_1 \cup T_2.\end{aligned}$$

Then $(S_1 \cup S_2, T_1 \cup T_2, \delta_1 \vee \delta_2)$ *is a valuated bimatroid.*

Proof. Let S_i' be a copy of S_i, and T_i' be a copy of T_i $(i = 1, 2)$, where $S_1' \cap S_2' = \emptyset$ and $T_1' \cap T_2' = \emptyset$. Consider a valuated bimatroid $(S_1' \cup S_2', T_1' \cup T_2', \delta_{12})$ defined by $\delta_{12}(X_1' \cup X_2', Y_1' \cup Y_2') = \delta_1(X_1, Y_1) + \delta_2(X_2, Y_2)$, where $X_i' \subseteq S_i'$ denotes the copy of $X_i \subseteq S_i$ and similarly for $Y_i' \subseteq T_i'$. Also consider valuated bimatroids $(S_1 \cup S_2, S_1' \cup S_2', \delta_S)$ and $(T_1' \cup T_2', T_1 \cup T_2, \delta_T)$, where $\delta_S(X, X_1' \cup X_2')$ is equal to 0 if $X_1 \cup X_2 = X$ and $X_1 \cap X_2 = \emptyset$, and to $-\infty$ otherwise; and similarly, $\delta_T(Y_1' \cup Y_2', Y)$ is equal to 0 if $Y_1 \cup Y_2 = Y$ and $Y_1 \cap Y_2 = \emptyset$, and to $-\infty$ otherwise. Then we have $\delta_1 \vee \delta_2 = \delta_S * \delta_{12} * \delta_T$, which is a valuated bimatroid by Theorem 5.2.23. ■

5.2.7 Characterizations

The objective of this section is to give other axioms that characterize a valuated matroid.

First we consider two seemingly weaker (but actually equivalent) exchange axioms for $\omega : \mathcal{B} \to \mathbf{R}$ or $\omega : 2^V \to \mathbf{R} \cup \{-\infty\}$ with $\mathcal{B} = \{B \subseteq V \mid \omega(B) \neq -\infty\}$.

(VM$_{\rm w}$) For $B, B' \in \mathcal{B}$ with $B \neq B'$, there exist $u \in B \setminus B'$ and $v \in B' \setminus B$ such that $B - u + v \in \mathcal{B}$, $B' + u - v \in \mathcal{B}$, and

$$\omega(B) + \omega(B') \leq \omega(B - u + v) + \omega(B' + u - v),$$

(VM$_{\rm loc}$) $\mathcal{B}$ satisfies (BM$_\pm$), and for $B, B' \in \mathcal{B}$ with $|B \setminus B'| = 2$, there exist $u \in B \setminus B'$ and $v \in B' \setminus B$ such that $B - u + v \in \mathcal{B}$, $B' + u - v \in \mathcal{B}$, and

$$\omega(B) + \omega(B') \leq \omega(B - u + v) + \omega(B' + u - v).$$

Note that (VM$_{\rm w}$) is a quantitative generalization of the self-dual exchange axiom (BM$_{\pm \rm w}$) of §2.3.4, and (VM$_{\rm loc}$) states a local exchange property.

These exchange axioms are in fact equivalent as follows (Dress–Wenzel [56], Murota [228]).

Theorem 5.2.25. *For a function $\omega : 2^V \to \mathbf{R} \cup \{-\infty\}$, the three conditions* (VM), (VM$_{\rm w}$), *and* (VM$_{\rm loc}$) *are equivalent, where $\mathcal{B} = \{B \subseteq V \mid \omega(B) \neq -\infty\}$.*

Proof. Since (VM$_{\rm w}$) $\Rightarrow$ (BM$_{\pm \rm w}$) and also (BM$_{\pm \rm w}$) $\Leftrightarrow$ (BM$_\pm$) by Theorem 2.3.14, we see (VM$_{\rm w}$) $\Rightarrow$ (VM$_{\rm loc}$), whereas (VM) $\Rightarrow$ (VM$_{\rm w}$) is obvious since $B \setminus B' \neq \emptyset$ for distinct $B, B' \in \mathcal{B}$. To prove (VM$_{\rm loc}$) $\Rightarrow$ (VM) we assume (VM$_{\rm loc}$). For $p : V \to \mathbf{R}$ we abbreviate $\omega[p]$ of (5.16) to ω_p, and define

$$\omega_p(B, u, v) = \omega_p(B - u + v) - \omega_p(B) \qquad (B \in \mathcal{B}),$$

consistently with $\omega(B, u, v)$ of (5.21), where $\omega_p(B, u, v) = -\infty$ if $B - u + v \notin \mathcal{B}$. For $B, B' \in \mathcal{B}$ and $u \in B \setminus B'$, $v \in B' \setminus B$, we have

$$\omega(B,u,v)+\omega(B',v,u)=\omega_p(B,u,v)+\omega_p(B',v,u). \tag{5.36}$$

If $B \in \mathcal{B}$, $B \setminus B' = \{u_0, u_1\}$, $B' \setminus B = \{v_0, v_1\}$ (with $u_0 \neq u_1$, $v_0 \neq v_1$), ($\mathrm{VM_{loc}}$) implies

$$\begin{aligned}\omega_p(B')-\omega_p(B) \le \max(&\omega_p(B,u_0,v_0)+\omega_p(B,u_1,v_1),\\ &\omega_p(B,u_0,v_1)+\omega_p(B,u_1,v_0)).\end{aligned} \tag{5.37}$$

Define

$$\begin{aligned}\mathcal{D}=\{(B,B') \mid B,B' \in \mathcal{B}, \exists u_* \in B \setminus B', \forall v \in B' \setminus B :\\ \omega(B)+\omega(B') > \omega(B-u_*+v)+\omega(B'+u_*-v)\},\end{aligned}$$

which denotes the set of pairs (B, B') for which the exchangeability in (VM) fails. We want to show $\mathcal{D} = \emptyset$.

Suppose, to the contrary, that $\mathcal{D} \neq \emptyset$, and take $(B, B') \in \mathcal{D}$ such that $|B' \setminus B|$ is minimum and let $u_* \in B \setminus B'$ be as in the definition of $\mathcal{D}$. We have $|B' \setminus B| > 2$. Define $p : V \to \mathbf{R}$ by

$$p(v)=\begin{cases}-\omega(B,u_*,v) & (v \in B' \setminus B, B-u_*+v \in \mathcal{B})\\ \omega(B',v,u_*)+\varepsilon & (v \in B' \setminus B, B-u_*+v \notin \mathcal{B}, B'+u_*-v \in \mathcal{B})\\ 0 & (\text{otherwise})\end{cases}$$

with some $\varepsilon > 0$ and consider ω_p.

Claim 1:

$$\omega_p(B,u_*,v)=0 \qquad \text{if } v \in B' \setminus B, B-u_*+v \in \mathcal{B}, \tag{5.38}$$

$$\omega_p(B',v,u_*)<0 \qquad \text{for } v \in B' \setminus B. \tag{5.39}$$

The inequality (5.39) can be shown as follows. If $B - u_* + v \in \mathcal{B}$, we have $\omega_p(B, u_*, v) = 0$ by (5.38) and

$$\omega_p(B,u_*,v)+\omega_p(B',v,u_*)=\omega(B,u_*,v)+\omega(B',v,u_*)<0$$

by (5.36) and the definition of u_*. Otherwise we have $\omega_p(B', v, u_*) = -\varepsilon$ or $-\infty$ according to whether $B' + u_* - v \in \mathcal{B}$ or not.

Claim 2: There exist $u_0 \in B \setminus B'$ and $v_0 \in B' \setminus B$ such that $u_0 \neq u_*$, $B' + u_0 - v_0 \in \mathcal{B}$, and

$$\omega_p(B',v_0,u_0) \ge \omega_p(B',v,u_0) \qquad (v \in B' \setminus B). \tag{5.40}$$

Since $|B \setminus B'| > 2$, there exists $u_0 \in B \setminus B'$ distinct from u_*. By ($\mathrm{BM_{\pm}}$) we have $B' + u_0 - v_0 \in \mathcal{B}$ for some $v_0 \in B' \setminus B$. We can further assume (5.40) by redefining v_0 to be the element $v \in B' \setminus B$ that maximizes $\omega_p(B', v, u_0)$.

Claim 3: $(B, B'') \in \mathcal{D}$ with $B'' = B' + u_0 - v_0$.

To prove this it suffices to show

$$\omega_p(B, u_*, v) + \omega_p(B'', v, u_*) < 0 \qquad (v \in B'' \setminus B).$$

We may restrict ourselves to v with $B - u_* + v \in \mathcal{B}$, since otherwise the first term $\omega_p(B, u_*, v)$ is equal to $-\infty$. For such v the first term is equal to zero by (5.38). For the second term it follows from (5.37), (5.39), and (5.40) that

$$\begin{aligned} &\omega_p(B'', v, u_*) \\ &= \omega_p(B' + \{u_0, u_*\} - \{v_0, v\}) - \omega_p(B' + u_0 - v_0) \\ &\leq \max\left[\omega_p(B', v_0, u_0) + \omega_p(B', v, u_*), \omega_p(B', v, u_0) + \omega_p(B', v_0, u_*)\right] \\ &\quad -\omega_p(B', v_0, u_0) \\ &< \max\left[\omega_p(B', v_0, u_0), \omega_p(B', v, u_0)\right] - \omega_p(B', v_0, u_0) \ = 0. \end{aligned}$$

Since $|B''\setminus B| = |B'\setminus B| - 1$, Claim 3 contradicts our choice of $(B, B') \in \mathcal{D}$. Therefore we conclude $\mathcal{D} = \emptyset$. ∎

It is also mentioned that another exchange axiom

(VM$_{\rm d}$) For $B, B' \in \mathcal{B}$ and $u \in B \setminus B'$, there exist $v \in B' \setminus B$ and $u' \in B \setminus B'$ such that $B - u + v \in \mathcal{B}$, $B' + u' - v \in \mathcal{B}$, and

$$\omega(B) + \omega(B') \leq \omega(B - u + v) + \omega(B' + u' - v)$$

is known to be equivalent to (VM) (see Murota [226]) .

A valuated matroid can be characterized as a family of matroids. For $p : V \to \mathbf{R}$ we define

$$\mathcal{B}_p = \{B \in \mathcal{B} \mid \omega[p](B) \geq \omega[p](B') \ (\forall\, B' \in \mathcal{B})\},$$

which denotes the set of the maximizers of $\omega[p]$ (see (5.16) for the notation $\omega[p]$). It is immediate from (VM) that $\mathcal{B}_p$ forms the basis family of a matroid, whereas Murota [227, 228] points out that the converse is also true.

Theorem 5.2.26. *Let $(V, \mathcal{B})$ be a matroid. A function $\omega : \mathcal{B} \to \mathbf{R}$ is a valuation of $(V, \mathcal{B})$ if and only if for any $p : V \to \mathbf{R}$ the family $\mathcal{B}_p$ of the maximizers of $\omega[p]$ forms the basis family of a matroid. If ω is integer-valued, we may restrict p to be integer-valued in the "if" part.*

Proof. We abbreviate $\omega[p]$ to ω_p.

The "only if" part is easy to see. Take $B, B' \in \mathcal{B}_p$ and $u \in B \setminus B'$. Since ω_p satisfies (VM), there exists $v \in B' \setminus B$ such that

$$2 \max \omega_p = \omega_p(B) + \omega_p(B') \leq \omega_p(B - u + v) + \omega_p(B' + u - v),$$

which shows $B - u + v \in \mathcal{B}_p$ and $B' + u - v \in \mathcal{B}_p$. That is, $\mathcal{B}_p$ satisfies (BM$_\pm$).

For the "if" part it suffices, by Theorem 5.2.25, to show the local exchange axiom (VM$_{\rm loc}$). Under the assumption of (BM$_\pm$) for $\mathcal{B}$, (VM$_{\rm loc}$) is equivalent to the following claim.

Claim 1: If $B, B' \in \mathcal{B}$, $B \setminus B' = \{u_0, u_1\}$, $B' \setminus B = \{v_0, v_1\}$ (with $u_0 \neq u_1$, $v_0 \neq v_1$), then

$$\omega(B') - \omega(B) \leq \max(\omega_{00} + \omega_{11}, \omega_{01} + \omega_{10}), \tag{5.41}$$

where $\omega_{ij} = \omega(B, u_i, v_j)$ for $i, j = 0, 1$.

This claim can be proven as follows. Denote by γ and μ the left hand side and the right hand side of (5.41), respectively. We consider a bipartite graph $G = (\{u_0, u_1\}, \{v_0, v_1\}; E)$ with $E = \{(u_i, v_j) \mid B - u_i + v_j \in \mathcal{B}\}$. The graph G has a perfect matching (of size 2) by (BM$_\pm$). In addition we associate ω_{ij} with edge (u_i, v_j) as the weight. Then μ is equal to the maximum weight of a perfect matching in G and, by a variant of Theorem 2.2.36, there exists $\hat{p} : \{u_0, u_1, v_0, v_1\} \to \mathbf{R}$ such that

$$\hat{p}(u_i) + \hat{p}(v_j) \geq \omega_{ij} \quad ((u_i, v_j) \in E), \qquad \sum_{i=0,1} \hat{p}(u_i) + \sum_{j=0,1} \hat{p}(v_j) = \mu.$$

To show $\gamma \leq \mu$, suppose, to the contrary, that $\gamma > \mu$. Then we can modify $\hat{p}$ to $\bar{p} : \{u_0, u_1, v_0, v_1\} \to \mathbf{R}$ such that

$$\bar{p}(u_i) + \bar{p}(v_j) \geq \omega_{ij} \quad ((u_i, v_j) \in E), \qquad \sum_{i=0,1} \bar{p}(u_i) + \sum_{j=0,1} \bar{p}(v_j) = \gamma.$$

Using $\bar{p}$ we define $p : V \to \mathbf{R}$ by

$$p(v) = \begin{cases} +\bar{p}(v) & (v \in B \setminus B') \\ -\bar{p}(v) & (v \in B' \setminus B) \\ +M & (v \in B \cap B') \\ -M & (v \in V \setminus (B' \cup B)) \end{cases}$$

where $M > 0$ is a sufficiently large number.

For this p we have $\{B, B'\} \subseteq \mathcal{B}_p$, i.e., $\omega_p(B) = \omega_p(B') \geq \omega_p(B'')$ ($\forall B'' \in \mathcal{B}$). In fact, this is immediate from the following relations:

$$\omega_p(B') - \omega_p(B) = [\omega(B') - \omega(B)] - \sum_{i=0,1} \bar{p}(u_i) - \sum_{j=0,1} \bar{p}(v_j) = 0,$$

$$\begin{aligned} &\omega_p(B - u_i + v_j) - \omega_p(B) \\ &\quad = [\omega(B - u_i + v_j) - \omega(B)] + [p(B - u_i + v_j) - p(B)] \\ &\quad = \omega_{ij} - \bar{p}(u_i) - \bar{p}(v_j) \ \leq 0, \end{aligned}$$

$$\omega_p(B'') - \omega_p(B) \leq \omega(B'') - \omega(B) + \sum_{i=0,1} |\bar{p}(u_i)| + \sum_{j=0,1} |\bar{p}(v_j)| - M \ \leq 0 \quad \text{unless} \quad B \cap B' \subseteq B'' \subseteq B \cup B'.$$

Since $B, B' \in \mathcal{B}_p$ and $(V, \mathcal{B}_p)$ satisfies (BM$_\pm$) by the assumption, there exists $j \in \{0, 1\}$ such that $\omega_p(B - u_0 + v_j) = \omega_p(B' + u_0 - v_j) = \omega_p(B)$. Putting $k = 1 - j$ and noting $B' + u_0 - v_j = B - u_1 + v_k$, we obtain

$$\begin{aligned}0 &= \omega_p(B - u_0 + v_j) + \omega_p(B - u_1 + v_k) - 2\omega_p(B) \\ &= \omega(B - u_0 + v_j) + \omega(B - u_1 + v_k) - 2\omega(B) - \sum_{i=0,1} \bar{p}(u_i) - \sum_{j=0,1} \bar{p}(v_j) \\ &= \omega_{0j} + \omega_{1k} - \gamma \leq \mu - \gamma,\end{aligned}$$

a contradiction to $\gamma > \mu$. This completes the proof of the claim.

If ω is integer-valued, we have $\omega_{ij} \in \mathbf{Z}$ for all (i, j), and consequently, $\hat{p}$, $\bar{p}$ and p can be chosen to be integer-valued. ■

Finally we consider "level sets" of $\omega : \mathcal{B} \to \mathbf{R}$ defined by

$$\mathcal{L}(\omega, \alpha) = \{B \in \mathcal{B} \mid \omega(B) \geq \alpha\} \tag{5.42}$$

for $\alpha \in \mathbf{R}$. A level set of a matroid valuation does not necessarily form the basis family of a matroid, as the following example shows.

Example 5.2.27. Consider a valuated matroid $(V, \mathcal{B}, \omega)$ defined by $V = \{1, 2, 3, 4\}$, $\mathcal{B} = \{\{1, 2\}, \{2, 3\}, \{3, 4\}, \{4, 1\}\}$, and $\omega(\{1, 2\}) = 1$, $\omega(\{2, 3\}) = 2$, $\omega(\{3, 4\}) = 1$, $\omega(\{4, 1\}) = 0$. Then $\mathcal{L}(\omega, 1) = \{\{1, 2\}, \{2, 3\}, \{3, 4\}\}$ does not satisfy the simultaneous basis exchange property ($\mathrm{BM}_{\pm}$). □

It follows from (VM), however, that the family $\mathcal{L} = \mathcal{L}(\omega, \alpha)$ satisfies the following (weaker) exchange properties:

(BL) For $B, B' \in \mathcal{L}$ and for $u \in B \setminus B'$, there exists $v \in B' \setminus B$ such that $B - u + v \in \mathcal{L}$ or $B' + u - v \in \mathcal{L}$,

($\mathrm{BL_w}$) For distinct $B, B' \in \mathcal{L}$, there exist $u \in B \setminus B'$ and $v \in B' \setminus B$ such that $B - u + v \in \mathcal{L}$ or $B' + u - v \in \mathcal{L}$.

To see this, observe that the inequality (5.14) for (VM) implies:

$$\omega(B) \geq \alpha, \omega(B') \geq \alpha \Longrightarrow \max\{\omega(B - u + v), \omega(B' + u - v)\} \geq \alpha.$$

For $\mathcal{L} \subseteq 2^V$ in general, it holds obviously that ($\mathrm{BM}_{\pm}$) $\Rightarrow$ (BL) $\Rightarrow$ ($\mathrm{BL_w}$), but the converse is not true. In fact, "($\mathrm{BM}_{\pm}$) $\not\Leftarrow$ (BL)" is demonstrated by $\mathcal{L} = \{\{1, 2\}, \{2, 3\}, \{3, 4\}\}$ and "(BL) $\not\Leftarrow$ ($\mathrm{BL_w}$)" is by $\mathcal{L} = \{\{1, 2, 3\}, \{2, 3, 4\}, \{3, 4, 5\}, \{4, 5, 6\}\}$.

The following theorem of Shioura [299] characterizes a valuated matroid in terms of the level sets of $\omega[p]$. See (5.16) for the notation $\omega[p]$.

Theorem 5.2.28. *Let $(V, \mathcal{B})$ be a matroid. For a function $\omega : \mathcal{B} \to \mathbf{R}$, the following three conditions are equivalent:*

(i) *$\omega : \mathcal{B} \to \mathbf{R}$ is a valuation of $(V, \mathcal{B})$,*

(ii) *For any $p : V \to \mathbf{R}$ and for any $\alpha \in \mathbf{R}$, $\mathcal{L}(\omega[p], \alpha)$ satisfies* (BL),

(iii) *For any $p : V \to \mathbf{R}$ and for any $\alpha \in \mathbf{R}$, $\mathcal{L}(\omega[p], \alpha)$ satisfies* ($\mathrm{BL_w}$).

Proof. [(i) ⇒ (ii) ⇒ (iii)] If ω is a valuation, so is $\omega[p]$. Hence the claim follows from the observations above.

[(iii) ⇒ (i)] By Theorem 5.2.25 it suffices to show the local exchange axiom ($\mathrm{VM_{loc}}$). Suppose that $\{B, B'\} \subseteq \mathcal{B}$, $B \setminus B' = \{u_0, u_1\}$, $B' \setminus B = \{v_0, v_1\}$ with $u_0 \neq u_1$, $v_0 \neq v_1$. Put $B_{ij} = B - u_i + v_j$ for $i, j = 0, 1$. By ($\mathrm{BM_\pm}$) for $\mathcal{B}$ it holds that $\{B_{00}, B_{11}\} \subseteq \mathcal{B}$ or $\{B_{01}, B_{10}\} \subseteq \mathcal{B}$. Hence we may assume $\{B_{00}, B_{11}\} \subseteq \mathcal{B}$ without loss of generality. Define p by

$$\begin{aligned} p(u_0) &= [\omega(B') - \omega(B) + \omega(B_{00}) - \omega(B_{11})]/2, \\ p(u_1) &= [\omega(B') - \omega(B) - \omega(B_{00}) + \omega(B_{11})]/2 + t, \\ p(v_1) &= t, \\ p(v) &= 0 \qquad (v \in V \setminus \{u_0, u_1, v_1\}), \end{aligned}$$

so that $\omega[p](B') = \omega[p](B)$ and $\omega[p](B_{00}) = \omega[p](B_{11})$, where $t \in \mathbf{R}$ is a parameter. Since $\{B, B'\} \subseteq \mathcal{L}(\omega[p], \alpha)$ for $\alpha = \omega[p](B') = \omega[p](B)$, the proof of ($\mathrm{VM_{loc}}$) is completed if

$$\{B_{00}, B_{11}\} \subseteq \mathcal{L}(\omega[p], \alpha) \quad \text{or} \quad \{B_{01}, B_{10}\} \subseteq \mathcal{L}(\omega[p], \alpha) \tag{5.43}$$

is shown for some t. If $\{B_{01}, B_{10}\} \subseteq \mathcal{B}$, take

$$t = [\omega(B_{00}) - \omega(B_{11}) + \omega(B_{10}) - \omega(B_{01})]/2$$

so that $\omega[p](B_{01}) = \omega[p](B_{10})$. Then (5.43) follows from ($\mathrm{BL_w}$). If $B_{01} \notin \mathcal{B}$, we can take t large enough for $\omega[p](B_{10}) < \alpha$, and then ($\mathrm{BL_w}$) implies $\{B_{00}, B_{11}\} \subseteq \mathcal{L}(\omega[p], \alpha)$. In the remaining case where $B_{10} \notin \mathcal{B}$, we can take t small enough for $\omega[p](B_{01}) < \alpha$, and then ($\mathrm{BL_w}$) implies $\{B_{00}, B_{11}\} \subseteq \mathcal{L}(\omega[p], \alpha)$. ∎

5.2.8 Further Exchange Properties

We shall establish a number of lemmas concerning basis exchange in a single valuated matroid. They will play key roles for the valuated matroid intersection problem, to be explained later.

For $B \in \mathcal{B}$ and $B' \subseteq V$ we consider the exchangeability graph $G(B, B')$ introduced in §2.3.4. $G(B, B') = (B \setminus B', B' \setminus B; A)$ is a bipartite graph having $(B \setminus B', B' \setminus B)$ as the vertex bipartition and

$$A = \{(u, v) \mid u \in B \setminus B', v \in B' \setminus B, B - u + v \in \mathcal{B}\} \tag{5.44}$$

as the arc set. The key properties of the exchangeability in a matroid have been formulated as the perfect-matching lemma (Lemma 2.3.16) and the unique-matching lemma (Lemma 2.3.18).

To capture the exchangeability with valuations, we need quantitative extensions of the perfect-matching lemma and the unique-matching lemma. To this end we attach the exchange gain $\omega(B,u,v)$ of (5.21) to each arc (u,v) as "arc weight," and denote by $\widehat{\omega}(B,B')$ the maximum weight of a perfect matching in $G(B,B')$ with respect to the arc weight $\omega(B,u,v)$, i.e.,

$$\widehat{\omega}(B,B') = \max\{ \sum_{(u,v)\in M} \omega(B,u,v) \mid M\text{: perfect matching in } G(B,B')\}. \tag{5.45}$$

The perfect-matching lemma is generalized as follows (Murota [224]).

Lemma 5.2.29 (Upper-bound lemma). *For $B, B' \in \mathcal{B}$,*

$$\omega(B') \leq \omega(B) + \widehat{\omega}(B,B'). \tag{5.46}$$

Proof. For any $u_1 \in B \setminus B'$ there exists $v_1 \in B' \setminus B$ with

$$\omega(B) + \omega(B') \leq \omega(B - u_1 + v_1) + \omega(B' + u_1 - v_1),$$

which can be rewritten as

$$\omega(B') \leq \omega(B, u_1, v_1) + \omega(B'_2)$$

with $B'_2 = B' + u_1 - v_1$. By the same argument applied to (B, B'_2) we obtain

$$\omega(B'_2) \leq \omega(B, u_2, v_2) + \omega(B'_3)$$

for some $u_2 \in (B\setminus B') - u_1$ and $v_2 \in (B'\setminus B) - v_1$, where $B'_3 = B'_2 + u_2 - v_2 = B' + \{u_1, u_2\} - \{v_1, v_2\}$. Hence

$$\omega(B') \leq \omega(B'_3) + \sum_{i=1}^{2} \omega(B, u_i, v_i).$$

Repeating this process we arrive at

$$\omega(B') \leq \omega(B) + \sum_{i=1}^{m} \omega(B, u_i, v_i) \leq \omega(B) + \widehat{\omega}(B,B'),$$

where $m = |B \setminus B'| = |B' \setminus B|$, $B \setminus B' = \{u_1, \cdots, u_m\}$, $B' \setminus B = \{v_1, \cdots, v_m\}$. ■

Remark 5.2.30. For a trivial valuation $\omega : 2^V \to \{0, -\infty\}$, the upper-bound lemma guarantees $\widehat{\omega}(B,B') \neq -\infty$, which shows the existence of a perfect matching in $G(B,B')$. Thus, the perfect-matching lemma is a special case of the upper-bound lemma. □

Remark 5.2.31. The upper-bound lemma gives an alternative proof for the optimality condition given in Theorem 5.2.7. The necessity of (5.22) is obvious. For sufficiency take any $B' \in \mathcal{B}$ and consider $G(B,B')$. The condition (5.22) is equivalent to all the arcs having nonpositive weights. Hence $\widehat{\omega}(B,B') \leq 0$. Then the upper-bound lemma implies $\omega(B') \leq \omega(B)$. □

In the upper-bound lemma it is natural to ask for a (sufficient) condition under which the bound (5.46) is tight. Comparison of the unique-matching lemma (Lemma 2.3.18) and the upper-bound lemma will suggest

[Unique-max condition]
There exists exactly one maximum-weight perfect matching in $G(B, B')$.

In what follows we shall show (in Lemma 5.2.35 below) that this is indeed a sufficient condition for the tightness in (5.46).

First we note the following fact, rephrasing the unique-max condition in terms of "potential" or "dual variable".

Lemma 5.2.32. *Let $B \in \mathcal{B}$ and $B' \subseteq V$ with $|B' \setminus B| = |B \setminus B'| = m$.*

(1) *$G(B, B')$ has a perfect matching if and only if there exist $\widehat{p} : (B \setminus B') \cup (B' \setminus B) \to \mathbf{R}$ and indexings of the elements of $B \setminus B'$ and $B' \setminus B$, say $B \setminus B' = \{u_1, \cdots, u_m\}$ and $B' \setminus B = \{v_1, \cdots, v_m\}$, such that*

$$\omega(B, u_i, v_j) - \widehat{p}(u_i) + \widehat{p}(v_j) \begin{cases} = 0 \ (1 \leq i = j \leq m) \\ \leq 0 \ (1 \leq i, j \leq m). \end{cases} \tag{5.47}$$

Then, $\widehat{\omega}(B, B') = \sum_{i=1}^{m} \widehat{p}(u_i) - \sum_{j=1}^{m} \widehat{p}(v_j)$.

(2) *The pair (B, B') satisfies the unique-max condition if and only if there exist $\widehat{p} : (B \setminus B') \cup (B' \setminus B) \to \mathbf{R}$ and indexings of the elements of $B \setminus B'$ and $B' \setminus B$, say $B \setminus B' = \{u_1, \cdots, u_m\}$ and $B' \setminus B = \{v_1, \cdots, v_m\}$, such that*

$$\omega(B, u_i, v_j) - \widehat{p}(u_i) + \widehat{p}(v_j) \begin{cases} = 0 \ (1 \leq i = j \leq m) \\ \leq 0 \ (1 \leq j < i \leq m) \\ < 0 \ (1 \leq i < j \leq m). \end{cases} \tag{5.48}$$

Proof. This follows from Theorem 2.2.36. ∎

Lemma 5.2.33. *Let $B \in \mathcal{B}$ and u, u°, v, v° be four distinct elements with $\{u, u^\circ\} \subseteq B$, $\{v, v^\circ\} \subseteq V \setminus B$, and let $B' = B - \{u, u^\circ\} + \{v, v^\circ\}$. Assume that $M = \{(u, v), (u^\circ, v^\circ)\}$ is the unique maximum-weight perfect matching in $G(B, B')$.*

(1) *$B' \in \mathcal{B}$ and $\omega(B') = \omega(B) + \widehat{\omega}(B, B')$.*

(2) *For $B^\circ = B - u^\circ + v^\circ$ we have*

$$\begin{aligned} \omega(B^\circ, u, v) &= \omega(B, u, v), \\ \omega(B^\circ, u, u^\circ) &= \omega(B, u, v^\circ) - \omega(B, u^\circ, v^\circ), \\ \omega(B^\circ, v^\circ, v) &= \omega(B, u^\circ, v) - \omega(B, u^\circ, v^\circ). \end{aligned}$$

Proof. (1) Putting $B^* = B - u + v$ we see

$$\omega(B^*) + \omega(B^\circ) = \omega(B, u, v) + \omega(B, u^\circ, v^\circ) + 2\omega(B) = \widehat{\omega}(B, B') + 2\omega(B). \tag{5.49}$$

By applying the exchange axiom (5.14) to (B°, B^*) with $u \in B^\circ \setminus B^*$ we have

$$\omega(B^*) + \omega(B^\circ) \leq \omega(B^* - v' + u) + \omega(B^\circ + v' - u)$$

for some $v' \in B^* \setminus B^\circ = \{u^\circ, v\}$. Combining this with (5.49) we obtain

$$\widehat{\omega}(B, B') + 2\omega(B) \leq \omega(B^* - v' + u) + \omega(B^\circ + v' - u). \tag{5.50}$$

Suppose that $v' = u^\circ$. Then

$$\begin{aligned} \text{RHS of (5.50)} &= \omega(B^* - u^\circ + u) + \omega(B^\circ + u^\circ - u) \\ &= \omega(B - u^\circ + v) + \omega(B - u + v^\circ) \\ &= \omega(B, u^\circ, v) + \omega(B, u, v^\circ) + 2\omega(B). \end{aligned}$$

This means that $M' = \{(u^\circ, v), (u, v^\circ)\}$ is also a maximum-weight perfect matching in $G(B, B')$, a contradiction to the uniqueness of M.

Therefore we have $v' = v$ in (5.50), and then

$$\text{RHS of (5.50)} = \omega(B^* - v + u) + \omega(B^\circ + v - u) = \omega(B) + \omega(B').$$

Hence follows $\omega(B) + \widehat{\omega}(B, B') \leq \omega(B')$. The reverse inequality has already been shown in the upper-bound lemma (Lemma 5.2.29). Note that $B' \in \mathcal{B}$ follows from $\omega(B') \neq -\infty$.

(2) By straightforward calculations as follows:

$$\begin{aligned} \omega(B^\circ, u, v) &= \omega(B - u^\circ + v^\circ - u + v) - \omega(B - u^\circ + v^\circ) \\ &= \omega(B') - \omega(B) - \omega(B, u^\circ, v^\circ) \\ &= \widehat{\omega}(B, B') - \omega(B, u^\circ, v^\circ) \\ &= \omega(B, u, v), \\ \omega(B^\circ, u, u^\circ) &= \omega(B - u + v^\circ) - \omega(B - u^\circ + v^\circ) \\ &= \omega(B, u, v^\circ) - \omega(B, u^\circ, v^\circ), \\ \omega(B^\circ, v^\circ, v) &= \omega(B - u^\circ + v) - \omega(B - u^\circ + v^\circ) \\ &= \omega(B, u^\circ, v) - \omega(B, u^\circ, v^\circ). \end{aligned}$$

■

Lemma 5.2.34. *Let $B \in \mathcal{B}$ and $B' \subseteq V$ with $|B'| = |B|$. If there exists exactly one maximum-weight perfect matching M in $G(B, B')$, then for any $(u^\circ, v^\circ) \in M$ the following hold true.*

(1) $B^\circ \equiv B - u^\circ + v^\circ \in \mathcal{B}$.

(2) There exists exactly one maximum-weight perfect matching in $G(B^\circ, B')$.

(3) $\widehat{\omega}(B^\circ, B') = \widehat{\omega}(B, B') - \omega(B, u^\circ, v^\circ)$.

Proof. (1) This is obvious.

(2) Using the notation in Lemma 5.2.32 we have $M = \{(u_i, v_i) \mid i = 1, \cdots, m\}$ and $(u^\circ, v^\circ) = (u_k, v_k)$ for some k. Put

$$B_{ij} = B^\circ - u_i + v_j = B - \{u_i, u^\circ\} + \{v_j, v^\circ\}$$

for $i \neq k$, $j \neq k$. It then follows from (5.46) and (5.48) that

$$\begin{aligned}
&\omega(B^\circ, u_i, v_j) \\
&= \omega(B_{ij}) - \omega(B^\circ) \\
&\leq \widehat{\omega}(B, B_{ij}) - \omega(B, u_k, v_k) \\
&= \max\,[\omega(B, u_k, v_k) + \omega(B, u_i, v_j), \omega(B, u_i, v_k) + \omega(B, u_k, v_j)] \\
&\quad -\omega(B, u_k, v_k) \\
&\leq [\widehat{p}(u_i) + \widehat{p}(u_k) - \widehat{p}(v_j) - \widehat{p}(v_k)] - [\widehat{p}(u_k) - \widehat{p}(v_k)] \\
&= \widehat{p}(u_i) - \widehat{p}(v_j),
\end{aligned}$$

where the second inequality is strict for $i < j$. For $i = j$, on the other hand, both inequalities are satisfied with equalities, since $G(B, B_{ii})$ has a unique maximum-weight perfect matching $\{(u_i, v_i), (u^\circ, v^\circ)\}$ and Lemma 5.2.33 implies $\omega(B^\circ, u_i, v_i) = \omega(B, u_i, v_i) = \widehat{p}(u_i) - \widehat{p}(v_i)$. Thus, the potential $\widehat{p}$ for (B, B') serves as a certificate of the unique-max condition also for (B°, B').

(3) $\widehat{\omega}(B^\circ, B') = \sum_{i \neq k} (\widehat{p}(u_i) - \widehat{p}(v_i)) = \widehat{\omega}(B, B') - \omega(B, u^\circ, v^\circ)$. ■

We are now in a position to state a main result of this section. The "unique-max lemma" below, due to Murota [224], is a quantitative extension of the unique-matching lemma (Lemma 2.3.18).

Lemma 5.2.35 (Unique-max lemma). *Let $B \in \mathcal{B}$ and $B' \subseteq V$ with $|B'| = |B|$. If there exists exactly one maximum-weight perfect matching in $G(B, B')$, then $B' \in \mathcal{B}$ and*

$$\omega(B') = \omega(B) + \widehat{\omega}(B, B'). \tag{5.51}$$

Proof. By induction on $m = |B \setminus B'|$. The case of $m = 1$ is obvious. So assume $m \geq 2$. Take any (u°, v°) contained in the unique maximum-weight perfect matching, and put $B^\circ = B - u^\circ + v^\circ$. (B°, B') satisfies the unique-max condition by Lemma 5.2.34(2), and we have

$$\omega(B') = \omega(B^\circ) + \widehat{\omega}(B^\circ, B')$$

by the induction hypothesis. By Lemma 5.2.34(3) we see

$$\widehat{\omega}(B^\circ, B') = \widehat{\omega}(B, B') - \omega(B, u^\circ, v^\circ)$$

while $\omega(B^\circ) = \omega(B) + \omega(B, u^\circ, v^\circ)$ by definition. Hence follows (5.51). ■

Remark 5.2.36. A remark is in order on the relation between "unique-matching condition" (= uniqueness of a perfect matching in $G(B, B')$) and "unique-max condition" (= uniqueness of the *maximum-weight* perfect matching in $G(B, B')$). Obviously the former implies the latter, and not conversely in general. However, for a separable valuation, induced from a linear weighting (Example 5.2.2), these two conditions are equivalent, and consequently the unique-max lemma reduces to the unique-matching lemma. See also Frank [76, Lemma 2] in this connection. □

Remark 5.2.37. There is an alternative proof of the unique-max lemma, suggested by A. Sebő, that makes use of the unique-matching lemma as well as the upper-bound lemma in contrast to the above proof. Let $\widehat{p} : (B \setminus B') \cup (B' \setminus B) \to \mathbf{R}$ be as in Lemma 5.2.32 and extend it to $p : V \to \mathbf{R}$ by defining

$$p(v) = \begin{cases} \widehat{p}(v) \ (v \in (B \setminus B') \cup (B' \setminus B)) \\ +M \ (v \in B \cap B') \\ -M \ (v \in V \setminus (B' \cup B)) \end{cases}$$

with a sufficiently large $M > 0$. We abbreviate $\omega[p]$ of (5.16) to ω_p. It follows from Theorem 5.2.26 ("only if" part) that the family of the maximizers of ω_p:

$$\mathcal{B}_p = \{B''' \in \mathcal{B} \mid \omega_p(B''') \geq \omega_p(B'') \ (B'' \in \mathcal{B})\}$$

forms the basis family of a matroid, say $\mathbf{M}_p = (V, \mathcal{B}_p)$. We claim that $B \in \mathcal{B}_p$. To see this, first note that

$$\omega_p(B'') - \omega_p(B) \leq \omega(B'') - \omega(B) - M/2 \leq 0$$

unless $B \cap B' \subseteq B'' \subseteq B \cup B'$. If $B \cap B' \subseteq B'' \subseteq B \cup B'$, on the other hand, we have $\omega_p(B'') - \omega_p(B) \leq 0$ by the upper-bound lemma and the inequality

$$\omega_p(B, u, v) = \omega(B, u, v) - \widehat{p}(u) + \widehat{p}(v) \leq 0 \qquad (u \in B \setminus B'', v \in B'' \setminus B).$$

We also claim that the exchangeability graph $G_p(B, B')$ in $\mathbf{M}_p$ has a unique perfect matching, since $B - u_i + v_i \in \mathcal{B}_p$ $(1 \leq i \leq m)$ and $B - u_i + v_j \notin \mathcal{B}_p$ $(1 \leq i < j \leq m)$ by (5.48). By applying the unique-matching lemma to the given pair (B, B') in the matroid $\mathbf{M}_p = (V, \mathcal{B}_p)$, we obtain $B' \in \mathcal{B}_p$, which means $\omega_p(B') = \omega_p(B)$, i.e.,

$$\omega(B') = \omega(B) + \sum_{i=1}^{m} \widehat{p}(u_i) - \sum_{i=1}^{m} \widehat{p}(v_i) = \omega(B) + \widehat{\omega}(B, B').$$

□

The following lemma will be used in §5.2.12.

Lemma 5.2.38. *Under the same assumption as in Lemma 5.2.35, let $\widehat{p}$, u_i, and v_j be as in Lemma 5.2.32. Then*

$$\omega(B', v_j, u_i) \leq \widehat{p}(v_j) - \widehat{p}(u_i) \qquad (1 \leq i, j \leq m).$$

Proof. Putting $B'_{ij} = B' - v_j + u_i$ and using Lemma 5.2.29, Lemma 5.2.35, and (5.48) we see

$$\begin{aligned} \omega(B', v_j, u_i) &= \omega(B'_{ij}) - \omega(B') \leq \widehat{\omega}(B, B'_{ij}) - \widehat{\omega}(B, B') \\ &\leq \left[\sum_{k \neq i} \widehat{p}(u_k) - \sum_{k \neq j} \widehat{p}(v_k)\right] - \left[\sum_{k=1}^{m} \widehat{p}(u_k) - \sum_{k=1}^{m} \widehat{p}(v_k)\right] \\ &= \widehat{p}(v_j) - \widehat{p}(u_i). \end{aligned}$$

■

5.2.9 Valuated Independent Assignment Problem

The independent matching problem (§2.3.5) is generalized in this section to a weighted version, called the valuated independent assignment problem, introduced by Murota [224, 225].

The problem we consider is the following:

[Valuated independent assignment problem (VIAP)]
Given a bipartite graph $G = (V^+, V^-; A)$, a pair of valuated matroids $\mathbf{M}^+ = (V^+, \mathcal{B}^+, \omega^+)$ and $\mathbf{M}^- = (V^-, \mathcal{B}^-, \omega^-)$, and a weight function $w : A \to \mathbf{R}$, find a matching M ($\subseteq A$) that maximizes

$$\Omega(M) \equiv w(M) + \omega^+(\partial^+ M) + \omega^-(\partial^- M) \tag{5.52}$$

subject to the constraint

$$\partial^+ M \in \mathcal{B}^+, \qquad \partial^- M \in \mathcal{B}^-, \tag{5.53}$$

where $w(M) = \sum\{w(a) \mid a \in M\}$, and $\partial^+ M$ (resp., $\partial^- M$) denotes the set of vertices in V^+ (resp., V^-) incident to M. A matching M satisfying the constraint (5.53) is called an *independent assignment*. Obviously, an independent assignment is an independent matching in the sense of §2.3.5.

The above problem reduces to the independent assignment problem of Iri–Tomizawa [133] if the valuations are trivial with $\omega^\pm = 0$ on $\mathcal{B}^\pm$:

[Independent assignment problem (IAP)]
Given a bipartite graph $G = (V^+, V^-; A)$, a pair of matroids $\mathbf{M}^+ = (V^+, \mathcal{B}^+)$ and $\mathbf{M}^- = (V^-, \mathcal{B}^-)$, and a weight function $w : A \to \mathbf{R}$, find a matching M ($\subseteq A$) that maximizes $w(M)$ subject to the constraint $\partial^+ M \in \mathcal{B}^+$, $\partial^- M \in \mathcal{B}^-$.

The special case of IAP where the matroids are free with $\mathcal{B}^+ = 2^{V^+}$ and $\mathcal{B}^- = 2^{V^-}$ coincides with the conventional assignment problem, and the further special case with $w \equiv 0$ is the problem of finding a perfect matching.

Another series of specializations of VIAP is obtained by choosing a very special underlying graph $G_\equiv = (V^+, V^-; A_\equiv)$ that represents a one-to-one correspondence between V^+ and V^-. In other words, given a pair of valuated matroids $\mathbf{M}_1 = (V, \mathcal{B}_1, \omega_1)$ and $\mathbf{M}_2 = (V, \mathcal{B}_2, \omega_2)$ defined on a common ground set V, and a weight function $w : V \to \mathbf{R}$, we consider a VIAP in which V^+ and V^- are disjoint copies of V and $A_\equiv = \{(v^+, v^-) \mid v \in V\}$, where $v^+ \in V^+$ and $v^- \in V^-$ denote the copies of $v \in V$, and $\mathbf{M}^+$ and $\mathbf{M}^-$ are isomorphic to $\mathbf{M}_1$ and $\mathbf{M}_2$, respectively. The VIAP in this case is equivalent to

[Valuated matroid intersection problem]
Given a pair of valuated matroids $\mathbf{M}_1 = (V, \mathcal{B}_1, \omega_1)$ and $\mathbf{M}_2 = (V, \mathcal{B}_2, \omega_2)$ and a weight function $w : V \to \mathbf{R}$, find a common base $B \in \mathcal{B}_1 \cap \mathcal{B}_2$ that maximizes $w(B) + \omega_1(B) + \omega_2(B)$,

where $w(B) = \sum_{v \in B} w(v)$. If the valuations are trivial with $\omega^{\pm} = 0$ on $\mathcal{B}^{\pm}$, this reduces to

> **[Optimal common base problem]**
> Given a pair of matroids $\mathbf{M}_1 = (V, \mathcal{B}_1)$ and $\mathbf{M}_2 = (V, \mathcal{B}_2)$ and a weight function $w : V \to \mathbf{R}$, find a common base $B \in \mathcal{B}_1 \cap \mathcal{B}_2$ that maximizes $w(B)$.

A further special case with $w \equiv 0$ is the problem of finding a common base of two matroids.

The following two problems also fall into the category of VIAP.

> **[Disjoint bases problem]**
> Given a pair of valuated matroids $\mathbf{M}_1 = (V, \mathcal{B}_1, \omega_1)$ and $\mathbf{M}_2 = (V, \mathcal{B}_2, \omega_2)$, find disjoint bases B_1 and B_2 (i.e., $B_1 \cap B_2 = \emptyset$, $B_1 \in \mathcal{B}_1$, and $B_2 \in \mathcal{B}_2$) that maximize $\omega_1(B_1) + \omega_2(B_2)$.

> **[Partition problem]**
> Given a pair of valuated matroids $\mathbf{M}_1 = (V, \mathcal{B}_1, \omega_1)$ and $\mathbf{M}_2 = (V, \mathcal{B}_2, \omega_2)$, find a partition $(B, V \setminus B)$ of V that maximizes $\omega_1(B) + \omega_2(V \setminus B)$.

The disjoint bases problem for more than two valuated matroids can also be formulated as a valuated independent assignment problem (on a bipartite graph similar to Fig. 2.11). The partition problem is an intersection problem in disguise, since it is the intersection problem for $\mathbf{M}_1$ and $(\mathbf{M}_2)^*$, the dual of $\mathbf{M}_2$.

In the ordinary independent matching problem (§2.3.5), the constraint imposed on a matching M is that $\partial^{\pm} M$ be independent in $\mathbf{M}^{\pm}$ rather than that $\partial^{\pm} M$ be a base in $\mathbf{M}^{\pm}$. This motivates us to consider the following extension of VIAP parametrized by an integer k:

> **[VIAP(k)]**
> Given a bipartite graph $G = (V^+, V^-; A)$, a pair of valuated matroids $\mathbf{M}^+ = (V^+, \mathcal{B}^+, \omega^+)$ and $\mathbf{M}^- = (V^-, \mathcal{B}^-, \omega^-)$, and a weight function $w : A \to \mathbf{R}$, find a matching M ($\subseteq A$) that maximizes
>
> $$\Omega(M, B^+, B^-) \equiv w(M) + \omega^+(B^+) + \omega^-(B^-)$$
>
> subject to the constraint that $|M| = k$ and
>
> $$\partial^+ M \subseteq B^+ \in \mathcal{B}^+, \qquad \partial^- M \subseteq B^- \in \mathcal{B}^-. \tag{5.54}$$

Obviously, VIAP(k) with $k = \max(r^+, r^-)$ agrees with the original VIAP, where $r^+ = \operatorname{rank} \mathbf{M}^+$ and $r^- = \operatorname{rank} \mathbf{M}^-$. Note that VIAP($k$) with $k = \max(r^+, r^-)$ is feasible only if $r^+ = r^-$ and the same is true for VIAP. A special case of VIAP(k) with trivial valuations $\omega^{\pm} = 0$ and trivial weighting vector $w = 0$ reads:

> **[Independent matching problem]**
> Given a bipartite graph $G = (V^+, V^-; A)$ and a pair of matroids $\mathbf{M}^+ = (V^+, \mathcal{I}^+)$ and $\mathbf{M}^- = (V^-, \mathcal{I}^-)$, find a matching M $(\subseteq A)$ such that $|M| = k$, $\partial^+ M \in \mathcal{I}^+$, and $\partial^- M \in \mathcal{I}^-$.

This problem may be regarded as a variant of the independent matching problem treated in §2.3.5. The VIAP(k) contains another well-studied important problem:

> **[Weighted matroid intersection problem]**
> Given a pair of matroids $\mathbf{M}_1 = (V, \mathcal{I}_1)$ and $\mathbf{M}_2 = (V, \mathcal{I}_2)$ and a weight function $w : V \to \mathbf{R}$, find a common independent set $I \in \mathcal{I}_1 \cap \mathcal{I}_2$ of size k that maximizes $w(I)$.

In the above we have demonstrated that VIAP(k) contains a host of important problems. Finally, we explain that we can formulate VIAP(k) as an instance of the original VIAP. This fact justifies our subsequent developments focusing on the original VIAP.

For an instance of VIAP(k), we consider a VIAP on a bipartite graph $G_k = (V_k^+, V_k^-; A_k)$ with valuated matroids $\mathbf{M}_k^+ = (V_k^+, \mathcal{B}_k^+, \omega_k^+)$ and $\mathbf{M}_k^- = (V_k^-, \mathcal{B}_k^-, \omega_k^-)$ defined as follows. Let r^+ and r^- denote the ranks of the given valuated matroids $\mathbf{M}^+$ and $\mathbf{M}^-$, respectively. The graph $G_k = (V_k^+, V_k^-; A_k)$ is defined by

$$\begin{aligned}
V_k^+ &= V^+ \cup U_k^+, \qquad U_k^+ \equiv \{u_i^+ \mid 1 \le i \le r^- - k\}, \\
V_k^- &= V^- \cup U_k^-, \qquad U_k^- \equiv \{u_i^- \mid 1 \le i \le r^+ - k\}, \\
A_k &= A \cup \{(u, u_i^-) \mid u \in V^+, u_i^- \in U_k^-\} \cup \{(u_i^+, u) \mid u \in V^-, u_i^+ \in U_k^+\}.
\end{aligned}$$

The valuated matroid $\mathbf{M}_k^+$ is the direct sum of $\mathbf{M}^+$ and the free matroid on U_k^+ with trivial valuation (which is zero), i.e., $\mathcal{B}_k^+ = \{B \cup U_k^+ \mid B \in \mathcal{B}^+\}$ and $\omega_k^+(B \cup U_k^+) = \omega^+(B)$ for $B \in \mathcal{B}^+$. Similarly for $\mathbf{M}_k^-$. Note that $\mathbf{M}_k^+$ and $\mathbf{M}_k^-$ have a common rank equal to $r^+ + r^- - k$. The weight $w_k : A_k \to \mathbf{R}$ is defined by

$$w_k(a) = \begin{cases} w(a) & (a \in A) \\ 0 & (a \in A_k \setminus A). \end{cases}$$

With an independent assignment M_k in G_k we can associate a feasible solution (M, B^+, B^-) for VIAP(k) by defining $M = M_k \cap A$, $B^+ = \partial^+ M_k \setminus U_k^+$, and $B^- = \partial^- M_k \setminus U_k^-$. Conversely, from (M, B^+, B^-) feasible for VIAP(k) we can construct an independent assignment M_k in G_k. Moreover, for the objective function $\Omega(M, B^+, B^-)$ of VIAP(k), we have $\Omega(M, B^+, B^-) = w_k(M_k) + \omega_k^+(\partial^+ M_k) + \omega_k^-(\partial^- M_k)$, in agreement with the objective function of the associated VIAP.

5.2.10 Optimality Criteria

In this section we establish two forms of optimality criteria for the valuated independent assignment problem (VIAP). The first criterion refers to a "potential" function and the second to "negative cycles." Both criteria are natural extensions of the well-established corresponding results for the independent assignment problem. Recall that the VIAP is given in terms of a bipartite graph $G = (V^+, V^-; A)$, a pair of valuated matroids $\mathbf{M}^+ = (V^+, \mathcal{B}^+, \omega^+)$ and $\mathbf{M}^- = (V^-, \mathcal{B}^-, \omega^-)$, and a weight function $w : A \to \mathbf{R}$.

Potential Criterion. The first optimality criterion is stated in the following theorem of Murota [224]. The formulation in (1) refers to the existence of a "potential" function, whereas its reformulation in (2) reveals its duality nature.

Theorem 5.2.39 (Potential criterion for VIAP).
(1) *An independent assignment M in G is optimal for the valuated independent assignment problem* (5.52)–(5.53) *if and only if there exists a "potential" function $p : V^+ \cup V^- \to \mathbf{R}$ such that*

(i) $w(a) - p(\partial^+ a) + p(\partial^- a) \begin{cases} \leq 0 \ (a \in A) \\ = 0 \ (a \in M), \end{cases}$

(ii) *$\partial^+ M$ is a maximum-weight base of $\mathbf{M}^+$ with respect to $\omega^+[p^+]$,*

(iii) *$\partial^- M$ is a maximum-weight base of $\mathbf{M}^-$ with respect to $\omega^-[-p^-]$,*

where $p^\pm$ is the restriction of p to $V^\pm$, and $\omega^+[p^+]$ (resp., $\omega^-[-p^-]$) is the similarity transformation defined in (5.16)*; namely,*

$$\omega^+[p^+](B^+) = \omega^+(B^+) + \sum\{p(u) \mid u \in B^+\} \qquad (B^+ \subseteq V^+),$$
$$\omega^-[-p^-](B^-) = \omega^-(B^-) - \sum\{p(u) \mid u \in B^-\} \qquad (B^- \subseteq V^-).$$

(2)
$$\max_M\{\Omega(M) \mid M\text{: independent assignment}\}$$
$$= \min_p\{\max(\omega^+[p^+]) + \max(\omega^-[-p^-]) \mid$$
$$w(a) - p(\partial^+ a) + p(\partial^- a) \leq 0 \ (a \in A)\}.$$

(3) *If ω^+, ω^-, and w are all integer-valued, the potential p in* (1) *and* (2) *can be taken to be integer-valued.*

(4) *Let p be a potential that satisfies* (i)–(iii) *in* (1) *for some (optimal) independent assignment $M = M_0$. An independent assignment M' is optimal if and only if it satisfies* (i)–(iii) *(with M replaced by M').*

Proof. The proof is given later. ∎

The optimality condition for the valuated matroid intersection problem deserves a separate statement in a form of weight splitting, though it is an immediate corollary of the above theorem. Recall that the intersection problem is to maximize $w(B) + \omega_1(B) + \omega_2(B)$ for a pair of valuated matroids $\mathbf{M}_1 = (V, \mathcal{B}_1, \omega_1)$ and $\mathbf{M}_2 = (V, \mathcal{B}_2, \omega_2)$ and a weight function $w : V \to \mathbf{R}$.

Theorem 5.2.40 (Weight splitting for valuated matroid intersection).

(1) *A common base B of $\mathbf{M}_1 = (V, \mathcal{B}_1, \omega_1)$ and $\mathbf{M}_2 = (V, \mathcal{B}_2, \omega_2)$ maximizes $w(B) + \omega_1(B) + \omega_2(B)$ if and only if there exist $w_1, w_2 : V \to \mathbf{R}$ such that*

(i) [*"weight splitting"*] $w(v) = w_1(v) + w_2(v) \quad (v \in V)$,

(ii) *B is a maximum-weight base of $\mathbf{M}_1$ with respect to $\omega_1[w_1]$,*

(iii) *B is a maximum-weight base of $\mathbf{M}_2$ with respect to $\omega_2[w_2]$,*

where $\omega_1[w_1]$ (resp., $\omega_2[w_2]$) is the similarity transformation defined in (5.16).

(2) $$\max_B \{w(B) + \omega_1(B) + \omega_2(B)\}$$
$$= \min_{w_1, w_2} \{\max(\omega_1[w_1]) + \max(\omega_2[w_2]) \mid w(v) = w_1(v) + w_2(v) \ (v \in V)\}.$$

(3) *If ω_1, ω_2, and w are all integer-valued, we may assume that $w_1, w_2 : V \to \mathbf{Z}$.*

Proof. Formulate the valuated matroid intersection problem as the VIAP on $G_{\equiv} = (V^+, V^-; A_{\equiv})$ as defined in §5.2.9, and apply Theorem 5.2.39 above. ∎

By putting $\omega^{\pm} = 0$ in the above theorems we can obtain the standard results for the independent assignment problem and the optimal common base problem, as well as for the related problems such as the weighted matroid intersection problem. For these results, see Bixby–Cunningham [13], Edmonds [68, 70], Faigle [74], Frank [76, 77], Fujishige [79, 82], Iri–Tomizawa [133], Lawler [170, 171], Welsh [333], and Zimmermann [352].

Remark 5.2.41. The optimality criterion in Theorem 5.2.40(2) can be reformulated as a *Fenchel-type duality* between a pair of matroid valuations and their conjugate functions, as reported in Murota [230]. It is also mentioned that Theorem 5.2.39 can be extended for the submodular flow problem with a certain nonlinear objective function, called *M-convex function* (see Murota [227, 231, 234]). □

Negative-cycle Criterion. To describe the second criterion for optimality in VIAP we need to introduce an auxiliary graph $\tilde{G}_M = (\tilde{V}, \tilde{A})$ associated with an independent assignment M. We put $B^+ = \partial^+ M$ and $B^- = \partial^- M$. The vertex set $\tilde{V}$ of $\tilde{G}_M$ is given by $\tilde{V} = V^+ \cup V^-$ and the arc set $\tilde{A}$ consists of four disjoint parts:

$$\tilde{A} = A^\circ \cup M^\circ \cup A^+ \cup A^-,$$

where

$$\begin{aligned}
A^\circ &= \{a \mid a \in A\} && \text{(copy of } A\text{)}, \\
M^\circ &= \{\overline{a} \mid a \in M\} && (\overline{a}\text{: reorientation of } a), \\
A^+ &= \{(u,v) \mid u \in B^+, v \in V^+ \setminus B^+, B^+ - u + v \in \mathcal{B}^+\}, \\
A^- &= \{(v,u) \mid u \in B^-, v \in V^- \setminus B^-, B^- - u + v \in \mathcal{B}^-\}.
\end{aligned}$$

In addition, arc length $\gamma_M(a)$ $(a \in \tilde{A})$ is defined by

$$\gamma_M(a) = \begin{cases} -w(a) & (a \in A^\circ) \\ w(\overline{a}) & (a = (u,v) \in M^\circ, \overline{a} = (v,u) \in M) \\ -\omega^+(B^+, u, v) & (a = (u,v) \in A^+) \\ -\omega^-(B^-, u, v) & (a = (v,u) \in A^-) \end{cases}$$

where $\omega^+(B^+, u, v)$ and $\omega^-(B^-, u, v)$ are defined according to (5.21). A directed cycle of negative length will be called a *negative cycle.*

The second criterion for optimality is stated in the following theorem of Murota [224].

Theorem 5.2.42 (Negative-cycle criterion for VIAP). *An independent assignment M in G is optimal for the valuated independent assignment problem* (5.52)–(5.53) *if and only if there exists in $\tilde{G}_M$ no negative cycle with respect to the arc length γ_M.*

Proof. The proof is given later. ∎

Remark 5.2.43. The arcs of A^+ or A^- represent the exchangeability in the respective matroids. In fact, the subgraphs (V^+, A^+) and (V^-, A^-) of $\tilde{G}_M$ can be identified respectively with the exchangeability graphs $G(B^+, V^+ \setminus B^+)$ for $\mathbf{M}^+$ and $G(B^-, V^- \setminus B^-)$ for $\mathbf{M}^-$ introduced in §5.2.8. Note, however, that the arc length in $\tilde{G}_M$ is the negative of the arc weight in $G(B^+, V^+ \setminus B^+)$ or $G(B^-, V^- \setminus B^-)$. □

Proof of the Optimality Criteria. The main body of the proof consists in proving the equivalence of the following three conditions for an independent assignment M:

(OPT) M is optimal,
(NNC) There is no negative cycle in $\tilde{G}_M$,
(POT) There exists a potential p with (i)–(iii) in Theorem 5.2.39(1).

We prove (OPT) ⇒ (NNC) ⇒ (POT) ⇒ (OPT). We abbreviate γ_M to γ whenever convenient.

(OPT) ⇒ (NNC): Suppose $\tilde{G}_M$ has a negative cycle. Let Q $(\subseteq \tilde{A})$ be the arc set of a negative cycle having the smallest number of arcs, and put

$$\overline{B}^+ = (B^+ \setminus \{\partial^+ a \mid a \in Q \cap A^+\}) \cup \{\partial^- a \mid a \in Q \cap A^+\}, \qquad (5.55)$$

$$\overline{B}^- = (B^- \setminus \{\partial^- a \mid a \in Q \cap A^-\}) \cup \{\partial^+ a \mid a \in Q \cap A^-\}, \qquad (5.56)$$

where $B^+ = \partial^+ M$ and $B^- = \partial^- M$ as before.

Lemma 5.2.44. $(B^+, \overline{B}^+)$ *and* $(B^-, \overline{B}^-)$ *satisfy the unique-max condition in* $\mathbf{M}^+$ *and* $\mathbf{M}^-$ *respectively.*

Proof. We prove the claim for $(B^+, \overline{B}^+)$ by adapting Fujishige's proof technique developed in Fujishige [79] for the independent assignment problem (see also Fujishige [82, Lemma 5.4]).

First note that the exchangeability graph $G(B^+, \overline{B}^+)$ for $\mathbf{M}^+$ has a perfect matching $Q \cap A^+$ under the correspondence in Remark 5.2.43. Take a maximum-weight perfect matching $M' = \{(u_i, v_i) \mid i = 1, \cdots, m\}$ in $G(B^+, \overline{B}^+)$, where $m = |B^+ \setminus \overline{B}^+|$, as well as the potential function $\widehat{p}$ in Lemma 5.2.32(1). Then M' is a subset of

$$A^* = \{(u, v) \mid u \in B^+ \setminus \overline{B}^+, v \in \overline{B}^+ \setminus B^+, \omega^+(B^+, u, v) - \widehat{p}(u) + \widehat{p}(v) = 0\}.$$

Put $Q' = (Q \setminus A^+) \cup M'$, regarding M' as a subset of A^+. Then Q' is a disjoint union of cycles in $\tilde{G}_M$ with its length

$$\gamma(Q') = \gamma(Q) + [\gamma(M') - \gamma(Q \cap A^+)]$$

being negative, since $-\gamma(M')$ is equal to the maximum weight of a perfect matching in $G(B^+, \overline{B}^+)$ and $Q \cap A^+$ is a perfect matching in $G(B^+, \overline{B}^+)$. The minimality of Q (with respect to the number of arcs) implies that Q' itself is a negative cycle having the smallest number of arcs.

Suppose, to the contrary, that $(B^+, \overline{B}^+)$ does not satisfy the unique-max condition. Since $(u_i, v_i) \in A^*$ for $i = 1, \cdots, m$, it follows from Lemma 5.2.32(2) that there are distinct indices i_k $(k = 1, \cdots, q; q \geq 2)$ such that $(u_{i_k}, v_{i_{k+1}}) \in A^*$ for $k = 1, \cdots, q$, where $i_{q+1} = i_1$. That is,

$$\omega^+(B^+, u_{i_k}, v_{i_{k+1}}) = \widehat{p}(u_{i_k}) - \widehat{p}(v_{i_{k+1}}) \qquad (k = 1, \cdots, q).$$

On the other hand we have

$$\omega^+(B^+, u_{i_k}, v_{i_k}) = \widehat{p}(u_{i_k}) - \widehat{p}(v_{i_k}) \qquad (k = 1, \cdots, q).$$

It then follows that

$$\sum_{k=1}^{q} \omega^+(B^+, u_{i_k}, v_{i_{k+1}}) = \sum_{k=1}^{q} \omega^+(B^+, u_{i_k}, v_{i_k}) \quad \left(= \sum_{k=1}^{q} [\widehat{p}(u_{i_k}) - \widehat{p}(v_{i_k})] \right)$$

i.e.,

$$\sum_{k=1}^{q} \gamma(u_{i_k}, v_{i_{k+1}}) = \sum_{k=1}^{q} \gamma(u_{i_k}, v_{i_k}). \tag{5.57}$$

For $k = 1, \cdots, q$, let $P'(v_{i_{k+1}}, u_{i_k})$ denote the path on Q' from $v_{i_{k+1}}$ to u_{i_k}, and let Q'_k be the directed cycle formed by arc $(u_{i_k}, v_{i_{k+1}})$ and path $P'(v_{i_{k+1}}, u_{i_k})$. Obviously,

$$\gamma(Q'_k) = \gamma(u_{i_k}, v_{i_{k+1}}) + \gamma(P'(v_{i_{k+1}}, u_{i_k})) \qquad (k = 1, \cdots, q). \tag{5.58}$$

A simple but crucial observation here is that

$$\left(\bigcup_{k=1}^{q} P'(v_{i_{k+1}}, u_{i_k})\right) \cup \{(u_{i_k}, v_{i_k}) \mid k = 1, \cdots, q\} = q' \cdot Q'$$

for some q' with $1 \leq q' < q$, where the union denotes the multiset union, and this expression means that each element of Q' appears q' times on the left hand side. Hence by adding (5.58) over $k = 1, \cdots, q$ we obtain

$$\begin{aligned}\sum_{k=1}^{q} \gamma(Q'_k) &= \sum_{k=1}^{q} \gamma(u_{i_k}, v_{i_{k+1}}) + \sum_{k=1}^{q} \gamma(P'(v_{i_{k+1}}, u_{i_k})) \\ &= \left[\sum_{k=1}^{q} \gamma(u_{i_k}, v_{i_{k+1}}) - \sum_{k=1}^{q} \gamma(u_{i_k}, v_{i_k})\right] + q' \cdot \gamma(Q') \\ &= q' \cdot \gamma(Q') \ < 0,\end{aligned}$$

where the last equality is due to (5.57). This implies that $\gamma(Q'_k) < 0$ for some k, while Q'_k has a smaller number of arcs than Q'. This contradicts the minimality of Q'. Therefore $(B^+, \overline{B}^+)$ satisfies the unique-max condition. Similarly for $(B^-, \overline{B}^-)$. ■

Lemma 5.2.45. *For a negative cycle Q in $\tilde{G}_M$ having the smallest number of arcs,*

$$\overline{M} = (M \setminus \{a \in M \mid \overline{a} \in Q \cap M^\circ\}) \cup (Q \cap A^\circ)$$

is an independent assignment with $\Omega(\overline{M}) \geq \Omega(M) - \gamma_M(Q)$ $(> \Omega(M))$.

Proof. Note that $\overline{B}^+ = \partial^+ \overline{M}$ and $\overline{B}^- = \partial^- \overline{M}$ for $\overline{B}^+$ and $\overline{B}^-$ defined in (5.55) and (5.56), and recall the notation $B^+ = \partial^+ M$ and $B^- = \partial^- M$. By Lemma 5.2.44 and the unique-max lemma (Lemma 5.2.35) we have

$$\begin{aligned}\omega^+(\overline{B}^+) &= \omega^+(B^+) + \widehat{\omega}^+(B^+, \overline{B}^+) \geq \omega^+(B^+) - \gamma(Q \cap A^+), \\ \omega^-(\overline{B}^-) &= \omega^-(B^-) + \widehat{\omega}^-(B^-, \overline{B}^-) \geq \omega^-(B^-) - \gamma(Q \cap A^-).\end{aligned}$$

Also we have

$$w(\overline{M}) = w(M) - \gamma(Q \cap (A^\circ \cup M^\circ)).$$

Addition of these inequalities yields $\Omega(\overline{M}) \geq \Omega(M) - \gamma(Q)$. ■

The above lemma shows "(OPT) $\Rightarrow$ (NNC)".

(NNC) $\Rightarrow$ (POT): By Theorem 2.2.35(1), (NNC) implies the existence of a function $p : V^+ \cup V^- \to \mathbf{R}$ such that

$$\gamma(a) + p(\partial^+ a) - p(\partial^- a) \geq 0 \qquad (a \in \tilde{A}). \tag{5.59}$$

This condition for $a \in A^\circ \cup M^\circ$ is equivalent to the condition (i) in Theorem 5.2.39(1). For $a = (u, v) \in A^+$ it means

$$\omega^+(B^+, u, v) - p(u) + p(v) \le 0.$$

Namely, $\omega^+[p^+](B^+, u, v) \le 0$ for all (u, v) with $B^+ + u - v \in \mathcal{B}^+$. This in turn implies the condition (ii) by Theorem 5.2.7. Similarly, the above condition for $a \in A^-$ implies the condition (iii). Thus "(NNC) $\Rightarrow$ (POT)" has been shown.

(POT) $\Rightarrow$ (OPT): For any independent assignment M and any function $p : V^+ \cup V^- \to \mathbf{R}$ we see

$$\begin{aligned} \Omega(M) &= \omega^+(\partial^+ M) + \omega^-(\partial^- M) + w(M) \\ &= \left[\omega^+(\partial^+ M) + \sum_{a\in M} p(\partial^+ a)\right] + \left[\omega^-(\partial^- M) - \sum_{a\in M} p(\partial^- a)\right] \\ &\quad + \sum_{a\in M} [w(a) - p(\partial^+ a) + p(\partial^- a)] \\ &= \omega^+[p^+](\partial^+ M) + \omega^-[-p^-](\partial^- M) + \sum_{a\in M} w_p(a), \end{aligned} \tag{5.60}$$

where $w_p(a) = w(a) - p(\partial^+ a) + p(\partial^- a)$.

Suppose M and p satisfy (i)–(iii) of Theorem 5.2.39(1), and take an arbitrary independent assignment M'. Then we have

$$\begin{aligned} \Omega(M') &= \omega^+[p^+](\partial^+ M') + \omega^-[-p^-](\partial^- M') + \sum_{a\in M'} w_p(a) \\ &\le \omega^+[p^+](\partial^+ M) + \omega^-[-p^-](\partial^- M) \\ &= \Omega(M). \end{aligned} \tag{5.61}$$

This shows that M is optimal, establishing "(POT) $\Rightarrow$ (OPT)". Thus we have shown the equivalence of the three conditions (OPT), (NNC), and (POT).

For the statement (2) of Theorem 5.2.39, the expression (5.60), valid for any M and p, implies that LHS $\le$ RHS, whereas Theorem 5.2.39(1) shows that the equality is attained. The integrality asserted in the statement (3) of Theorem 5.2.39 can be imposed in (5.59). Finally for the statement (4) of Theorem 5.2.39 we note in the inequality (5.61) that $\Omega(M') = \Omega(M)$ if and only if $\omega^+[p^+](\partial^+ M') = \omega^+[p^+](\partial^+ M)$, $\omega^-[-p^-](\partial^- M') = \omega^-[-p^-](\partial^- M)$, and $w_p(a) = 0$ for $a \in M'$.

We have completed the proofs of Theorem 5.2.39 and Theorem 5.2.42.

Extension to VIAP(k). The optimality criteria for VIAP(k) introduced in §5.2.9 are stated explicitly for later references.

Theorem 5.2.46.

(1) *A feasible solution (M, B^+, B^-) for VIAP(k) is optimal if and only if there exists a "potential" function $p : V^+ \cup V^- \to \mathbf{R}$ such that*

(i) $w(a) - p(\partial^+ a) + p(\partial^- a) \begin{cases} \le 0 & (a \in A) \\ = 0 & (a \in M) \end{cases}$

(ii) *B^+ is a maximum-weight base of $\mathbf{M}^+$ with respect to $\omega^+[p^+]$,*
(iii) *B^- is a maximum-weight base of $\mathbf{M}^-$ with respect to $\omega^-[-p^-]$,*
(iv) $p(u) \geq p(v) \qquad (u \in V^+,\ v \in B^+ \setminus \partial^+ M)$,
(v) $p(u) \leq p(v) \qquad (u \in V^-,\ v \in B^- \setminus \partial^- M)$.

(2) *If ω^+, ω^-, and w are all integer-valued, the potential p in* (1) *can be taken to be integer-valued.*

(3) *Let p be a potential that satisfies* (i)–(v) *in* (1) *for some (optimal) (M_0, B_0^+, B_0^-). Then (M, B^+, B^-) is optimal if and only if it satisfies* (i)–(v).

Proof. Formulate VIAP(k) as the VIAP on $G_k = (V_k^+, V_k^-; A_k)$ as defined in §5.2.9, and apply Theorem 5.2.39. ■

For a negative-cycle criterion of optimality we need to introduce an auxiliary graph $\tilde{G}_{(M,B^+,B^-)} = (\tilde{V}, \tilde{A})$ associated with (M, B^+, B^-), which is a slight modification of the one used for VIAP. The vertex set $\tilde{V}$ of $\tilde{G}_{(M,B^+,B^-)}$ is given by

$$\tilde{V} = V^+ \cup V^- \cup \{s^+, s^-\},$$

where s^+ and s^- are new vertices referred to as the source vertex and the sink vertex respectively. The arc set $\tilde{A}$ consists of eight disjoint parts:

$$\tilde{A} = (A^\circ \cup M^\circ) \cup (A^+ \cup F^+ \cup S^+) \cup (A^- \cup F^- \cup S^-),$$

where

$$\begin{aligned}
A^\circ &= \{a \mid a \in A\} \qquad &&\text{(copy of } A\text{)},\\
M^\circ &= \{\overline{a} \mid a \in M\} \qquad &&(\overline{a}\text{: reorientation of } a),\\
A^+ &= \{(u,v) \mid u \in B^+, v \in V^+ \setminus B^+, B^+ - u + v \in \mathcal{B}^+\}, &&\\
F^+ &= \{(u, s^+) \mid u \in V^+\}, &&\\
S^+ &= \{(s^+, v) \mid v \in B^+ \setminus \partial^+ M\}, &&\\
A^- &= \{(v,u) \mid u \in B^-, v \in V^- \setminus B^-, B^- - u + v \in \mathcal{B}^-\}, &&\\
F^- &= \{(s^-, u) \mid u \in V^-\}, &&\\
S^- &= \{(v, s^-) \mid v \in B^- \setminus \partial^- M\}. &&
\end{aligned} \tag{5.62}$$

The arc length $\gamma(a) = \gamma_{(M,B^+,B^-)}(a)$ $(a \in \tilde{A})$ is defined by

$$\gamma(a) = \begin{cases} -w(a) & (a \in A^\circ) \\ w(\overline{a}) & (a = (u,v) \in M^\circ, \overline{a} = (v,u) \in M) \\ -\omega^+(B^+, u, v) & (a = (u,v) \in A^+) \\ -\omega^-(B^-, u, v) & (a = (v,u) \in A^-) \\ 0 & (a \in F^+ \cup S^+ \cup F^- \cup S^-). \end{cases} \tag{5.63}$$

Theorem 5.2.47. *A feasible solution (M, B^+, B^-) for VIAP(k) is optimal if and only if there exists in $\tilde{G}_{(M,B^+,B^-)}$ no negative cycle with respect to the arc length $\gamma_{(M,B^+,B^-)}$.*

Proof. Formulate VIAP(k) as the VIAP on $G_k = (V_k^+, V_k^-; A_k)$ as defined in §5.2.9, and apply Theorem 5.2.42. ∎

Remark 5.2.48. The definitions of F^+ and F^- could be replaced by

$$F^+ = \{(u, s^+) \mid u \in \partial^+ M \cup (V^+ \setminus B^+)\},$$
$$F^- = \{(s^-, u) \mid u \in \partial^- M \cup (V^- \setminus B^-)\}$$

without affecting the above theorem. The present definition is more convenient for the algorithm to be developed later. □

Remark 5.2.49. When $k = r^+ = r^-$, the auxiliary graph $\tilde{G}_{(M,B^+,B^-)}$ contains the auxiliary graph $\tilde{G}_M$ as a subgraph. □

5.2.11 Application to Triple Matrix Product

Before going on to algorithms for the valuated independent assignment problem, we digress here into a possible application of the duality result in Theorem 5.2.39 to linear algebra. The connection to a *triple matrix product* explained here conforms with the historical development of the independent assignment problem explained in Remark 2.3.37.

The following fact is noted by Murota [233].

Theorem 5.2.50. *Assume that a matrix product* $P(s) = Q_1(s)T(s)Q_2(s)$ *is nonsingular, where* $Q_1(s)$ *(resp.,* $Q_2(s)$*) is a* $k \times m$ *(resp.,* $n \times k$*) rational matrix over a field* $\boldsymbol{K}$*, and* $T(s)$ *is an* $m \times n$ *rational matrix over an extension field* $\boldsymbol{F}$ $(\supseteq \boldsymbol{K})$ *such that the set of the coefficients is algebraically independent over* $\boldsymbol{K}$*. Then there exist* $k \times k$ *nonsingular rational matrices* $S_1(s)$*,* $S_2(s)$ *and diagonal matrices* $\mathrm{diag}\,(s;p) = \mathrm{diag}\,(s^{p_1}, \cdots, s^{p_m})$*,* $\mathrm{diag}\,(s;q) = \mathrm{diag}\,(s^{q_1}, \cdots, s^{q_n})$ *with* $p \in \mathbf{Z}^m$ *and* $q \in \mathbf{Z}^n$ *such that*

$$\deg_s \det P = \deg_s \det S_1 + \deg_s \det S_2$$

and the matrices

$$\bar{Q}_1(s) = S_1(s)^{-1}\, Q_1(s)\, \mathrm{diag}\,(s;p),$$
$$\bar{T}(s) = \mathrm{diag}\,(s;-p)\, T(s)\, \mathrm{diag}\,(s;-q),$$
$$\bar{Q}_2(s) = \mathrm{diag}\,(s;q)\, Q_2(s)\, S_2(s)^{-1}$$

are all proper. Note that $S_1(s)^{-1}P(s)S_2(s)^{-1} = \bar{Q}_1(s)\bar{T}(s)\bar{Q}_2(s)$*.*

Proof. Firstly, by the Cauchy–Binet formula (Proposition 2.1.6), we have

$$\det P = \sum_{|I|=|J|=k} \pm \det Q_1[R, I] \cdot \det T[I, J] \cdot \det Q_2[J, C],$$

where $R = \mathrm{Row}(Q_1)$ and $C = \mathrm{Col}(Q_2)$. There is no numerical cancellation in the summation above by virtue of the assumed algebraic independence of the coefficients in $T(s)$, and hence

$$\deg_s \det P = \max_{|I|=|J|=k} \{\deg_s \det Q_1[R,I] + \deg_s \det T[I,J] + \deg_s \det Q_2[J,C]\}.$$

Next, consider a valuated independent assignment problem defined as follows. The vertex sets V^+ and V^- are the row set and the column set of $T(s)$, respectively, and the arc set $A = \{(i,j) \mid T_{ij}(s) \neq 0\}$. The valuated matroids $\mathbf{M}^+ = (V^+, \omega^+)$ and $\mathbf{M}^- = (V^-, \omega^-)$ attached to V^+ and V^- respectively are those defined by $Q_1(s)$ and the transpose of $Q_2(s)$ as in (5.15), i.e.,

$$\omega^+(I) = \deg_s \det Q_1[R,I], \qquad \omega^-(J) = \deg_s \det Q_2[J,C]$$

and the weight w_{ij} of an edge $(i,j) \in A$ is defined by $w_{ij} = \deg_s T_{ij}(s)$. Note that the maximum value of $\sum_{(i,j)\in M} w_{ij}$ over all matchings M with $I = \partial^+ M$ and $J = \partial^- M$ is equal to $\deg_s \det T[I,J]$.

Then we see from the above expression of $\deg_s \det P$ that $\deg_s \det P$ is equal to the maximum value of the objective function $\Omega(M)$ of (5.52) over all independent assignment M. Let M be an optimal independent assignment and put $I = \partial^+ M$ and $J = \partial^- M$. Let $\hat{p} : V^+ \cup V^- \to \mathbf{Z}$ be the potential in Theorem 5.2.39, and define $p \in \mathbf{Z}^m$ and $q \in \mathbf{Z}^n$ by $p_i = \hat{p}_i$ for $i \in V^+$ and $q_j = -\hat{p}_j$ for $j \in V^-$. Define $S_1 = Q_1[R,I]\,\mathrm{diag}\,(s; p_I)$ and $S_2 = \mathrm{diag}\,(s; q_J)\,Q_2[J,C]$, where $p_I = (p_i \mid i \in I) \in \mathbf{Z}^I$ is the restriction of p to I and similarly for $q_J \in \mathbf{Z}^J$.

The conditions (i), (ii), and (iii) in Theorem 5.2.39(1), coupled with (5.24), imply the properness of $\bar{T}(s)$, $\bar{Q}_1(s)$, and $\bar{Q}_2(s)$, respectively. ■

Remark 5.2.51. The close relationship between the triple matrix product and the independent assignment problem through the Cauchy–Binet formula was first observed by Tomizawa–Iri [317, 318]. To be more precise, the rank of $P = Q_1 T Q_2$ was expressed in Tomizawa–Iri [317] as the maximum size of an independent matching (cf. Remark 2.3.37), whereas the degree of the determinant of $P(s) = Q_1 T(s) Q_2$ with constant matrices Q_i $(i = 1,2)$ was represented in Tomizawa–Iri [318] as the optimal value of an independent assignment. Theorem 5.2.50 gives an extension of this idea to the more general case with polynomial/rational matrices $Q_i(s)$ $(i = 1,2)$ by means of valuated matroids with an additional explicit statement concerning the transformation into proper matrices. □

5.2.12 Cycle-canceling Algorithms

This section describes a primal-type cycle-canceling algorithm for the valuated independent assignment problem, due to Murota [225]. The algorithm runs in strongly polynomial time with oracles for the valuations $\omega^\pm$. Another algorithm of primal-dual augmenting type will be given in §5.2.13.

Algorithms. Our cycle-canceling algorithm is based on the negative-cycle criterion (Theorem 5.2.42). It can be polished up to a strongly polynomial algorithm using the minimum-ratio-cycle strategy.

In Theorem 5.2.42 we have shown a negative-cycle criterion for the optimality with reference to an auxiliary graph $\tilde{G}_M = (\tilde{V}, \tilde{A})$ with $\tilde{V} = V^+ \cup V^-$ and $\tilde{A} = A^\circ \cup M^\circ \cup A^+ \cup A^-$, where

$$\begin{aligned} A^\circ &= \{a \mid a \in A\} \qquad \text{(copy of } A\text{)}, \\ M^\circ &= \{\overline{a} \mid a \in M\} \qquad (\overline{a}\text{: reorientation of } a), \\ A^+ &= \{(u,v) \mid u \in B^+, v \in V^+ \setminus B^+, B^+ - u + v \in \mathcal{B}^+\}, \\ A^- &= \{(v,u) \mid u \in B^-, v \in V^- \setminus B^-, B^- - u + v \in \mathcal{B}^-\}. \end{aligned}$$

Here $B^+ = \partial^+ M$ and $B^- = \partial^- M$. The arc length $\gamma_M(a)$ $(a \in \tilde{A})$ is defined by

$$\gamma_M(a) = \begin{cases} -w(a) & (a \in A^\circ) \\ w(\overline{a}) & (a = (u,v) \in M^\circ, \overline{a} = (v,u) \in M) \\ -\omega^+(B^+, u, v) & (a = (u,v) \in A^+) \\ -\omega^-(B^-, u, v) & (a = (v,u) \in A^-) \end{cases}$$

where $\omega^+(B^+, u, v)$ and $\omega^-(B^-, u, v)$ are as in (5.21).

Theorem 5.2.42 as well as its proof suggests the following algorithm for solving the valuated independent assignment problem.

Cycle-canceling algorithm

Starting from an arbitrary independent assignment M, repeat (i)–(ii) below while there exists a negative cycle in $\tilde{G}_M$:

(i) Find a negative cycle Q having the smallest number of arcs in the auxiliary graph $\tilde{G}_M$ (with respect to the arc length γ_M).

(ii) Modify the current independent matching along the cycle Q by

$$\overline{M} = (M \setminus \{a \in M \mid \overline{a} \in Q \cap M^\circ\}) \cup (Q \cap A^\circ).$$

The validity of this procedure follows from Theorem 5.2.42 and Lemma 5.2.45.

Remark 5.2.52. This is a straightforward extension of the primal algorithm of Fujishige [79] for the ordinary independent assignment problem, which extends the classical idea of Klein [160] and which is further extended later by Fujishige [80] and by Zimmermann [352] for the submodular flow problem (see also Fujishige [82] and the references therein). □

The above algorithm assumes an initial independent assignment M, which can be found by the algorithm for the independent matching problem treated in §2.3.5. For each M the graph $\tilde{G}_M$ can be constructed with $r^+(|V^+| - r^+)$ evaluations of ω^+ and $r^-(|V^-| - r^-)$ evaluations of ω^-, where r^+ and r^- are the ranks of $\mathbf{M}^+$ and $\mathbf{M}^-$ respectively (we have $r^+ = r^-$ for a feasible problem). When the valuated matroids are associated with polynomial/rational matrices as in Examples 5.2.3 and 5.2.10, $\omega^{\pm}(\cdot,\cdot,\cdot)$ can be determined by pivoting operations on the matrices if arithmetic operations on rational functions can be performed.

A negative cycle having the smallest number of arcs in (i) can be found easily by a variant of the standard shortest-path algorithm. It should however be worth noting that the minimality of the number of arcs is not really necessary, and in fact this observation adds more flexibility to the algorithm, as we will see soon. Recalling the notation

$$\overline{B}^{+} = (B^{+} \setminus \{\partial^{+}a \mid a \in Q \cap A^{+}\}) \cup \{\partial^{-}a \mid a \in Q \cap A^{+}\}, \qquad (5.64)$$

$$\overline{B}^{-} = (B^{-} \setminus \{\partial^{-}a \mid a \in Q \cap A^{-}\}) \cup \{\partial^{+}a \mid a \in Q \cap A^{-}\}, \qquad (5.65)$$

we call a cycle Q in $\tilde{G}_M$ *admissible* if both $(B^{+},\overline{B}^{+})$ and $(B^{-},\overline{B}^{-})$ satisfy the unique-max condition in $\mathbf{M}^{+}$ and $\mathbf{M}^{-}$ respectively. The admissibility of Q guarantees (by the unique-max lemma) that the modified matching $\overline{M}$ remains an independent assignment.

In the proof of Lemma 5.2.44 it has been shown that if a negative cycle Q is not admissible, a family of cycles, denoted Q'_k $(k = 1,\cdots,q)$ there, is naturally defined and that at least one of its members is a negative cycle. We call each Q'_k an *induced cycle*. The above observations lead to the following refinements of Lemma 5.2.44 and Lemma 5.2.45.

Lemma 5.2.53. *Let Q be a negative cycle in $\tilde{G}_M$. Then either Q is admissible or else it induces a negative cycle having a smaller number of arcs than Q. In particular, a negative cycle having the smallest number of arcs is admissible.* □

Lemma 5.2.54. *For an admissible cycle Q in $\tilde{G}_M$, $\overline{M}$ is an independent assignment with $\Omega(\overline{M}) \geq \Omega(M) - \gamma_M(Q)$.* □

The algorithm finds the optimal independent assignment in a finite number of steps since there exist a finite number of independent assignments in the given graph and the objective function value $\Omega(M)$ increases monotonically; we have seen

$$\Omega(\overline{M}) \geq \Omega(M) - \gamma_M(Q) \qquad (> \Omega(M)). \qquad (5.66)$$

However, the number of iterations of the loop (i)–(ii) is not bounded by a polynomial in the problem size, as is also the case with thc original form of the primal algorithm for the ordinary independent assignment problem.

Zimmermann [353] has shown (for the submodular flow problem) that a judicious choice of a negative cycle renders the number of iterations bounded by r^{+} $(= r^{-})$. The idea is to introduce an auxiliary weight function α on $\tilde{A}$ and to select a cycle Q of minimum ratio $\gamma_M(Q)/\alpha(Q)$ (satisfying some extra condition). In what follows we shall show that this idea carries over to our problem, making the number of iterations of the loop (i)–(ii) of our algorithm bounded by r^+ $(= r^-)$.

We maintain a subset $M^\bullet$ of $\tilde{A}$, called the *active arc set*, and define $\alpha : \tilde{A} \to \{0, 1\}$ by

$$\alpha(a) = \begin{cases} 1 & (a \in M^\bullet) \\ 0 & (a \in \tilde{A} \setminus M^\bullet). \end{cases}$$

An arc is said to be *active* if it belongs to $M^\bullet$. A cycle Q $(\subseteq \tilde{A})$ is called a *minimum-ratio cycle* with respect to (γ_M, α) if $\gamma_M(Q)/\alpha(Q)$ takes the minimum value among all cycles with $\alpha(Q) > 0$.

Cycle-canceling algorithm with minimum-ratio cycle
Starting from an arbitrary independent assignment M and active arc set defined by $M^\bullet = M^\circ$ $(\equiv \{\overline{a} \mid a \in M\})$, repeat (i)–(iii) below while there exists a negative cycle in $\tilde{G}_M$:

(i) Find an admissible minimum-ratio cycle Q in the auxiliary graph $\tilde{G}_M$ (with respect to (γ_M, α)).

(ii) Modify the current active arc set by

$$\overline{M^\bullet} = M^\bullet \setminus (Q \cap M^\circ)$$

and the function α accordingly.

(iii) Modify the current independent matching along the cycle Q by

$$\overline{M} = (M \setminus \{a \in M \mid \overline{a} \in Q \cap M^\circ\}) \cup (Q \cap A^\circ).$$

The following properties are maintained throughout the computation:

- Any negative cycle in $\tilde{G}_M$ contains an active arc (cf. Lemma 5.2.60).
- M is an independent assignment (i.e., $\partial^+ M \in \mathcal{B}^+$, $\partial^- M \in \mathcal{B}^-$).

Because of the first property, the minimum-ratio cycle in (i) is well-defined, as long as $\tilde{G}_M$ contains a negative cycle. In (ii), on the other hand, the active arc set $M^\bullet$ decreases monotonically, at least by one element in each iteration. This implies the termination of the algorithm in at most r^+ $(= r^-)$ iterations, whereas the obtained matching M is an optimal independent assignment by the second property and Theorem 5.2.42.

An admissible minimum-ratio cycle can be found in a polynomial time in the problem size as follows. By an algorithm of Megiddo [193] a minimum-ratio cycle Q can be generated in $\mathrm{O}(|\tilde{V}|^2 |\tilde{A}| \log |\tilde{V}|)$ time. We can test for the admissibility of Q on the basis of Lemma 5.2.32 by means of an algorithm for the weighted bipartite matching problem. This takes $\mathrm{O}(|\tilde{V}|^3)$ or less time. In case Q is not admissible, it induces at least one minimum-ratio cycle having a smaller number of arcs than Q, as will be shown later in Lemma 5.2.57. We pick up one of the induced minimum-ratio cycles, and repeat the above procedure. After repeating not more than $|\tilde{V}|$ times we are guaranteed to obtain an admissible minimum-ratio cycle.

Summarizing the above arguments we have the following theorem due to Murota [225].

Theorem 5.2.55. *The cycle-canceling algorithm with minimum-ratio cycle selection is a strongly polynomial time algorithm (modulo a polynomial number of evaluations of $\omega^{\pm}$).* □

Other algorithms for VIAP based on the negative-cycle criterion can be found in Murota [225] and Shigeno [296].

Validity of the Minimum-ratio Cycle Algorithm. We shall show the validity of the cycle-canceling algorithm using the minimum-ratio cycle selection. Basically we follow the arguments in Goldberg–Tarjan [96], Zimmermann [353] while establishing two lemmas (Lemma 5.2.57 and Lemma 5.2.59) specific to our problem. We abbreviate γ_M to γ for notational simplicity.

For $\epsilon \geq 0$ an independent assignment M is said to be *ϵ-optimal* (with respect to α) if there exists a function $p : \tilde{V} \to \mathbf{R}$ such that

$$\gamma_p(a) \equiv \gamma(a) + p(\partial^+ a) - p(\partial^- a) \geq -\epsilon\alpha(a) \qquad (a \in \tilde{A}). \tag{5.67}$$

Noting (5.67) is equivalent to saying that the modified arc length $\hat{\gamma}(a) = \gamma(a) + \epsilon\alpha(a)$ admits a function p such that

$$\hat{\gamma}(a) + p(\partial^+ a) - p(\partial^- a) \geq 0 \qquad (a \in \tilde{A}),$$

we see that the existence of p with (5.67) is also equivalent to

$$\gamma(Q) \geq -\epsilon\alpha(Q) \qquad (Q: \text{ negative cycle}).$$

This implies obviously that $\alpha(Q) > 0$ for any negative cycle Q; that is:

$$\text{any negative cycle in } \tilde{G}_M \text{ contains an active arc.} \tag{5.68}$$

Conversely suppose (5.68) is true and

$$\text{there exists a negative cycle.} \tag{5.69}$$

Then the "minimum cycle ratio"

$$\mu = \min\left\{\frac{\gamma(Q)}{\alpha(Q)} \mid Q: \text{ cycle with } \alpha(Q) > 0\right\} \tag{5.70}$$

is a well-defined negative number, and M is ϵ-optimal for $\epsilon = -\mu > 0$. Hence we have the following statement.

Lemma 5.2.56. *Condition* (5.68) *is satisfied if and only if M is ϵ-optimal for some $\epsilon \geq 0$.*

Proof. In addition to the above argument note that the case $\epsilon = 0$ corresponds to an optimal M, for which (5.68) is vacuously true due to Theorem 5.2.42. ■

Under the condition (5.68) we define $\epsilon(M)$ to be the minimum value of $\epsilon \geq 0$ for which M is ϵ-optimal. The above argument shows, under (5.69), that

$$\epsilon(M) = -\mu. \tag{5.71}$$

The following lemma substantiates the step (i) of the algorithm.

Lemma 5.2.57. *Assume* (5.68) *and* (5.69), *and let Q be a minimum-ratio cycle. Either Q is admissible or else it induces a minimum-ratio cycle having a smaller number of arcs than Q. In particular, a minimum-ratio cycle having the smallest number of arcs is admissible.*

Proof. We modify the proof of Lemma 5.2.44. Let $\overline{B}^{+}$ and $\overline{B}^{-}$ be defined by (5.64) and (5.65). Suppose that Q is not admissible, and assume without loss of generality that $(B^{+}, \overline{B}^{+})$ does not satisfy the unique-max condition. Take a maximum-weight perfect matching M' in $G(B^{+}, \overline{B}^{+})$ for $\mathbf{M}^{+}$. Put $Q' = (Q \setminus A^{+}) \cup M'$, which is a collection of disjoint cycles, say $Q' = \bigcup_{j=1}^{l} Q'_j$. Then $\alpha(Q') = \alpha(Q)$ (since $\alpha(M') = \alpha(Q \cap A^{+}) = 0$) and

$$\gamma(Q') = \gamma(Q) + [\gamma(M') - \gamma(Q \cap A^{+})] \tag{5.72}$$

holds. By the choice of M' we have $\gamma(Q') \leq \gamma(Q)$, which implies

$$\gamma(Q')/\alpha(Q') \leq \gamma(Q)/\alpha(Q) = \mu. \tag{5.73}$$

We claim that the equality holds in (5.73). In fact, (5.73) shows

$$\gamma(Q') = \sum_{j=1}^{l} \gamma(Q'_j) \ \leq \ \mu\, \alpha(Q') = \mu \sum_{j=1}^{l} \alpha(Q'_j),$$

whereas $\gamma(Q'_j) \geq \mu\alpha(Q'_j)$ for all j by (5.68) and (5.70). With the equality in (5.73) we obtain $\gamma(Q') = \gamma(Q)$ since $\alpha(Q') = \alpha(Q)$.

It then follows from (5.72) that

$$\gamma(Q \cap A^{+}) = \gamma(M') = -\widehat{\omega}^{+}(B^{+}, \overline{B}^{+}). \tag{5.74}$$

Hence, putting

$$M'' = Q \cap A^{+} = \{(u_i, v_i) \mid i = 1, \cdots, m\}$$

we have $M'' \subseteq A^{*}$, where

$$A^{*} = \{(u, v) \mid u \in B^{+} \setminus \overline{B}^{+}, v \in \overline{B}^{+} \setminus B^{+}, \omega^{+}(B^{+}, u, v) - \widehat{p}(u) + \widehat{p}(v) = 0\},$$

and $\widehat{p}$ is the potential function in Lemma 5.2.32(1).

Since $(B^+, \overline{B}^+)$ does not satisfy the unique-max condition, there exist distinct indices i_k $(k = 1, \cdots, q; q \geq 2)$ such that $(u_{i_k}, v_{i_{k+1}}) \in A^*$ for $k = 1, \cdots, q$, where $i_{q+1} = i_1$. Then

$$\omega^+(B^+, u_{i_k}, v_{i_{k+1}}) = \widehat{p}(u_{i_k}) - \widehat{p}(v_{i_{k+1}}) \qquad (k = 1, \cdots, q),$$
$$\omega^+(B^+, u_{i_k}, v_{i_k}) = \widehat{p}(u_{i_k}) - \widehat{p}(v_{i_k}) \qquad (k = 1, \cdots, q),$$
$$\sum_{k=1}^{q} \gamma(u_{i_k}, v_{i_{k+1}}) = \sum_{k=1}^{q} \gamma(u_{i_k}, v_{i_k})$$

hold true, where the second equation is due to $M'' \subseteq A^*$.

For $k = 1, \cdots, q$, let $P(v_{i_{k+1}}, u_{i_k})$ denote the path on Q from $v_{i_{k+1}}$ to u_{i_k}, and let Q_k be the directed cycle formed by arc $(u_{i_k}, v_{i_{k+1}})$ and path $P(v_{i_{k+1}}, u_{i_k})$. By a similar argument as in the proof of Lemma 5.2.44 we obtain

$$\sum_{k=1}^{q} (\gamma(Q_k) - \mu\alpha(Q_k)) = q'(\gamma(Q) - \mu\alpha(Q)) = 0$$

for some q' with $1 \leq q' < q$, which shows $\gamma(Q_k) - \mu\alpha(Q_k) = 0$ for each k. Therefore Q_k is a minimum-ratio cycle for k with $\alpha(Q_k) > 0$, while such k exists since $\sum_{k=1}^{q} \alpha(Q_k) = q'\alpha(Q) > 0$. ∎

Lemma 5.2.58. *Assume* (5.68) *and* (5.69), *and let Q be an admissible minimum-ratio cycle. Then $\overline{M}$ is an independent assignment with $\Omega(\overline{M}) = \Omega(M) - \gamma_M(Q)$.*

Proof. The same as the proof of Lemma 5.2.45, except that (5.74) is used. ∎

Lemma 5.2.59. *Assume* (5.68) *and* (5.69), *and let Q be an admissible minimum-ratio cycle. Then $\epsilon(\overline{M}) \leq \epsilon(M)$ for $\overline{M} = (M \setminus \{a \in M \mid \overline{a} \in Q \cap M^\circ\}) \cup (Q \cap A^\circ)$.*

Proof. Put $\epsilon = \epsilon(M)$, which is equal to $-\mu$ by (5.71). By the ϵ-optimality of M, we have

$$\gamma_p(a) \equiv \gamma(a) + p(\partial^+ a) - p(\partial^- a) \geq -\epsilon\alpha(a) \qquad (a \in \tilde{A})$$

for some p. Note that

$$\gamma_p(a) = -\epsilon\alpha(a) \qquad (a \in Q). \tag{5.75}$$

Denote by $\tilde{G}_{\overline{M}} = (\overline{V}, \overline{A})$ the auxiliary graph for $\overline{M}$; with obvious additional notations $\overline{A} = A^\circ \cup \overline{M}^\circ \cup \overline{A}^+ \cup \overline{A}^-$, $\overline{\gamma}$, and $\overline{\alpha}$. We will show

$$\overline{\gamma}_p(a) \equiv \overline{\gamma}(a) + p(\partial^+ a) - p(\partial^- a) \geq -\epsilon\overline{\alpha}(a) \qquad (a \in \overline{A}) \tag{5.76}$$

for the same p. This is obvious for $a \in \overline{M}^{\circ} \setminus M^{\circ}$ since $\overline{\alpha}(a) = 0$ and its reorientation $\overline{a} \in Q \cap A^{\circ}$ satisfies $\gamma_p(\overline{a}) = 0$.

In what follows we show (5.76) for $a \in \overline{A}^{+}$; the proof for the remaining case with $a \in \overline{A}^{-}$ is similar. We abbreviate ω^+, V^+, B^+ and $\overline{B}^{+}$ to ω, V, B and $\overline{B}$ respectively. Then (5.76) for $a \in \overline{A}^{+}$ can be written as

$$\omega(\overline{B}, u, v) \leq p(u) - p(v) \qquad (u \in \overline{B}, v \in V \setminus \overline{B}) \tag{5.77}$$

since $\omega(\overline{B}, u, v) = -\infty$ if $(u, v) \notin \overline{A}^{+}$.

Recalling the definition

$$\gamma(a) = -\omega(B, u, v) \qquad (a = (u, v) \in A^+)$$

and noting $\alpha(a) = 0$ $(a \in A^+)$ we see from (5.67) that

$$\omega(B, u, v) \leq p(u) - p(v) \qquad (u \in B, v \in V \setminus B). \tag{5.78}$$

[Note that $(u, v) \notin A^+$ implies $\omega(B, u, v) = -\infty$.] The equation (5.75) shows that this is satisfied with equality for $(u, v) \in Q \cap A^+$. Hence

$$\widehat{\omega}(B, \overline{B}) = \sum_{u \in B \setminus \overline{B}} p(u) - \sum_{v \in \overline{B} \setminus B} p(v). \tag{5.79}$$

For $u \in \overline{B}$ and $v \in V \setminus \overline{B}$ put $B' = \overline{B} - u + v$. It follows from the upper-bound lemma (Lemma 5.2.29), the unique-max lemma (Lemma 5.2.35), (5.78), and (5.79) that

$$\begin{aligned}
&\omega(\overline{B}, u, v) \\
&= \omega(B') - \omega(\overline{B}) \\
&\leq \widehat{\omega}(B, B') - \widehat{\omega}(B, \overline{B}) \\
&\leq \left[\sum_{u' \in B \setminus B'} p(u') - \sum_{v' \in B' \setminus B} p(v') \right] - \left[\sum_{u' \in B \setminus \overline{B}} p(u') - \sum_{v' \in \overline{B} \setminus B} p(v') \right] \\
&= p(u) - p(v).
\end{aligned}$$

Thus (5.77) is established. It may be remarked that the essence of (5.77) lies in Lemma 5.2.38. ■

Combining Lemma 5.2.56 and Lemma 5.2.59 we see that the condition (5.68) is preserved in updating an independent matching in the step (iii) of the algorithm. That is, we have the following.

Lemma 5.2.60. *Assume* (5.68) *and* (5.69), *and let* Q *be an admissible minimum-ratio cycle. Then the condition* (5.68) *is satisfied by* $\overline{M}$. □

We have justified all the claims about the cycle-canceling algorithm with minimum-ratio cycle selection.

5.2.13 Augmenting Algorithms

This section describes a primal-dual-type augmenting algorithm for the valuated independent assignment problem, due to Murota [225]. The algorithm is an extension of the well-established primal-dual algorithm for the ordinary independent assignment problem and the weighted matroid intersection problem. The algorithm will be used in §6.2.6 for an analysis of mixed polynomial matrices.

Algorithms. The augmenting algorithm for VIAP solves VIAP(k) for $k = 0, 1, 2, \cdots$ with the aid of the auxiliary graph $\tilde{G}_{(M,B^+,B^-)} = (\tilde{V}, \tilde{A})$ introduced in §5.2.10. The vertex set $\tilde{V}$ is given by

$$\tilde{V} = V^+ \cup V^- \cup \{s^+, s^-\},$$

where s^+ and s^- are new vertices referred to as the source vertex and the sink vertex respectively, and the arc set $\tilde{A}$ consists of eight disjoint components:

$$\tilde{A} = (A^\circ \cup M^\circ) \cup (A^+ \cup F^+ \cup S^+) \cup (A^- \cup F^- \cup S^-)$$

with components defined (cf. (5.62)) by

$$\begin{aligned}
A^\circ &= \{a \mid a \in A\} \qquad \text{(copy of } A\text{)},\\
M^\circ &= \{\overline{a} \mid a \in M\} \qquad (\overline{a}\text{: reorientation of } a),\\
A^+ &= \{(u,v) \mid u \in B^+, v \in V^+ \setminus B^+, B^+ - u + v \in \mathcal{B}^+\},\\
F^+ &= \{(u, s^+) \mid u \in V^+\},\\
S^+ &= \{(s^+, v) \mid v \in B^+ \setminus \partial^+ M\},\\
A^- &= \{(v,u) \mid u \in B^-, v \in V^- \setminus B^-, B^- - u + v \in \mathcal{B}^-\},\\
F^- &= \{(s^-, u) \mid u \in V^-\},\\
S^- &= \{(v, s^-) \mid v \in B^- \setminus \partial^- M\}.
\end{aligned}$$

The arc length $\gamma(a) = \gamma_{(M,B^+,B^-)}(a)$ $(a \in \tilde{A})$ is defined (cf. (5.63)) by

$$\gamma(a) = \begin{cases} -w(a) & (a \in A^\circ)\\ w(\overline{a}) & (a = (u,v) \in M^\circ, \overline{a} = (v,u) \in M)\\ -\omega^+(B^+, u, v) & (a = (u,v) \in A^+)\\ -\omega^-(B^-, u, v) & (a = (v,u) \in A^-)\\ 0 & (a \in F^+ \cup S^+ \cup F^- \cup S^-). \end{cases}$$

The following fact is most fundamental.

Lemma 5.2.61. *Let (M, B^+, B^-) be a feasible solution to VIAP(k). The problem VIAP($k+1$) has a feasible solution if and only if there exists a directed path from s^+ to s^- in $\tilde{G}_{(M,B^+,B^-)}$.*

Proof. First note that the graph $\tilde{G}_{(M,B^+,B^-)}$ does not depend on w nor on $\omega^{\pm}$, except for the arc length. By Theorem 2.3.33 it suffices to prove the claim that there exists a directed path from s^+ to s^- in $\tilde{G}_{(M,B^+,B^-)}$ if and only if there exists a directed path from S^+ to S^- in the auxiliary graph $\tilde{G}_M$ of §2.3.5. This claim follows from general facts (i) and (ii) below valid for $I \subseteq B \in \mathcal{B}$ in a matroid $(V, \mathcal{I}, \mathcal{B}, \mathrm{cl})$:

(i) For $v \in V \setminus B$: $\quad v \notin \mathrm{cl}(I) \iff \exists u \in B \setminus I : \ B - u + v \in \mathcal{B}$,

(ii) For $u \in I, v \in \mathrm{cl}(I) \setminus I$: $\quad I - u + v \in \mathcal{I} \iff B - u + v \in \mathcal{B}$. ■

Suppose that (M, B^+, B^-) is optimal for VIAP(k), and that VIAP($k+1$) is feasible. It follows from Lemma 5.2.61 that there is a (directed) path in $\tilde{G}_{(M,B^+,B^-)}$ from the source s^+ to the sink s^-, and from Theorem 5.2.47 that there is a shortest path from s^+ to s^- with respect to γ. Then the following theorem holds true (Murota [225]).

Theorem 5.2.62. *Let (M, B^+, B^-) be optimal for VIAP(k) and P be a shortest path, from the source s^+ to the sink s^- in $\tilde{G}_{(M,B^+,B^-)}$, having the smallest number of arcs. Then $(\overline{M}, \overline{B}^+, \overline{B}^-)$ defined by*

$$\overline{M} = (M \setminus \{a \in M \mid \overline{a} \in P \cap M^\circ\}) \cup (P \cap A^\circ), \tag{5.80}$$

$$\overline{B}^+ = (B^+ \setminus \{\partial^+ a \mid a \in P \cap A^+\}) \cup \{\partial^- a \mid a \in P \cap A^+\}, \tag{5.81}$$

$$\overline{B}^- = (B^- \setminus \{\partial^- a \mid a \in P \cap A^-\}) \cup \{\partial^+ a \mid a \in P \cap A^-\} \tag{5.82}$$

is optimal for VIAP($k+1$).

Proof. The proof is given later. ■

With this theorem, we obtain the following algorithm of augmenting type that solves VIAP(k) for $k = 0, 1, 2, \cdots$. At the beginning of the algorithm we set $M = \emptyset$ and find a maximum-weight base B^+ of $\mathbf{M}^+$ with respect to ω^+ and a maximum-weight base B^- of $\mathbf{M}^-$ with respect to ω^-. Obviously this choice gives the optimal solution to VIAP(0).

Augmenting algorithm (outline)
Starting from the empty matching M and maximum-weight bases B^+ and B^- of $\mathbf{M}^+$ and $\mathbf{M}^-$ with respect to ω^+ and ω^-, repeat (i)–(ii) below for $k = 0, 1, 2, \cdots$:

(i) Find a shortest path P having the smallest number of arcs from s^+ to s^- in $\tilde{G}_{(M,B^+,B^-)}$ with respect to the arc length $\gamma_{(M,B^+,B^-)}$.
[Stop if there is no path from s^+ to s^-.]

(ii) Update (M, B^+, B^-) to $(\overline{M}, \overline{B}^+, \overline{B}^-)$ by (5.80), (5.81), (5.82).

Remark 5.2.63. The above algorithm is a natural extension of the primal-dual algorithm for the ordinary independent assignment problem and the weighted matroid intersection problem due to Iri–Tomizawa [133] and Lawler [170, 171] (see also Frank [76]). □

The algorithm outlined above can be made more efficient by the explicit use of a potential function $p : \tilde{V} \to \mathbf{R}$, the use of which has been invented independently by Tomizawa [312] and by Edmonds–Karp [71] in the primal-dual algorithm for the ordinary minimum-cost flow problem.

Suppose again that (M, B^+, B^-) is optimal for VIAP(k). By Theorem 5.2.47 there is a potential $p : \tilde{V} \to \mathbf{R}$ such that

$$\gamma_p(a) \equiv \gamma(a) + p(\partial^+ a) - p(\partial^- a) \geq 0 \qquad (a \in \tilde{A}). \tag{5.83}$$

This condition is equivalent to the following set of conditions appearing in Theorem 5.2.46(1):

$$w(a) - p(\partial^+ a) + p(\partial^- a) \begin{cases} \leq 0 \ (a \in A) \\ = 0 \ (a \in M) \end{cases} \tag{5.84}$$

B^+ is a maximum-weight base of $\mathbf{M}^+$ with respect to $\omega^+[p^+]$, (5.85)

B^- is a maximum-weight base of $\mathbf{M}^-$ with respect to $\omega^-[-p^-]$, (5.86)

$$p(u) \geq p(v) \qquad (u \in V^+,\ v \in B^+ \setminus \partial^+ M), \tag{5.87}$$

$$p(u) \leq p(v) \qquad (u \in V^-,\ v \in B^- \setminus \partial^- M), \tag{5.88}$$

where $p^\pm$ denotes the restriction of p to $V^\pm$ and

$$\omega^+[p^+](B) = \omega^+(B) + \sum_{v \in B} p^+(v) = \omega^+(B) + \sum_{v \in B} p(v) \qquad (B \in \mathcal{B}^+),$$

$$\omega^-[-p^-](B) = \omega^-(B) - \sum_{v \in B} p^-(v) = \omega^-(B) - \sum_{v \in B} p(v) \qquad (B \in \mathcal{B}^-).$$

We maintain such a potential function p in addition to (M, B^+, B^-) and seek a shortest path with respect to the modified arc length γ_p, which is nonnegative by virtue of (5.83). At the beginning of the algorithm the potential p is chosen as

$$p(v) = \begin{cases} 0 & (v \in V^+ \cup \{s^+\}) \\ -\max_{a \in A} w(a) & (v \in V^- \cup \{s^-\}) \end{cases} \tag{5.89}$$

which is easily seen to be legitimate. In the general steps p is updated to

$$\overline{p}(v) = p(v) + \Delta p(v) \qquad (v \in \tilde{V}) \tag{5.90}$$

based on the length $\Delta p(v)$ of the shortest path from the source s^+ to v with respect to the modified arc length γ_p.

Augmenting algorithm (with potential)

(Step 0)

(i) Set $M = \emptyset$.

(ii) Define p by (5.89).

(iii) Find maximum-weight bases B^+ and B^- of $\mathbf{M}^+$ and $\mathbf{M}^-$ with respect to ω^+ and ω^-.

(Step 1) Repeat (i)–(iii) below for $k = 0, 1, 2, \cdots$:

(i) Find a shortest path P having the smallest number of arcs from s^+ to s^- in $\tilde{G}_{(M,B^+,B^-)}$ with respect to the modified arc length γ_p of (5.83).
[Stop if there is no path from s^+ to s^-.]

(ii) For each $v \in \tilde{V}$ compute the length $\Delta p(v)$ of the shortest path from s^+ to v in $\tilde{G}_{(M,B^+,B^-)}$ with respect to the modified arc length γ_p; Update p to $\overline{p}$ by (5.90).

(iii) Update (M, B^+, B^-) to $(\overline{M}, \overline{B}^+, \overline{B}^-)$ by (5.80), (5.81), (5.82).

Remark 5.2.64. In the description of the algorithm above, we have assumed that $\Delta p(v)$ takes a finite value for all v in order to focus on the main ideas. In actual implementations, however, this issue should be taken care of in an appropriate manner. □

Validity of the Augmenting Algorithm. We show that $(\overline{M}, \overline{B}^+, \overline{B}^-, \overline{p})$ satisfies the conditions (5.84)–(5.88). It then follows from Theorem 5.2.46 that $(\overline{M}, \overline{B}^+, \overline{B}^-)$ is optimal for VIAP$(k+1)$. Theorem 5.2.62 also follows from this.

First note that $\overline{M}$ is a matching of size $k+1$ and that

$$\begin{aligned} \gamma_{\overline{p}}(a) &\equiv \gamma(a) + \overline{p}(\partial^+ a) - \overline{p}(\partial^- a) \\ &= \gamma_p(a) + \Delta p(\partial^+ a) - \Delta p(\partial^- a) \geq 0 \qquad (a \in \tilde{A}) \end{aligned} \tag{5.91}$$

by the definition of Δp.

Lemma 5.2.65.

$$w(a) - \overline{p}(\partial^+ a) + \overline{p}(\partial^- a) \begin{cases} \leq 0 & (a \in A) \\ = 0 & (a \in \overline{M}). \end{cases}$$

Proof. The first follows from (5.91) for $a \in A^\circ$, while the second is due to

$$\gamma_p(a) + \Delta p(\partial^+ a) - \Delta p(\partial^- a) = \gamma(a) + \overline{p}(\partial^+ a) - \overline{p}(\partial^- a) = 0 \quad (a \in M \cup P).$$

■

Lemma 5.2.66.

$$\begin{aligned} \overline{p}(u) &\geq \overline{p}(v) \qquad (u \in V^+, v \in \overline{B}^+ \setminus \partial^+ \overline{M}), \\ \overline{p}(u) &\leq \overline{p}(v) \qquad (u \in V^-, v \in \overline{B}^- \setminus \partial^- \overline{M}). \end{aligned}$$

Proof. The inequality (5.91) for $a = (u, s^+), (v, s^+), (s^+, v)$ implies $\overline{p}(u) - \overline{p}(s^+) \geq 0$ and $\overline{p}(v) - \overline{p}(s^+) = 0$. The proof for the second claim is similar. ■

Let

$$\{(u_i^+, v_i^+) \mid i = 1, \cdots, l^+\} = P \cap A^+,$$
$$\{(v_i^-, u_i^-) \mid i = 1, \cdots, l^-\} = P \cap A^-,$$

where $l^+ = |P \cap A^+|$, $l^- = |P \cap A^-|$, and the indices are chosen so that $u_1^+, v_1^+, u_2^+, v_2^+, \cdots, u_{l^+}^+, v_{l^+}^+$ represents the order in which they appear on P, and similarly for $v_{l^-}^-, u_{l^-}^-, \cdots, v_2^-, u_2^-, v_1^-, u_1^-$. We see

$$\overline{B}^+ = B^+ - \{u_1^+, \cdots, u_{l^+}^+\} + \{v_1^+, \cdots, v_{l^+}^+\} \supseteq \partial^+ \overline{M},$$
$$\overline{B}^- = B^- - \{u_1^-, \cdots, u_{l^-}^-\} + \{v_1^-, \cdots, v_{l^-}^-\} \supseteq \partial^- \overline{M}.$$

Lemma 5.2.67.
(1) $(B^+, \overline{B}^+)$ *and* $(B^-, \overline{B}^-)$ *satisfy the unique-max condition in* $\mathbf{M}^+$ *and* $\mathbf{M}^-$ *respectively.*
(2)

$$\widehat{\omega}(B^+, \overline{B}^+) = \sum_{i=1}^{l^+} \left(\overline{p}(u_i^+) - \overline{p}(v_i^+)\right),$$
$$\widehat{\omega}(B^-, \overline{B}^-) = -\sum_{i=1}^{l^-} \left(\overline{p}(u_i^-) - \overline{p}(v_i^-)\right).$$

Proof. We prove the case "+" only and omit the superscript "+". By (5.91) for $a \in A^+$ we have

$$\omega(B, u_i, v_j) \leq \overline{p}(u_i) - \overline{p}(v_j) \qquad (1 \leq i, j \leq l).$$

Here we have an equality if $i = j$ and a strict inequality if $i < j$ by the definitions of $\overline{p}$ and P. Then the unique-max property and the expression in (2) follow from Lemma 5.2.32. ■

Lemma 5.2.68.

$$\omega^+[\overline{p}^+](\overline{B}^+) \geq \omega^+[\overline{p}^+](B_1^+) \qquad (B_1^+ \in \mathcal{B}^+),$$
$$\omega^-[-\overline{p}^-](\overline{B}^-) \geq \omega^-[-\overline{p}^-](B_1^-) \qquad (B_1^- \in \mathcal{B}^-).$$

Proof. Again we prove the case "+" only. By Lemma 5.2.7 it suffices to show

$$\omega[\overline{p}](\overline{B} - u + v) \leq \omega[\overline{p}](\overline{B}) \qquad (u \in \overline{B}, v \in V \setminus \overline{B}).$$

Note first that

$$\omega[\overline{p}](\overline{B} - u + v) - \omega[\overline{p}](\overline{B}) = \omega(\overline{B} - u + v) - \omega(\overline{B}) - \overline{p}(u) + \overline{p}(v). \quad (5.92)$$

Here we have

$$\omega(\overline{B} - u + v) - \omega(B) \leq \widehat{\omega}(B, \overline{B} - u + v) \leq \sum_{u' \in B} \overline{p}(u') - \sum_{v' \in \overline{B} - u + v} \overline{p}(v')$$

by the upper-bound lemma (Lemma 5.2.29) and (5.91) for $a \in A^+$, and

$$\omega(\overline{B}) - \omega(B) = \widehat{\omega}(B, \overline{B}) = \sum_{i=1}^{l} (\overline{p}(u_i) - \overline{p}(v_i))$$

by Lemma 5.2.67 and the unique-max lemma (Lemma 5.2.35). Therefore the RHS of (5.92) is bounded by

$$\sum_{u' \in B} \overline{p}(u') - \sum_{v' \in \overline{B} - u + v} \overline{p}(v') - \sum_{i=1}^{l} (\overline{p}(u_i) - \overline{p}(v_i)) - \overline{p}(u) + \overline{p}(v) = 0.$$

■

Thus we have shown (5.84) in Lemma 5.2.65, (5.85) and (5.86) in Lemma 5.2.68, (5.87) and (5.88) in Lemma 5.2.66. This completes the proof of Theorem 5.2.62.

Notes. The concept of valuated matroids has been obtained by Dress–Wenzel [54, 57] through a quantitative generalization of the exchange axiom of matroids. Duality results for a pair of valuated matroids are established by Murota [224, 230]. This direction is further pursued by Murota [227, 231, 234] to arrive at the concept of "M-convex functions." Besides the exchange axiom, the concept of matroids can also be defined in terms of submodular functions, and the equivalence between the exchange property and the submodularity is one of the most fundamental facts in matroid theory, described in §2.3.2. In harmony with the generalization of matroids to M-convex functions in terms of the exchange axiom, the concept of submodular functions has been generalized to that of "L-convex functions" by Murota [231]. Then the equivalence between the exchange property and the submodularity is generalized to the conjugacy between M-convex functions and L-convex functions. With these concepts a discrete analogue of convex analysis (Rockafellar [280, 281, 282]), called "discrete convex analysis," has been developed by Murota [231]. See Murota [232, 235, 236] for expositions on "discrete convex analysis" using M-convex and L-convex functions.

6. Theory and Application of Mixed Polynomial Matrices

This chapter is devoted to a study of the mathematical properties of mixed polynomial matrices with particular emphasis on applications to control theoretic problems. Mathematically, the analysis of mixed polynomial matrices relies heavily on the results in Chap. 4 and Chap. 5, in particular, the CCF of LM-matrices and the properties of valuated matroids.

6.1 Descriptions of Dynamical Systems

Mixed polynomial matrices arise naturally from the description of dynamical systems. The objective of this section is to collect relevant definitions and concepts for later reference.

6.1.1 Mixed Polynomial Matrix Descriptions

In §3.1.2 as well as in §1.2.2 we have compared two kinds of descriptions of linear time-invariant dynamical systems from the viewpoint of structural analysis using mixed polynomial matrices. The first kind is the *standard form*:

$$\frac{\mathrm{d}\boldsymbol{x}}{\mathrm{d}t} = A\boldsymbol{x} + B\boldsymbol{u}, \tag{6.1}$$

where $\boldsymbol{x} \in \mathbf{R}^n$ and $\boldsymbol{u} \in \mathbf{R}^m$, and the other the *descriptor form*:

$$F\frac{\mathrm{d}\boldsymbol{x}}{\mathrm{d}t} = A\boldsymbol{x} + B\boldsymbol{u}, \tag{6.2}$$

where $\boldsymbol{x} \in \mathbf{R}^n$, $\boldsymbol{u} \in \mathbf{R}^m$, and F is usually a square matrix. It has been argued by way of examples that the coefficient matrix $[A - sF \mid B]$ of the frequency domain representation of the descriptor form (6.2), with a suitable choice of variables, can often be modeled by a mixed polynomial matrix.

Let us recall the definition of a mixed polynomial matrix. A matrix $A(s)$ of polynomials in s over a field $\boldsymbol{F}$ is called a mixed polynomial matrix with respect to $(\boldsymbol{K}, \boldsymbol{F})$, where $\boldsymbol{K}$ is a subfield of $\boldsymbol{F}$, if $A(s)$ is split into two parts:

$$A(s) = Q(s) + T(s) \tag{6.3}$$

in such a way that

(MP-Q1) The coefficients in $Q(s)$ belong to $\boldsymbol{K}$, and
(MP-T) The collection $\mathcal{T}$ of nonzero coefficients in $T(s)$ is algebraically independent over $\boldsymbol{K}$.

A mixed polynomial matrix with respect to $(\boldsymbol{K}, \boldsymbol{F})$ is a mixed matrix with respect to $(\boldsymbol{K}(s), \boldsymbol{F}(s))$. On expressing

$$A(s) = \sum_{k=0}^{N} s^k A_k \ , \quad Q(s) = \sum_{k=0}^{N} s^k Q_k \ , \quad T(s) = \sum_{k=0}^{N} s^k T_k,$$

we have

$$A_k = Q_k + T_k \qquad (k = 0, 1, \cdots, N)$$

and for each k, A_k is a mixed matrix with respect to $(\boldsymbol{K}, \boldsymbol{F})$.

A subclass of mixed polynomial matrices has been identified with reference to the physical dimensional consistency. The subclass is characterized by replacing (MP-Q1) with a stronger condition:

(MP-Q2) Every nonvanishing subdeterminant of $Q(s)$ is a monomial over $\boldsymbol{K}$, i.e., of the form αs^p with $\alpha \in \boldsymbol{K}$ and an integer p.

Recall from Theorem 3.3.2 that (MP-Q2) holds if and only if

$$Q(s) = \mathrm{diag}\,[s^{r_1}, \cdots, s^{r_m}] \cdot Q(1) \cdot \mathrm{diag}\,[s^{-c_1}, \cdots, s^{-c_n}] \tag{6.4}$$

for some integers r_i $(i = 1, \cdots, m)$ and c_j $(j = 1, \cdots, n)$.

An *LM-polynomial matrix* with respect to $(\boldsymbol{K}, \boldsymbol{F})$ will mean a mixed polynomial matrix with respect to $(\boldsymbol{K}, \boldsymbol{F})$ which is an LM-matrix with respect to $(\boldsymbol{K}(s), \boldsymbol{F}(s))$. Namely, an LM-polynomial matrix $A(s)$ with respect to $(\boldsymbol{K}, \boldsymbol{F})$ can be expressed as

$$A(s) = \begin{pmatrix} Q(s) \\ T(s) \end{pmatrix} \tag{6.5}$$

in such a way that (MP-Q1) and (MP-T) are satisfied. A subclass of LM-polynomial matrices can be identified by imposing the condition (MP-Q2).

A *generic polynomial matrix* with respect to $(\boldsymbol{K}, \boldsymbol{F})$ will mean a matrix $A(s)$ of polynomials in s over $\boldsymbol{F}$ such that the collection of nonzero coefficients is algebraically independent over $\boldsymbol{K}$. In other words, a generic polynomial matrix is a mixed polynomial matrix $A(s) = Q(s) + T(s)$ with $Q(s) = O$, which trivially satisfies (MP-Q2).

6.1.2 Relationship to Other Descriptions

In the literature of structural approach in control theory a number of different mathematical frameworks have been proposed to cope with the problem of "parameter dependency." In this section we consider to what extent the mixed polynomial matrix description is general in comparison with other frameworks.

Firstly, the present framework includes the graph-theoretic approach based on the standard form, in which all the nonvanishing entries of the matrices A and B of (6.1) are assumed to be algebraically independent parameters, since the decomposition

$$[A - sI_n \mid B] = [-sI_n \mid O] + [A \mid B]$$

satisfies (MP-Q2) and (MP-T) with $Q(s) = [-sI_n \mid O]$ and $T(s) = [A \mid B]$. Similarly, the graph-theoretic approach based on the descriptor form, in which all the nonvanishing entries of the matrices F, A, and B of (6.2) are assumed to be algebraically independent parameters, also fits the present setting.

Corfmat–Morse [42] considered the standard form (6.1) with the coefficients A and B "linearly parametrized" as

$$A = A_0 + \sum_{i=1}^{k} B_i P_i C_i, \quad B = B_0 + \sum_{i=1}^{k} B_i P_i D_i, \tag{6.6}$$

where P_i $(i = 1, \cdots, k)$ are matrices such that all the entries are independent parameters, and B_i, C_i, D_i $(i = 1, \cdots, k)$ as well as A_0, B_0 are fixed constant matrices. The linear parametrization (6.6) for the standard form (6.1) may be extended to that for the descriptor system (6.2) by assuming the following forms of the coefficients:

$$F = F_0 + \sum_{i=1}^{k} B_i P_i F_i, \quad A = A_0 + \sum_{i=1}^{k} B_i P_i C_i, \quad B = B_0 + \sum_{i=1}^{k} B_i P_i D_i, \tag{6.7}$$

where P_i $(i = 1, \cdots, k)$ are matrices such that all the nonvanishing entries are independent parameters, and F_i, B_i, C_i, D_i $(i = 1, \cdots, k)$ as well as F_0, A_0, B_0 are fixed constant matrices. The case of P_i being scalars is considered by Hosoe–Hayakawa–Aoki [114].

A mixed polynomial matrix description for (6.7) can be obtained by setting

$$\overline{F} = \begin{bmatrix} F_1 \\ \vdots \\ F_k \end{bmatrix} \quad \overline{C} = \begin{bmatrix} C_1 \\ \vdots \\ C_k \end{bmatrix} \quad \overline{D} = \begin{bmatrix} D_1 \\ \vdots \\ D_k \end{bmatrix} \quad \overline{P} = \begin{bmatrix} P_1 & & \\ & \ddots & \\ & & P_k \end{bmatrix}$$

and $\overline{B} = [B_1 \mid \cdots \mid B_k]$, and introducing auxiliary variables

$$\overline{\boldsymbol{v}} = (\overline{C} - s\overline{F})\boldsymbol{x} + \overline{D}\boldsymbol{u}, \qquad \overline{\boldsymbol{w}} = \overline{P}\overline{\boldsymbol{v}}.$$

Then the descriptor system (6.2) with (6.7) can be equivalently rewritten into another descriptor system in descriptor-vector $(\boldsymbol{x}, \overline{\boldsymbol{v}}, \overline{\boldsymbol{w}})$ and input-vector $\boldsymbol{u}$. The coefficient matrix is given by

$$\left[\begin{array}{ccc|c} A_0 - sF_0 & O & \overline{B} & B_0 \\ \overline{C} - s\overline{F} & -I & O & \overline{D} \\ O & \overline{P} & -I & O \end{array}\right],$$

which is a mixed polynomial matrix $Q(s) + T(s)$ with

$$Q(s) = \left[\begin{array}{ccc|c} A_0 - sF_0 & O & \overline{B} & B_0 \\ \overline{C} - s\overline{F} & -I & O & \overline{D} \\ O & O & -I & O \end{array}\right], \qquad T(s) = \left[\begin{array}{ccc|c} O & O & O & O \\ O & O & O & O \\ O & \overline{P} & O & O \end{array}\right].$$

Thus, the linear parametrization as extended above is included in the present framework. The stronger condition (MP-Q2) need not be satisfied from the mathematical point of view, though it is likely to be the case for the physical reason.

A special class of linearly parametrized systems is considered by Hayakawa–Hosoe–Hayashi–Ito [106]. This class includes the so-called *compartmental systems* (see, e.g., Hayakawa–Hosoe–Hayashi–Ito [106, 107], Zazworsky–Knudsen [351] for compartmental systems). Denote by $\boldsymbol{a}_i$ $(i = 1, \cdots, n)$ and $\boldsymbol{b}_j$ $(j = 1, \cdots, m)$, respectively, the column-vectors of the matrices A and B in the standard form (6.1), namely,

$$A = [\boldsymbol{a}_1, \cdots, \boldsymbol{a}_n], \qquad B = [\boldsymbol{b}_1, \cdots, \boldsymbol{b}_m].$$

It is assumed that the column vectors are expressed as

$$\boldsymbol{a}_i = \overline{A}_i \boldsymbol{p}_i \quad (i = 1, \cdots, n), \qquad \boldsymbol{b}_j = \overline{B}_j \boldsymbol{r}_j \quad (j = 1, \cdots, m)$$

with vectors $\boldsymbol{p}_i$ and $\boldsymbol{r}_j$ of independent parameters and fixed constant matrices $\overline{A}_i$ and $\overline{B}_j$. By introducing auxiliary variables

$$\boldsymbol{w}_i = x_i \boldsymbol{p}_i \quad (i = 1, \cdots, n), \qquad \boldsymbol{v}_j = u_j \boldsymbol{r}_j \quad (j = 1, \cdots, m),$$

we obtain a descriptor system in descriptor-vector $(\boldsymbol{x}, \overline{\boldsymbol{w}}, \overline{\boldsymbol{v}})$ and input-vector $\boldsymbol{u}$, where $\overline{\boldsymbol{w}} = ({\boldsymbol{w}_1}^{\mathrm{T}}, \cdots, {\boldsymbol{w}_n}^{\mathrm{T}})^{\mathrm{T}}$ and $\overline{\boldsymbol{v}} = ({\boldsymbol{v}_1}^{\mathrm{T}}, \cdots, {\boldsymbol{v}_m}^{\mathrm{T}})^{\mathrm{T}}$. The coefficient matrix is then given by

$$\left[\begin{array}{ccc|c} -sI_n & \overline{A} & \overline{B} & O \\ \overline{P} & -I & O & O \\ O & O & -I & \overline{R} \end{array}\right], \tag{6.8}$$

where

$$\overline{A} = [\overline{A}_1 \mid \cdots \mid \overline{A}_n], \qquad \overline{B} = [\overline{B}_1 \mid \cdots \mid \overline{B}_m],$$

$$\overline{P} = \begin{bmatrix} \boldsymbol{p}_1 & & \\ & \ddots & \\ & & \boldsymbol{p}_n \end{bmatrix}, \qquad \overline{R} = \begin{bmatrix} \boldsymbol{r}_1 & & \\ & \ddots & \\ & & \boldsymbol{r}_m \end{bmatrix}.$$

The matrix in (6.8) is a mixed polynomial matrix $Q(s) + T(s)$ with

$$Q(s) = \left[\begin{array}{ccc|c} -sI_n & \overline{A} & \overline{B} & O \\ O & -I & O & O \\ O & O & -I & O \end{array}\right], \qquad T(s) = \left[\begin{array}{ccc|c} O & O & O & O \\ \overline{P} & O & O & O \\ O & O & O & \overline{R} \end{array}\right].$$

The matrix $Q(s)$ meets the stronger condition (MP-Q2).

Anderson–Hong [6] considered the standard form (6.1) with A and B expressed in the form of "matrix nets":

$$A = A_0 + \sum_{i=1}^{k} \mu_i A_i, \quad B = B_0 + \sum_{i=1}^{k} \mu_i B_i,$$

where μ_i $(i = 1, \cdots, k)$ are scalar independent parameters, and A_i and B_i $(i = 0, 1, \cdots, k)$ are fixed constant matrices. This does not seem to fit in the present framework of mixed polynomial matrices.

6.2 Degree of Determinant of Mixed Polynomial Matrices

6.2.1 Introduction

In this section we investigate combinatorial characterizations of the highest degree of a minor of order k:

$$\delta_k(A) = \max_{I,J}\{\deg_s \det A[I, J] \mid |I| = |J| = k\} \tag{6.9}$$

for a mixed polynomial matrix $A(s)$ with a view to laying a theoretical foundation for the structural analysis of dynamical systems by means of mixed polynomial matrices. Recall from §5.1.2 and §5.1.3 that the sequence of $\delta_k(A)$ $(k = 1, 2, \cdots)$ determines the Smith–McMillan form at infinity and also the structural indices of the Kronecker form.

We have already encountered the function $\delta(I, J) = \deg_s \det A[I, J]$ as an example of the abstract concept of a valuated bimatroid (see Example 5.2.15). In particular, Theorem 5.2.13 shows the concavity of the sequence $\delta_1(A)$, $\delta_2(A)$, $\cdots$, in the sense of

$$\delta_{k-1}(A) + \delta_{k+1}(A) \leq 2\delta_k(A) \qquad (1 \leq k \leq r - 1), \tag{6.10}$$

where $\delta_0(A) = 0$ and $r = \operatorname{rank} A$. It is emphasized that $A(s)$ is not restricted to a mixed polynomial matrix in this inequality.

The framework of the valuated independent assignment problem introduced in §5.2.9 will play the major role in the investigation of $\delta_k(A)$ for a mixed polynomial matrix $A(s)$.

Remark 6.2.1. In parallel to $\delta_k(A)$ it is often meaningful (see Example 5.2.16) to consider

$$o_k(A) = \min_{I,J}\{\operatorname{ord}_s \det A[I, J] \mid |I| = |J| = k\}, \tag{6.11}$$

where ord_s denotes the minimum degree of a nonzero term in a polynomial in s. Since $o_k(A)$ is equal to $-\delta_k(B)$ for $B(s) = A(1/s)$, all the results for $\delta_k(A)$ can be translated to those for $o_k(A)$. For instance, (6.10) yields

$$o_{k-1}(A) + o_{k+1}(A) \geq 2o_k(A) \qquad (1 \leq k \leq r-1),$$

where $o_0(A) = 0$ and $r = \mathrm{rank}\, A$. □

6.2.2 Graph-theoretic Method

Let us start with a graph-theoretic characterization of $\delta_k(A)$ for a polynomial matrix $A(s)$. We consider a bipartite graph $G(A) = (R, C; E)$, where $R = \mathrm{Row}(A)$, $C = \mathrm{Col}(A)$, and $E = \{(i,j) \mid i \in R, j \in C, A_{ij}(s) \neq 0\}$. To arc $(i,j) \in E$ is attached a weight $w_{ij} = \deg_s A_{ij}(s)$, and the weight of $M \subseteq E$ is defined by $w(M) = \sum_{(i,j)\in M} w_{ij}$. The weighted matching problem treated in §2.2.5 is closely related to $\delta_k(A)$.

Theorem 6.2.2. *Let $A(s)$ be a polynomial matrix.*

(1) $\delta_k(A) \leq \max\{w(M) \mid M\text{: } k\text{-matching in } G(A)\}$,
where the right-hand side is equal to $-\infty$ if no k-matching exists.

(2) *The equality holds if $A(s)$ is a generic polynomial matrix, i.e., if the nonzero coefficients in $A(s)$ are algebraically independent.*

Proof. Consider the defining expansion of the determinant of a submatrix $A[I,J]$ of order $|I| = |J| = k$:

$$\det A[I,J] = \sum_{\sigma: I \to J} \mathrm{sgn}\sigma \prod_{i\in I} A_{i\sigma(i)}(s), \tag{6.12}$$

where σ runs over all one-to-one correspondences from I to J, and $\mathrm{sgn}\sigma$ is defined with reference to a fixed one-to-one correspondence. Put $\hat{\delta}_k(A) = \max\{w(M) \mid M\text{: } k\text{-matching}\}$. It is easy to see that the highest degree of a nonzero term $\prod_{i\in I} A_{i\sigma(i)}(s)$ is equal to $\hat{\delta}_k(A)$, i.e.,

$$\max_{|I|=|J|=k}\ \max_{\sigma: I\to J} \deg_s \prod_{i\in I} A_{i\sigma(i)}(s) = \max_{|I|=|J|=k}\ \max_{\sigma: I\to J} \sum_{i\in I} w_{i\sigma(i)} = \hat{\delta}_k(A).$$

This expression, when combined with the definition (6.9) of $\delta_k(A)$, shows that $\delta_k(A) \leq \hat{\delta}_k(A)$. The equality holds if $A(s)$ is a generic polynomial matrix since no cancellation occurs on the right-hand side of (6.12). ■

The above theorem is most fundamental among the graph-theoretic methods for degree of determinant as well as for structure at infinity. The graph-theoretic method for the DAE-index problem based on this theorem has been fully demonstrated in §1.1. For graph-theoretic methods for structure at infinity and related topics, see Commault–Dion–Hovelaque [38], Commault–Dion–Perez [39], Dion–Commault [48], Linnemann [174], Svaricek [307], van der Woude [326, 327], and van der Woude–Murota [328].

Remark 6.2.3. Theorem 6.2.2 can be translated for $o_k(A)$ defined by (6.11). Namely,

$$o_k(A) = \min\{w(M) \mid M\text{: } k\text{-matching in } G(A)\} \tag{6.13}$$

for a generic polynomial matrix $A(s)$. Henceforth no explicit statements will be made on such translations for $o_k(A)$. □

6.2.3 Basic Identities

We present basic identities concerning the degree of the determinant of (layered) mixed polynomial matrices.

Theorem 6.2.4. *For a square mixed polynomial matrix $A(s) = Q(s)+T(s)$,*

$$\deg_s \det A = \max_{\substack{|I|=|J| \\ I\subseteq R, J\subseteq C}} \{\deg_s \det Q[I,J] + \deg_s \det T[R\setminus I, C\setminus J]\}. \tag{6.14}$$

(It is implied that the right-hand side is equal to $-\infty$ for a singular matrix A.) In other words, the valuated bimatroid associated with $A(s)$ by (5.29) is the union of the valuated bimatroids defined by $Q(s)$ and $T(s)$.

Proof. It follows from the defining expansion (2.2) of determinant that

$$\det A = \sum_{|I|=|J|} \pm \det Q[I,J] \cdot \det T[R\setminus I, C\setminus J].$$

Since the degree of a sum is bounded by the maximum degree of a summand, we obtain

$$\begin{aligned}\deg_s \det A &\leq \max_{|I|=|J|} \deg_s(\det Q[I,J] \cdot \det T[R\setminus I, C\setminus J]) \\ &= \max_{|I|=|J|} \{\deg_s \det Q[I,J] + \deg_s \det T[R\setminus I, C\setminus J]\},\end{aligned}$$

where the inequality turns into an equality provided the highest-degree terms do not cancel one another. The algebraic independence of the nonzero coefficients in $T(s)$ ensures this. ■

The above theorem immediately yields a similar identity for an LM-polynomial matrix $A(s) = \binom{Q(s)}{T(s)}$. Recall the notations $R_Q = \mathrm{Row}(Q)$, $R_T = \mathrm{Row}(T)$, $C = \mathrm{Col}(A)$, $m_Q = |R_Q|$, $m_T = |R_T|$, and $n = |C|$.

Theorem 6.2.5. *For a square LM-polynomial matrix $A(s) = \binom{Q(s)}{T(s)}$,*

$$\deg_s \det A = \max_{J\subseteq C, |J|=|R_Q|} \{\deg_s \det Q[R_Q, J] + \deg_s \det T[R_T, C\setminus J]\}. \tag{6.15}$$

(It is implied that the right-hand side is equal to $-\infty$ for a singular matrix A.) □

In what follows we focus on an LM-polynomial matrix and consider a variant of δ_k. Namely, for $A(s) = \binom{Q(s)}{T(s)}$ we define[1]

$$\delta_k^{\mathrm{LM}}(A) = \max_{I,J}\{\deg_s \det A[R_Q \cup I, J] \mid I \subseteq R_T, J \subseteq C, |I| = k, |J| = m_Q + k\}, \qquad (6.16)$$

where $0 \leq k \leq \min(m_T, n - m_Q)$. It should be clear that $\delta_k^{\mathrm{LM}}(A)$ designates the highest degree of a minor of order $m_Q + k$ with row set containing R_Q. By convention, $\delta_k^{\mathrm{LM}}(A) = -\infty$ if there exists no (I, J) that satisfies the conditions on the right-hand side of (6.16). By substituting (6.15) into (6.16) we obtain

$$\delta_k^{\mathrm{LM}}(A) = \max_{I,J,B}\{\deg_s \det Q[R_Q, B] + \deg_s \det T[I, J \setminus B] \mid I \subseteq R_T, B \subseteq J \subseteq C, |I| = k, |J| = m_Q + k, |B| = m_Q\}. \qquad (6.17)$$

We prefer to work with δ_k^{LM} for an LM-polynomial matrix rather than to deal directly with δ_k for a mixed polynomial matrix. This is because (i) any algorithm for δ_k^{LM} can be used to compute δ_k for a general mixed polynomial matrix (as explained below), and (ii) our algorithm description is much simpler for δ_k^{LM}.

The reduction of δ_k to δ_k^{LM} is as follows. Given an $m \times n$ mixed polynomial matrix $A(s) = Q(s) + T(s)$ we consider a $(2m) \times (m + n)$ LM-polynomial matrix

$$\tilde{A}(s) = \begin{pmatrix} \tilde{Q}(s) \\ \tilde{T}(s) \end{pmatrix} = \begin{pmatrix} \mathrm{diag}\,[s^{d_1}, \cdots, s^{d_m}] & Q(s) \\ -\mathrm{diag}\,[t_1 s^{d_1}, \cdots, t_m s^{d_m}] & T(s) \end{pmatrix} \qquad (6.18)$$

using "new" variables $t_1, \cdots, t_m$ and exponents (integers)

$$d_i = \max_{j \in C_A} \deg_s Q_{ij}(s) \qquad (i \in R_A), \qquad (6.19)$$

where $R_A = \mathrm{Row}(A)$ and $C_A = \mathrm{Col}(A)$.

Lemma 6.2.6. *Let $A(s)$ be an $m \times n$ mixed polynomial matrix and $\tilde{A}(s)$ be the associated LM-polynomial matrix defined by* (6.18) *and* (6.19)*. Then*

$$\delta_k(A) = \delta_k^{\mathrm{LM}}(\tilde{A}) - \sum_{i=1}^{m} d_i.$$

Proof. Define

$$\hat{A}(s) = \begin{array}{c} \\ R_Q \\ R_T \end{array} \begin{array}{c} \begin{array}{cc} \quad R_A & \qquad\qquad C_A \end{array} \\ \begin{pmatrix} \mathrm{diag}\,(s^{d_1}, \cdots, s^{d_m}) & Q(s) \\ -\mathrm{diag}\,(s^{d_1}, \cdots, s^{d_m}) & T(s) \end{pmatrix} \end{array},$$

[1] The notation $\delta_k^{\mathrm{LM}}(A)$ is defined also for a rational matrix $A(s)$ by (6.16).

where $R_Q = \mathrm{Row}(Q)$ and $R_T = \mathrm{Row}(T)$ have a natural one-to-one correspondence with R_A. The matrix $\hat{A}(s)$ is obtained from $\tilde{A}(s)$ by dividing the $(m+i)$th row by t_i and redefining $T_{ij}(s)/t_i$ to be $T_{ij}(s)$ ($j \in C_A$) for $i = 1, \cdots, m$. The latter fact implies $\delta_k^{\mathrm{LM}}(\tilde{A}) = \delta_k^{\mathrm{LM}}(\hat{A})$, where $\delta_k^{\mathrm{LM}}(\hat{A})$ is defined similarly to (6.16), though $\hat{A}(s)$ is not an LM-polynomial matrix.

If $J \supseteq R_A$, we have

$$\deg_s \det \hat{A}[R_Q \cup I, J] = \deg_s \det A[I, C_A \cap J] + \sum_{i=1}^{m} d_i \qquad (I \subseteq R_T). \quad (6.20)$$

Hence, taking the maximum of this expression over all I and J with $|I| = |J| - m = k$ and $J \supseteq R_A$, we see that $\delta_k(A) + \sum_{i=1}^{m} d_i$ is equal to

$$\max\{\deg_s \det \hat{A}[R_Q \cup I, J] \mid I \subseteq R_T,\ R_A \subseteq J \subseteq C, |I| = k, |J| = m + k\}.$$

It remains to be shown that the extra constraint "$J \supseteq R_A$" can be removed without affecting the maximum value. Fix $I \subseteq R_T$ and let $J \subseteq R_A \cup C_A$ be a maximizer of $\deg_s \det \hat{A}[R_Q \cup I, J]$ satisfying $J \supseteq R_A$. We claim that $\hat{A}[R_Q \cup I, J]^{-1} \hat{A}[R_Q \cup I, C_A \setminus J]$ is a proper rational matrix. Then, by (5.24), J is an optimum solution to the maximization problem without the constraint "$J \supseteq R_A$".

The claim can be proven as follows. Denoting by I_Q and I_A the copies of I in R_Q and R_A, respectively, we partition the matrix $\hat{A}[R_Q \cup I, R_A \cup C_A]$ as

$$\hat{A}[R_Q \cup I, R_A \cup C_A] = \begin{array}{c} \\ R_Q \cap I_Q \\ R_Q \setminus I_Q \\ R_T \cap I \end{array} \begin{array}{c} \begin{array}{cccc} R_A \cap I_A & R_A \setminus I_A & C_A \cap J & C_A \setminus J \end{array} \\ \left(\begin{array}{cccc} D_1 & O & Q_{11} & Q_{12} \\ O & D_2 & Q_{21} & Q_{22} \\ -D_1 & O & T_{11} & T_{12} \end{array} \right) \end{array}$$

with the obvious short-hand notations D_1, Q_{11}, T_{11}, etc. for the relevant submatrices of $\mathrm{diag}\,(s^{d_1}, \cdots, s^{d_m})$, $Q(s)$, $T(s)$, etc. By row transformations we obtain

$$\begin{pmatrix} D_1 & O & Q_{11} & Q_{12} \\ O & D_2 & Q_{21} & Q_{22} \\ -D_1 & O & T_{11} & T_{12} \end{pmatrix} \Rightarrow \begin{pmatrix} I & O & O & D_1{}^{-1}[Q_{12} - Q_{11} A_{11}{}^{-1} A_{12}] \\ O & I & O & D_2{}^{-1}[Q_{22} - Q_{21} A_{11}{}^{-1} A_{12}] \\ O & O & I & A_{11}{}^{-1} A_{12} \end{pmatrix},$$

where $A_{ij} = Q_{ij} + T_{ij}$. This shows $\hat{A}[R_Q \cup I, J]^{-1} \hat{A}[R_Q \cup I, C_A \setminus J] = \begin{pmatrix} B_1(s) \\ B_2(s) \end{pmatrix}$ with

$$B_1(s) = \mathrm{diag}\,(s^{-d_1}, \cdots, s^{-d_m})\{Q[R_Q, C_A \setminus J] - Q[R_Q, C_A \cap J] B_2(s)\},$$
$$B_2(s) = A[I, C_A \cap J]^{-1} A[I, C_A \setminus J].$$

Here $B_2(s)$ is a proper rational matrix by the choice of J (cf. (6.20) and (5.24)), and $\mathrm{diag}\,(s^{-d_1}, \cdots, s^{-d_m}) Q[R_Q, C_A]$ is also proper by the definition (6.19) of d_i. Therefore, $\hat{A}[R_Q \cup I, J]^{-1} \hat{A}[R_Q \cup I, C_A \setminus J]$ is a proper rational matrix. ∎

Example 6.2.7. Consider a 2×3 mixed polynomial matrix:

$$A(s) = \begin{array}{c|ccc|} & c_1 & c_2 & c_3 \\ \hline r_1 & s^3+1 & s^2+\alpha_1 & \alpha_2 s+1 \\ r_2 & s^2+\alpha_3 & s & 0 \\ \hline \end{array}$$

with respect to $(\boldsymbol{K}, \boldsymbol{F}) = (\mathbf{Q}, \mathbf{Q}(\alpha_1, \alpha_2, \alpha_3))$, where $\{\alpha_1, \alpha_2, \alpha_3\}$ is assumed to be algebraically independent over $\mathbf{Q}$. We have $A(s) = Q(s) + T(s)$ with

$$Q(s) = \begin{array}{|ccc|} \hline s^3+1 & s^2 & 1 \\ s^2 & s & 0 \\ \hline \end{array}, \qquad T(s) = \begin{array}{|ccc|} \hline 0 & \alpha_1 & \alpha_2 s \\ \alpha_3 & 0 & 0 \\ \hline \end{array}.$$

The associated LM-polynomial matrix is given by

$$\tilde{A}(s) = \begin{array}{c|cc|ccc|} & r_1 & r_2 & c_1 & c_2 & c_3 \\ \hline r_{Q1} & s^3 & & s^3+1 & s^2 & 1 \\ r_{Q2} & & s^2 & s^2 & s & 0 \\ \hline r_{T1} & -t_1 s^3 & & 0 & \alpha_1 & \alpha_2 s \\ r_{T2} & & -t_2 s^2 & \alpha_3 & 0 & 0 \\ \hline \end{array},$$

where $d_1 = 3$ and $d_2 = 2$ in (6.19). It is easy to see by inspection that

$$\begin{aligned} \delta_1(A) &= \deg_s \det A[r_1, c_1] = 3, \\ \delta_2(A) &= \deg_s \det A[\{r_1, r_2\}, \{c_1, c_3\}] = 3, \\ \delta_1^{\mathrm{LM}}(\tilde{A}) &= \deg_s \det \tilde{A}[\{r_{Q1}, r_{Q2}, r_{T1}\}, \{r_1, r_2, c_1\}] = 3 + 5, \\ \delta_2^{\mathrm{LM}}(\tilde{A}) &= \deg_s \det \tilde{A}[\{r_{Q1}, r_{Q2}, r_{T1}, r_{T2}\}, \{r_1, r_2, c_1, c_3\}] = 3 + 5, \end{aligned}$$

which verify the relation $\delta_k^{\mathrm{LM}}(\tilde{A}) = \delta_k(A) + (d_1 + d_2)$ in Lemma 6.2.6. □

6.2.4 Reduction to Valuated Independent Assignment

We describe how the computation of $\delta_k^{\mathrm{LM}}(A)$ for an LM-polynomial matrix $A(s) = \binom{Q(s)}{T(s)}$ can be reduced to solving a valuated independent assignment problem of §5.2.9. Assuming $Q(s)$ to be of full-row rank, we denote by $\mathbf{M}_Q = (C_Q, \mathcal{B}_Q, \omega_Q)$ the valuated matroid associated with $Q(s)$ (see Example 5.2.3); namely,

$$\mathcal{B}_Q = \{B \subseteq C_Q \mid \det Q[R_Q, B] \neq 0\}, \tag{6.21}$$

$$\omega_Q(B) = \deg_s \det Q[R_Q, B] \qquad (B \in \mathcal{B}_Q). \tag{6.22}$$

Here and henceforth $C_Q = \{j_Q \mid j \in C\}$ denotes a disjoint copy of the column set C of A (with $j_Q \in C_Q$ denoting the copy of $j \in C$), whereas R_Q and R_T mean, as before, the row sets of $Q(s)$ and $T(s)$, respectively, with $|R_Q| = m_Q$, $|R_T| = m_T$ and $|C| = n$.

We consider a valuated independent assignment problem defined on a bipartite graph $G = (V^+, V^-; E)$ with $V^+ = R_T \cup C_Q$, $V^- = C$, and $E = E_T \cup E_Q$, where

$$E_T = \{(i,j) \mid i \in R_T, j \in C, T_{ij}(s) \neq 0\}, \quad E_Q = \{(j_Q, j) \mid j \in C\}.$$

The valuated matroids $\mathbf{M}_k^+ = (V^+, \mathcal{B}_k^+, \omega_k^+)$ and $\mathbf{M}_k^- = (V^-, \mathcal{B}_k^-, \omega_k^-)$ attached to V^+ and V^- are defined by

$$\begin{aligned} \mathcal{B}_k^+ &= \{B^+ \subseteq V^+ \mid B^+ \cap C_Q \in \mathcal{B}_Q, |B^+ \cap R_T| = k\}, \\ \mathcal{B}_k^- &= \{B^- \subseteq V^- \mid |B^-| = m_Q + k\} \end{aligned}$$

and

$$\begin{aligned} &\omega_k^+(B^+) = \omega_Q(B^+ \cap C_Q) \qquad (B^+ \in \mathcal{B}_k^+), \\ &\omega_k^-(B^-) = 0 \qquad (B^- \in \mathcal{B}_k^-). \end{aligned}$$

The weight w_{ij} of an arc $(i,j) \in E$ is defined by

$$w_{ij} = \begin{cases} \deg_s T_{ij}(s) & ((i,j) \in E_T) \\ 0 & ((i,j) \in E_Q). \end{cases} \tag{6.23}$$

The value of an independent assignment M is given by

$$\begin{aligned} \Omega_k(M) &= w(M) + \omega_k^+(\partial^+ M) + \omega_k^-(\partial^- M) \\ &= \sum_{(i,j) \in M \cap E_T} \deg_s T_{ij}(s) + \deg_s \det Q[R_Q, \partial^+(M \cap E_Q)]. \end{aligned}$$

We then have the following characterization of $\delta_k^{\mathrm{LM}}(A)$ in terms of the optimal value of the valuated independent assignment problem.

Theorem 6.2.8. *For an LM-polynomial matrix $A(s) = \binom{Q(s)}{T(s)}$ with $Q(s)$ of full-row rank and an integer k with $0 \leq k \leq \min(m_T, n - m_Q)$, $\delta_k^{\mathrm{LM}}(A)$ of* (6.16) *coincides with the optimal value of the valuated independent assignment problem defined above. That is,*

$$\delta_k^{\mathrm{LM}}(A) = \max\{\Omega_k(M) \mid M\text{: independent assignment}\},$$

where the right-hand side is defined to be $-\infty$ if there exists no independent assignment M.

Proof. Define

$$\Delta(I, J, B) = \deg_s \det Q[R_Q, B] + \deg_s \det T[I, J \setminus B],$$

which is the function to be maximized in the expression (6.17) for $\delta_k^{\mathrm{LM}}(A)$. By virtue of the algebraic independence of the nonzero coefficients in $T(s)$, the second term, $\deg_s \det T[I, J \setminus B]$, is equal to the maximum weight (with respect to w_{ij}) of a matching of size $|I| = |J \setminus B|$ in the bipartite graph $(R_T, C; E_T)$ that covers I and $J \setminus B$ (see Theorem 6.2.2). Given (I, J, B) with $|I| = k$ and $\Delta(I, J, B) > -\infty$, we can construct an independent assignment M such that

$$I = \partial^+(M \cap E_T), \quad J = \partial^- M, \quad B = \partial^+(M \cap E_Q), \qquad (6.24)$$

and that $M \cap E_T$ is a maximum weight k-matching in the graph $(R_T, C; E_T)$ that covers I and $J \setminus B$. Note that $\det Q[R_Q, B] \neq 0$ and $|I| = k$ if and only if $B \cup I \in \mathcal{B}_k^+$. Moreover, $\omega_k^+(B \cup I) = \deg_s \det Q[R_Q, B]$ by the definition, and therefore we have $\Delta(I, J, B) = \Omega_k(M)$. Conversely, an independent assignment M with $\Omega_k(M)$ maximum determines (I, J, B), as above, for which $\Delta(I, J, B) = \Omega_k(M)$ holds true. Hence the maximum value of $\Delta(I, J, B)$ is equal to that of $\Omega_k(M)$. ■

Example 6.2.9. The valuated independent assignment problem associated with a 4×5 LM-polynomial matrix

$$A(s) = \begin{array}{c|ccccc|} & x_1 & x_2 & x_3 & x_4 & x_5 \\ \hline & s^3 & 0 & s^3+1 & s^2 & 1 \\ & 0 & s^2 & s^2 & s & 0 \\ \hline f_1 & -t_1 s^3 & 0 & 0 & \alpha_1 & \alpha_2 s \\ f_2 & 0 & -t_2 s^2 & \alpha_3 & 0 & 0 \\ \hline \end{array} \qquad (6.25)$$

with $k = 2$ is illustrated in Fig. 6.1. This matrix is essentially the same as $\tilde{A}(s)$ in Example 6.2.7, but the columns and the rows are now indexed as $C = \{x_1, x_2, x_3, x_4, x_5\}$ and $R_T = \{f_1, f_2\}$; accordingly $C_Q = \{x_{1Q}, x_{2Q}, x_{3Q}, x_{4Q}, x_{5Q}\}$. An optimal independent assignment

$$M = \{(f_1, x_5), (f_2, x_2), (x_{1Q}, x_1), (x_{3Q}, x_3)\}$$

is marked by ○ in Fig. 6.1. We have $I = \partial^+(M \cap E_T) = \{f_1, f_2\}$, $J = \partial^- M = \{x_1, x_2, x_3, x_5\}$, $B = \partial^+(M \cap E_Q) = \{x_{1Q}, x_{3Q}\} \in \mathcal{B}_Q$, $\omega_Q(B) = 5$, $w(M) = 1 + 2 = 3$, and therefore $\Omega_2(M) = 5 + 3 = 8$, which agrees with $\delta_2^{\mathrm{LM}}(A) = 8$. □

Theorem 6.2.8 enables us to design an efficient algorithm to compute δ_k^{LM} by specializing the general algorithmic scheme for valuated independent assignment problems given in §5.2.13. This will be described in detail in §6.2.6.

Remark 6.2.10. When the stronger condition (MP-Q2) may be assumed for the matrix $Q(s)$ of an LM-polynomial matrix $A(s)$, the valuated independent assignment problem reduces to a linearly-weighted independent assignment problem. By Theorem 3.3.2, (MP-Q2) implies $Q(s) = \mathrm{diag}\,[s^{r_1}, \cdots, s^{r_m}] \cdot$

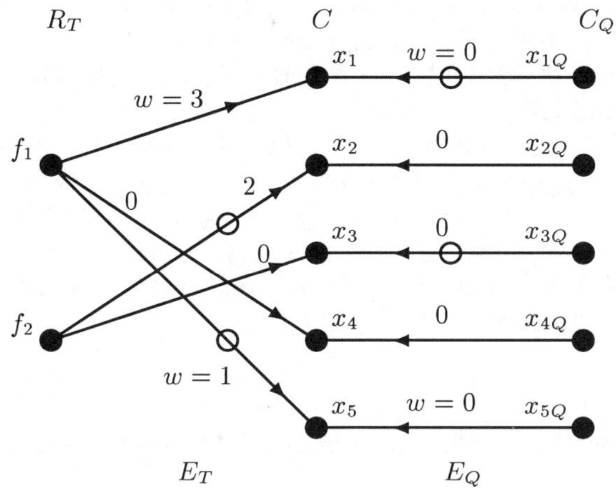

Fig. 6.1. Valuated independent assignment problem for $\delta_2^{\mathrm{LM}}(A)$ of Example 6.2.9 (○: arc in M, $B = \{x_{1Q}, x_{3Q}\}$)

$Q(1) \cdot \mathrm{diag}\,[s^{-c_1}, \cdots, s^{-c_n}]$ for some integers r_i ($i = 1, \cdots, m$) and c_j ($j = 1, \cdots, n$). Hence $\omega_Q(B) = \sum_{i \in R_Q} r_i - \sum_{j \in B} c_j$, which is a separable valuation (cf. Example 5.2.2). In place of the valuated independent assignment problem we may consider an independent assignment problem on the same bipartite graph $G = (V^+, V^-; E)$ and the same (non-valuated) matroids $\mathbf{M}_k^+ = (V^+, \mathcal{B}_k^+)$ and $\mathbf{M}_k^- = (V^-, \mathcal{B}_k^-)$, but with the arc weight redefined as

$$w_{ij} = \begin{cases} \deg_s T_{ij}(s) & ((i,j) \in E_T) \\ -c_j & ((i,j) = (j_Q, j) \in E_Q). \end{cases}$$

Then we have

$$\delta_k^{\mathrm{LM}}(A) = \max\{w(M) \mid M\text{: independent assignment}\} + \sum_{i \in R_Q} r_i.$$

This formulation under (MP-Q2) was introduced first by Murota [200] in characterizing the dynamical degree (see also Murota [204, §27]), and then applied to the problem of disturbance rejection by Murota–van der Woude [242]. □

6.2.5 Duality Theorems

The basic identities on the degree of subdeterminants presented in §6.2.3 are recast here into novel identities of duality nature. They are obtained from the duality result (Theorem 5.2.39) on the valuated independent assignment problem.

Consider the valuated independent assignment problem for $\delta_k^{\mathrm{LM}}(A)$. Let M be an optimal independent assignment and (I, J, B) be defined by $I = \partial^+(M \cap E_T)$, $J = \partial^- M$, and $B = \partial^+(M \cap E_Q)$ (cf. (6.24)), where $|I| = k$, $|J| = m_Q + k$, and $|B| = m_Q$.

Let $\hat{p} : R_T \cup C \cup C_Q \to \mathbf{Z}$ be the potential function in Theorem 5.2.39. We may assume that $\hat{p}(j_Q) = \hat{p}(j)$ for $j \in C$, where $j_Q \in C_Q$ denotes the copy of $j \in C$. To see this, first note that $\hat{p}(j_Q) \geq \hat{p}(j)$ for $j \in C$ and the equality holds if $(j_Q, j) \in M$. For $j \in C$ with $(j_Q, j) \notin M$, we can redefine $\hat{p}(j_Q)$ to be equal to $\hat{p}(j)$ without violating the conditions (i) and (ii) in Theorem 5.2.39(1). Define $q \in \mathbf{Z}^{R_T}$ and $p \in \mathbf{Z}^C$ by

$$q_i = \hat{p}(i) \quad (i \in R_T), \qquad p_j = -\hat{p}(j) \quad (j \in C). \tag{6.26}$$

The conditions (i)–(iii) in Theorem 5.2.39(1) are expressed as follows:

$$\deg_s T_{ij}(s) \leq q_i + p_j \quad ((i,j) \in E_T), \tag{6.27}$$
$$\deg_s T_{ij}(s) = q_i + p_j \quad ((i,j) \in M \cap E_T), \tag{6.28}$$
$$\omega_Q[-p](B) = \max_{B' \in \mathcal{B}_Q} \omega_Q[-p](B'), \tag{6.29}$$
$$q(I) = \max_{|I'|=k} q(I'), \tag{6.30}$$
$$p(J) = \max_{|J'|=m_Q+k} p(J'), \tag{6.31}$$

where $q(I) = \sum_{i \in I} q_i$ and $p(J) = \sum_{j \in J} p_j$. These conditions imply

$$\begin{aligned}
\delta_k^{\mathrm{LM}}(A) &= \deg_s \det Q[R_Q, B] + \deg_s \det T[I, J \setminus B] \\
&= \omega_Q(B) + q(I) + p(J \setminus B) \\
&= \omega_Q[-p](B) + q(I) + p(J) \\
&= \max_{B' \in \mathcal{B}_Q} \omega_Q[-p](B') + \max_{|I'|=k} q(I') + \max_{|J'|=m_Q+k} p(J').
\end{aligned} \tag{6.32}$$

Thus we obtain the following theorem of Murota [233].

Theorem 6.2.11. *For an LM-polynomial matrix* $A(s) = \binom{Q(s)}{T(s)}$ *and an integer* k *such that* $\delta_k^{\mathrm{LM}}(A) > -\infty$, *the following identity holds true:*

$$\delta_k^{\mathrm{LM}}(A) = \min_{q_i + p_j \geq \deg_s T_{ij}} \left[\max_{|I|=k} q(I) + \max_{|J|=m_Q+k} p(J) + \max_{B \in \mathcal{B}_Q} \omega_Q[-p](B) \right],$$

where the minimum is taken over all $q \in \mathbf{R}^{R_T}$ *and* $p \in \mathbf{R}^C$ *satisfying* $q_i + p_j \geq \deg_s T_{ij}$ *for all* (i, j), *and the minimum is attained by integer vectors* $q \in \mathbf{Z}^{R_T}$ *and* $p \in \mathbf{Z}^C$.

Proof. Let (I, J, B) be associated with an optimal M as above. For any (q', p') with $q'_i + p'_j \geq \deg_s T_{ij}$ $(\forall (i,j))$, we have

$$\begin{aligned}\delta_k^{\rm LM}(A) &= \deg_s \det Q[R_Q, B] + \deg_s \det T[I, J \setminus B] \\ &\le \omega_Q(B) + q'(I) + p'(J \setminus B) \\ &= \omega_Q[-p'](B) + q'(I) + p'(J) \\ &\le \max_{B' \in \mathcal{B}_Q} \omega_Q[-p'](B') + \max_{|I'|=k} q'(I') + \max_{|J'|=m_Q+k} p'(J'),\end{aligned}$$

whereas the inequalities turn into equalities for $(q', p') = (q, p)$, as in (6.32). ∎

With p and q above, we can transform the matrix $A(s)$ to another LM-polynomial matrix that is somehow canonical with respect to $\delta_k^{\rm LM}$.

Theorem 6.2.12. *For an LM-polynomial matrix $A(s) = \binom{Q(s)}{T(s)}$ and an integer k such that $\delta_k^{\rm LM}(A) > -\infty$, there exist $p \in \mathbf{Z}^C$ and $q \in \mathbf{Z}^{R_T}$ such that*

$$\bar{A}(s) = \begin{pmatrix} \bar{Q}(s) \\ \bar{T}(s) \end{pmatrix} = \begin{pmatrix} I_{m_Q} & O \\ O & \mathrm{diag}\,(s; -q) \end{pmatrix} \cdot \begin{pmatrix} Q(s) \\ T(s) \end{pmatrix} \cdot \mathrm{diag}\,(s; -p)$$

satisfy

$$\delta_k^{\rm LM}(\bar{A}) = \max_{|B|=m_Q} \deg_s \det \bar{Q}[R_Q, B] + \max_{|I|=|J|=k} \deg_s \det \bar{T}[I, J]. \tag{6.33}$$

We may additionally impose either

$$\delta_k^{\rm LM}(A) = \delta_k^{\rm LM}(\bar{A}) + \max_{|I|=k} q(I) + \max_{|J|=m_Q+k} p(J) \tag{6.34}$$

or that $\bar{A}(s)$ be a polynomial matrix.

Proof. Let (I, J, B) be associated with an optimal M, and p and q be defined by (6.26). Put $S(s) = (Q[R_Q, B] \cdot \mathrm{diag}\,(s; -p_B))^{-1}$, where p_B is the restriction of p to B, and define

$$\hat{A}(s) = \begin{pmatrix} \hat{Q}(s) \\ \hat{T}(s) \end{pmatrix} = \begin{pmatrix} S(s) & O \\ O & \mathrm{diag}\,(s; -q) \end{pmatrix} \cdot A(s) \cdot \mathrm{diag}\,(s; -p). \tag{6.35}$$

The conditions (6.27)–(6.29) mean that

$$\hat{A}(s) = \begin{array}{c} \\ R_Q \\ I \\ R_T \setminus I \end{array} \begin{array}{c} \begin{array}{ccc} B & J \setminus B & C \setminus J \end{array} \\ \begin{pmatrix} I_{m_Q} & Q_2'(s) & Q_3'(s) \\ T_1'(s) & T_2^\bullet(s) & T_3'(s) \\ T_1''(s) & T_2''(s) & T_3''(s) \end{pmatrix} \end{array}$$

is a proper rational matrix, in which $T_2^\bullet(s)$ admits a transversal consisting of entries of degree zero. Obviously,

$$\delta_k^{\rm LM}(\hat{A}) = \max_{|B'|=m_Q} \deg_s \det \hat{Q}[R_Q, B'] + \max_{|I'|=|J'|=k} \deg_s \det \hat{T}[I', J'],$$

in which all the three terms are equal to zero. This implies the first identity (6.33). The second identity (6.34) is due to (6.32) combined with $\delta_k^{\rm LM}(\bar{A}) = \delta_k^{\rm LM}(\hat{A}) + \deg_s \det S^{-1} = \omega_Q[-p](B)$. To make $\bar{A}(s)$ a polynomial matrix, replace p by $p - \alpha\mathbf{1}$ with a sufficiently large $\alpha \in \mathbf{Z}$, where $\mathbf{1}$ is the vector of all components equal to one. Note that this change does not affect (6.33). ■

Example 6.2.13. We illustrate the above argument for the LM-polynomial matrix $A(s)$ of (6.25) with $k = 2$. The vectors $p \in \mathbf{Z}^C$ and $q \in \mathbf{Z}^{R_T}$ of (6.26) are given by $p = (-1, -1, -3, -4, -3)$ and $q = (4, 3)$. Accordingly we have

$$\bar{A}(s) = \begin{array}{c} \\ \\ \\ f_1 \\ f_2 \end{array} \begin{array}{ccccc} x_1 & x_2 & x_3 & x_4 & x_5 \\ \hline s^4 & 0 & s^6+s^3 & s^6 & s^3 \\ 0 & s^3 & s^5 & s^5 & 0 \\ \hline -t_1 & 0 & 0 & \alpha_1 & \alpha_2 \\ 0 & -t_2 & \alpha_3 & 0 & 0 \\ \hline \end{array},$$

for which (6.33) holds true with $\delta_2^{\rm LM}(\bar{A}) = 9 = 9 + 0$. All the entries of $\bar{A}(s)$ are polynomials, and the identity (6.34) also holds true, though these two may not be compatible in general. Recall from Example 6.2.9 that $I = \{f_1, f_2\}$, $J = \{x_1, x_2, x_3, x_5\}$, $B = \{x_{1Q}, x_{3Q}\}$. The matrix $\hat{A}(s)$ of (6.35) is equal to

$$\hat{A}(s) = \begin{array}{c} \\ \\ \\ f_1 \\ f_2 \end{array} \begin{array}{cc|c|c|c} x_1 & x_3 & x_5 & x_2 & x_4 \\ \hline 1 & 0 & \frac{1}{s} & \frac{-s^3-1}{s^3} & \frac{-1}{s} \\ 0 & 1 & 0 & \frac{1}{s^2} & 1 \\ \hline -t_1 & 0 & \alpha_2 & 0 & \alpha_1 \\ 0 & \alpha_3 & 0 & -t_2 & 0 \\ \hline \end{array}.$$

In §6.2.6 we will come back to this example and explain how the vectors p and q can be found (see the variable p in Fig. 6.4, in particular). □

As corollaries to the above theorem we obtain the following two theorems on the degree of the whole determinant. The first theorem shows that an LM-polynomial matrix can be transformed so that the maximization on the right-hand side of (6.15) in Theorem 6.2.5 may be done separately for Q- and T-parts. The second is a similar statement for a mixed polynomial matrix treated in Theorem 6.2.4.

Theorem 6.2.14. *For a nonsingular LM-polynomial matrix* $A(s) = \binom{Q(s)}{T(s)}$ *there exists* $p \in \mathbf{Z}^C$ *such that*

$$\bar{A}(s) = \begin{pmatrix} \bar{Q}(s) \\ \bar{T}(s) \end{pmatrix} = \begin{pmatrix} Q(s) \\ T(s) \end{pmatrix} \cdot \mathrm{diag}\,(s; -p)$$

satisfies

$$\deg_s \det \bar{A} = \max_{|B|=|R_Q|} \deg_s \det \bar{Q}[R_Q, B] + \max_{|J|=|R_T|} \deg_s \det \bar{T}[R_T, J].$$

An additional condition that $\bar{A}(s)$ be a polynomial matrix may be imposed.

Proof. Apply Theorem 6.2.12 with $k = m_T = n - m_Q$ to obtain $p \in \mathbf{Z}^C$. The row transformation by $\mathrm{diag}\,(s; -q)$ is not necessary in the case of $k = m_T$. To make $\bar{A}(s)$ a polynomial matrix, replace p by $p - \alpha\mathbf{1}$ with a sufficiently large $\alpha \in \mathbf{Z}$, where $\mathbf{1}$ is the vector of all components equal to one. ∎

Theorem 6.2.15. *For a nonsingular mixed polynomial matrix $A(s) = Q(s) + T(s)$, there exist $p_R \in \mathbf{Z}^R$ and $p_C \in \mathbf{Z}^C$ such that*

$$\bar{A}(s) = \mathrm{diag}\,(s; -p_R) \cdot A(s) \cdot \mathrm{diag}\,(s; p_C)$$

satisfies

$$\deg_s \det \bar{A} = \max_{\substack{|I|=|J| \\ I \subseteq R, J \subseteq C}} \deg_s \det \bar{Q}[I, J] + \max_{\substack{|I|=|J| \\ I \subseteq R, J \subseteq C}} \deg_s \det \bar{T}[R \setminus I, C \setminus J], \quad (6.36)$$

where

$$\bar{Q}(s) = \mathrm{diag}\,(s; -p_R) \cdot Q(s) \cdot \mathrm{diag}\,(s; p_C),$$
$$\bar{T}(s) = \mathrm{diag}\,(s; -p_R) \cdot T(s) \cdot \mathrm{diag}\,(s; p_C).$$

An additional condition (i) $p_R \geq \mathbf{0}$, $p_C \geq \mathbf{0}$, *or* (ii) $p_R \leq \mathbf{0}$, $p_C \leq \mathbf{0}$, *may be imposed on p_R and p_C.*

Proof. Apply Theorem 6.2.14 to the associated LM-polynomial matrix (6.18) to obtain $\hat{p} \in \mathbf{Z}^{R \cup C}$. Denote by $\hat{p}_R$ and $\hat{p}_C$ the restrictions of $\hat{p}$ to R and to C, respectively. Then put $p_R = d - \hat{p}_R$ and $p_C = -\hat{p}_C$, where $d \in \mathbf{Z}^R$ is the vector of exponents in (6.18). For the additional condition, replace p_R with $p_R + \alpha\mathbf{1}$ and p_C with $p_C + \alpha\mathbf{1}$ using a suitable $\alpha \in \mathbf{Z}$. ∎

Example 6.2.16. The matrix $\bar{A}(s)$ in Theorem 6.2.15 may not be a polynomial matrix. Consider, for example, a 2×2 mixed matrix $A(s) = Q(s) + T(s)$ with

$$A(s) = \begin{bmatrix} s & s+1+t_1 s \\ s+1+t_2 s & t_3 s \end{bmatrix}, \quad Q(s) = \begin{bmatrix} s & s+1 \\ s+1 & 0 \end{bmatrix}, \quad T(s) = \begin{bmatrix} 0 & t_1 s \\ t_2 s & t_3 s \end{bmatrix}.$$

We may take $p_R = (1, 1)$ and $p_C = (0, 0)$ to obtain

$$\bar{A}(s) = \begin{bmatrix} 1 & 1+1/s+t_1 \\ 1+1/s+t_2 & t_3 \end{bmatrix}, \quad \bar{Q}(s) = \begin{bmatrix} 1 & 1+1/s \\ 1+1/s & 0 \end{bmatrix}, \quad \bar{T}(s) = \begin{bmatrix} 0 & t_1 \\ t_2 & t_3 \end{bmatrix},$$

for which (6.36) holds true. However, it can be verified that $\bar{A}(s)$ cannot be a polynomial matrix in (6.36). □

See (1.37) for another example of Theorem 6.2.15.

6.2.6 Algorithm

In §6.2.4 we have explained how to reduce the computation of $\delta_k^{\mathrm{LM}}(A)$ to solving a valuated independent assignment problem. Here we will provide an algorithm for $\delta_k^{\mathrm{LM}}(A)$ by adapting the augmenting algorithm of §5.2.13 for a general valuated independent assignment problem.

The associated valuated independent assignment problem is defined on the bipartite graph $G = (V^+, V^-; E) = (R_T \cup C_Q, C; E_T \cup E_Q)$, where C_Q is a disjoint copy of C (with $j_Q \in C_Q$ denoting the copy of $j \in C$), and

$$E_T = \{(i,j) \mid i \in R_T, j \in C, T_{ij}(s) \neq 0\}, \quad E_Q = \{(j_Q, j) \mid j \in C\}.$$

To V^+ and V^- are attached the valuated matroids $\mathbf{M}^+ = (V^+, \mathcal{B}^+, \omega^+)$ and $\mathbf{M}^- = (V^-, \mathcal{B}^-, \omega^-)$ defined by

$$\begin{aligned} \mathcal{B}^+ &= \{B^+ \subseteq V^+ \mid B^+ \supseteq R_T, B^+ \cap C_Q \in \mathcal{B}_Q\}, & \mathcal{B}^- &= \{V^-\}, \\ \omega^+(B^+) &= \omega_Q(B^+ \cap C_Q) \quad (B^+ \in \mathcal{B}^+), & \omega^-(B^-) &= 0 \quad (B^- \in \mathcal{B}^-), \end{aligned}$$

where $\mathcal{B}_Q$ and ω_Q are given in (6.21) and (6.22). The arc weight w is the same as in (6.23).

The algorithm solves VIAP$(m_Q + k)$ for $k = 0, 1, 2, \cdots, k_{\max}$ by the augmenting algorithm of §5.2.13 to compute the value of $\delta_k^{\mathrm{LM}}(A)$ successively for $k = 0, 1, 2, \cdots, k_{\max}$, where $k_{\max}$ is the maximum k with $\delta_k^{\mathrm{LM}}(A) > -\infty$. Namely, the algorithm maintains a pair (M, B) of a matching $M \subseteq E_T \cup E_Q$ and a base $B \in \mathcal{B}_Q$ $(\subseteq 2^{C_Q})$ that maximizes

$$\Omega''(M, B) \equiv w(M) + \omega_Q(B) = w(M \cap E_T) + \omega_Q(B) \qquad (6.37)$$

subject to the constraint that $\partial^+(M \cap E_Q) = B$ and M is of a specified size. We put

$$M_T = M \cap E_T, \qquad M_Q = M \cap E_Q.$$

With reference to (M, B) it constructs an auxiliary directed graph $\tilde{G} = \tilde{G}_{(M,B)} = (\tilde{V}, \tilde{E})$ with vertex set $\tilde{V} = R_T \cup C_Q \cup C$ and arc set $\tilde{E} = E_T \cup E_Q \cup E^+ \cup M^\circ$, where

$$\begin{aligned} E^+ &= \{(i_Q, j_Q) \mid i_Q \in B, j_Q \in C_Q \setminus B, B - i_Q + j_Q \in \mathcal{B}_Q\}, \\ M^\circ &= \{\overline{a} \mid a \in M\} \qquad (\overline{a}\text{: reorientation of } a). \end{aligned}$$

It should be emphasized that the arcs in E^+ have both ends in C_Q and that the arcs in M° are directed from C to $R_T \cup C_Q$, i.e., $\partial^+ M^\circ \subseteq C$ and $\partial^- M^\circ \subseteq R_T \cup C_Q$. We put

$$\begin{aligned} M_T^\circ &= \{a \in M^\circ \mid \partial^- a \in R_T\} = \{\overline{a} \mid a \in M_T\}, \\ M_Q^\circ &= \{a \in M^\circ \mid \partial^- a \in C_Q\} = \{\overline{a} \mid a \in M_Q\}. \end{aligned}$$

We define the entrance $S^+ \subseteq \tilde{V}$ and the exit $S^- \subseteq \tilde{V}$ by

$$S^+ = R_T \setminus \partial^+ M_T = R_T \setminus \partial^- M_T^\circ, \qquad S^- = C \setminus \partial^- M = C \setminus \partial^+ M^\circ.$$

Note that no vertex in C_Q belongs to the entrance S^+.

We define the arc length $\gamma = \gamma_{(M,B)} : \tilde{E} \to \mathbf{Z}$ by

$$\gamma_{(M,B)}(a) = \begin{cases} -\deg_s T_{ij}(s) & (a = (i,j) \in E_T) \\ \quad \deg_s T_{ij}(s) & (a = (j,i) \in M_T^\circ) \\ -\omega_Q(B, i_Q, j_Q) & (a = (i_Q, j_Q) \in E^+) \\ 0 & (a \in E_Q \cup M_Q^\circ) \end{cases} \tag{6.38}$$

where $\omega_Q(B, i_Q, j_Q) = \omega_Q(B - i_Q + j_Q) - \omega_Q(B)$, compatibly with the notation (5.21). By (5.23) we can compute $\omega_Q(B, i_Q, j_Q)$ by means of pivoting operations on $Q(s)$, namely, for $P(s) = S(s)Q(s)$ with $S(s) = Q[R_Q, B]^{-1}$ we have $\omega_Q(B, i_Q, j_Q) = \deg_s P_{ij}(s)$.

Suppose there is a shortest path in $\tilde{G}_{(M,B)}$ from the entrance S^+ to the exit S^- with respect to the arc length γ, and let L be (the set of arcs on) a shortest path from S^+ to S^- having the smallest number of arcs. Then we can update (M, B) to $(\overline{M}, \overline{B})$ by

$$\overline{M} = M - \{a \in M \mid \overline{a} \in L \cap M^\circ\} + (L \cap (E_T \cup E_Q)), \tag{6.39}$$

$$\overline{B} = B - \{\partial^+ a \mid a \in L \cap E^+\} + \{\partial^- a \mid a \in L \cap E^+\}. \tag{6.40}$$

In fact, $\overline{M}$ is obviously a matching with $\partial^+(\overline{M} \cap E_Q) = \overline{B}$ and $|\overline{M}| = |M| + 1$, and furthermore, Theorem 5.2.62 shows that $\overline{B} \in \mathcal{B}_Q$ and $(\overline{M}, \overline{B})$ maximizes $\Omega''(\overline{M}, \overline{B})$ under these constraints.

Our algorithm for $\delta_k^{\mathrm{LM}}(A)$ repeats finding a shortest path and updating (M, B) as follows.

Outline of the algorithm

Starting from a maximum-weight base $B \in \mathcal{B}_Q$ with respect to ω_Q and the corresponding matching $M = \{(j_Q, j) \mid j_Q \in B\}$, repeat (i)–(ii) below:

(i) Find a shortest path L having the smallest number of arcs from S^+ to S^- in $\tilde{G}_{(M,B)}$ with respect to the arc length $\gamma_{(M,B)}$ of (6.38). [Stop if there is no path from S^+ to S^-.]

(ii) Update (M, B) according to (6.39) and (6.40).

An initial base B of maximum value of ω_Q can be found by the greedy algorithm described in §5.2.4. At each stage of this algorithm it holds that $\delta_k^{\mathrm{LM}}(A) = \Omega''(M, B)$ for $k = |M| - m_Q$ and that (I, J, B) defined by (6.24) gives the maximum in the expression (6.17) of $\delta_k^{\mathrm{LM}}(A)$.

As has been explained in §5.2.13, the above algorithm can be made more efficient by the explicit use of a potential function on the auxiliary graph $\tilde{G} = (\tilde{V}, \tilde{E})$. To this end we maintain $p : \tilde{V} \to \mathbf{Z}$ that satisfies

$$-\deg_s T_{ij}(s) + p(i) - p(j) \geq 0 \qquad ((i,j) \in E_T), \tag{6.41}$$
$$-\deg_s T_{ij}(s) + p(i) - p(j) = 0 \qquad ((i,j) \in M_T), \tag{6.42}$$
$$p(j_Q) - p(j) \geq 0 \qquad (j \in C), \tag{6.43}$$
$$p(j_Q) - p(j) = 0 \qquad ((j_Q, j) \in M_Q), \tag{6.44}$$
$$-\omega_Q(B, i_Q, j_Q) + p(i_Q) - p(j_Q) \geq 0 \qquad ((i_Q, j_Q) \in E^+), \tag{6.45}$$
$$p(i) - p(k) \geq 0 \qquad (i \in R_T, k \in S^+), \tag{6.46}$$
$$p(k) - p(j) \geq 0 \qquad (j \in C, k \in S^-). \tag{6.47}$$

It is remarked that the existence of such p implies the optimality of (M, B) with respect to Ω'' of (6.37). In fact, for any (M', B') with $|M'| = |M|$ and $\partial^+(M' \cap E_Q) = B'$ we have

$$\begin{aligned} w(M') + \omega_Q(B') &= w_p(M') + \omega_Q[p_Q](B') + p(\partial^+ M'_T) - p(\partial^- M') \\ &\leq \omega_Q[p_Q](B) + p(\partial^+ M_T) - p(\partial^- M) \\ &= w(M) + \omega_Q(B), \end{aligned}$$

where $M'_T = M' \cap E_T$, $w_p(a) = w(a) - p(\partial^+ a) + p(\partial^- a)$, and p_Q denotes the restriction of p to C_Q. Note that $w_p(M') \leq w_p(M) = 0$ by (6.41)–(6.44), $\omega_Q[p_Q](B') \leq \omega_Q[p_Q](B)$ by (6.45) and Theorem 5.2.7, $p(\partial^+ M'_T) \leq p(\partial^+ M_T)$ by (6.46), and $p(\partial^- M') \geq p(\partial^- M)$ by (6.47).

Initially, we have $M_T = \emptyset$ and $\omega_Q(B, i_Q, j_Q) \leq 0$ for all $(i_Q, j_Q) \in E^+$, and therefore we can put

$$p(i) = \max_{k \in R_T, j \in C} \deg_s T_{kj}(s) \quad (i \in R_T), \qquad p(j) = p(j_Q) = 0 \quad (j \in C) \tag{6.48}$$

to meet the conditions (6.41)–(6.47). In general steps, p is updated to

$$\overline{p}(v) = p(v) + \Delta p(v) \qquad (v \in \tilde{V}) \tag{6.49}$$

based on the length $\Delta p(v)$ of the shortest path from S^+ to v with respect to the modified arc length

$$\gamma_p(a) = \gamma(a) + p(\partial^+ a) - p(\partial^- a) \geq 0 \qquad (a \in \tilde{E}), \tag{6.50}$$

where the nonnegativity of γ_p is due to (6.41)–(6.47). Then $\overline{p}$ satisfies the conditions (6.41)–(6.47) (see Lemmas 5.2.65, 5.2.66, and 5.2.68).

To compute $\omega_Q(B, i_Q, j_Q)$ we use two matrices (or two-dimensional arrays) $P = P(s)$ and $S = S(s)$, as well as two vectors (or one-dimensional arrays) *base* and p. The array P represents an $m_Q \times n$ matrix of rational functions in s over $\boldsymbol{K}$, where $P = Q$ at the beginning of the algorithm (Step 1 below). The other array S is an $m_Q \times m_Q$ matrix of rational functions in s over $\boldsymbol{K}$, which is set to the unit (identity) matrix in Step 1. The variable *base* is a vector of size m_Q, which represents a mapping (correspondence):

$R_Q \to C \cup \{0\}$. The vector p, indexed by $R_T \cup C \cup C_Q$, represents the potential function satisfying (6.41)–(6.47). We also use a scalar (integer-valued) variable δ_Q to compute $\omega_Q(B)$.

The following algorithm computes $\delta_k^{\mathrm{LM}}(A)$ for $k = 0, 1, 2, \cdots, k_{\max}$, as well as the value of $k_{\max}$, where $k_{\max} = -1$ by convention, if $\operatorname{rank} Q(s) < m_Q$.

Algorithm for $\delta_k^{\mathrm{LM}}(A)$ ($k = 0, 1, 2, \cdots, k_{\max}$)

Step 1: [Initialize]
  $M := \emptyset$; $B := \emptyset$; $\delta_Q := 0$;
  $\mathit{base}[i] := 0$ $(i \in R_Q)$; $P[i,j] := Q_{ij}$ $(i \in R_Q, j \in C)$;
  $S :=$ unit matrix of order m_Q;
  $p[i] := \max_{k \in R_T, j \in C} \deg_s T_{kj}$ $(i \in R_T)$; $p[j] := p[j_Q] := 0$ $(j \in C)$. [cf. (6.48)]
Step 2: [Find $B \in \mathcal{B}_Q$ that maximizes ω_Q]
  While $|B| < m_Q$ do
    {Find (h,j) that maximizes $\deg_s P[h,j]$
      subject to $\mathit{base}[h] = 0$, $j_Q \notin B$, and $P[h,j] \neq 0$;
    If there exists no such (h,j), then stop with $k_{\max} := -1$;
    $B := B + j_Q$; $\delta_Q := \delta_Q + \deg_s P[h,j]$; $M := M + (j_Q, j)$;
    $\mathit{base}[h] := j$; $w := 1/P[h,j]$;
    $P[h,l] := w \times P[h,l]$ $(l \in C)$; $S[h,l] := w \times S[h,l]$ $(l \in R_Q)$;
    $P[m,l] := P[m,l] - P[m,j] \times P[h,l]$ $(m \in R_Q \setminus \{h\}, l \in C \setminus \{j\})$;
    $S[m,l] := S[m,l] - P[m,j] \times S[h,l]$ $(m \in R_Q \setminus \{h\}, l \in R_Q)$;
    $P[m,j] := 0$ $(m \in R_Q \setminus \{h\})$ };
  $k := 0$.
Step 3: [Construct the auxiliary graph $\tilde{G}_{(M,B)}$]
  $\delta_k^{\mathrm{LM}}(A) := \delta_Q + \sum_{(i,j) \in M \cap E_T} \deg_s T_{ij}$;
  $S^+ := R_T \setminus \partial^+(M \cap E_T)$; $S^- := C \setminus \partial^- M$; $M^\circ := \{\overline{a} \mid a \in M\}$;
  $E^+ := \{(i_Q, j_Q) \mid h \in R_Q, j_Q \notin B, P[h,j] \neq 0, i = \mathit{base}[h]\}$;

$$\gamma(a) := \begin{cases} -\deg_s T_{ij}(s) & (a = (i,j) \in E_T) \\ \deg_s T_{ij}(s) & (a = (j,i) \in M_T^\circ) \\ -\deg_s P[h,j] & (a = (i_Q, j_Q) \in E^+, \mathit{base}[h] = i) \\ 0 & (a \in E_Q \cup M_Q^\circ) \end{cases}$$

[cf. (6.38)]

  where $M_T^\circ = \{\overline{a} \mid a \in M \cap E_T\}$, $M_Q^\circ = \{\overline{a} \mid a \in M \cap E_Q\}$;
  $\gamma_p(a) := \gamma(a) + p[\partial^+ a] - p[\partial^- a]$ $(a \in \tilde{E})$. [cf. (6.50)]
Step 4: [Augment M along a shortest path]
  For each $v \in \tilde{V}$ compute the length $\Delta p(v)$ of the shortest path from S^+ to v in $\tilde{G}_{(M,B)}$ with respect to the modified arc length γ_p;
  If there is no path from S^+ to S^- (including the case where $S^+ = \emptyset$ or $S^- = \emptyset$), then stop with $k_{\max} := k$;
  Let L ($\subseteq \tilde{E}$) be (the set of arcs on) a shortest path, having the smallest number of arcs, from S^+ to S^- with respect to the modified arc length γ_p;

$M := M - \{a \in M \mid \overline{a} \in L \cap M^\circ\} + (L \cap (E_T \cup E_Q))$; $\quad k := k + 1$;
$p[v] := p[v] + \Delta p(v) \quad (v \in \tilde{V})$; [cf. (6.49)]
For all $(i_Q, j_Q) \in L \cap E^+$ (in the order from S^+ to S^- along L) do the following:
  {Find h such that $i = base[h]$;
  $B := B - i_Q + j_Q; \quad \delta_Q := \delta_Q + \deg_s P[h,j]$;
  $base[h] := j; \quad w := 1/P[h,j]$;
  $P[h,l] := w \times P[h,l] \quad (l \in C); \qquad S[h,l] := w \times S[h,l] \quad (l \in R_Q)$;
  $P[m,l] := P[m,l] - P[m,j] \times P[h,l] \quad (m \in R_Q \setminus \{h\}, l \in C \setminus \{j\})$;
  $S[m,l] := S[m,l] - P[m,j] \times S[h,l] \quad (m \in R_Q \setminus \{h\}, l \in R_Q)$;
  $P[m,j] := 0 \quad (m \in R_Q \setminus \{h\})$ };
Go to Step 3. □

Step 2 for finding a maximum-weight base $B \in \mathcal{B}_Q$ is justified by the greedy algorithm for a valuated bimatroid given in §5.2.5 (see also Example 5.2.15).

For the updates of P in Steps 2 and 4, the algorithm assumes the availability of arithmetic operations on rational functions in a single variable s over the subfield $\boldsymbol{K}$. It is emphasized that no arithmetic operations are done on the T-part, so that no rational function operations involving coefficients in T (which are independent symbols) are needed.

The updates of P are the standard pivoting operations on rational functions in s over $\boldsymbol{K}$, the total number of which is bounded by $\mathrm{O}(|R|^2|C|\,k_{\max})$. Note that pivoting operations are required for each arc $(i_Q, j_Q) \in L \cap E^+$ (see Step 4). The sparsity of P should be taken into account in actual implementations of the algorithm.

The matrix $S(s)$ gives the inverse of $Q[R_Q, B]$, which is often useful (see, e.g., the proof of Theorem 6.2.12). When $S(s)$ is not needed, it may simply be eliminated from the computation without any side effect.

The shortest path in Step 4 can be found in time linear in the size of the graph $\tilde{G}$, which is $\mathrm{O}((|R|+|C|)^2)$, by means of the standard graph algorithms; see, e.g., Aho–Hopcroft–Ullman [1, 2].

Remark 6.2.17. The above algorithm can be used to compute $\delta_k(A)$ for a mixed polynomial matrix $A(s)$ by considering the associated LM-mixed polynomial matrix on the basis of Lemma 6.2.6. This is what is called "Algorithm D" in §1.3.2. □

Example 6.2.18. The algorithm above is illustrated here for the 4×5 LM-polynomial matrix $A(s)$ of (6.25):

$$
A(s) = \begin{array}{c|ccccc|}
 & x_1 & x_2 & x_3 & x_4 & x_5 \\
\hline
 & s^3 & 0 & s^3+1 & s^2 & 1 \\
 & 0 & s^2 & s^2 & s & 0 \\
\hline
f_1 & -t_1 s^3 & 0 & 0 & \alpha_1 & \alpha_2 s \\
f_2 & 0 & -t_2 s^2 & \alpha_3 & 0 & 0 \\
\hline
\end{array} .
$$

We work with a 2×5 matrix $P(s)$, a 2×2 matrix $S(s)$, a vector *base* of size 2, and another vector p of size 12.

The flow of computation is traced below.

Step 1: $M := \emptyset$; $B := \emptyset$; $\delta_Q := 0$;

$$(base, P, S) := \begin{array}{c} \\ r_1 \\ r_2 \end{array}\!\begin{array}{c} \\ \boxed{\begin{array}{c} 0 \\ 0 \end{array}} \end{array},\ \begin{array}{c} \begin{array}{ccccc} x_1 & x_2 & x_3 & x_4 & x_5 \end{array} \\ \boxed{\begin{array}{ccccc} s^3 & 0 & s^3+1 & s^2 & 1 \\ 0 & s^2 & s^2 & s & 0 \end{array}} \end{array},\ \boxed{\begin{array}{cc} 1 & 0 \\ 0 & 1 \end{array}};$$

$$p := \begin{array}{cc|ccccc|ccccc} f_1 & f_2 & x_1 & x_2 & x_3 & x_4 & x_5 & x_{1Q} & x_{2Q} & x_{3Q} & x_{4Q} & x_{5Q} \\ \hline 3 & 3 & 0 & 0 & 0 & 0 & 0 & 0 & 0 & 0 & 0 & 0 \end{array}.$$

Step 2: $(h, j) := (r_1, x_1)$; $B := \{x_{1Q}\}$, $\delta_Q := 3$; $M := \{(x_{1Q}, x_1)\}$;

$$(base, P, S) := \begin{array}{c} \\ r_1 \\ r_2 \end{array}\!\begin{array}{c} \\ \boxed{\begin{array}{c} x_1 \\ 0 \end{array}} \end{array},\ \begin{array}{c} \begin{array}{ccccc} x_1 & x_2 & x_3 & x_4 & x_5 \end{array} \\ \boxed{\begin{array}{ccccc} 1 & 0 & \frac{s^3+1}{s^3} & \frac{1}{s} & \frac{1}{s^3} \\ 0 & s^2 & s^2 & s & 0 \end{array}} \end{array},\ \boxed{\begin{array}{cc} \frac{1}{s^3} & 0 \\ 0 & 1 \end{array}};$$

$(h, j) := (r_2, x_2)$; $B := \{x_{1Q}, x_{2Q}\}$, $\delta_Q := 5$; $M := \{(x_{1Q}, x_1), (x_{2Q}, x_2)\}$;

$$(base, P, S) := \begin{array}{c} \\ r_1 \\ r_2 \end{array}\!\begin{array}{c} \\ \boxed{\begin{array}{c} x_1 \\ x_2 \end{array}} \end{array},\ \begin{array}{c} \begin{array}{ccccc} x_1 & x_2 & x_3 & x_4 & x_5 \end{array} \\ \boxed{\begin{array}{ccccc} 1 & 0 & \frac{s^3+1}{s^3} & \frac{1}{s} & \frac{1}{s^3} \\ 0 & 1 & 1 & \frac{1}{s} & 0 \end{array}} \end{array},\ \boxed{\begin{array}{cc} \frac{1}{s^3} & 0 \\ 0 & \frac{1}{s^2} \end{array}};$$

$k := 0$.

Step 3: $\delta_0^{\mathrm{LM}}(A) := 5$; $S^+ := \{f_1, f_2\}$; $S^- := \{x_3, x_4, x_5\}$;
$M^\circ := \{(x_1, x_{1Q}), (x_2, x_{2Q})\}$;
$E^+ := \{(x_{1Q}, x_{3Q}), (x_{1Q}, x_{4Q}), (x_{1Q}, x_{5Q}), (x_{2Q}, x_{3Q}), (x_{2Q}, x_{4Q})\}$;
γ and γ_p are given in $\tilde{G}^{(0)}$ of Fig. 6.2. [See $\tilde{G}^{(0)}$ in Fig. 6.2]

Step 4:

$$\Delta p := \begin{array}{cc|ccccc|ccccc} f_1 & f_2 & x_1 & x_2 & x_3 & x_4 & x_5 & x_{1Q} & x_{2Q} & x_{3Q} & x_{4Q} & x_{5Q} \\ \hline 0 & 0 & 0 & 1 & 0 & 1 & 2 & 0 & 1 & 0 & 1 & 3 \end{array};$$

There exists a path from S^+ to S^-;
$L := \{(f_1, x_1), (x_1, x_{1Q}), (x_{1Q}, x_{3Q}), (x_{3Q}, x_3)\}$;
$M := \{(f_1, x_1), (x_{2Q}, x_2), (x_{3Q}, x_3)\}$; $k := 1$;

$$p := \begin{array}{cc|ccccc|ccccc} f_1 & f_2 & x_1 & x_2 & x_3 & x_4 & x_5 & x_{1Q} & x_{2Q} & x_{3Q} & x_{4Q} & x_{5Q} \\ \hline 3 & 3 & 0 & 1 & 0 & 1 & 2 & 0 & 1 & 0 & 1 & 3 \end{array};$$

$(i_Q, j_Q) := (x_{1Q}, x_{3Q}) \in L \cap E^+$; $h := r_1$; $B := \{x_{3Q}, x_{2Q}\}$, $\delta_Q := 5$;

$$(base, P, S) := \begin{array}{c} \\ r_1 \\ r_2 \end{array}\!\begin{array}{c} \\ \boxed{\begin{array}{c} x_3 \\ x_2 \end{array}} \end{array},\ \begin{array}{c} \begin{array}{ccccc} x_1 & x_2 & x_3 & x_4 & x_5 \end{array} \\ \boxed{\begin{array}{ccccc} \frac{s^3}{s^3+1} & 0 & 1 & \frac{s^2}{s^3+1} & \frac{1}{s^3+1} \\ -\frac{s^3}{s^3+1} & 1 & 0 & \frac{1}{s(s^3+1)} & \frac{-1}{s^3+1} \end{array}} \end{array},\ \boxed{\begin{array}{cc} \frac{1}{s^3+1} & 0 \\ \frac{-1}{s^3+1} & \frac{1}{s^2} \end{array}}.$$

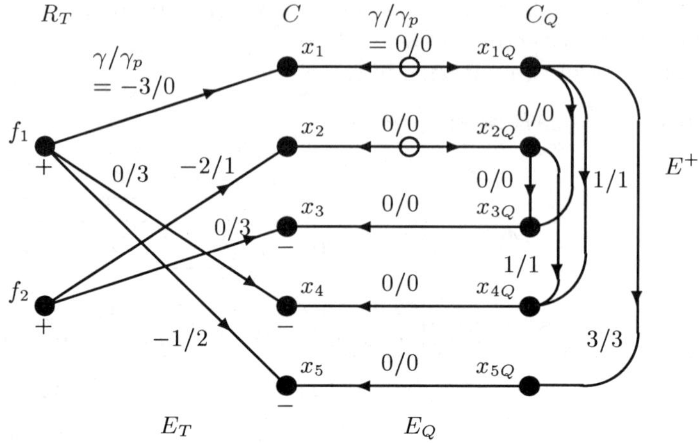

v	f_1	f_2	x_1	x_2	x_3	x_4	x_5	x_{1Q}	x_{2Q}	x_{3Q}	x_{4Q}	x_{5Q}
p	3	3	0	0	0	0	0	0	0	0	0	0
Δp	0	0	0	1	0	1	2	0	1	0	1	3

Fig. 6.2. Graph $\tilde{G}^{(0)}$ ($\bigcirc$: arc in M, $B = \{x_{1Q}, x_{2Q}\}$, $S^+ = \{f_1, f_2\}$, $S^- = \{x_3, x_4, x_5\}$)

Step 3: $\delta_1^{\rm LM}(A) := 5 + 3 = 8$; $S^+ := \{f_2\}$; $S^- := \{x_4, x_5\}$;
$M^\circ := \{(x_1, f_1), (x_2, x_{2Q}), (x_3, x_{3Q})\}$;
$E^+ := \{(x_{2Q}, x_{1Q}), (x_{2Q}, x_{4Q}), (x_{2Q}, x_{5Q}), (x_{3Q}, x_{1Q}), (x_{3Q}, x_{4Q}), (x_{3Q}, x_{5Q})\}$;
γ and γ_p are given in $\tilde{G}^{(1)}$ of Fig. 6.3. [See $\tilde{G}^{(1)}$ in Fig. 6.3]

Step 4:

$$\Delta p := \begin{array}{cc|ccccc|ccccc} f_1 & f_2 & x_1 & x_2 & x_3 & x_4 & x_5 & x_{1Q} & x_{2Q} & x_{3Q} & x_{4Q} & x_{5Q} \\ \hline 1 & 0 & 1 & 0 & 3 & 3 & 1 & 1 & 0 & 3 & 3 & 1 \end{array};$$

There exists a path from S^+ to S^-;
$L := \{(f_2, x_2), (x_2, x_{2Q}), (x_{2Q}, x_{1Q}), (x_{1Q}, x_1), (x_1, f_1), (f_1, x_5)\}$;
$M := \{(f_1, x_5), (f_2, x_2), (x_{1Q}, x_1), (x_{3Q}, x_3)\}$; $k := 2$;

$$p := \begin{array}{cc|ccccc|ccccc} f_1 & f_2 & x_1 & x_2 & x_3 & x_4 & x_5 & x_{1Q} & x_{2Q} & x_{3Q} & x_{4Q} & x_{5Q} \\ \hline 4 & 3 & 1 & 1 & 3 & 4 & 3 & 1 & 1 & 3 & 4 & 4 \end{array};$$

$(i_Q, j_Q) := (x_{2Q}, x_{1Q}) \in L \cap E^+$; $h := r_2$; $B := \{x_{3Q}, x_{1Q}\}$, $\delta_Q := 5$;

$$(base, P, S) := \begin{array}{c|c|} r_1 & x_3 \\ r_2 & x_1 \end{array}, \quad \begin{array}{ccccc} x_1 & x_2 & x_3 & x_4 & x_5 \\ \hline 0 & 1 & 1 & \frac{1}{s} & 0 \\ 1 & \frac{-s^3-1}{s^3} & 0 & \frac{-1}{s^4} & \frac{1}{s^3} \end{array}, \quad \begin{array}{|cc|} 0 & \frac{1}{s^2} \\ \frac{1}{s^3} & \frac{-s^3-1}{s^5} \end{array}.$$

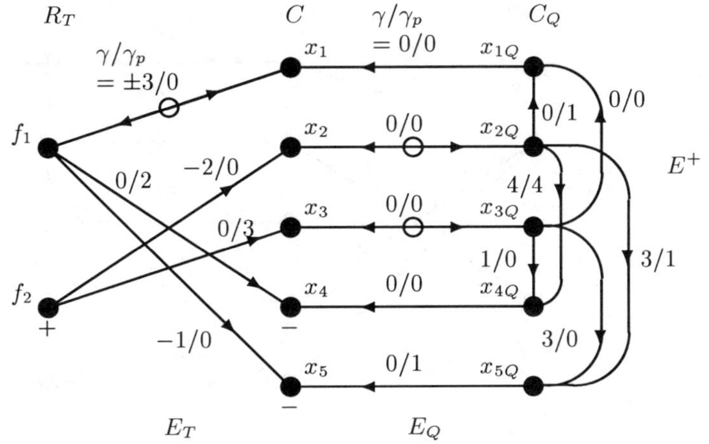

v	f_1	f_2	x_1	x_2	x_3	x_4	x_5	x_{1Q}	x_{2Q}	x_{3Q}	x_{4Q}	x_{5Q}
p	3	3	0	1	0	1	2	0	1	0	1	3
Δp	1	0	1	0	3	3	1	1	0	3	3	1

Fig. 6.3. Graph $\tilde{G}^{(1)}$ (○: arc in M, $B = \{x_{2Q}, x_{3Q}\}$, $S^+ = \{f_2\}$, $S^- = \{x_4, x_5\}$)

Step 3: $\delta_2^{\mathrm{LM}}(A) := 5 + 3 = 8$; $S^+ := \emptyset$; $S^- := \{x_4\}$;
$M^\circ := \{(x_5, f_1), (x_2, f_2), (x_1, x_{1Q}), (x_3, x_{3Q})\}$;
$E^+ := \{(x_{1Q}, x_{2Q}), (x_{1Q}, x_{4Q}), (x_{1Q}, x_{5Q}), (x_{3Q}, x_{2Q}), (x_{3Q}, x_{4Q})\}$;
γ and γ_p are given in $\tilde{G}^{(2)}$ of Fig. 6.4. [See $\tilde{G}^{(2)}$ in Fig. 6.4]
Step 4: There exists no path from S^+ $(= \emptyset)$ to S^-;
Stop with $k_{\max} := 2$. □

Notes. This section is based mostly on Murota [233]. In particular, Theorems 6.2.8, 6.2.11, 6.2.14, and 6.2.15 are found in Murota [233], whereas Theorem 6.2.4 is given in Murota [200]. The problem of computing the degree of determinant will be considered again in §7.1.

6.3 Smith Form of Mixed Polynomial Matrices

The Smith normal form of a mixed polynomial matrix $A(s) = Q(s) + T(s)$ is investigated. It is shown that all the invariant factors except for the last are polynomials in s free from the coefficients in $T(s)$, and the last invariant factor can be expressed in terms of the CCF of an associated LM-matrix.

6.3.1 Expression of Invariant Factors

Let $A(s) = Q(s) + T(s)$ be an $m \times n$ mixed polynomial matrix of rank r with respect to $(\boldsymbol{K}, \boldsymbol{F})$ satisfying, by definition, (MP-Q1) and (MP-T) in §6.1.1.

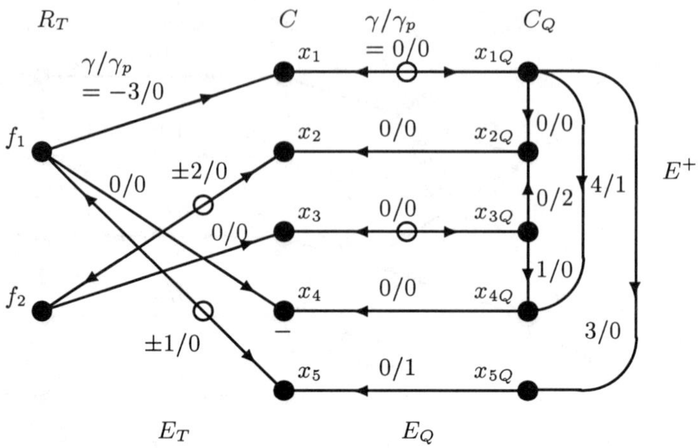

v	f_1	f_2	x_1	x_2	x_3	x_4	x_5	x_{1Q}	x_{2Q}	x_{3Q}	x_{4Q}	x_{5Q}
p	4	3	1	1	3	4	3	1	1	3	4	4

Fig. 6.4. Graph $\tilde{G}^{(2)}$ (○: arc in M, $B = \{x_{1Q}, x_{3Q}\}$, $S^+ = \emptyset$, $S^- = \{x_4\}$)

Regarding $A(s)$ as a polynomial matrix in s over $\boldsymbol{F}$, we define (cf. §5.1.1) the kth determinantal divisor $d_k(s) \in \boldsymbol{F}[s]$ by

$$d_k(s) = \gcd{}_{\boldsymbol{F}[s]}\{\det A[I, J] \mid |I| = |J| = k\} \qquad (k = 1, \cdots, r) \tag{6.51}$$

and the kth invariant factor $e_k(s) \in \boldsymbol{F}[s]$ by

$$e_k(s) = \frac{d_k(s)}{d_{k-1}(s)} \qquad (k = 1, \cdots, r), \tag{6.52}$$

where $d_k(s)$ and $e_k(s)$ are chosen to be monic in $\boldsymbol{F}[s]$ for $k = 1, \cdots, r$. Then the Smith form of $A(s)$ is given (cf. Theorem 5.1.1) by

$$\Sigma_A(s) = \mathrm{diag}\,(e_1(s), \cdots, e_r(s), 0, \cdots, 0).$$

Note that the coefficients of $d_k(s)$ and $e_k(s)$ are, in general, rational functions in $\mathcal{T}$ over $\boldsymbol{K}$, where $\mathcal{T}$ ($\subseteq \boldsymbol{F}$) denotes the set of the coefficients in the entries of $T(s)$.

Example 6.3.1. Consider a 2×2 mixed polynomial matrix $A(s) = Q(s) + T(s)$ with respect to $(\boldsymbol{K}, \boldsymbol{F}) = (\mathbf{Q}, \mathbf{Q}(\tau_1, \tau_2, \tau_3))$ given by

$$A(s) = \begin{pmatrix} s + \tau_1 & s + \tau_3 \\ 0 & \tau_2 s + 1 \end{pmatrix}, \quad Q(s) = \begin{pmatrix} s & s \\ 0 & 1 \end{pmatrix}, \quad T(s) = \begin{pmatrix} \tau_1 & \tau_3 \\ 0 & \tau_2 s \end{pmatrix}.$$

The Smith form of $A(s)$ is given by $\Sigma_A(s) = \mathrm{diag}\,[1,\ (s+\tau_1)(s+1/\tau_2)]$, which is true under the condition of algebraic independence of $\mathcal{T} = \{\tau_1, \tau_2, \tau_3\}$.

This expression, however, is no longer valid if specific numerical values are given to the parameters in $\mathcal{T}$. Namely, it can be verified easily that $\Sigma_A(s) = \mathrm{diag}\,[s+\tau_1,\ s+\tau_1]$ if $\tau_1 = 1/\tau_2 = \tau_3$, and otherwise the first expression is valid. It is emphasized that the theorems of this section deal with the generic situation where no algebraic relation exists among the parameters in $\mathcal{T}$. □

Before entering into the Smith form it is in order to point out a remarkable implication of the stronger condition on $Q(s)$:

(MP-Q2) Every nonvanishing subdeterminant of $Q(s)$ is a monomial over $\boldsymbol{K}$, i.e., of the form αs^p with $\alpha \in \boldsymbol{K}$ and an integer p,

introduced for physical-dimensional consistency. If $A(s) = Q(s) + T(s)$ is nonsingular, its determinant is a nonvanishing polynomial in $(s, \mathcal{T})$ over $\boldsymbol{K}$. The following lemma claims that $\det A(s)$ contains no (nonmonomial) polynomial in s free from $\mathcal{T}$ as an irreducible factor in $\boldsymbol{K}[s,\mathcal{T}]$. This fact affords a rich structure to the class of mixed polynomial matrices with the condition (MP-Q2) (see Remark 6.2.10 for another implication of (MP-Q2)).

Lemma 6.3.2. *For a nonsingular mixed polynomial matrix* $A(s) = Q(s) + T(s)$ *satisfying the stronger condition* (MP-Q2), *the decomposition of* $\det A(s)$ *into irreducible factors in* $\boldsymbol{K}[s,\mathcal{T}]$ *is expressed as*

$$\det A(s) = \alpha s^p \cdot \prod_k \psi_k(s,\mathcal{T}),$$

where $\alpha \in \boldsymbol{K} \setminus \{0\}$, *$p$ is a nonnegative integer, and* $\psi_k(s,\mathcal{T}) \in \boldsymbol{K}[s,\mathcal{T}] \setminus \boldsymbol{K}[s]$ *and* $\psi_k(s,\mathcal{T})$ *is irreducible in* $\boldsymbol{K}[s,\mathcal{T}]$ *for each* k. *Hence,* $\det A(z) = 0$ *for* z *algebraic over* $\boldsymbol{K}(\mathcal{T})$ *implies either* $z = 0$ *or* z *is transcendental over* $\boldsymbol{K}$.

Proof. By (6.4) we have

$$A(s) = \mathrm{diag}\,[s^{r_1}, \cdots, s^{r_n}] \cdot (Q(1) + \tilde{T}(s)) \cdot \mathrm{diag}\,[s^{-c_1}, \cdots, s^{-c_n}],$$

where $\tilde{T}(s) = \mathrm{diag}\,[s^{-r_1}, \cdots, s^{-r_n}] \cdot T(s) \cdot \mathrm{diag}\,[s^{c_1}, \cdots, s^{c_n}]$. For any nonzero number, say z, that is algebraic over $\boldsymbol{K}$, $Q(1) + \tilde{T}(z)$ is a mixed matrix with respect to $(\boldsymbol{K}, \boldsymbol{F}(z))$. Applications of Theorem 4.2.8 to $Q(1)+\tilde{T}(s)$ and $Q(1)+\tilde{T}(z)$ show that the nonsingularity of $A(s)$ implies that of $A(z)$. This means that $\det A(s)$ has no factor in $\boldsymbol{K}[s]$ except for a monomial in s. ■

The properties of the Smith form of $A(s)$ are stated in Theorems 6.3.3 and 6.3.4 below. The former refers to $e_k(s)$ for $k = 1, \cdots, r-1$, whereas the latter to $e_r(s)$. Recall the notation $\mathrm{ord}_s(\cdot)$ for the lowest degree of a nonvanishing term of a polynomial (cf. §2.1.1).

Theorem 6.3.3. *Let* $A(s) = Q(s) + T(s)$ *be a mixed polynomial matrix of rank* r *with respect to* $(\boldsymbol{K}, \boldsymbol{F})$. *For* $k = 1, \cdots, r-1$, *the* k*th monic determinantal divisor* $d_k(s)$ *and the* k*th monic invariant factor* $e_k(s)$ *of* $A(s)$ *contain* *no elements of* $\mathcal{T}$, *that is,* $d_k(s) \in \boldsymbol{K}[s]$ *and* $e_k(s) \in \boldsymbol{K}[s]$. *Moreover, if* $Q(s)$ *satisfies the stronger condition* (MP-Q2), *then they are monomials:*

$$d_k(s) = s^{p_k}, \quad e_k(s) = s^{p_k - p_{k-1}} \qquad (k = 1, \cdots, r-1) \tag{6.53}$$

with exponents given by

$$p_k = \min\{\mathrm{ord}_s \det A[I, J] \mid |I| = |J| = k\} \qquad (k = 1, \cdots, r-1), \tag{6.54}$$

where $p_0 = 0$ *by convention.*

Proof. The proof is given in §6.3.2. ■

To determine the last invariant factor $e_r(s)$ we associate with $A(s)$ an augmented $(2m) \times (m+n)$ matrix

$$\tilde{A}(s) = \tilde{A}(s;t) = \begin{pmatrix} I_m & Q(s) \\ -\mathrm{diag}\,(t_1, \cdots, t_m) & T(s) \end{pmatrix} = \begin{pmatrix} \tilde{Q}(s) \\ \tilde{T}(s;t) \end{pmatrix}, \tag{6.55}$$

where $t_1, \cdots, t_m$ are new indeterminates, $t = (t_1, \cdots, t_m)$, and

$$\tilde{Q}(s) = [I_m \mid Q(s)], \quad \tilde{T}(s;t) = [-\mathrm{diag}\,(t_1, \cdots, t_m) \mid T(s)].$$

Since

$$\begin{pmatrix} I_m & O \\ I_m & I_m \end{pmatrix} \begin{pmatrix} I_m & Q(s) \\ -I_m & T(s) \end{pmatrix} \begin{pmatrix} I_m & -Q(s) \\ O & I_n \end{pmatrix} = \begin{pmatrix} I_m & O \\ O & A(s) \end{pmatrix},$$

the Smith form of $\tilde{A}(s;1) = \tilde{A}(s;t)\Big|_{t_1 = \cdots = t_m = 1}$ is given as

$$\begin{pmatrix} I_m & O \\ O & \Sigma_A(s) \end{pmatrix} \tag{6.56}$$

in terms of the Smith form $\Sigma_A(s)$ of $A(s)$. In particular, the last invariant factor of $\tilde{A}(s;1)$ is equal to $e_r(s)$.

The augmented matrix $\tilde{A}(s;t)$ in (6.55) is an LM-matrix with respect to $(\boldsymbol{K}(s), \boldsymbol{F}(s,t))$. It can also be viewed as an LM-matrix with respect to $(\boldsymbol{K}[s], \boldsymbol{F}(s,t))$ for the integral domain $\boldsymbol{D} = \boldsymbol{K}[s]$ in the sense of §4.4.7, i.e., $\tilde{A}(s;t) \in \mathrm{LM}(\boldsymbol{K}[s], \boldsymbol{F}(s,t))$, since $\tilde{Q}(s)$ is a polynomial matrix over $\boldsymbol{K}$. The CCF of such an LM-matrix has been considered in Theorem 4.4.19.

Let $\bar{A} = \bar{A}(s;t)$ be the CCF of $\tilde{A}(s;t)$ in Theorem 4.4.19 and denote by $\{\bar{A}_l(s;t) \mid l = 0, 1, \cdots, b, \infty\}$ the family of the LM-irreducible diagonal blocks of $\bar{A}(s;t)$, where $\bar{A}_0 = \bar{A}_0(s;t)$ and $\bar{A}_\infty = \bar{A}_\infty(s;t)$ are the horizontal and the vertical tail, and $\bar{A}_l = \bar{A}_l(s;t)$ $(l = 1, \cdots, b)$ are the square LM-irreducible blocks; we put $R_l = \mathrm{Row}(\bar{A}_l)$ and $C_l = \mathrm{Col}(\bar{A}_l)$ for $l = 0, 1, \cdots, b, \infty$. Recall from Theorem 4.4.19 that there exists $\hat{A}(s;t)$ obtained from $\tilde{A}(s;t)$ through a unimodular transformation over $\boldsymbol{K}[s]$ such that $\hat{A}[R_l, C_j] = \bar{A}[R_l, C_j] = O$ for $l > j$ and $\hat{A}[R_l, C_l] = \bar{A}[R_l, C_l]$ for $l = 0, 1, \cdots, b, \infty$. In particular, the diagonal block $\bar{A}[R_l, C_l]$ consists of polynomials in $(s, t, \mathcal{T})$ over $\boldsymbol{K}$ for

$l = 0, 1, \cdots, b, \infty$. We denote by $\bar{d}_r^0(s)$ and $\bar{d}_r^\infty(s)$ the monic determinantal divisors in $\boldsymbol{F}(t)[s]$ of the largest order of $\bar{A}_0(s;t)$ and $\bar{A}_\infty(s;t)$, respectively.

The following theorem states that $e_r(s)$ is characterized by the diagonal blocks of $\bar{A}(s;t)$.

Theorem 6.3.4. *Let $A(s) = Q(s) + T(s)$ be a mixed polynomial matrix of rank r with respect to $(\boldsymbol{K}, \boldsymbol{F})$ and $\{\bar{A}_l(s;t) \mid l = 0, 1, \cdots, b, \infty\}$ be as above. The rth monic determinantal divisor $d_r(s)$ and the rth monic invariant factor $e_r(s)$ of $A(s)$ can be expressed as*

$$d_r(s) = \alpha_r \cdot d_r'(s) \cdot \prod_{l=1}^{b} \det \bar{A}_l(s;1), \quad e_r(s) = \alpha_r \cdot e_r'(s) \cdot \prod_{l=1}^{b} \det \bar{A}_l(s;1), \tag{6.57}$$

where $\alpha_r \in \boldsymbol{F}$, $d_r'(s) = \bar{d}_r^0(s) \cdot \bar{d}_r^\infty(s) \in \boldsymbol{K}[s]$, and $e_r'(s) \in \boldsymbol{K}[s]$. Moreover, if $Q(s)$ satisfies the stronger condition (MP-Q2), *it holds that*

$$\bar{d}_r^0(s) = s^{\bar{p}_r^0}, \qquad \bar{d}_r^\infty(s) = s^{\bar{p}_r^\infty}, \qquad e_r'(s) = s^{\bar{p}_r^0 + \bar{p}_r^\infty - p_{r-1}}, \tag{6.58}$$

where

$$\bar{p}_r^0 = \min\{\mathrm{ord}_s \det \bar{A}_0(s;t)[R_0, J] \mid |J| = |R_0|\}, \tag{6.59}$$
$$\bar{p}_r^\infty = \min\{\mathrm{ord}_s \det \bar{A}_\infty(s;t)[I, C_\infty] \mid |I| = |C_\infty|\}, \tag{6.60}$$

and p_{r-1} is given by (6.54).

Proof. The proof is given in §6.3.2. ■

As a corollary to this theorem we can identify those parameters in $\mathcal{T}$ which affect the Smith form of $A(s)$.

Theorem 6.3.5. *The rth monic invariant factor of $A(s)$ depends on $\tau \in \mathcal{T}$ if and only if τ is contained in some square LM-irreducible block $\bar{A}_l(s;t)$ of the CCF of $\tilde{A}(s;t)$ such that* $\det \bar{A}_l(s;t)$ *is not a monomial in s over $\boldsymbol{F}$.*

Proof. Recall the expression of $e_r(s)$ in Theorem 6.3.4. All the parameters of $\mathcal{T}$ contained in $\bar{A}_l(s;1)$ with $1 \le l \le b$ appear in $\det \bar{A}_l(s;1)$ by Theorem 4.5.4. They remain after the normalization by $\alpha_r \in \boldsymbol{F}$ if and only if $\det \bar{A}_l(s;1)$ is not a monomial in s over $\boldsymbol{F}$. ■

Remark 6.3.6. The monomiality of $\det \bar{A}_l(s;1)$ in Theorem 6.3.5 can be checked efficiently by computing $\deg_s \det \bar{A}_l(s;1)$ and $\mathrm{ord}_s \det \bar{A}_l(s;1)$ by the algorithm of §6.2, since $\det \bar{A}_l(s;1)$ is monomial in s if and only if $\deg_s \det \bar{A}_l(s;1) = \mathrm{ord}_s \det \bar{A}_l(s;1)$. See §6.4.4 for a concrete procedure for this idea with an additional improvement. □

Remark 6.3.7. In case $A(s)$ is an LM-matrix, there is no need to introduce the augmented LM-matrix $\tilde{A}(s;t)$. The claims in Theorems 6.3.4 and 6.3.5 remain valid when we redefine $\bar{A}$ to be the CCF of $A(s) \in \mathrm{LM}(\boldsymbol{K}[s], \boldsymbol{F}(s))$ such as in Theorem 4.4.19. □

Let us specialize Theorems 6.3.3 and 6.3.4 to a generic polynomial matrix $A(s)$ (of which all the nonzero coefficients are algebraically independent). Such $A(s)$ is a mixed polynomial matrix with $Q(s) = O$, satisfying the stronger assumption (MP-Q2) trivially.

Theorem 6.3.8. *Let $A(s)$ be a generic polynomial matrix of rank r, and $\{\bar{A}_l(s) \mid l = 0, 1, \cdots, b, \infty\}$ be the DM-components of $A(s)$. For $k = 1, \cdots, r-1$, the kth monic determinantal divisor $d_k(s)$ and the kth monic invariant factor $e_k(s)$ of $A(s)$ are monomials given by* (6.53) *and* (6.54). *The rth monic determinantal divisor $d_r(s)$ and the rth monic invariant factor $e_r(s)$ of $A(s)$ are given by* (6.57)–(6.60), *in which $\bar{A}_l(s;t)$ is replaced by $\bar{A}_l(s)$.*

Proof. In addition to Theorems 6.3.3 and 6.3.4, note Remark 6.3.7 and the fact that the CCF reduces to the DM-decomposition in this special case. ■

Remark 6.3.9. The present results on the Smith form, Theorems 6.3.4 and 6.3.8 in particular, will find direct applications in the argument on structural controllability in §6.4. □

Example 6.3.10. The theorems above are illustrated here for a 5×5 matrix

$$A(s) = \begin{array}{c} \\ w_1 \\ w_2 \\ w_3 \\ w_4 \\ w_5 \end{array} \begin{array}{c} \begin{array}{ccccc} x_1 & x_2 & x_3 & x_4 & x_5 \end{array} \\ \begin{pmatrix} 0 & 0 & 1+\tau_1 s & 3s & \tau_2 \\ s & 1 & 1 & 0 & \tau_3+\tau_4 s \\ 2s^2 & 2s & 2s & 0 & \tau_5 s \\ 0 & 0 & 0 & s^2 & \tau_6 \\ 2s^3 & 2s^2 & 2s^2 & 0 & s+\tau_7 s^2 \end{pmatrix} \end{array}.$$

This is a mixed polynomial matrix with respect to $(\boldsymbol{K}, \boldsymbol{F}) = (\mathbf{Q}, \mathbf{Q}(\mathcal{T}))$ for $\mathcal{T} = \{\tau_1, \tau_2, \tau_3, \tau_4, \tau_5, \tau_6, \tau_7\}$, admitting the decomposition $A(s) = Q(s) + T(s)$ with

$$Q(s) = \begin{pmatrix} 0 & 0 & 1 & 3s & 0 \\ s & 1 & 1 & 0 & 0 \\ 2s^2 & 2s & 2s & 0 & 0 \\ 0 & 0 & 0 & s^2 & 0 \\ 2s^3 & 2s^2 & 2s^2 & 0 & s \end{pmatrix}, \quad T(s) = \begin{pmatrix} 0 & 0 & \tau_1 s & 0 & \tau_2 \\ 0 & 0 & 0 & 0 & \tau_3+\tau_4 s \\ 0 & 0 & 0 & 0 & \tau_5 s \\ 0 & 0 & 0 & 0 & \tau_6 \\ 0 & 0 & 0 & 0 & \tau_7 s^2 \end{pmatrix}.$$

Note that $Q(s)$ satisfies the stronger condition (MP-Q2) with

$$(r_1, \cdots, r_5) = (1, 1, 2, 2, 3), \qquad (c_1, \cdots, c_5) = (0, 1, 1, 0, 2)$$

in (6.4). The augmented LM-matrix $\tilde{A}(s;t)$ of (6.55) is given by

$$
\tilde{A}(s;t) = \begin{array}{c}
\begin{array}{cccccccccc} w_1 & w_2 & w_3 & w_4 & w_5 & x_1 & x_2 & x_3 & x_4 & x_5 \end{array} \\
\left[\begin{array}{ccccc|ccccc}
1 & & & & & 0 & 0 & 1 & 3s & 0 \\
 & 1 & & & & s & 1 & 1 & 0 & 0 \\
 & & 1 & & & 2s^2 & 2s & 2s & 0 & 0 \\
 & & & 1 & & 0 & 0 & 0 & s^2 & 0 \\
 & & & & 1 & 2s^3 & 2s^2 & 2s^2 & 0 & s \\
\hline
-t_1 & & & & & 0 & 0 & \tau_1 s & 0 & \tau_2 \\
 & -t_2 & & & & 0 & 0 & 0 & 0 & \tau_3 + \tau_4 s \\
 & & -t_3 & & & 0 & 0 & 0 & 0 & \tau_5 s \\
 & & & -t_4 & & 0 & 0 & 0 & 0 & \tau_6 \\
 & & & & -t_5 & 0 & 0 & 0 & 0 & \tau_7 s^2
\end{array}\right]
\end{array}.
$$

The CCF $\bar{A}(s;t)$ of $\tilde{A}(s;t)$, of which the diagonal blocks are matrices over $\mathbf{Q}[s,t,\mathcal{T}]$, is given by

$$
\bar{A}(s;t) = \begin{array}{c}
\begin{array}{cc|cc|c|c|cccc} \multicolumn{2}{c}{C_0} & \multicolumn{2}{c}{C_1} & C_2 & C_3 & \multicolumn{4}{c}{C_\infty} \\ x_1 & x_2 & w_1 & x_3 & x_4 & w_4 & w_3 & w_5 & w_2 & x_5 \end{array} \\
\left[\begin{array}{cc|cc|c|c|cccc}
s & 1 & & 1 & & & & & 1 & \\
\hline
 & & 1 & 1 & & -3/s & & & & \\
 & & -t_1 & \tau_1 s & & & & & & \tau_2 \\
\hline
 & & & & s^2 & 1 & & & & \\
\hline
 & & & & & -t_4 & & & & \tau_6 \\
\hline
 & & & & & & 1 & 0 & -2s & 0 \\
 & & & & & & 0 & 1 & -2s^2 & s \\
 & & & & & & -t_3 & 0 & 0 & \tau_5 s \\
 & & & & & & 0 & -t_5 & 0 & \tau_7 s^2 \\
 & & & & & & 0 & 0 & -t_2 & \tau_3 + \tau_4 s
\end{array}\right]
\end{array}.
$$

Note that the diagonal blocks are polynomial matrices in s whereas a fraction "$-3/s$" is contained in an off-diagonal block. The CCF consists of nonempty tails and three square diagonal blocks. The CCF reveals that $r = \operatorname{rank} A(s) = 4$ (< 5). According to Theorem 6.3.4 we see that

$$d_4(s) = \alpha_4 \cdot s^{\bar{p}_4^0 + \bar{p}_4^\infty} \cdot (\tau_1 s + 1) \cdot s^2 \cdot (-1)$$

with $\bar{p}_4^0 = 0$ and $\bar{p}_4^\infty = 1$, where $\alpha_4 = -1/\tau_1$ to make $d_4(s)$ a monic polynomial. Therefore,

$$d_4(s) = s^3 \cdot (s + 1/\tau_1).$$

As for the other determinantal divisors, we obtain $d_1(s) = d_2(s) = d_3(s) = 1$ from (6.53) and (6.54). Hence the Smith form $\Sigma_A(s)$ of $A(s)$ is given by

$$\Sigma_A(s) = \operatorname{diag}[1, 1, 1, s^3(s + 1/\tau_1), 0].$$

Note that τ_1, contained in $\bar{A}_1(s;t)$, is the only member of $\mathcal{T}$ that appears in $\Sigma_A(s)$, as predicted by Theorem 6.3.5.

Finally, we mention the matrix $\hat{A}(s;t)$. We choose, for instance,

$$\{x_1, x_2\} \prec \{w_1, x_3\} \prec \{x_4\} \prec \{w_4\} \prec \{w_3, w_5, w_2, x_5\}$$

as a linear extension of the partial order in the CCF of $\tilde{A}(s;t)$. Then

$$\hat{A}(s;t) = P_{\rm r} \begin{pmatrix} L(s) & O \\ O & I_5 \end{pmatrix} \tilde{A}(s;t) P_{\rm c}, \quad \bar{A}(s;t) = P_{\rm r} \begin{pmatrix} U(s)L(s) & O \\ O & I_5 \end{pmatrix} \tilde{A}(s;t) P_{\rm c}$$

with

$$L(s) = \begin{pmatrix} 1 & & & & \\ & 1 & & & \\ & -2s & 1 & & \\ & & & 1 & \\ & -2s^2 & & & 1 \end{pmatrix}, \quad U(s) = \begin{pmatrix} 1 & & & -3/s & \\ & 1 & & & \\ & & 1 & & \\ & & & 1 & \\ & & & & 1 \end{pmatrix}$$

and permutation matrices $P_{\rm r}$ and $P_{\rm c}$. Note that $L(s)$ is a unimodular polynomial matrix in s over $\mathbf{Q}$. The block-triangular matrix $\hat{A}(s;t)$ is given by

$$\hat{A}(s;t) = \begin{array}{cc|cc|c|c|cccc} \multicolumn{2}{c}{C_0} & \multicolumn{2}{c}{C_1} & C_2 & C_3 & & & C_\infty & \\ x_1 & x_2 & w_1 & x_3 & x_4 & w_4 & w_3 & w_5 & w_2 & x_5 \\ \hline s & 1 & & 1 & & & & & 1 & \\ \hline & & 1 & 1 & 3s & & & & & \\ & & -t_1 & \tau_1 s & & & & & & \tau_2 \\ \hline & & & & s^2 & 1 & & & & \\ \hline & & & & & -t_4 & & & & \tau_6 \\ \hline & & & & & & 1 & 0 & -2s & 0 \\ & & & & & & 0 & 1 & -2s^2 & s \\ & & & & & & -t_3 & 0 & 0 & \tau_5 s \\ & & & & & & 0 & -t_5 & 0 & \tau_7 s^2 \\ & & & & & & 0 & 0 & -t_2 & \tau_3 + \tau_4 s \end{array} .$$

As claimed, the matrix $\hat{A}$ is a polynomial matrix in s and it agrees with $\bar{A}$ in the diagonal blocks. Also notice the difference between the zero/nonzero structures of $\bar{A}$ and $\hat{A}$. In particular, we can exchange the positions of the two blocks $\{w_1, x_3\}$ and $\{x_4\}$ in $\bar{A}$ without destroying the block-triangular structure if we accordingly exchange the corresponding rows, whereas these two blocks must be arranged in this order in $\hat{A}$ to put it in an explicit block-triangular form. In other words, the square diagonal blocks are partially ordered as $\{w_1, x_3\} \prec \{w_4\}$, $\{x_4\} \prec \{w_4\}$ with respect to the zero/nonzero structure in $\bar{A}$, whereas they are totally ordered as $\{w_1, x_3\} \prec \{x_4\} \prec \{w_4\}$ in $\hat{A}$. □

Remark 6.3.11. It is natural to ask whether the Smith form of $A(s) = Q(s) + T(s)$ can be computed efficiently. This problem has been solved in two special cases. If $T(s) = O$, then $A(s)$ is simply a polynomial matrix over $\boldsymbol{K}$, for which Kannan [154] proposes a polynomial time algorithm. If $Q(s)$ satisfies the stronger condition (MP-Q2), which is trivially true in the other extreme case of $Q(s) = O$, an efficient (polynomial-time) matroid-theoretic algorithm of §6.2 is available. □

6.3.2 Proofs

A minor (subdeterminant) of $A(s) = A(s, \mathcal{T})$, is a polynomial in s and $\mathcal{T}$ over $\boldsymbol{K}$. Let $d_k^*(s, \mathcal{T}) \in \boldsymbol{K}[s, \mathcal{T}]$ denote the kth determinantal divisor of A, i.e., the greatest common divisor of all minors of order k in A as polynomials in $(s, \mathcal{T})$ over $\boldsymbol{K}$. Theorem 6.3.3 follows from the following lemma as well as Lemma 6.3.2.

Lemma 6.3.12. *$d_{r-1}^*(s, \mathcal{T}) \in \boldsymbol{K}[s]$, that is, no $\tau \in \mathcal{T}$ appears in d_{r-1}^*.*

Proof. Since $r = \operatorname{rank} A$, there exists a nonsingular submatrix $A[I, J]$ with $|I| = |J| = r$. For $\tau \in \mathcal{T}$ let (i, j) denote the position at which τ appears in A. If τ does not appear in $\delta = \det A[I, J]$ $(\neq 0)$, then d_{r-1}^* is free from τ since d_{r-1}^* divides δ. If τ does appear in δ, then $i \in I$ and $j \in J$ and furthermore $\delta' = \det A[I \setminus \{i\}, J \setminus \{j\}] \neq 0$. Obviously, δ' does not contain τ and hence d_{r-1}^* is free from τ since d_{r-1}^* divides δ'. ■

We now turn to the proof of Theorem 6.3.4. Firstly, $\tilde{A}(s;t)$ and $\hat{A}(s;t)$ share the same Smith form, since they are connected by a unimodular transformation. Secondly, the Smith form $\Sigma_A(s)$ of $A(s)$ can be obtained from that of $\tilde{A}(s;1)$ by (6.56). The following lemma claims that $\tilde{A}(s;t)$ and $\tilde{A}(s;1)$ have essentially identical Smith forms. We write $\tilde{A}(s;t,\mathcal{T})$ for $\tilde{A}(s;t)$ to explicitly indicate its dependence on the coefficients $\mathcal{T}$ in $T(s)$.

Lemma 6.3.13. *The Smith form of $\tilde{A}(s;1,\mathcal{T})$, as a matrix over $\boldsymbol{F}[s]$, is obtained from that of $\tilde{A}(s;t,\mathcal{T})$, as a matrix over $\boldsymbol{F}(t)[s]$, by setting $t_1 = \cdots = t_m = 1$. Conversely, the Smith form of $\tilde{A}(s;t,\mathcal{T})$ is obtained from that of $\tilde{A}(s;1,\mathcal{T})$ by replacing $\tau \in \mathcal{T}$ with τ/t_i if τ is contained in the ith row.* □

This allows us to concentrate on the Smith form of $\hat{A}(s;t)$. Regarding $\hat{A}(s;t) = \hat{A}(s;t,\mathcal{T})$ as a matrix over the ring $\boldsymbol{K}[s,t,\mathcal{T}]$, we denote by $\hat{d}_{r+m}(s;t)$ $(\in \boldsymbol{K}[s,t,\mathcal{T}])$ the $(r+m)$th determinantal divisor of $\hat{A}(s;t)$. Then

$$d_r(s) = \alpha_r \cdot \hat{d}_{r+m}(s;1), \tag{6.61}$$

where $\alpha_r \in \boldsymbol{K}(\mathcal{T}) \subseteq \boldsymbol{F}$ is introduced for normalization to a monic polynomial in $\boldsymbol{F}[s]$. Since $\hat{A}$ is block-triangularized with full-rank diagonal blocks (cf. Theorem 4.4.19), a nonvanishing minor of $\hat{A}$ of order $r+m$ is expressed as

$$\begin{aligned} &\det \hat{A}[R_0, J] \cdot \det \hat{A}[I, C_\infty] \cdot \prod_{l=1}^{b} \det \hat{A}[R_l, C_l] \\ &= \det \bar{A}[R_0, J] \cdot \det \bar{A}[I, C_\infty] \cdot \prod_{l=1}^{b} \det \bar{A}[R_l, C_l] \end{aligned} \tag{6.62}$$

for $J \subseteq C_0$ and $I \subseteq R_\infty$. Then Theorem 6.3.4 follows from (6.61) and (6.62) and the lemma below, where $\gcd_{\boldsymbol{K}[s,t,\mathcal{T}]}\{\cdot\}$ denotes the greatest common divisor in the ring $\boldsymbol{K}[s,t,\mathcal{T}]$.

Lemma 6.3.14.

$$\gcd{}_{\boldsymbol{K}[s,t,\mathcal{T}]}\{\det \bar{A}[R_0, J] \mid |J| = |R_0|,\ J \subseteq C_0\} \in \boldsymbol{K}[s],$$
$$\gcd{}_{\boldsymbol{K}[s,t,\mathcal{T}]}\{\det \bar{A}[I, C_\infty] \mid |I| = |C_\infty|,\ I \subseteq R_\infty\} \in \boldsymbol{K}[s].$$

Proof. This follows from Theorem 4.5.8. ■

Notes. This section is based on Murota [213, 216].

6.4 Controllability of Dynamical Systems

Structural controllability of a control system is investigated using mixed polynomial matrices. A necessary and sufficient condition for structural controllability is given in terms of the CCF of an associated LM-matrix, along with an efficient algorithm for testing it. As a special case, the structural controllability of a descriptor system is expressed in terms of the Dulmage–Mendelsohn decomposition of an associated bipartite graph.

6.4.1 Controllability

Controllability is one of the most fundamental characteristics for a control system. A linear time-invariant dynamical system in the standard form

$$\frac{\mathrm{d}\boldsymbol{x}}{\mathrm{d}t} = A\boldsymbol{x} + B\boldsymbol{u}, \tag{6.63}$$

where $\boldsymbol{x} = \boldsymbol{x}(t) \in \mathbf{R}^n$ is the state-vector and $\boldsymbol{u} = \boldsymbol{u}(t) \in \mathbf{R}^m$ is the input-vector, is said to be *controllable* if any initial state $\boldsymbol{x}_0 = \boldsymbol{x}(0)$ can be brought to any prescribed final state $\boldsymbol{x}_f = \boldsymbol{x}(t_f)$ in a finite time t_f by suitably chosen input $\boldsymbol{u}(t)$ $(0 \le t \le t_f)$.

The following characterizations of *controllability* are well known (see Kailath [152], Rosenbrock [284], Wolovich [342], Wonham [343]).

Theorem 6.4.1. *The following five conditions are equivalent.*

(i) *The system* (6.63) *is controllable.*

(ii) $\operatorname{rank}[B \mid AB \mid A^2B \mid \cdots \mid A^{n-1}B] = n.$

(iii) *The* $n^2 \times n(n+m-1)$ *matrix*

$$\bar{D} = \begin{bmatrix} B & -I_n & & & & & & \\ O & A & B & -I_n & & & & \\ & & & A & B & -I_n & & \\ & & & & \ddots & \ddots & \ddots & \\ & & & & & A & B & -I_n & O \\ & & & & & & & A & B \end{bmatrix} \tag{6.64}$$

is of rank n^2.

(iv) $\operatorname{rank}[A - zI_n \mid B] = n$ *for any complex number* z.

(v) *The nth monic determinantal divisor of* $[A - sI_n \mid B]$ *is equal to 1.* □

The $n \times nm$ matrix $[B \mid AB \mid A^2B \mid \cdots \mid A^{n-1}B]$ in (ii) above is called the *controllability matrix*, while $[A - sI_n \mid B]$ in (v) the *modal controllability matrix*.

Controllability concept can be defined for a system of descriptor form

$$F \frac{\mathrm{d}\boldsymbol{x}}{\mathrm{d}t} = A\boldsymbol{x} + B\boldsymbol{u} \tag{6.65}$$

in a number of different ways (see, for example, Kodama–Ikeda [162], Pandolfi [263], Hayakawa–Hosoe–Ito [108], Verghese–Lévy–Kailath [330], Yip–Sincovec [349], Cobb [36]), where the matrix F is square ($n \times n$) but not necessarily nonsingular. We define the descriptor system (6.65) to be controllable if

$$\mathrm{rank}\,[A - zF \mid B] = n \quad \text{for any complex number } z. \tag{6.66}$$

It should be obvious that this is equivalent to saying that the nth monic determinantal divisor of $[A - sF \mid B]$ is equal to 1.

The significance of this definition of controllability can be understood with reference to the canonical decomposition of a descriptor system explained in Remark 5.1.9. It is easy to see from Theorem 6.4.1 that the descriptor system (6.65) is controllable if and only if the subsystem (5.11) derived from it is controllable in the ordinary sense. In other words, the present definition of controllability means the controllability of the exponential modes, agreeing with the notion of "R-controllability" of Yip–Sincovec [349].

For later references, we put together the relevant conditions:

(C1) $\det(A - sF) \neq 0$,
(C2) $\mathrm{rank}\,[A \mid B] = n$,
(C3) $\mathrm{rank}\,[A - zF \mid B] = n$ for any $z \in \mathbf{C}$, $z \neq 0$,

where it should be evident that (C2) and (C3) together are equivalent to the controllability condition (6.66). The condition (C2) is for the controllability of zero mode, whereas (C3) for nonzero modes. We often use the notation

$$D(s) = \left[\, A - sF \mid B \,\right]. \tag{6.67}$$

6.4.2 Structural Controllability

The notion of "structural controllability" was first introduced by Lin [173] along with its graphical condition for single-input systems, followed by subsequent extensions to multi-input systems by Shields–Pearson [295], Glover–Silverman [94], Davison [45], Hosoe–Matsumoto [115], and Maeda [184]. It also motivated many related works (e.g., Kobayashi–Yoshikawa [161], Maeda–Yamada [185], Hosoe [113], Linnemann [174], Murota–Poljak [240], Reinschke [279], Yamada–Foulds [345], Yamada–Luenberger [346, 347, 348]). This section is devoted to a sketch of the graph-theoretic approach to structural controllability with particular emphasis on the comparison of different graph representations, signal-flow graphs and dynamic graphs for systems in standard form, and bipartite graphs for systems in descriptor form. The reader is referred to Murota [204, Chap. 3], Reinschke [279, Chap. 1], and Šiljak [300, Chap. 1] for more details on the graph-theoretic approach to structural controllability.

Let us first consider the conventional case of the standard form (6.63). Associated with a particular instance of the system (6.63) with the entries of A and B being given concrete real numbers, we consider the *structured system*, which is described by the same state-space equations as the original system except that the nonvanishing entries of the coefficient matrices A and B are replaced by independent parameters (or indeterminates). Then the original system is said to be *structurally controllable* if the associated structured system is generically controllable with respect to those parameters, i.e., if it is controllable (in the sense of Theorem 6.4.1) for those parameter values which lie outside some proper algebraic variety in the parameter space.

It is easy to see that the structural controllability is equivalent to the condition that the generic-rank of the controllability matrix is equal to n, i.e.,

$$\text{generic-rank}\,[B \mid AB \mid A^2B \mid \cdots \mid A^{n-1}B] = n \tag{6.68}$$

with respect to those parameters. Note that the generic-rank of the controllability matrix of the structured system is equal to the rank of the controllability matrix with parameter values fixed to arbitrary transcendental numbers which are algebraically independent over the rational number field $\mathbf{Q}$, since each entry of the controllability matrix is a polynomial (with coefficients from $\mathbf{Q}$) in those parameters. Note also that the condition (6.68) is not equivalent to

$$\text{term-rank}\,[B \mid AB \mid A^2B \mid \cdots \mid A^{n-1}B] = n.$$

The following is the fundamental result (Lin [173], Shields–Pearson [295], Glover–Silverman [94], Maeda [184]) giving a graph-theoretic characterization of the structural controllability in terms of the signal-flow graph $G = (V, E)$ associated with (6.63). Recall from §2.2.1 that the vertex set V and the arc set E are defined by

$$\begin{aligned} V &= X \cup U, \qquad X = \{x_1, \cdots, x_n\}, \quad U = \{u_1, \cdots, u_m\}, \\ E &= \{(x_j, x_i) \mid A_{ij} \neq 0\} \cup \{(u_j, x_i) \mid B_{ij} \neq 0\}. \end{aligned}$$

By a *stem* we mean a directed path in G with its initial vertex belonging to U.

Theorem 6.4.2. *A system in the standard form* (6.63) *is structurally controllable if and only if the signal-flow graph G satisfies both* (a) *and* (b) *below:*

(a) *There exists a set of mutually disjoint cycles and stems such that all the vertices in X are covered,*

(b) *Any vertex* x_i $(\in X)$ *is reachable by a directed path from some* u_j $(\in U)$, *i.e.*, $u_j \overset{*}{\longrightarrow} x_i$ *on* G.

Proof. This theorem is derived later from a more general result in Remark 6.4.9. See also Shields–Pearson [295], Glover–Silverman [94], Davison [45], Hosoe–Matsumoto [115], and Maeda [184], as well as Murota [204, Chap. 3], Šiljak [300, Chap. 1], and Linnemann [176]. ∎

A system is said to be *reachable* if it satisfies the condition (b) above, namely, if any vertex x_i $(\in X)$ is reachable by a directed path from some u_j $(\in U)$ in the signal-flow graph G.

The structural controllability can be characterized also in terms of the dynamic graph, as is observed by Murota [204, Theorem 15.1]. Recall from §2.2.1 (see also Example 2.2.4) that for a system (6.63) the dynamic graph of time-span n is defined to be $G_0^n = (X_0^n \cup U_0^{n-1}, E_0^{n-1})$ with

$$\begin{aligned}
X_0^n &= \bigcup_{t=0}^{n} X^t, \quad X^t = \{x_i^t \mid i = 1, \cdots, n\} \quad (t = 0, 1, \cdots, n), \\
U_0^{n-1} &= \bigcup_{t=0}^{n-1} U^t, \quad U^t = \{u_j^t \mid j = 1, \cdots, m\} \quad (t = 0, 1, \cdots, n-1), \\
E_0^{n-1} &= \{(x_j^t, x_i^{t+1}) \mid A_{ij} \neq 0; t = 0, 1, \cdots, n-1\} \\
&\quad \cup \{(u_j^t, x_i^{t+1}) \mid B_{ij} \neq 0; t = 0, 1, \cdots, n-1\}.
\end{aligned}$$

Theorem 6.4.3. *A system in the standard form* (6.63) *is structurally controllable if and only if there exists in the dynamic graph* G_0^n *of time-span* n *a Menger-type vertex-disjoint linking of size* n *from* U_0^{n-1} *to* X^n.

Proof. This follows from Theorem 6.4.4 below. ∎

The *generic dimension of the controllable subspace* means the generic rank of the controllability matrix, i.e., $\mathrm{rank}\,[B \mid AB \mid \cdots \mid A^{n-1}B]$ when the nonzero entries of A and B are algebraically independent parameters. A system is structurally controllable if and only if the generic dimension of the controllable subspace is equal to n.

The following result of Poljak [271] is an extension of Theorem 6.4.3 above.

Theorem 6.4.4. *For a reachable system* (6.63), *the generic dimension of the controllable subspace is equal to the maximum size of a Menger-type vertex-disjoint linking from* U_0^{n-1} *to* X^n *in the dynamic graph* G_0^n *of time-span* n.

Proof. The proof is based on the max-flow min-cut theorem (Theorem 2.2.30) and Theorem 6.4.5 below. See Poljak [271] for the detail. ∎

The following theorem, due to Hosoe [113], is a fundamental result on the generic dimension of the controllable subspace. For $X' \subseteq X = \{x_1, \cdots, x_n\}$ we use the notation $[A[X', X'] \mid B[X', U]]$ to mean the $|X'| \times (|X'| + m)$

matrix formed with the submatrices $A[X',X']$ and $B[X',U]$, where $U = \{u_1,\cdots,u_m\}$.

Theorem 6.4.5. *For a reachable system* (6.63), *the generic dimension of the controllable subspace is equal to*

$$\max\{|X'| \mid \text{term-rank}\,([A[X',X'] \mid B[X',U]]) = |X'|,\ X' \subseteq X\}. \qquad \square$$

Remark 6.4.6. Under the reachability assumption, Theorem 6.4.4 implies

$$\text{term-rank}\,\bar{D} = \text{generic-rank}\,\bar{D}$$

for the matrix $\bar{D}$ in (6.64), where the generic-rank is defined with respect to the nonzero entries of A and B. Note that generic-rank $\bar{D} = n(n-1) +$ generic-rank $[B \mid AB \mid \cdots \mid A^{n-1}B]$ and that term-rank $\bar{D}$ equals to $n(n-1)$ plus the maximum size of a Menger-type vertex-disjoint linking from U_0^{n-1} to X^n in G_0^n. The former can be shown by row elimination on $\bar{D}$ and the latter by the linkage lemma (cf. Murota [204, Prop. 7.1], Welsh [333, Chap. 13, §1]). $\square$

In the remainder of this section we consider structural controllability for descriptor systems, following Murota [199]. As has been discussed in §1.2.2, the descriptor form (6.65) is a more elementary description than the standard form (6.63), and hence will be more suitable for structural analysis. We define a descriptor system (6.65) to be *structurally solvable* if the condition (C1) in §6.4.1 is satisfied under the assumption that the nonvanishing entries of the coefficient matrices F, A, and B are algebraically independent over $\mathbf{Q}$. A descriptor system (6.65) is said to be structurally controllable if, in addition, the conditions (C2) and (C3) in §6.4.1 are satisfied under the same assumption.

We shall derive a necessary and sufficient graph-theoretic condition for the structural controllability. As has been discussed in §2.2.1, the natural graph representation of a descriptor system is a bipartite graph, and accordingly, it will be reasonable to aim at establishing a graph-theoretic condition on the bipartite graph for the structural controllability.

Let $G = (V^+, V^-; E)$ be the bipartite graph associated with the descriptor system (6.65). Namely, $V^+ = X \cup U = \{x_1,\cdots,x_n\} \cup \{u_1,\cdots,u_m\}$, $V^- = \{e_1,\cdots,e_n\}$, and $E = E_A \cup E_F \cup E_B$ with $E_A = \{(x_j,e_i) \mid A_{ij} \neq 0\}$, $E_F = \{(x_j,e_i) \mid F_{ij} \neq 0\}$, and $E_B = \{(u_j,e_i) \mid B_{ij} \neq 0\}$. No parallel arcs are introduced even if $E_A \cap E_F \neq \emptyset$. We call an arc an s-*arc* if it belongs to E_F.

The first two conditions (C1) and (C2) are easy to handle. Let G_{A-sF} and $G_{[A|B]}$ denote the bipartite graphs associated with $A - sF$ and $[A \mid B]$, respectively. Namely, $G_{A-sF} = (X, V^-; E_A \cup E_F)$ and $G_{[A|B]} = (X \cup U, V^-; E_A \cup E_B)$. Then, by Proposition 2.2.25, we have

$$\text{(C1)} \iff \text{term-rank}\,(A - sF) = n \iff \nu(G_{A-sF}) = n, \qquad (6.69)$$

$$\text{(C2)} \iff \text{term-rank}\,[A \mid B] = n \iff \nu(G_{[A|B]}) = n, \qquad (6.70)$$

where $\nu(\cdot)$ denotes the size of a maximum matching in a bipartite graph. For the third condition (C3) we consider the DM-decomposition of G. Let $G_k = (V_k^+, V_k^-; E_k)$ $(k = 0, 1, \cdots, b, \infty)$ be the DM-components of $G = (V^+, V^-; E)$. In this notation, $k = 0$ and $k = \infty$ designate the horizontal tail and the vertical tail, respectively, though the vertical tail does not exist (is empty) under the condition (C1) or (C2).

A necessary and sufficient graph-theoretic condition for the structural controllability of a descriptor system is given as follows (Murota [199]).

Theorem 6.4.7. *A descriptor system* (6.65) *is structurally solvable if and only if*

(B1) $\nu(G_{A-sF}) = n$.

It is structurally controllable if and only if the following two conditions (B2) *and* (B3) *hold in addition to* (B1)*:*

(B2) $\nu(G_{[A|B]}) = n$,
(B3) *None of the consistent DM-components* G_k $(k = 1, \cdots, b)$ *of the bipartite graph* G *contain* s*-arcs.*

Proof. This follows from (6.69), (6.70), and Theorem 6.3.8 as well as Theorem 6.4.1(v). See Murota [204, §14.2] for an alternative proof. ■

In the particular case with nonsingular F, Theorem 6.4.7 reduces to the following, which makes no reference to the DM-decomposition.

Corollary 6.4.8. *A descriptor system* (6.65) *with* term-rank $F = n$ *is structurally controllable if and only if the following two conditions* (B2) *and* (B4) *hold:*

(B2) $\nu(G_{[A|B]}) = n$,
(B4) $\nu(G \setminus \{x_j\}) = n$ *for any* $x_j \in X$.

Proof. Since term-rank $F = n$, the condition (B3) is satisfied if and only if the whole graph G is the horizontal tail. The latter condition is equivalent to (B4) by Corollary 2.2.23(2). ■

Remark 6.4.9. Theorem 6.4.2 can be derived from Corollary 6.4.8. First observe that the system in the standard form (6.63) is structurally controllable if and only if so is the descriptor system (6.65) with the same A and B, and a nonsingular diagonal F. The condition (a) in Theorem 6.4.2 is easily seen to be equivalent to (B2). According to the algorithm for the DM-decomposition in §2.2.3, the condition (b) in Theorem 6.4.2 is equivalent to saying that the whole graph G is the horizontal tail, which is equivalent to (B4) by Corollary 2.2.23(2). □

The conditions given in Theorem 6.4.7 can be checked efficiently with $\mathrm{O}((m+n)^{5/2})$ graph manipulations as follows. (B1) and (B2) may be checked by finding maximum matchings in G_{A-sF} and $G_{[A|B]}$, respectively. Suppose (B1) is satisfied and there exists in G_{A-sF} a perfect matching, say M. It is also a maximum matching in G. Let $G_M = (V^+ \cup V^-, E \cup M^\circ)$ be the auxiliary graph associated with the matching M in G, where M° is the set of reorientations of the arcs in M, and define G' to be the subgraph of G_M which is obtained from G_M by deleting all the vertices reachable from U. Then (B3) is equivalent to the condition that none of the strong components of G' contain s-arcs.

Remark 6.4.10. Graph-theoretic conditions for the structural controllability of a descriptor system were first given, almost simultaneously, by Aoki–Hosoe–Hayakawa [7] and Matsumoto–Ikeda [187] with different expressions. However, both of these graph-theoretic conditions lack in the natural invariance, being expressed in terms of noninvariant graph representations as follows. Aoki–Hosoe–Hayakawa [7] uses a graph $\hat{G} = (V, E)$ that has vertex set $V = X \cup U$ and the arc set

$$E = \{(x_j, x_i) \mid (A - sF)_{ij} \neq 0\} \cup \{(u_j, x_i) \mid B_{ij} \neq 0\}.$$

The graph $\hat{G}$ thus defined is not unique in that $\hat{G}$ depends on the casual choice of ordering of the equations. In particular, the subgraph of $\hat{G}$ on X does not reflect the fact that $A - sF$ is subject not to similarity transformations, but to equivalence transformations. On the other hand, Matsumoto–Ikeda [187] employs a graph representation which can only be determined after a maximum matching on the bipartite graph associated with $A - sF$ is found. This representation is not unique, either, since it depends on the choice of the maximum matching.

Though the conclusions derived from the criteria of Aoki–Hosoe–Hayakawa [7] and Matsumoto–Ikeda [187] are known to be unaffected by the nonuniqueness of the graph representations, it would be preferable to express the controllability condition in such a way that the underlying invariance may be represented explicitly. The conditions (B1)–(B3) in Theorem 6.4.7 are invariant in this respect, since the DM-decomposition of the bipartite graph G remains the same (isomorphic) under the changes of ordering of equations. □

Example 6.4.11. The structural controllability criterion in Theorem 6.4.7 is illustrated for a descriptor system (6.65) with

$$F = \begin{pmatrix} f_1 & 0 & 0 \\ f_2 & f_3 & f_4 \\ 0 & 0 & 0 \end{pmatrix}, \quad A = \begin{pmatrix} a_1 & a_2 & 0 \\ a_3 & 0 & 0 \\ 0 & 0 & a_4 \end{pmatrix}, \quad B = \begin{pmatrix} 0 \\ b \\ 0 \end{pmatrix},$$

which is taken from Matsumoto–Ikeda [187]. We assume that the set of parameters $\{f_1, f_2, f_3, f_4; a_1, a_2, a_3, a_4; b\}$ is algebraically independent over

Q. The bipartite graph G is depicted in Fig. 6.5 together with its DM-decomposition, which consists of the horizontal tail $G_0 = (V_0^+, V_0^-; E_0)$ with $V_0^+ = \{x_1, x_2, u\}$ and $V_0^- = \{e_1, e_2\}$, and the only one consistent component $G_1 = (V_1^+, V_1^-; E_1)$ with $V_1^+ = \{x_3\}$, $V_1^- = \{e_3\}$, and $E_1 = \{a_4\}$. No s-arc is contained in G_1, in agreement with the condition (B3) of Theorem 6.4.7. The other two conditions, (B1) and (B2), are easily seen to be met. Thus this system has been shown to be structurally controllable. It should be noted that the s-arcs contained in G_0 do not affect the controllability. □

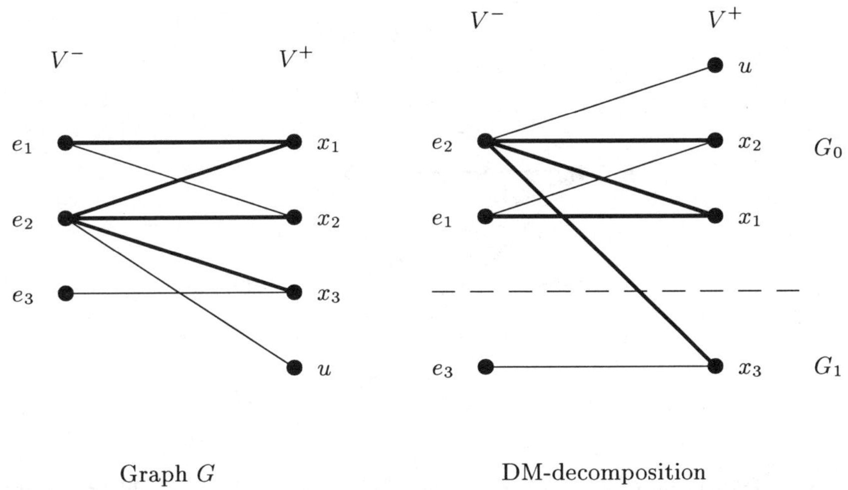

Fig. 6.5. Bipartite graph G of Example 6.4.11 and its DM-decomposition (bold line: s-arc)

Example 6.4.12. Modify the system of Example 6.4.11 by fixing $a_2 = 0$, following Matsumoto–Ikeda [187]. That is, we assume $\{f_1, f_2, f_3, f_4; a_1, a_3, a_4; b\}$ is the set of algebraically independent parameters. The two conditions (B1) and (B2) are still satisfied, whereas (B3) is not, as demonstrated in Fig. 6.6. The DM-decomposition of the modified bipartite graph yields a horizontal tail G_0 with $V_0^+ = \{x_2, u\}$ and $V_0^- = \{e_2\}$, and two consistent components G_1 with $V_1^+ = \{x_1\}$ and $V_1^- = \{e_1\}$, and G_2 with $V_2^+ = \{x_3\}$ and $V_2^- = \{e_3\}$. The component G_1 contains an s-arc. In fact, it is easy to verify that $\operatorname{rank} D(z) = 2 < 3$ for $z = a_1/f_1$. □

Example 6.4.13. Consider the descriptor system (6.65) given by

$$F = \begin{pmatrix} f_1 & 0 & f_2 & 0 \\ 0 & f_3 & 0 & f_4 \\ 0 & 0 & 0 & 0 \\ 0 & 0 & 0 & 0 \end{pmatrix}, \quad A = \begin{pmatrix} 0 & 0 & 0 & a_1 \\ 0 & a_2 & 0 & 0 \\ a_3 & 0 & a_4 & 0 \\ 0 & a_5 & 0 & a_6 \end{pmatrix}, \quad B = \begin{pmatrix} b \\ 0 \\ 0 \\ 0 \end{pmatrix},$$

where $\{f_i \mid i = 1, \cdots, 4\} \cup \{a_i \mid i = 1, \cdots, 6\} \cup \{b\}$ is assumed to be algebraically independent over $\mathbf{Q}$. (This example is taken from Matsumoto–Ikeda [187].) The conditions (B1) and (B2) are satisfied. The DM-decomposition of the bipartite graph G yields the horizontal tail G_0 with $V_0^+ = \{x_1, x_3, u\}$ and $V_0^- = \{e_1, e_3\}$, and one consistent component G_1 with $V_1^+ = \{x_2, x_4\}$ and $V_1^- = \{e_2, e_4\}$. The s-arcs corresponding to f_3 and f_4 are contained in G_1, causing this system to be uncontrollable. □

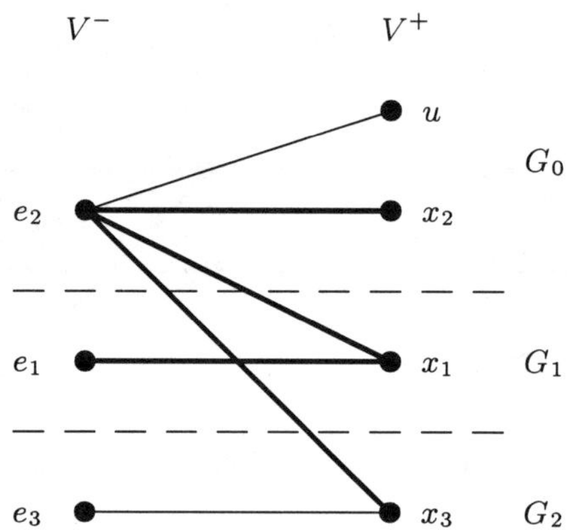

Fig. 6.6. DM-decomposition of the graph G of Example 6.4.12 (bold line: s-arc)

Example 6.4.14. Consider the descriptor system (6.65) with

$$F = \begin{pmatrix} 0 & 0 & 0 & 0 \\ f_1 & 0 & 0 & 0 \\ 0 & f_2 & 0 & 0 \\ f_3 & 0 & 0 & 0 \end{pmatrix}, \quad A = \begin{pmatrix} a_1 & 0 & 0 & a_2 \\ 0 & 0 & 0 & a_3 \\ 0 & a_4 & a_5 & 0 \\ 0 & a_6 & a_7 & 0 \end{pmatrix}, \quad B = \begin{pmatrix} b_1 & 0 \\ 0 & b_2 \\ 0 & 0 \\ 0 & 0 \end{pmatrix},$$

taken from Aoki–Hosoe–Hayakawa [7]. The conditions (B1) and (B2) are easily verified to hold. The third condition (B3) is trivially met, since the whole graph G constitutes the horizontal tail in the DM-decomposition. Therefore, this system is structurally controllable. □

6.4.3 Mixed Polynomial Matrix Formulation

Though the notion of structural controllability is quite appealing, it is often doubtful from the physical point of view to assume that all the nonvanishing entries of the coefficient matrices are independent parameters, since they usually do not stand for individual physical parameters, as has been discussed in

Chap. 3. In particular, some of the entries may be fixed or correlated numbers having mutual algebraic dependence. This observation has motivated a number of generalizations and refinements in the structural approach in control theory (see, e.g., Hayakawa–Hosoe–Hayashi–Ito [106, 107], Yamada–Luenberger [346]).

In this section we are concerned with a combinatorial characterization of structural controllability in the spirit of our physical observations in Chap. 3. In particular, we consider the descriptor system (6.65) in which F, A, and B are mixed matrices with ground field $\mathbf{Q}$:

$$F = Q_F + T_F, \quad A = Q_A + T_A, \quad B = Q_B + T_B, \tag{6.71}$$

such that the set $\mathcal{T}$ of the nonvanishing entries of $[T_F \mid T_A \mid T_B]$ is algebraically independent over $\mathbf{Q}$. This implies that $[A - sF \mid B] = Q(s) + T(s)$ is a mixed polynomial matrix. Furthermore, it is assumed that the matrix $Q(s)$ satisfies the stronger condition for the dimensional consistency.

It should be clear that assuming algebraic independence for $\mathcal{T}$ is equivalent to regarding the members of $\mathcal{T}$ as independent parameters, and therefore to considering a family of systems parametrized by those parameters in $\mathcal{T}$. A particular system in this family having algebraically independent parameter values is controllable if and only if almost all members of the family are controllable.

We formulate the above problem in more general terms for a mixed polynomial matrix, following Murota [203]. Let

$$A(s) = Q(s) + T(s) \tag{6.72}$$

be an $m \times n$ mixed polynomial matrix of rank m with respect to $(\boldsymbol{K}, \boldsymbol{F})$ such that $Q(s)$ satisfies the stronger assumption

(MP-Q2) Every nonvanishing subdeterminant of $Q(s)$ is a monomial over $\boldsymbol{K}$, i.e., of the form αs^p with $\alpha \in \boldsymbol{K}$ and an integer p.

In view of (6.66) we simply say that $A(s)$ is *structurally controllable* if the mth monic determinantal divisor of $A(s)$ is equal to 1. This condition is tantamount to saying that the Smith form of $A(s)$, as a polynomial matrix in s over $\boldsymbol{F}$, is equal to $[I_m \mid O]$. We put $R = \mathrm{Row}(A)$ and $C = \mathrm{Col}(A)$ and denote the mth monic determinantal divisor of $A(s)$ by $d_m(s)$. The roots of $d_m(s)$ will be called the *uncontrollable modes.*

We shall derive a necessary and sufficient condition for structural controllability together with an efficient algorithm for testing it. The proposed algorithm is suitable for practical applications in that it is free from numerical difficulty of rounding errors and is guaranteed to run in polynomial time in the size of the control system in question.

The existence of a zero uncontrollable mode is easy to characterize.

Proposition 6.4.15. *An $m \times n$ mixed polynomial matrix $A(s)$ of rank m satisfying* (MP-Q2) *is free from a zero uncontrollable mode if and only if there exists (I, J) such that* $\mathrm{rank}\, Q(0)[R \setminus I, C \setminus J] + \text{term-rank}\, T(0)[I, J] = m$.

Proof. $A(s)$ does not have a zero uncontrollable mode if and only if $\mathrm{rank}\, A(0) = m$. Then Theorem 4.2.8 establishes this. ∎

The multiplicity of the zero uncontrollable mode is obviously equal to $o_m(A) = \min\{\mathrm{ord}_s \det A[I, J] \mid |I| = |J| = m\}$, where ord_s denotes the minimum degree in s of a nonzero term in a polynomial. Then, by Remark 6.2.3, it can be characterized in terms of an independent assignment problem (see also Remark 6.2.10).

The nonzero uncontrollable modes can be treated by means of the CCF of an LM-polynomial matrix. This is based on the fact (Theorem 6.3.4) that the CCF corresponds to the decomposition of the determinantal divisor into irreducible factors.

To be specific, we consider, as in (6.55), an LM-polynomial matrix

$$\tilde{A}(s) = \tilde{A}(s; t) = \begin{pmatrix} I_m & Q(s) \\ -\mathrm{diag}\,(t_1, \cdots, t_m) & T(s) \end{pmatrix} = \begin{pmatrix} \tilde{Q}(s) \\ \tilde{T}(s; t) \end{pmatrix} \tag{6.73}$$

associated with $A(s)$, where $t_1, \cdots, t_m$ are new indeterminates and $t = (t_1, \cdots, t_m)$. Put $\tilde{C} = \mathrm{Col}(\tilde{A}) \simeq R \cup C$. With reference to (6.4) we define $\zeta : \tilde{C} \to \mathbf{Z}$ by

$$\zeta(j) = \begin{cases} -r_j & (j \in R) \\ -c_j & (j \in C) \end{cases} \tag{6.74}$$

as well as the usual convention $\zeta(J) = \sum_{j \in J} \zeta(j)$ for $J \subseteq \tilde{C}$.

Regarding $\tilde{A}(s)$ as an LM-matrix with respect to $(\boldsymbol{K}[s], \boldsymbol{F}(s,t))$ we may think of its block-triangular form ("CCF over a ring") in the sense of Theorem 4.4.19, which is obtained from $\tilde{A}(s)$ through a unimodular transformation over $\boldsymbol{K}[s]$. Let $\hat{A}(s)$ and $\bar{A}(s)$ be the block-triangular matrix and the CCF of $\tilde{A}(s)$ as in Theorem 4.4.19. Note that $\hat{A}(s)$ and $\bar{A}(s)$ have identical diagonal blocks, though they may differ in the upper-triangular part. The families of the row sets and the column sets in the CCF are denoted respectively by $\{\bar{R}_k \mid k = 0, 1, \cdots, b\}$ and $\{\bar{C}_k \mid k = 0, 1, \cdots, b\}$, and the diagonal blocks by

$$\bar{A}_k = \begin{pmatrix} \bar{Q}_k \\ \bar{T}_k \end{pmatrix} = \bar{A}[\bar{R}_k, \bar{C}_k], \qquad k = 0, 1, \cdots, b.$$

In this notation, $k = 0$ designates the horizontal tail, whereas the vertical tail does not exist (is empty) since $\mathrm{rank}\, \tilde{A} = 2m$ by $\mathrm{rank}\, A = m$. We define

$$\begin{aligned} \mathcal{J}_k = \{J \subseteq \bar{C}_k \mid\ & \bar{Q}_k[\mathrm{Row}(\bar{Q}_k), \bar{C}_k \setminus J]\text{: nonsingular}, \\ & \bar{T}_k[\mathrm{Row}(\bar{T}_k), J]\text{: term-nonsingular}\}, \quad k = 1, \cdots, b. \end{aligned}$$

For $J \subseteq \bar{C}_k$ such that $\bar{T}_k[\mathrm{Row}(\bar{T}_k), J]$ is term-nonsingular, we denote by $\xi_k(J)$ and $\eta_k(J)$ the highest and lowest degrees in s of a nonzero term in

$\det \bar{T}_k[\mathrm{Row}(\bar{T}_k), J]$. Note that $\xi_k(J)$ and $\eta_k(J)$ can be expressed in terms of weighted-matching problems (cf. §6.2.2).

Theorem 6.4.16. *For an $m \times n$ mixed polynomial matrix $A(s)$ of rank m satisfying* (MP-Q2), *the number of nonzero uncontrollable modes is given by*

$$\sum_{k=1}^{b} \left[\max\{\zeta(\bar{C}_k \setminus J) + \xi_k(J) \mid J \in \mathcal{J}_k\} - \min\{\zeta(\bar{C}_k \setminus J) + \eta_k(J) \mid J \in \mathcal{J}_k\}\right].$$

Hence there exist no nonzero uncontrollable modes if and only if

$$\max\{\zeta(\bar{C}_k \setminus J) + \xi_k(J) \mid J \in \mathcal{J}_k\} = \min\{\zeta(\bar{C}_k \setminus J) + \eta_k(J) \mid J \in \mathcal{J}_k\} \quad (6.75)$$

for each $k = 1, \cdots, b$.

Proof. The determinant of $\bar{A}_k(s;1) = \bar{A}_k(s;t)\big|_{t_1=\cdots=t_m=1}$ can be expressed as $\det \bar{A}_k(s;1) = \alpha_k s^{p_k} \cdot \bar{\psi}_k(s,\mathcal{T})$, where α_k is a nonzero constant, p_k is a nonnegative integer, and $\bar{\psi}_k(s,\mathcal{T}) \in \boldsymbol{K}[s,\mathcal{T}]$ is not divisible by s. We note

$$\deg_s \bar{\psi}_k(s) = \max\{\zeta(\bar{C}_k \setminus J) + \xi_k(J) \mid J \in \mathcal{J}_k\} - \min\{\zeta(\bar{C}_k \setminus J) + \eta_k(J) \mid J \in \mathcal{J}_k\},$$

which is a corollary of Theorem 6.2.5. Then the claim follows from Theorem 6.3.4. ■

Remark 6.4.17. See Murota [203] as well as Murota [204, Theorem 28.1] for an alternative formulation of the condition for structural controllability in the form of a weighted matroid partition problem. □

On the basis of the combinatorial characterization in Proposition 6.4.15 and Theorem 6.4.16, an efficient algorithm for testing the existence of zero/nonzero uncontrollable modes is designed in the next section.

Theorem 6.4.16 above includes many previously known results on the structural controllability as special cases. In particular, it is a direct generalization of Theorem 6.4.7 for the case where all the nonvanishing entries of the coefficients are taken for independent parameters. Note that the CCF used in Theorem 6.4.16 is a generalization of the DM-decomposition used in Theorem 6.4.7. Theorem 6.4.16 also implies the results of Hayakawa–Hosoe–Hayashi–Ito [106] (see Murota [204, §31.1] for detail).

6.4.4 Algorithm

In this section we describe an efficient algorithm to check for the existence of nonzero uncontrollable modes of $A(s) = Q(s) + T(s)$ on the basis of Theorem 6.4.16, whereas the algorithm of §4.2.4 for computing the rank of an LM-matrix can be utilized readily for the zero uncontrollable mode by Proposition 6.4.15.

Before describing the concrete procedure for detecting nonzero uncontrollable modes we will outline the basic idea in general terms. As shown in §4.2.3 and §4.4.4, the CCF of the LM-matrix $\tilde{A}(s)$ of (6.73) can be computed via the independent matching problem on a bipartite graph. The CCF can be obtained from the strong components of a subgraph of the auxiliary graph associated with a maximum independent matching. Moreover, the argument in §6.2 shows that $\max\{\zeta(\bar{C}_k \setminus J) + \xi_k(J)\}$ and $\min\{\zeta(\bar{C}_k \setminus J) + \eta_k(J)\}$, characterizing the existence of nonzero uncontrollable modes in Theorem 6.4.16, can be expressed in terms of independent assignment problems in the strong component corresponding to the block $\bar{A}_k$ of the CCF. In this way the existence of nonzero uncontrollable modes can be found by computing $\max\{\zeta(\bar{C}_k \setminus J) + \xi_k(J)\}$ and $\min\{\zeta(\bar{C}_k \setminus J) + \eta_k(J)\}$ separately by efficient algorithms that employ arithmetic operations on rational numbers only.

It is possible to design a faster algorithm by making use of a fundamental fact about the network flow problem. To be more specific, the condition (6.75) is equivalent to a graph-theoretic condition that there exists in the strong component for the block $\bar{A}_k$ no directed cycle of nonzero length with respect to an appropriately defined arc length. This latter condition is equivalent further to the existence of a potential function such that the length of an arc is the difference of the potentials of the end-vertices (see Theorem 2.2.35(2)). The existence of such a potential function is easy to check. The objective of this section is to describe this idea in concrete terms.

A concrete description of the algorithm for the condition (6.75) follows. We use an auxiliary network $N = (V, E, \gamma)$ with underlying graph $G = (V, E)$ and length function $\gamma : E \to \mathbf{Z}$, in a way consistent with the algorithm in §4.2.4 for a mixed matrix. The vertex set V is defined as

$$V = V_Q \cup V_T = (R_Q \cup C_Q) \cup (R_T \cup C_T),$$

where $R_Q = \mathrm{Row}(Q)$, $C_Q = \mathrm{Col}(Q)$, $R_T = \mathrm{Row}(T)$, $C_T = \mathrm{Col}(T)$, $V_Q = R_Q \cup C_Q$, and $V_T = R_T \cup C_T$. The arc set E consists of five disjoint parts,

$$E = E_{TQ} \cup E_{QT} \cup E_Q \cup E_T \cup E_M,$$

to be defined below. We denote by $\varphi_Q : R \cup C \to R_Q \cup C_Q$ and $\varphi_T : R \cup C \to R_T \cup C_T$ the obvious one-to-one correspondences.

Let $\hat{I} \subseteq R$ and $\hat{J} \subseteq C$ be such that $Q(1)[R \setminus \hat{I}, C \setminus \hat{J}]$ is nonsingular and $\text{term-rank}\, T[\hat{I}, \hat{J}] = |\hat{I}|$, where such $(\hat{I}, \hat{J})$ exists by Theorem 4.2.8 since $\mathrm{rank}\, A(1) = m$. We define

$$E_{TQ} = \{(\varphi_T(i), \varphi_Q(i)) \mid i \in \hat{I}\} \cup \{(\varphi_T(j), \varphi_Q(j)) \mid j \in C \setminus \hat{J}\},$$
$$E_{QT} = \{(\varphi_Q(i), \varphi_T(i)) \mid i \in R \setminus \hat{I}\} \cup \{(\varphi_Q(j), \varphi_T(j)) \mid j \in \hat{J}\}.$$

Let P be the pivotal transform of $Q = Q(1)$ with pivot $\hat{Q} \equiv Q[R \setminus \hat{I}, C \setminus \hat{J}]$. Namely,

$$P = \begin{array}{c} \\ C \setminus \hat{J} \\ \hat{I} \end{array} \begin{array}{c} \begin{array}{cc} R \setminus \hat{I} & \hat{J} \end{array} \\ \left(\begin{array}{cc} \hat{Q}^{-1} & \hat{Q}^{-1}Q[R \setminus \hat{I}, \hat{J}] \\ -Q[\hat{I}, C \setminus \hat{J}]\hat{Q}^{-1} & Q[\hat{I}, \hat{J}] - Q[\hat{I}, C \setminus \hat{J}]\hat{Q}^{-1}Q[R \setminus \hat{I}, \hat{J}] \end{array} \right) \end{array}, \tag{6.76}$$

where $\mathrm{Row}(P) = (C \setminus \hat{J}) \cup \hat{I}$ and $\mathrm{Col}(P) = (R \setminus \hat{I}) \cup \hat{J}$. Note that P is a constant matrix free from s. With reference to P we define

$$E_Q = \{(\varphi_Q(i), \varphi_Q(j)) \mid P_{ij} \neq 0, i \in (C \setminus \hat{J}) \cup \hat{I}, j \in (R \setminus \hat{I}) \cup \hat{J}\}.$$

The structure of T is represented by E_T and E_M. For each nonzero entry $T_{ij}(s)$ of $T(s)$ we consider a pair of parallel arcs a_{ij}^0 and a_{ij}^1 with $\partial^+ a_{ij}^0 = \partial^+ a_{ij}^1 = \varphi_T(i) \in R_T$ and $\partial^- a_{ij}^0 = \partial^- a_{ij}^1 = \varphi_T(j) \in C_T$. Putting

$$E_T^0 = \{a_{ij}^0 \mid T_{ij} \neq 0, i \in R, j \in C\}, \quad E_T^1 = \{a_{ij}^1 \mid T_{ij} \neq 0, i \in R, j \in C\},$$

we define $E_T = E_T^0 \cup E_T^1$. Since $\text{term-rank}\, T[\hat{I}, \hat{J}] = |\hat{I}|$, the bipartite graph $(R_T, C_T; E_T)$ with vertex set $R_T \cup C_T$ and arc set E_T has a matching M $(\subseteq E_T)$ such that $|M| = |\hat{I}|$, $\varphi_T(\hat{I}) = \partial^+ M$, and $\varphi_T(\hat{J}) \supseteq \partial^- M$. We define E_M as the set of reoriented arcs of M, i.e.,

$$E_M = \{\bar{a} \mid a \in M\},$$

where $\bar{a}$ denotes the reorientation of a.

The length function $\gamma : E \to \mathbf{Z}$ is defined with reference to r_i $(i = 1, \cdots, m)$ and c_j $(j = 1, \cdots, n)$ of (6.4) as

$$\gamma(a) = \begin{cases} -r_i & \text{if} \quad a = (\varphi_T(i), \varphi_Q(i)) \in E_{TQ},\ i \in \hat{I}, \\ -c_j & \text{if} \quad a = (\varphi_T(j), \varphi_Q(j)) \in E_{TQ},\ j \in C \setminus \hat{J}, \\ r_i & \text{if} \quad a = (\varphi_Q(i), \varphi_T(i)) \in E_{QT},\ i \in R \setminus \hat{I}, \\ c_j & \text{if} \quad a = (\varphi_Q(j), \varphi_T(j)) \in E_{QT},\ j \in \hat{J}, \\ 0 & \text{if} \quad a \in E_Q, \\ -\mathrm{ord}_s T_{ij}(s) & \text{if} \quad a \in E_T^0, \\ -\deg_s T_{ij}(s) & \text{if} \quad a \in E_T^1, \\ -\gamma(a') & \text{if} \quad a \in E_M \text{ is the reorientation of } a' \in M \subseteq E_T. \end{cases}$$

For a nonzero entry $T_{ij}(s)$ of $T(s)$ with $\mathrm{ord}_s T_{ij}(s) = \deg_s T_{ij}(s)$ (which is the case if $T_{ij}(s)$ is a monomial in s), the pair of arcs, having the same length, may be replaced by a single arc of the same length.

We are now ready to rephrase the condition (6.75) in terms of the network $N = (G, \gamma)$. Let V° be the set of vertices of G which are not reachable to the exit

$$S^- = \varphi_T(\hat{J}) \setminus \partial^- M \ \subseteq C_T \tag{6.77}$$

by a directed path. The subgraph of G induced on V° is denoted as G°. The strong components of G° correspond to the consistent diagonal blocks of the CCF of $\tilde{A}$, where it is noted that the vertical tail is empty. For each strong component of G°, say $\hat{G} = (\hat{V}, \hat{E})$, we consider the condition that the sum of the lengths $\gamma(a)$ along any directed cycle in $\hat{G}$ is equal to zero, i.e.,

$$\sum_{a\in\hat{C}} \gamma(a) = 0 \qquad (\forall \hat{C} : \text{directed cycle in } \hat{G}). \tag{6.78}$$

Since $\hat{G}$ is strongly connected, this condition is equivalent, by Theorem 2.2.35(2), to the existence of a potential function $\pi : \hat{V} \to \mathbf{Z}$ such that

$$\gamma(a) = \pi(\partial^- a) - \pi(\partial^+ a) \qquad (a \in \hat{E}). \tag{6.79}$$

Theorem 6.4.18. *An $m \times n$ mixed polynomial matrix $A(s)$ of rank m satisfying* (MP-Q2) *has no nonzero uncontrollable mode if and only if each strong component of the subgraph G° admits a potential function π such that* (6.79) *holds.*

Proof. For simplicity of notation let us assume that G itself is a strong component. We also assume for simplicity of argument that each $T_{ij}(s)$ is a monomial in s so that each pair of parallel arcs in E_T is replaced by a single arc. Consider the independent assignment problem as in §6.2 to compute $\deg_s \det \tilde{A}(s)$ for $\tilde{A}(s)$ of (6.73). Then $\max\{\zeta(\bar{C}_k \setminus J) + \xi_k(J)\}$ corresponds to the maximum weight of an independent assignment, whereas $\min\{\zeta(\bar{C}_k \setminus J) + \eta_k(J)\}$ to the minimum. Hence, the condition (6.75) is tantamount to saying that all the independent assignments have the same weight. This is the case if and only if the weight of an arbitrarily chosen independent assignment is the maximum and the minimum at the same time. By Theorem 5.2.42, the independent assignment associated with $(\hat{I}, \hat{J}, M)$ has the maximum weight if and only if there exists no negative cycle in the auxiliary network, whereas it has the minimum weight if and only if there exists no positive cycle. The network N employed above is essentially the same as the auxiliary network as defined in §5.2.10. Therefore, (6.78) is necessary and sufficient for (6.75). ■

Remark 6.4.19. The potential function π of (6.79), if it exists, can be constructed as follows. First fix a spanning tree $\hat{T} \subseteq \hat{E}$ and a vertex $u \in \hat{V}$. For each $v \in \hat{V}$, set $\pi(v)$ equal to the length of the path in $\hat{T}$ connecting u to v. Finally check for the validity of this π by verifying the condition (6.79) for each $a \in \hat{E} \setminus \hat{T}$. □

The overall computational complexity for testing for the existence of uncontrollable modes on the basis of Proposition 6.4.15 and Theorem 6.4.18 is dominated by that for the construction of the graph G and therefore bounded by $\mathrm{O}(n^3 \log n)$, where $m \leq n$ is assumed. Note that the decomposition of G into strong components can be done in $\mathrm{O}(|E|)$ time and the potential function of (6.79) for a strong component $\hat{G} = (\hat{V}, \hat{E})$, if it exists, can be found in time of $\mathrm{O}(|\hat{E}|)$ by the procedure of Remark 6.4.19. It should be emphasized here that the whole algorithm involves only pivoting operations on the matrix $Q(1)$, the entries of which are rational numbers (simple numbers such as ± 1 in practical applications).

Remark 6.4.20. When the above algorithm is applied to $[A - sF \mid B] = Q(s)+T(s)$ with nonsingular $A-sF$, we can choose $\hat{J}$ and M so that $\varphi_T(\hat{J}) \setminus \partial^- M = \varphi_T(U)$, where $U = \mathrm{Col}(B)$. Then the exit S^- defined in (6.77) coincides with $\varphi_T(U)$. □

Remark 6.4.21. The graph-theoretic criterion in Theorem 6.4.7 can be derived from Proposition 6.4.15 and Theorem 6.4.18 applied to the matrix $[A - sF \mid B] = Q(s) + T(s)$ with $Q(s) = O$. The derivation relies on the observation of Remark 6.4.20. Note that the graph G in Theorem 6.4.7 is identical with the subgraph $(R_T, C_T; E_T)$ in the network N, except that the arcs are reoriented. □

6.4.5 Examples

This section illustrates the algorithm of §6.4.4 as well as Theorem 6.4.16 by means of two examples.

Example 6.4.22. Recall again the mechanical system (Fig. 3.5) treated in Example 3.1.7. The matrix $D(s) = [A - sF \mid B]$ is given as

$$D(s) = \begin{array}{c} \\ w_1 \\ w_2 \\ w_3 \\ w_4 \\ w_5 \\ w_6 \end{array} \begin{array}{|cccccc|c|} \multicolumn{1}{c}{x_1} & x_2 & x_3 & x_4 & x_5 & \multicolumn{1}{c}{x_6} & \multicolumn{1}{c}{u} \\ \hline -s & 0 & 1 & 0 & 0 & 0 & 0 \\ 0 & -s & 0 & 1 & 0 & 0 & 0 \\ -k_1 & 0 & -sm_1 & 0 & -1 & 0 & 1 \\ 0 & -k_2 & 0 & -sm_2 & 1 & 0 & 0 \\ 0 & 0 & 0 & 0 & -1 & f & 0 \\ -s & s & 0 & 0 & 0 & 1 & 0 \\ \hline \end{array},$$

which can be expressed as $D(s) = Q(s) + T(s)$ with $Q(s)$ and $T(s)$ of (3.18) and (3.19). As we have seen in Examples 3.2.2 and 3.3.1 the stronger condition (MP-Q2) is satisfied with $(r_1, \cdots, r_6) = (1, 1, 2, 2, 2, 1)$ and $(c_1, \cdots, c_7) = (0, 0, 1, 1, 2, 1, 2)$. Hence the nonsingularity of $A - sF$ is equivalent to that of $A - F$, which can be verified by the algorithm of §4.2.4. It can also be verified that $\mathrm{rank}\, D(0) = 6$, which means, by Proposition 6.4.15, the controllability of the zero mode.

As for the controllability of nonzero modes, we may take $\hat{I} = \{w_3, w_4\}$, $\hat{J} = \{x_3, x_4, u\}$, and $M = \{(w_3^T, x_3^T), (w_4^T, x_4^T)\}$ in accordance with Remark 6.4.20. Then the matrix P of (6.76) is

$$P = \begin{array}{c} \\ x_1 \\ x_2 \\ x_5 \\ x_6 \\ w_3 \\ w_4 \end{array} \begin{array}{|cccc|ccc|} \multicolumn{1}{c}{w_1} & w_2 & w_5 & \multicolumn{1}{c}{w_6} & x_3 & x_4 & \multicolumn{1}{c}{u} \\ \hline -1 & 0 & 0 & 0 & -1 & 0 & 0 \\ 0 & -1 & 0 & 0 & 0 & -1 & 0 \\ 0 & 0 & -1 & 0 & 0 & 0 & 0 \\ -1 & 1 & 0 & 1 & -1 & 1 & 0 \\ \hline 0 & 0 & -1 & 0 & 0 & 0 & 1 \\ 0 & 0 & 1 & 0 & 0 & 0 & 0 \\ \hline \end{array} .$$

The auxiliary network $N = (G, \gamma) = (V, E, \gamma)$ is depicted in Fig. 6.7, where $x_i^T = \varphi_T(x_i)$, $x_i^Q = \varphi_Q(x_i)$, etc., and the associated length $\gamma(a)$ is

$$\gamma(a) = \begin{cases} -2 & (a = (x_5^T, x_5^Q), (w_3^T, w_3^Q), (w_4^T, w_4^Q)) \\ -1 & (a = (x_6^T, x_6^Q), (w_3^T, x_3^T), (w_4^T, x_4^T)) \\ 1 & (a = (x_3^T, w_3^T), (x_4^T, w_4^T), (x_3^Q, x_3^T), (x_4^Q, x_4^T), \\ & \quad (w_1^Q, w_1^T), (w_2^Q, w_2^T), (w_6^Q, w_6^T)) \\ 2 & (a = (u^Q, u^T), (w_5^Q, w_5^T)) \\ 0 & (\text{otherwise}). \end{cases}$$

All the vertices except those in $V^\circ = \{w_1^T, w_1^Q, w_2^T, w_2^Q, w_6^T, w_6^Q\}$ are reachable to $S^- = \{u^T\}$, and the subgraph G° consists of three (disconnected) arcs (w_1^Q, w_1^T), (w_2^Q, w_2^T) and (w_6^Q, w_6^T). Then condition in Theorem 6.4.18 is trivially met, and therefore this mechanical system is structurally controllable. □

Example 6.4.23. Consider a hypothetical descriptor system with

$$F = \begin{bmatrix} 0 & 0 & p_1 \\ 1 & 1 & p_2 \\ 0 & 0 & 0 \end{bmatrix}, \quad A = \begin{bmatrix} 1 & p_3 & 0 \\ 0 & 0 & 1 \\ -1 & -1 & p_4 \end{bmatrix}, \quad B = \begin{bmatrix} p_5 \\ 0 \\ 0 \end{bmatrix}, \tag{6.80}$$

where $\{p_i \mid i = 1, \cdots, 5\}$ is to be understood as independent parameters. The matrix $D(s) = [A - sF \mid B]$ is then a mixed matrix $D(s) = Q(s) + T(s)$ with

$$Q(s) = \left[\begin{array}{ccc|c} 1 & 0 & 0 & 0 \\ -s & -s & 1 & 0 \\ -1 & -1 & 0 & 0 \end{array}\right], \qquad T(s) = \left[\begin{array}{ccc|c} 0 & p_3 & -sp_1 & p_5 \\ 0 & 0 & -sp_2 & 0 \\ 0 & 0 & p_4 & 0 \end{array}\right],$$

where $\mathrm{Row}(D) = \{w_1, w_2, w_3\}$ and $\mathrm{Col}(D) = \{x_1, x_2, x_3, u\}$. Note that $Q(s)$ satisfies the property (MP-Q2), admitting the expression (6.4) with, e.g., $(r_1, r_2, r_3) = (0, 1, 0)$ and $(c_1, c_2, c_3, c_4) = (0, 0, 1, 0)$.

It is easy to see by inspection that $A - sF$ is nonsingular and the zero mode is controllable (i.e., $\mathrm{rank}\, D(0) = 3$). For the controllability of nonzero modes, we may take $\hat{I} = \{w_1, w_2\}$, $\hat{J} = \{x_2, x_3, u\}$, and $M = \{(w_1^T, x_2^T), (w_2^T, x_3^T)\}$ in accordance with Remark 6.4.20. Then the matrix P of (6.76) is

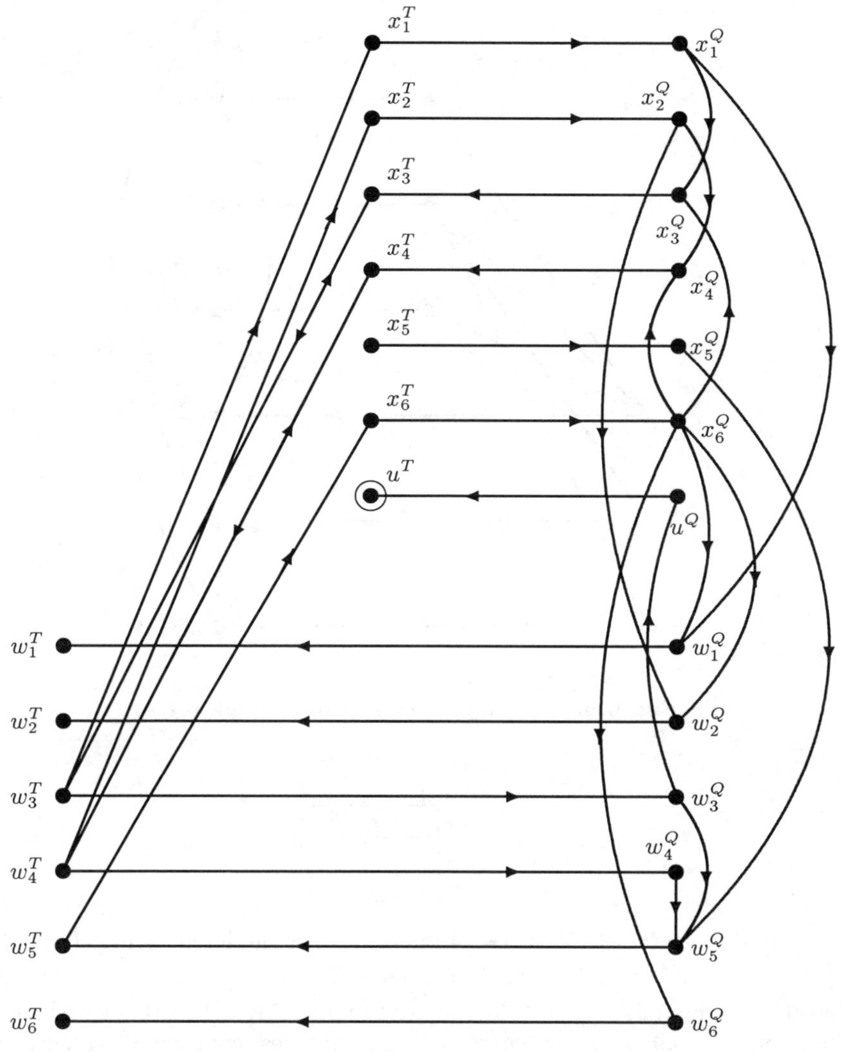

Fig. 6.7. Auxiliary network N for the mechanical system (Example 6.4.22)

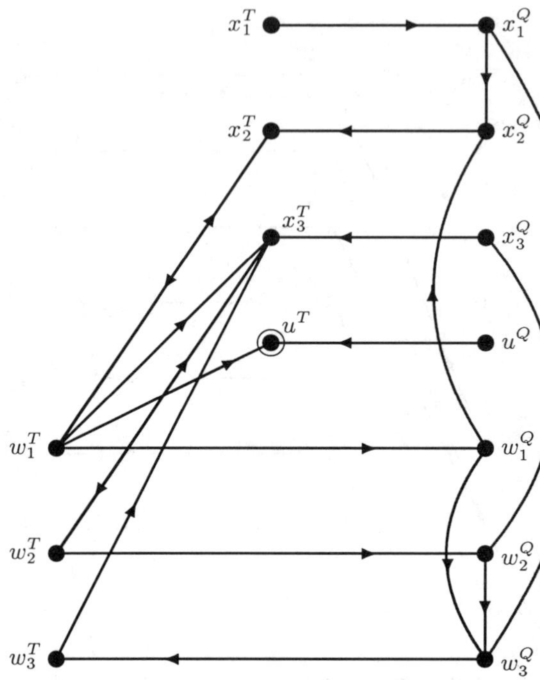

Fig. 6.8. Auxiliary network N in Example 6.4.23

$$P = \begin{array}{c|c|ccc|} & w_3 & x_2 & x_3 & u \\ \hline x_1 & -1 & 1 & 0 & 0 \\ \hline w_1 & 1 & -1 & 0 & 0 \\ \hline w_2 & -1 & 0 & 1 & 0 \\ \hline \end{array} .$$

The auxiliary network $N = (G, \gamma) = (V, E, \gamma)$ is depicted in Fig. 6.8, where $x_i^T = \varphi_T(x_i)$, $x_i^Q = \varphi_Q(x_i)$, etc. The exit is $S^- = \{u^T\}$, to which the vertices not in $V^\circ = \{x_3^T, x_3^Q, w_2^T, w_2^Q, w_3^T, w_3^Q\}$ are reachable. The subnetwork on G°, shown in Fig. 6.9 with the length γ in parentheses, contains directed cycles. The sum of the lengths along the cycle consisting of $\{w_2^T, w_2^Q, w_3^Q, w_3^T, x_3^T\}$ vanishes, whereas that of $\{w_2^T, w_2^Q, x_3^Q, x_3^T\}$ does not. Thus it is revealed, by Theorem 6.4.18, that this system has a nonzero uncontrollable mode. The graph-theoretic methods such as Theorem 6.4.7, treating the nonvanishing entries of F, A, and B of (6.80) as if they were independent, would fail to detect this fact.

The associated LM-polynomial matrix in (6.73) and its CCF are given respectively by

$$
\begin{array}{ccccccc}
w_1 & w_2 & w_3 & x_1 & x_2 & x_3 & u \\
\hline
1 & & & 1 & 0 & 0 & 0 \\
& 1 & & -s & -s & 1 & 0 \\
& & 1 & -1 & -1 & 0 & 0 \\
\hline
-t_1 & & & 0 & p_3 & -sp_1 & p_5 \\
& -t_2 & & 0 & 0 & -sp_2 & 0 \\
& & -t_3 & 0 & 0 & p_4 & 0
\end{array}
,\qquad
\begin{array}{cccc|ccc}
\multicolumn{4}{c|}{\bar{C}_0} & \multicolumn{3}{c}{\bar{C}_1} \\
u & w_1 & x_1 & x_2 & w_2 & w_3 & x_3 \\
\hline
0 & 1 & 0 & -1 & & 1 & \\
0 & 0 & 1 & 1 & & -1 & \\
p_5 & -t_1 & 0 & p_3 & & & -sp_1 \\
\hline
& & & & 1 & -s & 1 \\
& & & & -t_2 & 0 & -sp_2 \\
& & & & 0 & -t_3 & p_4
\end{array}
.
$$

The CCF has the horizontal tail with $\bar{C}_0 = \{u, w_1, x_1, x_2\}$ and one square block with $\bar{C}_1 = \{w_2, w_3, x_3\}$, while the vertical tail is empty. The determinant of the 3×3 diagonal block corresponding to $\bar{C}_1$ is equal to $-(t_3 p_2 + t_2 p_4)s + t_2 t_3$, which has the root $s = 1/(p_2 + p_4)$ when $t_i = 1$. This represents the uncontrollable mode. Thus the CCF reveals directly how the nonzero uncontrollable mode arises. Finally it is mentioned that the CCF above is obtained through a unimodular transformation by

$$
U = \begin{bmatrix} 1 & 0 & 1 \\ 0 & 0 & -1 \\ 0 & 1 & -s \end{bmatrix} .
$$

□

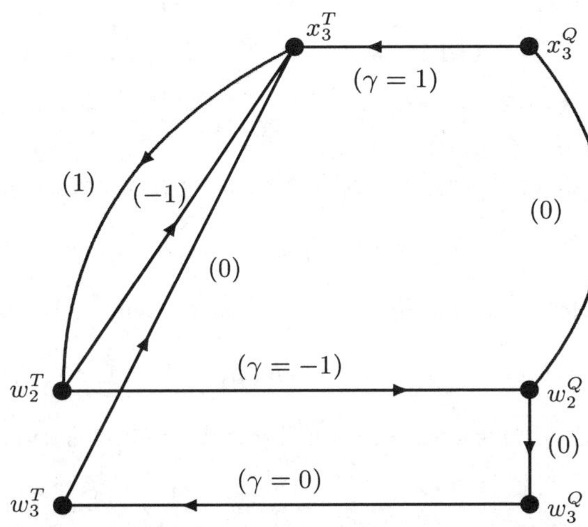

Fig. 6.9. Subnetwork on G° in Example 6.4.23

Notes. The mixed matrix formulation in §6.4.3 and the algorithm in §6.4.4 are taken from Murota [203].

6.5 Fixed Modes of Decentralized Systems

Structurally fixed modes of a decentralized control system are investigated using mixed polynomial matrices. A necessary and sufficient condition for the existence of structurally fixed modes is given along with an efficient algorithm for testing it.

6.5.1 Fixed Modes

The concept of (decentralized) fixed modes, introduced by Wang–Davison [332], is recognized as one of the fundamental concepts for the decentralized control (see Šiljak [300, §1.6], Trave–Titli–Tarras [320]). To be specific, consider a linear time-invariant dynamical system with ν local control stations described by

$$\dot{\boldsymbol{x}}(t) = A\boldsymbol{x}(t) + B\boldsymbol{u}(t), \qquad \boldsymbol{y}(t) = C\boldsymbol{x}(t), \tag{6.81}$$

where A, B, and C are real matrices, $\boldsymbol{x} \in \mathbf{R}^n$ is the state-vector, $\boldsymbol{u} = ({\boldsymbol{u}_1}^{\mathrm{T}}, \cdots, {\boldsymbol{u}_\nu}^{\mathrm{T}})^{\mathrm{T}} \in \mathbf{R}^m$ and $\boldsymbol{y} = ({\boldsymbol{y}_1}^{\mathrm{T}}, \cdots, {\boldsymbol{y}_\nu}^{\mathrm{T}})^{\mathrm{T}} \in \mathbf{R}^l$ are the input-vector and the output-vector, respectively, consisting of the input-vectors $\boldsymbol{u}_k$ ($k = 1, \cdots, \nu$) and the output-vectors $\boldsymbol{y}_k$ ($k = 1, \cdots, \nu$) of the local control stations. The matrices B and C are partitioned into ν blocks as

$$B = (B_1 \mid \cdots \mid B_\nu), \quad C = \begin{pmatrix} C_1 \\ \vdots \\ C_\nu \end{pmatrix}$$

in correspondence to the local stations.

The local output feedback is specified by a block-diagonal matrix

$$K = \mathrm{diag}\,[K_1, \cdots, K_\nu], \tag{6.82}$$

which represents the nondynamic decentralized output feedback

$$\boldsymbol{u}(t) = K\boldsymbol{y}(t), \quad \text{i.e.,} \quad \boldsymbol{u}_k = K_k\boldsymbol{y}_k \quad (k = 1, \cdots, \nu).$$

The local output feedback control with dynamic compensation is described by

$$\dot{\boldsymbol{z}}(t) = L\boldsymbol{z}(t) + M\boldsymbol{y}(t), \qquad \boldsymbol{u}(t) = N\boldsymbol{z}(t) + K\boldsymbol{y}(t) + P\boldsymbol{v}(t), \tag{6.83}$$

where $\boldsymbol{z} = ({\boldsymbol{z}_1}^{\mathrm{T}}, \cdots, {\boldsymbol{z}_\nu}^{\mathrm{T}})^{\mathrm{T}}$ and $\boldsymbol{v} = ({\boldsymbol{v}_1}^{\mathrm{T}}, \cdots, {\boldsymbol{v}_\nu}^{\mathrm{T}})^{\mathrm{T}}$ are the state-vector and the external input-vector, respectively, consisting of the kth feedback controller ($k = 1, \cdots, \nu$); the matrices L, M, N, and P are block-diagonal matrices of appropriate sizes.

Let $\mathcal{K}$ be the family of all matrices K of the form (6.82). The greatest common divisor of the characteristic polynomials of $A + BKC$, for all $K \in \mathcal{K}$, is called the *fixed polynomial* of (A, B, C) with respect to $\mathcal{K}$, and denoted by $\psi(s) = \psi(s; A, B, C, \mathcal{K})$. Namely,

$$\psi(s; A, B, C, \mathcal{K}) = \gcd\{\det(A + BKC - sI_n) \mid K \in \mathcal{K}\}. \tag{6.84}$$

A complex number $\lambda \in \mathbf{C}$ is called a *fixed mode* of (A, B, C) with respect to $\mathcal{K}$ if λ is an eigenvalue of $A + BKC$ for all $K \in \mathcal{K}$, or equivalently, if $\psi(\lambda; A, B, C, \mathcal{K}) = 0$.

The importance of the concept of fixed modes is demonstrated by the following facts due to Wang–Davison [332] and Corfmat–Morse [41]:

1. The system (6.81) is stabilizable by the decentralized dynamic output feedback (6.83) if and only if all the fixed modes of (A, B, C) have negative real parts, and
2. The spectrum of the closed-loop system (6.81) and (6.83) is freely assignable by means of $K \in \mathcal{K}$ if and only if there exist no fixed modes of (A, B, C).

The fixed polynomial and fixed modes can be defined by (6.84) with respect to an arbitrarily specified family (feedback structure) $\mathcal{K}$ of the matrices K, not necessarily of the form (6.82). A natural choice is to let $\mathcal{K}$ be a family of matrices K which are subject to an arbitrarily specified zero/nonzero structure (cf. Wang–Davison [332], Pichai–Sezer–Šiljak [269]). Namely, for an $m \times l$ matrix $\hat{K} = (\hat{K}_{ij})$ with $\hat{K}_{ij} \in \{0, 1\}$ we define

$$\mathcal{K} = \{K \mid K_{ij} = 0 \text{ if } \hat{K}_{ij} = 0\}. \tag{6.85}$$

We refer to the following fundamental result of Anderson–Clements [5] in a form extended to a general feedback structure and with a proof based on a rank identity for mixed matrices. We use the notation $X = \mathrm{Row}(A) \simeq \mathrm{Col}(A) \simeq \mathrm{Row}(B) \simeq \mathrm{Col}(C)$, $U = \mathrm{Col}(B)$, and $Y = \mathrm{Row}(C)$, where $|X| = n$, $|U| = m$, and $|Y| = l$; and also[2]

$$\mathcal{C}_{\mathcal{K}} = \{(I, J) \mid I \subseteq U, J \subseteq Y,\ \hat{K}[U \setminus I, Y \setminus J] = O\}. \tag{6.86}$$

Note that $(I, J) \in \mathcal{C}_{\mathcal{K}}$ is a cover of $K \in \mathcal{K}$ as defined in §2.2.3.

Theorem 6.5.1. *Let $\mathcal{K}$ and $\mathcal{C}_{\mathcal{K}}$ be defined by* (6.85) *and* (6.86). *For a complex number $\lambda \in \mathbf{C}$ and a nonnegative integer $d \in \mathbf{Z}$, we have*

$$\max_{K \in \mathcal{K}} \mathrm{rank}\,(A + BKC - \lambda I_n) \leq n - d$$

if and only if there exists $(I, J) \in \mathcal{C}_{\mathcal{K}}$ such that

$$\mathrm{rank} \begin{pmatrix} A - \lambda I_n & B[X, I] \\ C[J, X] & O \end{pmatrix} \leq n - d. \tag{6.87}$$

In particular, $\lambda \in \mathbf{C}$ is a fixed mode of (A, B, C) with respect to $\mathcal{K}$ if and only if (6.87) *with $d = 1$ holds for some $(I, J) \in \mathcal{C}_{\mathcal{K}}$.*

[2] It is understood that $(I, J) \in \mathcal{C}_{\mathcal{K}}$ if $I = U$ or $J = Y$.

Proof. First note that $\operatorname{rank}(A + BKC - \lambda I_n) = \operatorname{rank} D - (m + l)$ for

$$D = \begin{array}{c} \\ X \\ U \\ Y \end{array} \!\!\begin{array}{c} \begin{array}{ccc} X & U & Y \end{array} \\ \begin{pmatrix} A - \lambda I_n & B & O \\ O & -I_m & K \\ C & O & -I_l \end{pmatrix} \end{array}. \tag{6.88}$$

The maximum of $\operatorname{rank} D$ over all $K \in \mathcal{K}$ is attained when $K_{ij} \neq 0$ for all (i, j) with $\hat{K}_{ij} \neq 0$, and the nonzero entries of K are indeterminates. In this case, D is a mixed matrix $D = Q + T$ with

$$Q = \begin{pmatrix} A - \lambda I_n & B & O \\ O & -I_m & O \\ C & O & -I_l \end{pmatrix}, \qquad T = \begin{pmatrix} O & O & O \\ O & O & K \\ O & O & O \end{pmatrix}, \tag{6.89}$$

and an application of the rank identity (4.23) in Corollary 4.2.12 to D yields the desired result as follows. Let $(\tilde{I}, \tilde{J})$ be a minimizer on the right-hand side of (4.23), where $\tilde{I} \subseteq \operatorname{Row}(D)$ and $\tilde{J} \subseteq \operatorname{Col}(D)$, for which we have

$$\operatorname{rank} D = \operatorname{rank} Q[\tilde{I}, \tilde{J}] - |\tilde{I}| - |\tilde{J}| + 2(n + m + l)$$

and $\operatorname{rank} T[\tilde{I}, \tilde{J}] = 0$. The structure of T allows us to assume that $\tilde{I} \supseteq X \cup Y$ and $\tilde{J} \supseteq X \cup U$. Putting $I = U \setminus \tilde{I}$ and $J = Y \setminus \tilde{J}$, we have $(I, J) \in \mathcal{C}_{\mathcal{K}}$ and

$$\operatorname{rank} Q[\tilde{I}, \tilde{J}] = \operatorname{rank} \begin{pmatrix} A - \lambda I_n & B[X, I] \\ C[J, X] & O \end{pmatrix} + |U \setminus I| + |Y \setminus J|.$$

Therefore, we obtain

$$\operatorname{rank}(A + BKC - \lambda I_n) = \operatorname{rank} \begin{pmatrix} A - \lambda I_n & B[X, I] \\ C[J, X] & O \end{pmatrix}.$$

■

As the above proof reveals, the content of Theorem 6.5.1 lies in a min-max duality assertion that

$$\max_{K \in \mathcal{K}} \operatorname{rank}(A + BKC - \lambda I_n) = \min_{(I,J) \in \mathcal{C}_{\mathcal{K}}} \operatorname{rank} \begin{pmatrix} A - \lambda I_n & B[X, I] \\ C[J, X] & O \end{pmatrix}. \tag{6.90}$$

This identity is observed by Tanino–Takahashi [309] in the special case of K of the form (6.82).

We also mention the following result of Tarokh [311], with a simple proof using a basic fact about mixed matrices.

Theorem 6.5.2. *A complex number $\lambda \in \mathbf{C}$ is a fixed mode of (A, B, C) with respect to $\mathcal{K}$ of (6.85) if and only if $\begin{pmatrix} A - \lambda I_n & B[X, I] \\ C[J, X] & O \end{pmatrix}$ is singular for all (I, J) such that $K[I, J]$ is nonsingular.*

Proof. Application of Lemma 4.2.7 to the mixed matrix $D = Q + T$ in (6.88) and (6.89). ■

As a refinement of the above theorem, Tanino–Takahashi [309] showed

$$\max_{K \in \mathcal{K}} \operatorname{rank}(A + BKC - \lambda I_n) = \max_{I,J} \left\{ \operatorname{rank} \begin{pmatrix} A - \lambda I_n & B[X,I] \\ C[J,X] & O \end{pmatrix} - |I| \right\}, \tag{6.91}$$

where the maximum on the right-hand side is taken over all (I, J) such that $K[I, J]$ is nonsingular. This identity can be derived similarly from Theorem 4.2.8 applied to the mixed matrix $D = Q + T$ in (6.88) and (6.89).

Remark 6.5.3. Though both Theorem 6.5.1 and Theorem 6.5.2 are concerned with combinatorial characterizations of fixed modes, they are complementary in the following sense. The former guarantees the existence of a "certificate" (namely, (I, J) in the theorem) for λ being a fixed mode, whereas the latter (in its contraposition) for λ not being a fixed mode. □

Remark 6.5.4. In §6.4 we have discussed the controllability for (A, B). The fixed mode problem contains this as a special case. Given (A, B), consider a fixed mode problem with $C = I_n$ and $\mathcal{K}$ defined by $\hat{K}_{ij} = 1$ for all (i, j). Then, by a fundamental result in control theory (Wolovich [342]), (A, B) is controllable if and only if (A, B, C) has no fixed modes with respect to $\mathcal{K}$. □

6.5.2 Structurally Fixed Modes

The concept of a structurally fixed mode is proposed by Sezer–Šiljak [294] on the basis of the observation that some fixed modes stem from an accidental matching of numerical values of system parameters and others from the combinatorial structure of the system. For a system (A, B, C) we associate a family $\mathcal{S}$ of systems that are "structurally equivalent" to (A, B, C), where $(\hat{A}, \hat{B}, \hat{C})$ is said to be structurally equivalent to (A, B, C) if $\hat{A}$, $\hat{B}$, and $\hat{C}$ have respectively the same zero/nonzero structure as that of A, B, and C. A system (A, B, C) is said to have *structurally fixed modes* if every $(\hat{A}, \hat{B}, \hat{C}) \in \mathcal{S}$ has fixed modes. It is noted that considering the family $\mathcal{S}$ of structurally equivalent systems is algebraically tantamount to considering a single system in which all the nonzero entries of A, B, and C are algebraically independent. See Šiljak [300, §1.6] and Trave–Titli–Tarras [320] for more account on structurally fixed modes.

Example 6.5.5. For a scalar system $(A, B, C) = ((1), (0), (0))$ we associate a structured system $(\hat{A}, \hat{B}, \hat{C}) = ((a), (0), (0))$ with an independent parameter a. The system (A, B, C) has a fixed mode $\lambda = 1$ with respect to $\mathcal{K} = \{(k) \mid k \in \mathbf{R}\}$, and the structured system $(\hat{A}, \hat{B}, \hat{C})$ has a fixed mode $\lambda = a$. Accordingly, the system (A, B, C) has a structurally fixed mode. Note that the fixed mode of $(\hat{A}, \hat{B}, \hat{C})$ varies with a. □

The following theorem of Sezer–Šiljak [294] gives a combinatorial characterization of the existence of structurally fixed modes of a system (A, B, C). The feedback structure is represented by a family $\mathcal{K}$ of matrices K subject to an arbitrarily specified zero/nonzero structure.

Theorem 6.5.6. *A system* (A, B, C) *has structurally fixed modes with respect to* $\mathcal{K}$ *of* (6.85) *if and only if either of the following two conditions is satisfied, where* $\mathcal{C}_{\mathcal{K}}$ *is defined in* (6.86).

(i) *There exists* $(I, J) \in \mathcal{C}_{\mathcal{K}}$ *and a partition of* X *into disjoint subsets* X_1, X_2, X_3 *with* $X_2 \neq \emptyset$ *such that* $A[X_1, X_2 \cup X_3] = O$, $A[X_2, X_3] = O$, $B[X_1 \cup X_2, I] = O$, *and* $C[J, X_2 \cup X_3] = O$, *that is, such that*

$$A = \begin{array}{c} \\ X_1 \\ X_2 \\ X_3 \end{array} \!\!\begin{array}{c} \begin{array}{ccc} X_1 & X_2 & X_3 \end{array} \\ \begin{pmatrix} A_{11} & O & O \\ A_{21} & A_{22} & O \\ A_{31} & A_{32} & A_{33} \end{pmatrix} \end{array}, \qquad B = \begin{array}{c} \\ X_1 \\ X_2 \\ X_3 \end{array} \!\!\begin{array}{c} \begin{array}{cc} I & U \setminus I \end{array} \\ \begin{pmatrix} O & B_{12} \\ O & B_{22} \\ B_{31} & B_{32} \end{pmatrix} \end{array},$$

$$C = \begin{array}{c} \\ J \\ Y \setminus J \end{array} \!\!\begin{array}{c} \begin{array}{ccc} X_1 & X_2 & X_3 \end{array} \\ \begin{pmatrix} C_{11} & O & O \\ C_{21} & C_{22} & C_{23} \end{pmatrix} \end{array}.$$

(ii) *There exists* $(I, J) \in \mathcal{C}_{\mathcal{K}}$ *such that*

$$\text{term-rank} \begin{pmatrix} A & B[X, I] \\ C[J, X] & O \end{pmatrix} \leq n - 1. \tag{6.92}$$

Proof. This is proven later using Theorem 6.5.7 below. It may be noted that the necessity of (ii) follows from Theorem 6.5.1. ∎

The criteria given in the above theorem can be reformulated as follows (Linnemann [175], Pichai–Sezer–Šiljak [269]). We represent the structure of a system (A, B, C) with feedback K by a directed graph $G = (V, E)$. The vertex set V and the arc set E are defined by

$$\begin{aligned} &V = X \cup U \cup Y, \quad X = \{x_1, \cdots, x_n\}, \ U = \{u_1, \cdots, u_m\}, \ Y = \{y_1, \cdots, y_l\}, \\ &E = E_A \cup E_B \cup E_C \cup E_K, \\ &\qquad E_A = \{(x_j, x_i) \mid A_{ij} \neq 0\}, \quad E_B = \{(u_j, x_i) \mid B_{ij} \neq 0\}, \\ &\qquad E_C = \{(x_j, y_i) \mid C_{ij} \neq 0\}, \quad E_K = \{(y_j, u_i) \mid \hat{K}_{ij} \neq 0\}. \end{aligned}$$

Note that G is the graph associated with the matrix $\begin{pmatrix} A & B & O \\ O & O & K \\ C & O & O \end{pmatrix}$ as in §2.2.1.

Theorem 6.5.7. *A system* (A, B, C) *has no structurally fixed modes with respect to* $\mathcal{K}$ *of* (6.85) *if and only if both of the following two conditions are satisfied:*

(G1) *Each vertex of* X *is contained in a strong component of* G *which includes an arc of* E_K,

(G2) *There exists a set of mutually disjoint cycles in G that covers the vertices of X.*

Proof. This will be proven later as a corollary of a more general result in §6.5.3; see Remark 6.5.16. ■

The equivalence of Theorem 6.5.6 and Theorem 6.5.7 can be shown by a fairly easy graph-theoretic argument. First, the condition (i) of Theorem 6.5.6 is easily seen to be equivalent to the violation of (G1) of Theorem 6.5.7. Next, denote by D_0 the matrix D in (6.88) with $\lambda = 0$. The condition in (G2) of Theorem 6.5.7 is equivalent to the term-nonsingularity of D_0. Note that $(\tilde{I}, \tilde{J})$ is a cover of D_0 if and only if $Y \setminus \tilde{J} \subseteq \tilde{I}$, $U \setminus \tilde{I} \subseteq \tilde{J}$, $(I, J) \in \mathcal{C}_K$ and $(\tilde{I} \cap (X \cup Y), \tilde{J} \cap (X \cup U))$ is a cover of the matrix in (6.92) for $I = \tilde{I} \cap U$, $J = \tilde{J} \cap Y$. Then the König–Egerváry theorem (Theorem 2.2.15) shows the equivalence between the condition (ii) of Theorem 6.5.6 and the violation of (G2) of Theorem 6.5.7.

The two theorems are certainly equivalent as above, and moreover both show how to check for the existence of a structurally fixed mode efficiently using binary operations only. They are, however, complementary in the sense that Theorem 6.5.6 guarantees a "certificate" for the existence of a structurally fixed mode whereas Theorem 6.5.7 for the nonexistence. See also Remark 6.5.3.

Example 6.5.8. The conditions (G1) and (G2) in Theorem 6.5.7 do not discriminate the existence of zero and nonzero fixed modes. Consider, for example, a scalar system ($n = 1$) with $A = (0)$, $B = (b)$, $C = (c)$, and $K = (0)$. Obviously, this system has a (structurally) fixed mode at zero, and no nonzero fixed mode. Neither (G1) nor (G2) in Theorem 6.5.7 is satisfied. Note also that both of the conditions (i) and (ii) in Theorem 6.5.6 are satisfied. In contrast, the matroid-theoretic method to be developed in the next subsection will separate zero and nonzero fixed modes. □

Example 6.5.9. Consider a decentralized system ($n = 6$) with three local stations described by

$$A = \begin{pmatrix} 0 & a_1 & 0 & 0 & 0 & 0 \\ 0 & 0 & 0 & 0 & 0 & 0 \\ a_2 & 0 & 0 & 1 & 0 & 1 \\ 0 & 0 & 0 & 0 & 0 & 0 \\ 0 & 0 & 0 & 1 & 0 & 1 \\ 0 & 0 & 0 & 0 & 0 & 0 \end{pmatrix}, \quad B = \begin{pmatrix} 0 & 0 & 0 \\ b_1 & 0 & 0 \\ 0 & 0 & 0 \\ 0 & b_2 & 0 \\ 0 & 0 & 0 \\ 0 & 0 & b_3 \end{pmatrix}, \quad K = \begin{pmatrix} k_1 & k_2 & 0 & 0 & 0 & 0 \\ 0 & 0 & k_3 & k_4 & 0 & 0 \\ 0 & 0 & 0 & 0 & k_5 & k_6 \end{pmatrix},$$

and $C = I_6$, where $\{a_1, a_2, b_1, b_2, b_3\}$ is the set of independent parameters (see Reinschke [278, Example 4] and the references cited therein for the origin of this system). It can be verified that $\lambda = 0$ is a fixed mode of all the systems parametrized by $\{a_1, a_2, b_1, b_2, b_3\}$. In this sense, the family parametrized by $\{a_1, a_2, b_1, b_2, b_3\}$ has a structure that admits a fixed mode. This fact, however, cannot be captured in the present formulation of structurally fixed modes, since $\lambda = 0$ is not a fixed mode if the four entries of A having a constant value of one are replaced by free parameters. Accordingly, the graph-theoretic method of Theorem 6.5.7 leads us to the conclusion that this system has no structurally fixed mode. In contrast, the matroid-theoretic method to be developed in the next subsection is capable of detecting this kind of fixed mode, as will be explained in Example 6.5.19. □

6.5.3 Mixed Polynomial Matrix Formulation

Let us introduce a formulation of structurally fixed modes, due to Murota [209], which is more general and would be more realistic than the one described in §6.5.2.

Let

$$A(s) = Q(s) + T(s) \tag{6.93}$$

be an $n \times n$ mixed polynomial matrix with respect to $(\boldsymbol{K}, \boldsymbol{F}) = (\mathbf{Q}, \mathbf{R})$ such that $Q(s)$ satisfies the stronger assumption

(MP-Q2) Every nonvanishing subdeterminant of $Q(s)$ is a monomial over $\boldsymbol{K}$, i.e., of the form αs^p with $\alpha \in \boldsymbol{K}$ and an integer p.

We put $R = \mathrm{Row}(A)$ and $C = \mathrm{Col}(A)$. Let K be an $n \times n$ generic matrix, the nonzero entries of which are algebraically independent numbers in $\mathbf{R}$. Then

$$A_K(s) = A(s) + K = Q(s) + T(s) + K = Q(s) + T_K(s) \tag{6.94}$$

is a mixed polynomial matrix, where $T_K(s) = T(s) + K$. We assume throughout that $A_K(s)$ is nonsingular.

Denote by $\mathcal{K}$ the set of nonzero entries of K, and by $\mathcal{S}$ the set of nonzero coefficients in $T(s)$. Then $\mathcal{K} \cup \mathcal{S}$ is algebraically independent over $\mathbf{Q}$. It should be clear that assuming the algebraic independence of $\mathcal{S}$ is equivalent to regarding the members of $\mathcal{S}$ as independent parameters, and therefore to considering a family of systems parametrized by those parameters in $\mathcal{S}$. A particular system in this family having algebraically independent parameter values has a fixed mode with respect to $\mathcal{K}$ if and only if each system parametrized by $\mathcal{S}$ has a fixed mode with respect to $\mathcal{K}$. Note, however, the value of a fixed mode varies, in general, with the parameters in $\mathcal{S}$.

We define the *fixed polynomial* $\psi(s)$ as the greatest common divisor in $\mathbf{C}[s]$ of all $\det A_K(s)$, where arbitrary values are substituted into $\mathcal{K}$. Namely,

$$\psi(s) = \gcd\{\det A_K(s) \mid K \in \mathcal{K}\} \tag{6.95}$$

with the obvious understanding of the notation "$K \in \mathcal{K}$". Also we call a complex number $\lambda \in \mathbf{C}$ a *fixed mode* if $\psi(\lambda) = 0$.

Remark 6.5.10. The structurally fixed mode as formulated in §6.5.2 is a special case of the present formulation. To see this, note the identity

$$\det(A + BKC - sI_n) = (-1)^{m+l} \det \begin{pmatrix} A - sI_n & B & O \\ O & -I_m & K \\ C & O & -I_l \end{pmatrix}$$

and take

$$Q(s) = \begin{pmatrix} -sI_n & O & O \\ O & -I_m & O \\ O & O & -I_l \end{pmatrix}, \quad T(s) = \begin{pmatrix} A & B & O \\ O & O & O \\ C & O & O \end{pmatrix}, \quad \mathcal{K} = \begin{pmatrix} O & O & O \\ O & O & K \\ O & O & O \end{pmatrix} \qquad (6.96)$$

in the decomposition (6.94). Then $\mathcal{S}$ is equal to the set of nonzero entries of A, B, and C, and $\mathcal{K}$ to the set of nonzero entries of K. □

We shall derive a necessary and sufficient condition, of a combinatorial nature, for the existence of fixed modes with the aid of the CCF of LM-matrices. The derived condition can be tested efficiently. The proposed algorithm is suitable for practical applications in that it is free from numerical difficulty of rounding errors and is guaranteed to run in polynomial time in the size of the control system in question. The established criterion naturally reduces to the graph-theoretic criterion of §6.5.2.

Regarding $\det A_K(s)$ as a polynomial in $(s, \mathcal{S}, \mathcal{K})$ over $\mathbf{Q}$, we consider the decomposition into irreducible polynomials in $\mathbf{Q}[s, \mathcal{S}, \mathcal{K}]$. As a consequence of the assumption (MP-Q2), this is expressed (cf. Lemma 6.3.2) as

$$\det A_K(s) = \alpha s^p \cdot \prod_{k \in \Psi_1} \psi_k(s, \mathcal{S}) \cdot \prod_{k \in \Psi_2} \psi_k(s, \mathcal{S}, \mathcal{K}), \qquad (6.97)$$

where $\alpha \in \mathbf{Q} \setminus \{0\}$, p is a nonnegative integer, $\psi_k(s, \mathcal{S}) \in \mathbf{Q}[s, \mathcal{S}] \setminus \mathbf{Q}[s]$ for $k \in \Psi_1$, and $\psi_k(s, \mathcal{S}, \mathcal{K}) \in \mathbf{Q}[s, \mathcal{S}, \mathcal{K}] \setminus \mathbf{Q}[s, \mathcal{S}]$ for $k \in \Psi_2$. The index sets, Ψ_1 and Ψ_2, classify the irreducible factors according to whether they do not contain or do contain variables in $\mathcal{K}$.

Lemma 6.5.11. *The fixed polynomial* $\psi(s)$ *is given as*

$$\psi(s) = \alpha s^p \cdot \prod_{k \in \Psi_1} \psi_k(s, \mathcal{S}). \qquad (6.98)$$

Proof. This is easy to see. ■

The existence of a zero fixed mode is easy to characterize.

Theorem 6.5.12. *For nonsingular* $A_K(s)$ *of* (6.94) *satisfying* (MP-Q2), *the following conditions are equivalent.*

(i) $\lambda = 0$ *is not a fixed mode.*

(ii) *There exists* (I, J) *such that* $Q(0)[R \setminus I, C \setminus J]$ *is nonsingular and* $(T(0) + K)[I, J]$ *is term-nonsingular.*

(iii) *There exists no* (I, J) *such that* $T(0)[I, J] = O$, $K[I, J] = O$, *and* $\operatorname{rank} Q(0)[I, J] \leq |I| + |J| - n - 1$.

Proof. The equivalence of (i) to the nonsingularity of $A_K(0)$ is obvious. The equivalence to (ii) and (iii) are by Lemma 4.2.7 and by (4.23) in Corollary 4.2.12. ■

The multiplicity of the zero fixed mode, the exponent p in (6.97), is equal to $\mathrm{ord}_s \det A_K(s)$, which can be characterized in terms of an independent assignment problem (weighted matroid intersection problem) as in §6.2.4 (see Remark 6.2.3 and Remark 6.2.10). Recall the notation ord_s for the minimum degree in s of a nonzero term in a polynomial introduced in (2.1). To be specific, we consider, as in (6.55), an LM-polynomial matrix

$$\tilde{A}_K(s) = \tilde{A}_K(s;t) = \begin{pmatrix} I_n & Q(s) \\ -\mathrm{diag}\,(t_1, \cdots, t_n) & T_K(s) \end{pmatrix} = \begin{pmatrix} \tilde{Q}(s) \\ \tilde{T}_K(s;t) \end{pmatrix} \tag{6.99}$$

associated with $A_K(s)$, where $t_1, \cdots, t_n$ are new indeterminates and $t = (t_1, \cdots, t_n)$. Put $\tilde{C} = \mathrm{Col}(\tilde{A}_K) \simeq R \cup C$. With reference to (6.4) we define $\zeta : \tilde{C} \to \mathbf{Z}$ by

$$\zeta(j) = \begin{cases} -r_j & (j \in R) \\ -c_j & (j \in C) \end{cases} \tag{6.100}$$

as well as the usual convention $\zeta(J) = \sum_{j \in J} \zeta(j)$ for $J \subseteq \tilde{C}$. For $J \subseteq \tilde{C}$ such that $\tilde{T}_K[\mathrm{Row}(\tilde{T}_K), J]$ is term-nonsingular, we denote by $\eta(J)$ the lowest degree in s of a nonzero term in $\det \tilde{T}_K[\mathrm{Row}(\tilde{T}_K), J]$. Note that $\eta(J)$ admits a combinatorial expression, namely (cf. (6.13)),

$$\eta(J) = \min\{w(M) \mid M\text{: } n\text{-matching with } \partial^- M = J \text{ in } G_{\tilde{T}}\ \}, \tag{6.101}$$

where $G_{\tilde{T}} = (\mathrm{Row}(\tilde{T}_K), \tilde{C}; E_{\tilde{T}})$ is a bipartite graph with arc set $E_{\tilde{T}} = \{(i,j) \mid i \in \mathrm{Row}(\tilde{T}_K), j \in \tilde{C}, (\tilde{T}_K)_{ij} \neq 0\}$ and $w(M) = \sum_{(i,j) \in M} \mathrm{ord}_s (\tilde{T}_K)_{ij}$. Also we define

$$\begin{aligned} \mathcal{J} = \{J \subseteq \tilde{C} \mid\ & \tilde{Q}[\mathrm{Row}(\tilde{Q}), \tilde{C} \setminus J]\text{: nonsingular}, \\ & \tilde{T}_K[\mathrm{Row}(\tilde{T}_K), J]\text{: term-nonsingular}\}. \end{aligned} \tag{6.102}$$

Theorem 6.5.13. *For nonsingular $A_K(s)$ of* (6.94) *satisfying* (MP-Q2)*, the multiplicity p of the zero fixed mode is given by*

$$p = \min\{\zeta(\tilde{C} \setminus J) + \eta(J) \mid J \in \mathcal{J}\} - \zeta(R).$$

Proof. This is a straightforward adaptation of Theorem 6.2.5. ■

On the basis of this theorem, the multiplicity of the zero fixed mode can be computed efficiently by a variant of the algorithm in §6.2.6.

The nonzero fixed modes can be treated by means of the CCF of the LM-polynomial matrix $\tilde{A}_K(s)$. This is based on the fact (Theorem 4.5.9) that the CCF corresponds to the decomposition of the determinant into irreducible factors.

Regarding $\tilde{A}_K(s)$ as an LM-matrix with respect to $(\mathbf{Q}[s], \mathbf{R}(s))$ we may think of its block-triangular form ("CCF over a ring") in the sense of Theorem 4.4.19, which is obtained from $\tilde{A}_K(s)$ through a unimodular transformation over $\mathbf{Q}[s]$. Let $\hat{A}_K(s)$ and $\bar{A}_K(s)$ be the block-triangular matrix and the CCF of $\tilde{A}_K(s)$ as in Theorem 4.4.19. The families of the row sets and the column sets in the CCF are denoted respectively by $\{\bar{R}_k \mid k = 1, \cdots, b\}$ and $\{\bar{C}_k \mid k = 1, \cdots, b\}$, and the square diagonal blocks by

$$\bar{A}_k = \begin{pmatrix} \bar{Q}_k \\ \bar{T}_k \end{pmatrix} = \bar{A}_K[\bar{R}_k, \bar{C}_k], \qquad k = 1, \cdots, b.$$

Note that $\hat{A}_K(s)$ and $\bar{A}_K(s)$ have identical diagonal blocks, though they may differ in the upper-triangular part. Similarly to (6.102) we define

$$\mathcal{J}_k = \{J \subseteq \bar{C}_k \mid \bar{Q}_k[\mathrm{Row}(\bar{Q}_k), \bar{C}_k \setminus J]\text{: nonsingular}, \\ \bar{T}_k[\mathrm{Row}(\bar{T}_k), J]\text{: term-nonsingular}\}, \quad k = 1, \cdots, b.$$

For $J \subseteq \bar{C}_k$ such that $\bar{T}_k[\mathrm{Row}(\bar{T}_k), J]$ is term-nonsingular, we denote by $\xi_k(J)$ and $\eta_k(J)$ the highest and lowest degrees in s of a nonzero term in $\det \bar{T}_k[\mathrm{Row}(\bar{T}_k), J]$. Note that $\xi_k(J)$ and $\eta_k(J)$ can be expressed in terms of weighted-matching problems, just as (6.101).

To identify the nonzero fixed modes, we classify the diagonal blocks into three categories by defining three index sets

$$\bar{\Psi}_0 = \{k \mid \bar{A}_k \text{ contains no variable of } \mathcal{S} \cup \mathcal{K}\}, \tag{6.103}$$
$$\bar{\Psi}_1 = \{k \mid \bar{A}_k \text{ contains a variable of } \mathcal{S} \text{ and no variable of } \mathcal{K}\}, \tag{6.104}$$
$$\bar{\Psi}_2 = \{k \mid \bar{A}_k \text{ contains a variable of } \mathcal{K}\}. \tag{6.105}$$

Theorem 6.5.14. *For nonsingular $A_K(s)$ of* (6.94) *satisfying* (MP-Q2), *the number of nonzero fixed modes is given by*

$$\sum_{k \in \bar{\Psi}_1} \left[\max\{\zeta(\bar{C}_k \setminus J) + \xi_k(J) \mid J \in \mathcal{J}_k\} - \min\{\zeta(\bar{C}_k \setminus J) + \eta_k(J) \mid J \in \mathcal{J}_k\}\right].$$

Hence there exist no nonzero fixed modes if and only if

$$\max\{\zeta(\bar{C}_k \setminus J) + \xi_k(J) \mid J \in \mathcal{J}_k\} = \min\{\zeta(\bar{C}_k \setminus J) + \eta_k(J) \mid J \in \mathcal{J}_k\} \tag{6.106}$$

for each $k \in \bar{\Psi}_1$.

Proof. It follows from (6.99) that

$$\det A_K(s) = \det \tilde{A}_K(s; 1), \tag{6.107}$$

where $\tilde{A}_K(s;1) = \tilde{A}_K(s;t)\Big|_{t_1 = \cdots = t_n = 1}$. On the other hand, we have

$$\det \tilde{A}_K(s;t) = \det \hat{A}_K(s;t) = \det \bar{A}_K(s;t) = \prod_{k=1}^{b} \det \bar{A}_k(s;t), \qquad (6.108)$$

where the first equality may be assumed by the fact that $\hat{A}_K(s;t)$ is obtained form $\tilde{A}_K(s;t)$ through a unimodular transformation over $\mathbf{Q}[s]$.

According to Theorem 4.5.9 the expression (6.108) gives a decomposition into irreducible factors in the ring $\mathbf{Q}(s)[\mathcal{S}, \mathcal{K}, t]$. In view of Lemma 6.3.2 we see further that, for each k, $\det \bar{A}_k(s;t)$ is a product of a monomial in s and an irreducible polynomial in $\mathbf{Q}[s, \mathcal{S}, \mathcal{K}, t]$. This statement needs only a marginal modification even after the substitution of $t_1 = \cdots = t_n = 1$. Namely, we claim that, for $k = 1, \cdots, b$, we have

$$\det \bar{A}_k(s;1) = \rho_k(s) \cdot \bar{\psi}_k(s, \mathcal{S}, \mathcal{K}), \qquad (6.109)$$

where $\rho_k(s) \in \mathbf{Q}[s]$ is a monomial in s, and $\bar{\psi}_k(s, \mathcal{S}, \mathcal{K}) \in \mathbf{Q}[s, \mathcal{S}, \mathcal{K}] \setminus \mathbf{Q}[s]$ is irreducible in $\mathbf{Q}[s, \mathcal{S}, \mathcal{K}]$.

The proof of (6.109) is easy. Denote by $\mathcal{T}_i$ the set of elements of $\mathcal{T} \equiv \mathcal{S} \cup \mathcal{K}$ contained in the row of $\tilde{T}_K(s;t)$ corresponding to t_i $(i = 1, \cdots, n)$. Also denote $\det \bar{A}_k(s;t)$ by $f_k(\mathcal{T}_1, \cdots, \mathcal{T}_n; t_1, \cdots, t_n)$, which is irreducible in $\mathbf{Q}(s)[\mathcal{T}_1, \cdots, \mathcal{T}_n, t_1, \cdots, t_n]$ by Theorem 4.5.6. We see

$$f_k(\mathcal{T}_1, \cdots, \mathcal{T}_n; t_1, \cdots, t_n) = \left(\prod_{i \in \mathrm{Row}(\bar{T}_k)} t_i \right) \cdot f_k(\mathcal{T}_1/t_1, \cdots, \mathcal{T}_n/t_n; 1, \cdots, 1),$$

where $\mathcal{T}_i/t_i$ means substituting a/t_i for each indeterminate $a \in \mathcal{T}_i$. This expression implies that $\det \bar{A}_k(s;1)$ is irreducible in $\mathbf{Q}(s)[\mathcal{S}, \mathcal{K}]$, which, with Lemma 6.3.2, completes the proof of (6.109).

With reference to (6.97), a combination of (6.107), (6.108), and (6.109) shows

$$\prod_{k \in \Psi_1} \psi_k(s, \mathcal{S}) = \prod_{k \in \bar{\Psi}_1} \bar{\psi}_k(s, \mathcal{S}), \qquad \prod_{k \in \Psi_2} \psi_k(s, \mathcal{S}, \mathcal{K}) = \prod_{k \in \bar{\Psi}_2} \bar{\psi}_k(s, \mathcal{S}, \mathcal{K}).$$

In particular, the first expression above gives the nonmonomial part of the fixed polynomial in Lemma 6.5.11. Finally we note

$$\deg_s \bar{\psi}_k(s) = \max\{\zeta(\bar{C}_k \setminus J) + \xi_k(J) \mid J \in \mathcal{J}_k\} - \min\{\zeta(\bar{C}_k \setminus J) + \eta_k(J) \mid J \in \mathcal{J}_k\},$$

which is a corollary of Theorem 6.2.5. ∎

On the basis of the combinatorial characterization of fixed modes in Theorem 6.5.12 and Theorem 6.5.14, an efficient algorithm for testing the existence of zero/nonzero fixed modes is designed in the next section.

6.5.4 Algorithm

In this section we describe an efficient algorithm to check for the existence of nonzero fixed modes on the basis of Theorem 6.5.14, whereas the algorithm of §4.2.4 for computing the rank of an LM-matrix can be utilized readily for the zero fixed mode by Theorem 6.5.12. The basic idea of the algorithm for nonzero fixed modes is the same as that for nonzero uncontrollable modes in §6.4.4.

A concrete description of the algorithm for the condition (6.106) follows. We use an auxiliary network $N = (V, E, \gamma)$ with underlying graph $G = (V, E)$ and length function $\gamma : E \to \mathbf{Z}$, in a way consistent with §4.2.4. The vertex set V is defined as

$$V = V_Q \cup V_T = (R_Q \cup C_Q) \cup (R_T \cup C_T),$$

where $R_Q = \mathrm{Row}(Q)$, $C_Q = \mathrm{Col}(Q)$, $R_T = \mathrm{Row}(T)$, $C_T = \mathrm{Col}(T)$, $V_Q = R_Q \cup C_Q$, and $V_T = R_T \cup C_T$. The arc set E consists of six disjoint parts,

$$E = E_{TQ} \cup E_{QT} \cup E_Q \cup E_T \cup E_K \cup E_M,$$

to be defined below. We denote by $\varphi_Q : R \cup C \to R_Q \cup C_Q$ and $\varphi_T : R \cup C \to R_T \cup C_T$ the obvious one-to-one correspondences.

Let $\hat{I} \subseteq R$ and $\hat{J} \subseteq C$ be such that $Q(1)[R \setminus \hat{I}, C \setminus \hat{J}]$ is nonsingular and $T_K[\hat{I}, \hat{J}]$ is term-nonsingular, where such $(\hat{I}, \hat{J})$ exists by the nonsingularity of $A_K(1)$ and Lemma 4.2.7. We define

$$\begin{aligned} E_{TQ} &= \{(\varphi_T(i), \varphi_Q(i)) \mid i \in \hat{I}\} \cup \{(\varphi_T(j), \varphi_Q(j)) \mid j \in C \setminus \hat{J}\}, \\ E_{QT} &= \{(\varphi_Q(i), \varphi_T(i)) \mid i \in R \setminus \hat{I}\} \cup \{(\varphi_Q(j), \varphi_T(j)) \mid j \in \hat{J}\}. \end{aligned}$$

Let P be the pivotal transform of $Q = Q(1)$ with pivot $\hat{Q} \equiv Q[R \setminus \hat{I}, C \setminus \hat{J}]$ (cf. (6.76)), where $\mathrm{Row}(P) = (C \setminus \hat{J}) \cup \hat{I}$ and $\mathrm{Col}(P) = (R \setminus \hat{I}) \cup \hat{J}$. Note that P is a constant matrix free from s. With reference to P we define

$$E_Q = \{(\varphi_Q(i), \varphi_Q(j)) \mid P_{ij} \neq 0, i \in (C \setminus \hat{J}) \cup \hat{I}, j \in (R \setminus \hat{I}) \cup \hat{J}\}.$$

The structure of $T_K(s) = T(s) + K$ is represented by E_T, E_K, and E_M. For each nonzero entry $T_{ij}(s)$ of $T(s)$ we consider a pair of parallel arcs a_{ij}^0 and a_{ij}^1 with $\partial^+ a_{ij}^0 = \partial^+ a_{ij}^1 = \varphi_T(i) \in R_T$ and $\partial^- a_{ij}^0 = \partial^- a_{ij}^1 = \varphi_T(j) \in C_T$. Putting

$$E_T^0 = \{a_{ij}^0 \mid T_{ij} \neq 0, i \in R, j \in C\}, \quad E_T^1 = \{a_{ij}^1 \mid T_{ij} \neq 0, i \in R, j \in C\},$$

we define $E_T = E_T^0 \cup E_T^1$ and

$$E_K = \{(\varphi_T(i), \varphi_T(j)) \mid K_{ij} \neq 0, i \in R, j \in C\}.$$

Since $T_K[\hat{I}, \hat{J}]$ is term-nonsingular, the bipartite graph $(R_T, C_T; E_T \cup E_K)$ with vertex set $R_T \cup C_T$ and arc set $E_T \cup E_K$ has a matching M $(\subseteq E_T \cup E_K)$

such that $|M| = |\hat{I}|(= |\hat{J}|)$, $\varphi_T(\hat{I}) = \partial^+ M$, and $\varphi_T(\hat{J}) = \partial^- M$. We define E_M as the set of reoriented arcs of M, i.e.,

$$E_M = \{\bar{a} \mid a \in M\},$$

where $\bar{a}$ denotes the reorientation of a.

The length function $\gamma : E \to \mathbf{Z}$ is defined with reference to r_i and c_i $(i = 1, \cdots, n)$ of (6.4) as

$$\gamma(a) = \begin{cases} -r_i & \text{if} \quad a = (\varphi_T(i), \varphi_Q(i)) \in E_{TQ},\ i \in \hat{I}, \\ -c_j & \text{if} \quad a = (\varphi_T(j), \varphi_Q(j)) \in E_{TQ},\ j \in C \setminus \hat{J}, \\ r_i & \text{if} \quad a = (\varphi_Q(i), \varphi_T(i)) \in E_{QT},\ i \in R \setminus \hat{I}, \\ c_j & \text{if} \quad a = (\varphi_Q(j), \varphi_T(j)) \in E_{QT},\ j \in \hat{J}, \\ 0 & \text{if} \quad a \in E_Q \cup E_K, \\ -\mathrm{ord}_s T_{ij}(s) & \text{if} \quad a \in E_T^0, \\ -\deg_s T_{ij}(s) & \text{if} \quad a \in E_T^1, \\ -\gamma(a') & \text{if} \quad a \in E_M \text{ is the reorientation of } a' \in M. \end{cases}$$

For a nonzero entry $T_{ij}(s)$ of $T(s)$ with $\mathrm{ord}_s T_{ij}(s) = \deg_s T_{ij}(s)$ (which is the case if $T_{ij}(s)$ is a monomial in s), the pair of arcs, having the same length, may be replaced by a single arc of the same length.

We are now ready to rephrase the condition (6.106) in terms of the network $N = (G, \gamma)$. Note that the strong components of G correspond to diagonal blocks of the CCF of $\tilde{A}_K$. For each strong component of G, say $\hat{G} = (\hat{V}, \hat{E})$, we consider the condition that the sum of the lengths $\gamma(a)$ along any directed cycle in $\hat{G}$ is equal to zero (cf. (6.78)). Since $\hat{G}$ is strongly connected, this condition is equivalent, by Theorem 2.2.35(2), to the existence of a potential function $\pi : \hat{V} \to \mathbf{Z}$ such that

$$\gamma(a) = \pi(\partial^- a) - \pi(\partial^+ a) \qquad (\forall a \in \hat{E}). \tag{6.110}$$

Theorem 6.5.15. *For nonsingular* $A_K(s)$ *of* (6.94) *satisfying* (MP-Q2), *there exist no nonzero fixed modes if and only if each strong component of* G *either contains an arc of* E_K *or admits a potential function* π *such that* (6.110) *holds.*

Proof. For simplicity of notation let us assume that G itself is a strong component that does not contain an arc of E_K. We also assume for simplicity of argument that each $T_{ij}(s)$ is a monomial in s so that each pair of parallel arcs in E_T is replaced by a single arcs. Consider the independent assignment problem as in §6.2 to compute $\deg_s \det \tilde{A}_K(s)$ for $\tilde{A}_K(s)$ of (6.99). The rest of the proof is the same as that of Theorem 6.4.18. ∎

The overall computational complexity for testing for the existence of fixed modes on the basis of Theorem 6.5.12 and Theorem 6.5.15 is dominated by that for the construction of the graph G and therefore bounded by $\mathrm{O}(n^3 \log n)$. Note that the decomposition of G into strong components can be done in $\mathrm{O}(|E|)$ time and the potential function of (6.110) for a strong component $\hat{G} = (\hat{V}, \hat{E})$, if any, can be found in time of $\mathrm{O}(|\hat{E}|)$ by the procedure of Remark 6.4.19. It should be emphasized here that the whole algorithm involves only pivoting operations on the matrix $Q(1)$, the entries of which are rational numbers (simple numbers such as ± 1 in practical applications).

Remark 6.5.16. The graph-theoretic criterion in Theorem 6.5.7 can be derived from Theorem 6.5.12 and Theorem 6.5.15 applied to the matrix $D(s) = Q(s)+T(s)+K$ defined by (6.96) in Remark 6.5.10. Recall (cf. (6.88)) that both $\mathrm{Row}(D)$ and $\mathrm{Col}(D)$ have a natural one-to-one correspondence with $X \cup U \cup Y$, to be denoted by $\nu_R : \mathrm{Row}(D) \to X \cup U \cup Y$ and $\nu_C : \mathrm{Col}(D) \to X \cup U \cup Y$. Note also that $Q(s)$ satisfies (MP-Q2) with

$$r_i = \begin{cases} 1 \ (\nu_R(i) \in X) \\ 0 \ (\nu_R(i) \in U \cup Y), \end{cases} \qquad c_j = 0 \quad (j \in \mathrm{Col}(D)).$$

In considering nonzero fixed modes we may take $(\hat{I}, \hat{J}) = (\emptyset, \emptyset)$ since $Q(s)$ is nonsingular. The auxiliary network $N = (G, \gamma)$ for $(\hat{I}, \hat{J}) = (\emptyset, \emptyset)$ is easy to identify. We have

$$\begin{aligned} E_{TQ} &= \{(\varphi_T(j), \varphi_Q(j)) \mid j \in \mathrm{Col}(D)\}, \\ E_{QT} &= \{(\varphi_Q(i), \varphi_T(i)) \mid i \in \mathrm{Row}(D)\}, \\ E_Q &= \{(\varphi_Q(j), \varphi_Q(i)) \mid \nu_R(i) = \nu_C(j), i \in \mathrm{Row}(D), j \in \mathrm{Col}(D)\}, \\ E_T &= \{(\varphi_T(i), \varphi_T(j)) \mid T_{ij} \neq 0\}, \\ E_K &= \{(\varphi_T(i), \varphi_T(j)) \mid K_{ij} \neq 0\}, \\ E_M &= \emptyset, \end{aligned}$$

and $\gamma(a) = 1$ if $a = (\varphi_Q(i), \varphi_T(i)) \in E_{QT}$ with $\nu_R(i) \in X$, and $\gamma(a) = 0$ otherwise. Let us call an arc a with $\gamma(a) = 1$ a *critical arc*. A critical arc corresponds to an element of X. It is easy to see that a strong component of G (of N) containing a critical arc cannot admit a potential function.

We mean by (M1) the condition that each strong component of G either contains an arc of E_K or admits a potential function π such that (6.110) holds, and by (M2) the condition of nonsingularity of $D(0)$. We also refer to the conditions (G1) and (G2) in Theorem 6.5.7.

The equivalence of (G2) and (M2) is easy to see. Also the implication, (G1) $\Longrightarrow$ (M1), is easy to see. The converse is not always true (see Example 6.5.18). Under the condition (M2), however, every critical arc is contained in a strong component of G, and consequently, the converse, (M1) $\Longrightarrow$ (G1), is also true. Thus we have shown [(M2) $\Longleftrightarrow$ (G2)], [(G1) $\Longrightarrow$ (M1)], and [(M1), (M2) $\Longrightarrow$ (G1)], proving [(M1), (M2) $\Longleftrightarrow$ (G1), (G2)]. It is emphasized, however, that (G1) alone does not correspond to the nonexistence of nonzero fixed modes, as we have seen in Example 6.5.8. □

6.5.5 Examples

Example 6.5.17. The algorithm described in §6.5.4 as well as the derivation in §6.5.3 is illustrated here by means of an example. Consider a 9×9 mixed polynomial matrix $A_K(s) = Q(s) + T(s) + K$ of (6.94) with $Q(s)$ and $T(s)$ given by

$$
\begin{array}{c|ccccccccc|}
 & x_1 & x_2 & x_3 & x_4 & x_5 & x_6 & x_7 & x_8 & x_9 \\ \hline
w_1 & 1 & 0 & 1 & 0 & 0 & 0 & 0 & 0 & 0 \\
w_2 & 0 & 0 & 1 & 0 & 0 & 0 & 0 & 0 & 0 \\
w_3 & 0 & 0 & 0 & 0 & 0 & 0 & 0 & 0 & 0 \\
w_4 & -1 & 0 & -1 & 0 & 0 & 0 & 0 & 0 & 0 \\
w_5 & 0 & 0 & 0 & 0 & 1 & s & 0 & s & 0 \\
w_6 & 1 & 0 & 1 & 0 & -1 & -s & s & 0 & 0 \\
w_7 & 0 & 0 & 0 & 0 & 0 & 0 & 0 & 0 & 0 \\
w_8 & -s & 0 & -s & 0 & 0 & 0 & 0 & 0 & 1 \\
w_9 & 0 & 0 & 0 & 0 & 0 & 0 & 0 & 0 & 0 \\ \hline
\end{array}
\; , \quad
\begin{array}{|ccccccccc|}
x_1 & x_2 & x_3 & x_4 & x_5 & x_6 & x_7 & x_8 & x_9 \\ \hline
0 & 0 & 0 & 0 & 0 & 0 & 0 & 0 & 0 \\
0 & sf_1 & 0 & 0 & 0 & 0 & a_1 & 0 & 0 \\
0 & a_2 & sf_2 & a_3 & a_4 & 0 & 0 & 0 & 0 \\
0 & 0 & 0 & 0 & 0 & 0 & 0 & 0 & 0 \\
0 & 0 & 0 & 0 & 0 & 0 & 0 & 0 & a_5 \\
0 & 0 & 0 & 0 & 0 & 0 & 0 & a_6 & 0 \\
0 & 0 & 0 & 0 & 0 & 0 & 0 & a_7 & a_8 \\
0 & 0 & 0 & 0 & 0 & 0 & 0 & 0 & 0 \\
0 & 0 & 0 & 0 & a_9 & sf_3 & 0 & 0 & 0 \\ \hline
\end{array}
\; ,
$$

and

$$
K = \begin{array}{c|ccccccccc|}
 & x_1 & x_2 & x_3 & x_4 & x_5 & x_6 & x_7 & x_8 & x_9 \\ \hline
w_1 & 0 & 0 & 0 & 0 & 0 & 0 & 0 & 0 & k_1 \\
w_2 & 0 & 0 & 0 & 0 & 0 & 0 & 0 & 0 & 0 \\
w_3 & 0 & 0 & 0 & 0 & 0 & 0 & 0 & 0 & 0 \\
w_4 & 0 & 0 & 0 & k_2 & 0 & 0 & 0 & 0 & 0 \\
w_5 & 0 & 0 & 0 & 0 & 0 & 0 & k_3 & 0 & 0 \\
w_6 & 0 & 0 & 0 & 0 & 0 & 0 & 0 & 0 & 0 \\
w_7 & 0 & 0 & 0 & 0 & 0 & 0 & k_4 & 0 & 0 \\
w_8 & 0 & 0 & 0 & 0 & 0 & 0 & 0 & 0 & 0 \\
w_9 & 0 & 0 & 0 & 0 & 0 & 0 & 0 & 0 & 0 \\ \hline
\end{array} \; .
$$

The assumption (MP-Q2) is satisfied, where (6.4) holds true with

$$(r_1, \cdots, r_9) = (0,0,0,0,0,0,0,1,0), \;\; (c_1, \cdots, c_9) = (0,0,0,0,0,-1,-1,-1,1).$$

Note that $\mathcal{S} = \{a_1, \cdots, a_9\} \cup \{f_1, \cdots, f_3\}$ and $\mathcal{K} = \{k_1, \cdots, k_4\}$.

By direct calculation we obtain

$$
\begin{aligned}
\det A_K(s) = [s] \times & \left[(a_9 - f_3)(f_1 f_2 s^2 - a_2)\right] \\
& \times \left[k_2(k_1 s + 1)(a_7 s - k_4 s + a_7 k_3 - a_6 k_4)\right],
\end{aligned}
$$

where the brackets [ ] correspond to the three parts in (6.97). It follows from Lemma 6.5.11 that the fixed polynomial is given by $\psi(s) = (a_9 - f_3) \cdot s \cdot (f_1 f_2 s^2 - a_2)$, where $\alpha = (a_9 - f_3)$ and $p = 1$ in (6.98).

The associated LM-polynomial matrix $\tilde{A}_K(s)$ of (6.99) is given by

w_1	w_2	w_3	w_4	w_5	w_6	w_7	w_8	w_9	x_1	x_2	x_3	x_4	x_5	x_6	x_7	x_8	x_9
1									1	0	1	0	0	0	0	0	0
	1								0	0	1	0	0	0	0	0	0
		1							0	0	0	0	0	0	0	0	0
			1						-1	0	-1	0	0	0	0	0	0
				1					0	0	0	0	1	s	0	s	0
					1				1	0	1	0	-1	$-s$	s	0	0
						1			0	0	0	0	0	0	0	0	0
							1		$-s$	0	$-s$	0	0	0	0	0	1
								1	0	0	0	0	0	0	0	0	0
$-t_1$									0	0	0	0	0	0	0	0	k_1
	$-t_2$								0	sf_1	0	0	0	0	a_1	0	0
		$-t_3$							0	a_2	sf_2	a_3	a_4	0	0	0	0
			$-t_4$						0	0	0	k_2	0	0	0	0	0
				$-t_5$					0	0	0	0	0	0	k_3	0	a_5
					$-t_6$				0	0	0	0	0	0	0	a_6	0
						$-t_7$			0	0	0	0	0	0	k_4	a_7	a_8
							$-t_8$		0	0	0	0	0	0	0	0	0
								$-t_9$	0	0	0	0	a_9	sf_3	0	0	0

The CCF $\bar{A}_K(s)$, being identical with the block-triangular matrix $\hat{A}_K(s)$ in this example, is given by

$\bar{C}_1$	$\bar{C}_2$			$\bar{C}_3$	$\bar{C}_4$	$\bar{C}_5$	$\bar{C}_6$		$\bar{C}_7$	$\bar{C}_8$				$\bar{C}_9$	$\bar{C}_{10}$		$\bar{C}_{11}$
x_1	w_2	x_2	x_3	w_3	x_4	w_4	x_5	x_6	w_9	w_5	w_6	x_7	x_8	w_7	w_1	x_9	w_8
1			1												1		
	1	0	1														
	$-t_2$	sf_1	0									a_1					
	0	a_2	sf_2	$-t_3$	a_3		a_4										
				1											1		
					k_2	$-t_4$											
						1											
							-1	$-s$			1	s			-1		
							a_9	sf_3	$-t_9$								
									1								
										1	1	s	s		-1		
										0	$-t_6$	0	a_6				
										$-t_5$	0	k_3	0			a_5	
										0	0	k_4	a_7	$-t_7$		a_8	
														1			
															s	1	1
															$-t_1$	k_1	
																	$-t_8$

The CCF has 11 blocks of column sets: $\bar{C}_1 = \{x_1\}$, $\bar{C}_2 = \{w_2, x_2, x_3\}$, $\bar{C}_3 = \{w_3\}$, $\bar{C}_4 = \{x_4\}$, $\bar{C}_5 = \{w_4\}$, $\bar{C}_6 = \{x_5, x_6\}$, $\bar{C}_7 = \{w_9\}$, $\bar{C}_8 = \{w_5, w_6, x_7, x_8\}$, $\bar{C}_9 = \{w_7\}$, $\bar{C}_{10} = \{w_1, x_9\}$, $\bar{C}_{11} = \{w_8\}$. The index sets of (6.103)–(6.105) are given by $\bar{\Psi}_0 = \{1,3,5,7,9,11\}$, $\bar{\Psi}_1 = \{2,6\}$, and $\bar{\Psi}_2 = \{4,8,10\}$, and $\bar{\psi}_k(s)$ of (6.109) for $k \in \bar{\Psi}_1 \cup \bar{\Psi}_2$ are:

$$
\begin{aligned}
&\bar{\psi}_2(s) = (f_1 f_2 s^2 - a_2), \quad \bar{\psi}_6(s) = (a_9 - f_3); \\
&\bar{\psi}_4(s) = k_2, \quad \bar{\psi}_8(s) = (a_7 s - k_4 s + a_7 k_3 - a_6 k_4), \quad \bar{\psi}_{10}(s) = (k_1 s + 1).
\end{aligned}
$$

Note also that $\det \bar{A}_6 = s \cdot \bar{\psi}_6$.

We now illustrate the algorithm of §6.5.4. Suppose we have found $\hat{I} = \{w_2, w_3, w_4, w_5, w_7, w_9\}$, $\hat{J} = \{x_2, x_3, x_4, x_6, x_7, x_8\}$, and a matching $M = \{a_7, f_1, f_2, f_3, k_2, k_3\}$ by applying the algorithm of §4.2.4 to $A_K(1)$ above. Then the matrix P of (6.76) is given by

$$P = \begin{array}{c|ccc|cccccc|} & w_1 & w_6 & w_8 & x_2 & x_3 & x_4 & x_6 & x_7 & x_8 \\ \hline x_1 & 1 & 0 & 0 & 0 & 1 & 0 & 0 & 0 & 0 \\ x_5 & 1 & -1 & 0 & 0 & 0 & 0 & 1 & -1 & 0 \\ x_9 & 1 & 0 & 1 & 0 & 0 & 0 & 0 & 0 & 0 \\ \hline w_2 & 0 & 0 & 0 & 0 & 1 & 0 & 0 & 0 & 0 \\ w_3 & 0 & 0 & 0 & 0 & 0 & 0 & 0 & 0 & 0 \\ w_4 & 1 & 0 & 0 & 0 & 0 & 0 & 0 & 0 & 0 \\ w_5 & -1 & 1 & 0 & 0 & 0 & 0 & 0 & 1 & 1 \\ w_7 & 0 & 0 & 0 & 0 & 0 & 0 & 0 & 0 & 0 \\ w_9 & 0 & 0 & 0 & 0 & 0 & 0 & 0 & 0 & 0 \\ \hline \end{array}$$

and the auxiliary network $N = (G, \gamma) = (V, E, \gamma)$ is depicted in Fig. 6.10, where $x_i^T = \varphi_T(x_i)$, $x_i^Q = \varphi_Q(x_i)$, etc. The associated length $\gamma(a)$ is as follows:

$$\gamma(a) = \begin{cases} -1 & (a = (x_6^Q, x_6^T), (x_7^Q, x_7^T), (x_8^Q, x_8^T), (x_9^T, x_9^Q), \\ & \quad (w_2^T, x_2^T), (w_3^T, x_3^T), (w_9^T, x_6^T)) \\ 1 & (a = (w_8^Q, w_8^T), (x_2^T, w_2^T), (x_3^T, w_3^T), (x_6^T, w_9^T)) \\ 0 & (\text{otherwise}). \end{cases}$$

The diagonal blocks in the CCF are determined from the strong components of G. In particular, the diagonal blocks with indices in $\bar{\Psi}_1 = \{2, 6\}$ correspond respectively to $\hat{G}_2$ consisting of $\{w_2^T, w_2^Q, w_3^T, x_2^T, x_3^T, x_3^Q\}$ and $\hat{G}_6$ of $\{w_9^T, x_5^T, x_5^Q, x_6^T, x_6^Q\}$. These two strong components are extracted in Fig. 6.11, where the length $\gamma(a)$ is attached in parentheses to each arc a.

Theorem 6.5.15 reveals that $\hat{G}_2$ brings about nonzero fixed modes since it contains a directed cycle of nonzero length. On the other hand, $\hat{G}_6$ has no directed cycle of nonzero length, introducing no nonzero fixed modes. Accordingly, $\hat{G}_6$ admits a potential function π such that $\pi(x_6^T) = -1$, $\pi(w_9^T) = \pi(x_5^T) = \pi(x_5^Q) = \pi(x_6^Q) = 0$. We also see by the equivalence of (i) and (iii) in Theorem 6.5.12 that $\lambda = 0$ is a fixed mode, since $T[I, J](0) = O$, $K[I, J] = O$, and $\operatorname{rank} Q(0)[I, J] \le |I| + |J| - 10$ for $I = \{w_1, \cdots, w_9\}$ and $J = \{x_6\}$. Furthermore, by Theorem 6.5.13, $\lambda = 0$ is simple (i.e., with multiplicity one). □

Example 6.5.18. The present method successfully discriminates the existence of zero and nonzero fixed modes. Consider again the problem of Example 6.5.8. It has a zero fixed mode and no nonzero fixed mode, while neither graph-theoretic condition in Theorem 6.5.7 is satisfied. In line with Remark 6.5.16 we apply the present method to

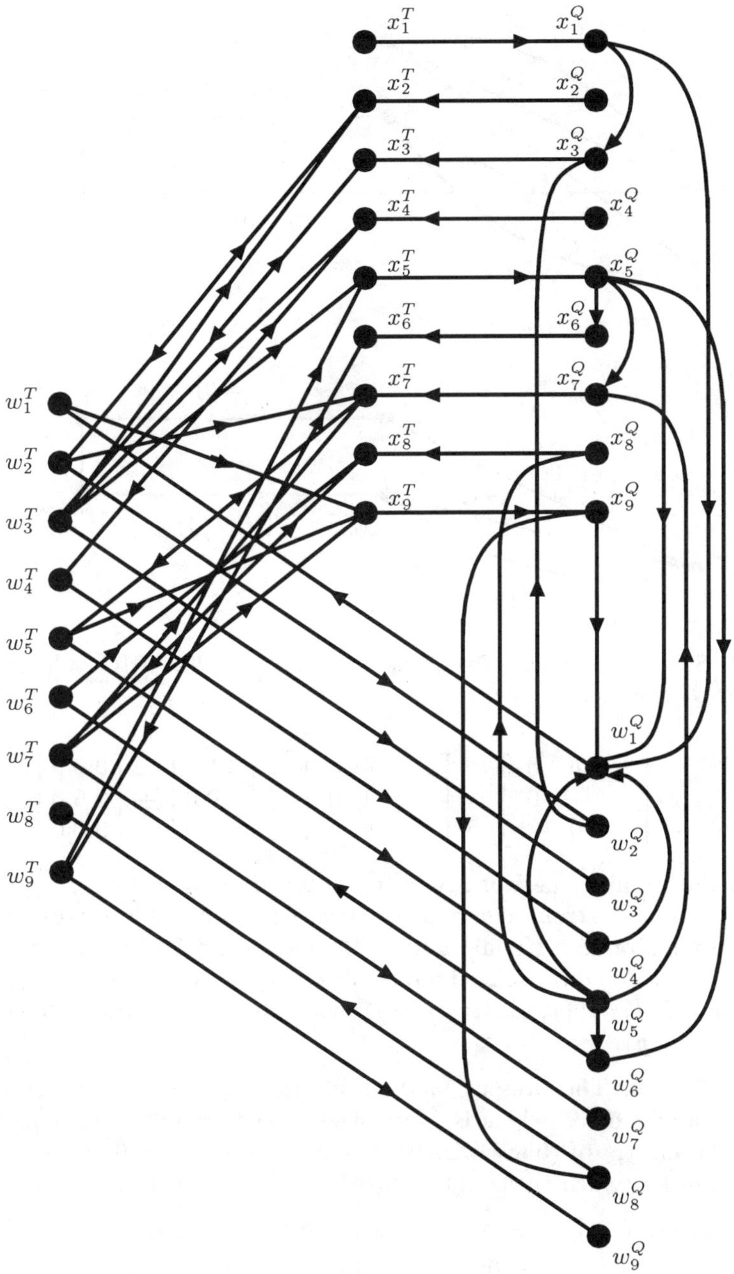

Fig. 6.10. Auxiliary network N in Example 6.5.17

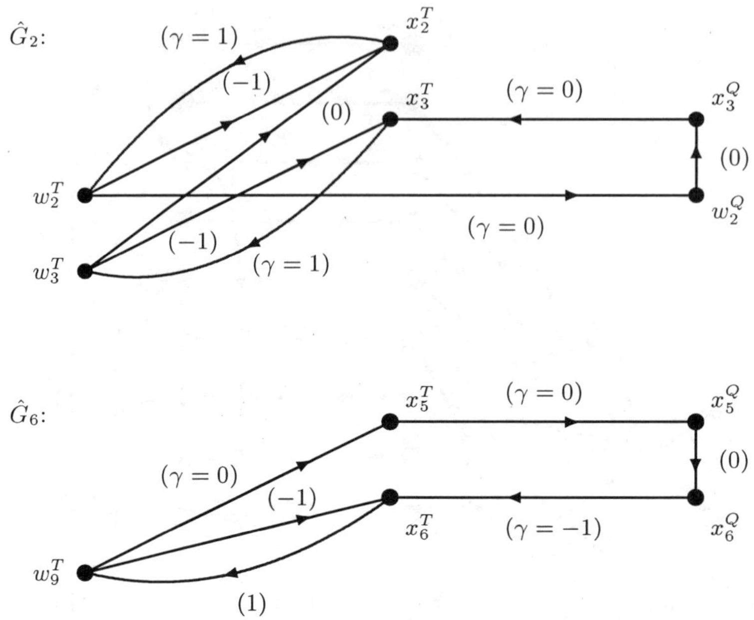

Fig. 6.11. Strong components free from E_K in Example 6.5.17 ($\hat{G}_6$ admits a potential and $\hat{G}_2$ does not)

$$D(s) = \begin{bmatrix} -s & b & 0 \\ 0 & -1 & 0 \\ c & 0 & -1 \end{bmatrix} = \begin{bmatrix} -s & 0 & 0 \\ 0 & -1 & 0 \\ 0 & 0 & -1 \end{bmatrix} + \begin{bmatrix} 0 & b & 0 \\ 0 & 0 & 0 \\ c & 0 & 0 \end{bmatrix} + \begin{bmatrix} 0 & 0 & 0 \\ 0 & 0 & 0 \\ 0 & 0 & 0 \end{bmatrix}$$

to obtain the auxiliary network $N = (G, \gamma)$ for $(\hat{I}, \hat{J}) = (\emptyset, \emptyset)$. The graph G is acyclic, and each strong component, consisting of a single vertex, admits a potential function in a trivial manner. By Theorem 6.5.15 we can conclude that there exists no nonzero fixed mode. The existence of a zero fixed mode is exhibited by $(I, J) = (\{x, u\}, \{x, y\})$ in Theorem 6.5.12(iii), where $\mathrm{Row}(D) = \mathrm{Col}(D) = \{x, u, y\}$. □

Example 6.5.19. The present method successfully detects the zero fixed mode in Example 6.5.9, which is overlooked by the graph-theoretic method. We take the matrix of (6.96) with $n = 6$, $m = 3$, and $l = 6$ as the matrix $A_K(s)$ of size 15, which turns out to satisfy (MP-Q2) with

$$(r_1, \cdots, r_{15}) = (1, 1, 1, 2, 1, 2; 0, 0, 0; 0, 0, 0, 1, 0, 1),$$
$$(c_1, \cdots, c_{15}) = (0, 0, 0, 1, 0, 1; 0, 0, 0; 0, 0, 0, 1, 0, 1).$$

Then Theorem 6.5.12 reveals the existence of the zero fixed mode. □

Notes. The mixed matrix formulation in §6.5.3 and the algorithm in §6.5.4 are taken from Murota [209].

7. Further Topics

This chapter introduces three supplementary, mutually independent, topics: the combinatorial relaxation algorithm, combinatorial system theory, and mixed skew-symmetric matrices.

7.1 Combinatorial Relaxation Algorithm

The "combinatorial relaxation" approach to algebraic computation initiated by Murota [212] is described here for the problem of computing the maximum degree of subdeterminants of a polynomial/rational matrix, the problem treated in §6.2 by graph-theoretic and valuated matroid-theoretic methods. A purely combinatorial algorithm, whether graph-theoretic or matroid-theoretic, is based on a genericity assumption, and hence can possibly fail when the assumed genericity is not satisfied by specific input data. An algorithm of "combinatorial relaxation" type is a remedy for this. It always returns the correct answer, while sharing the spirit of generic approach. It is efficient, behaving as a combinatorial algorithm in most cases, and at the same time it is reliable, coping with nongeneric cases where numerical cancellation does affect the answer.

7.1.1 Outline of the Algorithm

Let $A(s) = (A_{ij}(s))$ be an $m \times n$ rational function matrix with $A_{ij}(s)$ being a rational function in s with coefficients from a certain field $\boldsymbol{F}$ (typically the real number field $\mathbf{R}$). We shall present an algorithm for computing the highest degree of a minor of order k of $A(s)$:

$$\delta_k(A) = \max\{\deg_s \det A[I, J] \mid |I| = |J| = k\}. \tag{7.1}$$

As a combinatorial counterpart of $\delta_k(A)$ we consider the maximum weight of a k-matching in the associated bipartite graph $G(A) = (V, E)$ introduced in §6.2.2; each arc of $G(A)$ corresponds to a nonzero entry $A_{ij}(s)$ and is given a weight $w_{ij} = \deg_s A_{ij}(s)$. We then define

$$\hat{\delta}_k(A) = \max\{w(M) \mid M \text{ is a } k\text{-matching in } G(A)\}, \tag{7.2}$$

where $\hat{\delta}_k(A) = -\infty$ if no k-matching exists.

As has been discussed in §6.2.2 (Theorem 6.2.2 in the case of a polynomial matrix), the combinatorial value $\hat{\delta}_k(A)$ is an upper bound on $\delta_k(A)$ and it is generically tight. Recall that the word "generic" refers to an algebraic assumption that the nonzero coefficients in $A(s)$ are subject to no algebraic relations, whereas its practical interpretation would be "so long as no accidental numerical cancellation occurs." To make this statement more precise, we define an $m \times n$ constant matrix $A^\circ = (A^\circ_{ij})$ by

$$A^\circ_{ij} = \begin{cases} \lim_{s\to\infty} s^{-w_{ij}} A_{ij}(s) & \text{if } A_{ij}(s) \neq 0 \\ 0 & \text{if } A_{ij}(s) = 0. \end{cases} \tag{7.3}$$

Let us call A°_{ij} the *leading coefficient* of $A_{ij}(s)$, since, when $A_{ij}(s)$ is a polynomial, A°_{ij} is equal to the coefficient of the highest-degree term in $A_{ij}(s)$.

Theorem 7.1.1. *Let $A(s)$ be a rational function matrix.*

(1) $\delta_k(A) \le \hat{\delta}_k(A)$.

(2) *The equality holds generically, i.e., if the set of nonzero leading coefficients $\{A^\circ_{ij} \mid A^\circ_{ij} \neq 0\}$ is algebraically independent (over a subfield of $\boldsymbol{F}$).* □

We say that $A(s)$ is *upper-tight* (for k) if $\delta_k(A) = \hat{\delta}_k(A)$. Note that genericity is sufficient and not necessary for the upper-tightness.

The algorithm, outlined below, takes advantage of two facts:

(i) $\hat{\delta}_k(A)$ is generically equal to $\delta_k(A)$, and

(ii) $\hat{\delta}_k(A)$ can be computed efficiently by a combinatorial algorithm.

The algorithm first computes $\hat{\delta}_k(A)$, instead of $\delta_k(A)$, by solving a weighted-matching problem using an efficient combinatorial algorithm (Phase 1), and then checks whether $\hat{\delta}_k(A)$ equals $\delta_k(A)$ (Phase 2). The algorithm invokes an exception-handling algebraic elimination routine to modify A only when it detects discrepancy between $\hat{\delta}_k(A)$ and $\delta_k(A)$ due to numerical cancellation (Phase 3). In Phase 3, where $\delta_k(A) \le \hat{\delta}_k(A) - 1$, the matrix A is modified to another matrix A' such that $\delta_k(A') = \delta_k(A)$ and $\hat{\delta}_k(A') \le \hat{\delta}_k(A) - 1$.

Algorithm for computing $\delta_k(A)$ (outline)

Phase 1 : Compute $\hat{\delta}_k(A)$ by solving the weighted-matching problem in $G(A)$ using an efficient combinatorial algorithm (cf. Ahuja–Magnanti–Orlin [3], Cook–Cunningham–Pulleyblank–Schrijver [40], Lawler [171]).

Phase 2 : Test whether $\delta_k(A) = \hat{\delta}_k(A)$ or not (without computing $\delta_k(A)$). If so, output $\hat{\delta}_k(A)$ and stop.

Phase 3 : Modify A to another matrix A' such that $\delta_k(A') = \delta_k(A)$ and $\hat{\delta}_k(A') \le \hat{\delta}_k(A) - 1$. Put $A := A'$ and go to Phase 1. □

The test in Phase 2 for the upper-tightness can be reduced to computing the ranks of four constant matrices (see Theorem 7.1.9). The modification algorithm of Phase 3, to be described in detail in §7.1.3, makes essential use of dual variables based on the duality theorem for the polyhedral description of matchings. Since numerical cancellation occurs only rarely (or nongenerically) the above algorithm is combinatorial in almost all cases and hence suitable for large scale problems.

Remark 7.1.2. In more general terms an algorithm of "combinatorial relaxation" type consists of the following three distinct phases:

Phase 1: Consider a relaxation (or an easier problem) of a combinatorial nature to the original problem and find a solution to the relaxed problem.
Phase 2: Test for the validity of this solution to the original problem (without computing the solution to the original problem).
Phase 3 (In case of invalid solution): Modify the relaxation so that the invalid solution is eliminated.

It is crucial for computational efficiency that the relaxed problem can be solved efficiently and that the modification of the relaxation in Phase 3 need not be invoked many times. □

Example 7.1.3. Some technical issues of the combinatorial relaxation algorithm above are illustrated here for a specific example. Consider a polynomial matrix over $\boldsymbol{F} = \mathbf{R}$ $(m = n = 4)$:

$$A(s) = \begin{array}{c} \\ r_1 \\ r_2 \\ r_3 \\ r_4 \end{array} \begin{array}{c} \begin{array}{cccc} c_1 & c_2 & c_3 & c_4 \end{array} \\ \begin{pmatrix} \alpha s^4 & s^5 & 0 & 2s^3 \\ s^5 & s^6+1 & s^4 & s^2 \\ s^4+s & s^5 & -s^3 & 0 \\ 2s^2 & s & 0 & s+2 \end{pmatrix} \end{array}$$

with a nonzero parameter α introduced for an illustrative purpose. The associated bipartite graph $G = G(A)$, shown in Fig. 7.1, has 8 vertices and 13 arcs. The leading coefficient matrix (7.3) is given by

$$A^\circ = \begin{pmatrix} \alpha & 1 & 0 & 2 \\ 1 & 1 & 1 & 1 \\ 1 & 1 & -1 & 0 \\ 2 & 1 & 0 & 1 \end{pmatrix}. \tag{7.4}$$

Let us consider the minors of order $k = 3$, and assume $\alpha \neq 1$ as the first case. We may take, for example, $M^{(1)} = \{(r_1, c_4), (r_2, c_2), (r_3, c_1)\}$ as a matching of size 3 of maximum weight $w(M^{(1)}) = 3 + 6 + 4 = 13$. The corresponding submatrix

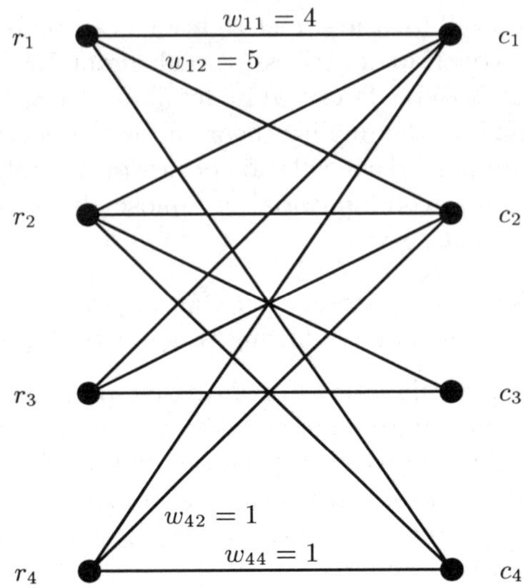

Fig. 7.1. Bipartite graph $G(A)$ (Example 7.1.3)

$$A[\partial^+ M^{(1)}, \partial^- M^{(1)}] = \begin{array}{c} \\ r_1 \\ r_2 \\ r_3 \end{array} \begin{array}{c} \begin{array}{ccc} c_1 & c_2 & c_4 \end{array} \\ \left(\begin{array}{ccc} \alpha s^4 & s^5 & 2s^3 \\ s^5 & s^6+1 & s^2 \\ s^4+s & s^5 & 0 \end{array} \right) \end{array}$$

has another matching $M^{(2)} = \{(r_1, c_4), (r_2, c_1), (r_3, c_2)\}$ of weight $w(M^{(2)}) = 3 + 5 + 5 = 13$. In the determinant expansion of this minor, the two terms of degree 13 arising from $M^{(1)}$ and $M^{(2)}$ cancel each other, and

$$\det A[\{r_1, r_2, r_3\}, \{c_1, c_2, c_4\}] = (1 - \alpha)s^{11} - 2s^{10} + s^8 - 2s^7 - 2s^4.$$

Therefore,

$$\begin{aligned} &\delta_3(A[\{r_1, r_2, r_3\}, \{c_1, c_2, c_4\}]) = 11 \\ &< 13 = w(M^{(1)}) = \hat{\delta}_3(A[\{r_1, r_2, r_3\}, \{c_1, c_2, c_4\}]). \end{aligned}$$

Nevertheless, we have $\delta_3(A) = 13 = \hat{\delta}_3(A)$, provided $\alpha \neq 1$, because of the existence of another minor of degree 13. Consider a third matching $M^{(3)} = \{(r_1, c_2), (r_2, c_1), (r_3, c_3)\}$ of weight $w(M^{(3)}) = 5 + 5 + 3 = 13$. The corresponding submatrix

$$A[\partial^+ M^{(3)}, \partial^- M^{(3)}] = \begin{array}{c} \\ r_1 \\ r_2 \\ r_3 \end{array} \begin{array}{c} \begin{array}{ccc} c_1 & c_2 & c_3 \end{array} \\ \left(\begin{array}{ccc} \alpha s^4 & s^5 & 0 \\ s^5 & s^6+1 & s^4 \\ s^4+s & s^5 & -s^3 \end{array} \right) \end{array}$$

admits four matchings (of size 3) of weight 13, and has the determinant

$$\det A[\{r_1, r_2, r_3\}, \{c_1, c_2, c_3\}] = 2(1-\alpha)s^{13} + s^{10} - \alpha s^7.$$

Hence $\delta_3(A) = 13 = \hat{\delta}_3(A)$, provided $\alpha \neq 1$.

The phenomenon observed above illustrates a challenging complication: after we have found a matching M of maximum weight in Phase 1, we must look globally for a $k \times k$ minor of degree $w(M)$ before we can decide in Phase 2 whether $w(M)$ is equal to $\delta_k(A)$ or not.

In case $\alpha = 1$ we can verify by inspection that there exists no 3×3 minor of degree equal to 13, whereas $\deg_s \det A[\{r_1, r_2, r_3\}, \{c_2, c_3, c_4\}] = 12$. Accordingly we conclude

$$\hat{\delta}_3(A) = 13, \qquad \delta_3(A) = \begin{cases} 13 & \text{if } \alpha \neq 1 \\ 12 & \text{if } \alpha = 1. \end{cases}$$

The values of $\hat{\delta}_k(A)$ and $\delta_k(A)$ for $k = 1, 2, 4$ can be found similarly:

$$\hat{\delta}_1(A) = 6, \ \delta_1(A) = 6,$$
$$\hat{\delta}_2(A) = 10, \ \delta_2(A) = \begin{cases} 10 & \text{if } \alpha \neq 1 \\ 9 & \text{if } \alpha = 1, \end{cases}$$
$$\hat{\delta}_4(A) = 14, \ \delta_4(A) = \begin{cases} 14 & \text{if } \alpha \neq 5 \\ 13 & \text{if } \alpha = 5. \end{cases}$$

□

7.1.2 Test for Upper-tightness

This section describes a procedure for Phase 2 which tests for the upper-tightness $\delta_k(A) = \hat{\delta}_k(A)$ of $A(s)$ without computing $\delta_k(A)$. The procedure makes use of the standard duality result for bipartite matchings, which follows from the integrality of the associated linear programs.

We consider the following primal-dual pair of linear programs (Chvátal [35], Lawler [171], Lovász–Plummer [181], Schrijver [292]):

$$\begin{aligned} \text{PLP}(k)\text{: Maximize } & \sum_{e \in E} w_e \xi_e, \\ \text{subject to } & \sum_{\partial e \ni i} \xi_e \leq 1 \quad (i \in V), \\ & \sum_{e \in E} \xi_e = k, \\ & \xi_e \geq 0 \quad (e \in E); \end{aligned} \tag{7.5}$$

$$\begin{aligned} \text{DLP}(k)\text{: Minimize } & \sum_{i \in V} p_i + kq \quad (\equiv \pi(p,q)), \\ \text{subject to } & p_i + p_j + q \geq w_{ij} \quad ((i,j) \in E), \\ & p_i \geq 0 \quad (i \in V). \end{aligned} \tag{7.6}$$

Note that $\boldsymbol{\xi} = (\xi_e \mid e \in E) \in \mathbf{R}^E$ is the primal variable and $p = (p_i \mid i \in V) = p_{\mathrm{R}} \oplus p_{\mathrm{C}} = (p_{\mathrm{R}i} \mid i \in R) \oplus (p_{\mathrm{C}j} \mid j \in C) \in \mathbf{R}^V$ and $q \in \mathbf{R}$ are the dual variables.

As is well known, these linear programs enjoy the integrality property.

Lemma 7.1.4.

(1) *PLP(k) has an integral optimal solution with* $\xi_e \in \{0, 1\}$ $(e \in E)$.

(2) *If* w_e *is integer for* $e \in E$, *DLP(k) has an integral optimal solution with* $p_i \in \mathbf{Z}$ $(i \in V)$ *and* $q \in \mathbf{Z}$.

Proof. The coefficient matrix is seen to be totally unimodular by Camion's criterion (Lawler [171, Th.16.4], Schrijver [292, Th.19.3(vi)]). ■

By virtue of the linear programming duality as well as the primal integrality we have

$$\hat{\delta}_k(A) = \min\{\pi(p,q) \mid (p,q) \text{ is feasible to DLP}(k)\}. \tag{7.7}$$

By the dual integrality we henceforth assume that the dual variables are integer-valued.

The optimality of a k-matching is expressed as follows. For $e = (i,j) \in E$, the reduced weight is defined by

$$\tilde{w}_e = \tilde{w}_{ij} = w_{ij} - p_i - p_j - q. \tag{7.8}$$

Then (p,q) is (dual) feasible if and only if $\tilde{w}_e \le 0$ $(e \in E)$ and $p_i \ge 0$ $(i \in V)$. An arc $e \in E$ is said to be *tight* (with respect to (p,q)) if $\tilde{w}_e = 0$. We put

$$E^* = E^*(p,q) = \{e \in E \mid \tilde{w}_e = 0\}, \tag{7.9}$$

which is the set of tight arcs, and define a subgraph $G^* = G^*(p,q) = (V, E^*(p,q))$. A vertex $i \in V$ is said to be *active* (with respect to p) if $p_i > 0$, and we put

$$V^* = V^*(p) = \{i \in V \mid p_i > 0\}, \tag{7.10}$$

$$I^* = I^*(p) = \{i \in R \mid p_i > 0\} = V^* \cap R, \tag{7.11}$$

$$J^* = J^*(p) = \{j \in C \mid p_j > 0\} = V^* \cap C. \tag{7.12}$$

We call I^* and J^* *active* rows and columns, respectively. The complementary slackness condition yields the following optimality criterion. Note that this is essentially the same as Theorem 2.2.36.

Lemma 7.1.5. *Let* M *be a* k*-matching in* $G(A)$ *and* (p,q) *be a dual feasible solution. Then both* M *and* (p,q) *are optimal (i.e.,* $w(M) = \pi(p,q)$*) if and only if* $M \subseteq E^*(p,q)$ *and* $\partial M \supseteq V^*(p)$. □

The following corollary is important for our algorithm. Note that $G^*(p,q)$ depends on the choice of (p,q).

Lemma 7.1.6. *Let (p, q) be an optimal dual solution. Then M is an optimal k-matching in G if and only if M is a k-matching in $G^*(p, q)$ such that $\partial M \supseteq V^*(p)$.* □

To derive a necessary and sufficient condition for the upper-tightness we extract the "tight part" from $A(s)$ which is composed of the entries that can potentially contribute to the coefficient of $s^{\hat{\delta}_k(A)}$ in a minor of order k. For a dual feasible (p, q) we define an $m \times n$ constant matrix

$$\mathcal{T}(A; p, q) = A^* = (A^*_{ij})$$

by

$$A^*_{ij} = \lim_{s \to \infty} s^{-p_i - p_j - q} A_{ij}(s) = \begin{cases} A^\circ_{ij} & \text{if } (i,j) \in E^*(p,q) \\ 0 & \text{otherwise.} \end{cases} \tag{7.13}$$

We call A^* the *tight coefficient matrix* (with respect to (p, q)). Note that $\mathcal{T}(A; p, q) = A^*$ varies with the choice of (p, q), not unique even for optimal (p, q). The tight coefficient matrix A^* can also be defined by

$$A_{ij}(s) = s^{p_i + p_j + q}(A^*_{ij} + \mathrm{o}(1)), \tag{7.14}$$

where $\mathrm{o}(1)$ denotes an expression (rational function) that tends to zero as $s \to \infty$. In a matrix form we can also write this as

$$A(s) = s^q \cdot \mathrm{diag}\,(s; p_{\mathrm{R}}) \cdot (A^* + \mathrm{o}(1)) \cdot \mathrm{diag}\,(s; p_{\mathrm{C}}) \tag{7.15}$$

using the notation

$$\mathrm{diag}\,(s; r) = \mathrm{diag}\,(s^{r_1}, s^{r_2}, \cdots) \tag{7.16}$$

for a diagonal matrix with diagonal entries $s^{r_1}, s^{r_2}, \cdots$, where $r = (r_1, r_2, \cdots)$.

In terms of the tight coefficient matrix A^*, Lemma 7.1.6 can be rephrased as follows. It should be clear that $A^*_{ij} \neq 0$ if and only if $(i, j) \in E^*$.

Lemma 7.1.7. *Let (p, q) be an optimal dual solution and assume $|I| = |J| = k$ for $I \subseteq R$ and $J \subseteq C$. Then $\hat{\delta}_k(A[I, J]) = \hat{\delta}_k(A)$ if and only if $I \supseteq I^*$, $J \supseteq J^*$, and*

$$\text{term-rank}\, A^*[I, J] = |I| = |J| = k. \tag{7.17}$$

In particular, there exist such $I \subseteq R$ and $J \subseteq C$. □

For $I \subseteq R$ and $J \subseteq C$ with $|I| = |J| = k$ it follows from (7.14) that

$$\det A[I, J] = s^{p(I \cup J) + kq} \left(\det A^*[I, J] + \mathrm{o}(1) \right),$$

where $p(I \cup J) = \sum_{i \in I \cup J} p_i$. If (p, q) is optimal and if $I \supseteq I^*$ and $J \supseteq J^*$, we have $p(I \cup J) + kq = p(V) + kq = \pi(p, q) = \hat{\delta}_k(A)$, and therefore

$$\det A[I, J] = s^{\hat{\delta}_k(A)} \left(\det A^*[I, J] + \mathrm{o}(1) \right).$$

This yields the following criterion for the upper-tightness. Note that "term-rank" in (7.17) of Lemma 7.1.7 is replaced with "rank" in (7.18) below.

Lemma 7.1.8. *Let (p, q) be an optimal dual solution. Then $\delta_k(A) = \hat{\delta}_k(A)$ if and only if there exist $I \supseteq I^*$ and $J \supseteq J^*$ such that*

$$\operatorname{rank} A^*[I, J] = |I| = |J| = k. \tag{7.18}$$

□

The above criterion, involving existential quantifiers, is not readily checked efficiently. It can, however, be rewritten in a form suitable for straightforward verification. In fact, the following theorem shows that the upper-tightness is equivalent to a set of rank conditions for four constant matrices.

Theorem 7.1.9. *Let (p, q) be an optimal dual solution, I^* and J^* be the active rows and columns defined by* (7.11) *and* (7.12), *and A^* be the tight coefficient matrix defined by* (7.13). *Then $\delta_k(A) = \hat{\delta}_k(A)$ if and only if the following four conditions are satisfied:*

(R1) $\operatorname{rank} A^*[R, C] \geq k$,
(R2) $\operatorname{rank} A^*[I^*, C] = |I^*|$,
(R3) $\operatorname{rank} A^*[R, J^*] = |J^*|$,
(R4) $\operatorname{rank} A^*[I^*, J^*] \geq |I^*| + |J^*| - k$.

Proof. This follows from Lemma 7.1.8 and Theorem 2.3.46 for $\lambda(I, J) = \operatorname{rank} A^*[I, J]$. ■

The following similar theorem for term-rank will be used later.

Theorem 7.1.10. *Let (p, q) be a dual feasible solution, and I^* and J^* be defined by* (7.11) *and* (7.12), *and A^* by* (7.13). *Then the following three conditions* (i)–(iii) *are equivalent.*

(i) *(p, q) is optimal.*
(ii) *There exist $I \supseteq I^*$ and $J \supseteq J^*$ such that*
$\quad$ $\text{term-rank}\, A^*[I, J] = |I| = |J| = k$.
(iii) *The following four conditions are satisfied:*
(T1) $\text{term-rank}\, A^*[R, C] \geq k$,
(T2) $\text{term-rank}\, A^*[I^*, C] = |I^*|$,
(T3) $\text{term-rank}\, A^*[R, J^*] = |J^*|$,
(T4) $\text{term-rank}\, A^*[I^*, J^*] \geq |I^*| + |J^*| - k$.

Proof. This follows from Lemma 7.1.5, Lemma 7.1.7 and Theorem 2.3.46 for $\lambda(I, J) = \text{term-rank}\, A^*[I, J]$. ■

Example 7.1.11 (Continued from Example 7.1.3). First we consider the case of $k = 3$. As the optimal dual variables we may take

$$p_{r_1} = 2, \quad p_{r_2} = 3, \quad p_{r_3} = 2, \quad p_{r_4} = 0; \tag{7.19}$$
$$p_{c_1} = 1, \quad p_{c_2} = 2, \quad p_{c_3} = 0, \quad p_{c_4} = 0; \tag{7.20}$$

and $q = 1$. We have

$$\hat{\delta}_3(A) = \pi(p,q) = \sum_{i \in V} p_i + kq = 13.$$

Those variables and the reduced weights $\tilde{w}_e$ of (7.8) are illustrated in Fig. 7.2.

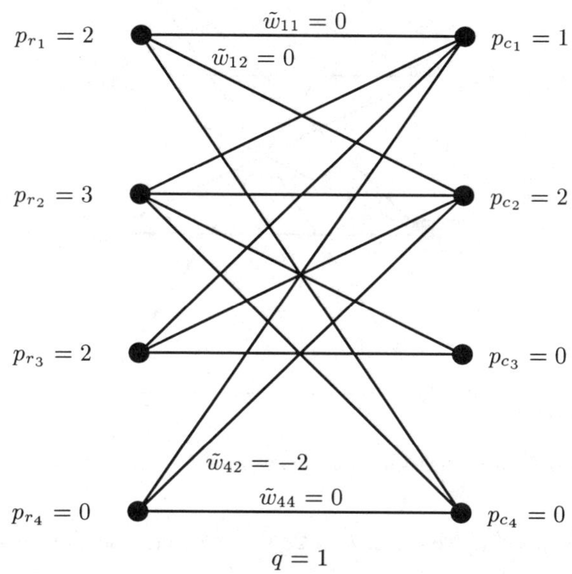

Fig. 7.2. Dual variables and reduced weights for $G(A)$ (Example 7.1.11, $k = 3$)

According to (7.13) we have the tight coefficient matrix

$$\mathcal{T}(A;p,q) = A^* = \begin{array}{c} \\ \bullet \\ \bullet \\ \bullet \\ \\ \end{array} \!\!\! \begin{array}{c} \begin{array}{cccc} \bullet & \bullet & & \end{array} \\ \begin{pmatrix} \alpha & 1 & 0 & 2 \\ 1 & 1 & 1 & 0 \\ 1 & 1 & -1 & 0 \\ 2 & 0 & 0 & 1 \end{pmatrix} \end{array}, \tag{7.21}$$

which should be compared with A° of (7.4); A^* contains a smaller number of nonzero entries. The symbol $\bullet$ denotes active rows and columns. The graph G^* consisting of the tight arcs is shown in Fig. 7.3. Noting $I^*(p) = \{r_1, r_2, r_3\}$, $J^*(p) = \{c_1, c_2\}$, we see that the conditions (R1)–(R3) in Theorem 7.1.9 are satisfied for all values of α. On the other hand, the last condition (R4) is violated when $\alpha = 1$. Hence by Theorem 7.1.9 we see

$$\delta_3(A) \quad \begin{cases} = \hat{\delta}_3(A) = 13 & \text{if } \alpha \neq 1 \\ \leq \hat{\delta}_3(A) - 1 = 12 & \text{if } \alpha = 1. \end{cases}$$

In the case of $\alpha = 1$ we cannot tell how small $\delta_3(A)$ is, though we know for sure that $\delta_3(A)$ is strictly smaller than $\hat{\delta}_3(A) = 13$.

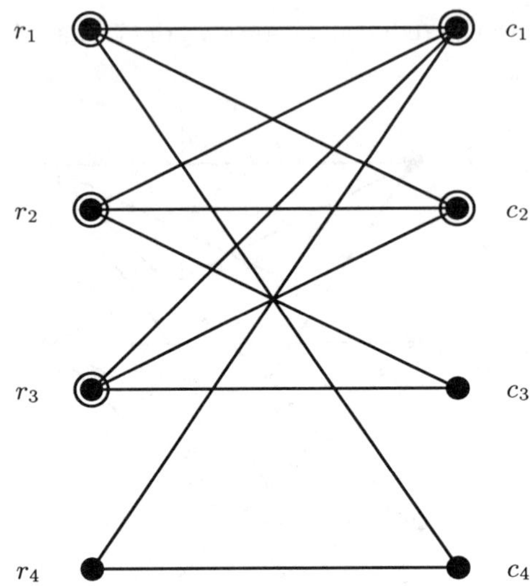

Fig. 7.3. Graph $G^*(p,q)$ of tight arcs (Example 7.1.11, $k = 3$) (○: active vertex)

Next we consider the case of $k = 2$. We may choose the following optimal dual variables:

$$p'_{r_1} = 0,\ p'_{r_2} = 1,\ p'_{r_3} = 0,\ p'_{r_4} = 0;\ p'_{c_1} = 0,\ p'_{c_2} = 1,\ p'_{c_3} = 0,\ p'_{c_4} = 0;$$

and $q' = 4$, from which we see $\hat{\delta}_2(A) = \pi(p', q') = 10$. The tight coefficient matrix is given by

$$\mathcal{T}(A; p', q') = A^* = \begin{array}{c} \\ \bullet \\ \\ \\ \end{array}\!\!\begin{array}{c} \quad\quad\bullet \\ \begin{pmatrix} \alpha & 1 & 0 & 0 \\ 1 & 1 & 0 & 0 \\ 1 & 1 & 0 & 0 \\ 0 & 0 & 0 & 0 \end{pmatrix} \end{array},$$

where the symbol $\bullet$ indicates the active row and column. Noting $I^*(p') = \{r_2\}$, $J^*(p') = \{c_2\}$, we see that the conditions (R2)–(R4) in Theorem 7.1.9 are satisfied for all values of α, whereas the first condition (R1) is violated when $\alpha = 1$. Hence by Theorem 7.1.9 we see

$$\delta_2(A) \begin{cases} = \hat{\delta}_2(A) = 10 & \text{if } \alpha \neq 1 \\ \leq \hat{\delta}_2(A) - 1 = 9 & \text{if } \alpha = 1. \end{cases}$$

The exact values of $\delta_3(A)$ and $\delta_2(A)$ when $\alpha = 1$ will be determined later in Example 7.1.14. □

7.1.3 Transformation Towards Upper-tightness

When the matrix $A(s)$ is not upper-tight, the combinatorial quantity $\hat{\delta}_k(A)$ gives only an upper bound on $\delta_k(A)$. We will show how to transform $A(s)$ efficiently to an upper-tight matrix through repeated biproper transformations (see §5.1.2 for the definition of biproper transformations). Note that the Smith–McMillan form at infinity (cf. §5.1.2) guarantees the existence of such an upper-tight matrix, though this fact is not used below. For $S = \boldsymbol{F}(s)$, $\boldsymbol{F}[s]$, $\boldsymbol{F}[1/s]$, or $\boldsymbol{F}[s, 1/s]$, we denote by $\mathcal{M}(S)$ the set of matrices with entries in S.

Given $A(s) \in \mathcal{M}(\boldsymbol{F}(s))$ with $\delta_k(A) < \hat{\delta}_k(A)$, we are to modify $A(s)$ to another matrix $A'(s) = (A'_{ij}(s))$ such that

(P1) $A'(s) = U(s)A(s)V(s)$, where $U(s), V(s) \in \mathcal{M}(\boldsymbol{F}[1/s])$ and $\det U(s), \det V(s) \in \boldsymbol{F} \setminus \{0\}$, and

(P2) $\hat{\delta}_k(A') \leq \hat{\delta}_k(A) - 1$.

In particular, $U(s)$ and $V(s)$ are biproper matrices, and therefore $\delta_k(A') = \delta_k(A)$ by (5.5). When $A(s) \in \mathcal{M}(\boldsymbol{F}[s, 1/s])$, the modified matrix $A'(s)$ remains in $\mathcal{M}(\boldsymbol{F}[s, 1/s])$ by the condition that $U(s), V(s) \in \mathcal{M}(\boldsymbol{F}[1/s])$. It should be obvious that we can get an upper-tight matrix by repeatedly applying this transformation.

Recall that a constant matrix A^*, called the tight coefficient matrix, is derived from $A(s)$ with reference to an optimal dual variable (p, q), which is assumed to be integer-valued. Since $A(s)$ is not upper-tight, at least one of the four rank conditions (R1)–(R4) in Theorem 7.1.9 is violated, whereas the term-rank conditions (T1)–(T4) in Theorem 7.1.10 are satisfied. We consider the modification algorithm for each case.

If (R1) is violated: We have $\operatorname{rank} A^*[R, C] < k \leq \text{term-rank}\, A^*[R, C]$. Then there exists a nonsingular constant matrix $U = (U_{hi})$ (see the construction below) such that

$$\text{term-rank}\,(UA^*) \leq k - 1. \tag{7.22}$$

We transform A to A' by

$$A'(s) = U(s)A(s), \tag{7.23}$$

where $U(s) = (U_{hi}(s))$ is given by

$$U_{hi}(s) = U_{hi}\ s^{\sigma(h,i)}, \quad \sigma(h, i) = p_{\mathrm{R}h} - p_{\mathrm{R}i} \tag{7.24}$$

with reference to the dual variable $p_{\mathrm{R}} = (p_{\mathrm{R}i} \mid i \in R)$ associated with the rows. The transformation matrix $U(s)$ can be written also as

$$U(s) = \operatorname{diag}(s; p_{\mathrm{R}}) \cdot U \cdot \operatorname{diag}(s; -p_{\mathrm{R}}) \tag{7.25}$$

using the notation (7.16).

We claim that the property (P2) is satisfied.

Lemma 7.1.12. (P2) *is implied by* (7.22).

Proof. The dual variable (p,q) is optimal for A. First we claim that (p,q) is feasible for the modified matrix A'. By (7.15) and (7.25) we have

$$\begin{aligned} & s^{-q} \cdot \mathrm{diag}\,(s; -p_{\mathrm{R}}) \cdot A'(s) \cdot \mathrm{diag}\,(s; -p_{\mathrm{C}}) \\ &= U \cdot s^{-q} \cdot \mathrm{diag}\,(s; -p_{\mathrm{R}}) \cdot A(s) \cdot \mathrm{diag}\,(s; -p_{\mathrm{C}}) \\ &= U(A^* + \mathrm{o}(1)) = UA^* + \mathrm{o}(1), \end{aligned}$$

which shows that $w'_{ij} - p_i - p_j - q \le 0$ for $w'_{ij} = \deg_s A'_{ij}(s)$.

Then (7.22) and Theorem 7.1.10 imply that (p,q) is not optimal for the modified matrix A', whereas it is for the original matrix A. Hence we have

$$\hat{\delta}_k(A') < \pi(p,q) = \hat{\delta}_k(A).$$

By the integrality we finally obtain $\hat{\delta}_k(A') \le \hat{\delta}_k(A) - 1$. ∎

As to the property (P1) we easily see the following.

Lemma 7.1.13. (P1) *is satisfied if* $[\, U_{hi} \neq 0 \implies p_{\mathrm{R}h} \le p_{\mathrm{R}i} \,]$. □

The above statement says that U should be in a triangular form with respect to the orderings of rows and columns determined by the dual variable $p_{\mathrm{R}} = (p_{\mathrm{R}i} \mid i \in R)$ associated with the rows of $A(s)$. Such U can be constructed as follows.

Let $\tau : \{1, \cdots, m\} \to R$ be a one-to-one correspondence such that $[i \le h \Rightarrow p_{\mathrm{R}\tau(i)} \ge p_{\mathrm{R}\tau(h)}]$. We denote by $\boldsymbol{a}_i^* \in \boldsymbol{F}^n$ the row vector of A^* indexed by $i \in R$. Let $\{\boldsymbol{a}_i^* \mid i \in B\}$ be the basis of row vectors of A^* that is constructed by picking up the independent vectors from the sequence $\boldsymbol{a}_{\tau(1)}^*$, $\boldsymbol{a}_{\tau(2)}^*$, $\cdots$, considered in this order. Also let $\{\boldsymbol{a}_i^* \mid i \in D\}$ be the set of vectors consisting of the first

$$d = m - k + 1 \tag{7.26}$$

vectors in this sequence that do not belong to the basis. Hence $B \subseteq R$, $D \subseteq R$, $B \cap D = \emptyset$, $|B| = \mathrm{rank}\, A^*[R,C] < k$, and $|D| = d$.

The row vector $(U_{hi} \mid i \in R)$ of the matrix U is defined for $h \in D$ by the relation

$$-\boldsymbol{a}_h^* = \sum_{i \in B} U_{hi} \boldsymbol{a}_i^* \qquad (h \in D), \tag{7.27}$$

where $U_{hh} = 1$, and $U_{hi} = 0$ if $i \in R \setminus (B \cup \{h\})$. For $h \in R \setminus D$, it is defined to be the unit vector corresponding to h (i.e, $U_{hi} = \delta_{hi}$), where δ_{hi} denotes the Kronecker delta ($\delta_{hi} = 1$ for $h = i$ and $= 0$ otherwise). Then we have $U_{hi} = 0$ if $\tau^{-1}(i) > \tau^{-1}(h)$, which guarantees the condition in Lemma 7.1.13.

The row vector of UA^* indexed by $h \in D$ is zero by (7.27), and therefore term-rank $(UA^*) \le m - d = k - 1$, as desired in (7.22).

If (R2) is violated: We have $\operatorname{rank} A^*[I^*, C] < |I^*| = \text{term-rank}\, A^*[I^*, C]$. Then there exists a nonsingular constant matrix $U = (U_{hi})$ such that

$$\text{term-rank}\,(UA^*)[I^*, C] \leq |I^*| - 1. \tag{7.28}$$

As in the case of (R1) we transform A to A' by (7.23) with $U(s)$ defined by (7.24). Lemma 7.1.13 remains true, insuring the property (P1). On the other hand, (P2) is implied by (7.28). The proof is similar for Lemma 7.1.12; the feasibility of (p, q) for A' is shown in the same manner, while the nonoptimality follows from (7.28) and Theorem 7.1.10.

The constant matrix U is constructed similarly as in the case (R1). However, in the sequence $\boldsymbol{a}^*_{\tau(1)}$, $\boldsymbol{a}^*_{\tau(2)}$, $\cdots$, we consider only those terms $\boldsymbol{a}^*_{\tau(i)}$ which are the row vectors of the submatrix $A^*[I^*, C]$ (i.e., those vectors with $\tau(i) \in I^*$). The independent vectors are chosen from this subsequence so that $\{\boldsymbol{a}^*_i \mid i \in B\}$ may constitute the basis of row vectors of the submatrix $A^*[I^*, C]$. Also we put $d = 1$ instead of (7.26). Namely, we pick up the first dependent vector in the subsequence. Then the row vector of UA^* indexed by $h \in D$ is zero by (7.27), and therefore $\text{term-rank}\,(UA^*)[I^*, C] \leq |I^*| - 1$, as required by (7.28).

If (R3) is violated: We have $\operatorname{rank} A^*[R, J^*] < |J^*| = \text{term-rank}\, A^*[R, J^*]$. The modification in this case should be obvious from the one for the case (R2). Just exchange the roles of the rows and the columns. This means in particular that the matrix A is modified to A' by means of a transformation of the form $A'(s) = A(s)V(s)$ with

$$V(s) = \operatorname{diag}(s; -p_C) \cdot V \cdot \operatorname{diag}(s; p_C), \tag{7.29}$$

where V is a nonsingular constant matrix such that

$$\text{term-rank}\,(A^*V)[R, J^*] \leq |J^*| - 1. \tag{7.30}$$

If (R4) is violated: We have $\operatorname{rank} A^*[I^*, J^*] < |I^*| + |J^*| - k \leq \text{term-rank}\, A^*[I^*, J^*]$. Then there exists a nonsingular constant matrix $U = (U_{hi})$ such that

$$\text{term-rank}\,(UA^*)[I^*, J^*] \leq |I^*| + |J^*| - k - 1. \tag{7.31}$$

As in the case of (R1) we transform A to A' by (7.23) with $U(s)$ defined by (7.24). Lemma 7.1.13 remains true, insuring the property (P1). On the other hand, (P2) is implied by (7.31). The proof is similar for Lemma 7.1.12; the feasibility of (p, q) for A' is shown in the same manner, while the nonoptimality follows from (7.31) and Theorem 7.1.10.

The constant matrix U is constructed similarly as in the case (R1). However, we order the row vectors $\{\boldsymbol{a}^*_i[J^*] \mid i \in I^*\}$ of the submatrix $A^*[I^*, J^*]$ into a sequence $(\boldsymbol{a}^*_{\tau(i)}[J^*] \mid i = 1, 2, \cdots)$, and consider its subsequence consisting of the vectors with $\tau(i) \in I^*$. The independent vectors are chosen from this subsequence so that $\{\boldsymbol{a}_i^*[J^*] \mid i \in B\}$ may constitute the basis of row vectors of the submatrix $A^*[I^*, J^*]$. Also we put $d = k + 1 - |J^*|$ instead of (7.26), and define $D \subseteq I^*$ to be the set of the first d indices of the dependent row vectors $\boldsymbol{a}_i^*[J^*]$.

The row vector $(U_{hi} \mid i \in R)$ of the matrix U is defined similarly except that (7.27) is replaced by

$$-\boldsymbol{a}_h^*[J^*] = \sum_{i \in B} U_{hi} \boldsymbol{a}_i^*[J^*] \qquad (h \in D).$$

Then the row vector of $(UA^*)[I^*, J^*]$ indexed by $h \in D$ is zero, and therefore term-rank $(UA^*)[I^*, J^*] \leq |I^*| - d = |I^*| + |J^*| - k - 1$, as required by (7.31).

Example 7.1.14 (Continued from Example 7.1.11). Consider the case of $k = 3$, $\alpha = 1$, where the matrix $A(s)$ of Example 7.1.3 is not upper-tight, since, as we have seen in Example 7.1.11, the tight coefficient matrix A^* of (7.21),

$$\mathcal{T}(A; p, q) = A^* = \begin{array}{c} \\ \bullet \\ \bullet \\ \bullet \\ \\ \end{array} \!\!\! \begin{array}{c} \begin{array}{cccc} \bullet & \bullet & & \end{array} \\ \begin{pmatrix} 1 & 1 & 0 & 2 \\ 1 & 1 & 1 & 0 \\ 1 & 1 & -1 & 0 \\ 2 & 0 & 0 & 1 \end{pmatrix} \end{array},$$

does not satisfy the condition (R4) of Theorem 7.1.9. Namely, we have $\operatorname{rank} A^*[I^*, J^*] = 1 < |I^*| + |J^*| - 3 = 2$ where $I^*(p) = \{r_1, r_2, r_3\}$, $J^*(p) = \{c_1, c_2\}$ (indicated by $\bullet$), and the optimal dual variable (p, q) is given by (7.19) and (7.20).

We take the second row as the basis of the row vectors of $A^*[I^*, J^*]$ (i.e., $B = \{r_2\}$, $D = \{r_1, r_3\}$, $d = 2$) to get

$$U = \begin{pmatrix} 1 & -1 & 0 & 0 \\ 0 & 1 & 0 & 0 \\ 0 & -1 & 1 & 0 \\ 0 & 0 & 0 & 1 \end{pmatrix},$$

which, together with the dual variable p_{R} of (7.19), yields

$$U(s) = \operatorname{diag}(s^2, s^3, s^2, s^0) \cdot U \cdot \operatorname{diag}(s^{-2}, s^{-3}, s^{-2}, s^0) = \begin{pmatrix} 1 & -s^{-1} & 0 & 0 \\ 0 & 1 & 0 & 0 \\ 0 & -s^{-1} & 1 & 0 \\ 0 & 0 & 0 & 1 \end{pmatrix}.$$

Then the matrix A is modified to

$$A'(s) = U(s)A(s) = \begin{array}{c} \\ r_1 \\ r_2 \\ r_3 \\ r_4 \end{array} \!\!\! \begin{array}{c} \begin{array}{cccc} c_1 & c_2 & c_3 & c_4 \end{array} \\ \begin{pmatrix} 0 & -s^{-1} & -s^3 & 2s^3 - s \\ s^5 & s^6 + 1 & s^4 & s^2 \\ s & -s^{-1} & -2s^3 & -s \\ 2s^2 & s & 0 & s + 2 \end{pmatrix} \end{array}.$$

This matrix turns out to be upper-tight for $k = 3$ with $\hat{\delta}_3(A') = \delta_3(A') = 12$, which is verified as follows.

We find an optimal matching $M^{(4)} = \{(r_1, c_4), (r_2, c_2), (r_3, c_3)\}$ of size 3 of weight $w(M^{(4)}) = 3 + 6 + 3 = 12$ as well as the optimal dual variables

$$p'_{r_1} = 1,\ p'_{r_2} = 3,\ p'_{r_3} = 0,\ p'_{r_4} = 0;\ p'_{c_1} = 0,\ p'_{c_2} = 1,\ p'_{c_3} = 1,\ p'_{c_4} = 0;$$

and $q' = 2$, which shows $\hat{\delta}_3(A') = \pi(p', q') = 12$. The tight coefficient matrix

$$\mathcal{T}(A'; p', q') = (A')^* = \begin{matrix} \\ \bullet \\ \bullet \\ \\ \\ \end{matrix} \begin{matrix} & \bullet & \bullet & \\ \end{matrix} \begin{pmatrix} 0 & 0 & 0 & 2 \\ 1 & 1 & 0 & 0 \\ 0 & 0 & -2 & 0 \\ 2 & 0 & 0 & 0 \end{pmatrix}$$

satisfies the four conditions (R1)–(R4) in Theorem 7.1.9, where $I^*(p') = \{r_1, r_2\}$, $J^*(p') = \{c_2, c_3\}$ (indicated by $\bullet$). Hence $\hat{\delta}_3(A') = \delta_3(A')$. □

7.1.4 Algorithm Description

Combining the procedures given above we obtain the following algorithm for computing $\delta_k(A)$ for $A(s) \in \mathcal{M}(\boldsymbol{F}(s))$. For

$$w_{\max}(A) = \max_{i \in R, j \in C} \deg_s A_{ij}(s), \tag{7.32}$$

$$w_{\min}(A) = -\max_{j \in C} \sum_{i \in R} (\text{degree of the denominator of } A_{ij}(s)), \tag{7.33}$$

we have $k \cdot w_{\min}(A) \leq \delta_k(A) \leq \hat{\delta}_k(A) \leq k \cdot w_{\max}(A)$ if $\delta_k(A) > -\infty$. Hence the number of modifications of the matrix is bounded by

$$\hat{\delta}_k(A^{(0)}) - \delta_k(A^{(0)}) \leq k(w_{\max}(A^{(0)}) - w_{\min}(A^{(0)})),$$

where $A^{(0)}$ denotes the input matrix.

Algorithm for computing $\delta_k(A)$

Step 0

Define $w_{\min}(A)$ as (7.33).

Step 1

(1) : Find a maximum weight k-matching M and integer-valued optimal dual variables, $p_{\mathrm{R}i}$ $(i \in R)$, $p_{\mathrm{C}j}$ $(j \in C)$ and q, for $G(A)$;
$\hat{\delta}_k(A) := w(M)$ $\quad (\hat{\delta}_k(A) = -\infty$ if no k-matching exists).

(2) : If $\hat{\delta}_k(A) < k \cdot w_{\min}$, then stop with $\delta_k(A) = -\infty$.

Step 2

(1) : $A^*_{ij} := \lim_{s \to \infty} s^{-p_{\mathrm{R}i} - p_{\mathrm{C}j} - q} A_{ij}(s) \quad (i \in R, j \in C)$. [cf.(7.13)]

(2) : If the four conditions (R1)–(R4) in Theorem 7.1.9 are satisfied, then stop with $\delta_k(A) = \hat{\delta}_k(A)$.

Step 3 [A is not upper-tight]

(1) : Modify the matrix $A(s)$ as described in §7.1.3, according to which of (R1)–(R4) is violated. (If (R1) is violated, for example, let U be defined by (7.27) and $A(s) := \mathrm{diag}\,(s; p_{\mathrm{R}}) \cdot U \cdot \mathrm{diag}\,(s; -p_{\mathrm{R}}) A(s))$.

(2) : Go to Step 1. □

The stopping criterion in Step 1 (2) is to cope with the case of $\delta_k(A^{(0)}) = -\infty$. In Step 2, we need row/column elimination operations on A^*. Though it requires $\mathrm{O}(\max(m^3, n^3))$ arithmetic operations in $\boldsymbol{F}$ in the worst case, it can be done much faster since A^* is usually very sparse in practical applications.

Finally let us mention the probabilistic behavior of the algorithm. As already noted in Theorem 7.1.1, $\hat{\delta}_k(A)$ differs from $\delta_k(A)$ only because of accidental numerical cancellation. Let us fix the structure (i.e., the position of the nonzero coefficients) of the input matrix $A = A^{(0)}$ and regard the numerical values of coefficients in $\mathbf{R} = \boldsymbol{F}$ as real-valued random variables with continuous distributions. Then we have $\hat{\delta}_k(A) = \delta_k(A)$ with probability one, which means that Step 3 is performed only with null probability. Since the worst-case time complexity for the weighted-matching problem is bounded by $\mathrm{O}((m+n)^3)$, we obtain the following statement, indicating the practical efficiency of the proposed algorithm: The average time complexity (in the above sense) of the proposed algorithm for a fixed k is bounded by a polynomial in $m+n$ (e.g., $(m+n)^3$).

Notes. The idea of the combinatorial relaxation algorithm was proposed first by Murota [212] for the Newton diagram of determinantal equations, which was followed by Murota [219] (the degree of the determinant of (skew-symmetric) polynomial matrices), Murota [220] (the degree of subdeterminants), Iwata–Murota–Sakuta [147] (primal-dual type algorithm together with comparative computational results), Iwata–Murota [146] (mixed polynomial matrix formulation using valuated matroids), and Iwata [141] (matrix pencil using strict equivalence). This section is based on Murota [220].

7.2 Combinatorial System Theory

This section is devoted to the description of a combinatorial analogue of the dynamical system theory of Murota [202, 206, 211]. The theory is developed in a matroid-theoretic framework by replacing the matrices A and B in the state-space equations $\dot{\boldsymbol{x}} = A\boldsymbol{x} + B\boldsymbol{u}$ or $\boldsymbol{x}_{k+1} = A\boldsymbol{x}_k + B\boldsymbol{u}_k$ with bimatroids. The main objective is to extend the combinatorial mathematical aspects in the arguments of structural controllability, rather than to pursue physical faith in the structural approach. Combinatorial counterparts are given to a number of fundamental concepts such as controllability in the conventional system theory, revealing the combinatorial nature of some fundamental results in the conventional system theory. However, the present theory is not a direct generalization of the existing "generic" approach, mainly because of the possible discrepancy between the matrix multiplication and bimatroid multiplication. Here the discrepancy means the failure of the relation $\mathbf{L}(A) * \mathbf{L}(B) = \mathbf{L}(A \cdot B)$ for matrices A and B, where $\mathbf{L}(\cdot)$ denotes the bimatroid defined by a matrix.

7.2.1 Definition of Combinatorial Dynamical Systems

A combinatorial analogue of the dynamical system

$$\boldsymbol{x}_{k+1} = A\boldsymbol{x}_k + B\boldsymbol{u}_k \tag{7.34}$$

is obtained by replacing the matrices A and B with two bimatroids. To be more precise, a *combinatorial dynamical system* (to be abbreviated as *CDS*) is a pair $(\mathbf{A}, \mathbf{B})$ of bimatroids such that $\mathrm{Row}(\mathbf{A}) = \mathrm{Col}(\mathbf{A}) = \mathrm{Row}(\mathbf{B})$ $(\equiv S)$ and that S and $\mathrm{Col}(\mathbf{B})$ $(\equiv P)$ are mutually disjoint: $\mathbf{A} = (S, S, \Lambda(\mathbf{A}), \alpha)$, $\mathbf{B} = (S, P, \Lambda(\mathbf{B}), \beta)$, where α and β are birank functions. The set S is called the *state set*, whereas P is the *input set*. A bimatroid $\mathbf{F}$ is called a *state feedback* if $\mathrm{Row}(\mathbf{F}) = P$ and $\mathrm{Col}(\mathbf{F}) = S$.

As a counterpart of (7.34), we consider

$$(X_{k+1}, X_k \cup U_k) \in \Lambda(\mathbf{A} \vee \mathbf{B}), \qquad X_k, X_{k+1} \subseteq S,\ U_k \subseteq P. \tag{7.35}$$

By definition, this means that X_{k+1} can be partitioned into two disjoint parts, say X'_{k+1} and X''_{k+1}, such that $(X'_{k+1}, X_k) \in \Lambda(\mathbf{A})$ and $(X''_{k+1}, U_k) \in \Lambda(\mathbf{B})$. See Fig. 7.4 and compare it with the dynamic graph introduced in §2.2.1 (see also Fig. 2.5). The equation (7.35) will be referred to as the *state-space equation* for the CDS.

An *input* is a sequence $(U_k)_{k=0}^{K-1} = (U_k \mid k = 0, 1, \cdots, K-1)$, $U_k \subseteq P$. We say that a sequence $(X_k)_{k=0}^{K} = (X_k \mid k = 0, 1, \cdots, K), X_k \subseteq S$, is a trajectory *compatible* with $(U_k)_{k=0}^{K-1}$ if (7.35) holds for $k = 0, 1, \cdots, K-1$. An input is said to be *admissible* for (X', X) if there exists a trajectory $(X_k)_{k=0}^{K}$ compatible with it such that $X_0 = X'$ and $X_K = X$. Put

$$\mathrm{RS}_k(X_0) = \{X_k \subseteq S \mid \text{some } (U_i)_{i=0}^{k-1} \text{ is admissible for } (X_0, X_k)\}, \tag{7.36}$$

$$\mathrm{RS}(X_0) = \bigcup_{k=0}^{\infty} \mathrm{RS}_k(X_0). \tag{7.37}$$

Then we say that X $(\subseteq S)$ is *reachable* at time k (≥ 0) from X_0 $(\subseteq S)$ if $X \in \mathrm{RS}_k(X_0)$, and that a CDS is *reachable* if $\{x\} \in \mathrm{RS}(\emptyset)$ for $\forall x \in S$. We also say that a CDS is *controllable* if $\mathrm{RS}(\emptyset) = 2^S$.

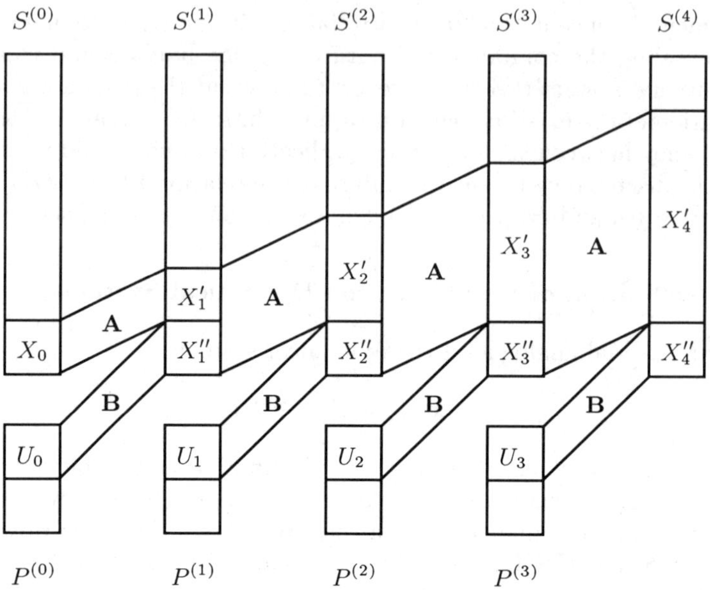

Fig. 7.4. Combinatorial dynamical system

7.2.2 Power Products

We first consider an "autonomous" system in which input bimatroid $\mathbf{B}$ is trivial (of rank zero). In this case the state-space equation (7.35) reduces to

$$(X_{k+1}, X_k) \in \Lambda(\mathbf{A}), \qquad X_k, X_{k+1} \subseteq S. \tag{7.38}$$

In other words, we investigate the power products of a single bimatroid $\mathbf{A}$ such that $\mathrm{Row}(\mathbf{A}) = \mathrm{Col}(\mathbf{A})$.

Since $\mathrm{Row}(\mathbf{A}) = \mathrm{Col}(\mathbf{A})$, we can think of the product of $\mathbf{A}$ with itself. We define $\mathbf{A}^k$ recursively by $\mathbf{A}^k = \mathbf{A}^{k-1} * \mathbf{A} = \mathbf{A} * \mathbf{A}^{k-1}$ for $k = 1, 2, \cdots$, where, for convenience, we put $\mathbf{A}^0 = (S, S, \Lambda(\mathbf{A}^0))$ with $\Lambda(\mathbf{A}^0) = \{(X, X) \mid X \subseteq S\}$. For an autonomous system, $\bigcup_{X_0 \subseteq S} \mathrm{RS}_k(X_0)$ agrees with the family of independent sets of $\mathbf{RM}(\mathbf{A}^k)$, which we are interested in.

Remark 7.2.1. Even in the special case where $\mathbf{A}$ arises from a generic matrix A with independent nonzero entries, the power products of $\mathbf{A} = \mathbf{L}(A)$ do not agree with the bimatroids associated with the power products of A, i.e., $\mathbf{L}(A^k) \neq (\mathbf{L}(A))^k$. In fact, for

$$A = \begin{array}{c|ccccc|} & x_1 & x_2 & x_3 & x_4 & x_5 \\ \hline x_1 & 0 & 0 & t_1 & 0 & 0 \\ x_2 & t_2 & 0 & t_3 & 0 & 0 \\ x_3 & t_4 & 0 & 0 & 0 & 0 \\ x_4 & 0 & t_5 & 0 & 0 & 0 \\ x_5 & 0 & 0 & 0 & t_6 & 0 \\ \hline \end{array},$$

where t_i $(i = 1, \cdots, 6)$ are indeterminates, the 2×2 submatrix of A^3 with row set $\{x_2, x_5\}$ and column set $\{x_1, x_3\}$ is singular, i.e., $(\{x_2, x_5\}, \{x_1, x_3\}) \notin \Lambda(\mathbf{L}(A^3))$, whereas $(\{x_2, x_5\}, \{x_1, x_3\}) \in \Lambda(\mathbf{L}(A)^3)$. □

In spite of such discrepancy between $\mathbf{L}(A^k)$ and $\mathbf{L}(A)^k$ in the families of the linked pairs, they share the same rank (=the maximum size of linked pairs), as is stated in the following theorem of Poljak [270].

Theorem 7.2.2. $\mathrm{rank}\,(\mathbf{L}(A^k)) = \mathrm{rank}\,(\mathbf{L}(A)^k)$ *for a square generic matrix* A.

Proof. See Poljak [270]. ■

Fundamental properties of the power products of a bimatroid are stated in the following two theorems shown by Murota [211]. The first theorem reveals that the sequence $r(\mathbf{A}^k) = \mathrm{rank}\,(\mathbf{A}^k)$, $k = 0, 1, 2, \cdots$, is convex and nonincreasing.

Theorem 7.2.3. *Let* $\mathbf{A}$ *be a bimatroid with* $\mathrm{Row}(\mathbf{A}) = \mathrm{Col}(\mathbf{A}) = S$.

(1) $\quad r(\mathbf{A}^{k-1}) - r(\mathbf{A}^k) \geq r(\mathbf{A}^k) - r(\mathbf{A}^{k+1}), \qquad k = 1, 2, \cdots.$

(2) *There exists* $\tau = \tau(\mathbf{A})$ $(0 \leq \tau \leq |S|)$ *such that*

$$r(\mathbf{A}^0) > r(\mathbf{A}^1) > \cdots > r(\mathbf{A}^{\tau-1}) > r(\mathbf{A}^\tau) = r(\mathbf{A}^k), \quad k = \tau+1, \tau+2, \cdots.$$

Proof. (1) This follows from Theorem 2.3.55 with $\mathbf{L}_1 = \mathbf{L}_3 = \mathbf{A}$ and $\mathbf{L}_2 = \mathbf{A}^{k-1}$.

(2) First note the obvious relation: $r(\mathbf{A}^k) \geq r(\mathbf{A}^{k+1})$. Let τ be the smallest k such that the equality holds, where $\tau \leq |S|$ since $0 \leq r(\mathbf{A}^\tau) \leq r(\mathbf{A}^0) - \tau = |S| - \tau$. Then (1) implies that the equality must hold for all $k \geq \tau$. ■

The integer $\tau = \tau(\mathbf{A})$ above is called the *transition index* of $\mathbf{A}$. We symbolically write $r(\mathbf{A}^\infty)$ for $r(\mathbf{A}^\tau)$ though the limit of $\mathbf{A}^k$ (as $k \to \infty$) may not exist.

The ranks of the associated row and column matroids, $\mathbf{RM}(\mathbf{A}^k)$ and $\mathbf{CM}(\mathbf{A}^k)$, satisfy similar inequalities, since $\mathrm{rank}\,(\mathbf{A}^k) = \mathrm{rank}\,(\mathbf{RM}(\mathbf{A}^k)) = \mathrm{rank}\,(\mathbf{CM}(\mathbf{A}^k))$. The following theorem establishes a stronger assertion than the inequality of Theorem 7.2.3(2) above. It claims that the sequences of $\{\mathbf{RM}(\mathbf{A}^k)\}_k$ and $\{\mathbf{CM}(\mathbf{A}^k)\}_k$ have nice nesting structures and consequently their limits do exist, which will be denoted by $\mathbf{RM}(\mathbf{A}^\infty)$ and $\mathbf{CM}(\mathbf{A}^\infty)$.

Theorem 7.2.4. *Let τ be the transition index of a bimatroid $\mathbf{A}$ with* $\mathrm{Row}(\mathbf{A}) = \mathrm{Col}(\mathbf{A})$. *Then*

$$\mathbf{RM}(\mathbf{A}^0) \underset{\neq}{\to} \mathbf{RM}(\mathbf{A}^1) \underset{\neq}{\to} \cdots \underset{\neq}{\to} \mathbf{RM}(\mathbf{A}^{\tau-1}) \underset{\neq}{\to} \mathbf{RM}(\mathbf{A}^{\tau}) = \mathbf{RM}(\mathbf{A}^k), \quad k \geq \tau+1,$$

$$\mathbf{CM}(\mathbf{A}^0) \underset{\neq}{\to} \mathbf{CM}(\mathbf{A}^1) \underset{\neq}{\to} \cdots \underset{\neq}{\to} \mathbf{CM}(\mathbf{A}^{\tau-1}) \underset{\neq}{\to} \mathbf{CM}(\mathbf{A}^{\tau}) = \mathbf{CM}(\mathbf{A}^k), \quad k \geq \tau+1,$$

where $\mathbf{M}_1 \underset{\neq}{\to} \mathbf{M}_2$ means that $\mathbf{M}_2$ is a strong quotient of, and not isomorphic to, $\mathbf{M}_1$.

Proof. It follows from Theorem 2.3.59 that $\mathbf{RM}(\mathbf{A}^k) \to \mathbf{RM}(\mathbf{A}^{k+1})$ and $\mathbf{CM}(\mathbf{A}^k) \to \mathbf{CM}(\mathbf{A}^{k+1})$. Combining these with Theorem 7.2.3(2) and Lemma 2.3.1, we establish the theorem. ■

Based on Theorem 7.2.3 we can define a set of characteristic indices $(\omega_0; \omega_1, \omega_2, \cdots)$ of $\mathbf{A}$ by

$$\begin{aligned} \omega_0 &= r(\mathbf{A}^{\infty}), \\ \omega_k &= r(\mathbf{A}^{k+1}) + r(\mathbf{A}^{k-1}) - 2r(\mathbf{A}^k), \quad k = 1, 2, \cdots, \end{aligned}$$

where $\omega_k \geq 0$ $(k = 0, 1, \cdots)$. Note that $\omega_k = 0$ for $k > \tau$ and that

$$\omega_0 + \sum_{k=1}^{\tau} k\omega_k = |S|.$$

In the particular case where $\mathbf{A} = \mathbf{L}(A)$ with a square generic matrix A having independent nonzero entries, this set of indices $(\omega_0; \omega_1, \omega_2, \cdots)$ determines the Jordan canonical form of the matrix A. Motivated by this we call $(\omega_0; \omega_1, \omega_2, \cdots)$ the *Jordan type* of a bimatroid $\mathbf{A}$.

Remark 7.2.5. Theorems 7.2.3 and 7.2.4 are motivated by similar phenomena for matrix products A^k, $k = 0, 1, 2, \cdots$. The same inequalities as in Theorem 7.2.3 hold true for the sequence $\mathrm{rank}\, A^k$, $k = 0, 1, 2, \cdots$, with τ being the maximum size of a Jordan block for a zero eigenvalue. The strong map sequence of Theorem 7.2.4 corresponds to the nesting structure of subspaces:

$$\mathrm{Im}(\mathbf{A}^0) \underset{\neq}{\supseteq} \mathrm{Im}(\mathbf{A}^1) \underset{\neq}{\supseteq} \cdots \underset{\neq}{\supseteq} \mathrm{Im}(\mathbf{A}^{\tau-1}) \underset{\neq}{\supseteq} \mathrm{Im}(\mathbf{A}^{\tau}) = \mathrm{Im}(\mathbf{A}^k), \quad k \geq \tau + 1.$$

Recall from Example 2.3.8 that the strong map relation may be interpreted as a combinatorial abstraction of the nesting of linear subspaces. □

7.2.3 Eigensets and Recurrent Sets

In this section we study eigensets and recurrent sets of a bimatroid, introduced by Murota [202, 211]. A subset $X \subseteq S$ is said to be an *eigenset* of a bimatroid $\mathbf{A} = (S, S, \Lambda(\mathbf{A}))$ if $(X, X) \in \Lambda(\mathbf{A})$, and a *recurrent set* of $\mathbf{A}$ if $(X, X) \in \Lambda(\mathbf{A}^k)$ for some $k \geq 1$. By definition, an eigenset is a recurrent set, but the converse is not true in general.

Eigensets of a bimatroid may be regarded as a combinatorial analogue of eigenvectors of a matrix as follows. Let A be a matrix and $\mathbf{A}$ the corresponding bimatroid, both having S as the row set and the column set. Just as we think of A as a mapping in $\mathbf{R}^S$ we may regard $\mathbf{A}$ as a (multivalued) mapping in 2^S, where it is kept in mind that $X \in 2^S$ can be expressed alternatively by its characteristic vector $\chi_X \in \mathbf{R}^S$. A vector $\boldsymbol{x} \in \mathbf{R}^S$ is an eigenvector of A if it is invariant in direction when transformed by A. In parallel, a subset $X \in 2^S$ or its characteristic vector χ_X is an eigenset of $\mathbf{A}$ if it can be kept invariant when transformed by $\mathbf{A}$.

We denote by EIG($\mathbf{A}$) the family of eigensets of $\mathbf{A}$, and by max-EIG($\mathbf{A}$) that of maximum-sized eigensets of $\mathbf{A}$. Similarly, we denote by REC($\mathbf{A}$) the family of recurrent sets of $\mathbf{A}$, and by max-REC($\mathbf{A}$) that of maximum-sized recurrent sets of $\mathbf{A}$.

Example 7.2.6. For illustration, let us consider the bimatroid $\mathbf{A}$ defined by a matrix

	x_1	x_2	x_3
x_1	1	1	1
x_2	1	1	2
x_3	1	1	1

Writing $X \to Y$ instead of $(X,Y) \in \Lambda(\mathbf{A})$, we have $\{x_1,x_2\} \to \{x_1,x_3\}$, $\{x_1,x_2\} \to \{x_2,x_3\}$, $\{x_2,x_3\} \to \{x_1,x_3\}$, $\{x_2,x_3\} \to \{x_2,x_3\}$; $\{x_i\} \to \{x_j\}$ for $i,j = 1,2,3$, and $\emptyset \to \emptyset$. Hence,

$$\begin{aligned}\mathrm{EIG}(\mathbf{A}) = \mathrm{REC}(\mathbf{A}) &= \{\{x_2,x_3\},\{x_1\},\{x_2\},\{x_3\},\emptyset\},\\ \text{max-EIG}(\mathbf{A}) = \text{max-REC}(\mathbf{A}) &= \{\{x_2,x_3\}\}.\end{aligned}$$

This example shows that a maximal recurrent set is not necessarily a maximum-sized recurrent set, nor is a maximal eigenset a maximum eigenset. In fact, the singleton set $\{x_1\}$ is maximal and not maximum. □

We first consider recurrent sets of maximum size.

Theorem 7.2.7. *Let* $\mathbf{A}$ *be a bimatroid with* Row($\mathbf{A}$) = Col($\mathbf{A}$) *and* $k \geq \tau(\mathbf{A})$ *(=the transition index). Then*

$$\text{max-REC}(\mathbf{A}) = \text{max-EIG}(\mathbf{A}^k) = \max \mathbf{RM}(\mathbf{A}^\infty) \cap \max \mathbf{CM}(\mathbf{A}^\infty),$$

where $\max \mathbf{RM}(\mathbf{A}^\infty)$ *denotes the family of bases (=maximum-sized independent sets) of* $\mathbf{RM}(\mathbf{A}^\infty)$*, and similarly for* $\max \mathbf{CM}(\mathbf{A}^\infty)$*; accordingly, the right-most term means the family of common bases of* $\mathbf{RM}(\mathbf{A}^\infty)$ *and* $\mathbf{CM}(\mathbf{A}^\infty)$.

Proof. Suppose $X \in$ REC($\mathbf{A}$). Then $(X,X) \in \Lambda(\mathbf{A}^k)$ for some $k \geq 1$. This implies that $(X,X) \in \Lambda(\mathbf{A}^{km})$ for any m. Choosing m so that $km \geq \tau$, we see X is independent both in $\mathbf{RM}(\mathbf{A}^\infty)$ and $\mathbf{CM}(\mathbf{A}^\infty)$, cf. Theorem 7.2.4.

If X is a common base of $\mathbf{RM}(\mathbf{A}^\infty)$ and $\mathbf{CM}(\mathbf{A}^\infty)$ and $k \geq \tau$, then $(X, Y) \in \Lambda(\mathbf{A}^k)$ and $(Z, X) \in \Lambda(\mathbf{A}^k)$ for some Y and Z. It then follows from the property (L-3) of a bimatroid that $(X', X'') \in \Lambda(\mathbf{A}^k)$ for some $X' \supseteq X$ and $X'' \supseteq X$. We must have $X' = X'' = X$ since $|X| = \operatorname{rank}(\mathbf{A}^k)$. Hence $X \in \operatorname{EIG}(\mathbf{A}^k) \subseteq \operatorname{REC}(\mathbf{A})$.

The claim of the theorem follows from the above argument. ■

Theorem 7.2.8. REC(**A**) *is a hereditary family. That is, if* $Y \subseteq X \in \operatorname{REC}(\mathbf{A})$, *then* $Y \in \operatorname{REC}(\mathbf{A})$.

Proof. The proof consists of three steps (i)–(iii). The assertion of the theorem follows from the claim in (iii) by induction.

(i) First we claim: $x \in X \in \operatorname{EIG}(\mathbf{A}) \Rightarrow \{x\} \in \operatorname{REC}(\mathbf{A})$. By Theorem 2.3.44(1), there is a permutation σ of X such that $(\{x\}, \{\sigma(x)\}) \in \Lambda(\mathbf{A})$. For each $x \in X$ there exists $k \geq 1$ such that $\sigma^k(x) = x$. This implies that $(\{x\}, \{x\}) \in \Lambda(\mathbf{A}^k)$. Hence $\{x\} \in \operatorname{REC}(\mathbf{A})$.

(ii) Next we claim: $x \in X \in \operatorname{EIG}(\mathbf{A}) \Rightarrow X \setminus \{x\} \in \operatorname{REC}(\mathbf{A})$. Since $(X, X) \in \Lambda(\mathbf{A})$, $\mathbf{A}[X, X]$ is a nonsingular bimatroid. Put $\mathbf{A}' = \mathbf{A}[X, X]^{-1}$ (the inverse bimatroid; see §2.3). Then $X \in \operatorname{EIG}(\mathbf{A}')$. The first claim implies that, for each $x \in X$ there exists a sequence $x_0\ (= x), x_1, \cdots, x_k\ (= x)$ in X such that $(\{x_i\}, \{x_{i-1}\}) \in \Lambda(\mathbf{A}')$ for $i = 1, \cdots, k$. This is equivalent to $(X_{i-1}, X_i) \in \Lambda(\mathbf{A}[X, X])$, with $X_i = X \setminus \{x_i\}$ for $i = 1, \cdots, k$. Hence $(X \setminus \{x\}, X \setminus \{x\}) \in \Lambda(\mathbf{A}^k)$, establishing the claim.

(iii) We finally claim: $x \in X \in \operatorname{REC}(\mathbf{A}) \Rightarrow X \setminus \{x\} \in \operatorname{REC}(\mathbf{A})$. Since $X \in \operatorname{EIG}(\mathbf{A}^k)$ for some $k \geq 1$, we see $X \setminus \{x\} \in \operatorname{REC}(\mathbf{A}^k)$ by the second claim above. This implies $X \setminus \{x\} \in \operatorname{REC}(\mathbf{A})$. ■

Remark 7.2.9. By the proof of Theorem 7.2.7, a recurrent set of **A** is a common independent set of $\mathbf{RM}(\mathbf{A}^\infty)$ and $\mathbf{CM}(\mathbf{A}^\infty)$. The converse, however, is not true in general, since an arbitrary common independent set cannot always be augmented to a common base. For a concrete instance, consider the bimatroid **A** defined by a matrix

	x_1	x_2	x_3
x_1	1	1	0
x_2	0	0	1
x_3	0	0	1

and the singleton set $\{x_2\}$. This is not a recurrent set, though it is independent both in $\mathbf{RM}(\mathbf{A}^\infty)$ and $\mathbf{CM}(\mathbf{A}^\infty)$. □

Remark 7.2.10. The family of eigensets is not necessarily hereditary. For the bimatroid **A** defined by a matrix

	x_1	x_2
x_1	0	1
x_2	1	0

we have $\mathrm{EIG}(\mathbf{A}) = \{\{x_1, x_2\}, \emptyset\}$. □

Our next result shows that the maximum size of a recurrent set coincides with that of an eigenset. Since an eigenset is a "static" concept that does not involve dynamics, we may say that the formula below gives a "static" characterization of the limit of the dynamics, $\lim_{k\to\infty} r(\mathbf{A}^k)$.

Theorem 7.2.11. *For a bimatroid* $\mathbf{A}$ *with* $\mathrm{Row}(\mathbf{A}) = \mathrm{Col}(\mathbf{A})$,

$$\max\{|X| \mid X \in \mathrm{EIG}(\mathbf{A})\} = \max\{|X| \mid X \in \mathrm{REC}(\mathbf{A})\} = r(\mathbf{A}^\infty).$$

Proof. By Theorem 7.2.7 as well as the obvious inclusion $\mathrm{EIG}(\mathbf{A}) \subseteq \mathrm{REC}(\mathbf{A})$, we see

$$\max\{|X| \mid X \in \mathrm{EIG}(\mathbf{A})\} \le \max\{|X| \mid X \in \mathrm{REC}(\mathbf{A})\} = r(\mathbf{A}^\infty).$$

The inequality here is an equality by the following lemma. ■

Lemma 7.2.12. *If X is a recurrent set of maximum size and $x \in X$, then there exists an eigenset Y such that $x \in Y$ and $|X| = |Y|$.*

Proof. We denote by S^- and S^+ two disjoint copies of $S = \mathrm{Row}(\mathbf{A}) = \mathrm{Col}(\mathbf{A})$. For $X \subseteq S$ we compatibly denote by X^- and X^+ the copies of X in S^- and S^+, respectively.

Consider two matroids $\mathbf{M}_\mathbf{A} = (S^- \cup S^+, \mathcal{B}_\mathbf{A})$ and $\mathbf{M}_0 = (S^- \cup S^+, \mathcal{B}_0)$ with the base families

$$\mathcal{B}_\mathbf{A} = \{(S^-\setminus X^-)\cup Y^+ \mid (X,Y) \in \Lambda(\mathbf{A})\}, \quad \mathcal{B}_0 = \{(S^-\setminus X^-)\cup Y^+ \mid X = Y\},$$

where $\mathbf{M}_\mathbf{A}$ is the matroid associated with $\mathbf{A}$ by (2.86). It is easy to see that

$$\begin{aligned} X \in \mathrm{EIG}(\mathbf{A}) &\iff (S^- \setminus X^-) \cup X^+ \in \mathcal{B}_\mathbf{A} \\ &\iff \exists H \in \mathcal{B}_\mathbf{A} \cap \mathcal{B}_0 :\ X^+ = H \cap S^+. \end{aligned}$$

Let $B_\mathbf{A} \subseteq \mathbf{R}^{S^-\cup S^+}$ denote the convex hull of the characteristic vectors χ_H of $H \in \mathcal{B}_\mathbf{A}$, and define $B_0 \subseteq \mathbf{R}^{S^-\cup S^+}$ similarly from $\mathcal{B}_0$.

For $X \in$ max-REC$(\mathbf{A})$, there exist X_i $(i = 0, 1, \cdots, k)$ with $X_0 = X_k = X$ such that $(X_{i-1}, X_i) \in \Lambda(\mathbf{A})$ for $i = 1, \cdots, k$. This means that

$$[(\chi_S)^- - (\chi_{X_{i-1}})^-] \oplus (\chi_{X_i})^+ \in B_\mathbf{A}, \qquad i = 1, 2, \cdots, k,$$

where for a vector $\xi \in \mathbf{R}^S$ in general we write ξ^- and ξ^+ for the corresponding vectors in $\mathbf{R}^{S^-}$ and $\mathbf{R}^{S^+}$, respectively. Taking the average of these expressions and putting $\mu = \sum_{i=1}^k \chi_{X_i}/k$, we obtain

$$\bar{\boldsymbol{h}} = [(\chi_S)^- - \mu^-] \oplus \mu^+ \in B_\mathbf{A} \cap B_0.$$

We also have $\bar{h}(S^+) = |X| = r(\mathbf{A}^\infty)$. This shows that $\max\{h(S^+) \mid \boldsymbol{h} \in B_\mathbf{A} \cap B_0\} = r(\mathbf{A}^\infty)$ and that $\bar{\boldsymbol{h}}$ attains the maximum.

By the integrality of matroid intersection, $B_{\mathbf{A}} \cap B_0$ is the convex hull of the characteristic vectors of common bases of $\mathbf{M}_{\mathbf{A}}$ and $\mathbf{M}_0$. In particular, we can express $\bar{\boldsymbol{h}}$ as a convex combination

$$\bar{\boldsymbol{h}} = \sum_{j=1}^{m} c_j \boldsymbol{h}_j, \qquad \sum_{j=1}^{m} c_j = 1, \quad c_j > 0 \ (j = 1, \cdots, m)$$

of such incidence vectors $\boldsymbol{h}_j = \chi_{H_j}$ $(j = 1, \cdots, m)$ with $h_j(S^+) = r(\mathbf{A}^\infty)$. Note that $H_j \in \mathcal{B}_{\mathbf{A}} \cap \mathcal{B}_0$ is equivalent to $H_j = (S^- \setminus {Z_j}^-) \cup {Z_j}^+$ for some $Z_j \in \mathrm{EIG}(\mathbf{A})$, and that $h_j(S^+) = |Z_j| = r(\mathbf{A}^\infty)$. Since $x \in X$, we have $\bar{h}(x^+) > 0$. This implies that $h_j(x^+) > 0$ for some j, i.e., $x \in Z_j$ for some j. This Z_j gives the desired Y. ■

Remark 7.2.13. A recurrent set of maximum size is not necessarily an eigenset, i.e., max-REC($\mathbf{A}$) $\neq$ max-EIG($\mathbf{A}$). For the bimatroid $\mathbf{A}$ defined by a matrix

	x_1	x_2	x_3	x_4	x_5
x_1	0	0	1	0	0
x_2	0	0	0	1	0
x_3	0	1	0	0	1
x_4	0	0	1	0	0
x_5	1	0	0	0	0

it can be verified that max-EIG($\mathbf{A}$) $= \{\{x_1, x_3, x_5\}, \{x_2, x_3, x_4\}\}$ and max-REC($\mathbf{A}$) $=$ max-EIG($\mathbf{A}$) $\cup \{\{x_1, x_2, x_3\}, \{x_3, x_4, x_5\}\}$. □

Remark 7.2.14. Theorem 7.2.11 does not imply

$$\max\{|X| \mid \mathrm{rank}\, A[X, X] = |X|, X \subseteq S\} = \lim_{k\to\infty} \mathrm{rank}\,(A^k) \tag{7.39}$$

for a numerical matrix A with Row(A) $=$ Col(A) $= S$ because of the possible discrepancy between matrix multiplication and bimatroid multiplication (cf. Remark 2.3.54). In fact, $A = \begin{pmatrix} 1 & 1 \\ -1 & -1 \end{pmatrix}$ serves as a counterexample to (7.39). The equality (7.39) is true, however, if the nonzero entries of A are algebraically independent, which fact follows from the combination of Theorem 7.2.11 with Theorem 7.2.2. □

7.2.4 Controllability of Combinatorial Dynamical Systems

This section gives a number of controllability criteria for a CDS $(\mathbf{A}, \mathbf{B})$. Before stating the general results, it is worth while explaining the relation between the structural controllability and the controllability of a CDS in a typical situation.

Consider a conventional dynamical system (7.34) described by matrices A and B, with which we can associate a CDS $(\mathbf{A}, \mathbf{B})$ under the correspondence:

$\mathbf{A} = \mathbf{L}(A)$, $\mathbf{B} = \mathbf{L}(B)$. In the special case where the matrices A and B are generic ("structured" in the sense of §6.4.2), the associated bimatroids $\mathbf{A}$ and $\mathbf{B}$ can be represented by matchings in bipartite graphs. In this case the associated CDS $(\mathbf{A}, \mathbf{B})$ is controllable if and only if there exists in the dynamic graph G_0^n of the system (7.34) a Menger-type vertex-disjoint linking of size n from U_0^{n-1} to X^n (see §2.2.1 or §6.4.2 for the definitions of G_0^n, U_0^{n-1}, and X^n). Theorem 6.4.3 then reveals that a "structured" system (A, B) is controllable in the ordinary sense if and only if the associated CDS $(\mathbf{A}, \mathbf{B})$ is controllable. It is noted at the same time that the controllability of the associated CDS $(\mathbf{A}, \mathbf{B})$ is only necessary for general numerical matrices A and B.

For the controllability criteria we shall investigate the structure of the sequence $\{\mathrm{RS}_k(\emptyset)\}_k$, i.e., the sequence of those sets which are reachable at time k from the empty set (cf. (7.36)). As the following theorem claims, $\mathrm{RS}_k(\emptyset)$ forms the family of independent sets of a matroid, denoted by $\mathbf{R}_k$, and the sequence $\{\mathbf{R}_k\}_k$ is determined by a recurrence relation. The matroid $\mathbf{R}_k$ will be referred to as the *reachability matroid.*

Theorem 7.2.15. *For each k (≥ 0), $\mathrm{RS}_k(\emptyset)$ of (7.36) forms the family of independent sets of a matroid $\mathbf{R}_k$. The matroids are determined by*

$$\mathbf{R}_k = (\mathbf{A} * \mathbf{R}_{k-1}) \vee \mathbf{RM}(\mathbf{B}), \qquad k = 1, 2, \cdots,$$

where $\mathbf{R}_0$ is a trivial matroid (of rank zero), $\mathbf{A} * \mathbf{R}_{k-1}$ is the matroid induced from $\mathbf{R}_{k-1}$ by $\mathbf{A}$, and $\vee$ means the matroid union.

Proof. We prove the claims by induction on k. Assume that $\mathrm{RS}_{k-1}(\emptyset)$ defines a matroid $\mathbf{R}_{k-1}$. It follows from the state-space equation (7.35) that $X_k \in \mathrm{RS}_k(\emptyset)$ if and only if $X_k = X_k' \cup X_k''$, $X_k' \cap X_k'' = \emptyset$, $(X_k', X_{k-1}) \in \Lambda(\mathbf{A})$, $(X_k'', U_{k-1}) \in \Lambda(\mathbf{B})$, and $X_{k-1} \in \mathrm{RS}_{k-1}(\emptyset)$ for some X_k', X_k'', X_{k-1}, and U_{k-1}. The latter condition can be rephrased that $X_k = X_k' \cup X_k''$, $X_k' \cap X_k'' = \emptyset$ for some independent set X_k' of $\mathbf{A} * \mathbf{R}_{k-1}$ and some independent set X_k'' of $\mathbf{RM}(\mathbf{B})$. Note that, by the induction assumption, the notation $\mathbf{A} * \mathbf{R}_{k-1}$ makes sense to mean the matroid induced from $\mathbf{R}_{k-1}$ by $\mathbf{A}$. Hence $\mathrm{RS}_k(\emptyset)$ forms the family of independent sets of the union of $\mathbf{A} * \mathbf{R}_{k-1}$ and $\mathbf{RM}(\mathbf{B})$. ■

Theorem 7.2.15 motivates us to investigate a sequence of matroids subject to a recurrence relation. The following general theorem proven in Murota [206] reveals some fundamental properties of the reachability matroids, where we choose $\mathbf{N} = \mathbf{RM}(\mathbf{B})$.

Theorem 7.2.16. *Let $\mathbf{A} = (S, S, \alpha)$ be a bimatroid with birank function α, and $\mathbf{N} = (S, \nu)$ be a matroid with rank function ν. Define a sequence of matroids $\mathbf{R}_k$ by the recurrence relation*

$$\mathbf{R}_k = (\mathbf{A} * \mathbf{R}_{k-1}) \vee \mathbf{N}, \quad k = 1, 2, \cdots, \tag{7.40}$$

starting with a trivial matroid $\mathbf{R}_0$ *on* S.

(1) $r(\mathbf{R}_k) - r(\mathbf{R}_{k-1}) \geq r(\mathbf{R}_{k+1}) - r(\mathbf{R}_k), \quad k = 1, 2, \cdots.$

(2) *There exists* $\kappa = \kappa(\mathbf{A}, \mathbf{N})$ (≥ 0) *such that*

$$r(\mathbf{R}_0) < r(\mathbf{R}_1) < \cdots < r(\mathbf{R}_{\kappa-1}) < r(\mathbf{R}_\kappa) = r(\mathbf{R}_k), \quad k = \kappa+1, \kappa+2, \cdots.$$

$$(3) \qquad \mathbf{R}_0 \overset{\leftarrow}{\neq} \mathbf{R}_1 \overset{\leftarrow}{\neq} \cdots \overset{\leftarrow}{\neq} \mathbf{R}_{\kappa-1} \overset{\leftarrow}{\neq} \mathbf{R}_\kappa = \mathbf{R}_k, \quad k = \kappa+1, \kappa+2, \cdots,$$

where $\mathbf{R}_k \overset{\leftarrow}{\neq} \mathbf{R}_{k+1}$ *means that* $\mathbf{R}_k$ *is a strong quotient of, and not isomorphic to,* $\mathbf{R}_{k+1}$.

Proof. Let $S^{(i)}$ $(i = 0, 1, 2, \cdots)$ be disjoint copies of S, and denote by $\mathbf{A}^{(i)} = (S^{(i)}, S^{(i-1)}, \alpha^{(i)})$ and $\mathbf{N}^{(i)} = (S^{(i)}, \nu^{(i)})$ $(i = 0, 1, 2, \cdots)$ the bimatroids and matroids that are isomorphic to $\mathbf{A}$ and $\mathbf{N}$, respectively. Denote by $\mathbf{M}^{(i)}$ the dual of the matroid associated with $\mathbf{A}^{(i)}$ by (2.86); $S^{(i)} \cup S^{(i-1)}$ is the ground set of $\mathbf{M}^{(i)}$, and $(X, Y) \in \Lambda(\mathbf{A}^{(i)})$ if and only if $X \cup (S^{(i-1)} \setminus Y)$ is a base of $\mathbf{M}^{(i)}$. Furthermore, define a matroid

$$\mathbf{G}_{j,k} = \left(\bigvee_{i=j+1}^{k} \mathbf{M}^{(i)} \right) \vee \left(\bigvee_{i=j}^{k} \mathbf{N}^{(i)} \right),$$

the ground set of which is $\bigcup_{i=j}^{k} S^{(i)}$. It is straightforward to verify that $\mathbf{R}_k \simeq (\mathbf{G}_{1,k})_{S^{(k)}}$ (=the contraction of $\mathbf{G}_{1,k}$ to $S^{(k)}$) and that

$$r(\mathbf{R}_k) = r(\mathbf{G}_{1,k}) - (k-1)|S|, \tag{7.41}$$

since $\bigcup_{i=1}^{k-1} S^{(i)}$, being a base of $\bigvee_{i=2}^{k} \mathbf{M}^{(i)}$, is independent in $\mathbf{G}_{1,k}$.

Proposition 2.3.40 applied to a triple $(\mathbf{M}^{(2)} \vee \mathbf{N}^{(1)}, \mathbf{G}_{2,k}, \mathbf{M}^{(k+1)} \vee \mathbf{N}^{(k+1)})$ yields

$$r(\mathbf{G}_{1,k+1}) + r(\mathbf{G}_{2,k}) \leq r(\mathbf{G}_{1,k}) + r(\mathbf{G}_{2,k+1}).$$

Noting $\mathbf{G}_{2,k} \simeq \mathbf{G}_{1,k-1}$ and $\mathbf{G}_{2,k+1} \simeq \mathbf{G}_{1,k}$ and using (7.41), we obtain the desired inequality in (1).

The strong map relation $\mathbf{R}_k \leftarrow \mathbf{R}_{k+1}$ follows from the expressions

$$\mathbf{R}_k \simeq (\mathbf{G}_{2,k+1})_{S^{(k+1)}}, \qquad \mathbf{R}_{k+1} \simeq ([\mathbf{M}^{(2)} \vee \mathbf{N}^{(1)}] \vee \mathbf{G}_{2,k+1})_{S^{(k+1)}},$$

and Theorem 2.3.41. Hence, $r(\mathbf{R}_k) \leq r(\mathbf{R}_{k+1})$, $k = 0, 1, 2, \cdots$. Let κ be the smallest k (> 0) such that the equality holds. Then (1) implies that the equality must hold for all $k \geq \kappa$, establishing (2). Finally recall Lemma 2.3.1 to obtain the assertion of (3). ■

The strong map relation for $\mathbf{R}_k$ shows that the reachable part grows with time to a matroid $\mathbf{R}_\kappa$, denoted also as $\mathbf{R}_\infty$. Compare this with the similar phenomenon for the conventional system (7.34) that the controllable subspaces $R_k = \mathrm{Im}[B \mid AB \mid A^2B \mid \cdots \mid A^{k-1}B]$ increase with k, where A and B are numerical matrices. In this connection recall from Example 2.3.8 that the strong map relation may be interpreted as a combinatorial abstraction of the nesting of linear subspaces.

The ultimate rank $r(\mathbf{R}_\kappa)$, i.e., $r(\mathbf{R}_\infty)$, of the sequence $\{\mathbf{R}_k\}_k$ defined by the recurrence relation (7.40) admits a "static" characterization, as follows. Recall that $\mathbf{A}[X,X]$ denotes the restriction of $\mathbf{A}$ to (X,X), and $\mathbf{N}[X]$ the restriction of $\mathbf{N}$ to X.

Theorem 7.2.17. *Let* $\mathbf{A} = (S,S,\alpha)$, $\mathbf{N} = (S,\nu)$, *and* $\{\mathbf{R}_k\}_k$ *be as in Theorem 7.2.16. If*

$$\alpha(X, S \setminus X) + \nu(X) \geq 1, \qquad \emptyset \neq \forall X \subseteq S,$$

it holds that

$$\mathrm{rank}\,(\mathbf{R}_\infty) = \max\{|X| \mid \mathrm{rank}\,\big(\mathbf{A}[X,X] \vee \mathbf{N}[X]\big) = |X|,\ X \subseteq S\}.$$

Proof. See Murota [206]. ■

By specializing the above theorem to the case of $\mathbf{N} = \mathbf{RM}(\mathbf{B})$, we obtain a formula for the ultimate rank of the reachability matroid of a CDS $(\mathbf{A},\mathbf{B})$, which is called the *controllable dimension.* This result is a generalization of Theorem 6.4.5.

Corollary 7.2.18. *If a CDS* $(\mathbf{A},\mathbf{B})$ *is reachable (i.e.,* $\{x\} \in \mathrm{RS}(\emptyset)$ *for all* $x \in S$*), the controllable dimension is given by*

$$\mathrm{rank}\,(\mathbf{R}_\infty) = \max\{|X| \mid \mathrm{rank}\,\big(\mathbf{A}[X,X] \vee \mathbf{B}[X,P]\big) = |X|,\ X \subseteq S\}. \quad (7.42)$$

Proof. With Theorem 7.2.17 it suffices to show that the reachability of $(\mathbf{A},\mathbf{B})$ is equivalent to

$$\alpha(X, S \setminus X) + \beta(X,P) \geq 1, \qquad \emptyset \neq \forall X \subseteq S,$$

where α and β are the birank functions of $\mathbf{A}$ and $\mathbf{B}$. ■

The formula (7.42) yields the following controllability criteria shown in Murota [206].

Theorem 7.2.19. *For a CDS* $(\mathbf{A},\mathbf{B})$, *the following three conditions are equivalent:*

(i) $(\mathbf{A},\mathbf{B})$ *is controllable;*

(ii) $(\mathbf{A},\mathbf{B})$ *is reachable, and* $\mathbf{RM}(\mathbf{A}) \vee \mathbf{RM}(\mathbf{B})$ *is the free matroid;*

(iii) $(\mathbf{A},\mathbf{B})$ *is reachable, and there exists a state feedback* $\mathbf{F}$ *such that* $\mathbf{A} \vee (\mathbf{B} * \mathbf{F})$ *is a nonsingular bimatroid.*

Proof. First note that reachability is necessary for controllability; so we assume reachability. By definition, controllability is equivalent to $\mathrm{rank}\,(\mathbf{R}_\infty) = |S|$. On the other hand, the expression of $\mathrm{rank}\,(\mathbf{R}_\infty)$ in Corollary 7.2.18 shows that $\mathrm{rank}\,(\mathbf{R}_\infty) = |S|$ if and only if $\mathrm{rank}\,(\mathbf{A} \vee \mathbf{B}) = |S|$. Hence follows the equivalence of (i) and (ii), since $\text{rank}\,(\mathbf{A} \vee \mathbf{B}) = \text{rank}\,(\mathbf{RM}(\mathbf{A}) \vee \mathbf{RM}(\mathbf{B}))$. The equivalence of (ii) and (iii) is easy to show. Assume (iii), i.e., that $(S, S) \in \Lambda(\mathbf{A} \vee (\mathbf{B} * \mathbf{F}))$ for some $\mathbf{F}$. This is equivalent to saying that there exist $X, X' \subseteq S$, and $U \subseteq P$ such that $(X', X) \in \Lambda(\mathbf{A})$, $(S \setminus X', U) \in \Lambda(\mathbf{B})$, and $(U, S \setminus X) \in \Lambda(\mathbf{F})$. Hence $S = X' \cup (S \setminus X')$ is independent in $\mathbf{RM}(\mathbf{A}) \vee \mathbf{RM}(\mathbf{B})$, showing (ii). Similarly we can show that (ii) implies (iii), by choosing $\mathbf{F}$ to be the "free" bimatroid, in which every pair is a linked pair. ■

Remark 7.2.20. Compare the second criterion (ii) in Theorem 7.2.19 with the structural controllability theorem (Theorem 6.4.2). Note that, in case of $\mathbf{A} = \mathbf{L}(A)$, $\mathbf{B} = \mathbf{L}(B)$ with generic matrices A and B, $\mathbf{RM}(\mathbf{A}) \vee \mathbf{RM}(\mathbf{B})$ is the free matroid if and only if $\text{term-rank}\,[A \mid B] = n$. □

The integer $\kappa = \kappa(\mathbf{A}, \mathbf{B})$ $(0 \leq \kappa \leq n = |S|)$ in Theorem 7.2.16 for $\mathbf{N} = \mathbf{RM}(\mathbf{B})$ is called the *controllability index* of the CDS $(\mathbf{A}, \mathbf{B})$. The inequalities in (1) and (2) show that $\Delta r_k = r(\mathbf{R}_k) - r(\mathbf{R}_{k-1})$, $k = 1, 2, \cdots$, form a nonnegative and nonincreasing sequence that vanishes for $k > \kappa$. This enables us to define a set of nonnegative indices $\{\kappa_i\}$ by

$$\kappa_i = |\{k \mid \Delta r_k \geq i\}|, \qquad i = 1, 2, \cdots,$$

just as in the conventional dynamical system theory. Note that $\kappa_1 = \kappa(\mathbf{A}, \mathbf{B})$. The indices $\{\kappa_i\}$ are called the *controllability indices* of $(\mathbf{A}, \mathbf{B})$.

For a controllable system we can find a nicely nested input sequence with reference to the controllability indices as follows.

Theorem 7.2.21. *Suppose $(\mathbf{A}, \mathbf{B})$ is controllable. There exists an input $(U_k \mid k = 0, 1, \cdots, K - 1)$ with $K \leq |S|$ admissible for $(\emptyset, S)$ such that*

$$U_0 \subseteq U_1 \subseteq \cdots \subseteq U_{K-1}$$

and that

$$\{\tilde{\kappa}_u \mid u \in P\} = \{\kappa_i \mid i = 1, \cdots, |P|\},$$

where $\tilde{\kappa}_u = |\{k \mid U_k \ni u, 0 \leq k \leq K - 1\}|$ indicates how many times $u \in P$ is used.

Proof. See Murota [206]. ■

Notes. Sections 7.2.2 and 7.2.3 are based on Murota [211], whereas §7.2.4 is on Murota [206].

7.3 Mixed Skew-symmetric Matrix

7.3.1 Introduction

A square matrix $A = (A_{ij})$ over a field is said to be *skew-symmetric* if $A^{\mathrm{T}} = -A$ and all the diagonal entries are equal to zero, where the second condition is implied by the first if the characteristic of the underlying field is distinct from two. The definition of skew-symmetry presupposes that the row set and the column set of A are indexed by a common finite set, say $V = \{1, \cdots, n\}$.

Skew-symmetric matrices enjoy rich combinatorial structures. Among others, the rank of a skew-symmetric matrix A is equal to the maximum size of a matching in the associated undirected graph $G = (V, E)$ with vertex set V and arc set $E = \{(i,j) \mid A_{ij} \neq 0\}$, provided that the nonzero entries of the matrix are independent parameters except for the obvious constraint $A_{ij} + A_{ji} = 0$ due to skew-symmetry. This fact was exploited by Tutte [322] in deriving a fundamental duality result for maximum matchings (cf. Theorem 7.3.9). Combinatorial properties of a skew-symmetric matrix with numerical data are abstracted as delta-matroids by Bouchet [15]. A delta-matroid is an abstract discrete structure defined in terms of an exchange axiom (to be explained in §7.3.3).

Skew-symmetric matrices are also important in applications. Electrical network theory, in particular, employs an ideal element called a gyrator, which is described by a skew-symmetric matrix of order two. The solvability of RCG networks (electrical networks involving gyrators as well as resistors and capacitors) has been analyzed successfully with the aid of the results on the matroid parity problem (Recski [276, 277], Ueno–Kajitani [323]).

In this section we introduce the skew-symmetric version of a mixed matrix as a further mathematical tool for systems analysis by means of matroid-theoretic combinatorial methods. A mixed skew-symmetric matrix is a matrix A expressed as $A = Q + T$, where Q is a "constant" skew-symmetric matrix and T is a "generic" skew-symmetric matrix in the sense that the nonzero entries of T are independent parameters except for the obvious constraint due to skew-symmetry.

A formal definition of mixed skew-symmetric matrix reads as follows. Let $\boldsymbol{F}$ be a field, and $\boldsymbol{K}$ be a subfield of $\boldsymbol{F}$. A skew-symmetric matrix A over $\boldsymbol{F}$ (i.e., $A_{ij} = -A_{ji} \in \boldsymbol{F}$, $A_{ii} = 0$) is called a *mixed skew-symmetric matrix* with respect to $(\boldsymbol{K}, \boldsymbol{F})$ if

$$A = Q + T, \tag{7.43}$$

where

(MS-Q) Q is a skew-symmetric matrix over $\boldsymbol{K}$ (i.e., $Q_{ij} \in \boldsymbol{K}$), and
(MS-T) T is a skew-symmetric matrix over $\boldsymbol{F}$ (i.e., $T_{ij} \in \boldsymbol{F}$) such that the set $\mathcal{T} = \{T_{ij} \mid T_{ij} \neq 0, i < j\}$ of its nonzero entries in the upper-triangular part is algebraically independent over $\boldsymbol{K}$.

Example 7.3.1. Here is a small example of a mixed skew-symmetric matrix $A = Q + T$:

$$\begin{bmatrix} 0 & -1 & 1 & t_1 \\ 1 & 0 & 2 & 0 \\ -1 & -2 & 0 & t_2 \\ -t_1 & 0 & -t_2 & 0 \end{bmatrix} = \begin{bmatrix} 0 & -1 & 1 & 0 \\ 1 & 0 & 2 & 0 \\ -1 & -2 & 0 & 0 \\ 0 & 0 & 0 & 0 \end{bmatrix} + \begin{bmatrix} 0 & 0 & 0 & t_1 \\ 0 & 0 & 0 & 0 \\ 0 & 0 & 0 & t_2 \\ -t_1 & 0 & -t_2 & 0 \end{bmatrix},$$

where $\mathcal{T} = \{t_1, t_2\}$ is assumed to be algebraically independent over $\mathbf{Q}$. This is a mixed skew-symmetric matrix with respect to $(\boldsymbol{K}, \boldsymbol{F})$ for $\boldsymbol{K} = \mathbf{Q}$ and $\boldsymbol{F} = \mathbf{Q}(t_1, t_2)$. □

The objective of this section is to present major mathematical results on mixed skew-symmetric matrices and to treat the solvability of electrical networks with gyrators using those results. The mathematical analysis of mixed skew-symmetric matrices leads to an extension of the matroid parity problem, called the delta-matroid parity problem, introduced by Geelen–Iwata–Murota [93]. In particular, the rank formula for a mixed skew-symmetric matrix takes the form of a novel min-max formula for a pair of linear delta-matroids, which is an extension of the duality result for the linear matroid parity problem due to Lovász [177, 179, 180].

The table below indicates how the combinatorial tools used in the analysis of mixed matrices are generalized for mixed skew-symmetric matrices.

	Mixed matrix	Mixed skew-symmetric matrix
T-part	bipartite matching	nonbipartite matching
Q-part	linear matroid	linear delta-matroid
Combination	matroid union/intersection	delta-matroid covering/parity

Remark 7.3.2. Given a matroid $\mathbf{M} = (V, \mathcal{B}, \rho)$ on ground set V with basis family $\mathcal{B}$ and rank function ρ, and also a partition Π of V into pairs, called *lines*, the *matroid parity problem* is to find a base containing the maximum number of lines (or equivalently to find a largest independent set consisting of lines). We denote by $\nu(\mathbf{M}, \Pi)$ the optimal value of the matroid parity problem (the maximum number of lines contained in a base). The matroid parity problem was introduced by Lawler (cf. Lawler [171], Lovász–Plummer [181]).

The matroid parity problem is polynomially unsolvable in general, as pointed out by Jensen–Korte [151] and Lovász [179]. (This statement is independent of the P≠NP conjecture.) It is solvable, however, if the matroid in question is represented by a matrix. This special case is called the *linear matroid parity problem.* Lovász [177, 179, 180] showed a polynomial-time algorithm for finding an optimal base for the linear matroid parity problem. This algorithm has been followed by more efficient algorithms: an augmenting path algorithm of Gabow–Stallmann [83] and an algorithm of Orlin–Vande Vate [258]. Those algorithms are based on a min-max theorem for $\nu(\mathbf{M}, \Pi)$, due to Lovász [177, 179, 180], which states that, for a linear matroid $\mathbf{M}$ representable over a field $\boldsymbol{F}$, we have

$$\nu(\mathbf{M},\Pi) = \min_{\mathbf{M}\to\mathbf{M}^\circ,\{V_i\}} \left\{ \rho(V) - \rho^\circ(V) + \sum_i \left\lfloor \frac{\rho^\circ(V_i)}{2} \right\rfloor \right\}, \tag{7.44}$$

where the minimum is taken over all matroids $\mathbf{M}^\circ = (V, \mathcal{B}^\circ, \rho^\circ)$ that are strong quotients of $\mathbf{M}$ (see (2.65)) and all partitions $\{V_i\}$ of V that are compatible with the partition Π (that is, each V_i is a union of lines), and the minimum is attained by a linear matroid $\mathbf{M}^\circ$ representable over $\boldsymbol{F}$. □

7.3.2 Skew-symmetric Matrix

A matrix $A = (A_{ij})$ over a field $\boldsymbol{F}$ is said to be *skew-symmetric* if $A_{ij} = -A_{ji}$ for all (i,j) and $A_{ii} = 0$ for all i, where the second condition is implied by the first if the characteristic of $\boldsymbol{F}$ is distinct from two. The definition of skew-symmetry presupposes that the row set and the column set of A are indexed by a common finite set, say $V = \{1,\cdots,n\}$. The *support graph* of A is an undirected graph $G = (V,E)$ with vertex set V and arc set $E = \{(i,j) \mid A_{ij} \neq 0\}$. For $I \subseteq V$, $A[I]$ designates $A[I,I]$, the principal submatrix of A indexed by I. We also denote by $I \triangle J$ the *symmetric difference* of I and J, namely, $I \triangle J = (I \cup J) \setminus (I \cap J)$.

For a skew-symmetric A of an even order the *Pfaffian* of A is defined by

$$\mathrm{pf}\, A = \sum_P a_P, \tag{7.45}$$

where the summation is taken over all partitions $P = \{\{i_1,j_1\},\cdots,\{i_\nu,j_\nu\}\}$ ($\nu = n/2$) of V into unordered pairs and

$$a_P = \mathrm{sgn}\begin{pmatrix} 1 & 2 & \cdots & 2\nu-1 & 2\nu \\ i_1 & j_1 & \cdots & i_\nu & j_\nu \end{pmatrix} \prod_{k=1}^{\nu} A_{i_k j_k}.$$

Note that a nonzero term a_P in the Pfaffian corresponds to a perfect matching in the support graph G, since $\prod_{k=1}^{\nu} A_{i_k j_k} \neq 0$ if and only if $(i_k, j_k) \in E$ for $k = 1,\cdots,\nu$. We define $\mathrm{pf}\, A = 0$ if n is odd.

The following fact is well known.

Proposition 7.3.3. *For a skew-symmetric A we have* $\det A = (\mathrm{pf}\, A)^2$.

Proof. For n odd, both $\det A$ and $\mathrm{pf}\, A$ vanish, since $\det A = \det((-A)^{\mathrm{T}}) = (-1)^n \det A$. The content of this identity lies in the case of even n; see Muir [195] for the proof. ■

Pfaffians enjoy an identity similar to the Grassmann–Plücker identity for determinants (Proposition 2.1.4). We use the notation $\mathrm{pf}_A(i_1,\cdots,i_K)$

for $\mathrm{pf}(A[\{i_1,\cdots,i_K\}])$, where $\mathrm{pf}_A(i_2,i_1,i_3,\cdots,i_K) = -\mathrm{pf}_A(i_1,i_2,i_3,\cdots,i_K)$, etc., and, in particular, $\mathrm{pf}_A(i_1,\cdots,i_K) = 0$ if the indices $i_1,\cdots,i_K$ are not distinct. We also use the notation $\widehat{i}$ for the omission of an index i, that is, $(i_1,i_2,\cdots,\widehat{i_k},\cdots,i_K) = (i_1,i_2,\cdots,i_{k-1},i_{k+1},\cdots,i_K)$.

Proposition 7.3.4 (Grassmann–Plücker identity for Pfaffians). *For a skew-symmetric matrix A, it holds that*

$$\sum_{l=1}^{L}(-1)^l\cdot \mathrm{pf}_A(j_l,i_1,i_2,\cdots,i_K)\cdot \mathrm{pf}_A(j_1,j_2,\cdots,\widehat{j_l},\cdots,j_L)$$
$$+\sum_{k=1}^{K}(-1)^k\cdot \mathrm{pf}_A(i_1,i_2,\cdots,\widehat{i_k},\cdots,i_K)\cdot \mathrm{pf}_A(i_k,j_1,j_2,\cdots,j_L) = 0. \quad (7.46)$$

In particular, for $i \in I\triangle J$, it holds that

$$\mathrm{pf}_A(I)\cdot \mathrm{pf}_A(J) = \sum_{j\in(I\triangle J)\setminus\{i\}} \sigma_j\cdot \mathrm{pf}_A(I\triangle\{i,j\})\cdot \mathrm{pf}_A(J\triangle\{i,j\}) \quad (7.47)$$

with appropriately chosen sign $\sigma_j = \pm 1$ depending on the ordering of the elements of I, J, etc.

Proof. By the definition of the Pfaffian we have

$$\mathrm{pf}_A(j_l,i_1,i_2,\cdots,i_K) = \sum_{k=1}^{K}(-1)^{k-1}\cdot \mathrm{pf}_A(j_l,i_k)\cdot \mathrm{pf}_A(i_1,i_2,\cdots,\widehat{i_k},\cdots,i_K),$$
$$\mathrm{pf}_A(i_k,j_1,j_2,\cdots,j_L) = \sum_{l=1}^{L}(-1)^{l-1}\cdot \mathrm{pf}_A(i_k,j_l)\cdot \mathrm{pf}_A(j_1,j_2,\cdots,\widehat{j_l},\cdots,j_L),$$

as well as $\mathrm{pf}_A(j_l,i_k)+\mathrm{pf}_A(i_k,j_l) = 0$. Substitution of these into the left-hand side of (7.46) establishes the identity. For (7.47), take $I = \{j_1,i_1,i_2,\cdots,i_K\}$, $J = \{j_2,\cdots,j_L\}$, $i = j_1$ in (7.46). ■

Remark 7.3.5. The Grassmann–Plücker identity for Pfaffians is known to physicists as Wick's theorem. The expression (7.47) using symmetric difference is found in Wenzel [334] and Dress–Wenzel [55]. The present proof of (7.46) is taken from Ohta [253]. □

Proposition 7.3.6. *The rank of a skew-symmetric matrix is equal to the maximum size of a nonsingular principal submatrix.*

Proof. Let (I, J) be such that $\mathrm{rank}\,A = \mathrm{rank}\,A[I,J] = |I| = |J|$. Then Proposition 2.1.9(2) with $(I_1,J_1) = (I,I)$, $(I_2,J_2) = (I\cup J, I\cup J)$ shows

$$2r \geq \mathrm{rank}\,A[I,I]+\mathrm{rank}\,A[I\cup J,I\cup J] \geq \mathrm{rank}\,A[I\cup J,I]+\mathrm{rank}\,A[I,I\cup J] \geq 2r,$$

where $r = \operatorname{rank} A$. This implies $|I| = \operatorname{rank} A[I, I] = \operatorname{rank} A$. ■

For a matrix A, a *pivotal transform* means the matrix A' resulting from the transformation

$$A = \begin{bmatrix} A_{11} & A_{12} \\ A_{21} & A_{22} \end{bmatrix} \quad \mapsto \quad A' = \begin{bmatrix} {A_{11}}^{-1} & {A_{11}}^{-1} A_{12} \\ -A_{21} {A_{11}}^{-1} & A_{22} - A_{21} {A_{11}}^{-1} A_{12} \end{bmatrix}$$

for a nonsingular submatrix A_{11}. The pivotal transformation preserves skew-symmetry. We denote by $A * I$ the pivotal transform of a skew-symmetric matrix A with respect to a nonsingular principal submatrix $A[I]$. The following is a fundamental identity due to Tucker [321].

Proposition 7.3.7. *For a skew-symmetric matrix A and $I \subseteq V$ such that $A[I]$ is nonsingular we have*

$$\det(A * I)[J] = \det A[I \triangle J] / \det A[I] \qquad (J \subseteq V).$$

Proof. First note the relation

$$\begin{bmatrix} {A_{11}}^{-1} & O \\ -A_{21} {A_{11}}^{-1} & I_2 \end{bmatrix} \begin{bmatrix} A_{11} & O & I_1 & A_{12} \\ A_{21} & I_2 & O & A_{22} \end{bmatrix} = \left[\begin{array}{cc|cc} I_1 & O & {A_{11}}^{-1} & {A_{11}}^{-1} A_{12} \\ O & I_2 & -A_{21} {A_{11}}^{-1} & A_{22} - A_{21} {A_{11}}^{-1} A_{12} \end{array}\right]$$

where I_1 and I_2 denote unit matrices and $A_{11} = A[I]$. The claim can be proven by considering the determinant of the relevant submatrix. ■

A skew-symmetric matrix $A = (A_{ij})$ is said to be *generic* if the set $\{A_{ij} \mid A_{ij} \neq 0, i < j\}$ of its nonzero entries in the upper-triangular part is algebraically independent. A generic skew-symmetric matrix is in fact a combinatorial object, since it does not carry numerical information. Its structure is fully represented by the support graph G, and conversely, for any graph G we may associate a generic skew-symmetric matrix (sometimes called the *Tutte matrix* of G) such that its support graph coincides with G. For a graph G we denote by $\nu(G)$ the maximum size of a matching in G and by $\operatorname{odd}(G)$ the number of odd components of G, where an *odd component* means a connected component having an odd number of vertices.

For the rank of a generic skew-symmetric matrix we have the following graph-theoretic characterizations.

Proposition 7.3.8. *The rank of a generic skew-symmetric matrix A is equal to the maximum size of a matching in the support graph of A.*

Proof. The algebraic independence implies that $\operatorname{pf} A$ is nonzero if and only if there exists at least one nonzero term a_P in the definition (7.45). Hence, A is nonsingular if and only if the support graph of A has a perfect matching. Application of this argument to principal submatrices establishes the claim by Proposition 7.3.6. ■

Theorem 7.3.9 (Tutte–Berge formula). *For a graph $G = (V, E)$ we have*

$$2\nu(G) = \min\{|V| + |U| - \mathrm{odd}(G \setminus U) \mid U \subseteq V\}, \tag{7.48}$$

where $G \setminus U$ means the graph obtained from G by deleting the vertices of U.

Proof. This will be derived from a more general theorem below. Standard combinatorial proofs can be found in Lovász–Plummer [181] and Cook–Cunningham–Pulleyblank–Schrijver [40]. ∎

The rank of a principal submatrix $A[I, I]$ of a generic skew-symmetric matrix A can be expressed in terms of matchings by Proposition 7.3.8 above, and hence Theorem 7.3.9 gives a formula:

$$\mathrm{rank}\, A[I, I] = \min\{2|I| - |I'| - \mathrm{odd}(G[I']) \mid I' \subseteq I\},$$

where $G[I']$ means the subgraph of G induced on I'. This formula can be generalized for a general submatrix $A[I, J]$ in the following form, which is ascribed to L. Lovász in Cunningham–Geelen [44].

Theorem 7.3.10. *For a generic skew-symmetric matrix A we have*

$$\mathrm{rank}\, A[I, J] = \min\{|I \setminus I'| + |J \setminus J'| + |I' \cap J'| - \mathrm{odd}(G[I' \cap J']) \mid (I', J') \in D(I, J)\}, \tag{7.49}$$

where $D(I, J) = \{(I', J') \mid A[I' \setminus J', J'] = O, A[I', J' \setminus I'] = O, I' \subseteq I, J' \subseteq J\}$.

Proof.[1] The inequality

$$\mathrm{rank}\, A[I, J] \le |I \setminus I'| + |J \setminus J'| + |I' \cap J'| - \mathrm{odd}(G[I' \cap J']) \tag{7.50}$$

is valid for any $(I', J') \in D(I, J)$, since

$$\mathrm{rank}\, A[I, J] \le |I \setminus I'| + |J \setminus J'| + \mathrm{rank}\, A[I', J'],$$
$$\mathrm{rank}\, A[I', J'] = \mathrm{rank}\, A[I' \cap J'] \le |I' \cap J'| - \mathrm{odd}(G[I' \cap J']).$$

By Theorem 2.3.47 there exist $I^* \subseteq I$ and $J^* \subseteq J$ such that
(i) $|I^*| + |J^*| - \mathrm{rank}\, A[I^*, J^*] = |I| + |J| - \mathrm{rank}\, A[I, J]$,
(ii) $\mathrm{rank}\, A[I^* \setminus \{i\}, J^* \setminus \{j\}] = \mathrm{rank}\, A[I^*, J^*]$, $\forall i \in I^*$, $\forall j \in J^*$.
We claim that (ii) implies $(I^*, J^*) \in D(I, J)$. Suppose, to the contrary, that $A_{ij} \neq 0$ for some (i, j) with $i \in I^* \setminus J^*$, $j \in J^*$ or $i \in I^*$, $j \in J^* \setminus I^*$. Since $\mathrm{rank}\, A[I^* \setminus \{i\}, J^* \setminus \{j\}] = \mathrm{rank}\, A[I^*, J^*]$ $(=: r)$, there exist $I'' \subseteq I^* \setminus \{i\}$ and $J'' \subseteq J^* \setminus \{j\}$ such that $\mathrm{rank}\, A[I'', J''] = |I''| = |J''| = r$. Consider the Laplace expansion of $\det A[I'' \cup \{i\}, J'' \cup \{j\}]$. It contains a nonvanishing term $A_{ij} \cdot \det A[I'', J'']$, which is not cancelled out by virtue of

[1] This proof as well as the derivation of Theorem 7.3.9 from this theorem is taken from Geelen [90]. Compare this proof with Remark 4.2.15.

the unique occurrence of the variable A_{ij}. This implies a contradiction that $r = \operatorname{rank} A[I^*, J^*] \geq \operatorname{rank} A[I'' \cup \{i\}, J'' \cup \{j\}] = |I''| + 1 = r + 1$.

Since $(I^*, J^*) \in D(I, J)$, we have $\operatorname{rank} A[I^*, J^*] = \operatorname{rank} A[I^* \cap J^*]$, and also $\operatorname{rank} A[I^* \cap J^*] = \operatorname{rank} A[(I^* \cap J^*) \setminus \{i\}]$ for all $i \in I^* \cap J^*$ by (ii). The latter implies, by Gallai's lemma below, that $\operatorname{rank} A[I^* \cap J^*] = |I^* \cap J^*| - \operatorname{odd}(G[I^* \cap J^*])$. Substitution of these into (i) shows that (7.50) holds with equality for $(I', J') = (I^*, J^*)$. ■

The following fundamental fact, used in the proof above, deserves to be stated as a theorem.

Theorem 7.3.11 (Gallai's lemma). *Let $G = (V, E)$ be a connected graph. If $\nu(G \setminus \{v\}) = \nu(G)$ for all $v \in V$, then $2\nu(G \setminus \{v\}) = |V| - 1$ for all $v \in V$.*

Proof.[2] Let $\mathbf{M}$ be the matching matroid (see Example 2.3.7) defined by G. The assumption means that $\mathbf{M}$ has no coloops. If (u, v) is an arc, then there is no maximum matching missing both u and v, that is, u and v are in series. Since series pairs are transitive (cf. §2.3.2) and G is connected, every pair of vertices is in series. This implies that a maximum matching misses just one vertex, i.e., $2\nu(G) = |V| - 1$. Hence, $2\nu(G \setminus \{v\}) = 2\nu(G) = |V| - 1$ for all $v \in V$. ■

Theorem 7.3.9 is now proven from Theorem 7.3.10. For a minimizer $(I^*, J^*) \in D(V, V)$ in (7.49) we have

$$\begin{aligned}\operatorname{rank} A &= |V \setminus I^*| + |V \setminus J^*| + |I^* \cap J^*| - \operatorname{odd}(G[I^* \cap J^*]) \\ &\geq |V| + |V \setminus (I^* \cup J^*)| - \operatorname{odd}(G[I^* \cup J^*]).\end{aligned}$$

On the other hand, for any $U \subseteq V$ it holds that

$$\begin{aligned}\operatorname{rank} A &\leq \operatorname{rank} A[V \setminus U] + \operatorname{rank} A[U, V \setminus U] + \operatorname{rank} A[V, U] \\ &\leq (|V \setminus U| - \operatorname{odd}(G \setminus U)) + |U| + |U|.\end{aligned}$$

Hence follows (7.48), since $\operatorname{rank} A = 2\nu(G)$.

Remark 7.3.12. It is shown by Cunningham–Geelen [44] that the rank of a general submatrix $A[I, J]$ can be characterized in terms of "path-matching" (a generalization of matching) and that it can be computed in polynomial time, though the algorithm is not really combinatorial. □

Remark 7.3.13. For a numerically specified skew-symmetric matrix, the maximum size of a matching in the support graph is only an upper bound on the rank (due to accidental numerical cancellation). A systematic procedure has been given by Geelen [91] that assigns integers in the range of $\{1, \cdots, n\}$ (n: the size of the matrix) to the nonzero entries of a generic skew-symmetric matrix so that the rank of the resulting (numerical) skew-symmetric matrix attains this upper bound. □

[2] This proof, communicated by J. Geelen, reveals the matroid-theoretic nature of the proof given in Lovász–Plummer [181].

Remark 7.3.14. A canonical decomposition representing the structure of maximum matchings of nonbipartite graphs is known as the Gallai–Edmonds decomposition (see Lovász–Plummer [181]). As a refinement of this decomposition, a canonical block-triangularization of skew-symmetric matrices is given by Iwata [140]. This is a generalization of the Dulmage–Mendelsohn decomposition expounded in §2.2.3. □

7.3.3 Delta-matroid

The concept of a delta-matroid was introduced by Bouchet [15] as a generalization of a matroid. Essentially equivalent combinatorial structures were proposed independently by Chandrasekaran–Kabadi [30] and by Dress–Havel [50]; see also Bouchet–Cunningham [19] and Bouchet–Dress–Havel [20].

A *delta-matroid* is a pair $(V, \mathcal{F})$ of a finite set V and a nonempty family $\mathcal{F}$ of its subsets, called *feasible sets*, that satisfy the *symmetric exchange axiom*:

(DM) For $F, F' \in \mathcal{F}$ and $u \in F \triangle F'$, there exists $v \in F \triangle F'$ such that $F \triangle \{u, v\} \in \mathcal{F}$.

A delta-matroid is said to be an *even delta-matroid* if $|F \triangle F'|$ is even for all $F, F' \in \mathcal{F}$. As is easily seen, $(V, \mathcal{F})$ is an even delta-matroid if and only if it satisfies

($\mathrm{DM_{even}}$) For $F, F' \in \mathcal{F}$ and $u \in F \triangle F'$, there exists $v \in (F \triangle F') \setminus \{u\}$ such that $F \triangle \{u, v\} \in \mathcal{F}$.

It is also known (cf. Wenzel [335], Duchamp [58]) that an even delta-matroid is characterized by a stronger exchange axiom (*simultaneous exchange axiom*):

($\mathrm{DM_{\pm}}$) For $F, F' \in \mathcal{F}$ and $u \in F \triangle F'$, there exists $v \in (F \triangle F') \setminus \{u\}$ such that $F \triangle \{u, v\} \in \mathcal{F}$ and $F' \triangle \{u, v\} \in \mathcal{F}$.

A number of operations can be defined for a delta-matroid $\mathbf{M} = (V, \mathcal{F})$ with respect to a subset $X \subseteq V$. The *twisting* of $\mathbf{M}$ by X is a delta-matroid $\mathbf{M} \triangle X = (V, \mathcal{F} \triangle X)$, where $\mathcal{F} \triangle X = \{F \triangle X \mid F \in \mathcal{F}\}$. Two delta-matroids are said to be *equivalent* if they are transformed to each other by twisting. The delta-matroid $\mathbf{M}^* = \mathbf{M} \triangle V$ is called the *dual* of $\mathbf{M}$. The *deletion* of X from $\mathbf{M}$ means a delta-matroid $\mathbf{M} \setminus X = (V \setminus X, \mathcal{F} \setminus X)$ defined by $\mathcal{F} \setminus X = \{F \mid F \in \mathcal{F},\ F \subseteq V \setminus X\}$, where $\mathcal{F} \setminus X$ is assumed to be nonempty. The *contraction* of $\mathbf{M}$ by X, denoted $\mathbf{M}/X$, is defined as $(\mathbf{M} \triangle X) \setminus X$. Note that these operations preserve evenness.

A skew-symmetric matrix defines an even delta-matroid (Bouchet [16]). Let A be a skew-symmetric matrix over a field and V be its row/column set. The family of the nonsingular principal submatrices

$$\mathcal{F}(A) = \{X \subseteq V \mid \operatorname{rank} A[X] = |X|\}$$

satisfies the simultaneous exchange axiom ($\mathrm{DM_{\pm}}$) by (7.47) in Proposition 7.3.4, and hence $\mathbf{M}(A) = (V, \mathcal{F}(A))$ forms an even delta-matroid, in which the empty set is feasible. A delta-matroid $\mathbf{M}$ that can be expressed as $\mathbf{M} = \mathbf{M}(A)\triangle X$ for some skew-symmetric matrix A over a field $\boldsymbol{F}$ and a subset $X \subseteq V$ (not necessarily feasible in $\mathbf{M}(A)$) is called a *linear delta-matroid representable* over $\boldsymbol{F}$. If the matrix A and the subset X are given explicitly, $\mathbf{M}$ is said to be *represented* over $\boldsymbol{F}$. In case X is feasible in $\mathbf{M}(A)$, we have $\mathbf{M}(A)\triangle X = \mathbf{M}(A * X)$ by Proposition 7.3.7.

A matroid $\mathbf{M} = (V, \mathcal{B})$ given in terms of the basis family $\mathcal{B}$ is an even delta-matroid, since (BM$_\pm$) in §2.3.4 implies (DM$_\pm$) (or alternatively, since (BM$_+$) and (BM$_-$) together imply (DM$_{\rm even}$)). Moreover, a linear matroid is a linear delta-matroid, as follows. Let $\mathbf{M} = (V, \mathcal{B})$ be represented by a matrix with column set V in the sense that $\mathcal{B}$ is the family of column bases of the matrix (see Example 2.3.8). For any base B of $\mathbf{M}$ we may assume that the matrix is in the following form:

$$\begin{array}{cc} & \begin{array}{cc} B & V\setminus B \end{array} \\ B & \begin{pmatrix} I & \quad D \end{pmatrix}, \end{array}$$

where I is an identity matrix. Define a skew-symmetric matrix A by

$$A = \begin{array}{c} \\ B \\ V\setminus B \end{array} \!\!\begin{array}{c} \begin{array}{cc} B & \quad V\setminus B \end{array} \\ \begin{pmatrix} O & D \\ -D^{\mathrm{T}} & O \end{pmatrix} \end{array}. \tag{7.51}$$

Then, for $X \subseteq V$, $A[X\triangle B]$ is nonsingular if and only if $D[B \setminus X, X \setminus B]$ is nonsingular, which in turn is equivalent to $X \in \mathcal{B}$. Hence we have $\mathbf{M} = \mathbf{M}(A)\triangle B$, which shows that $\mathbf{M}$ is indeed a linear delta-matroid.

For a pair of delta-matroids $\mathbf{M}_1 = (V_1, \mathcal{F}_1)$ and $\mathbf{M}_2 = (V_2, \mathcal{F}_2)$ with $V_1 \cap V_2 = \emptyset$, their *direct sum* $\mathbf{M}_1 \oplus \mathbf{M}_2 = (V_1 \cup V_2, \mathcal{F}_1 \oplus \mathcal{F}_2)$ is a delta-matroid with

$$\mathcal{F}_1 \oplus \mathcal{F}_2 = \{F_1 \cup F_2 \mid F_1 \in \mathcal{F}_1, F_2 \in \mathcal{F}_2\}.$$

If $\mathbf{M}_1 = \mathbf{M}(A_1)\triangle X_1$ and $\mathbf{M}_2 = \mathbf{M}(A_2)\triangle X_2$, we have $\mathbf{M}_1 \oplus \mathbf{M}_2 = \mathbf{M}(A)\triangle X$ for $X = X_1 \cup X_2$ and

$$A = \begin{array}{c} \\ V_1 \\ V_2 \end{array} \!\!\begin{array}{c} \begin{array}{cc} V_1 & V_2 \end{array} \\ \begin{pmatrix} A_1 & O \\ O & A_2 \end{pmatrix} \end{array}.$$

For a pair of delta-matroids $\mathbf{M}_1 = (V, \mathcal{F}_1)$ and $\mathbf{M}_2 = (V, \mathcal{F}_2)$, we define their *union* by $\mathbf{M}_1 \vee \mathbf{M}_2 = (V, \mathcal{F}_1 \vee \mathcal{F}_2)$ with

$$\mathcal{F}_1 \vee \mathcal{F}_2 = \{F_1 \cup F_2 \mid F_1 \cap F_2 = \emptyset, F_1 \in \mathcal{F}_1, F_2 \in \mathcal{F}_2\}. \tag{7.52}$$

It is known (Bouchet [17]) that $\mathbf{M}_1 \vee \mathbf{M}_2$ is a delta-matroid.

The *delta-covering problem*, posed by Bouchet [18], is to find $F_1 \in \mathcal{F}_1$ and $F_2 \in \mathcal{F}_2$ maximizing $|F_1 \triangle F_2|$ for a given pair of delta-matroids $\mathbf{M}_1 = (V, \mathcal{F}_1)$ and $\mathbf{M}_2 = (V, \mathcal{F}_2)$. The delta-covering problem contains the following decision problems as special cases:

[Partition problem]
Given a pair of delta-matroids $\mathbf{M}_1 = (V, \mathcal{F}_1)$ and $\mathbf{M}_2 = (V, \mathcal{F}_2)$, does there exist a partition (F_1, F_2) of V such that $F_1 \in \mathcal{F}_1$ and $F_2 \in \mathcal{F}_2$?

[Intersection problem]
Given a pair of delta-matroids $\mathbf{M}_1$ and $\mathbf{M}_2$, does there exist a common feasible set?

Note that the intersection problem for $\mathbf{M}_1$ and $\mathbf{M}_2$ is the partition problem for $\mathbf{M}_1$ and ${\mathbf{M}_2}^*$.

The following is an observation, to be used in §7.3.4, that the optimal value of the delta-covering problem is equal to the maximum size of a feasible set in the sum $\mathbf{M}_1 \vee \mathbf{M}_2$ if the empty set is feasible in one of the given delta-matroids.

Lemma 7.3.15. *For a pair of delta-matroids $\mathbf{M}_1 = (V, \mathcal{F}_1)$ and $\mathbf{M}_2 = (V, \mathcal{F}_2)$ such that $\emptyset \in \mathcal{F}_1$, it holds that*

$$\begin{aligned} &\max\{|F_1 \triangle F_2| \mid F_1 \in \mathcal{F}_1, F_2 \in \mathcal{F}_2\} \\ &= \max\{|F_1 \cup F_2| \mid F_1 \cap F_2 = \emptyset, F_1 \in \mathcal{F}_1, F_2 \in \mathcal{F}_2\}. \end{aligned}$$

Proof. Take $F_1 \in \mathcal{F}_1$ and $F_2 \in \mathcal{F}_2$ with maximum $|F_1 \triangle F_2|$. For $u \in F_1 \cap F_2$, if any, there exists $v \in F_1 = F_1 \triangle \emptyset$ such that $F_1' = F_1 \setminus \{u, v\} \in \mathcal{F}_1$. The maximality of $|F_1 \triangle F_2|$ implies $v \in F_1 \setminus F_2$. Hence $|F_1' \triangle F_2| = |F_1 \triangle F_2|$ and $|F_1' \cap F_2| = |F_1 \cap F_2| - 1$. Repeated transformation from (F_1, F_2) to (F_1', F_2) leads to a disjoint pair (F_1, F_2). ∎

A min-max relation is known for the delta-covering problem in the case of linear delta-matroids. The min-max relation refers to a distance between two delta-matroids and the number of odd components with respect to a pair of even delta-matroids.

The *distance* between two delta-matroids $\mathbf{M} = (V, \mathcal{F})$ and $\mathbf{M}^\circ = (V, \mathcal{F}^\circ)$, denoted $\mathrm{dist}(\mathbf{M}, \mathbf{M}^\circ)$, is defined to be the minimum cardinality of Z such that $\mathbf{M} = \mathbf{M}^+ \setminus Z$ and $\mathbf{M}^\circ = \mathbf{M}^+ / Z$ for some delta-matroid $\mathbf{M}^+ = (V \cup Z, \mathcal{F}^+)$, where $\mathrm{dist}(\mathbf{M}, \mathbf{M}^\circ) = +\infty$ if no such $\mathbf{M}^+$ exists. Note that $\mathrm{dist}(\mathbf{M}, \mathbf{M}^\circ) = \mathrm{dist}(\mathbf{M}^\circ, \mathbf{M})$ since $\mathbf{M}^+ \setminus Z = (\mathbf{M}^+ \triangle Z)/Z$ and $\mathbf{M}^+/Z = (\mathbf{M}^+ \triangle Z) \setminus Z$.

For a pair of even delta-matroids $\mathbf{M}_1 = (V, \mathcal{F}_1)$ and $\mathbf{M}_2 = (V, \mathcal{F}_2)$, let $(V_1, \cdots, V_k)$ be the finest partition of V that simultaneously gives direct-sum decompositions of both $\mathbf{M}_1$ and $\mathbf{M}_2$. We say that V_i is an *odd component* with respect to $(\mathbf{M}_1, \mathbf{M}_2)$ if $|F_1 \cap V_i| + |F_2 \cap V_i| - |V_i|$ is odd for $F_1 \in \mathcal{F}_1$ and $F_2 \in \mathcal{F}_2$. Denoting by $\mathrm{odd}(\mathbf{M}_1, \mathbf{M}_2)$ the number of odd components with respect to $(\mathbf{M}_1, \mathbf{M}_2)$, we have $|F_1 \triangle F_2| \le |V| - \mathrm{odd}(\mathbf{M}_1, \mathbf{M}_2)$ for any $F_1 \in \mathcal{F}_1$ and $F_2 \in \mathcal{F}_2$.

The min-max relation, due to Geelen–Iwata–Murota [93], reads as follows.

Theorem 7.3.16. *For a pair of linear delta-matroids $\mathbf{M}_1 = (V, \mathcal{F}_1)$ and $\mathbf{M}_2 = (V, \mathcal{F}_2)$ representable over fields $\boldsymbol{F}_1$ and $\boldsymbol{F}_2$, respectively, we have*

$$\begin{aligned}&\max\{|F_1 \triangle F_2| \mid F_1 \in \mathcal{F}_1, F_2 \in \mathcal{F}_2\}\\&= \min\{\mathrm{dist}(\mathbf{M}_1, \mathbf{M}_1^\circ) + \mathrm{dist}(\mathbf{M}_2, \mathbf{M}_2^\circ) - \mathrm{odd}(\mathbf{M}_1^\circ, \mathbf{M}_2^\circ)\\&\qquad \mid \mathbf{M}_1^\circ, \mathbf{M}_2^\circ\text{: } \textit{even delta-matroids}\} + |V|,\end{aligned}$$

and the minimum is attained by $\mathbf{M}_1^\circ$ *and* $\mathbf{M}_2^\circ$ *representable over* $\boldsymbol{F}_1$ *and* $\boldsymbol{F}_2$*, respectively.* □

The *delta-matroid parity problem* (or *delta-parity problem*) has been introduced by Geelen–Iwata–Murota [93] as a natural generalization of the matroid parity problem (see Remark 7.3.2 for the matroid parity problem). Let $\mathbf{M} = (V, \mathcal{F})$ be a delta-matroid on V with $|V|$ even, and Π be a partition of V into pairs, called *lines*. For $F \subseteq V$, we denote by $\delta_\Pi(F)$ the number of lines exactly one element of which belongs to F. In other words,

$$\delta_\Pi(F) = |\{v \in V \mid v \in F, \bar{v} \in V \setminus F\}|, \tag{7.53}$$

where, for $v \in V$, $\bar{v}$ denotes the element such that $\{v, \bar{v}\}$ is a line. The delta-parity problem is to find a feasible set $F \in \mathcal{F}$ that minimizes $\delta_\Pi(F)$. We denote by $\delta(\mathbf{M}, \Pi)$ the optimal value of this problem, that is,

$$\delta(\mathbf{M}, \Pi) = \min\{\delta_\Pi(F) \mid F \in \mathcal{F}\}. \tag{7.54}$$

An obvious lower bound exists on $\delta(\mathbf{M}, \Pi)$ in the case of an even delta-matroid $\mathbf{M}$. Let $\mathbf{M} = \mathbf{M}_1 \oplus \cdots \oplus \mathbf{M}_k$ be the finest direct sum decomposition of $\mathbf{M}$ which is compatible with the partition Π (that is, the ground set of each $\mathbf{M}_i$ is a union of lines). Then each component $\mathbf{M}_i$ is also an even delta-matroid. A component $\mathbf{M}_i$ is called an *odd component* if every feasible set of $\mathbf{M}_i$ is of odd cardinality. Denoting the number of such odd components by $\mathrm{odd}(\mathbf{M}, \Pi)$, we have $\delta(\mathbf{M}, \Pi) \geq \mathrm{odd}(\mathbf{M}, \Pi)$.

In the linear case, we can tighten this lower bound by considering another delta-matroid $\mathbf{M}^\circ$ as follows (Geelen–Iwata–Murota [93]).

Theorem 7.3.17. *For a linear delta-matroid* $\mathbf{M}$ *representable over a field* $\boldsymbol{F}$*, we have*

$$\delta(\mathbf{M}, \Pi) = \max\{\mathrm{odd}(\mathbf{M}^\circ, \Pi) - \mathrm{dist}(\mathbf{M}, \mathbf{M}^\circ) \mid \mathbf{M}^\circ\text{: } \textit{even delta-matroid}\},$$

and the maximum is attained by $\mathbf{M}^\circ$ *representable over* $\boldsymbol{F}$. □

Remark 7.3.18. The delta-parity problem and the delta-covering problem are equivalent. Given an instance of the delta-parity problem, $\mathbf{M} = (V, \mathcal{F})$ and Π, let $\mathcal{L}$ denote the family of subsets of V that can be represented as a union of lines. Then $\mathbf{M}_\Pi = (V, \mathcal{L})$ forms an even delta-matroid, and $\delta_\Pi(F) = |V| - \max\{|F \triangle L| \mid L \in \mathcal{L}\}$ holds for $F \subseteq V$. Hence the delta-parity problem is a delta-covering problem for $(\mathbf{M}, \mathbf{M}_\Pi)$. Conversely, given a pair of delta-matroids $(V, \mathcal{F}_1)$ and $(V, \mathcal{F}_2)$ for the delta-covering problem, denote their copies by $\mathbf{M}_1 = (V_1, \mathcal{F}_1')$ and $\mathbf{M}_2 = (V_2, \mathcal{F}_2')$, respectively. Let

$\mathbf{M} = (V_1 \cup V_2, \mathcal{F})$ be the direct sum of $\mathbf{M}_1$ and $\mathbf{M}_2^*$, the dual of $\mathbf{M}_2$, and Π be the partition of the ground set $V_1 \cup V_2$ into the pairs of the corresponding copies. For a pair of feasible sets $F_1 \in \mathcal{F}_1$ and $F_2 \in \mathcal{F}_2$, it is easy to see that $\delta_\Pi(F) = |V| - |F_1 \triangle F_2|$ holds for $F = F_1' \cup (V_2 \setminus F_2') \in \mathcal{F}$, where F_1' and F_2' are the copies of F_1 and F_2. Therefore the delta-covering problem on $(V, \mathcal{F}_1)$ and $(V, \mathcal{F}_2)$ is reduced to the delta-parity problem for $\mathbf{M}$ with the partition Π. □

An augmenting path algorithm is given by Geelen–Iwata–Murota [93] for solving the delta-parity problem on linearly represented delta-matroids. The algorithm consists of $\mathrm{O}(n)$ augmentations, each augmentation involving $\mathrm{O}(n^3)$ elementary pivoting operations. Hence the time complexity of the algorithm is $\mathrm{O}(n^4)$ in total, where the bound can be reduced slightly with the use of the so-called fast matrix multiplications. This algorithm can be adapted to solve the delta-covering problem. Note in this connection that the delta-parity problem, as well as the delta-covering problem, for a pair of general delta-matroids is polynomially unsolvable, since it contains the matroid-parity problem as a special case (see Remark 7.3.2 and Remark 7.3.19).

Remark 7.3.19. The delta-parity problem is a natural generalization of the matroid parity problem, which has been explained in Remark 7.3.2. For a matroid $\mathbf{M} = (V, \rho)$ with rank function ρ and a partition Π of V into lines, let $\nu(\mathbf{M}, \Pi)$ denote the optimal value of the matroid parity problem, and $\delta(\mathbf{M}, \Pi)$ be the optimal value of the delta-parity problem when $\mathbf{M}$ is regarded as a delta-matroid. Then it is obvious that $2\nu(\mathbf{M}, \Pi) = \mathrm{rank}\,\mathbf{M} - \delta(\mathbf{M}, \Pi)$. This shows that the matroid parity problem is a special case of the delta-matroid parity problem. Moreover, the representation indicated in (7.51) shows that the linear matroid parity problem is a linear delta-matroid parity problem.

The augmenting path algorithm of Geelen–Iwata–Murota [93] for the linear delta-parity problem is based on the idea in the algorithm of Gabow–Stallmann [83] for the linear matroid parity problem.

Also the min-max theorem (Theorem 7.3.17) for the linear delta-matroid parity problem is closely related to the Lovász min-max theorem (7.44) for the linear matroid parity problem. To see this, first rewrite (7.44) to

$$\begin{aligned}\delta(\mathbf{M}, \Pi) &= \mathrm{rank}\,\mathbf{M} - 2\nu(\mathbf{M}, \Pi)\\ &= \max_{\mathbf{M}\to\mathbf{M}^\circ, \{V_i\}} \left[\left(\rho^\circ(V) - 2\sum_i \left\lfloor \frac{\rho^\circ(V_i)}{2} \right\rfloor \right) - (\rho(V) - \rho^\circ(V)) \right], \end{aligned} \tag{7.55}$$

where the maximum is taken over all matroids $\mathbf{M}^\circ = (V, \rho^\circ)$ that are strong quotients of $\mathbf{M}$ and all partitions $\{V_i\}$ of V that are compatible with the partition Π. For the second term in the maximization (7.55) we can show

$$\rho(V) - \rho^\circ(V) = \mathrm{dist}(\mathbf{M}, \mathbf{M}^\circ),$$

where the proof for "$\geq$" relies on the factorization theorem for strong maps (cf. Kung [168, §8.2.B], Welsh [333, §17.4]) and that for "$\leq$" is straightforward using (DM). The first term in the maximization (7.55) corresponds to $\mathrm{odd}(\mathbf{M}^\circ, \Pi)$ in the sense that, if $\{V_i\}$ runs over direct sum decompositions of $\mathbf{M}^\circ$, we have

$$\max_{\{V_i\}} \left(\rho^\circ(V) - 2\sum_i \left\lfloor \frac{\rho^\circ(V_i)}{2} \right\rfloor \right) = \mathrm{odd}(\mathbf{M}^\circ, \Pi).$$

This follows easily from $\rho^\circ(V) = \sum_i \rho^\circ(V_i)$ and the fact that $\rho^\circ(V_i) - 2\left\lfloor \frac{\rho^\circ(V_i)}{2} \right\rfloor$ is equal to 1 or 0 according to whether V_i is an odd component or not.

It should be emphasized, however, that the two min-max formulas, the expression (7.55) and Theorem 7.3.17 (specialized to the matroid parity problem), are not identical. To be specific, we cannot assume the partition $\{V_i\}$ in (7.55) to be a direct sum decomposition of $\mathbf{M}^\circ$, nor can we assume the $\mathbf{M}^\circ$ in Theorem 7.3.17 to be a strong quotient of $\mathbf{M}$. This subtle point is demonstrated by the matroid parity problem defined by a linear matroid $\mathbf{M}$ associated with the matrix

$$\begin{array}{cc|cc|cc|cc}
v_1 & v_2 & v_3 & v_4 & v_5 & v_6 & v_7 & v_8 \\ \hline
1 & 1 & 0 & 0 & 0 & 0 & 0 & 1 \\
0 & 0 & 1 & 1 & 0 & 0 & 0 & 1 \\
0 & 0 & 0 & 0 & 1 & 1 & 0 & 1 \\
0 & 0 & 0 & 0 & 0 & 0 & 1 & 1
\end{array}$$

over $\boldsymbol{F} = \mathrm{GF}(2)$, and a partition $\Pi = \{\{v_1, v_2\}, \{v_3, v_4\}, \{v_5, v_6\}, \{v_7, v_8\}\}$ of $V = \{v_1, \cdots, v_8\}$. We have $\mathrm{rank}\, \mathbf{M} = 4$, $\nu(\mathbf{M}, \Pi) = 1$, and $\delta(\mathbf{M}, \Pi) = 2$. In (7.55) we can take $\mathbf{M}$ for $\mathbf{M}^\circ$ and Π for $\{V_i\}$; then $\rho^\circ(\{v_1, v_2\}) = \rho^\circ(\{v_3, v_4\}) = \rho^\circ(\{v_5, v_6\}) = 1$ and $\rho^\circ(\{v_7, v_8\}) = 2$. Note that Π does not give a direct sum decomposition of $\mathbf{M}^\circ = \mathbf{M}$. For Theorem 7.3.17 let $\mathbf{M}^+$ be the linear delta-matroid defined by a skew-symmetric matrix

$$A^+ = \begin{array}{c|cccc|cccc|c|}
 & v_1 & v_3 & v_5 & v_7 & v_2 & v_4 & v_6 & v_8 & z \\ \hline
v_1 & & & & & 1 & 0 & 0 & 1 & 0 \\
v_3 & & & & & 0 & 1 & 0 & 1 & 0 \\
v_5 & & & & & 0 & 0 & 1 & 1 & 0 \\
v_7 & & & & & 0 & 0 & 0 & 1 & 0 \\ \hline
v_2 & 1 & 0 & 0 & 0 & & & & & 0 \\
v_4 & 0 & 1 & 0 & 0 & & & & & 0 \\
v_6 & 0 & 0 & 1 & 0 & & & & & 0 \\
v_8 & 1 & 1 & 1 & 1 & & & & & 1 \\ \hline
z & 0 & 0 & 0 & 0 & 0 & 0 & 0 & 1 & \\ \hline
\end{array}$$

over GF(2). Fixing a base $B = \{v_1, v_3, v_5, v_7\}$ of the matroid $\mathbf{M}$, we can identify the matroid $\mathbf{M}$ with the delta-matroid $(\mathbf{M}^+ \setminus \{z\}) \Delta B$. For $\mathbf{M}^\circ =$

$(\mathbf{M}^+/\{z\})\triangle B$, we have: $\mathrm{odd}(\mathbf{M}^\circ, \Pi) - \mathrm{dist}(\mathbf{M}, \mathbf{M}^\circ) = 3 - 1 = 2 = \delta(\mathbf{M}, \Pi)$. Note that the family of feasible sets of $\mathbf{M}^\circ$ is given by $\{\{v_i, v_j, v_k, v_7, v_8\} \mid i \in \{1,2\}, j \in \{3,4\}, k \in \{5,6\}\}$, and that Π itself gives the finest direct sum decomposition of $\mathbf{M}^\circ$ compatible with Π, where $\{v_1, v_2\}$, $\{v_3, v_4\}$, and $\{v_5, v_6\}$ are the odd components. It should be emphasized that $\mathbf{M}^\circ$ can be identified with a matroid, which, however, is not a strong quotient of $\mathbf{M}$. □

7.3.4 Rank of Mixed Skew-symmetric Matrices

The rank of a mixed skew-symmetric matrix $A = Q + T$ can be treated successfully by means of the delta-covering problem for the associated linear delta-matroids.

The following identity is most fundamental, where it is recalled that the nonsingularity of a principal submatrix of T is characterized by graph-theoretic terms (see Proposition 7.3.8).

Lemma 7.3.20. *A mixed skew-symmetric matrix $A = Q + T$ is nonsingular if and only if both $Q[I]$ and $T[V \setminus I]$ are nonsingular for some $I \subseteq V$.*

Proof. By the definition of Pfaffians we see

$$\mathrm{pf}\, A = \sum_{I \subseteq V} \pm \mathrm{pf}\, Q[I] \cdot \mathrm{pf}\, T[V \setminus I]. \tag{7.56}$$

No cancellation can occur among terms with distinct I by virtue of the algebraic independence of the nonzero entries of T in the upper-triangular part. Hence $\mathrm{pf}\, A \neq 0$ if and only if $\mathrm{pf}\, Q[I] \neq 0$ and $\mathrm{pf}\, T[V \setminus I] \neq 0$ for some $I \subseteq V$. ■

The above statement can be rephrased in terms of the union of delta-matroids as follows.

Theorem 7.3.21. *For a mixed skew-symmetric matrix $A = Q+T$, the delta-matroid defined by A is the union of the delta-matroids defined by Q and T, that is, $\mathbf{M}(A) = \mathbf{M}(Q) \vee \mathbf{M}(T)$.*

Proof. This follows from the definition (7.52) of the union and Lemma 7.3.20 applied to principal submatrices of A. ■

Theorem 7.3.22. *For a mixed skew-symmetric matrix $A = Q + T$,*

$$\begin{aligned} \mathrm{rank}\, A &= \max\{\mathrm{rank}\, Q[I] + \mathrm{rank}\, T[V \setminus I] \mid I \subseteq V\} && (7.57)\\ &= \max\{|F_Q \triangle F_T| \mid F_Q \in \mathcal{F}_Q, F_T \in \mathcal{F}_T\}, && (7.58) \end{aligned}$$

where $\mathbf{M}(Q) = (V, \mathcal{F}_Q)$ and $\mathbf{M}(T) = (V, \mathcal{F}_T)$ are the linear delta-matroids defined respectively by Q and T.

Proof. The first identity follows from Lemma 7.3.20, whereas the second is obtained from Theorem 7.3.21 with Lemma 7.3.15. ■

The rank formula (7.58) enables us to compute the rank of $A = Q + T$ by solving the delta-covering problem for $(\mathbf{M}(Q), \mathbf{M}(T))$. This can be done in polynomial time ($\mathrm{O}(n^4)$ to be specific) using arithmetic operations in $\boldsymbol{K}$ by adapting the algorithm for delta-covering problem for a pair of linear delta-matroids.

Remark 7.3.23. The linear matroid parity problem can be reduced to the problem of computing the rank of a mixed skew-symmetric matrix. Given a pair of matrices $B = (\boldsymbol{b}_i \mid i = 1, \cdots, N)$ and $C = (\boldsymbol{c}_i \mid i = 1, \cdots, N)$, the matroid parity problem (cf. Remark 7.3.2) is to find $I \subseteq V = \{1, \cdots, N\}$ of maximum cardinality such that the column vectors $\{\boldsymbol{b}_i, \boldsymbol{c}_i \mid i \in I\}$ are linearly independent. We denote by ν the optimal value ($= \max |I|$) of the matroid parity problem.

Defining a mixed skew-symmetric matrix

$$A = \left[\begin{array}{c|cc|c|cc} O & \boldsymbol{b}_1 & \boldsymbol{c}_1 & \cdots & \boldsymbol{b}_N & \boldsymbol{c}_N \\ \hline -\boldsymbol{b}_1^{\mathrm{T}} & 0 & t_1 & & 0 & 0 \\ -\boldsymbol{c}_1^{\mathrm{T}} & -t_1 & 0 & & 0 & 0 \\ \hline \vdots & & & \ddots & & \\ \hline -\boldsymbol{b}_N^{\mathrm{T}} & 0 & 0 & & 0 & t_N \\ -\boldsymbol{c}_N^{\mathrm{T}} & 0 & 0 & & -t_N & 0 \end{array}\right] \tag{7.59}$$

using indeterminates $t_1, \cdots, t_N$, we have

$$\operatorname{rank} A = 2(N + \nu). \tag{7.60}$$

Proof of (7.60): The rank identity (7.57) yields

$$\operatorname{rank} A = 2 \max_I (f(I) + |V \setminus I|), \quad f(I) = \operatorname{rank} (\boldsymbol{b}_i, \boldsymbol{c}_i \mid i \in I).$$

Hence it suffices to show ν coincides with $\hat{\nu} = \max_I (f(I) - |I|)$. For an optimal I of the matroid parity problem we have $\nu = |I| = f(I) - |I| \leq \hat{\nu}$. Conversely, let I be a maximizer of $f(I) - |I|$ that is minimal with respect to set inclusion. Then $f(I) = 2|I|$, since otherwise there exists $i \in I$ such that $f(I \setminus \{i\}) \geq f(I) - 1$, which implies that $I \setminus \{i\}$ is also a maximizer. Hence $\nu \geq |I| = f(I) - |I| = \hat{\nu}$.

The formula (7.60) is equivalent to a well-known identity due to Lovász [178] (cf. Lovász–Plummer [181, Theorem 11.1.2]), which reads

$$\operatorname{rank} \sum_{i=1}^{N} x_i (\boldsymbol{b}_i \wedge \boldsymbol{c}_i) = 2\nu, \tag{7.61}$$

where $\boldsymbol{b} \wedge \boldsymbol{c} = \boldsymbol{b}\boldsymbol{c}^{\mathrm{T}} - \boldsymbol{c}\boldsymbol{b}^{\mathrm{T}}$ (called *wedge product*) and x_i $(i = 1, \cdots, N)$ are indeterminates. The equivalence between (7.60) and (7.61) can be shown easily by considering a Schur complement of A and using the identity

$$[\boldsymbol{b}\ \boldsymbol{c}] \begin{bmatrix} 0 & t \\ -t & 0 \end{bmatrix}^{-1} \begin{bmatrix} \boldsymbol{b}^{\mathrm{T}} \\ \boldsymbol{c}^{\mathrm{T}} \end{bmatrix} = -\frac{1}{t}(\boldsymbol{b} \wedge \boldsymbol{c})$$

(see Proposition 2.1.7 for Schur complement). □

7.3.5 Electrical Network Containing Gyrators

When the branch characteristics of an electrical network are given in terms of self- and mutual admittances Y, the network can be described by a matrix A of the form

$$A = \begin{array}{|cc|} \hline D & O \\ O & R \\ \hline -I & Y \\ \hline \end{array}\,, \tag{7.62}$$

where D is a fundamental cutset matrix and R is a fundamental circuit matrix of the underlying graph. Recall from §4.7.3 the convention that the above system of equations describes the "free" network that is obtained after the branches of voltage sources are contracted and those of current sources are deleted. Since $\ker D = (\ker R)^{\perp}$, the matrix A is nonsingular if and only if DYD^{T} is nonsingular (see Lemma 4.7.11). That is, the unique solvability of the network is equivalent to the nonsingularity of DYD^{T}.

Under the genericity assumption that the set of the nonvanishing entries of Y is algebraically independent over $\mathbf{Q}$, the matrix A above is a mixed matrix. This makes it possible to formulate the unique solvability of the network in terms of an independent matching problem (see also Remark 2.3.37 for a variant of this formulation for the nonsingularity of DYD^{T}). This genericity assumption, though fairly reasonable in many cases, is not always justified.

An ideal element called a *gyrator* is commonly employed in electrical network theory. It is a two-port element, the element characteristic of which is represented as

$$\begin{bmatrix} \xi \\ \bar{\xi} \end{bmatrix} = \begin{bmatrix} 0 & g \\ -g & 0 \end{bmatrix} \begin{bmatrix} \eta \\ \bar{\eta} \end{bmatrix} \tag{7.63}$$

for the current-voltage pairs (ξ, η), $(\bar{\xi}, \bar{\eta})$ at the ports, where $g \neq 0$. Note that the admittance matrix of a gyrator is a skew-symmetric matrix of order two. Accordingly, it is not reasonable to impose the above-mentioned genericity assumption on Y when the electrical network in question contains gyrators. Gyrators are certainly ideal or artificial elements, but they play a pivotal role in electrical network theory (cf. Rohrer [283], Saito [287]). For example, any passive network is known to be "equivalent" to an *RCG network*, which is, by definition, a network consisting of resistors, capacitors, and gyrators (and possibly, sources).

Example 7.3.24. Consider the electrical network in Fig. 7.5, taken from Ueno–Kajitani [323], which consists of five elements: two gyrators (two pairs of branches $\{1, \bar{1}\}$ and $\{2, \bar{2}\}$), two capacitors C_3 and C_4 (branches 3 and 4), and one resistor of conductance g_5 (branch 5). It is understood that voltage sources and current sources are already contracted and deleted, respectively. The matrix A describing this network is given by

$$
A = \begin{array}{c}
\begin{array}{ccccccc|ccccccc}
\xi^1 & \xi^{\bar{1}} & \xi^2 & \xi^{\bar{2}} & \xi^3 & \xi^4 & \xi^5 & \eta_1 & \eta_{\bar{1}} & \eta_2 & \eta_{\bar{2}} & \eta_3 & \eta_4 & \eta_5 \\ \hline
1 & 0 & -1 & 0 & 0 & 1 & 0 & & & & & & & \\
0 & 1 & 1 & 1 & 0 & -1 & 0 & & & & & & & \\
0 & 0 & 1 & 0 & 1 & -1 & 1 & & & & & & & \\ \hline
 & & & & & & & 1 & -1 & 1 & 0 & -1 & 0 & 0 \\
 & & & & & & & 0 & -1 & 0 & 1 & 0 & 0 & 0 \\
 & & & & & & & -1 & 1 & 0 & 0 & 1 & 1 & 0 \\
 & & & & & & & 0 & 0 & 0 & 0 & -1 & 0 & 1 \\ \hline
-1 & & & & & & & & g_1 & & & & & \\
 & -1 & & & & & & -g_1 & & & & & & \\
 & & -1 & & & & & & & & g_2 & & & \\
 & & & -1 & & & & & & -g_2 & & & & \\
 & & & & -1 & & & & & & & sC_3 & & \\
 & & & & & -1 & & & & & & & sC_4 & \\
 & & & & & & -1 & & & & & & & g_5
\end{array}
\end{array}
\qquad (7.64)
$$

and the underlying graph G is depicted in the right of Fig. 7.5. □

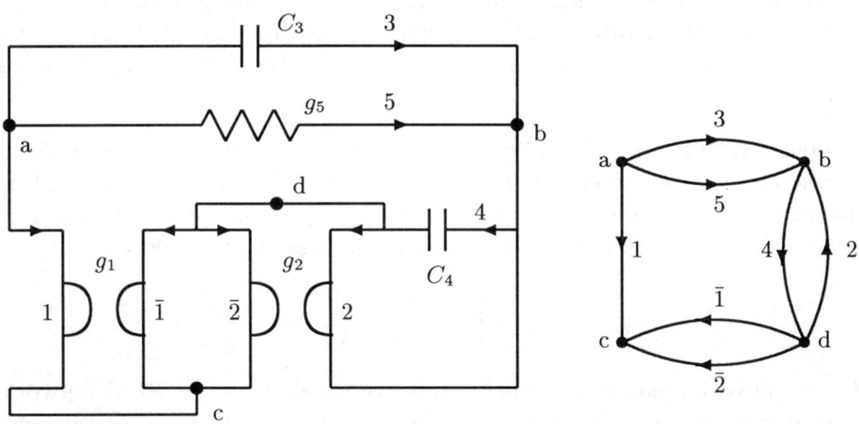

Fig. 7.5. An electrical network with gyrators (Example 7.3.24)

Thus we are motivated to consider matrices of the form (7.62) such that Y is a direct sum of a generic skew-symmetric matrix Y_s and a generic diagonal matrix Y_d. Namely, we consider a matrix

$$A = \left[\begin{array}{cccc} D_s & D_d & O & O \\ O & O & R_s & R_d \\ \hline -I_s & O & Y_s & O \\ O & -I_d & O & Y_d \end{array}\right], \tag{7.65}$$

in which Y_s is a generic skew-symmetric matrix, Y_d is a generic diagonal matrix, D_s, D_d, R_s and R_d are matrices over a field such that $[D_s\ D_d]$ and $[R_s\ R_d]$ are of full-row rank and

$$\ker[D_s\ D_d] = (\ker[R_s\ R_d])^{\perp}, \tag{7.66}$$

and I_s and I_d are unit matrices of appropriate dimensions. In Example 7.3.24, for instance, we have

$$Y_s = \left[\begin{array}{cc|cc} & g_1 & & \\ -g_1 & & & \\ \hline & & & g_2 \\ & & -g_2 & \end{array}\right], \qquad Y_d = \left[\begin{array}{ccc} sC_3 & & \\ & sC_4 & \\ & & g_5 \end{array}\right]$$

under the reasonable assumption that $\{g_1, g_2, C_3, C_4, g_5\}$ is algebraically independent. It is emphasized, however, that Y_s and Y_d are not assumed to be nonsingular, and that Y_s is not restricted to a block-diagonal matrix consisting of 2×2 blocks.

The objective of this subsection is to show the equivalence of the nonsingularity of A to that of a certain mixed skew-symmetric matrix associated with A. This implies by Theorem 7.3.22 that the nonsingularity of A, and hence the unique solvability of an electrical network described by A, can be tested by the efficient algorithm developed for the delta-parity/covering problem.

Remark 7.3.25. Though any passive network is "equivalent" to an RCG network, this does not mean that the present framework is general enough to treat an arbitrary passive network under the reasonable genericity assumption. Recall, for example, that an ideal *transformer* is described as

$$\begin{bmatrix} \eta \\ \xi \end{bmatrix} = \begin{bmatrix} t & 0 \\ 0 & -1/t \end{bmatrix} \begin{bmatrix} \bar{\eta} \\ \bar{\xi} \end{bmatrix}.$$

When a network containing transformers are rewritten as an RCG network, the genericity of the element t is not translated nicely to fit in our present formulation. □

First we observe a linear algebraic fact, independent of the genericity of Y_s and Y_d. Define a skew-symmetric matrix $\bar{A}$ by

$$\bar{A} = \left[\begin{array}{cc|cccc} O & O & R_s & R_d & O & O \\ O & O & O & O & -R_s & -R_d \\ \hline -R_s{}^{\mathrm{T}} & O & Y_s & O & O & O \\ -R_d{}^{\mathrm{T}} & O & O & O & O & -Y_d \\ O & R_s{}^{\mathrm{T}} & O & O & -Y_s & O \\ O & R_d{}^{\mathrm{T}} & O & Y_d{}^{\mathrm{T}} & O & O \end{array}\right]. \tag{7.67}$$

Lemma 7.3.26. *Let A and $\bar{A}$ be defined by* (7.65) *and* (7.67), *respectively, where $[D_s\ D_d]$ and $[R_s\ R_d]$ are of full-row rank and* (7.66) *is assumed.*[3] *Then,* $\det \bar{A} = c^2 \cdot (\det A)^2$ *for some $c \neq 0$.*

Proof. First assume that Y_s and Y_d are nonsingular and put $Z_s = Y_s{}^{-1}$ and $Z_d = Y_d{}^{-1}$. Define

$$B = \begin{bmatrix} R_s & R_d \end{bmatrix} \begin{bmatrix} Z_s & O \\ O & Z_d \end{bmatrix} \begin{bmatrix} R_s{}^{\mathrm{T}} \\ R_d{}^{\mathrm{T}} \end{bmatrix} = R_s Z_s R_s{}^{\mathrm{T}} + R_d Z_d R_d{}^{\mathrm{T}}.$$

Taking the Schur complement (Proposition 2.1.7) we see $\det \bar{A} = (\det Y_s \cdot \det Y_d)^2 \cdot \det S$, where

$$S = \begin{bmatrix} R_s Z_s R_s{}^{\mathrm{T}} & -R_d Z_d R_d{}^{\mathrm{T}} \\ R_d Z_d R_d{}^{\mathrm{T}} & -R_s Z_s R_s{}^{\mathrm{T}} \end{bmatrix}.$$

On the other other hand, $\det S = (\det B)^2$, since

$$\det \begin{bmatrix} M & -N \\ N & -M \end{bmatrix} = \det[N + M] \cdot \det[N - M]$$

for two square matrices M and N of the same size. Finally, $\det B = c \cdot (\det Y_s \cdot \det Y_d)^{-1} \cdot \det A$ for some $c \neq 0$ by Lemma 4.7.11. Therefore, $\det \bar{A} = c^2 \cdot (\det A)^2$ with nonzero c independent of Y_s and Y_d. This identity makes sense regardless of the nonsingularity of Y_s and Y_d. ■

With the matrix A of (7.65) we associate a mixed skew-symmetric matrix $\hat{A}$ defined by

$$\hat{A} = \left[\begin{array}{cc|cccc} O & O & R_s & R_d & O & O \\ O & O & O & O & -R_s & -R_d \\ \hline -R_s{}^{\mathrm{T}} & O & \hat{Y}_s & O & O & O \\ -R_d{}^{\mathrm{T}} & O & O & O & O & -Y_d \\ O & R_s{}^{\mathrm{T}} & O & O & -Y_s & O \\ O & R_d{}^{\mathrm{T}} & O & Y_d{}^{\mathrm{T}} & O & O \end{array}\right], \tag{7.68}$$

where $\hat{Y}_s$ is a copy of Y_s but with a new indeterminate for each independent entry of Y_s. The matrix $\hat{A}$ is almost the same as $\bar{A}$, but $\hat{A}$ is a mixed skew-symmetric matrix while $\bar{A}$ is not because of the repeated occurrence of Y_s.

[3] In this lemma Y_d can be any symmetric matrix and no genericity assumption on Y_s and Y_d is needed.

Lemma 7.3.27. *$\bar{A}$ is nonsingular if and only if $\hat{A}$ is nonsingular.*

Proof. Obviously, the nonsingularity of $\bar{A}$ implies that of $\hat{A}$. To show the converse, we use Lemma 7.3.20. Denote by $W_1 \cup W_2 \cup E_{s1} \cup E_{d1} \cup E_{s2} \cup E_{d2}$ the column set of $\hat{A}$ in the natural order with reference to (7.68). This serves also as the index set for $\bar{A}$. We have a natural correspondences $\varphi_s : E_{s1} \to E_{s2}$ and $\varphi_d : E_{d1} \to E_{d2}$. By the special structure of $\hat{A}$ we can take I in Lemma 7.3.20 so that $I \supseteq W_1 \cup W_2$, $\varphi_s(I \cap E_{s1}) = I \cap E_{s2}$ and $\varphi_d(I \cap E_{d1}) = I \cap E_{d2}$. Consider now the expansion of the Pfaffian of $\bar{A}$ in the form of (7.56). The term corresponding to the I above has no similar terms, and cannot be cancelled out. Hence $\mathrm{pf}\,\bar{A} \neq 0$, i.e., $\bar{A}$ is nonsingular. ■

The following theorem due to Iwata [142] gives a combinatorial characterization of the nonsingularity of A in (7.65). Put $E_s = \mathrm{Col}(R_s)$ and $E_d = \mathrm{Col}(R_d)$.

Theorem 7.3.28. *The following three conditions,* (i) *to* (iii), *are equivalent for the matrix A in* (7.65), *where Y_s is a generic skew-symmetric matrix, Y_d is a generic diagonal matrix, and the orthogonality* (7.66) *is assumed.*

(i) *A is nonsingular.*

(ii) *There exists $J \subseteq E_s \cup E_d$ such that $Y_s[J \cap E_s]$ is nonsingular, $Y_d[J \cap E_d]$ is nonsingular, and the submatrix of $[R_s\ R_d]$ with columns in $(E_s \cup E_d) \setminus J$ and all rows is nonsingular.*

(iii) *The associated mixed skew-symmetric matrix $\hat{A}$ defined by* (7.68) *is nonsingular.*

Proof. The equivalence of (i) and (iii) is due to Lemma 7.3.26 and Lemma 7.3.27. The equivalence of (ii) and (iii) is implicit in the proof of Lemma 7.3.27. ■

The above theorem has a number of implications. First, the equivalence of (i) and (iii) enables us to test for the nonsingularity of A by using the algorithm for the delta-parity/covering problem. Second, the equivalence of (i) and (ii) implies as an immediate corollary the unique solvability criterion for electrical networks, which is explained below.

Let us consider an RCG network, though the following argument is valid for an electrical network consisting of gyrators and other elements free from mutual couplings. After the branches of voltage sources are contracted and those of current sources are deleted, the network is described by a matrix A of the form (7.65) under the genericity assumption that the element characteristics are independent of one another. In this case, Y_s is a direct sum of generic skew-symmetric matrices of order two and Y_d is a generic diagonal matrix; both Y_s and Y_d are nonsingular. Moreover, $[D_s\ D_d]$ is a fundamental cutset matrix and $[R_s\ R_d]$ is a fundamental circuit matrix of the underlying graph, say $G = (V, E)$. Note that $E = E_s \cup E_d$ for $E_s = \mathrm{Col}(R_s)$ and $E_d = \mathrm{Col}(R_d)$, and that E_s is partitioned into pairs according to the block structure of Y_s.

In the literature of electrical network theory a tree in G is called a *proper tree* if each pair in E_s is either contained in the tree or disjoint from the tree.[4] Similarly, a spanning forest in G is said to be *proper* if each pair in E_s is either contained in it or disjoint from it.

The following solvability criterion for an RCG network, essentially due to Milić [194] (see also Recski [277], Ueno–Kajitani [323]), can be derived as an immediate consequence of Theorem 7.3.28.

Theorem 7.3.29. *An RCG network is uniquely solvable (under the genericity assumption) if and only if there exists a proper spanning forest.*

Proof. This follows from the equivalence of (i) and (ii) in Theorem 7.3.28. Note that the submatrix of $[R_s\ R_d]$ with columns in $(E_s \cup E_d) \setminus J$ and all rows is nonsingular if and only if J is a spanning forest, whereas the nonsingularity of $Y_s[J \cap E_s]$ imposes the properness on the spanning forest. ■

The connection of the solvability condition above to the matroid parity problem was pointed out first by Recski [276]. In the special case where the network consists of gyrators only, the associated mixed skew-symmetric matrix $\hat{A}$ takes the form of (7.59), and therefore, testing for the nonsingularity of $\hat{A}$ can be reduced to a matroid parity problem, as is shown in (7.60). It is indicated by Ueno–Kajitani [323] and Recski [277] that the solvability in the general case can be reduced to solving polynomially many matroid parity problems; at most $(|V| + |E_d|)|E|^2$ problems by Ueno–Kajitani [323] and at most $|E_d|$ problems by Recski [277]. Recently, it is observed by Iwata [142] that a single matroid parity problem suffices, as follows.

Given a graph $G = (V, E_s \cup E_d)$ with E_s partitioned into pairs, we make a copy of G, denoted $G' = (V', E_s' \cup E_d')$, and consider the direct sum of G and G', denoted $\hat{G} = (\hat{V}, \hat{E})$, where $\hat{V} = V \cup V'$, $\hat{E} = E_s \cup E_d \cup E_s' \cup E_d'$. The arc set $\hat{E}$ is partitioned into pairs as follows: $\{a, b\}$ is a pair in $\hat{E}$ if (i) $\{a, b\} \subseteq E_s$ and it is a pair in E_s, (ii) $\{a, b\} \subseteq E_s'$ and it is the copy of a pair in E_s, or (iii) $a \in E_d$, $b \in E_d'$ and they are the copies of each other. We denote this partition of $\hat{E}$ by $\hat{\Pi}$ and call $\hat{G}$ the *duplication* of G.

The observation of Iwata [142] reads as follows. Recall the notation $\nu(\cdot)$ for the maximum number of pairs contained in a base.

Theorem 7.3.30. *A graph G has a proper spanning forest if and only if $\nu(\mathbf{M}(\hat{G}), \hat{\Pi}) = r$ for the duplication $\hat{G}$ of G, where r denotes the number of arcs in a spanning forest of G. Hence, the unique solvability of an RCG network can be determined by solving a single matroid parity problem for a graphic matroid.*

---

[4] In the literature (e.g., Recski [277]) "normal tree" sometimes used as a synonym for "proper tree." A normal tree in an RCG network, however, often means a proper tree that contains as many capacitors as possible.

Proof. The existence of a proper spanning forest in G is obviously equivalent to the existence of a proper spanning forest in $\hat{G}$. The latter condition can be stated in terms of a single matroid parity problem for the graphic matroid $\mathbf{M}(\hat{G})$ defined by $\hat{G}$ and the partition $\hat{\Pi}$ above. ■

Example 7.3.31. The solvability conditions above are illustrated for the RCG network (Fig. 7.5) of Example 7.3.24. There exists a proper tree, e.g., $\{1, \bar{1}, 3\}$, in G. Therefore, this network is uniquely solvable by Theorem 7.3.29. The duplication $\hat{G}$ has eight vertices and seven parity pairs of arcs: $\{\{1, \bar{1}\}, \{2, \bar{2}\}, \{1', \bar{1}'\}, \{2', \bar{2}'\}, \{3, 3'\}, \{4, 4'\}, \{5, 5'\}\}$. The corresponding proper spanning forest in $\hat{G}$ is given by $\{1, \bar{1}, 1', \bar{1}', 3, 3'\}$. We have $\nu(\mathbf{M}(\hat{G}), \Pi) = 3 = r$ in the notation of Theorem 7.3.30. □

References

1. A. V. Aho, J. E. Hopcroft, and J. D. Ullman: *The Design and Analysis of Computer Algorithms*, Addison-Wesley, Reading, Mass., 1974.
2. A. V. Aho, J. E. Hopcroft, and J. D. Ullman: *Data Structures and Algorithms*, Addison-Wesley, Reading, Mass., 1983.
3. R. K. Ahuja, T. L. Magnanti, and J. B. Orlin: *Network Flows — Theory, Algorithms and Applications*, Prentice-Hall, Englewood Cliffs, 1993.
4. M. Aigner: *Combinatorial Theory*, Springer-Verlag, Berlin, 1979.
5. B. D. O. Anderson and D. J. Clements: Algebraic characterization of fixed modes in decentralized control, *Automatica*, **17** (1981), 703–712.
6. B. D. O. Anderson and H.-M. Hong: Structural controllability and matrix nets, *Inter. J. Control*, **35** (1982), 397–416.
7. T. Aoki, S. Hosoe, and Y. Hayakawa: Structural controllability for linear systems in descriptor form (in Japanese), *Trans. Soc. Instr. Control Engin.*, **19** (1983), 628–635.
8. C. Ashcraft and J. W. H. Liu: Applications of the Dulmage–Mendelsohn decomposition and network flow to graph bisection improvement, *SIAM J. Matrix Anal. Appl.*, **19** (1998), 325–354.
9. R. B. Bapat: König's theorem and bimatroids, *Linear Algebra Appl.*, **212/213** (1994), 353–365.
10. F. L. Bauer: Computational graphs and rounding error, *SIAM J. Numer. Anal.*, **11** (1974), 87–96.
11. S. P. Bhattacharyya: Transfer function conditions for output feedback disturbance rejection, *IEEE Trans. Automat. Control*, **AC-27** (1982), 974–977.
12. G. Birkhoff: *Lattice Theory*, 3rd ed., Amer. Math. Soc., 1995.
13. R. E. Bixby and W. H. Cunningham: Matroid optimization and algorithms, in: *Handbook of Combinatorics, Vol. I* (R. L. Graham, M. Grötschel, and L. Lovász, eds.), Elsevier, Amsterdam, 1995, Chapter 11, 551–609.
14. A. Björner, M. Las Vergnas, B. Sturmfels, N. White, and G. M. Ziegler: *Oriented Matroids*, Cambridge University Press, Cambridge, 1993.
15. A. Bouchet: Greedy algorithm and symmetric matroids, *Math. Programming*, **38** (1987), 147–159.
16. A. Bouchet: Representability of Δ-matroids, in: *Combinatorics* (A. Hajnal, L. Lovász, and V. T. Sós, eds.), The Janos Bolyai Mathematical Society, Budapest, 1988, 167–182.
17. A. Bouchet: Matchings and Δ-matroids, *Disc. Appl. Math.*, **24** (1989), 55–62.
18. A. Bouchet: Coverings and delta-coverings, in: *Integer Programming and Combinatorial Optimization* (E. Balas and J. Clausen, eds.), Springer-Verlag, Berlin, 1995, 228–243.
19. A. Bouchet and W. H. Cunningham: Delta-matroids, jump systems, and bisubmodular polyhedra, *SIAM J. Disc. Math.*, **8** (1995), 17–32.

20. A. Bouchet, A. W. M. Dress, and T. Havel: Δ-matroids and metroids, *Advances Math.*, **91** (1992), 136–142.
21. K. E. Brenan, S. L. Campbell, and L. R. Petzold: *Numerical Solution of Initial-Value Problems in Differential-Algebraic Equations*, North-Holland, New York, 1989; 2nd ed., SIAM, Philadelphia, 1996.
22. R. A. Brualdi: Term rank of the direct product of matrices, *Canad. J. Math.*, **18** (1966), 126–138.
23. R. A. Brualdi: Comments on bases in dependence structures, *Bull. Austral. Math. Soc.*, **1** (1969), 161–167.
24. R. A. Brualdi and H. J. Ryser: *Combinatorial Matrix Theory*, Cambridge University Press, London, 1991.
25. J. Bruno and L. Weinberg: The principal minors of a matroid, *Linear Algebra Appl.*, **4** (1971), 17–54.
26. P. Bujakiewicz: *Maximum Weighted Matching for High Index Differential Algebraic Equations*, Doctor's dissertation, Delft University of Technology, 1994.
27. P. Bujakiewicz and P. van den Bosch: Determination of perturbation index of a DAE with maximum weighted matching algorithm, *Proc. IEEE/IFAC Joint Symp. Computer-Aided Contr. Syst. Design* (S. E. Mattson, J. O. Gray, F. E. Cellier, eds.), Tucson, Arizona, March 1994.
28. F. E. Cellier: *Continuous System Modeling*, Springer-Verlag, Berlin, 1991.
29. F. E. Cellier and H. Elmqvist: Automated formula manipulation supports object-oriented continuous-system modeling, *IEEE Control Systems*, **13** (1993), 28–38.
30. R. Chandrasekaran and S. N. Kabadi: Pseudomatroids, *Disc. Math.*, **71** (1988), 205–217.
31. S. F. Chang and S. T. McCormick: A hierarchical algorithm for making sparse matrices sparser, *Math. Programming*, **56** (1992), 1–30.
32. S. F. Chang and S. T. McCormick: Computational results for the hierarchical algorithm for making sparse matrices sparser, *ACM Trans. Math. Software*, **19** (1993), 419–441.
33. Ch.-T. Chen: *Linear System Theory and Design*, 2nd ed., Holt, Rinehart and Winston, New York, 1970.
34. W.-K. Chen: *Applied Graph Theory — Graphs and Electrical Networks*, 2nd ed., North-Holland, Amsterdam, 1976.
35. V. Chvátal: *Linear Programming*, Freeman, New York, 1983.
36. D. Cobb: Controllability, observability, and duality in singular systems, *IEEE Trans. Automat. Control*, **AC-29** (1984), 1076–1082.
37. C. Commault and J.-M. Dion: Structure at infinity of linear multivariable systems — A geometric approach, *IEEE Trans. Automat. Control*, **AC-27** (1982), 693–696.
38. C. Commault, J.-M. Dion, and V. Hovelaque: A geometric approach for structured systems — Application to disturbance decoupling, *Automatica*, **33** (1997), 403–409.
39. C. Commault, J.-M. Dion, and A. Perez: Disturbance rejection for structured systems, *IEEE Trans. Automat. Control*, **AC-36** (1991), 884–887.
40. W. Cook, W. H. Cunningham, W. R. Pulleyblank, and A. Schrijver: *Combinatorial Optimization*, John Wiley, New York, 1998.
41. J. P. Corfmat and A. S. Morse: Decentralized control of linear multivariable systems, *Automatica*, **12** (1976), 479–495.
42. J. P. Corfmat and A. S. Morse: Structurally controllable and structurally canonical systems, *IEEE Trans. Automat. Control*, **AC-21** (1976), 129–131.
43. W. H. Cunningham: Improved bounds for matroid partition and intersection algorithms, *SIAM J. Comput.*, **15** (1986), 948–957.

44. W. H. Cunningham and J. F. Geelen: The optimal path-matching problem, *Combinatorica*, **17** (1997), 315–337.
45. E. J. Davison: Connectability and structural controllability of composite systems, *Automatica*, **13** (1977), 109–113.
46. F. J. de Jong: *Dimensional Analysis for Economists*, Contributions to Economic Analysis 50, North-Holland, Amsterdam, 1967.
47. J. Descusse and J.-M. Dion: On the structure at infinity of linear square decoupled systems, *IEEE Trans. Automat. Control*, **AC-27** (1982), 971–974.
48. J.-M. Dion and C. Commault: Feedback decoupling of structured systems, *IEEE Trans. Automat. Control*, **AC-38** (1993), 1132–1135.
49. G. Doetsch: *Introduction to the Theory and Application of the Laplace Transformation*, Springer-Verlag, New York, 1974 .
50. A. W. M. Dress and T. Havel: Some combinatorial properties of discriminants in metric vector spaces, *Advances Math.*, **62** (1986), 285–312.
51. A. W. M. Dress and W. Terhalle: Well-layered maps and the maximum-degree $k \times k$-subdeterminant of a matrix of rational functions, *Appl. Math. Lett.*, **8** (1995), 19–23.
52. A. W. M. Dress and W. Terhalle: Well-layered maps — A class of greedily optimizable set functions, *Appl. Math. Lett.*, **8** (1995), 77–80.
53. A. W. M. Dress and W. Terhalle: Rewarding maps — On greedy optimization of set functions, *Advances Appl. Math.*, **16** (1995), 464–483.
54. A. W. M. Dress and W. Wenzel: Valuated matroid: A new look at the greedy algorithm, *Appl. Math. Lett.*, **3** (1990), 33–35.
55. A. W. M. Dress and W. Wenzel: A greedy-algorithm characterization of valuated Δ-matroids, *Appl. Math. Lett.*, **4** (1991), 55–58.
56. A. W. M. Dress and W. Wenzel: Perfect matroids, *Advances Math.*, **91** (1992), 158–208.
57. A. W. M. Dress and W. Wenzel: Valuated matroids, *Advances Math.*, **93** (1992), 214–250.
58. A. Duchamp: A strong symmetric exchange axiom for delta-matroids, preprint, 1995.
59. I. S. Duff, A. M. Erisman, and J. K. Reid: *Discrete Methods for Sparse Matrices*, Clarendon Press, Oxford, 1986.
60. I. S. Duff and C. W. Gear: Computing the structural index, *SIAM J. Alg. Disc. Meth.*, **7** (1986), 594–603.
61. I. S. Duff, R. G. Grimes, and J. G. Lewis: Sparse matrix test problems, *ACM Trans. Math. Software*, **15** (1989), 1–14.
62. I. S. Duff, R. G. Grimes, and J. G. Lewis: *User's Guide for the Harwell–Boeing Sparse Matrix Collection (Release I)*, TR/PA/92/86, 1992.
63. A. L. Dulmage and N. S. Mendelsohn: Coverings of bipartite graphs, *Canad. J. Math.*, **10** (1958), 517–534.
64. A. L. Dulmage and N. S. Mendelsohn: A structure theory of bipartite graphs of finite exterior dimension, *Trans. Roy. Soc. Canada*, Section III, **53** (1959), 1–13.
65. A. L. Dulmage and N. S. Mendelsohn: On the inversion of sparse matrices, *Math. Comp.*, **16** (1962), 494–496.
66. A. L. Dulmage and N. S. Mendelsohn: Two algorithms for bipartite graphs, *J. Soc. Indust. Appl. Math.*, **11** (1963), 183–194.
67. J. Edmonds: Systems of distinct representatives and linear algebra, *J. National Bureau of Standards*, **71B** (1967), 241–245.
68. J. Edmonds: Submodular functions, matroids and certain polyhedra, in: *Combinatorial Structures and Their Applications* (R. Guy, H. Hanani, N. Sauer, and J. Schönheim, eds.), Gordon and Breach, New York, 1970, 69–87.

69. J. Edmonds: Matroid and the greedy algorithm, *Math. Programming*, **1** (1971), 127–136.
70. J. Edmonds: Matroid intersection, in: *Discrete Optimization* (P. L. Hammer, E. L. Johnson, and B. H. Korte, eds.), Ann. Disc. Math., **4** , North-Holland, Amsterdam, 1979, 39–49.
71. J. Edmonds and R. M. Karp: Theoretical improvements in algorithmic efficiency for network flow problems, *J. ACM*, **19** (1972), 248–264.
72. H. Elmqvist, M. Otter, and F. Cellier: Inline integration: A new mixed symbolic/numeric approach for solving differential-algebraic equation systems, *Proc. European Simulation Multiconference*, Prague, June 1995.
73. A. M. Erisman, R. G. Grimes, J. G. Lewis, W. G. Poole Jr., and H. D. Simon: Evaluation of orderings for unsymmetric sparse matrices, *SIAM J. Sci. Stat. Comput.*, **8** (1987), 600–624.
74. U. Faigle: Matroids in combinatorial optimization, in: *Combinatorial Geometries* (N. White, ed.), Cambridge University Press, London, 1987, 161–210.
75. L. R. Ford Jr. and D. R. Fulkerson: *Flows in Networks*, Princeton University Press, Princeton, 1962.
76. A. Frank: A weighted matroid intersection algorithm, *J. Algorithms*, **2** (1981), 328–336.
77. A. Frank: An algorithm for submodular functions on graphs, in: *Bonn Workshop on Combinatorial Optimization* (A. Bachem, M. Grötschel, and B. Korte, eds.), Ann. Disc. Math., **16**, North-Holland, Amsterdam, 1982, 97–120.
78. G. Frobenius: Über zerlegbare Determinanten, *Sitzungsber. Preuss. Akad. Wiss. Berlin*, 1917, 274–277. (*Gesammelte Abhandlungen*, **3** (1968), Springer-Verlag, New York, 701–704.)
79. S. Fujishige: A primal approach to the independent assignment problem, *J. Oper. Res. Soc. Japan*, **20** (1977), 1–15.
80. S. Fujishige: Algorithms for solving the independent-flow problems, *J. Oper. Res. Soc. Japan*, **21** (1978), 189–204.
81. S. Fujishige: Principal structure of submodular systems, *Disc. Appl. Math.*, **2** (1980), 77–79.
82. S. Fujishige: *Submodular Functions and Optimization*, North-Holland, Amsterdam, 1991.
83. H. N. Gabow and M. Stallmann: An augmenting path algorithm for linear matroid parity, *Combinatorica*, **6** (1986), 123–150.
84. H. N. Gabow and Y. Xu: Efficient theoretic and practical algorithms for linear matroid intersection problems, *J. Comput. Syst. Sci.*, **53** (1996), 129–147.
85. P. Gabriel and A. V. Roiter: *Algebra VIII, Representations of Finite-Dimensional Algebras*, Springer-Verlag, Berlin, 1992.
86. R. Gani and I. T. Cameron: Modelling for dynamic simulation of chemical processes — The index problem, *Chem. Engin. Sci.*, **47** (1992), 1311–1315.
87. F. R. Gantmacher: *The Theory of Matrices*, Chelsea, New York, 1959.
88. C. W. Gear: Differential-algebraic equation index transformations, *SIAM J. Sci. Stat. Comput.*, **9** (1988), 39–47.
89. C. W. Gear: Differential algebraic equations, indices, and integral algebraic equations, *SIAM J. Numer. Anal.*, **27** (1990), 1527–1534.
90. J. F. Geelen: *Matroids, Matchings, and Unimodular Matrices*, Ph. D. Thesis, University of Waterloo, 1995.
91. J. F. Geelen: An algebraic matching algorithm, preprint, 1997.
92. J. F. Geelen: Maximum rank matrix completion, *Linear Algebra Appl.*, **288** (1999), 211–217.
93. J. F. Geelen, S. Iwata, and K. Murota: The linear delta-matroid parity problem, *RIMS Preprint* No. 1149, Kyoto University, July 1997.

94. K. Glover and L. M. Silverman: Characterization of structural controllability, *IEEE Trans. Autom. Control*, **AC-21** (1976), 534–537.
95. I. Gohberg, P. Lancaster, and L. Rodman: *Matrix Polynomials*, Academic Press, New York, 1982.
96. A. V. Goldberg and R. E. Tarjan: Finding minimum-cost circulations by canceling negative cycles, *J. ACM*, **36** (1989), 873–886.
97. G. H. Golub and C. F. Van Loan: *Matrix Computations*, 3rd ed., The Johns Hopkins University Press, Baltimore, 1996.
98. M. Günther and U. Feldmann: The DAE-index in electric circuit simulation, *Math. Comput. Simulation*, **39** (1995), 573–582.
99. M. Günther and P. Rentrop: The differential-algebraic index concept in electric circuit simulation, *Zeitschrift für angewandte Mathematik und Mechanik*, **76** (1996), S1, 91–94.
100. E. Hairer, C. Lubich, and M. Roche: *The Numerical Solution of Differential-Algebraic Systems by Runge-Kutta Methods*, Springer-Verlag, Berlin, 1989.
101. E. Hairer and G. Wanner: *Solving Ordinary Differential Equations II*, Springer-Verlag, Berlin, 1991.
102. D. J. Hartfiel and R. Loewy: A determinantal version of the Frobenius–König theorem, *Linear Multilinear Algebra*, **16** (1984), 155–165.
103. M. L. J. Hautus: The formal Laplace transform for smooth linear systems, in: *Mathematical Systems Theory* (G. Marchesini and S. K. Mitter, eds.), Springer-Verlag, Berlin, 1976, 29–47.
104. M. L. J. Hautus and M. Heymann: New results on linear feedback decoupling, in: *Analysis and Optimization of Systems* (A. Bensoussan and J. L. Lions, eds.), Springer-Verlag, Berlin, 1980, 562–577.
105. M. L. J. Hautus and M. Heymann: Linear feedback decoupling — transfer function analysis, *IEEE Trans. Automat. Control*, **AC-28** (1983), 823–832.
106. Y. Hayakawa, S. Hosoe, M. Hayashi, and M. Ito: Controllability of structured systems with some linear dependence among variable elements (in Japanese), *Proc. 24th Conf. Japan Joint Automatic Control*, **1030** (1981), 59–60.
107. Y. Hayakawa, S. Hosoe, M. Hayashi, and M. Ito: On the structural controllability of compartmental systems, *IEEE Trans. Automat. Control*, **AC-29** (1984), 17–24.
108. Y. Hayakawa, S. Hosoe, and M. Ito: Dynamical degree and controllability for linear systems with intermediate standard form (in Japanese), *Trans. Inst. Electr. Comm. Engin. Japan*, **J64A** (1981), 752–759.
109. E. Hellerman and D. Rarick: Reinversion with the preassigned pivot procedure, *Math. Prog.*, **1** (1971), 195–216.
110. E. Hellerman and D. Rarick: The partitioned preassigned pivot procedure (P^4), in: *Sparse Matrices and Their Applications* (D. J. Rose and R. A. Willoughby, eds.), Plenum Press, New York, 1972, 67–76.
111. A. J. Hoffman: Some recent applications of the theory of linear inequalities to extremal combinatorial analysis, *Proc. of Symposia in Applied Mathematics*, **10** (1960), 113–127.
112. A. J. Hoffman and S. T. McCormick: A fast algorithm that makes matrices optimally sparse, in: *Progress in Combinatorial Optimization* (W. R. Pulleyblank, ed.), Academic Press, New York, 1984, 185–196.
113. S. Hosoe: Determination of generic dimension of controllable subspaces and its applications, *IEEE Trans. Automat. Control*, **AC-25** (1980), 1192–1196.
114. S. Hosoe, Y. Hayakawa, and T. Aoki: Structural controllability analysis for linear systems in linearly parameterized descriptor form, *IFAC World Congr.*, July 1984, Budapest, Hungary, 14.3/D6, 115–119.

115. S. Hosoe and K. Matsumoto: On the irreducibility condition in the structural controllability theorem, *IEEE Trans. Automat. Control*, **AC-24** (1979), 963–966.

116. H. E. Huntley: *Dimensional Analysis*, Macdonald, London, 1952.

117. M. Ichikawa: *An Application of Matroid Theory to Systems Analysis* (in Japanese), Graduation Thesis, Dept. Math. Eng. Instr. Phys., University of Tokyo, 1983.

118. Information-Technology Promotion Agency: *DPS User's Manual* (in Japanese), 1974.

119. Information-Technology Promotion Agency and Institute of the Union of Japanese Scientists and Engineers: *DPS-V2, Dynamic Process Simulation V2, User's Manual* (in Japanese), 1980.

120. Information-Technology Promotion Agency and Institute of the Union of Japanese Scientists and Engineers: *DPS-V3, Dynamic Process Simulation V3, User's Manual* (in Japanese), 1987.

121. Institute of the Union of Japanese Scientists and Engineers: *JUSE-L-GIFS User's Manual* (in Japanese), Ver. 3, 1976.

122. M. Iri: A min-max theorem for the ranks and term-ranks of a class of matrices — An algebraic approach to the problem of the topological degrees of freedom of a network (in Japanese), *Trans. Inst. Electr. Comm. Engin. Japan*, **51A** (1968), 180-187. (English translation in: *Electr. Comm. Japan*, **51** (1968), 18–25.)

123. M. Iri: *Network Flow, Transportation and Scheduling — Theory and Algorithms*, Academic Press, New York, 1969.

124. M. Iri: The maximum-rank minimum-term rank theorem for the pivotal transforms of a matrix, *Linear Algebra Appl.*, **2** (1969), 427–446.

125. M. Iri: Combinatorial canonical form of a matrix with applications to the principal partition of a graph (in Japanese), *Trans. Inst. Electr. Comm. Engin. Japan*, **54A** (1971), 30-37. (English translation in: *Electr. Comm. Japan*, **54A** (1971), 30–37.)

126. M. Iri: A review of recent work in Japan on principal partitions of matroids and their applications, *Ann. New York Acad. Sci.*, **319** (1979), 306–319.

127. M. Iri: Application of matroid theory to engineering systems problems, *Proc. Sixth Conf. Prob. Theory* (B. Bereanu et al., eds.), Editura Academiei Republicii Socialiste Romania, 1981, 107–127.

128. M. Iri: Applications of matroid theory, in: *Mathematical Programming — The State of the Art* (A. Bachem, M. Grötschel, and B. Korte, eds.), Springer-Verlag, Berlin, 1983, 158–201.

129. M. Iri: Structural theory for the combinatorial systems characterized by submodular functions, in: *Progress in Combinatorial Optimization* (W. R. Pulleyblank, ed.), Academic Press, New York, 1984, 197–219.

130. M. Iri and S. Fujishige: Use of matroid theory in operations research, circuits and systems theory, *Inter. J. Syst. Sci.*, **12** (1981), 27–54.

131. M. Iri and N. Tomizawa: A practical criterion for the existence of the unique solution in a linear electrical network with mutual couplings (in Japanese), *Trans. Inst. Electr. Comm. Engin. Japan*, **57A** (1974), 599–605.

132. M. Iri and N. Tomizawa: A unifying approach to fundamental problems in network theory by means of matroids (in Japanese), *Trans. Inst. Electr. Comm. Engin. Japan*, **58A** (1975), 33–40. (English translation in: *Electr. Comm. Japan*, **58A** (1975), 28–35.)

133. M. Iri and N. Tomizawa: An algorithm for finding an optimal "independent assignment", *J. Oper. Res. Soc. Japan*, **19** (1976), 32–57.

134. M. Iri, J. Tsunekawa, and K. Murota: Graph-theoretic approach to large-scale systems — Structural solvability and block-triangularization (in Japanese), *Trans. Infor. Process. Soc. Japan*, **23** (1982), 88–95. (English translation in: *Research Memorandum*, RMI 81-05, Dept. Math. Eng. Instr. Phys., University of Tokyo, 1981.)
135. M. Iri, J. Tsunekawa, and K. Yajima: The graphical techniques used for a chemical process simulator "JUSE GIFS", *Information Processing 71* (Proc. IFIP Congr. 71), Vol. 2 (Appl.), 1972, 1150–1155.
136. H. Ito: *An Algebraic Study on Discrete Stochastic Systems*, Doctor's dissertation, Dept. Math. Eng. Infor. Phys., University of Tokyo, 1992.
137. H. Ito, S. Amari, and K. Kobayashi: Identifiability of hidden Markov information sources and their minimum degrees of freedom, *IEEE Trans. Infor. Theory*, **IT-38** (1992), 324–333.
138. H. Ito, S. Iwata, and K. Murota: Block-triangularization of partitioned matrices under similarity/equivalence transformations, *SIAM J. Matrix Anal. Appl.*, **15** (1994), 1226–1255.
139. S. Iwata: Principal structure of submodular systems and Hitchcock-type independent flows, *Combinatorica*, **15** (1996), 515–532.
140. S. Iwata: Block-triangularization of skew-symmetric matrices, *Linear Algebra Appl.*, **273** (1998), 215–226.
141. S. Iwata: Computing the maximum degree of minors in matrix pencils via combinatorial relaxation, *Proc. Tenth Annual ACM-SIAM Symp. Disc. Algorithms*, (1999), 476–483.
142. S. Iwata: Structural analysis of electric networks with gyrators by matroid matchings, *Discrete Mathematics and Systems Science Research Report*, Osaka University, in preparation.
143. S. Iwata and K. Murota: A theorem on the principal structure for independent matchings, *Disc. Appl. Math.*, **61** (1995), 229–244.
144. S. Iwata and K. Murota: A minimax theorem and a Dulmage–Mendelsohn type decomposition for a class of generic partitioned matrices, *SIAM J. Matrix Anal. Appl.*, **16** (1995), 719–734.
145. S. Iwata and K. Murota: Horizontal principal structure of layered mixed matrices — Decomposition of discrete systems by design-variable selections, *SIAM J. Disc. Math.*, **9** (1996), 71–86.
146. S. Iwata and K. Murota: Combinatorial relaxation algorithm for mixed polynomial matrices, *RIMS Preprint*, No. 1113, Kyoto University, 1996.
147. S. Iwata, K. Murota, and I. Sakuta: Primal-dual combinatorial relaxation algorithms for the maximum degree of subdeterminants, *SIAM J. Sci. Comput.*, **17** (1996), 4, 993–1012.
148. N. Jacobson: *Lectures in Abstract Algebra, III — Theory of Fields and Galois Theory*, Van Nostrand, Princeton, 1964.
149. N. Jacobson: *Basic Algebra I*, 2nd ed., Freeman, 1985.
150. N. Jacobson: *Basic Algebra II*, 2nd ed., Freeman, 1989.
151. P. M. Jensen and B. Korte: Complexity of matroid property algorithms, *SIAM J. Comput.*, **11** (1982), 184–190.
152. T. Kailath: *Linear Systems*, Prentice-Hall, Englewood Cliffs, 1980.
153. R. E. Kalman: Mathematical description of linear dynamical systems, *J. Soc. Indust. Appl. Math., Ser.A, On control*, **1** (1963), 152–192.
154. R. Kannan: Solving systems of linear equations over polynomials, *Theor. Comput. Sci.*, **39** (1985), 69–88.
155. T. Katayama: *Optimal Control of Linear Systems — Introduction to Descriptor Systems* (in Japanese), Kindai-Kagakusha, Tokyo, 1999.

156. A. Kelmans: *Introduction to the matroid theory*, Lectures at the All-Union Conference on Graph Theory and Algorithms, Odessa, September 1973.
157. G. Kishi and Y. Kajitani: Maximally distinct trees in a linear graph (in Japanese), *Trans. Inst. Electr. Comm. Engin. Japan*, **51A** (1968), 196–203. (English translation in *Electr. Comm. Japan*, **51** (1968), 35–42.)
158. G. Kishi and Y. Kajitani: Maximally distant trees and principal partition of a linear graph, *Trans. Circuit Theory*, **CT–16** (1969), 323–330.
159. G. Kishi and Y. Kajitani: Topological considerations on minimal sets of variables describing an LCR network, *Trans. Circuit Theory*, **CT–20** (1973), 335–340.
160. M. Klein: A primal method for minimal cost flows, *Management Science*, **14** (1967), 205–220.
161. H. Kobayashi and T. Yoshikawa: Graph-theoretic approach to controllability and localizability of decentralized control, *IEEE Trans. Automat. Control*, **AC-27** (1982), 1096–1108.
162. S. Kodama and M. Ikeda: On representations of linear dynamical systems (in Japanese), *Trans. Inst. Electr. Comm. Engin. Japan*, **J56-D** (1973), 553–560.
163. B. Korte, L. Lovász, and R. Schrader: *Greedoids*, Springer-Verlag, Berlin, 1991.
164. S. Krogdahl: The dependence graph for bases in matroids, *Disc. Math.*, **19** (1977), 47–59.
165. J. P. S. Kung: Bimatroids and invariants, *Advances Math.*, **30** (1978), 238–249.
166. J. P. S. Kung: *A Source Book in Matroid Theory*, Birkhäuser, Boston, 1986.
167. J. P. S. Kung: Basis-exchange properties, in: *Theory of Matroids* (N. White, ed.), Cambridge University Press, London, 1986, Chapter 4, 62–75.
168. J. P. S. Kung: Strong maps, in: *Theory of Matroids* (N. White, ed.), Cambridge University Press, London, l986, Chapter 8, 224–253.
169. H. L. Langhaar: *Dimensional Analysis and Theory of Models*, John Wiley, New York, 1951.
170. E. L. Lawler: Matroid intersection algorithms, *Math. Programming*, **9** (1975), 31–56.
171. E. L. Lawler: *Combinatorial Optimization — Networks and Matroids*, Holt, Rinehart and Winston, New York, 1976.
172. J. Lee and J. Ryan: Matroid applications and algorithms, *ORSA J. Comput.*, **4** (1992), 70–98.
173. C.-T. Lin: Structural controllability, *IEEE Trans. Automat. Control*, **AC-19** (1974), 201–208.
174. A. Linnemann: Decoupling of structured systems, *Syst. Control Lett.*, **1** (1981), 79–86.
175. A. Linnemann: Graph-theoretic characterization of fixed modes in parametrized systems, *Inter. J. Control*, **38** (1983), 319–335.
176. A. Linnemann: A further simplification in the proof of the structural controllability theorem, *IEEE Trans. Automat. Control*, **AC-31** (1986), 638–639.
177. L. Lovász: The matroid matching problem, in: *Algebraic Methods in Graph Theory*, Vol. 2 (L. Lovász and V. T. Sós, eds.), The Janos Bolyai Mathematical Society, Budapest, 1978, 495–517.
178. L. Lovász: On determinants, matching and random algorithms, in: *Fundamentals of Computation Theory, FCT ' 79* (Proc. Conf. Algebraic, Arithmetic, and Categorial Methods in Computation Theory, Berlin/Wendisch-Rietz (GDR), September 1979) (L. Budach, ed.), Akademie-Verlag, Berlin, 1979.
179. L. Lovász: Matroid matching and some applications, *J. Combin. Theory*, **B28** (1980), 208–236.
180. L. Lovász: Selecting independent lines from a family of lines in a space, *Acta Sci. Math.*, **42** (1980), 121–131.

181. L. Lovász and M. Plummer: *Matching Theory*, North-Holland, Amsterdam, 1986.
182. D. G. Luenberger: Dynamic equations in descriptor form, *IEEE Trans. Automat. Control*, **AC-22** (1977), 312–321.
183. D. G. Luenberger: Time-invariant descriptor systems, *Automatica*, **14** (1978), 473–480.
184. H. Maeda: On structural controllability theorem, *IEEE Trans. Automat. Control*, **AC-26** (1981), 795–798.
185. H. Maeda and T. Yamada: Strong structural controllability, *SIAM J. Control Opt.*, **17** (1979), 123–138.
186. M. Marcus and H. Minc: Disjoint pairs of sets and incidence matrices, *Illinois J. Math.*, **7** (1963), 137–147.
187. T. Matsumoto and M. Ikeda: Structural controllability based on intermediate standard forms (in Japanese), *Trans. Soc. Instr. Control Engin.*, **19** (1983), 601–606.
188. S. E. Mattsson and G. Söderlind: Index reduction in differential-algebraic equations using dummy derivatives, *SIAM J. Sci. Comput.*, **14** (1993), 677–692.
189. S. B. Maurer: The maximum-rank minimum-term-rank theorem for matroids, *Linear Algebra Appl.*, **10** (1975), 129–237.
190. S. T. McCormick: A combinatorial approach to some sparse matrix problems, *Tech. Rept. SOL*, 83-5, Dept. Operations Research, Stanford University, 1983.
191. S. T. McCormick: Making sparse matrices sparser — Computational results, *Math. Programming*, **49** (1990), 91–111.
192. S. T. McCormick and S. F. Chang: The weighted sparsity problem — Complexity and algorithms, *SIAM J. Disc. Math.*, **6** (1993), 57–69.
193. N. Megiddo: Combinatorial optimization with rational objective functions, *Math. Oper. Res.*, **4** (1979), 414–424.
194. M. M. Milić: General passive networks — Solvability, degeneracies, and order of complexity, *IEEE Trans. Circuits Syst.*, **CAS-21** (1974), 177–183.
195. T. Muir: *The Theory of Determinants*, MacMillan, London, 1906.
196. K. Murota: Decomposition of a graph based on the Menger-type linkings on it (in Japanese), *Trans. Infor. Process. Soc. Japan*, **23** (1982), 280–287.
197. K. Murota: Structural analysis of a large-scale system of equations by means of the M-decomposition of a graph (in Japanese), *Trans. Infor. Process. Soc. Japan*, **23** (1982), 480–486.
198. K. Murota: LU-decomposition of a matrix with entries of different kinds, *Linear Algebra Appl.*, **49** (1983), 275–283.
199. K. Murota: Structural controllability of a system in descriptor form expressed in terms of bipartite graphs (in Japanese), *Trans. Soc. Instr. Control Engin.*, **20** (1984), 272–274.
200. K. Murota: Use of the concept of physical dimensions in the structural approach to systems analysis, *Japan J. Appl. Math.*, **2** (1985), 471–494.
201. K. Murota: Combinatorial canonical form of layered mixed matrices and block-triangularization of large-scale systems of linear/nonlinear equations, *Discussion Paper Series*, 257, Inst. Socio-Economic Planning, University of Tsukuba, 1985.
202. K. Murota: Combinatorial dynamical system theory, *Research Memorandum*, RMI 86-02, Dept. Math. Eng. Infor. Phys., University of Tokyo, 1986.
203. K. Murota: Refined study on structural controllability of descriptor systems by means of matroids, *SIAM J. Control Opt.*, **25** (1987), 967–989.
204. K. Murota: *Systems Analysis by Graphs and Matroids — Structural Solvability and Controllability*, Springer-Verlag, Berlin, 1987.

205. K. Murota: Menger-decomposition of a graph and its application to the structural analysis of a large-scale system of equations, *Disc. Appl. Math.*, **17** (1987), 107–134.
206. K. Murota: Combinatorial dynamical system theory: general framework and controllability criteria, *Disc. Appl. Math.*, **22** (1988/1989), 241–265.
207. K. Murota: On the irreducibility of layered mixed matrices, *Linear and Multilinear Algebra*, **24** (1989), 273–288.
208. K. Murota: Some recent results in combinatorial approaches to dynamical systems, *Linear Algebra Appl.*, **122/123/124** (1989), 725–759.
209. K. Murota: A matroid-theoretic approach to structurally fixed modes of control systems, *SIAM J. Control Opt.*, **27** (1989), 1381–1402.
210. K. Murota: Principal structure of layered mixed matrices, *Disc. Appl. Math.*, **27** (1990), 221–234.
211. K. Murota: Eigensets and power products of a bimatroid, *Advances Math.*, **80** (1990), 78–91.
212. K. Murota: Computing Puiseux-series solutions to determinantal equations via combinatorial relaxation, *SIAM J. Comput.*, **19** (1990), 1132–1161.
213. K. Murota: On the Smith normal form of structured polynomial matrices, *SIAM J. Matrix Anal. Appl.*, **12** (1991), 747–765.
214. K. Murota: A mathematical framework for combinatorial/structural analysis of linear dynamical systems by means of matroids, in: *Symbolic and Numerical Computation for Artificial Intelligence* (B. R. Donald, D. Kapur, and J. L. Mundy, eds.), Academic Press, London, 1992, Chapter 9, 221–244.
215. K. Murota: Matroids and systems analysis (in Japanese), in: *Discrete Structures and Algorithms, Vol. I* (S. Fujishige, ed.), Kindai-Kagakusha, Tokyo, 1992, Chapter 2, 57–109.
216. K. Murota: On the Smith normal form of structured polynomial matrices, II, *SIAM J. Matrix Anal. Appl.*, **14** (1993), 1103–1111.
217. K. Murota: Hierarchical decomposition of symmetric discrete systems by matroid and group theories, *Math. Programming*, **A59** (1993), 377–404.
218. K. Murota: Mixed matrices — Irreducibility and decomposition, in: *Combinatorial and Graph-Theoretical Problems in Linear Algebra* (R. A. Brualdi, S. Friedland, and V. Klee, eds.), Springer-Verlag, Berlin, 1993, 39–71.
219. K. Murota: Computing the degree of determinants via combinatorial relaxation, *SIAM J. Comput.*, **24** (1995), 765–796.
220. K. Murota: Combinatorial relaxation algorithm for the maximum degree of subdeterminants — Computing Smith-McMillan form at infinity and structural indices in Kronecker form, *Appl. Algebra Engin. Comm. Comput.*, **6** (1995), 251–273.
221. K. Murota: Finding optimal minors of valuated bimatroids, *Appl. Math. Lett.*, **8** (1995), 37–42.
222. K. Murota: Two algorithms for valuated delta-matroids, *Appl. Math. Lett.*, **9** (1996), 67–71.
223. K. Murota: Structural approach in systems analysis by mixed matrices — An exposition for index of DAE, *ICIAM 95 (Proc. Third Inter. Congr. Indust. Appl. Math.,* Hamburg, Germany, July 1995) (K. Kirchgässner, O. Mahrenholtz, and R. Mennicken, eds.), Akademie-Verlag, Berlin, 1996, 257–279.
224. K. Murota: Valuated matroid intersection, I: optimality criteria, *SIAM J. Disc. Math.*, **9** (1996), 545–561.
225. K. Murota: Valuated matroid intersection, II: algorithms, *SIAM J. Disc. Math.*, **9** (1996), 562–576.

226. K. Murota: On exchange axioms for valuated matroids and valuated delta-matroids, *Combinatorica*, **16** (1996), 591–596.
227. K. Murota: Convexity and Steinitz's exchange property, *Advances Math.*, **124** (1996), 272–311.
228. K. Murota: Characterizing a valuated delta-matroid as a family of delta-matroids, *J. Oper. Res. Soc. Japan*, **40** (1997), 565–578.
229. K. Murota: Matroid valuation on independent sets, *J. Combin. Theory*, **B69** (1997), 59–78.
230. K. Murota: Fenchel-type duality for matroid valuations, *Math. Programming*, **82** (1998), 357–375.
231. K. Murota: Discrete convex analysis, *Math. Programming*, **83** (1998), 313–371.
232. K. Murota: Discrete convex analysis (in Japanese), in: *Discrete Structure and Algorithms, Vol. V* (S. Fujishige, ed.), Kindai-Kagaku-sha, Tokyo, 1998, Chapter 2, 51–100.
233. K. Murota: On the degree of mixed polynomial matrices, *SIAM J. Matrix Anal. Appl.*, **20** (1999), 196–227.
234. K. Murota: Submodular flow problem with a nonseparable cost function, *Combinatorica*, **19** (1999), 87–109.
235. K. Murota: Discrete convex analysis — Exposition on conjugacy and duality, in: *Graph Theory and Combinatorial Biology* (L. Lovász, A. Gyarfas, G. O. H. Katona, A. Recski, and L. Szekely, eds.) (Proc. Inter. Colloquium on Combinatorics and Graph Theory, Balatonlelle, Hungary, July 1996), The Janos Bolyai Mathematical Society, Budapest, 1999, 253–278.
236. K. Murota: L-convex functions and M-convex functions, in: *Encyclopaedia of Optimization* (P. M. Pardalos and C. A. Floudas, eds.), Kluwer Academic Publishers, Boston, 2000, to appear.
237. K. Murota and M. Iri: Matroid-theoretic approach to the structural solvability of a system of equations (in Japanese), *Trans. Infor. Process. Soc. Japan*, **24** (1983), 157–164.
238. K. Murota and M. Iri: Structural solvability of systems of equations — A mathematical formulation for distinguishing accurate and inaccurate numbers in structural analysis of systems, *Japan J. Appl. Math.*, **2** (1985), 247–271.
239. K. Murota, M. Iri, and M. Nakamura: Combinatorial canonical form of layered mixed matrices and its application to block-triangularization of systems of equations, *SIAM J. Alg. Disc. Meth.*, **8** (1987), 123–149.
240. K. Murota and S. Poljak: Note on a graph-theoretic criterion for structural output controllability, *IEEE Trans. Automat. Control*, **AC-35** (1990), 939–942.
241. K. Murota and M. Scharbrodt: Computing the combinatorial canonical form of a layered mixed matrix, *Opt. Meth. Software*, **10** (1998), 373–391.
242. K. Murota and J. W. van der Woude: Structure at infinity of structured descriptor systems and its applications, *SIAM J. Control Opt.*, **29** (1991), 878–894.
243. M. Nakamura: *Mathematical Analysis of Discrete Systems and Its Applications* (in Japanese), Doctor's dissertation, Dept. Math. Eng. Instr. Phys., University of Tokyo, 1982.
244. M. Nakamura: Analysis of discrete systems and its applications (in Japanese), *Trans. Inst. Electr. Comm. Engin. Japan*, **J66A** (1983), 368–373.
245. M. Nakamura: Structural theorems for submodular functions, polymatroids and polymatroid intersections, *Graphs and Combinatorics*, **4** (1988), 257–284.
246. M. Nakamura and M. Iri: Fine structures of matroid intersections and their applications, *Proc. Int. Symp. Circuits Syst.*, Tokyo, 1979, 996–999.

247. M. Nakamura and M. Iri: A structural theory for submodular functions, polymatroids and polymatroid intersections, *Research Memorandum*, RMI 81-06 (1981), Dept. Math. Eng. Instr. Phys., University of Tokyo.
248. H. Narayanan: *Submodular Functions and Electrical Networks*, Elsevier, Amsterdam, 1997.
249. H. Narayanan and M. N. Vartak: An elementary approach to the principal partition of a matroid, *Trans. Inst. Electr. Comm. Engin. Japan*, **E64** (1981), 227–234.
250. G. L. Nemhauser, A. H. G. Rinnooy Kan, and M. J. Todd, eds.: *Optimization*, Handbooks in Operations Research and Management Science, Vol. 1, Elsevier, Amsterdam, 1989.
251. G. L. Nemhauser and L. A. Wolsey: *Integer and Combinatorial Optimization*, John Wiley, New York, 1988.
252. M. Newman: *Integral Matrices*, Academic Press, London, 1972.
253. Y. Ohta: *Bilinear Theory of Soliton*, Doctor's dissertation, University of Tokyo, 1992.
254. T. Ohtsuki, Y. Ishizaki, and H. Watanabe: Network analysis and topological degrees of freedom (in Japanese), *Trans. Inst. Electr. Comm. Engin. Japan*, **51A** (1968), 238–245.
255. J. O'Neil and D. B. Szyld: A block ordering method for sparse matrices, *SIAM J. Sci. Stat. Comput.*, **11** (1990), 811–823.
256. O. Ore: Graphs and matching theorems, *Duke Math. J.*, **22** (1955), 625–639.
257. O. Ore: Studies on directed graphs, I, *Annals of Math.*, **63** (1956), 383–406.
258. J. B. Orlin and J. H. Vande Vate: Solving the linear matroid parity problem as a sequence of matroid intersection problems, *Math. Programming*, **47** (1990), 81–106.
259. J. G. Oxley: *Matroid Theory*, Oxford University Press, Oxford, 1992.
260. T. Ozawa: Common trees and partition of two-graphs (in Japanese), *Trans. Inst. Electr. Comm. Engin. Japan*, **57A** (1974), 383–390.
261. T. Ozawa: Topological conditions for the solvability of linear active networks, *Int. J. Circuit Theory and Appl.*, **4** (1976), 125–136.
262. T. Ozawa: Structure of 2-graphs (in Japanese), *Trans. Inst. Electr. Comm. Engin. Japan*, **J59A** (1976), 262–263.
263. L. Pandolfi: Controllability and stabilization for linear systems of algebraic and differential equations, *J. Opt. Theory Appl.*, **30** (1980), 601–620.
264. C. C. Pantelides: The consistent initialization of differential-algebraic systems, *SIAM J. Sci. Stat. Comput.*, **9** (1988), 213–231.
265. C. H. Papadimitriou and K. Steiglitz: *Combinatorial Optimization: Algorithms and Complexity*, Prentice-Hall, Englewood Cliffs, 1982.
266. H. Perfect: A generalization of Rado's theorem on independent transversals, *Proc. Cambridge Philos. Soc.*, **66** (1969), 513–515.
267. B. Petersen: Investigating solvability and complexity of linear active networks by means of matroids, *IEEE Trans. Circuits Syst.*, **CAS-26** (1979), 330–342.
268. J. C. Picard and M. Queyranne: On the structure of all minimum cuts in a network and applications, *Math. Programming Study*, **13** (1980), 8–16.
269. V. Pichai, M. E. Sezer, and D. D. Šiljak: A graph-theoretic characterization of structurally fixed modes, *Automatica*, **20** (1984), 247–250.
270. S. Poljak: Maximum rank of powers of a matrix of a given pattern, *Proc. Amer. Math. Soc.*, **106** (1989), 1137–1144.
271. S. Poljak: On the generic dimension of controllable subspaces, *IEEE Trans. Automat. Control*, **AC-35** (1990), 367–369.
272. J. W. Ponton and P. J. Gawthrop: Systematic construction of dynamic models for phase equilibrium processes, *Comput. Chem. Engin.*, **15** (1991), 803–808.

273. A. Pothen and C.-J. Fan: Computing the block triangular form of a sparse matrix, *ACM Trans. Math. Software*, **16** (1990), 303–324.
274. R. Rado: A theorem on independence relations, *Quarterly J. Math. Oxford Ser.*, **13** (1942), 83–89.
275. A. Recski: Unique solvability and order of complexity of linear networks containing memoryless n-ports, *Int. J. Circuit Theory and Appl.*, **7** (1979), 31–42.
276. A. Recski: Sufficient conditions for the unique solvability of linear memoryless 2-ports, *Int. J. Circuit Theory and Appl.*, **8** (1980), 95–103.
277. A. Recski: *Matroid Theory and Its Applications in Electric Network Theory and in Statics*, Springer-Verlag, Berlin, 1989.
278. K. J. Reinschke: Graph-theoretic characterization of fixed modes in centralized and decentralized control, *Inter. J. Control*, **39** (1984), 715–729.
279. K. J. Reinschke: *Multivariable Control: A Graph-theoretic Approach*, Springer-Verlag, Berlin, 1988.
280. R. T. Rockafellar: *Convex Analysis*, Princeton University Press, Princeton, 1970.
281. R. T. Rockafellar: *Conjugate Duality and Optimization*, SIAM, Philadelphia, 1974.
282. R. T. Rockafellar: *Network Flows and Monotropic Optimization*, John Wiley, New York, 1984.
283. R. A. Rohrer: *Circuit Theory: An Introduction to the State Space Variable Approach*, McGraw-Hill, New York, 1970.
284. H. H. Rosenbrock: *State-space and Multivariable Theory*, Nelson, London, 1970.
285. H. J. Ryser: Indeterminates and incidence matrices, *Linear and Multilinear Algebra*, **1** (1973), 149–157.
286. H. J. Ryser: The formal incidence matrix, *Linear and Multilinear Algebra*, **3** (1975), 99–104.
287. M. Saito: *Newtork Analysis* (in Japanese), Korona-sha, Tokyo, 1974.
288. H. Schneider: The concepts of irreducibility and full indecomposability of a matrix in the works of Frobenius, König and Markov, *Linear Algebra Appl.*, **18** (1977), 139–162.
289. J. A. Schouten: *Tensor Analysis for Physicists*, 2nd ed., Clarendon Press, Oxford, 1954 (corrected printing, 1959; Dover edition, 1989).
290. A. Schrijver: *Matroids and Linking Systems*, Mathematics Centre Tracts, 88, Amsterdam, 1978.
291. A. Schrijver: Matroids and linking systems, *J. Combin. Theory*, **B26** (1979), 349–369.
292. A. Schrijver: *Theory of Linear and Integer Programming*, John Wiley, Chichester, 1986.
293. D. J. G. Sebastian, R. G. Noble, R. K. M. Thambynayagam, and R. K. Wood: DPS — A unique tool for process simulation, *Proc. 2nd World Congr. Chem. Engin.*, Montreal, V, 1981, 473–480.
294. M. E. Sezer and D. D. Šiljak: Structurally fixed modes, *Syst. Control Lett.*, **1** (1981), 60–64.
295. R. W. Shields and J. B. Pearson: Structural controllability of multiinput linear systems, *IEEE Trans. Automat. Control*, **AC-21** (1976), 203–212.
296. M. Shigeno: *A Dual Approximation Approach to Matroid Optimization Problems*, Doctor's dissertation, Tokyo Institute of Technology, 1996.
297. A. Shioura: A constructive proof for the induction of M-convex functions through networks, *Disc. Appl. Math.*, **82** (1998), 271–278.
298. A. Shioura: Minimization of an M-convex function, *Disc. Appl. Math.*, **84** (1998), 215–220.

299. A. Shioura: Level set characterization of M-convex functions, *Research Report*, No. 21, Dept. Mech. Eng., Sophia University, 1998.
300. D. D. Šiljak: *Decentralized Control of Complex Systems*, Academic Press, Boston, 1991.
301. D. Simson: *Linear Representations of Partially Ordered Sets and Vector Space Categories*, Gordon and Breach,1992.
302. M. Spivak: *Calculus on Manifolds*, Benjamin, New York, 1965.
303. N. Suda: *Linear Systems Theory* (in Japanese), Asakura, Tokyo, 1993.
304. K. Sugihara: Detection of structural inconsistency in systems of equations with degree of freedom and its applications, *Disc. Appl. Math.*, **10** (1985), 297–312.
305. K. Sugihara: *Machine Interpretation of Line Drawings*, MIT Press, Cambridge, Mass., 1986.
306. F. Svaricek: An improved graph theoretic algorithm for computing the structure at infinity of linear systems, *Proc. 29th Conf. Decision and Control*, Honolulu, Hawaii, December 1990, 2923–2924.
307. F. Svaricek: *Zuverlässige numerische Analyse linearer Regelungssysteme*, Teubner, Stuttgart, 1995.
308. T. Takamatsu, I. Hashimoto, and S. Tomita: Structural analysis of algebraic equations with the use of directed graphical representation (in Japanese), *Kagaku Kogaku Ronbunshu*, **8** (1982), 500–506.
309. T. Tanino and N. Takahashi: Degrees of fixed modes in linear control systems with constrained control structures (in Japanese), *Trans. Soc. Instr. Contr. Engin. Japan*, **24** (1988), 1150–1157.
310. R. E. Tarjan: *Data Structures and Network Algorithms*, SIAM, Philadelphia, 1983.
311. M. Tarokh: Fixed modes in multivariable systems using constrained controllers, *Automatica*, **21** (1985), 495–497.
312. N. Tomizawa: On some techniques useful for solution of transportation network problems, *Networks*, **1** (1971), 173–194.
313. N. Tomizawa: Strongly irreducible matroids and principal partition of a matroid into strongly irreducible minors (in Japanese), *Trans. Inst. Electr. Comm. Engin. Japan*, **J59A** (1976), 83–91.
314. N. Tomizawa: On a self-dual base axiom for matroids (in Japanese), *Papers of the Technical Group on Circuit and System Theory*, CST77–110, Inst. Electr. Comm. Engin. Japan, 1977.
315. N. Tomizawa and S. Fujishige: Theory of hyperspace (XIV) — Principal decompositions and principal structures of metric lattices with respect to supermodular functions (in Japanese), *Papers of the Technical Group on Circuit and System Theory*, CAS 82-2, Inst. Electr. Comm. Engin. Japan, 1982.
316. N. Tomizawa and S. Fujishige: Historical survey of extensions of the concept of principal partition and their unifying generalization to hypermatroids, *Syst. Sci. Res. Report*, No. 5, Dept. System Sci., Tokyo Institute of Technology, 1982.
317. N. Tomizawa and M. Iri: An algorithm for determining the rank of a triple matrix product AXB with application to the problem of discerning the existence of the unique solution in a network, *Trans. Inst. Electr. Comm. Engin. Japan*, **57A** (1974), 834–841. (English translation in *Electr. Comm. Japan*, **57A** (1974), 50–57).
318. N. Tomizawa and M. Iri: An algorithm for solving the "independent assignment" problem with application to the problem of determining the order of complexity of a network (in Japanese), *Trans. Inst. Electr. Comm. Engin. Japan*, **57A** (1974), 627–629.

319. D. M. Topkis: Minimizing a submodular function on a lattice, *Oper. Res.*, **26** (1978), 305–321.
320. L. Trave, A. Titli, and A. Tarras: *Large Scale Systems: Decentralization, Structure Constraints and Fixed Modes*, Springer-Verlag, Berlin, 1989.
321. A. W. Tucker: A combinatorial equivalence of matrices, *Proc. Symp. Appl. Math.*, **10** (1960), 129–140.
322. W. T. Tutte: The factorization of linear graphs, *J. London Math. Soc.*, **22** (1947), 107–111.
323. S. Ueno and Y. Kajitani: The unique solvability and state variables of an RCG network (in Japanese), *Trans. Inst. Electr. Comm. Engin. Japan*, **J68-A** (1985), 859–866.
324. J. Ungar, A. Kröner, and W. Marquardt: Structural analysis of differential-algebraic equation systems — Theory and application, *Comput. Chem. Engin.*, **19** (1995), 867–882.
325. B. L. van der Waerden: *Algebra*, Springer-Verlag, Berlin, 1955.
326. J. W. van der Woude: On the structure at infinity of a structured system, *Linear Algebra Appl.*, **148** (1991), 145–169.
327. J. W. van der Woude: The generic number of invariant zeros of a structured linear system, 1998.
328. J. van der Woude and K. Murota: Disturbance decoupling with pole placement for structured systems: a graph-theoretic approach, *SIAM J. Matrix Anal. Appl.*, **16** (1995), 922–942.
329. G. C. Verghese and T. Kailath: Rational matrix structure, *IEEE Trans. Automat. Control*, **AC-26** (1981), 434–439.
330. G. C. Verghese, B. C. Lévy, and T. Kailath: A generalized state-space for singular systems. *IEEE Trans. Automat. Control*, **AC-26** (1981), 811–831.
331. M. Vidyasagar: *Control System Synthesis: A Factorization Approach*, MIT Press, Cambridge, Mass., 1985.
332. S. H. Wang and E. J. Davison: On the stabilization of decentralized control systems, *IEEE Trans. Automat. Control*, **AC-18** (1973), 473–478.
333. D. J. A. Welsh: *Matroid Theory*, Academic Press, London, 1976.
334. W. Wenzel: Pfaffian forms and Δ-matroids, *Disc. Math.*, **115** (1993), 253–266.
335. W. Wenzel: Δ-matroids with the strong exchange conditions, *Appl. Math. Lett.*, **6** (1993), 67–70.
336. N. White, ed.: *Theory of Matroids*, Cambridge University Press, London, 1986.
337. N. White, ed.: *Combinatorial Geometries*, Cambridge University Press, London, 1987.
338. N. White, ed.: *Matroid Applications*, Cambridge University Press, London, 1992.
339. N. White: The Coxeter matroids of Gelfand et al., in: *Matroid Theory* (J. E. Bonin, J. G. Oxley, and B. Servatius, eds.), Amer. Math. Soc., Providence, R. I., 1996, 401–409.
340. H. Whitney: On the abstract properties of linear dependence, *Amer. J. Math.*, **57** (1935), 509–533.
341. D. V. Widder: *The Laplace Transform*, Princeton University Press, Princeton, 1941.
342. W. A. Wolovich: *Linear Multivariable Systems*, Springer-Verlag, New York, 1974.
343. W. M. Wonham: *Linear Multivariable Control: A Geometric Approach*, 3rd ed., Springer-Verlag, New York, 1985.
344. K. Yajima, J. Tsunekawa, and S. Kobayashi: On equation-based dynamic simulation, *Proc. 2nd World Congr. Chemical Eng.*, Montreal, V, 1981.

345. T. Yamada and L. R. Foulds: A graph-theoretic approach to investigate structural and qualitative properties of systems: a survey, *Networks*, **20** (1990), 427–452.
346. T. Yamada and D. G. Luenberger: Generic properties of column-structured matrices, *Linear Algebra Appl.*, **65** (1985), 189–206.
347. T. Yamada and D. G. Luenberger: Generic controllability theorem for descriptor systems, *IEEE Trans. Automat. Control*, **AC-30** (1985), 144–152.
348. T. Yamada and D. G. Luenberger: Algorithms to verify generic causality and controllability of descriptor systems, *IEEE Trans. Automat. Control*, **AC-30** (1985), 874–880.
349. E. L. Yip and R. F. Sincovec: Solvability, controllability, and observability of continuous descriptor systems, *IEEE Trans. Automat. Control*, **AC-26** (1981), 702–707.
350. L. A. Zadeh and C. A. Desoer: *Linear System Theory*, McGraw-Hill, New York, 1963.
351. R. M. Zazworsky and H. K. Knudsen: Controllability and observability of linear time-invariant compartmental models, *IEEE Trans. Automat. Control*, **AC-23** (1978), 872–877.
352. U. Zimmermann: Minimization on submodular flows, *Disc. Appl. Math.*, **4** (1982), 303–323.
353. U. Zimmermann: Negative circuits for flows and submodular flows, *Disc. Appl. Math.*, **36** (1992), 179–189.

Notation Table

Chapter 1

ξ^i : current in branch i §1.1.2
η_i : voltage across branch i §1.1.2
$\nu(A)$: index of polynomial matrix A (1.2)
A_{str} : structured matrix associated with polynomial matrix A (1.5)
$\nu_{\text{str}}(A)$: structural index of polynomial matrix A (1.6)
A-Q1 : assumption on Q-part of mixed polynomial matrix §1.2.1
A-T : assumption on T-part of mixed polynomial matrix §1.2.1
A-Q2 : stronger assumption on Q-part of mixed polynomial matrix §1.2.3
$\boldsymbol{F}$: field §1.3.1
$\boldsymbol{K}$: ground field, subfield of $\boldsymbol{F}$ §1.3.1
$A = Q + T$: mixed matrix (1.32)
M-Q : assumption on Q-part of mixed matrix §1.3.1
M-T : assumption on T-part of mixed matrix §1.3.1
$\mathcal{T}$: set of independent parameters §1.3.1
$A(s) = Q(s) + T(s)$: mixed polynomial matrix (1.33)
MP-Q1 : assumption on Q-part of mixed polynomial matrix §1.3.1
MP-T : assumption on T-part of mixed polynomial matrix §1.3.1
MP-Q2 : stronger assumption on Q-part of mixed polynomial matrix §1.3.1

Chapter 2

$\boldsymbol{F}$: field §2.1.1
$\boldsymbol{K}$: ground field, subfield of $\boldsymbol{F}$ §2.1.1
$\mathbf{Q}$: field of rational numbers §2.1.1
$\mathbf{R}$: field of real numbers §2.1.1
$\boldsymbol{K}[X]$: ring of polynomials in X over $\boldsymbol{K}$ §2.1.1
$\deg p$: degree of polynomial p §2.1.1
$\boldsymbol{K}(X)$: field of rational functions in X over $\boldsymbol{K}$ §2.1.1
$\boldsymbol{K}[X, 1/X]$: ring of Laurent polynomials in X over $\boldsymbol{K}$ §2.1.1
$\text{ord} f$: order of Laurent polynomial f §2.1.1
$\boldsymbol{K}(\mathcal{Y})$: field adjunction of $\mathcal{Y}$ to $\boldsymbol{K}$ §2.1.1
$\boldsymbol{K}[\mathcal{Y}]$: ring adjunction of $\mathcal{Y}$ to $\boldsymbol{K}$ §2.1.1

$\dim_{\boldsymbol{K}} \boldsymbol{F}$: degree of transcendency of $\boldsymbol{F}$ over $\boldsymbol{K}$	§2.1.1
$\mathrm{Row}(A)$: row set of matrix A	§2.1.2
$\mathrm{Col}(A)$: column set of matrix A	§2.1.2
A_{ij} : (i,j)-entry of matrix A	§2.1.2
$A[I,J]$: submatrix of A with row set I and column set J	§2.1.2
$\det A$: determinant of matrix A	(2.2)
$\mathrm{GL}(n, \boldsymbol{F})$: set of nonsingular matrices of order n over $\boldsymbol{F}$	§2.1.2
$\mathrm{BM}_{\pm}$: simultaneous exchange property of matroids	§2.1.2
VM : axiom of valuated matroids	§2.1.2
OM : axiom of oriented matroids	§2.1.2
$\mathrm{rank}\, A$: rank of matrix A	§2.1.3
$\text{term-rank}\, A$: term-rank of matrix A	§2.1.3
$G = (V, A)$: graph with vertex set V and arc set A	§2.2.1
$\partial^+ a$: initial vertex of arc a	§2.2.1
$\partial^- a$: terminal vertex of arc a	§2.2.1
∂a : set of vertices incident to arc a	§2.2.1
$\delta^+ v$: set of arcs leaving vertex v	§2.2.1
$\delta^- v$: set of arcs entering vertex v	§2.2.1
δv : set of arcs incident to vertex v	§2.2.1
$G \setminus U$: graph obtained from G by deleting vertices in U	§2.2.1
$u \xrightarrow{*} v$: directed path exists from u to v	§2.2.1
$\sim$: equivalence relation by reachability	§2.2.1
$\preceq$: partial order among strong components	§2.2.1
$G = (V^+, V^-; A)$: bipartite graph with bipartition (V^+, V^-) of vertex set and arc set A	§2.2.1
G_0^k : dynamic graph of time-span k	§2.2.1
$\mathcal{L}$: sublattice of 2^V	(2.18)
$\mathcal{L}_{\min}(f)$: family of the minimizers of f	(2.21)
$\mathcal{P}(\mathcal{L})$: partition determined by sublattice $\mathcal{L}$	(2.25)
$\mathcal{L}(\mathcal{P})$: sublattice determined by partition $\mathcal{P}$	(2.27)
$\Lambda(V; V_0, V_\infty)$: collection of sublattices of 2^V with minimum V_0 and maximum $V \setminus V_\infty$	(2.28)
$\Pi(V; V_0, V_\infty)$: collection of pairs of a partition of V with two distinguished subsets V_0 and V_∞ and a partial order $\preceq$	(2.29)
$\prec$: $\preceq$ and $\neq$	(2.30)
$\prec\!\cdot$: "covered by" relation with respect to a partial order	(2.31)
$\langle\ \rangle$: set of elements below with respect to a partial order	(2.32)
$\mathcal{L} = (S, \vee, \wedge)$: lattice with join $\vee$ and meet $\wedge$	§2.2.2
M : matching	§2.2.3
$\partial^+ M$: set of vertices in V^+ incident to arcs in M	§2.2.3
$\partial^- M$: set of vertices in V^- incident to arcs in M	§2.2.3
∂M : set of vertices incident to arcs in M	§2.2.3
$\nu(G)$: size of a maximum matching in bipartite graph G	§2.2.3
(U^+, U^-) : cover of bipartite graph G	§2.2.3

$\mathcal{C}(G)$: family of minimum covers of bipartite graph G	§2.2.3
Γ : set of adjacent vertices	(2.36)
γ : number of adjacent vertices	(2.37)
p_0 : surplus function	(2.39)
$N = (V, A, c; s^+, s^-)$: network with vertex set V, arc set A, capacity c, source s^+, and sink s^-	§2.2.4
φ : flow	§2.2.4
$\partial\varphi$: boundary of flow φ	(2.47)
$\mathrm{val}(\varphi)$: value of flow φ	§2.2.4
$\mathcal{S}$: family of S with $s^+ \in S$, $s^- \in V \setminus S$	(2.48)
$C(S)$: cut corresponding to S	(2.49)
$\kappa(S)$: capacity of cut S	(2.50)
$G = (V, A; X, Y)$: graph with vertex set V, arc set A, entrance X, and exit Y	§2.2.4
$N = (V, A, \overline{c}, \underline{c}, \gamma)$: network with vertex set V, arc set A, upper capacity $\overline{c}$, lower capacity $\underline{c}$, and cost γ	§2.2.5
$\mathrm{cost}(\varphi)$: cost of flow φ	§2.2.5
$w(M)$: weight of matching M	§2.2.5
$\mathbf{M} = (V, \mathcal{I})$: matroid on V with family of independent sets $\mathcal{I}$	§2.3.2
$\mathbf{M} = (V, \mathcal{B}, \mathcal{I}, \rho)$: matroid on V with family of bases $\mathcal{B}$, family of independent sets $\mathcal{I}$, and rank function ρ	§2.3.2
BM_- : (one-sided) basis exchange property	§2.3.2
$\mathrm{rank}\,\mathbf{M}$: rank of matroid $\mathbf{M}$	§2.3.2
$\mathrm{cl}(X)$: closure of subset X	§2.3.2
$\mathbf{M}^*$: dual of matroid $\mathbf{M}$	§2.3.2
BM_+ : dual exchange property	§2.3.2
$\mathbf{M}^U$: restriction of matroid $\mathbf{M}$ to U	§2.3.2
$\mathbf{M}_U$: contraction of matroid $\mathbf{M}$ to U	§2.3.2
$\mathbf{M}_1 \oplus \mathbf{M}_2$: direct sum of matroids $\mathbf{M}_1$ and $\mathbf{M}_2$	§2.3.2
$\mathbf{M}_1 \to \mathbf{M}_2$: strong map for matroids $\mathbf{M}_1$ and $\mathbf{M}_2$	§2.3.2
$\mathbf{M}(A)$: linear matroid defined by matrix A	§2.3.3
$\mathbf{M}\{U\}$: linear matroid defined by subspace U	§2.3.3
$\ker$: kernel of a matrix	§2.3.3
$\mathrm{BM}_\pm$: simultaneous exchange property	§2.3.4
$\mathrm{BM}_{+\mathrm{loc}}$: local exchange property	§2.3.4
$G(B, B')$: exchangeability graph for a pair of bases (B, B')	(2.66)
$\kappa(U)$: cut capacity of U	(2.71)
Γ : set of adjacent vertices	(2.73)
$\mathbf{M}_1 \vee \mathbf{M}_2$: union of matroids $\mathbf{M}_1$ and $\mathbf{M}_2$	§2.3.6
$\mathbf{L} = (S, T, \Lambda)$: bimatroid with row set S, column set T, and family of linked pairs Λ	§2.3.7
$\mathrm{Row}(\mathbf{L})$: row set of bimatroid $\mathbf{L}$	§2.3.7
$\mathrm{Col}(\mathbf{L})$: column set of bimatroid $\mathbf{L}$	§2.3.7
$\mathbf{RM}(\mathbf{L})$: row matroid of bimatroid $\mathbf{L}$	§2.3.7

$\mathbf{CM}(\mathbf{L})$: column matroid of bimatroid $\mathbf{L}$ §2.3.7
$G(\mathbf{L})$: underlying bipartite graph of bimatroid $\mathbf{L}$ §2.3.7
$\mathbf{L} \setminus Z$: deletion of Z from bimatroid $\mathbf{L}$ §2.3.7
$(D(\mathbf{L}), A(\mathbf{L}), C(\mathbf{L}))$: canonical partition of bimatroid $\mathbf{L}$ §2.3.7
$\mathbf{L}[X, Y]$: restriction of bimatroid $\mathbf{L}$ to (X, Y) §2.3.7
$\mathbf{L}^*$: dual of bimatroid $\mathbf{L}$ §2.3.7
$\mathbf{L}^{-1}$: inverse of bimatroid $\mathbf{L}$ §2.3.7
$\mathbf{L}_1 \vee \mathbf{L}_2$: union of bimatroids $\mathbf{L}_1$ and $\mathbf{L}_2$ §2.3.7
$\mathbf{L}_1 * \mathbf{L}_2$: product of bimatroids $\mathbf{L}_1$ and $\mathbf{L}_2$ §2.3.7
$\mathbf{L} * \mathbf{M}$: matroid induced from matroid $\mathbf{M}$ by bimatroid $\mathbf{L}$ §2.3.7

Chapter 3

$\mathcal{D}$: (multi)set of numbers characterizing a physical system (3.8)
$\mathcal{Q}$: (multi)set of accurate numbers (3.9)
$\mathcal{T}$: (multi)set of inaccurate numbers (3.9)
GA1 : first generality assumption §3.1.1
GA2 : second generality assumption §3.1.1
GA3 : third generality assumption §3.1.1
$A = Q + T$: mixed matrix (3.13)
M-Q : assumption on Q-part of mixed matrix §3.1.2
M-T : assumption on T-part of mixed matrix §3.1.2
$\mathrm{MM}(\boldsymbol{K}, \boldsymbol{F}; m, n)$: set of $m \times n$ mixed matrices with respect to $(\boldsymbol{K}, \boldsymbol{F})$ §3.1.2
$\mathrm{MM}(\boldsymbol{K}, \boldsymbol{F})$: set of mixed matrices with respect to $(\boldsymbol{K}, \boldsymbol{F})$ §3.1.2
$A(s) = Q(s) + T(s)$: mixed polynomial matrix (3.20)
MP-Q1 : assumption on Q-part of mixed polynomial matrix §3.1.2
MP-T : assumption on T-part of mixed polynomial matrix §3.1.2
$\mathcal{D}(\boldsymbol{F}; m, n; Z_1, \cdots, Z_d)$: set of $m \times n$ dimensioned matrices with ground field $\boldsymbol{F}$ and fundamental quantities $Z_1, \cdots, Z_d$ §3.2.2
$\mathcal{D}(\boldsymbol{F}; Z_1, \cdots, Z_d)$: set of dimensioned matrices with ground field $\boldsymbol{F}$ and fundamental quantities $Z_1, \cdots, Z_d$ §3.2.2
D_{r} : diagonal matrix representing physical dimensions of rows (3.27)
D_{c} : diagonal matrix representing physical dimensions of columns (3.28)
$\mathcal{U}(R; m, n)$: set of $m \times n$ totally unimodular matrices over ring R §3.2.3
$\mathcal{U}(R)$: set of totally unimodular matrices over ring R §3.2.3
$\boldsymbol{F}\langle Z_1, \cdots, Z_d \rangle$: ring generated over $\boldsymbol{F}$ by formal fractional powers of $Z_1, \cdots, Z_d$ (3.31)
MP-Q2 : stronger assumption on Q-part of mixed polynomial matrix §3.3.2

Chapter 4

$A = Q + T$: mixed matrix (4.1)
M-Q : assumption on Q-part of mixed matrix §4.1
M-T : assumption on T-part of mixed matrix §4.1
$\mathrm{MM}(\boldsymbol{K}, \boldsymbol{F}; m, n)$: set of $m \times n$ mixed matrices with respect to $(\boldsymbol{K}, \boldsymbol{F})$ §4.1

MM($\boldsymbol{K}, \boldsymbol{F}$) : set of mixed matrices with respect to $(\boldsymbol{K}, \boldsymbol{F})$ §4.1
$A = \begin{pmatrix} Q \\ T \end{pmatrix}$: LM-matrix (4.2)
L-Q : assumption on Q-part of LM-matrix §4.1
L-T : assumption on T-part of LM-matrix §4.1
LM($\boldsymbol{K}, \boldsymbol{F}; m_Q, m_T, n$) : set of $(m_Q + m_T) \times n$ LM-matrices with respect to $(\boldsymbol{K}, \boldsymbol{F})$ §4.1
LM($\boldsymbol{K}, \boldsymbol{F}$) : set of LM-matrices with respect to $(\boldsymbol{K}, \boldsymbol{F})$ §4.1
τ : term-rank of T-part (4.7)
Γ : set of nonzero rows of T-part (4.8)
γ : number of nonzero rows of T-part (4.9)
ρ : rank of Q-part (4.13)
p : LM-surplus function (4.16)
$J(\boldsymbol{x}, \boldsymbol{u})$: Jacobian matrix with respect to $\boldsymbol{x}$ and $\boldsymbol{u}$ (4.28)
GA1 : first generality assumption §4.3.2
GA2 : second generality assumption §4.3.3
GA3 : third generality assumption §4.3.3
S : nonsingular matrix in LM-admissible transformation (4.35)
P_{r} : row permutation matrix in LM-admissible transformation (4.35)
P_{c} : column permutation matrix in LM-admissible transformation (4.35)
$\boldsymbol{D}$: integral domain §4.4.7
d_k : kth determinantal divisor §4.5.1
p_τ : function characterizing the rank of LM-matrix (4.103)
LC($\boldsymbol{K}, \boldsymbol{F}_0, \boldsymbol{F}; m_Q, m_T, n$) : set of matrices §4.7.2
$\underset{\mathrm{pv}}{\sim}$: equivalence with respect to pivotal transformation §4.7.2
D : fundamental cutset matrix §4.7.3
R : fundamental circuit matrix §4.7.3
Y : admittance matrix §4.7.3
ker : kernel of a matrix §4.7.3
S_{r} : row transformation matrix in PE-equivalence (4.115)
S_{c} : column transformation matrix in PE-equivalence (4.115)
$\Pi = \{\Pi_\alpha\}_{\alpha=1}^{\mu}$: family of projection matrices §4.8.1
$\Gamma = \{\Gamma_\beta\}_{\beta=1}^{\nu}$: family of projection matrices §4.8.1
(A, Π, Γ) : partitioned matrix §4.8.1
$\mathcal{W}$: family of subspaces of V compatible with Γ (4.119)
p_{PE} : PE-surplus function (4.120)
$\mathcal{L}_{\min}(p_{\mathrm{PE}})$: family of minimizers of PE-surplus function p_{PE} (4.124)
$\mathcal{L}(A, \Pi, \Gamma)$: family of subspaces of V with property (4.126) §4.8.1
$\mathcal{P}(\tilde{A})$: partially ordered set determined by $\tilde{A}$ §4.8.1
$\mathcal{D}(\tilde{A})$: distributive lattice of order ideals of $\mathcal{P}(\tilde{A})$ §4.8.1
$\psi(J, S_{\mathrm{c}})$: subspace determined by (J, S_{c}) (4.127)
$\mathcal{W}^\circ$: family of subspaces of V compatible with Γ (4.128)
$\mathcal{Y}^\circ$: family of subspaces of U compatible with Π §4.8.4
p_{GP} : GP-surplus function (4.129)

λ : GP-birank function (4.130)
$\mathcal{L}$: lattice §4.9.2
f : submodular function §4.9.2
$\preceq$: partial order in $\mathcal{L}$ §4.9.2
$\mathcal{L}_{\min}(f;X)$: sublattice of minimizers of f not smaller than X (4.134)
$D(f;X)$: minimum element of $\mathcal{L}_{\min}(f;X)$ (4.135)
$\mathcal{K}_{\rm PS}(f)$: principal structure of $(\mathcal{L},f)$ (4.136)
$\mathcal{L}_{\rm PS}(f)$: principal sublattice of $(\mathcal{L},f)$ §4.9.2
$\mathcal{L}_{\min}(f)$: family of minimizers of f (4.137)
$\mathcal{L}_{\min}(f;v)$: family of minimizers of f containing v (4.138)
$D(f;v)$: minimum element of $\mathcal{L}_{\min}(f;v)$ §4.9.2
$\mathcal{B}_{\rm row}$: family of row-bases of a matrix (4.139)
$\mathcal{P}_{\rm DM}(I,C)$: partition in the DM-decomposition of $A[I,C]$ §4.9.3
$\mathcal{P}_{\rm CCF}(I,C)$: partition in the CCF of $A[I,C]$ §4.9.4
$\mathcal{L}_{\rm CCF}(I,C)$: sublattice corresponding to $\mathcal{P}_{\rm CCF}(I,C)$ §4.9.4
$\mathcal{B}_{\rm col}$: family of column-bases of a matrix (4.151)
q : surplus function for horizontal principal structure (4.153)
$\mathcal{L}_{\rm CCF}(R,J)$: sublattice corresponding to the CCF of $A[R,J]$ §4.9.5

Chapter 5

d_k : kth determinantal divisor (5.1)
e_k : kth invariant factor (invariant polynomial) (5.2)
δ_k : highest degree of a minor of order k (5.3)
t_k : contents at infinity (5.4)
$\mathbf{M}=(V,\omega)$: valuated matroid on V with valuation ω §5.2.1
$\mathbf{M}=(V,\mathcal{B},\omega)$: valuated matroid on V with family of bases $\mathcal{B}$ and valuation ω §5.2.1
VM : exchange axiom of valuated matroids §5.2.1
$\mathbf{M}[p]=(V,\mathcal{B},\omega[p])$: similarity transformation of valuated matroid $\mathbf{M}$ (5.16)
$\mathbf{M}^*=(V,\mathcal{B}^*,\omega^*)$: dual of valuated matroid $\mathbf{M}$ §5.2.3
$\mathbf{M}_I^U=(V,\mathcal{B}^U,\omega_I^U)$: restriction of valuated matroid $\mathbf{M}$ §5.2.3
$\mathbf{M}_U^J=(V,\mathcal{B}_U,\omega_U^J)$: contraction of valuated matroid $\mathbf{M}$ §5.2.3
$\mathbf{M}_{k,S_0}=(V,\mathcal{B}_k,\omega_{k,S_0})$: truncation of valuated matroid $\mathbf{M}$ (5.19)
$\mathbf{M}^{l,I_0}=(V,\mathcal{B}^l,\omega^{l,I_0})$: elongation of valuated matroid $\mathbf{M}$ (5.20)
$\omega(B,u,v)$: exchange gain (5.21)
VB-1, VB-2 : exchange axioms of valuated bimatroids §5.2.5
(S,T,δ) : valuated bimatroid §5.2.5
(S,T,Λ,δ) : valuated bimatroid §5.2.5
$\mathbf{M}_1\vee\mathbf{M}_2$: union of valuated matroids $\mathbf{M}_1$ and $\mathbf{M}_2$ §5.2.6
$\mathrm{VM_w}$: weak exchange axiom of valuated matroids §5.2.7
$\mathrm{VM_{loc}}$: local exchange axiom of valuated matroids §5.2.7
$\mathrm{VM_d}$: variant of exchange axiom of valuated matroids §5.2.7
$\mathcal{B}_p$: set of maximizers of $\omega[p]$ §5.2.7
$\mathcal{L}(\omega,\alpha)$: level set (5.42)

BL : exchange property of level sets §5.2.7
$\mathrm{BL_w}$: weaker exchange property of level sets §5.2.7
$G(B, B')$: exchangeability graph (5.44)
$\widehat{\omega}(B, B')$: maximum weight of a perfect matching in $G(B, B')$ (5.45)
VIAP : valuated independent assignment problem §5.2.9
$\Omega(M)$: objective function of VIAP (5.52)
VIAP(k) : valuated independent k-assignment problem §5.2.9
$\Omega(M, B^+, B^-)$: objective function of VIAP(k) (5.52)
$\mathrm{diag}\,(s; p)$: diagonal matrix with diagonal entries s^{p_i} §5.2.11

Chapter 6

$A(s) = Q(s) + T(s)$: mixed polynomial matrix (6.3)
MP-Q1 : assumption on Q-part of mixed polynomial matrix §6.1.1
MP-T : assumption on T-part of mixed polynomial matrix §6.1.1
MP-Q2 : stronger assumption on Q-part of mixed polynomial matrix §6.1.1
$A(s) = \begin{pmatrix} Q(s) \\ T(s) \end{pmatrix}$: LM-polynomial matrix (6.5)
δ_k : highest degree of a minor of order k (6.9)
o_k : lowest order of a minor of order k (6.11)
δ_k^{LM} : highest degree of a minor of order $m_Q + k$ for LM-matrix (6.16)
d_k : kth determinantal divisor (6.51)
e_k : kth invariant factor (invariant polynomial) (6.52)
Σ_A : Smith form of A §6.3.1
$D(s) = [A - sF \mid B]$: modal controllability matrix (6.67)
G_0^n : dynamic graph of time-span n §6.4.2
ζ : weight function for Q-part (6.74)
ξ_k : highest degree of a nonzero term in $\det \bar{T}_k[\mathrm{Row}(\bar{T}_k), J]$ §6.4.2
η_k : lowest degree of a nonzero term in $\det \bar{T}_k[\mathrm{Row}(\bar{T}_k), J]$ §6.4.2
$\psi(s; A, B, C, \mathcal{K})$: fixed polynomial of (A, B, C) with respect to $\mathcal{K}$ (6.84)
$\mathcal{K}$: feedback structure (6.85)
$\mathcal{C}_{\mathcal{K}}$: family of covers of feedback structure $\mathcal{K}$ (6.86)
$\mathcal{K}$: set of nonzero entries of K §6.5.3
$\mathcal{S}$: set of nonzero coefficients in $T(s)$ §6.5.3
$\psi(s)$: fixed polynomial (6.95)
ζ : weight function for Q-part (6.100)
$\eta(J)$: lowest degree of a nonzero term in $\det \tilde{T}_K[\mathrm{Row}(\tilde{T}_K), J]$ (6.101)
$\bar{\Psi}_0$: index set (6.103)
$\bar{\Psi}_1$: index set (6.104)
$\bar{\Psi}_2$: index set (6.105)

Chapter 7

δ_k : highest degree of a minor of order k (7.1)
$\hat{\delta}_k$: combinatorial counterpart of δ_k (7.2)
$A^\circ = (A_{ij}^\circ)$: leading coefficient matrix (7.3)

PLP(k) : primal linear program	(7.5)
DLP(k) : dual linear program	(7.6)
$\boldsymbol{\xi}$: primal variable	§7.1.2
$p = p_{\mathrm{R}} \oplus p_{\mathrm{C}}$: dual variable	§7.1.2
q : dual variable	§7.1.2
V^* : set of active vertices	(7.10)
I^* : set of active rows	(7.11)
J^* : set of active columns	(7.12)
$\mathcal{T}(A; p, q) = A^*$: tight coefficient matrix	(7.13)
$\mathrm{RS}_k(X_0)$: family of reachable sets at time k	(7.36)
$\mathrm{RS}(X_0)$: family of reachable sets	(7.37)
$\tau(\mathbf{A})$: transition index of bimatroid $\mathbf{A}$	§7.2.2
$\mathbf{RM}(\mathbf{A}^{\infty})$: limit of $\mathbf{RM}(\mathbf{A}^k)$	§7.2.2
$\mathbf{CM}(\mathbf{A}^{\infty})$: limit of $\mathbf{CM}(\mathbf{A}^k)$	§7.2.2
$(\omega_0; \omega_1, \omega_2, \cdots)$: Jordan type	§7.2.2
$\mathrm{EIG}(\mathbf{A})$: family of eigensets of bimatroid $\mathbf{A}$	§7.2.3
max-$\mathrm{EIG}(\mathbf{A})$: family of maximum-sized eigensets of bimatroid $\mathbf{A}$	§7.2.3
$\mathrm{REC}(\mathbf{A})$: family of recurrent sets of bimatroid $\mathbf{A}$	§7.2.3
max-$\mathrm{REC}(\mathbf{A})$: family of maximum-sized recurrent sets of bimatroid $\mathbf{A}$	§7.2.3
$\mathbf{R}_k$: reachability matroid	§7.2.4
$\mathbf{R}_{\infty}$: ultimate reachability matroid	§7.2.4
$r(\mathbf{R}_{\infty})$: controllable dimension	§7.2.4
$\kappa(\mathbf{A}, \mathbf{B})$: controllability index of CDS $(\mathbf{A}, \mathbf{B})$	§7.2.4
$\{\kappa_i\}$: controllability indices	§7.2.4
$A = Q + T$: mixed skew-symmetric matrix	(7.43)
MS-Q : assumption on Q-part of mixed skew-symmetric matrix	§7.3.1
MS-T : assumption on T-part of mixed skew-symmetric matrix	§7.3.1
$\nu(\mathbf{M}, \Pi)$: optimal value of the matroid parity problem $(\mathbf{M}, \Pi)$	§7.3.1
$A[I]$: principal submatrix of A indexed by I	§7.3.2
$I \triangle J$: symmetric difference of sets I and J	§7.3.2
$\mathrm{pf}\, A$: Pfaffian of skew-symmetric matrix A	(7.45)
$A * I$: pivotal transform of A with respect to principal submatrix $A[I]$	§7.3.2
$\nu(G)$: maximum size of a matching in graph G	§7.3.2
$\mathrm{odd}(G)$: number of odd components of graph G	§7.3.2
$G \setminus U$: graph obtained from G by deleting vertices of U	§7.3.2
$G[U]$: subgraph of G induced on U	§7.3.2
$\mathbf{M} = (V, \mathcal{F})$: delta-matroid on V with family of feasible sets $\mathcal{F}$	§7.3.3
DM : symmetric exchange axiom of delta-matroids	§7.3.3
$\mathrm{DM}_{\mathrm{even}}$: exchange axiom of even delta-matroids	§7.3.3
$\mathrm{DM}_{\pm}$: simultaneous exchange axiom of delta-matroids	§7.3.3
$\mathbf{M} \triangle X = (V, \mathcal{F} \triangle X)$: twisting of delta-matroid $\mathbf{M}$ by X	§7.3.3
$\mathbf{M}^*$: dual of delta-matroid $\mathbf{M}$	§7.3.3
$\mathbf{M} \setminus X = (V \setminus X, \mathcal{F} \setminus X)$: deletion of X from delta-matroid $\mathbf{M}$	§7.3.3

$\mathbf{M}/X$: contraction of delta-matroid $\mathbf{M}$ by X §7.3.3
$\mathbf{M}(A)$: delta-matroid defined by skew-symmetric matrix A §7.3.3
$\mathbf{M}_1 \oplus \mathbf{M}_2$: direct sum of delta-matroids $\mathbf{M}_1$ and $\mathbf{M}_2$ §7.3.3
$\mathbf{M}_1 \vee \mathbf{M}_2$: union of delta-matroids $\mathbf{M}_1$ and $\mathbf{M}_2$ §7.3.3
$\mathrm{dist}(\mathbf{M}_1, \mathbf{M}_2)$: distance between delta-matroids $\mathbf{M}_1$ and $\mathbf{M}_2$ §7.3.3
$\mathrm{odd}(\mathbf{M}_1, \mathbf{M}_2)$: number of odd components with respect to $(\mathbf{M}_1, \mathbf{M}_2)$ §7.3.3
Π : partition of V into pairs (lines) §7.3.3
$\delta_\Pi(F)$: number of lines exactly one of which belongs to F (7.53)
$\delta(\mathbf{M}, \Pi)$: optimal value of the delta-parity problem $(\mathbf{M}, \Pi)$ (7.54)
$\mathrm{odd}(\mathbf{M}, \Pi)$: number of odd components of $\mathbf{M}$ with respect to Π §7.3.3
$\boldsymbol{b} \wedge \boldsymbol{c}$: wedge product of vectors $\boldsymbol{b}$ and $\boldsymbol{c}$ §7.3.4
ker : kernel of a matrix §7.3.5
$\hat{G}$: duplication of graph G §7.3.5

Index

Printing: Saladruck, Berlin
Binding: Buchbinderei Lüderitz & Bauer, Berlin